Groundwater Hydrology

Groundwater Hydrology

Third Edition

David Keith Todd
University of California, Berkeley
and
Todd Engineers

Larry W. Mays
Arizona State University

WILEY
John Wiley & Sons, Inc.

Executive Editor	*Bill Zobrist*
Project Editor	*Jennifer Welter*
Senior Production Editor	*Valerie A. Vargas*
Marketing Manager	*Jennifer Powers*
Senior Designer	*Dawn L. Stanley*
New Media Editor	*Thomas Kulesa*
Cover Design	*David Levy*
Cover Image	*Photograph by Robert Morris, U.S. Geological Survey*
Production Management Services	*Argosy Publishing*

This book was set in 10/12 Times by Argosy Publishing and printed and bound by Hamilton Printing. The cover was printed by Phoenix Color Corporation.

On the Cover: Comal Springs. The highly productive Edwards aquifer, the first aquifer to be designated as a sole source aquifer under the Safe Drinking Water Act, is the source of water for more than 1 million people in San Antonio, Texas, some military bases and small towns, and for south-central Texas farmers and ranchers. The aquifer also supplies water to sustain threatened and endangered species habitat associated with natural springs in the region and supplies surface water to users downstream from the major springs. These various uses are in direct competition with ground-water development and have created challenging issues of ground-water management in the region.

Photo and description from the USGS website:
http://water.usgs.gov/pubs/circ/circ1186/html/gw_effect.html

This book is printed on acid-free paper. ∞

ISBN 0-471-05937-4 (cloth)
ISBN 0-471-45254-8 (WIE)

Printed in the United States of America

10 9 8 7 6 5 4 3 2 1

PREFACE to the 3rd Edition

Water is essential for life on Earth, and the largest source of fresh water lies under the Earth's surface. Increased demands for water have stimulated efforts to further develop, and in many cases over develop, our groundwater resources. In addition we have created pollution problems resulting from over development of groundwater resources and the mismanagement of wastes. As a result investigations of the occurrence and movement of groundwater have advanced, drilling equipment has improved, new models have been developed, techniques for managing groundwater have advanced, and research has extended our understanding of the resource.

Intended Audience and Introduction

This book is intended for a groundwater hydrology course in civil and environmental engineering, geology, or hydrogeology. It is appropriate for either a one or two term course at the junior, senior, or graduate level.

The first edition of this book by David K. Todd was published in 1959 and the second edition was published in 1980. This third edition of *Groundwater Hydrology* has the same purpose as the previous two editions, to present the fundamentals of groundwater hydrology in a manner understandable to those most concerned with such knowledge. Few people specialize in the subject, yet, because groundwater is a major natural resource, the subject becomes important for students and professionals in many fields: water supply, the environment, agriculture, economics, mining, and the law, to name only the obvious ones. Although it is impossible to present a subject fitted to such a diversity of students, the common need is an understanding of the fundamental principles, methods, and problems in the groundwater field. Thus, this book endeavors to make available a unified presentation of groundwater hydrology.

Since the earlier editions of this book, the groundwater resource field has made tremendous strides. Awareness of the environment, concerns and competition for water supplies, contamination of groundwater, and enhanced regulation of water resources have all focused attention on the subject. As a result educational programs, research funding, and professionals actively involved in developing and managing groundwater have grown dramatically. The National Ground Water Association together with journals such as Ground Water and Water Resources Research has been instrumental in advancing groundwater knowledge.

New to this edition

We have attempted to develop this book, not only as a stand-alone work, but in partnership with the Internet to serve as a portal to the vast resources on groundwater hydrology that now exist on the Internet. Hopefully this book will help guide the student, the professional, and the researcher to the general documents and other publications, program plans, field project details, software, and other information found on the Internet.

Internet References

Most of the web sites in the book are maintained by government agencies and nongovernmental organizations so that they will they continue in the future. Several end-of-chapter problems throughout the book are based upon Internet exercises.

- **Example Problems, Case Studies, Problem Sets** The most significant change in this edition has been the addition of example problems and case studies in the text, and the addition of problem sets at the end of each chapter. For educational purposes these problems enable a student to apply material in the text to realistic everyday situations.

- **Modeling with MODFLOW** The second most noticeable change has been an entirely new chapter on modeling (Chapter 9) featuring the U. S. Geological Survey MODFLOW model, which has become the standard modeling tool in the groundwater field. Chief among the new developments in the groundwater field is the role of computers, not only for organizing data and solving problems but also in managing groundwater resources on a basin-wide basis for known or anticipated inputs and outputs.

- **New Figures and Photos** We have included over 300 new figures and photos. Many of the new figures and tables within the book have been taken from various Web pages, in particular the U. S. Geological Survey. These items not only provide specific information but serve as samples of graphics and tabulations available on the Internet making the style of this book more variable than that of most textbooks.

- **References** References have been updated and selected on the basis of significance and availability.

- **Metric and U.S. Units** Both metric and U. S. customary units have been used in the example problems. Metric units are used in keeping with their growing acceptance in the United States as well as in recognition of the wide use of this book throughout the World. U. S. customary units have also been used because of the continued use by many in the United States.

Student and Instructor Resources

The website for the book, located at www.wiley.com/college/todd, includes resources for both students and faculty: corrections to the book, updates on changes in web addresses and new web pages used in the book, and other updates on material in the book that are of value to students and faculty. These resources are also included on the author's website at www.public.asu.edu/~lwmays/. Larry Mays can be contacted at mays@asu.edu

For instructors who have adopted the book for their course, a complete Solutions Manual for all homework problems in the text is available for download. Selected figures from the text, in PowerPoint format, are also available for easy creation of lecture slides. Visit the Instructor Companion Site portion of the book's website to register for a password for these assets available only to instructors.

For instructors who have adopted the book for their course, a complete Solutions Manual for all homework problems in the text is available for download. Selected figures from the text, in PowerPoint format, are also available for easy creation of lecture slides. Visit the Instructor Companion Site portion of the book's website to register for a password for these assets available only to instructors.

Acknowledgements

We are indebted to personnel of the U. S. Geological Survey for their numerous excellent publications on so many aspects of groundwater from which we have borrowed freely. We are deeply indebted to Sukru Ozger who prepared many of the problems and their solutions for the solutions manual. Students at Arizona State University have been exposed to advanced drafts of this new edition and offered several constructive comments. We want to thank the reviewers Scott Wolcott of Rochester Institute of Technology, Rameshwar Singh of San Jose State University, Rao S. Govindaraju of Purdue University, Albert J. Valocchi of the University of Illinois at Urbana-Champaign, Jeffrey D. Caulfield of the University of Missouri at Rolla, and Mark Widdowson of Virginia Tech. for their helpful comments and suggestions. David W. Abbott, Maureen Reilly, and Dan Rothman, P.E. provided useful reviews of selected chapters.

David Keith Todd
Piedmont, California

Larry W. Mays
Scottsdale, Arizona

Contents

6   Groundwater Levels and Environmental Influences   279

9   Groundwater Flow Modeling Techniques   413

Chapter 1

Introduction

$\mathbf{G}$*roundwater hydrology* may be defined as the science of the occurrence, distribution, and movement of water below the surface of the earth. *Geohydrology* has an identical connotation, and *hydrogeology* differs only by its greater emphasis on geology. Utilization of groundwater dates from ancient times, although an understanding of the occurrence and movement of subsurface water as part of the hydrologic cycle is recent.

The U.S. National Research Council (1991) presented the following definition of hydrology:

> Hydrology is the science that treats the waters of the Earth, their occurrence, circulation, and distribution, their chemical and physical properties, and their reaction with the environment, including the relation to living things. The domain of hydrology embraces the full life history of water on Earth.

Section 1.5 describes in further detail the concepts of the hydrologic cycle.

The importance of groundwater (hydrology) in the hydrologic cycle has been the subject of extensive technical research and publishing by many investigators over the past decades. Many of these publications are introduced in this book. Also, many books written on the subject of the fate of water have caught the attention of the general public, especially those interested in saving our resources. These include books by Carson[18] and de Villiers,[28] among others.

1.1 SCOPE

Groundwater (referred to without further specification) is commonly understood to mean water occupying all the voids within a geologic stratum. This *saturated zone* is to be distinguished from an *unsaturated,* or *aeration, zone* where voids are filled with water and air. Water contained in saturated zones is important for engineering works, geologic studies, and water supply developments; consequently, the occurrence of water in these zones will be emphasized here. Unsaturated zones are usually found above saturated zones and extend upward to the ground surface; because water here includes soil moisture within the root zone, it is a major concern of agriculture, botany, and soil science. No rigid demarcation of waters between the two zones is possible, for they possess an interdependent boundary, and water can move from zone to zone in either direction. The interrelationships are described more fully in Chapter 2.

Groundwater plays an important part in petroleum engineering. Two-fluid systems, involving oil and water, and three-fluid systems, involving gas, oil, and water, occur frequently in development of petroleum. Although the same hydrodynamic laws govern flows of these systems and groundwater, the distinctive nature of water in petroleum reservoirs sets it apart from other groundwater. Major differences exist in water quality, depth of occurrence, and methods of development and utilization, all of which contribute to a separation of interests and applications. Therefore, groundwater in petroleum reservoirs will not be treated specifically in this book. It should be noted, however, that groundwater hydrology has gained immeasurably from research conducted by the petroleum industry.

1.2   HISTORICAL BACKGROUND

1.2.1   *Qanats*

Groundwater development dates from ancient times.[15, 47*] The Old Testament contains numerous references to groundwater, springs, and wells. Other than dug wells, groundwater in ancient times was supplied from horizontal wells known as *qanats*.[†] These persist to the present day and can be found in a band across the arid regions of Southwestern Asia and North Africa extending from Afghanistan to Morocco. A cross section along a qanat is shown in Figure 1.2.1. Typically, a gently sloping tunnel dug through alluvial material leads water by gravity flow from beneath the water table at its upper end to a ground surface outlet and irrigation canal at its lower end.[13] Vertical shafts dug at closely spaced intervals provide access to the tunnel.[81] Qanats are laboriously hand constructed by skilled workers employing techniques that date back 3,000 years.[‡]

Iran possesses the greatest concentration of qanats; here some 22,000 qanats supply 75 percent of all water used in the country. Lengths of qanats extend up to 30 km, but most are less than 5 km.[13] The depth of the qanat mother well (see Figure 1.2.1) is normally less than 50 m, but instances of depths exceeding 250 m have been reported. Discharges of qanats vary seasonally with water table fluctuations and seldom exceed 100 m³/hr. Indicative of the density of qanats is the map in Figure 1.2.2. Based on aerial photographs of the Varamin Plain, located 40 km southeast of Tehran, this identifies 266 qanats within an area of 1,300 km².

1.2.2   Groundwater Theories

Utilization of groundwater greatly preceded understanding of its origin, occurrence, and movement. The writings of Greek and Roman philosophers to explain origins of springs and ground-

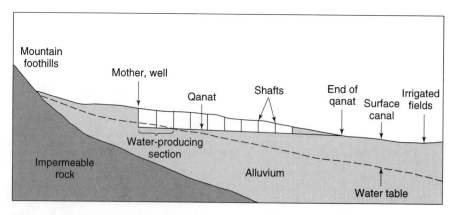

Figure 1.2.1. Vertical cross section along a qanat (after Beaumont[12]).

---

[*] Superscript numbers refer to references at the end of the chapter.

[†] *Qanat* comes from a Semitic word meaning "to dig." There are several variants of the name, including *karez, foggara,* and *falaj,* depending on location; in addition, there are numerous differences in spelling.[26]

[‡] Illustrative of the tremendous human effort expended to construct a qanat is a calculation by Beaumont.[13] The longest qanat near Zarand, Iran, is 29 km long with a mother well depth of 96 m and with 966 shafts along its length; the total volume of material excavated is estimated at 75,400 m³.

water contain theories ranging from fantasy to nearly correct accounts.[1, 6] As late as the seventeenth century it was generally assumed that water emerging from springs could not be derived from rainfall, for it was believed that the quantity was inadequate and the earth too impervious to permit penetration of rainwater far below the surface. Thus, early Greek philosophers such as Homer, Thales, and Plato hypothesized that springs were formed by seawater conducted through subterranean channels below the mountains, then purified and raised to the surface. Aristotle suggested that air enters cold dark caverns under the mountains where it condenses into water and contributes to springs.

The Roman philosophers, including Seneca and Pliny, followed the Greek ideas and contributed little to the subject. An important step forward, however, was made by the Roman architect Vitruvius. He explained the now-accepted infiltration theory that the mountains receive large amounts of rain that percolate through the rock strata and emerge at their base to form streams.

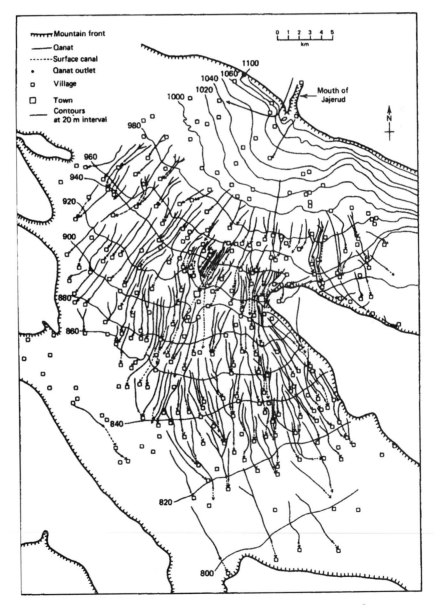

Figure 1.2.2. Map of qanats on the Varamin Plain, Iran (after Beaumont[9]).

The Greek theories persisted through the Middle Ages with no advances until the end of the Renaissance. The French potter and philosopher Bernard Palissy (c. 1510–1589) reiterated the infiltration theory in 1580, but his teachings were generally ignored. The German astronomer Johannes Kepler (1571–1630) was a man of strong imagination who likened the earth to a huge animal that takes in water of the ocean, digests and assimilates it, and discharges the end products of these physiological processes as groundwater and springs. The seawater theory of the Greeks, supplemented by the ideas of vaporization and condensation processes within the earth, was restated by the French philosopher René Descartes (1596–1650).

A clear understanding of the hydrologic cycle was achieved by the latter part of the seventeenth century. For the first time, theories were based on observations and quantitative data. Three Europeans made notable contributions, although others contributed to and supported these advances. Pierre Perrault[*] (1611–1680) measured rainfall during three years and estimated runoff of the upper Seine River drainage basin. He reported in 1674 that precipitation on the basin was about six times the river discharge, thereby demonstrating as false the early assumption of inadequate rainfall.[57] The French physicist Edme Mariotte (c. 1620–1684) made measurements of the Seine at Paris and confirmed Perrault's work. His publications appeared in 1686, after his death, and contained factual data strongly supporting the infiltration theory. Meinzer[54] once stated, "Mariotte . . . probably deserves more than any other man the distinction of being regarded as the founder of groundwater hydrology, perhaps I should say of the entire science of hydrology." The third contribution came from the English astronomer Edmund Halley (1656–1742), who reported in 1693 on measurements of evaporation, demonstrating that sea evaporation was sufficient to account for all springs and stream flow.

1.2.3   Recent Centuries

During the eighteenth century, fundamentals in geology were established that provided a basis for understanding the occurrence and movement of groundwater. During the first half of the nineteenth century many artesian wells were drilled in France, stimulating interest in groundwater. The French hydraulic engineer Henry Darcy (1803–1858) studied the movement of water through sand. His treatise of 1856 defined the relation, now known as Darcy's law, governing groundwater flow in most alluvial and sedimentary formations. Later European contributions of the nineteenth century emphasized the hydraulics of groundwater development. Significant contributions were made by J. Boussinesq, G. A. Daubrée, J. Dupuit, P. Forchheimer, and A. Thiem. In the twentieth century, increased activity in all phases of groundwater hydrology occurred. Many Europeans have participated with publications of either specialized or comprehensive works. There are too many people to mention them all, but R. Dachler, E. Imbeaux, K. Keilhack, W. Koehne, J. Kozeny, E. Prinz, H. Schoeller, and G. Thiem are best known in the United States.

---

* Pierre Perrault was a lawyer by profession and held administrative and financial positions in the French government; hence he is not well known in scientific circles. His interest in groundwater, leading to publication of *De l'Origine des Fontaines* in 1674, can be traced to the stimulus of the Dutch mathematician, astronomer, and physicist, Christiaan Huygens, who was then living in Paris and to whom the book is dedicated. Also, Pierre Perrault is often overshadowed by his four distinguished brothers: Jean (c. 1610–1669), a lawyer; Nicolas (1624–1662), a noted theologian; Claude (1613–1688), a physician, architect, and scientist, who is regarded as one of the most eminent French scholars of his time; and Charles (1628–1703), author and critic, who is best known for his Mother Goose fairy tales.

American contributions to groundwater hydrology date from near the end of the nineteenth century. In the past 110 years, tremendous advances have been made. Important early theoretical contributions were made by A. Hazen, F. H. King, and C. S. Slichter, while detailed field investigations were begun by men such as T. C. Chamberlin, N. H. Darton, W. T. Lee, and W. C. Mendenhall. O. E. Meinzer, through his consuming interest in groundwater and his dynamic leadership of groundwater activities of the U.S. Geological Survey, stimulated many individuals in the quest for groundwater knowledge. In recent decades the publications of M. S. Hantush, C. E. Jacob, G. B. Maxey, C. L. McGuinness, and R. W. Stallman are noteworthy. Within the last 40 years the surge in university research on groundwater problems, the establishment of professional consulting firms specializing in groundwater, the advent of the digital computer, and the extensive development of computer software have jointly produced a competence for development and management of groundwater resources.

1.3   TRENDS IN WATER WITHDRAWALS AND USE

The U.S. Geological Survey[65] estimated the total fresh and saline withdrawals in the United States during 1995 to have been 402,000 million gallons per day (Mgal/d) for all off-stream water-use categories (public supply, domestic, commercial, irrigation, livestock, industrial, mining, thermoelectric power). This estimate is nearly two percent less than the withdrawal estimate for 1990. Table 1.3.1 and Figure 1.3.1 provide a comparison of total water withdrawals by water-resources region. This comparison indicates that the California, South Atlantic–Gulf, and Mid-Atlantic regions account for one-third of the total water withdrawn in the United States.

National water-use compilations began in 1950 and are conducted at five-year intervals.[65] Estimates in Table 1.3.2 and Figure 1.3.2 summarize the water use—withdrawals, source of water, reclaimed wastewater, consumptive use, and in-stream use (hydroelectric power)—at five-year intervals. Figure 1.3.3 illustrates the trends in water withdrawals by water-use category and total withdrawals for 1960–1995. Table 1.3.2 also illustrates the percentage change in the 1990 and 1995 summary estimates. Estimates indicate that the general increase in water use from 1950 to 1980 and the decrease from 1980 to 1995 can be attributed in part to the following major factors:[65]

- Most of the increases in water use from 1950 to 1980 were the result of expansion of irrigation systems and increases in energy development.
- The development of center-pivot irrigation systems and the availability of plentiful and inexpensive groundwater resources supported the expansion of irrigation systems.
- Higher energy prices in the 1970s, and large drawdown in groundwater levels in some areas increased the cost of irrigation water. In the 1980s, improved application techniques, increased competition for water, and a downturn in farm economy reduced demands for irrigation water.
- The transition from water-supply management to water-demand management encouraged more efficient use of water.
- New technologies in the industrial sector that require less water, improved plant efficiencies, increased water recycling, higher energy prices, and changes in laws and regulations to reduce the discharge of pollutants resulted in decreased water use and less water being returned to natural systems after use.
- The enhanced awareness by the general public of water resources and the active conservation programs in many states have contributed to reduced water demands.

Table 1.3.1 Total Offstream Water Use by Water-Resources Region, 1995

Region	Population (thousands)	Per capita use, freshwater (gal/d)	Withdrawals (Mgal/d) (includes irrigation conveyance losses) By source and type									Reclaimed wastewater (Mgal/d)	Conveyance losses (Mgal/d)	Consumptive use, freshwater (Mgal/d)
			Groundwater			Surface water			Total					
			Fresh	Saline	Total	Fresh	Saline	Total	Fresh	Saline	Total			
New England	12,849	289	725	0	726	2,980	8,800	11,800	3,710	8,800	12,600	0	0	388
Mid-Atlantic	42,412	509	2,690	1	2,690	18,900	20,300	39,200	21,600	20,300	41,900	72	1.9	1,170
South Atlantic–Gulf	37,845	848	7,110	16	7,120	25,000	12,700	37,700	32,100	12,700	44,800	237	33	5,570
Great Lakes	21,836	1,500	1,510	4.6	1,520	31,100	6.5	31,100	32,700	11	32,700	0	.1	1,580
Ohio	22,631	1,330	1,980	22	2,000	28,100	.6	28,100	30,100	23	30,100	1.1	.7	1,870
Tennessee	4,198	2,140	258	0	258	8,730	0	8,730	8,980	0	8,980	.3	0	289
Upper Mississippi	22,268	1,050	2,570	4.2	2,570	20,700	0	20,700	23,300	4.2	23,300	11	0	1,660
Lower Mississippi	7,324	2,720	9,180	0	9,180	10,800	0	10,800	20,000	0	20,000	.7	553	7,740
Souris–Red–Rainy	693	364	115	0	115	138	0	138	253	0	253	0	1.8	122
Missouri Basin	10,664	3,380	9,320	38	9,360	26,700	0	26,700	36,000	38	36,100	22	7,840	14,200
Arkansas–White–Red	8,931	1,800	7,490	284	7,780	8,590	0	8,590	16,100	284	16,400	37	944	8,190
Texas–Gulf	16,755	1,050	5,960	324	6,280	11,700	4,860	16,600	17,700	5,190	22,900	71	390	7,340
Rio Grande	2,566	2,600	1,930	61	1,990	4,740	0	4,740	6,670	61	6,730	7.2	1,360	2,960
Upper Colorado	714	10,400	116	14	130	7,310	0	7,310	7,420	14	7,440	1.7	1,940	2,520
Lower Colorado	5,318	1,500	3,000	12	3,010	4,970	2.3	4,970	7,960	14	7,980	187	1,090	4,520
Great Basin	2,405	2,510	1,610	56	1,660	4,420	143	4,560	6,030	199	6,230	33	1,140	3,260
Pacific Northwest	9,948	3,220	5,500	0	5,500	26,500	38	26,500	32,000	38	32,000	.1	8,050	10,600
California	32,060	1,140	14,600	185	14,800	21,900	9,450	31,300	36,500	9,640	46,100	330	1,860	25,300
Alaska	604	350	58	75	132	154	43	196	211	117	329	0	.1	25
Hawaii	1,187	853	515	16	531	497	906	1,400	1,010	922	1,930	6.2	98	542
Caribbean	3,858	152	156	.2	156	433	2,450	2,880	588	2,450	3,040	0	15	189
Total	267,068	1,280	76,400	1,110	77,500	264,000	59,700	324,000	341,000	60,800	402,000	1,020	25,300	100,000

Source: Solley[65]

Figures may not add to totals because of independent rounding. Mgal/d = million gallons per day; gal/d = gallons per day

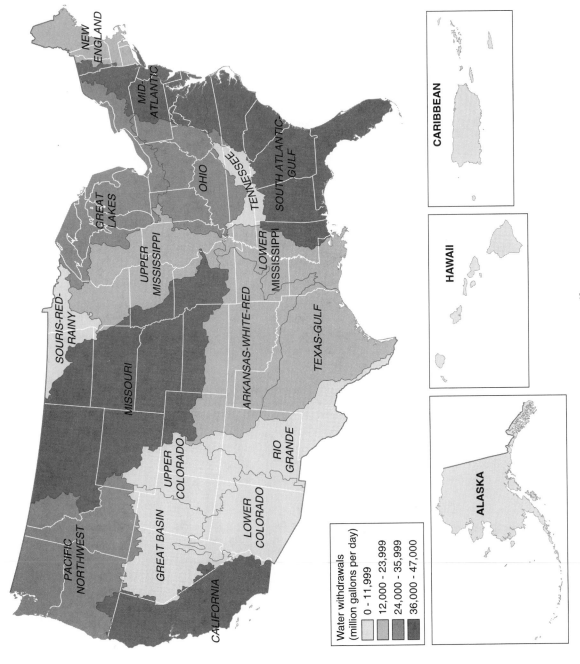

Figure 1.3.1. Total water withdrawals by water-resources region, 1995.[65]

Water withdrawals
(million gallons per day)

- 0 - 11,999
- 12,000 - 23,999
- 24,000 - 35,999
- 36,000 - 47,000

Table 1.3.2 Trends of Estimated Water Use in the United States, 1950–95

	1950[1]	1955[1]	1960[2]	1965[2]	1970[3]	1975[4]	1980[4]	1985[4]	1990[4]	1995[4]	Percentage change 1990–95
Population (millions)	150.7	164.0	179.3	193.8	205.9	216.4	229.6	242.4	252.3	267.1	+6
Offstream use											
Total withdrawals	180	240	270	310	370	420	440[5]	399	408	402	–2
Public supply	14	17	21	24	27	29	34	36.5	38.5	40.2	+4
Rural domestic and livestock	3.6	3.6	3.6	4.0	4.5	4.9	5.6	7.79	7.89	8.89	+13
Irrigation	89	110	110	120	130	140	150	137	137	134	–2
Industrial											
Thermoelectric power use	40	72	100	130	170	200	210	187	195	190	–3
Other industrial use	37	39	38	46	47	45	45	30.5	29.9	29.1	–3
Source of water											
Ground											
Fresh	34	47	50	60	68	82	83[5]	73.2	79.4	76.4	–4
Saline	([6])	.6	.4	.5	1	1	.9	.652	1.22	1.11	–9
Surface											
Fresh	140	180	190	210	250	260	290	265	259	264	+2
Saline	10	18	31	43	53	69	71	59.6	68.2	59.7	–12
Reclaimed wastewater	([6])	.2	.6	.7	.5	.5	.5	.579	.750	1.02	+36
Consumptive use	([6])	([6])	61	77	87[7]	96[7]	100[7]	92.3[7]	94[7]	100[7]	+6
Instream use											
Hydroelectric power	1,100	1,500	2,000	2,300	2,800	3,300	3,300	3,050	3,290	3,160	–4

[1] 48 States and District of Columbia

[2] 50 States and District of Columbia

[3] 50 States and District of Columbia, and Puerto Rico

[4] 50 States and District of Columbia, Puerto Rico, and Virgin Islands

[5] Revised

[6] Data not available.

[7] Freshwater only

Source: Solley[65]

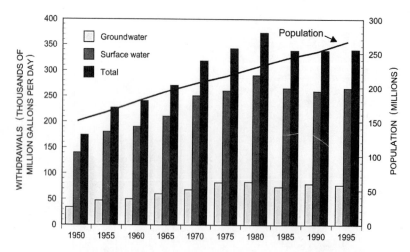

Figure 1.3.2. Trends in fresh groundwater and surface-water withdrawals, and population, 1950–95.[65]

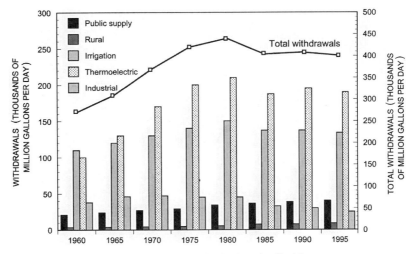

Figure 1.3.3. Trends in water withdrawals (fresh and saline) by water-use category and total (fresh and saline) withdrawals, 1960–95.[65]

1.4 UTILIZATION OF GROUNDWATER

Groundwater is an important source of water supply throughout the world. Its use in irrigation, industries, municipalities, and rural homes continues to increase. Figure 1.4.1 strikingly illustrates the dependence of an Indian village on its only water source—groundwater from a single dug well. Cooling and air conditioning have made heavy demands on groundwater because of its characteristic uniformity in temperature. Shortages of groundwater in areas where excessive withdrawals have occurred emphasize the need for accurate estimates of the available subsurface resources and the importance of proper planning to ensure the continued availability of water supplies.

There is a tendency to think of groundwater as being the primary water source in arid regions and of surface water in humid regions. But a study of groundwater use in the United States, for example, reveals that groundwater serves as an important resource in all climatic

Figure 1.4.1. Villagers laboriously lifting and carrying water from a deep dug well in northern India (photo by David K. Todd).

zones.[55] Reasons for this include its convenient availability near the point of use, its excellent quality (which typically requires little treatment), and its relatively low cost of development. Furthermore, in humid locales such as Barbados, Jamaica, and Hawaii, groundwater predominates as the water source because the high infiltration capacity of the soils sharply reduces surface runoff.

Figure 1.4.2 illustrates the relative proportion of water source and disposition and the general distribution of water from source to disposition for 1995. Table 1.4.1 lists the total off-stream water use by state for 1995—breaking down the withdrawals into groundwater and surface water and subdividing these into fresh and saline water. The total groundwater withdrawal was 77,500 Mgal/d and the total surface-water withdrawal was 324,000 Mgal/d. Table 1.4.2 lists the groundwater withdrawals by water-use category and water-resources region for 1995. The significant proportion of groundwater used for irrigation purposes is clearly indicated by the fact that 49,000 Mgal/d of the total 77,500 Mgal/d is used for irrigation.

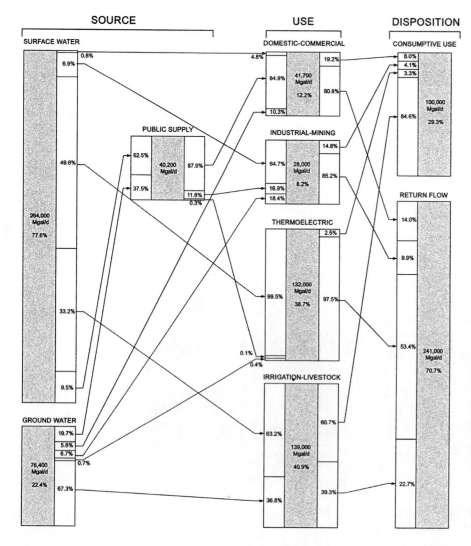

Figure 1.4.2. Source, use, and disposition of freshwater in the United States, 1995. For each water-use category, this diagram shows the relative proportion of water source and disposition and the general distribution of water from source to disposition. The lines and arrows indicate the distribution of water from source to disposition for each category; for example, surface water was 77.6 percent of total freshwater withdrawn, and going from "Source" to "Use" columns, the line from the surface-water block to the domestic and commercial block indicates that 0.8 percent of all surface water withdrawn was the source for 4.8 percent of total water (self-supplied withdrawals, public-supply deliveries) for domestic and commercial purposes. In addition, going from the "Use" to "Disposition" columns, the line from the domestic and commercial block to the consumptive use block indicates that 19.2 percent of the water for domestic and commercial purposes was consumptive use; this represents 8.0 percent of total consumptive use by all water-use categories.[65]

Table 1.4.1 Total Offstream Water Use by State, 1995

Withdrawals (Mgal/d) (includes irrigation conveyance losses)

Region	Population (thousands)	Per capita use freshwater (gal/d)	By source and type									Reclaimed wastewater (Mgal/d)	Conveyance losses (Mgal/d)	Consumptive use, freshwater (Mgal/d)
			Groundwater			Surface water			Total					
			Fresh	Saline	Total	Fresh	Saline	Total	Fresh	Saline	Total			
Alabama	4,253	1,670	436	9.1	445	6,650	0	6,650	7,090	9.1	7,100	0.1	0	532
Alaska	604	350	58	75	132	154	43	196	211	117	329	0	.1	25
Arizona	4,218	1,620	2,830	12	2,840	3,980	2.3	3,990	6,820	14	6,830	180	1,030	3,830
Arkansas	2,484	3,530	5,460	0	6,460	3,310	0	3,310	8,770	0	8,770	0	416	4,760
California	32,063	1,130	14,500	185	14,700	21,800	9,450	31,300	36,300	9,640	45,900	334	1,670	25,500
Colorado	3,747	3,690	2,260	17	2,270	11,600	0	11,600	13,800	17	13,800	11	3,770	5,230
Connecticut	3,275	389	166	0	166	1,110	3,180	4,290	1,280	3,180	4,450	0	0	97
Delaware	717	1,050	110	0	110	642	743	1,390	752	743	1,500	0	0	71
D.C.	554	18	.5	0	.6	9.7	0	9.7	10	0	10	0	0	15
Florida	14,166	509	4,340	4.6	4,340	2,880	11,000	13,800	7,210	11,000	18,200	236	32	2,780
Georgia	7,201	799	1,190	0	1,190	4,560	64	4,630	5,750	64	5,820	.6	0	1,170
Hawaii	1,187	853	515	16	531	497	906	1,400	1,010	922	1,930	6.2	98	542
Idaho	1,163	13,000	2,830	0	2,830	12,300	0	12,300	15,100	0	15,100	0	5,480	4,340
Illinois	11,830	1,680	928	25	953	19,000	0	19,000	19,900	25	19,900	2.0	0	857
Indiana	5,803	1,570	709	0	709	8,430	0	8,430	9,140	0	9,140	0	0	505
Iowa	2,842	1,070	528	0	528	2,510	0	2,510	3,030	0	3,030	0	0	290
Kansas	2,565	2,040	3,510	0	3,510	1,720	0	1,720	5,240	0	5,240	6.8	143	3,620
Kentucky	3,860	1,150	226	0	226	4,190	0	4,190	4,420	0	4,420	0	.5	318
Louisiana	4,342	2,270	1,350	0	1,350	8,500	0	8,500	9,850	0	9,850	0	166	1,930
Maine	1,241	178	80	0	80	141	105	248	221	105	326	0	0	48
Maryland	5,042	289	246	0	246	1,210	6,270	7,480	1,460	6,270	7,730	70	0	150
Massachusetts	6,074	189	351	0	351	795	4,370	5,160	1,150	4,370	5,510	0	0	180
Michigan	9,549	1,260	858	4.4	862	11,200	0	11,200	12,100	4.4	12,100	0	0	667
Minnesota	4,610	736	714	0	714	2,680	0	2,680	3,390	0	3,390	0	0	417
Mississippi	2,697	1,140	2,590	0	2,590	502	112	614	3,090	112	3,200	0	17	1,570
Missouri	5,324	1,320	891	0	891	6,140	0	6,140	7,030	0	7,030	11	0	692
Montana	870	10,200	204	13	217	8,640	0	8,640	8,850	13	8,860	0	4,410	1,960
Nebraska	1,637	6,440	6,200	4.7	6,200	4,350	0	4,350	10,500	4.7	10,500	2.0	906	7,020
Nevada	1,530	1,480	855	42	896	1,400	0	1,400	2,260	42	2,300	24	473	1,340
New Hampshire	1,148	388	81	0	81	364	877	1,240	446	877	1,320	0	0	35

(continues)

Table 1.4.1 (continued) Total Offstream Water Use by State, 1995

Region	Population (thousands)	Per capita use freshwater (gal/d)	Withdrawals (Mgal/d) (includes irrigation conveyance losses)									Reclaimed wastewater (Mgal/d)	Conveyance losses (Mgal/d)	Consumptive use, freshwater (Mgal/d)
			By source and type											
			Groundwater			Surface water			Total					
			Fresh	Saline	Total	Fresh	Saline	Total	Fresh	Saline	Total			
New Jersey	7,945	269	580	0	580	1,560	3,980	5,530	2,140	3,980	6,110	1.1	0	210
New Mexico	1,686	2,080	1,700	0	1,700	1,800	0	1,800	3,510	0	3,510	0	628	1,980
New York	18,136	567	1,010	1.5	1,010	9,270	6,500	15,800	10,300	6,500	16,800	0	0	469
North Carolina	7,195	1,070	535	2.1	535	7,200	1,550	8,750	7,730	1,560	9,290	1.0	0	713
North Dakota	641	1,750	122	0	122	1,000	0	1,000	1,120	0	1,120	0	5.1	181
Ohio	11,151	944	905	0	905	9,620	0	9,620	10,500	0	10,500	0	.2	791
Oklahoma	3,278	543	959	259	1,220	822	0	822	1,780	259	2,040	0	4.9	716
Oregon	3,140	2,520	1,050	0	1,050	6,860	0	6,860	7,910	0	7,910	0	1,300	3,210
Pennsylvania	12,072	802	860	0	860	8,820	0	8,820	9,680	0	9,680	1.1	0	565
Rhode Island	990	138	27	0	27	109	275	383	136	275	411	0	0	19
South Carolina	3,673	1,690	322	0	322	5,880	0	5,880	6,200	0	6,200	0	0	321
South Dakota	729	631	187	0	187	273	0	273	460	0	460	0	54	249
Tennessee	5,256	1,920	435	0	435	9,640	0	9,640	10,100	0	10,100	.5	0	233
Texas	18,724	1,300	8,370	411	8,780	16,000	4,860	20,800	24,300	5,280	29,600	109	540	10,500
Utah	1,951	2,200	776	14	790	3,530	143	3,670	4,300	157	4,460	14	612	2,200
Vermont	585	967	50	0	50	515	0	515	565	0	565	0	0	24
Virginia	6,618	826	358	0	358	5,110	2,800	7,900	5,470	2,800	8,260	0	2.9	218
Washington	5,431	1,620	1,760	0	1,760	7,060	38	7,100	8,820	38	8,860	0	1,090	3,080
West Virginia	1,828	2,530	146	.5	146	4,470	0	4,470	4,620	.5	4,620	0	0	352
Wisconsin	5,102	1,420	759	0	759	6,490	0	6,490	7,250	0	7,250	0	0	443
Wyoming	480	14,700	317	18	335	6,720	0	6,720	7,040	18	7,060	9.1	2,470	2,800
Puerto Rico	3,755	154	155	0	155	422	2,260	2,680	576	2,260	2,840	0	15	187
Virgin Islands	103	113	.5	.2	.7	11	190	201	12	190	202	0	0	1.9
Total	267,068	1,280	76,400	1,110	77,500	264,000	59,700	324,000	341,000	60,800	402,000	1,020	25,300	100,000

Figures may not add to totals because of independent rounding. Mgal/d = million gallons per day; gal/d = gallons per day

Source: Solley[65]

Table 1.4.2 Groundwater Withdrawals by Water-Use Category and Water-Resources Region, 1995

Region	Public supply	Domestic	Commer-cial	Irrigation	Live-stock	Industrial		Mining		Thermo-electric	Total	
	Fresh	Fresh	Fresh	Fresh	Fresh	Fresh	Saline	Fresh	Saline	Fresh	Fresh	Saline
New England	335	168	64	47	6.4	53	0	2.9	0	48	725	0
Mid-Atlantic	1,270	485	217	128	79	344	0	159	1.0	11	2,690	1.0
South Atlantic–Gulf	2,760	719	114	2,280	188	787	0	177	9.1	79	7,110	16
Great Lakes	585	354	44	170	50	270	3.6	34	1.0	7.6	1,510	4.6
Ohio	880	323	91	61	60	379	0	115	22	70	1,980	22
Tennessee	125	64	3.6	8.7	19	35	0	3.7	0	0	258	0
Upper Mississippi	1,150	311	94	430	216	328	0	22	4.2	24	2,570	4.2
Lower Mississippi	741	73	15	6,930	740	611	0	3.1	0	69	9,180	0
Souris–Red–Rainy	34	17	.2	45	17	1.7	0	.4	0	0	115	0
Missouri Basin	643	137	19	8,030	253	102	0	104	38	30	9,320	38
Arkansas–White–Red	378	105	16	6,660	190	78	0	30	284	37	7,490	284
Texas–Gulf	978	115	34	4,370	82	214	.5	118	324	50	5,960	324
Rio Grande	356	25	17	1,420	27	10	0	53	60	16	1,930	61
Upper Colorado	35	11	5.6	38	4.2	2.4	0	20	14	0	116	14
Lower Colorado	476	44	22	2,210	33	42	0	126	12	45	3,000	12
Great Basin	350	13	10	1,090	9.2	60	.1	71	19	2.6	1,610	56
Pacific Northwest	917	253	37	4,030	44	215	0	6.5	0	.5	5,500	0
California	2,730	112	77	10,900	231	522	10	16	151	3.6	14,600	185
Alaska	30	8.3	11	.1	.1	3.8	0	0	75	4.2	58	75
Hawaii	200	2.4	45	173	7.5	19	.9	.5	0	67	515	16
Caribbean	95	6.4	1.3	33	4.5	10	.2	3.4	0	2.2	156	.2
Total	15,100	3,350	939	49,000	2,260	4,090	15	1,070	1,010	565	76,400	1,110

Figures may not add to totals because of independent rounding. All values in million gallons per day

Source: Solley[65]

The largest single demand for groundwater is irrigation, amounting to 67.3% percent of all groundwater used in 1995. More than 90 percent of this water is pumped in the western states, where arid and semiarid conditions have fostered extensive irrigation development.

1.5   GROUNDWATER IN THE HYDROLOGIC CYCLE

1.5.1   Hydrologic Cycle

The central focus of hydrology is the *hydrologic cycle* consisting of the continuous processes shown in Figure 1.5.1. Water evaporates from the oceans and land surfaces to become water vapor that is carried over the earth by atmospheric circulation. The *water vapor* condenses and *precipitates* on the land and oceans. The precipitated water may be *intercepted* by vegetation, become overland flow over the ground surface, *infiltrate* into the ground, flow through the soil as *subsurface flow,* or discharge as *surface runoff.* Evaporation from the land surface comprises evaporation directly from soil and vegetation surfaces, and *transpiration* through plant leaves. Collectively these processes are called *evapotranspiration.* Infiltrated water may percolate deeper to recharge groundwater and later become *springflow* or *seepage into streams to also become streamflow.*

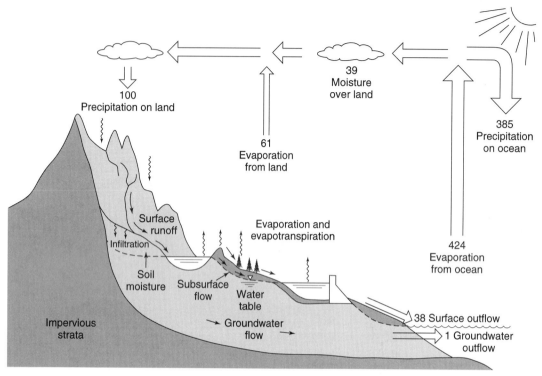

Figure 1.5.1. Hydrologic cycle with global annual average water balance given in units relative to a value of 100 for the rate of precipitation on land.[22]

A *hydrologic system* is defined as a structure or volume in space, surrounded by a boundary, that accepts water and other inputs, operates on them internally, and produces them as outputs.[21, 22] The structure (for surface or subsurface flow) or volume in space (for atmospheric moisture flow) is the totality of the flow paths through which the water may pass as throughput from the point it enters the system to the point it leaves. The boundary is a continuous surface defined in three dimensions enclosing the volume or structure. A *working medium* enters the system as input, interacts with the structure and other media, and leaves as output. Physical, chemical, and biological processes operate on the working media within the system; the most common working media involved in hydrologic analysis are water, air, and heat energy.

The *global hydrologic cycle* can be represented as a system containing three subsystems: the *atmospheric water system*, the *surface water system*, and the *subsurface water system* as shown in Figure 1.5.2. Another example is the storm–rainfall–runoff process on a watershed, which can be represented as a hydrologic system. The input is rainfall distributed in time and space over the watershed and the output is streamflow at the watershed outlet. The boundary is defined by the watershed divide and extends vertically upward and downward to horizontal planes.

Drainage basins, catchments, and *watersheds* are three synonymous terms that refer to the topographic area that collects and discharges surface streamflow through one outlet or mouth. Catchments are typically referred to as small drainage basins but no specific area limits have been established. The drainage basin divide, watershed divide, or catchment divide is the line dividing land whose drainage flows toward the given stream from land whose drainage flows away from that stream. Think of drainage basin sizes ranging from the Mississippi River drainage basin to a small urban drainage basin in your local community or some small valley in the countryside near you.

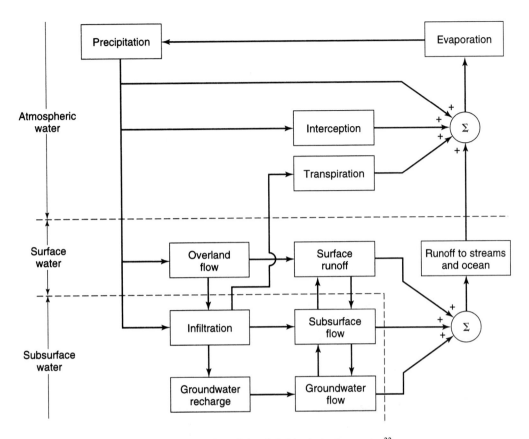

Figure 1.5.2. Block-diagram representation of the global hydrologic system.[22]

1.5.2   The Groundwater System in the Hydrologic Cycle

Groundwater constitutes one portion of the earth's water circulatory system known as the hydrologic cycle. Figure 1.5.1 illustrates some of the many facets involved in this cycle. Water-bearing formations of the earth's crust act as conduits for transmission and as reservoirs for storage of water. Water enters these formations from the ground surface or from bodies of surface water, after which it travels slowly for varying distances until it returns to the surface by action of natural flow, plants, or humans. The storage capacity of groundwater reservoirs combined with small flow rates provide large, extensively distributed sources of water supply. Groundwater emerging into surface stream channels aids in sustaining streamflow when surface runoff is low or nonexistent. Similarly, water pumped from wells represents the sole water source in many regions during much of every year.

Practically all groundwater originates as surface water. Principal sources of natural recharge include precipitation, streamflow, lakes, and reservoirs. Other contributions, known as artificial recharge, occur from excess irrigation, seepage from canals, and water purposely applied to augment groundwater supplies. Even seawater can enter underground along coasts where hydraulic gradients slope downward in an inland direction. Water within the ground moves downward through the unsaturated zone under the action of gravity, whereas in the saturated zone it moves in a direction determined by the surrounding hydraulic situation.

Discharge of groundwater occurs when water emerges from underground. Most natural discharge occurs as flow into surface water bodies, such as streams, lakes, and oceans; flow to the surface appears as a spring. Groundwater near the surface may return directly to the atmosphere by evaporation from within the soil and by transpiration from vegetation. Pumpage from wells constitutes the major artificial discharge of groundwater.

In this section we will discuss in general some of the aspects of the movement of groundwater in the hydrologic cycle. *The groundwater flow system* comprises the subsurface water, the geologic (porous) media containing the water, the flow boundaries, the sources (outcrop areas, streams for recharge to the aquifer), and the sinks (springs, interaquifer flow, and wells for flow from the aquifer). Water flows through and is stored within the groundwater system. Under natural conditions, the *travel time* of groundwater can range from less than a day to more than a million years. The *age* of the water can range from recent precipitation to water trapped in sediments that were deposited in geologic time. Chapter 3, Groundwater Movement, discusses the mechanics of groundwater movement.

The water that is withdrawn from a groundwater system initially comes from storage. The effects of the withdrawal from storage are propagated through the system, over time, as *water heads* (the *water table*) decrease at greater distances from the point of withdrawal. Ultimately, the effect of the withdrawal reaches a boundary such as a stream. At the stream, either increased *recharge* (*water added*) to the groundwater system occurs or increased discharge from the system occurs. Figure 1.5.3 shows the sources of water supplying pumpage from ten major regional aquifer systems in the United States. The figure illustrates the variability of aquifer response to long-term pumping and the extent to which changes in recharge and discharge can exceed changes in storage. It is important to quantify recharge, despite the difficulty of that undertaking.

Typically, most water from precipitation that infiltrates does not become recharge, but instead is stored in the soil zone and is eventually returned to the atmosphere by evaporation and plant transpiration.[2] The percentage of precipitation that becomes diffuse recharge is highly variable and depends upon many factors, such as depths to the water table, properties of surface soils, aquifer properties, and many other factors.

Interactions of surface-water systems with groundwater systems depend upon many factors, including positions of the surface-water systems relative to the groundwater systems; characteristics of the surface-water systems and their underlying materials; and the climate setting.[2, 80] Figure 1.5.4 illustrates the effect of *transient recharge* from precipitation on the configuration of a water table and the associated groundwater flow. The exchange of water across the interface between surface water and groundwater can result from downstream movement of water in and out of streambeds and banks, as illustrated in Figure 1.5.5. Other exchanges result from tides, wave action, filling or draining of reservoirs, and transpiration from vegetation at the edges of wetlands and other surface water sources. Most studies of exchanges have focused on streams.

Flows within groundwater systems can be on a local, intermediate, and regional basis, as illustrated in Figure 1.5.6. The recharge and discharge areas in a *local system of groundwater flow* are adjacent to each other. The recharge and discharge in an *intermediate groundwater flow system* are separated by one or more topographic high and low. In *regional groundwater flow* systems, recharge areas are along groundwater divides and discharge areas are located at the bottom of major drainage divides. Not every aquifer has each of these types of flow systems.[33, 71] In an aquifer system, the largest amount of groundwater flow is commonly in the local flow systems which are mostly affected by seasonal variations in recharge. Recharge areas of these local systems make up the largest part of the surface of a drainage basin, are relatively shallow, and have transient conditions. Regional groundwater flow systems are less transient than local and intermediate flow systems.

A *conceptual model of an aquifer system,* as illustrated in Figure 1.5.6 for the Midwestern Basins and Arches aquifer system (see Figure 1.5.7 for location), is a simplified qualitative

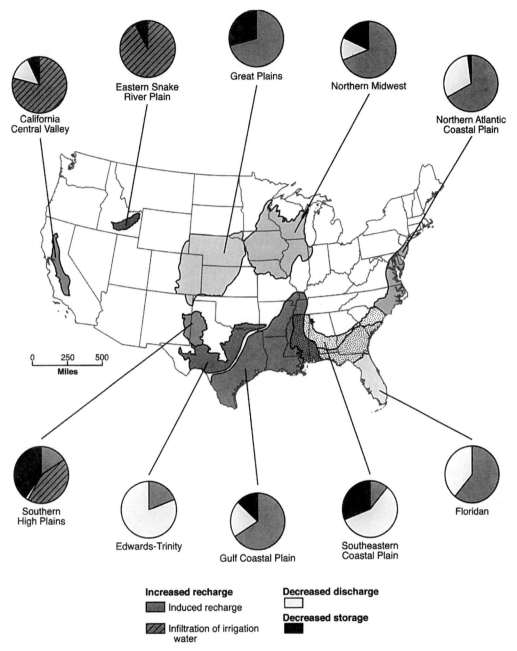

Figure 1.5.3. Sources of water that supply withdrawals from major aquifer systems in the United States are highly variable, as shown by these results from model simulations for various periods (Johnston, 1997). The Floridan and Edwards–Trinity aquifer systems, which equilibrate rapidly after pumping, were simulated as steady-state with no long-term change in storage. In contrast, the Southern High Plains (with most natural discharge occurring far from pumping wells) and the deeply buried Great Plains aquifer system have had substantial changes in groundwater storage. The distinction between changes in recharge and changes in discharge is a function of how the system was defined (i.e., a gain to one system may result in a loss from an adjoining system). For example, groundwater withdrawals from confined aquifers (Northern Atlantic Coastal Plain, Gulf Coastal Plain) can cause flow to be diverted (recharged) into the deeper regional flow regime that would otherwise discharge to streams in the outcrop areas or cause vertical leakage across confining units. Groundwater recharge in a region can be increased as a result of human modifications, such as return flow of excess irrigation water (California Central Valley). Note that the areal extent of the Southeastern Coastal Plain aquifer system overlaps the areal extents of the Floridan and Gulf Coastal Plain aquifer systems.[2]

description of the physical system.[33] Conceptual models may include a description of the aquifers and confining units that make up the aquifer system, the boundary conditions, flow regimes, sources and sinks of water, and general directions of the groundwater flow.

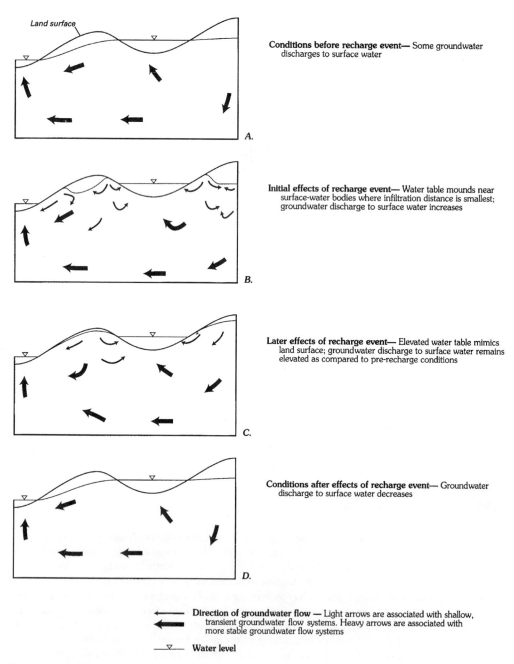

Conditions before recharge event— Some groundwater discharges to surface water

A.

Initial effects of recharge event— Water table mounds near surface-water bodies where infiltration distance is smallest; groundwater discharge to surface water increases

B.

Later effects of recharge event— Elevated water table mimics land surface; groundwater discharge to surface water remains elevated as compared to pre-recharge conditions

C.

Conditions after effects of recharge event— Groundwater discharge to surface water decreases

D.

Direction of groundwater flow — Light arrows are associated with shallow, transient groundwater flow systems. Heavy arrows are associated with more stable groundwater flow systems

▽ **Water level**

Figure 1.5.4. Diagrams showing the effect of transient recharge from precipitation on the configuration of a water table and associated groundwater flow.[33]

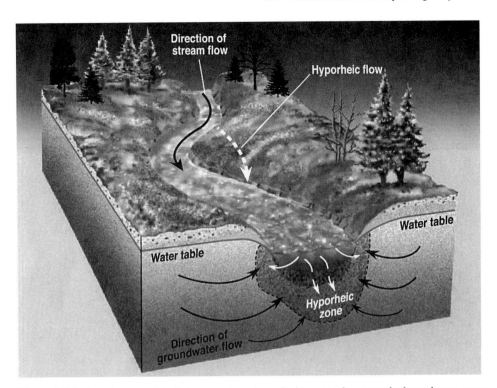

Figure 1.5.5. Local geomorphic features such as streambed topography, streambed roughness, meandering, and heterogeneities in sediment hydraulic conductivities can give rise to localized flow systems within streambeds and banks. The near-stream subsurface environment with active exchange between surface water and groundwater is commonly referred to as the *hyporheic zone*, although the transition between groundwater and surface water represents a hydrologic continuum, preventing a precise separation.[2]

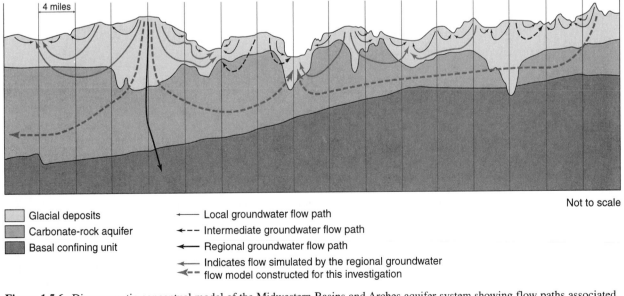

Glacial deposits
Carbonate-rock aquifer
Basal confining unit

Local groundwater flow path
Intermediate groundwater flow path
Regional groundwater flow path
Indicates flow simulated by the regional groundwater flow model constructed for this investigation

Not to scale

Figure 1.5.6. Diagrammatic conceptual model of the Midwestern Basins and Arches aquifer system showing flow paths associated with local, intermediate, and regional flow systems[71] and flow systems simulated by the regional groundwater flow model.[33]

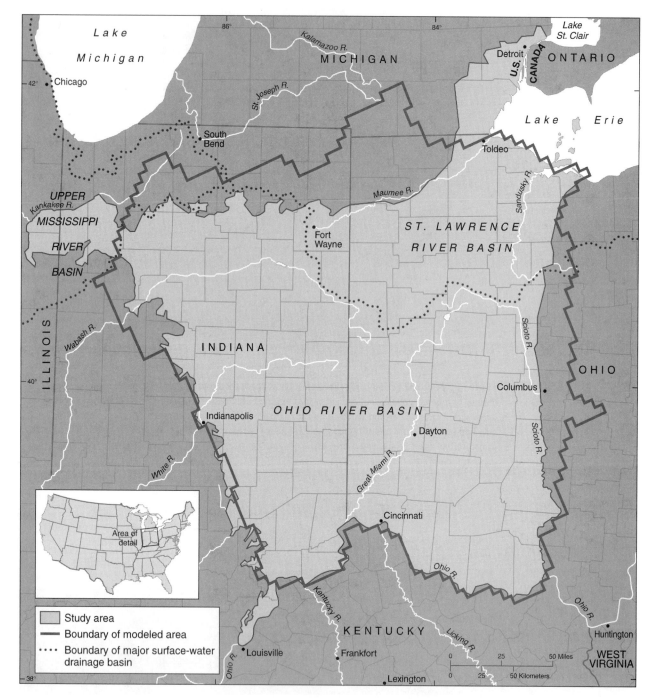

Figure 1.5.7. Midwestern Basin and Arches aquifer system in Parts of Indiana, Ohio, Michigan, and Illinois.[33]

1.6 HYDROLOGIC BUDGET

A *hydrologic budget, water budget,* or *water balance* is a measurement of continuity of the flow of water, which holds true for any time interval and applies to any size area ranging from local-scale areas to regional-scale areas or from any drainage area to the earth as a whole. The hydrologists usually must consider an open system, for which the quantification of the hydro-

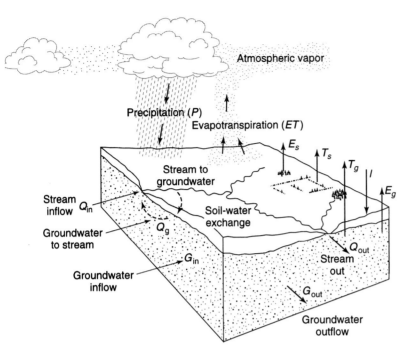

Figure 1.6.1. Components of hydrologic cycle in an open system: the major inflows and outflows of water from a parcel of land. *Source:* W.M. Marsh and J. Dozier, *Landscape: An Introduction to Physical Geography.* Copyright © 1986 by Wiley. Reprinted by permission of John Wiley & Sons, Inc.

logic cycle for that system becomes a mass balance equation in which the change of storage of water (dS/dt) with respect to time within that system is equal to the inputs (I) to the system minus the outputs (O) from the system.

Considering the open system in Figure 1.6.1, the water balance equation can be expressed for the surface water system and the groundwater system in units of volume per unit time separately, or for a given time period and area, in depth.

Surface Water System Hydrologic Budget

$$P + Q_{in} - Q_{out} + Q_g - E_s - T_s - I = \Delta S_s, \tag{1.6.1}$$

where P is the precipitation, Q_{in} is the surface water flow into the system, Q_{out} is the surface water flow out of the system, Q_g is the groundwater flow into the stream, E_s is the surface evaporation, T_s is the transpiration, I is the infiltration, and ΔS_s is the change in water storage of the surface water system.

Groundwater System Hydrologic Budget

$$I + G_{in} - G_{out} - Q_g - E_g - T_g = \Delta S_g, \tag{1.6.2}$$

where G_{in} is the groundwater flow into the system, G_{out} is the groundwater flow out of the system, and ΔS_g is the change in groundwater storage. The evaporation, E_g, and the transpiration, T_g, can be significant if the water table is near the ground surface.

System Hydrologic Budget

The system hydrologic budget is developed by adding the above two budgets together:

$$P - (Q_{out} - Q_{in}) - (E_s + E_g) - (T_s + T_g) - (G_{out} - G_{in}) = \Delta(S_s + S_g) \tag{1.6.3}$$

Using net mass exchanges, the above system hydrologic budget can be expressed as

$$P - Q - G - E - T = \Delta S \tag{1.6.4}$$

(a)

Figure 1.6.2. Diagrams illustrating water budgets for a groundwater system for predevelopment and development conditions.[3] (a) Predevelopment water-budget diagram illustrating that inflow equals outflow. (b) Water-budget diagram showing changes in flow for a groundwater system being pumped. The sources of water for the pumpage are changes in recharge, discharge, and the amount of water stored. The initial predevelopment values do not directly enter the budget calculation.

(b)

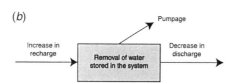

Hydrologic budgets can be used for numerous studies related to groundwater including

- Estimating groundwater exchange with lakes.[32, 39, 46, 60, 63, 64, 72]
- Estimating surface water and groundwater interaction.[38, 42, 62]
- Computing recharge from a well-hydrograph data.[39]

A diagram illustrating water budgets for a groundwater system for predevelopment and development conditions is presented in Figure 1.6.2. A groundwater budget for part of Nassau and Suffolk Counties, Long Island, New York, is shown in Figure 1.6.3. Both of these water budgets assume equilibrium conditions with little or no change in storage.

EXAMPLE 1.6.1

During 1996, the water budget terms for Lake Annie in Florida[60] included precipitation (P) of 43 inch/yr, evaporation (E) of 53 inch/yr, surface water inflow (Q_{in}) of 1 inch/yr, surface outflow (Q_{out}) of 173 inch/yr, and change in lake volume (ΔS) of −2 inch/yr. Determine the net groundwater flow (the groundwater inflow minus the groundwater outflow).

SOLUTION

Assuming $T_g = 0$, the water budget equation (1.6.4) to define the net groundwater flow for the lake is

$$G = \Delta S - P + E - Q_{in} + Q_{out}$$

$$= -2 - 43 + 53 - 1 + 173$$

$$= 180 \text{ inch/yr}$$ ∎

EXAMPLE 1.6.2

During January 1996, the water-budget terms for Lake Annie in Florida[60] included precipitation (P) of 1.9 inch, evaporation (E) of 1.5 inch, surface water inflow (Q_{in}) of 0 inch, surface outflow (Q_{out}) of 17.4 inch, and change in lake volume (ΔS) of 0 inch. Determine the net groundwater flow for January 1996 (the groundwater inflow minus the groundwater outflow).

SOLUTION

The water budget equation to define the net groundwater flow for the lake is

$$G = \Delta S - P + E - Q_{in} + Q_{out} = 0 - 1.9 + 1.5 - 0 + 17.4 = 17 \text{ inch for January 1996}$$ ∎

OVERALL PREDEVELOPMENT WATER-BUDGET ANALYSIS

INFLOW TO LONG ISLAND HYDROLOGIC SYSTEM	CUBIC FEET PER SECOND
1. Precipitation	2,475
OUTFLOW FROM LONG ISLAND HYDROLOGIC SYSTEM	
2. Evapotranspiration of precipitation	1,175
3. Ground-water discharge to sea	725
4. Streamflow discharge to sea	525
5. Evapotranspiration of ground water	25
6. Spring flow	25
Total outflow	2,475

GROUND-WATER PREDEVELOPMENT WATER-BUDGET ANALYSIS

INFLOW TO LONG ISLAND GROUND-WATER SYSTEM	CUBIC FEET PER SECOND
7. Ground-water recharge	1,275
OUTFLOW FROM LONG ISLAND GROUND-WATER SYSTEM	
8. Ground-water discharge to streams	500
9. Ground-water discharge to sea	725
10. Evapotranspiration of ground water	25
11. Spring flow	25
Total outflow	1,275

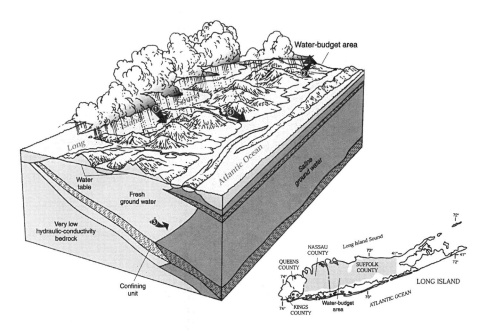

Figure 1.6.3. Groundwater budget for part of Nassau and Suffolk Counties, Long Island, New York.[3] Block diagram of Long Island, New York, and tables listing the overall water budget and groundwater budget under predevelopment conditions. Both water budgets assume equilibrium conditions with little or no change in storage.

The components of hydrologic budgets are either measured, calculated, or estimated. Evaporation, for example, may be obtained from measured pan evaporation data or calculated from the energy balance method, the aerodynamic method (such as the Thornthwaite–Holzman equation), or a combination of these methods (such as the Priestley–Taylor evaporation method). Detailed discussions of these methods are presented in References 22, 49, and 52. Precipitation data are measured or is obtained from recorded data, and in some cases are estimated based upon recorded data for other time periods of interest. Depending upon the time period of the budget, average monthly values could also be used. In the case when the hydrologic budgets are of lakes, lake volume changes are needed and they can be computed using actual lake-stage measurements and relationships between lake stage and lake volume. Using estimates of lake volume changes, evaporation, and precipitation, estimates of net groundwater flow to a lake can be made. The individual components of net groundwater flow are the groundwater inflow to the lake and the leakage from the lake, which can be determined through groundwater simulation models. Groundwater simulation models (Chapter 9), however, need to be calibrated, which requires the monitoring of groundwater levels (hydraulic heads, Chapter 2), lithographic data (Chapter 12), results (hydraulic conductivities) from pump tests (Chapter 4), or slug tests (Chapter 5). With a calibrated groundwater flow model, the groundwater flow into a lake and the leakage from a lake can be determined using simulated groundwater flow fields.

CASE STUDY *Lake Five-O, Florida*

Grubbs[39] described the hydrologic budgetary analysis that was performed for Lake Five-O (a seepage lake), located in Bay County in northwestern Florida (see Figures 1.6.4 and 1.6.5). This hydrologic budget (Table 1.6.1 and Figure 1.6.6) was determined using both collected data and model simulation results (see Chapter 9). Lake Five-O is located in an area characterized by coastal terrace deposits that have been modified by extensive karst development. Soils in the area are deep, excessively drained, and consist of very permeable, Lakeland series sands. Maximum lake depth ranged from 13.5 m to 15.4 m. Surface area ranged from 10.4 to 11.3 hectares, and the lake volume ranged from 9.09×10^5 to 1.11×10^6 m^3.

The net groundwater flow (groundwater inflow minus leakage) was determined by the water balance approach utilizing measurements of precipitation, lake evaporation was determined using the energy budget method, and lake volume changes were estimated from lake-stage measurements and a relationship between lake stage and lake volume. The analysis was utilized to make qualitative assessments of the significance of lake–groundwater exchanges during the study period. A groundwater simula-

tion model was used to determine the groundwater inflows and seepage from the lake. The simulation model was developed using lithographic data to define the three geohydrologic units, a network of monitoring wells to define the hydraulic heads over the time of the hydrologic budget, previously published data, and limited slug tests to help determine hydraulic conductivities. The analysis made it possible to develop quantitative estimates of minimum groundwater inflow and leakage rates not only during the study period, but also for long-term average conditions.

The hydrologic budget for Lake Five-O is expressed as

$$\Delta S = P - E + Q_{in} - Q_{out}$$

The hydrologic budget by Grubbs[39] showed that the groundwater inflow to the lake and leakage from the lake to the groundwater system are the dominant components, respectively, in total inflow (precipitation plus groundwater inflow) and total outflow (evaporation plus leakage) budgets of the lake. The groundwater movement, including the head distribution and groundwater flow near Lake Five-O, is discussed in Section 3.6.6.

LOCATION OF BAY COUNTY

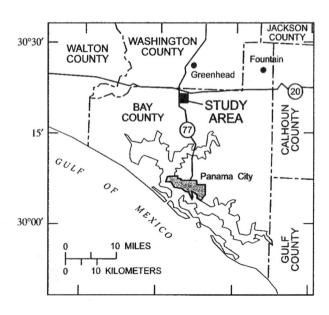

Figure 1.6.4. Location of Lake Five-O study area.[39]

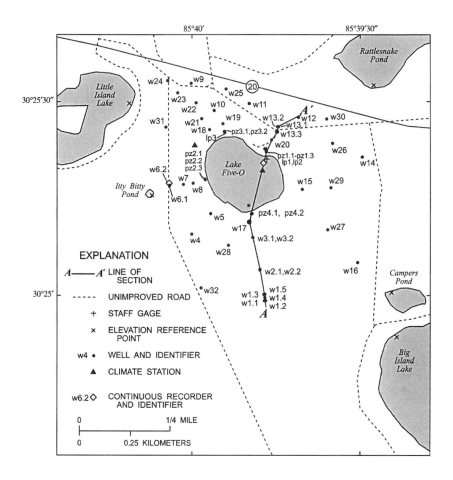

Figure 1.6.5. Location of data-collection sites in the Lake Five-O study area.[39]

Table 1.6.1 Monthly Net Groundwater Flow to Lake Five-O, 1989–90[39]

Month	Average lake volume	Change in lake volume	Precipitation	Evaporation	Net ground-water flow	Standard deviation	Standard deviation (percent of net ground-water flow)
				1989			
Jan.	989,700	−26,100	3,800	3,200	−26,700	2,300	9
Feb.	964,000	−21,900	6,900	4,100	−24,700	2,600	10
Mar.	943,300	−13,900	17,700	7,900	−23,700	3,300	14
Apr.	935,200	−8,700	10,600	11,000	−8,300	2,800	34
May	922,500	−14,700	8,900	13,800	−9,900	2,900	29
June	945,600	69,100	56,800	12,000	24,400	9,300	38
July	1,024,800	66,700	26,300	13,300	53,700	5,400	10
Aug.	1,068,400	26,200	20,900	12,300	17,600	3,700	21
Sept.	1,091,800	22,300	21,500	16,800	17,700	3,700	21
Oct.	1,103,400	3,100	15,100	13,000	1,000	2,700	270
Nov.	1,108,200	4,100	13,800	9,400	−300	2,300	770
Dec.	1,102,700	1,000	13,500	7,900	−4,600	2,300	50

(*continues*)

Table 1.6.1 (*continued*) Monthly Net Groundwater Flow to Lake Five-O, 1989–90

	Average lake volume	Change in lake volume	Precipitation	Evaporation	Net ground-water flow	Standard deviation	Standard deviation (percent of net ground-water flow)
Month							
			1990				
Jan.	1,107,000	−3,100	3,800	3,300	−3,500	1,500	43
Feb.	1,103,000	−4,800	5,700	4,800	−5,800	1,300	22
Mar.	1,092,000	−15,500	9,600	8,600	−16,500	2,000	12
Apr.	1,076,100	−17,800	12,000	11,100	−18,700	2,400	13
May	1,058,600	−22,000	9,400	15,100	−16,300	2,400	15
June	1,041,300	−9,100	23,700	14,300	−18,500	4,200	23
July	1,046,400	17,900	31,300	15,000	1,600	5,300	330
Aug.	1,055,600	−2,400	16,000	15,600	−2,800	3,600	130
Sept.	1,033,300	−35,600	3,900	17,900	−21,600	2,900	13
Oct.	998,600	−35,100	5,500	15,400	−25,200	2,800	11
Nov.	963,200	−35,600	800	11,300	−25,100	2,600	10
Dec.	932,000	−28,900	7,000	7,400	−28,500	2,600	9

All units are in cubic meters, unless otherwise noted. Standard deviation is the error component of the net groundwater flow estimate. Negative values of net groundwater flow indicate that leakage exceeded groundwater inflow.

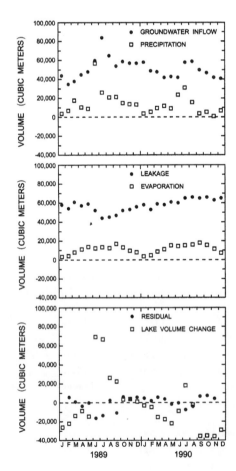

Figure 1.6.6. Monthly hydrologic budget components for Lake Five-O, 1989–90.[39]

The components of a hydrologic budget, whether they are measured or calculated, have associated errors. These errors are based on the degree of uncertainty of the measurements, limitations of methods, and the assumptions made to calculate the values. In many situations information on the rainfall, surface outflow, and withdrawal for supply may be the most reliable. Calculated values of lake evaporation and lake storage may be less reliable because off-site pan evaporation and estimated pan coefficients are used. Lake stage data may be less reliable because of the uncertainty in assessing the surface area of the lake. Groundwater flow is typically the least reliable. When the measured or calculated components are used in the budget calculations, imbalances between the inflow and outflow components, called *residuals*, occur. The residual term in a hydrologic budget is an accumulation of all the errors in the components of the budget. The previous hydrologic budget equations do not reflect residual terms; however, the analysis by Grubbs[39] describes the residuals in detail.

1.7   PUBLICATION SOURCES

1.7.1   Internet Resources

The Internet has changed the availability of sources of information on groundwater hydrology. Now we can access many types of data and publications very rapidly through the use of a computer. Appendix A provides a list of U.S. government and nongovernmental organizations' Web sites.

Much of this book has been developed to serve as a portal to the vast resources on groundwater hydrology that now exist on the Internet. Hopefully this book will help guide the student, the professional, and the researcher to the general documents, program plans, field project details, software, and other information found on the Internet. Many of the new figures and tables in this edition have been taken from various Web sites, in particular the U.S. Geological Survey site. These items not only provide specific information but also serve as samples of graphics and tabulations available on the Internet, making the style of this book more variable than that of most textbooks. The majority of Web sites referenced in this book are maintained by government agencies and established organizations, and therefore should be accessible for many years to come. Several end-of-chapter problems throughout the book are based upon Internet exercises.

1.7.2   U.S. Geological Survey Publications

In the United States, a majority of the field measurements and investigations of groundwater have been conducted by the U.S. Geological Survey (USGS). Most work has been on a cooperative basis with individual states. Results are published by the Survey as *circulars,* digital data series, fact sheets, hydrologic atlases, open-file reports, techniques of water-resources, water data reports, water-resources investigation reports, *professional papers,* and *water-supply papers.* Table 1.7.1 describes the various types of U.S. Geological Survey reports, and Table 1.7.2 describes how to find and reference on-line USGS water resources reports. Since 1935 records of groundwater measurements in key observation wells have been published in *water-supply papers* under the title *Groundwater Levels in the United States.* The U.S. Geological Survey publishes at irregular intervals other papers on the geology and groundwater resources of local areas. Invariably, these intensive investigations concern areas containing important groundwater problems and are carried out in cooperation with local agencies.

Table 1.7.1 Types of U.S. Geological Survey Reports

Circulars—Purpose: To present to general or scientific audiences short summaries or articles of short-term, popular, or local interest.

Digital Data Series—The Digital Data Series encompasses a broad range of digital data, including computer programs, interpreted results of investigations, comprehensive reviewed databases, spatial data sets, digital images and animation, and multimedia presentations that are not intended for printed release. Scientific reports in this series cover a wide variety of subjects and facets of U.S. Geological Survey investigations and research that are of lasting scientific interest and value. Releases in the Digital Data Series offer access to scientific information that is available in digital form; the information is primarily for viewing, processing, and (or) analyzing by computer.

Fact Sheets—Purpose: To describe Water Resources Division (WRD) programs, projects, products, and water-resources topics to either a general or professional audience. Water Fact Sheets are concise and timely publications that increase the understanding and visibility of WRD activities and accomplishments.

Hydrologic Atlases—Purpose: To present reports on hydrology or geohydrology in map format to a wide range of hydrologically oriented audiences.

Open-File Reports—Purpose: To make available (1) data reports, (2) reports preliminary findings that would be of interest to few persons other than the cooperating agency, (3) reports and maps pending publication elsewhere but requiring immediate release, and (4) timely information describing programs, projects, products, and water-resources topics.

Professional Papers—Purpose: To present comprehensive or topical reports on any field in the earth sciences. This series is commonly used for summaries of wide popular, scientific, or geographic interest, and for significant scientific contributions—generally on topics other than hydrology.

Techniques of Water-Resources—Purpose: To present to technically oriented audiences reports on methods and techniques used in collecting, analyzing, and processing hydrologic data.

Water Data Reports—A series of annual reports that document hydrologic data gathered from the U.S. Geological Survey's and cooperating agencies' surface and groundwater data-collection networks in each state, Puerto Rico, and Trust Territories. These records of streamflow, groundwater levels, and water quality provide the hydrologic information needed by state, local, and federal agencies, and the private sector for developing and managing our Nation's land and water resources.

Water-Resources Investigations Reports—Purpose: To (a) present to interdisciplinary audiences comprehensive or topical interpretive reports and maps that are mainly of local or short-term interest; (b) provide a medium of release for reports and maps that would not be feasible in any other series or journal or that would be published quickly.

Water-Supply Papers—Purpose: To present significant interpretive results of hydrologic investigations of broader than local interests.

Water Errata Sheets—Changes made to reports after publication.

Source: http://water.usgs.gov/pub.html

Table 1.7.2 How to Find and Reference Online USGS Water Resources Reports

Many USGS reports on water resources topics are now being served online. You can access them by their series and number. For example, to see Circular 1123, you should enter CIR 1123 in the search box at the following URL: http://water.usgs.gov/pubs.

Constructing a reference

Use one of these prefixes for the report series, followed by the report number. Dashes and underlines are permitted (and ignored), but blanks are not allowed. Case is insensitive.

Report Series	Preferred Prefix	Alternate Prefix(es) Allowed
Fact Sheet	FAC	FS FACT FS_
Open-File Report	OFR	OF
Water-Resources Investigation	WRI	WR WRIR
Professional Paper	PRO	PROF PP
Water-Data Report	WDR	WD DATA
Circular	CIR	CIRC C
Water-Supply Paper	WSP	WS
Bulletin	BUL	BULL
Techniques of Water-Resources Investigations	TWRI	TWRI

Table 1.7.2 (continued) How to Find and Reference Online USGS Water Resources Reports

Referencing parts of a document

Using only the series and number will reference the home page (index.html) of the document. You also can reference a specific part of a document in html format, even if it's not on the home page. For example, to find the section named *HRD4* in a file of Circular 1123 called *overview.html*, use this URL: http://pubs.water.usgs.gov/cir1123/overview.html#HDR4. This technique will work for any sub-page or figure of the html document.

The "pubs.water.usgs.gov" reference is persistent!

The USGS is committed to supporting this referencing system for the indefinite future. This means you can safely incorporate a "pubs.water.usgs.gov" reference in your Web pages and even in your printed documents and it will still work many years later.

When viewing a document, your browser's location may show you another URL that corresponds to the current physical location of the document. Do not use this physical location as a "persistent" reference! As our system grows, these locations will change but the "pubs.water.usgs.gov" reference will not.

Source: http://water.usgs.gov/pubs/referencing.html

1.7.3   Publications

The following journals provide articles on various topics of groundwater:

Environmental Science and Technology, American Chemical Society

Ground Water, National Ground Water Association

Groundwater Management, Water Well Journal Publishing Co.

Ground Water Monitoring and Remediation, Groundwater Publishing Company

Hydrological Science and Technology, American Institute of Hydrology

Journal of the American Water Resources Association, American Water Resources Association

Journal of Contaminant Hydrology, Elsevier Scientific Publishers

Journal of Hydraulics, American Society of Civil Engineers

Journal of Hydrology, Elsevier Scientific Publishers

Journal of Water Resources Planning and Management, American Society of Civil Engineers

There have been many books published on the subject of groundwater. Earlier books include References 4, 5, 7, 17, 19, 23, 26, 29–31, 40, 41, 43, 56, 58, 59, 61, 67, 70, 73, and 74.

Over the past three decades there are several previous books that have been published on groundwater including *Applied Hydrogeology*[35]; *Aquifer Hydraulics*[8]; *Contaminant Hydrogeology*[34]; *Groundwater*[36]; *Ground Water Contamination: Transport and Remediation*[14]; *Groundwater Engineering*[45]; *Groundwater Hydrology*[16]; *Groundwater Hydrology*[69]; *Groundwater Hydraulics and Pollutant Transport*[20]; *Groundwater Mechanics*[66]; *Groundwater Science*[36]; *Groundwater Systems Planning and Management*[77]; *The Handbook of Groundwater Engineering*[27]; *Hydraulics of Groundwater*[10]; *Manual of Applied Field Hydrogeology*[76]; *Modeling Groundwater Flow and Pollution*[11]; *Principles of Groundwater Engineering*[75]; and *Quantitative Hydrogeology.*[51]

1.8   DATA SOURCES

Table 1.8.1 lists the principal types of data and data compilations that are required for the analysis of groundwater systems. The lists are for the physical framework, hydrologic budgets and stresses, and the chemical framework. One of the sources of data used frequently by groundwater hydrologists is the U.S. Geological Survey NWIS system.

Table 1.8.1 Principal Types of Data and Data Compilations Required for Analysis of Groundwater Systems

Physical framework
Topographic maps showing the stream drainage network, surface-water bodies, landforms, cultural features, and locations of structures and activities related to water
Geologic maps of surficial deposits and bedrock
Hydrogeologic maps showing extent and boundaries of aquifers and confining units
Maps of tops and bottoms of aquifers and confining units
Saturated-thickness maps of unconfined (water-table) and confined aquifers
Average hydraulic conductivity maps for aquifers and confining units and transmissivity maps for aquifers
Maps showing variations in storage coefficient for aquifers
Estimates of age of groundwater at selected locations in aquifers

Hydrologic budgets and stresses
Precipitation data
Evaporation data
Streamflow data, including measurements of gain and loss of streamflow between gaging stations
Maps of the stream drainage network showing extent of normally perennial flow, normally dry channels, and normally seasonal flow
Estimates of total groundwater discharge to streams
Measurements of spring discharge
Measurements of surface-water diversions and return flows
Quantities and locations of interbasin diversions
History and spatial distribution of pumping rates in aquifers
Amount of groundwater consumed for each type of use and spatial distribution of return flows
Well hydrographs and historical head (water-level) maps for aquifers
Location of recharge areas (areal recharge from precipitation, losing streams, irrigated areas, recharge basins and recharge wells), and estimates of recharge

Chemical framework
Geochemical characteristics of earth materials and naturally occurring groundwater in aquifers and confining units
Spatial distribution of water quality in aquifers, both areally and with depth
Temporal changes in water quality, particularly for contaminated or potentially vulnerable unconfined aquifers
Sources and types of potential contaminants
Chemical characteristics of artificially introduced waters or waste liquids
Maps of land cover/land use at different scales, depending on study needs
Streamflow quality (water-quality sampling in space and time), particularly during periods of low flow

Source: Alley et al.[3]

1.8.1   NWISWeb Data for the Nation

The U.S. Geological Survey is the principal federal water data agency in the United States. The USGS collects and disseminates about 70 percent of the water data currently being used by numerous state, local, private, and other federal agencies to develop and manage water resources. The National WATer Data STOrage and REtrieval System (WATSTORE) was established in 1972 to provide an effective and efficient means for the processing and maintenance of water data collected through the USGS and to facilitate release of the data to the public. In 1976, the USGS opened WATSTORE to the public for direct access.

The U.S.G.S. National Water Information System (NWIS) has replaced WATSTORE and is referred to as NWISWeb (http://waterdata.usgs.gov/nwis). NWIS is a very large collection of data and information on the water resources of the United States. This database contains current and historical water data from more than 1.5 million locations across the nation. The data categories are real-time data, site information, surface water data, groundwater data, and water quality. Real-time data includes data transmitted from selected groundwater sites. Site information includes descriptive site information, including latitude, longitude, well depth, aquifer, and site use. The groundwater database includes groundwater site inventory, groundwater level data, and water-quality data. The groundwater site inventory consists of more than 850,000 records of wells, springs, test holes, tunnels, drains, and excavations in the United States. The groundwater data can be obtained at http://waterdata.usgs.gov/nwis/gw.

1.8.2 Real-Time Data

Real-time groundwater data include data that are automatically collected, transmitted, and made available to the public at least once a day according to the U.S.G.S. (Fact Sheet 090–01, December 2001). These data can be transmitted by land-lined telephone, cellular telephone, land-based *radio frequency* (RF) technology, satellite telemetry, or a combination of these technologies. Within the U.S.G.S., satellite telemetry is the most common method for real-time data transmission. Water levels are the most common data transmitted in real time by the USGS. Figure 1.8.1 illustrates a real-time data collection and transmission system. With this method, water-level data are recorded by a *data-collection platform* (DCP) (see Figure 1.8.2) and transmitted, often on a four-hour schedule, by satellite telemetry to a U.S.G.S. ground station. The data are then displayed at http://water.usgs.gov/nwis/gw.

Real-time data have many inherent advantages over data collected and distributed by traditional means, including timeliness, data quality, data availability, and cost. Additional information on real-time groundwater data can be obtained at http://water.usgs.gov/nwis/gw or from the following address: Office of Groundwater, U.S. Geological Survey, 411 National Center, 12201 Sunrise Valley Drive, Reston, Virginia 20192, 703–648–5001.

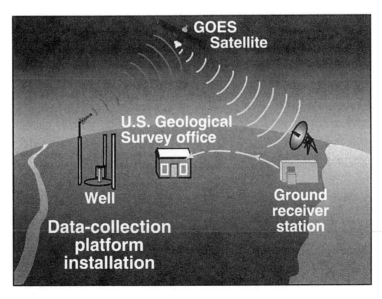

Figure 1.8.1. Real-time data collection and transmission system.[25]

Figure 1.8.2. Multiple sensor data-collection platform (DCP) installation in Kansas.[25]

PROBLEMS

1.6.1 Using the data for Lake Five-O in Table 1.6.1, show that the net groundwater flow for December 1989 is –4,600 cubic meters.

1.6.2 Using the data for Lake Five-O in Table 1.6.1, show that the net groundwater flow for December 1990 is –28,500 cubic meters.

1.6.3 Determine cumulative net groundwater flow for Lake Five-O during 1989.

1.6.4 Determine cumulative net groundwater flow for Lake Five-O during 1990.

1.7.1 Perform a search of the U.S. Geological Survey online publications (including circulars, fact sheets, open-file reports, professional papers, water-resources investigation reports, and water supply papers) to determine what studies, if any, have been performed on the regional aquifer system closest to where you live.

1.7.2 Develop an inventory of wells in the county where you live using the USGS data sources for your state. Select a well that has a time history of water levels and print the hydrograph.

1.7.3 Perform a search of U.S.G.S. publications for the topic "hydrologic budget and water budget." To perform the search, go to http://usgs-georef.cos.com. How many publications are listed?

1.7.4 Perform a search of U.S.G.S. publications for the High Plains Aquifer. To perform the search, go to http://usgs-geo ref.cos.com. How many publications are listed?

1.7.5 Perform a search of U.S.G.S. publications for the Edwards Aquifer. To perform the search, go to http://usgs-georef.cos.com. How many publications are listed?

1.7.6 Perform a search of U.S.G.S. publications for the topic of karst terrains. To perform the search go to http://usgs-geo ref.cos.com. How many publications are listed?

1.7.7 Go to the site http://water.usgs.gov/software and obtain a list of the groundwater software that the U.S.G.S. has available.

1.7.8 Write a description of the U.S.G.S. Ground-Water Resources Program. Use the site http://water.usgs.gov/ogw/GWRP.html.

REFERENCES

1. Adams, F. D., Origin of springs and rivers-an historical review, *Fennia*, v. 50, no. 1, 16 pp., 1928.

2. Alley, W. M., R. W. Healy, J. W. LaBaugh, and T. E. Reilly, Flow and storage in groundwater systems, *Science,* v. 206, June 14, 2002.

3. Alley, W. M., T. E. Reilly, and O. L. Franke, *Sustainability of Groundwater Resources*, U.S. Geological Survey circular 1186, http://water.usgs.gov/pubs/circ/circ1186, U.S. Geological Survey, Denver, CO, 1999.

4. Amer. Soc. Civil Engrs. (ASCE), *Ground Water Management*, Manual Engrng. Practice 40, New York, 216 pp., 1972.

5. American Water Works Assoc. (AWWA), *Ground Water*, AWWA Manual M21, New York, 130 pp., 1973.

6. Baker, M. N., and R. E. Horton, Historical development of ideas regarding the origin of springs and ground-water, *Trans. Amer. Geophysical Union*, v. 17, pp. 395–400, 1936.

7. Baldwin, G. V., and C. L. McGuinness, *A Primer on Ground Water*, *U.S. Geological Survey*, 26 pp., 1963.

8. Batu, V., *Aquifer Hydraulics,* Wiley Interscience, New York, 1998.

9. Bear, J., *Dynamics of Fluids in Porous Media*, Amer. Elsevier, New York, 1972.

10. Bear, J., *Hydraulics of Groundwater,* McGraw-Hill, New York, 1979.

11. Bear, J. and A. Verruijt, *Modeling Groundwater Flow and Pollution,* Reidel, Dordrecht, The Netherlands, 1987.

12. Beaumont, P., Qanats on the Varamin Plain, Iran, *Trans. Inst. British Geographers*, Publ. no. 45, pp. 169–179, 1968.

13. Beaumont, P., Qanat systems in Iran, *Bull. Intl. Assoc. Sci. Hydrology*, v. 16, pp. 39–50, 1971.

14. Bedient, P. B., H. S. Rifai, and C. J. Newell, *Ground Water Contamination: Transport and Remediation,* Prentice Hall, Englewood Cliffs, NJ, 1994.

15. Biswas, A. K., *History of Hydrology*, Amer. Elsevier, New York, 348 pp., 1970.

16. Bouwer, H., *Groundwater Hydrology*, McGraw-Hill, New York, 1978.

17. Bureau of Reclamation, *Ground Water Manual*, U.S. Dept. Interior, 480 pp., 1977.

18. Carson, R., *The Silent Spring,* Houghton Mifflin, Boston, 1962.

19. Cedergren, H. R., *Seepage, Drainage, and Flow Nets*, 2nd ed., John Wiley & Sons, New York, 534 pp., 1977.

20. Charbeneau, R. J., *Groundwater Hydraulics and Pollutant Transport,* Prentice Hall, Upper Saddle River, NJ, 2000.

21. Chow, V. T. (ed.), *Handbook of Applied Hydrology*, McGraw-Hill, New York, 1453 pp., 1964.

22. Chow, V. T., D. R. Maidment, and L. W. Mays, *Applied Hydrology,* McGraw-Hill, New York, 1988.

23. Collins, R. E., *Flow of Fluids through Porous Materials*, Reinhold, New York, 270 pp., 1961.

24. Cressey, G. B., Qanats, karez and foggaras, *Geogr. Review*, v. 48, pp. 27–44, 1958.

25. Cunningham, W.L., Real-Time Ground-Water Data for the Nation, U.S. Geological Fact Sheet 090-01, Dec. 2001.

26. Davis, S. N., and R. J. M. DeWiest, *Hydrogeology*, John Wiley & Sons, New York, 463 pp., 1966.

27. Delleur, J. W., ed., *The Handbook of Groundwater Engineering,* CRC Press, Boca Raton, FL, 1999.

28. De Villiers, M., *Water: The Fate of Our Most Precious Resource,* Mariner Books, Houghton Mifflin, Boston, 2000.

29. DeWiest, R. J. M., *Geohydrology*, John Wiley & Sons, New York, 366 pp., 1965.

30. DeWiest, R. J. M. (ed.), *Flow through Porous Media*, Academic, New York, 530 pp., 1969.

31. Domenico, P. A., *Concepts and Models in Groundwater Hydrology*, McGraw-Hill, New York, 405 pp., 1972.

32. Duwelius, R. F., Hydrologic data and hydrologic budget for Summit Lake Reservoir, Henry County, east–central Indiana, water years 1989 and 1990, 1993.

33. Eberts, S. M., and L. L. George, *Regional Groundwater Flow and Geochemistry in the Midwestern Basins and Arches Aquifer System in Parts of Indiana, Ohio, Michigan, and Illinois, U.S. Geological Survey Professional Paper* 1423-C, 2000.

34. Fetter, C. W., *Contaminant Hydrogeology,* 2nd edition, Prentice Hall, Upper Saddle River, NJ, 1999.

35. Fetter, C. W., *Applied Hydrogeology,* Prentice Hall, Upper Saddle River, NJ, 2001.

36. Fitts, C. R., *Groundwater Science,* Academic Press, San Diego, 2002.

37. Freeze, R. A., and J. A. Cherry, *Groundwater*, Prentice Hall, Englewood Cliffs, NJ, 604 pp., 1979.

38. Gronberg, J. A. M., and K. R. Belitz, *Estimation of a Water Budget for the Central Part of the Western San Joaquin Valley, California, U.S. Geological Survey Water-Resources Investigation*, WRI-91-4192, 1992.

39. Grubbs, J. W., *Evaluation of Groundwater Flow and Hydrologic Budget for Lake Five-O, A Seepage Lake in Northwestern Florida, U.S. Geological Survey Water-Resources Investigations Report* 94-4145, 1995.

40. Harr, M. E., *Groundwater and Seepage*, McGraw-Hill, New York, 315 pp., 1962.

41. Heath, R. C., and F. W. Trainer, *Introduction to Groundwater Hydrology*, John Wiley & Sons, New York, 284 pp., 1968.

42. Hedman, E. R., and Jorgenson, *Surface- and Ground-water Interaction and Hydrologic Budget of the Missouri River Valley Aquifer Between Yankton, South Dakota and St. Louis, Missouri, U.S. Geological Survey Hydrologic Investigations Atlas*, HA-0721, 1990.

43. Huisman, L., *Groundwater Recovery*, Winchester Press, New York, 336 pp., 1972.

44. Johnston, R. H., *Hydrologic Budgets of Regional Aquifer Systems of the United States—Predevelopment and Development Conditions, U.S. Geological Survey Professional Paper* No. 1425, 1997.

45. Kashef, A. A. I., *Groundwater Engineering,* McGraw-Hill, New York, 1986.

46. Lee, T. M., and A. Swancar, *The Influence of Evaporation, Groundwater and Uncertainty in the Hydrologic Budget of Lake Lucerne, A Seepage Lake in Polk County, Florida, U.S. Geological Survey Water-Supply Paper* 2439, 1997.

47. Lightfoot, D. R., The origin and diffusion of qanats in Arabia: New evidence from the northern and southern peninsula, *Geogr. Jour.*, v. 166, pp. 215–226, 2000.

48. MacKichan, K. A., Estimated use of water in the United States, 1955, *Jour. Amer. Water Works Assoc.*, v. 49, pp. 369–391, 1957.

49. Maidment, D. R., ed., *Handbook of Hydrology,* McGraw-Hill, New York, 1993.

50. Marsh, W. M., and J. Dozier, *Landscape: An Introduction to Physical Geography,* John Wiley & Sons, New York, 1986.

51. Marsily, de G., *Quantitative Hydrogeology,* Academic Press, New York, 1986.

52. Mays, L. W. (ed.), *Water Resources Handbook,* McGraw-Hill, New York, 1996.

53. Mays, L. W., *Water Resources Engineering,* John Wiley & Sons, New York, 2001.

54. Meinzer, O. E., The history and development of groundwater hydrology, *Jour. Washington Acad. Sci.,* v. 24, pp. 6–32, 1934.

55. Murray, C. R., and E. B. Reeves, *Estimated Use of Water in the United States in 1975, U.S. Geological Survey Circular* 765, 39 pp., 1977.

56. Muskat, M., *The Flow of Homogeneous Fluids through Porous Media,* McGraw-Hill, New York, 763 pp., 1937.

57. Perrault, P., *On the Origin of Springs,* trans. by A. LaRocque, Hafner, New York, 209 pp., 1957.

58. Polubarinova-Kochina, P. Y., *Theory of Groundwater Movement,* Princeton Univ. Press, Princeton, NJ, 613 pp., 1962.

59. Raudkivi, A. J., and R. A. Callander, *Analysis of Groundwater Flow,* John Wiley & Sons, New York, 214 pp., 1976.

60. Sacks, L. A., A. Swancar, and T. M. Lee, *Estimating Groundwater Exchange with Lakes Using Water-Budget and Chemical Mass-Balance Approaches for Ten Lakes in Ridge Areas of Polk and Highlands Counties, Florida, U.S. Geological Survey Water-Resources Investigations Report,* WRI-98-4133, 1998.

61. Scheidegger, A. E., *The Physics of Flow through Porous Media,* 3rd ed., Univ. of Toronto, Toronto, 353 pp., 1974.

62. Shade, P. J., *Water Budget for the Lahaina District, Island of Maui, Hawaii, U.S. Geological Survey Water-Investigations Report,* WRI-96-4238, 1996.

63. Skrobialowski, S. C. and M. J. Focazio, *Hydrologic Characteristics and Water Budgets for Swift Creek Reservoir, Virginia, 1996, U.S. Geological Survey Open-File Report* 97-0229, 1997.

64. Skrobialowski, S. C., *Hydrologic Characteristics and Water Budget for Swift Creek Reservoir, Virginia, 1997, U.S. Geological Survey Water-Resources Investigations,* WRI-98-4122, 1998.

65. Solley, W. B., *Preliminary Estimates of Water Use in the United States, 1995, U.S. Geological Survey Open-File Report* 97-0645, Reston, VA, 1997.

66. Strack, O. D. L., *Groundwater Mechanics,* Prentice Hall, Englewoods Cliffs, NJ, 1989.

67. Thomas, H. E., *The Conservation of Ground Water,* McGraw-Hill, New York, 327 pp., 1951.

68. Thurner, A., *Hydrogeologie,* Springer, Vienna, 350 pp., 1967.

69. Todd, D. K., *Groundwater Hydrology,* 2nd ed, John Wiley & Sons, New York, 1980.

70. Tolman, C. F., *Ground Water,* McGraw-Hill, New York, 593 pp., 1937.

71. Toth, J., A theoretical analysis of groundwater in small drainage basins, *Jour. Geophys. Res.,* v. 68, pp. 4795–4812, 1963.

72. Trommer, J. T., M. J. DelCharco, and B. R. Lewelling, *Water Budget and Water Quality of Ward Lake, Flow and Water-Quality Characteristics of the Braden River Estuary, and the Effects of Ward Lake on the Hydrologic System, West-Central Florida, U.S. Geological Survey Water-Resources Investigations Report* 98-4251, 1999.

73. Verruijt, A., *Theory of Groundwater Flow,* Gordon and Breach, New York, 190 pp., 1970.

74. Walton, W. C., *Groundwater Resource Evaluation,* McGraw-Hill, New York, 664 pp., 1970.

75. Walton, W. C., *Principles of Groundwater Engineering,* Lewis Publishers, Chelsea, MI, 1991.

76. Weight, W. D., and J. L. Sonderegger, *Manual of Applied Field Hydrogeology,* McGraw-Hill, New York, 2001.

77. Willis, R., and W. W-G. Yeh, *Groundwater Systems Planning and Management,* Prentice Hall, Englewood Cliffs, NJ, 1987.

78. Winter, T. C., Uncertainties in estimating the water balance of lakes, *Water Resources Bull.,* v. 17, pp. 82–115, 1981.

79. Winter, T. C., The interactions of lakes with variably saturated porous media, *Water Resources Research,* v. 19, pp. 1203–1218, 1985.

80. Winter, T. C., J. W. Harvey, O. L. Franke, and W. M. Alley, *Groundwater and Surface Water: A Single Source, U.S. Geological Survey Circular* 1139, http://water.usgs.gov/pubs/circ/circ1139, Denver, CO, 1998.

81. Wulff, H. E., The qanats of Iran, *Sci. Amer.,* v. 218, pp. 94–100, 105, 1968.

EXAMPLE PUBLICATIONS OF ORGANIZATIONS AND GOVERNMENT AGENCIES

American Water Works Association, *AWWA Standard for Disinfection of Water Wells,* ANSI/AWWA C654, Denver, CO.

American Water Works Association, *AWWA Standard for Water Wells,* ANSI/AWWA A100, Denver, CO.

American Water Works Association, Manual 21, *Groundwater,* Denver, CO, 1989.

Borch, M. A., S. A. Smith, and L. N. Noble, *Evaluation and Restoration of Water Supply Wells,* American Water Works Association and American Water Works Association Research Foundation, Denver, CO, 1993.

U.S. Environmental Protection Agency, Office of Drinking Water, *Local Financing for Wellhead Protection,* Washington, D. C., 1989.

U.S. Environmental Protection Agency, Office of Drinking Water, *Citizen's Guide to Ground-Water Protection,* Washington, DC, 1990.

U.S. Environmental Protection Agency, Office of Drinking Water, *Guide to Ground-Water Supply Contingency Planning for Local and State Governments,* Washington, DC, 1991.

U.S. Environmental Protection Agency, Office of Drinking Water, *Protecting Local Ground-Water Supplies Through Wellhead Protection,* Washington, DC, 1991.

Chapter 2

Occurrence of Groundwater

To describe the occurrence of groundwater necessitates a review of where and how groundwater exists; subsurface distribution, in both vertical and areal extents, needs to be considered. The geologic zones important to groundwater must be identified, as well as their structure in terms of water-holding and water-yielding capabilities. If hydrologic conditions furnish water to the underground zone, the subsurface strata govern its distribution and movement; hence the important role of geology in groundwater hydrology cannot be overemphasized. Springs, hydrothermal phenomena, and water in permanently frozen ground constitute special groundwater occurrences.

2.1 ORIGIN AND AGE OF GROUNDWATER

Almost all groundwater can be thought of as a part of the hydrologic cycle, including surface and atmospheric (meteoric) waters. Relatively minor amounts of groundwater may enter this cycle from other origins.

Water that has been out of contact with the atmosphere for at least an appreciable part of a geologic period is termed *connate water;* essentially, it consists of fossil interstitial water that has migrated from its original burial location.[62] This water may have been derived from oceanic or freshwater sources and, typically, is highly mineralized.[20] *Magmatic water* is water derived from magma; where the separation is deep, the term *plutonic water* is applied, while *volcanic water* designates water from relatively shallow depths (perhaps 3 to 5 km).[61] New water of magmatic or cosmic origin that has not previously been a part of the hydrosphere is referred to as *juvenile water.* And finally, *metamorphic water* is water that is or has been associated with rocks during their metamorphism. The diagram in Figure 2.1.1 illustrates the interrelations of these genetic types of groundwater.

The residence time of water underground has always been a topic of considerable speculation. But with the advent of radioisotopes, determinations of the age of groundwater have become possible. Hydrogen-3 (tritium) and carbon-14 are the two isotopes that have proved most useful. Tritium with a half-life of 12.33 years is produced in the upper atmosphere by cosmic radiation; carried to earth by rainfall and hence underground, this natural level of tritium begins to decay as a function of time, such that

$$A = A_o\, e^{-\lambda t} \tag{2.1.1}$$

where A is the observed radioactivity, A_o is the activity at the time the water entered the aquifer, λ is the decay constant, and t is the age of the water. Carbon-14 has a half-life of 5,730 years and is also produced at an established constant level in the atmosphere. This isotope is present in groundwater as dissolved bicarbonate originating from the biologically active layers of the

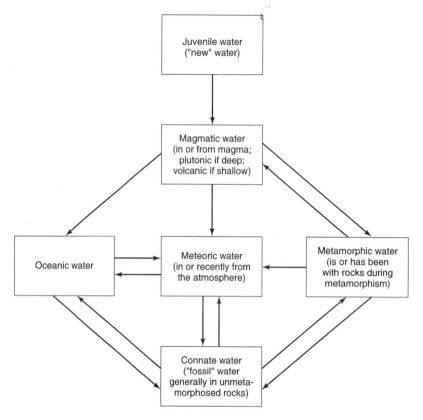

Figure 2.1.1. Diagram illustrating relationships of genetic types of water[62] (courtesy The Geological Society of America, 1957).

soil where CO_2 is generated by root respiration and the decay of humus.[59, 66] Tritium is applicable for estimating groundwater residence times of up to 50 years, while carbon-14 spans the age bracket of a several hundred to about 50,000 years.

Applications of the age-dating techniques have revealed groundwaters ranging in age from a few years or less to many thousand years.[*] Measurements of water samples taken from deep wells in deserts of the United Arab Republic and Saudi Arabia indicate ages of 20,000 to 30,000 years.[35] This period is compatible with the Wisconsin ice age, when these desert areas last possessed a high rainfall capable of recharging the underlying major aquifers.

2.2   ROCK PROPERTIES AFFECTING GROUNDWATER

2.2.1   Aquifers

Groundwater occurs in many types of geologic formations; those known as aquifers are of most importance. An *aquifer* may be defined as a formation that contains sufficient saturated permeable material to yield significant quantities of water to wells and springs.[35] This implies an ability to store and to transmit water; unconsolidated sands and gravels are a typical example. Furthermore, it is generally understood that an aquifer includes the unsaturated portion of the permeable unit. Synonyms frequently employed include *groundwater reservoir* and *water-bearing formation.* Aquifers are generally areally extensive and may be overlain or underlain

*The fallout of bomb tritium and C-14 in precipitation since the advent of nuclear weapon testing in 1952 has greatly complicated much of the dating of groundwater because recent levels greatly exceed the prebomb level.

by a *confining bed,* which may be defined as a relatively impermeable material stratigraphically adjacent to one or more aquifers. Clearly, there are various types of confining beds; the following types are well established in the literature:

1. *Aquiclude*—A saturated but relatively impermeable material that does not yield appreciable quantities of water to wells; clay is an example.

2. *Aquifuge*—A relatively impermeable formation neither containing nor transmitting water; solid granite belongs in this category.

3. *Aquitard*—A saturated but poorly permeable stratum that impedes groundwater movement and does not yield water freely to wells, that may transmit appreciable water to or from adjacent aquifers and, where sufficiently thick, may constitute an important groundwater storage zone; sandy clay is an example.[45]*

2.2.2   Porosity

Those portions of a rock or soil not occupied by solid mineral matter can be occupied by groundwater. These spaces are known as *voids, interstices, pores,* or *pore space*. Because interstices serve as water conduits, they are of fundamental importance to the study of groundwater. Typically, they are characterized by their size, shape, irregularity, and distribution. Original interstices were created by geologic processes governing the origin of the geologic formation and are found in sedimentary and igneous rocks. Secondary interstices developed after the rock was formed; examples include joints, fractures, solution openings, and openings formed by plants and animals. With respect to size, interstices may be classed as capillary, supercapillary, and subcapillary. Capillary interstices are sufficiently small that surface tension forces will hold water within them; supercapillary interstices are those larger than capillary ones; and subcapillary interstices are so small that water is held primarily by adhesive forces. Depending on the connection of interstices with others, they may be classed as communicating or isolated.

The *porosity* of a rock or soil is a measure of the contained interstices or voids expressed as the ratio of the volume of interstices to the total volume. If α is the porosity, then

$$\alpha = \frac{V_v}{V} = \frac{V_t - V_s}{V_t} \tag{2.2.1}$$

where V_v is the volume of interstices (voids), V_s is the volume of solids, and V is the total volume (bulk volume). Porosity may also be expressed by

$$\alpha = \frac{\rho_m - \rho_d}{\rho_m} = 1 - \frac{\rho_d}{\rho_m} \tag{2.2.2}$$

where ρ_m is the density of mineral particles (grain density)[†] and ρ_d is the bulk density.

The term *effective porosity* refers to the amount of interconnected pore space available for fluid flow and is expressed as a ratio of interconnected interstices to total volume. For unconsolidated porous media and for many consolidated rocks, the two porosities are identical. Porosity may also be expressed as a percentage by multiplying the right-hand side of Equations 2.2.1 or 2.2.2 by 100. The terms *primary* and *secondary porosity* are associated with original and secondary interstices, respectively.

---

*The word *aquifer* can be traced to its Latin origin. *Aqui-* is a combining form of *aqua* ("water") and *-fer* comes from *ferre* ("to bear"). Hence, an aquifer is a water-bearer. The suffix *-clude* of aquiclude is derived from the Latin *claudere* ("to shut or close"). Similarly, the suffix *-fuge* of aquifuge comes from *fugere* ("to drive away"), while the suffix *-tard* of aquitard follows from the Latin *tardus* ("slow").

†The density of solid rock varies with the type of mineral. For alluvium where quartz is the predominant mineral, a value of 2.65 g/cm^3 is typical; limestone and granite fall in the range 2.7–2.8 g/cm^3, and basalt can approach 3.0 g/cm^3.

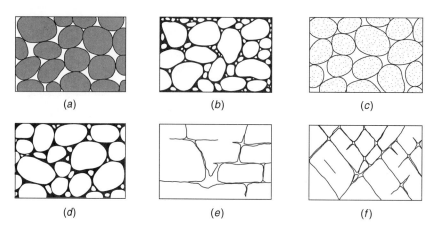

Figure 2.2.1. Examples of rock interstices and the relation of rock texture to porosity. (*a*) Well-sorted sedimentary deposit having high porosity. (*b*) Poorly sorted sedimentary deposit having low porosity. (*c*) Well-sorted sedimentary deposit consisting of pebbles that are themselves porous, so that the deposit as a whole has a very high porosity. (*d*) Well-sorted sedimentary deposit whose porosity has been diminished by the deposition of mineral matter in the interstices. (*e*) Rock rendered porous by solution. (*f*) Rock rendered porous by fracturing.[42]

Figure 2.2.1 shows several types of interstices and their relation to porosity. In terms of groundwater supply, granular sedimentary deposits are of major importance. Porosities in these deposits depend on the shape and arrangement of individual particles, distribution by size, and degree of cementation and compaction. In consolidated formations, removal of mineral matter by solution and degree of fracture are also important. Porosities range from near zero to more than 50 percent, depending on the above factors and the type of material. Representative porosity values for various geologic materials are listed in Table 2.2.1. It should be recognized that porosities for a particular soil or rock can vary considerably from these values.

In sedimentary rocks subject to compaction, measurements show that porosity decreases with depth of burial.[37] Thus, a typical relation has the form

$$\alpha_z = \alpha_o \, e^{-az} \tag{2.2.3}$$

Table 2.2.1 Representative Values of Porosity (after Morris and Johnson[45])

Material	Porosity, percent	Material	Porosity, percent
Gravel, coarse	28[a]	Loess	49
Gravel, medium	32[a]	Peat	92
Gravel, fine	34[a]	Schist	38
Sand, coarse	39	Siltstone	35
Sand, medium	39	Claystone	43
Sand, fine	43	Shale	6
Silt	46	Till, predominantly silt	34
Clay	42	Till, predominantly sand	31
Sandstone, fine grained	33	Tuff	41
Sandstone, medium grained	37	Basalt	17
Limestone	30	Gabbro, weathered	43
Dolomite	26	Granite, weathered	45
Dune sand	45		

[a]These values are for repacked samples; all others are undisturbed.

where α_z is the porosity at depth z, α_o is the porosity at the surface, a is a constant, and e is the base of Naperian logarithms.

EXAMPLE 2.2.1

An undisturbed sample of a medium sand weighs 484.68 g. The core of the undisturbed sample is 6 cm in diameter and 10.61 cm high. The sample is oven-dried for 24 hr at 110°C to remove the water content. At the end of the 24 hr, the core sample weighs 447.32 g. Determine the bulk density, void ratio, water content, porosity, and saturation percentage of the sample.

SOLUTION:

The dry weight of the sample is $W_d = 447.32$ g and the total weight is $W_T = 484.68$ g. The total volume of the undisturbed sample is

$$V_t = \pi r^2 h = \pi(3\text{ cm})^2(10.61\text{ cm}) = 300\text{ cm}^3$$

The *bulk density* is defined as the density of solids and voids together, after drying. Thus,

$$\rho_d = \frac{W_d}{V_t} = \frac{447.32\text{ g}}{300\text{ cm}^3} = 1.491\text{ g/cm}^3$$

Assuming quartz is the predominant mineral in the sample, then $\rho_m = 2.65$ g/cm^3
Thus, the volume V_s of the solid phase of the sample is

$$V_s = \frac{W_d}{\rho_m} = \frac{447.32\text{ g}}{2.65\text{ g/cm}^3} = 168.8\text{ cm}^3$$

Thus, the *total volume of voids* in the sample is

$$V_v = V_t - V_s = 300\text{ cm}^3 - 168.8\text{ cm}^3 = 131.2\text{ cm}^3$$

With this information, we can calculate the *void ratio e* of the sample is

$$e = \frac{V_v}{V_s} = \frac{131.2\text{ cm}^3}{168.8\text{ cm}^3} = 0.777$$

The *volumetric water content* of a sample is the volume of the water divided by the volume of the sample

$$\theta_v = \frac{V_{water}}{V_t} = \frac{(W_T - W_d)/\rho_{water}}{V_t} = \frac{484.68\text{ g} - 447.32\text{ g}}{300\text{ cm}^3}/1\text{g}/\text{cm}^3 = 0.1245\text{ g/cm}^3 = 0.125$$

where W_w is the total weight of the undisturbed sample before drying.
The *gravimetric water content* of the sample is

$$\theta_w = \frac{W_T - W_d}{W_d} \times 100 = \frac{484.68\text{ g} - 447.32\text{ g}}{447.32\text{ g}} \times 100 = 8.35\%$$

The porosity of the sample is

$$\alpha = \frac{V_t - V_s}{V_t} \times 100 = \frac{300\text{ cm}^3 - 168.8\text{ cm}^3}{300\text{ cm}^3} \times 100 = 43.73\%$$

Finally, the *saturation percentage of a sample* is defined as the percentage of the pore space that is filled by water,

$$\frac{\theta_v}{\alpha} \times 100 = \frac{(0.1245)}{(0.4373)} \times 100 = 28.47\%$$

<table>
<tr><td>

EXAMPLE 2.2.2

</td><td>

The void ratio of an unconsolidated clay sample is 1.19. Determine the porosity of the sample.

</td></tr>
<tr><td>

SOLUTION

</td><td>

Using the definition of the void ratio of an undisturbed sample, $e = \dfrac{V_v}{V_s}$, and substituting $V_v = V_t - V_s$, then the void ratio is $e = \dfrac{V_t - V_s}{V_s} = \dfrac{V_t}{V_s} - 1 \rightarrow \dfrac{V_t}{V_s} = 1 + e$.

Substituting this into the porosity equation, we obtain

$$\alpha = \frac{V_t - V_s}{V_t} \times 100 = \left[1 - \frac{V_s}{V_t}\right] \times 100 = \left[1 - \frac{1}{1+e}\right] \times 100 = \frac{e}{1+e} \times 100$$

Thus, the porosity of the sample is

$$\alpha = \frac{e}{1+e} \times 100 = \frac{1.19}{1+1.19} \times 100 = 54.34\%$$

■

</td></tr>
</table>

<table>
<tr><td>

EXAMPLE 2.2.3

</td><td>

The porosity of a quartz sand sample is 38.41%. Determine the bulk density of the sample.

</td></tr>
<tr><td>

SOLUTION

</td><td>

The bulk density and porosity of an undisturbed sample are defined as $\rho_d = \dfrac{W_d}{V_t}$ and $\alpha = \dfrac{V_t - V_s}{V_t} \times 100$, respectively.

Substituting the dry weight of a sample $W_d = \rho_m V_s$ into the bulk density expression, we have $\rho_d = \dfrac{W_d}{V_t} = \dfrac{\rho_m V_s}{V_t}$ and the porosity is

$$\alpha = \frac{V_t - V_s}{V_t} \times 100 = \left[1 - \frac{V_s}{V_t}\right] \times 100 \rightarrow \frac{V_s}{V_t} = 1 - \frac{\alpha}{100}$$

Using the bulk density expression then yields $\rho_d = \dfrac{\rho_m V_s}{V_t} = \rho_m \left[1 - \dfrac{\alpha}{100}\right]$.

For quartz sand, $\rho_m = 2.65$ g/cm^3, the bulk density is

$$\rho_d = \rho_m \left[1 - \frac{\alpha}{100}\right] = \left(2.65 \text{ g/cm}^3\right)\left[1 - \frac{38.41}{100}\right] = 1.63 \text{ g/cm}^3$$

■

</td></tr>
</table>

2.2.3 Soil Classification

Unconsolidated geologic materials are normally classified according to their size and distribution. A commonly employed system based on particle, or grain, size is listed in Table 2.2.2. Evaluation of the distribution of sizes is accomplished by mechanical analysis. This involves sieving particles coarser than 0.05 mm and measuring rates of settlement for smaller particles in suspension. Results are plotted on a particle-size distribution graph such as that shown in Figure 2.2.2. The percentage finer scale on the ordinate shows the percentage of material smaller than that of a given size particle on a dry-weight basis.

The effective particle size is the 10 percent finer than value (d_{10}). The distribution of particles is characterized by the *uniformity coefficient* U_c as

$$U_c = d_{60}/d_{10} \tag{2.2.4}$$

where d_{60} is the 60 percent finer than value. A *uniform material* has a low uniformity coefficient (the dune sand in Figure 2.2.2), while a *well-graded material* has a high uniformity coefficient (the alluvium).

Table 2.2.2 Soil Classification Based on Particle
Size (after Morris and Johnson[45])

Material	Particle size, mm
Clay	<0.004
Silt	0.004 – 0.062
Very fine sand	0.062 – 0.125
Fine sand	0.125 – 0.25
Medium sand	0.25 – 0.5
Coarse sand	0.5 – 1.0
Very coarse sand	1.0 – 2.0
Very fine gravel	2.0 – 4.0
Fine gravel	4.0 – 8.0
Medium gravel	8.0 – 16.0
Coarse gravel	16.0 – 32.0
Very coarse gravel	32.0 – 64.0

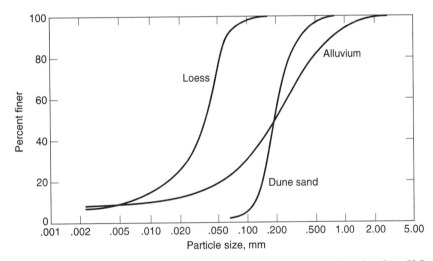

Figure 2.2.2. Particle-size distribution graph for three geologic samples (data from U.S. Geological Survey).

The *texture* of a soil is defined by the relative proportions of sand, silt, and clay present in the particle-size analysis. This can be expressed by the soil-textural triangle in Figure 2.2.3. Note, for example, that a soil composed of 30 percent clay, 60 percent silt, and 10 percent sand constitutes a silty clay loam.

2.2.4   Porosity and Representative Elementary Volume

We can define porosity in a spatial field as a function, $\alpha(x)$, at any point x using spatial averaging over a *representative elementary volume* (REV). Consider a porous medium with different sizes of averaging volumes, V^*, centered at point $\tilde{x}$. The point value of porosity (volumetric porosity) is associated with an averaging volume centered at that point, expressed as[14]

$$\alpha(\tilde{x}) = \frac{1}{V^*} \int_{V^*(x)} X(\tilde{x} - \tilde{y})\, d\tilde{y} \tag{2.2.4}$$

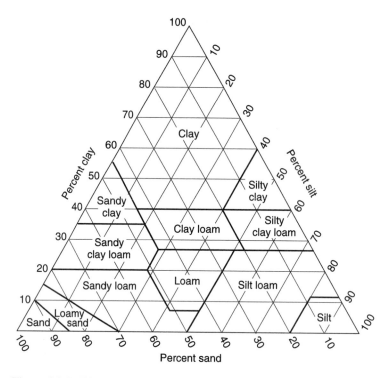

Figure 2.2.3. Triangle of soil textures for describing various combinations of sand, silt, and clay (after Soil Survey Staff[50]).

where $X(\tilde{x})$ is an indicator function for void space: $X(\tilde{x}) = 0$ if point $\tilde{x}$ is located in the solid space or $X(\tilde{x}) = 1$ if point x is located in the void space. V^* is the averaging volume centered on the point $\tilde{x}$, with the estimated porosity value as a function of averaging volume size. If the volume is too small, the average value is not well defined and the limit approaches either 0 or 1. There may be a range of volumes surrounding point $\tilde{x}$ for which the average is well defined, and if the averaging volume is too large, then soils with different textures may be included, resulting in a deviation from the average.

2.2.5   Specific Surface

The water retentive property of a soil or rock is markedly influenced by its surface area. This area depends on particle size and shape and on the type of clay minerals present. The term *specific surface* refers to the area per unit weight of the material, usually expressed as m²/g. Relative methods for measuring specific surface are based on retention of a polar organic molecule such as ethylene glycol; these have been related to absolute values derived from statistical calculations of surface area.[6] Clay particles contribute the greatest amount of surface area in unconsolidated formations. Nonswelling clays such as kaolinite have only an external surface and exhibit specific surfaces in the range of 10–30 m²/g; however, swelling clays such as montmorillonite and vermiculite have internal and external surfaces that yield specific surface values near 800 m²/g.

An illustration of the importance of particle size to specific surface is presented in Table 2.2.3. Here, considering only uniform spheres, it can be seen that when a given volume is transformed into 100 small spheres totaling the same volume, the specific surface increases by a factor of 100. Furthermore, it can be shown that when the volume is deformed into rod, disk, or plate shapes, specific surface increases even more.[4]

Table 2.2.3 Relation of Surface Area to Particle Size for Uniform Spheres

Diameter of particle, mm	Soil classification	Number of particles per cm^3	Total surface area, cm^2
10	Medium gravel	1	3.14
1	Coarse sand	1×10^3	31.4
0.1	Very fine sand	1×10^6	314
0.02	Silt	1.25×10^8	1,570
0.002	Clay	1.25×10^{11}	15,700

Note: Rectangular packing is assumed in a cubic container 1 cm on a side so that the total volume, and weight, of spheres remains constant at $\pi/6$ cm^3.

EXAMPLE 2.2.4

Using the tabulated results of a grain size distribution test on a field sample, perform the following tasks:

(a) Prepare a grain size distribution curve for this sample.

(b) Is this a well-graded or poorly graded sample?

(c) Classify the sample using Table 2.2.2.

(d) What would be reasonable porosity values for this sample?

U.S. Standard Sieve Number	Mass retained (g)
3/8	49.95
4	26.70
8	25.29
16	50.58
30	72.57
40	25.50
100	33.60
200	7.53
Pan (passes through #200 sieve)	8.28
Total sample weight	300.00

SOLUTION

(a) The given data are analyzed as shown in the table below. Note that the particle size (sieve opening) corresponding to each U.S. Standard Sieve number is given in the table. The results yield the grain-size distribution curve shown in Figure 2.2.4.

Sieve	Grain size (mm)	Mass retained (g)	Percent finer by mass
3/8	9.5	49.95	83.35
4	4.75	26.70	74.45
8	2.36	25.29	66.02
16	1.18	50.58	49.16
30	0.6	72.57	24.97
40	0.425	25.50	16.47
100	0.15	33.60	5.27
200	0.075	7.53	2.76
Pan	<0.075	8.28	
Total sample weight		300	

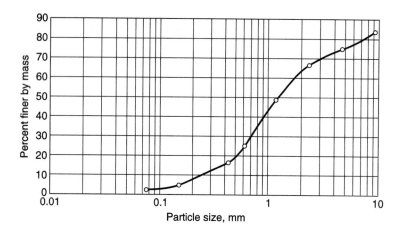

Figure 2.2.4.
Grain-size distribution curve for Example 2.2.4.

(b) From the grain-size distribution curve:

$$d_{60} \cong 1.6 \text{ mm} \qquad \text{and} \qquad d_{10} \cong 0.23 \text{ mm}$$

From Equation 2.2.4, the uniformity coefficient is

$$U_c = \frac{d_{60}}{d_{10}} = \frac{1.6 \text{ mm}}{0.23 \text{ mm}} \approx 7$$

Since $U_c > 6$, the sample can be described as well graded (i.e., low uniformity).

(c) The percentage of clay and silt in the sample is approximately 2–3 percent, while about 60 percent of the sample is sand. The remaining 37–38 percent is composed of very fine to coarse gravel.

(d) The porosity of the sample could be somewhere between 20 and 35 percent based on our classification in part (c). ∎

EXAMPLE 2.2.5

The following data are obtained from a hydrometer test to determine the gradation of a sample of fine sediments.

(a) Prepare a grain size distribution curve for this sample.
(b) Is this a well-graded or poorly graded sample?
(c) Classify the sample using Table 2.2.2.
(d) What would be reasonable porosity values for this sample?

Particle grain size (mm)	Percent finer than
1.000	100.00
0.700	98.90
0.500	88.50
0.350	49.40
0.250	10.30
0.175	0.90
0.125	0.00
0.088	0.00
0.0625	0.00

SOLUTION

(a) Using the given data, the grain size distribution curve is plotted for this sample in Figure 2.2.5.

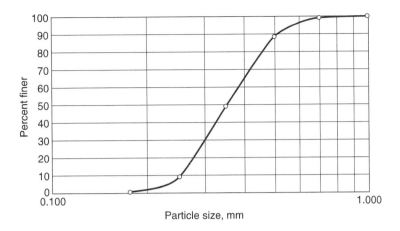

Figure 2.2.5. Grain-size distribution curve for Example 2.2.5.

(b) From Figure 2.2.5, $d_{60} \cong 0.36$ mm and $d_{10} \cong 0.25$ mm, the uniformity coefficient is given by Equation 2.2.4:

$$U_c = \frac{d_{60}}{d_{10}} = \frac{0.36 \text{ mm}}{0.25 \text{ mm}} \approx 1.44$$

Since $U_c < 4$, the sample can be classified as poorly graded (i.e., highly uniform).

(c) From the given data and the soil classifications of Table 2.2.2, the sample consists of 11.5 percent coarse sand, 78.2 percent medium sand, and 10.3 percent fine sand. Thus, the sample can be classified as a medium sand with some proportions of coarse and fine sand.

(d) Based on the classification in part (c), the porosity of the sample would be in the range of 35 to 45 percent. ∎

2.3   VERTICAL DISTRIBUTION OF GROUNDWATER

The subsurface occurrence of groundwater may be divided into zones of aeration and saturation. The *zone of aeration* consists of interstices occupied partially by water and partially by air. In the *zone of saturation*, all interstices are filled with water under hydrostatic pressure. On most of the land masses of the earth, a single zone of aeration overlies a single zone of saturation and extends upward to the ground surface, as shown in Figure 2.3.1.

In the zone of aeration, *vadose water*[*] occurs. This general zone may be further subdivided into the *soil water zone*, the *intermediate vadose zone*, and the *capillary zone* (Figure 2.3.1).[14]

The saturated zone extends from the upper surface of saturation down to underlying impermeable rock. In the absence of overlying impermeable strata, the *water table*, or *phreatic surface*,[†] forms the upper surface of the zone of saturation. This is defined as the surface of atmospheric pressure and appears as the level at which water stands in a well penetrating the aquifer. Actually, saturation extends slightly above the water table due to capillary attraction; however, water is held there at less than atmospheric pressure. Water occurring in the zone of saturation is commonly referred to simply as *groundwater,* but the term *phreatic water* is also employed.

---

[*]*Vadose* is derived from the Latin *vadosus* ("shallow").

[†]*Phreatic* is derived from the Greek *phrear*, *-atos* ("a well").

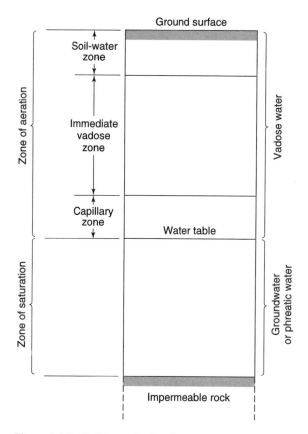

Figure 2.3.1. Divisions of subsurface water.

2.4 ZONE OF AERATION

2.4.1 Soil–Water Zone

Water in the soil–water zone exists at less than saturation except temporarily when excessive water reaches the ground surface as from rainfall or irrigation. The zone extends from the ground surface down through the major root zone. Its thickness varies with soil type and vegetation. Because of the agricultural importance of soil water in supplying moisture to roots, agriculturists and soil scientists have studied soil moisture distribution and movement extensively.

The amount of water present in the soil–water zone depends primarily on the recent exposure of the soil to moisture. Under hot, arid conditions, a water-vapor equilibrium tends to become established between the ambient air and the surfaces of fine-grained soil particles. As a result, only thin films of moisture—known as *hygroscopic water*—remain adsorbed on the surfaces. For coarse-grained materials and where additional moisture is available, water also forms liquid rings surrounding contacts between grains, as sketched in Figure 2.4.1. This water is held by surface tension forces and is sometimes referred to as *capillary water*. Temporarily, the soil–water zone may contain water in excess of capillary water from rainfall or irrigation; this *gravitational water* drains through the soil under the influence of gravity.

2.4.2 Intermediate Vadose Zone

The intermediate vadose zone extends from the lower edge of the soil–water zone to the upper limit of the capillary zone (Figure 2.3.1). The thickness may vary from zero, where the bound-

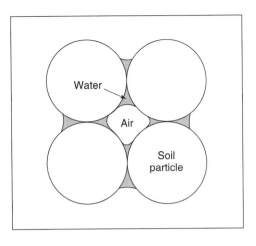

Figure 2.4.1. Illustration of vadose water held at contact points of particles in the unsaturated zone.

ing zones merge with a high water table approaching ground surface, to more than 100 m under deep water table conditions. The zone serves primarily as a region connecting the zone near ground surface with that near the water table through which water moving vertically downward must pass. Nonmoving vadose water is held in place by hygroscopic and capillary forces. Temporary excesses of water migrate downward as gravitational water.

2.4.3   Capillary Zone

The *capillary zone* (or *capillary fringe*) extends from the water table up to the limit of capillary rise of water. If a pore space could be idealized to represent a capillary tube, the *capillary rise* h_c (Figure 2.4.2) can be derived from an equilibrium between surface tension of water and the weight of water raised. Thus,

$$h_c = \frac{2\tau}{r\gamma} \cos \lambda \qquad (2.4.1)$$

where τ is surface tension, γ is the specific weight of water, r is the tube radius, and λ is the angle of contact between the meniscus and the wall of the tube. For pure water in a clean glass, $\lambda = 0$, and at 20° C, $\tau = 0.074$ g/cm and $\gamma = 1$ g/cm^3, so that for r in cm, the capillary rise (in cm) is

$$h_c = \frac{0.15}{r} \qquad (2.4.2)$$

It follows from Equation 2.4.2 that the thickness of the capillary zone will vary inversely with the pore size of a soil or rock. Measurements of capillary rise in unconsolidated materials shown in Table 2.4.1 bear out this relationship. Furthermore, for a material containing innumerable pores of a wide range in size, the upper boundary of the zone will form a jagged limit when studied microscopically. Taken macroscopically, however, a gradual decrease in water content results with height. That is, just above the water table almost all pores contain capillary water; higher, only the smaller connected pores contain water; and still higher, only the few smallest connected pores contain water lifted above the water table. This distribution of water above the water table is shown in Figure 2.4.3 from a drainage test on a sand. The visual capillary rise is invariably less than the actual capillary zone as defined in Figure 2.4.3.

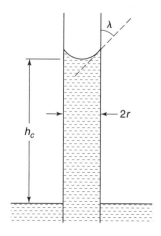

Figure 2.4.2. Rise of water in a capillary tube.

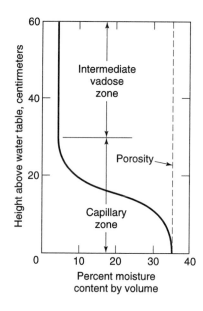

Figure 2.4.3. Distribution of water in a coarse sand above the water table after drainage (after Prill[47]).

Table 2.4.1 Capillary Rise in Samples of Unconsolidated Materials (after Lohman[34])

Material	Grain size (mm)	Capillary rise (cm)
Fine gravel	5–2	2.5
Very coarse sand	2–1	6.5
Coarse sand	1–0.5	13.5
Medium sand	0.5–0.2	24.6
Fine sand	0.2–0.1	42.8
Silt	0.1–0.05	105.5
Silt	0.05–0.02	200[a]

Note: Capillary rise measured after 72 days; all samples have virtually the same porosity of 41 percent.

[a]Still rising after 72 days.

EXAMPLE 2.4.1

Assuming uniform spherical grains of cubic packing, approximate the capillary rise in a soil as a function of the grain diameter. Using this relationship, estimate the capillary rise for each material in Table 2.4.1.

SOLUTION

The accompanying figure shows a typical arrangement of cubical packing with spherical grains of equal diameter. It can be shown, using geometry, that the radius of the pore space between the grains is 0.2 times the grain diameter.

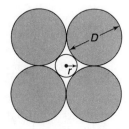

Assuming that this pore space could be idealized to represent a capillary tube, $r = 0.20D$ can be substituted into Equation 2.4.2 yielding $h_c = \dfrac{0.15}{r} = \dfrac{0.15}{0.20D} = \dfrac{3}{4D}$

Material	Grain size (mm)	Estimated capillary rise (cm)	Actual capillary rise (cm)
Fine gravel	5–2	1.5–3.75	2.5
Very coarse sand	2–1	3.75–7.5	6.5
Coarse sand	1–0.5	7.5–15	13.5
Medium sand	0.5–0.2	15–37.5	24.6
Fine sand	0.2–0.1	37.5–75	42.8
Silt	0.1–0.05	75–150	105.5
Silt	0.05–0.02	150–375	200	■

2.4.4   Measurement of Water Content

Determination of the water content of soils can be accomplished by various direct methods based on removal of the water from a sample by evaporation, leaching, or chemical reaction, followed by measurement of the amount removed.[1, 24] Thus, the *gravimetric method* involves weighing a wet soil sample, removing the water by oven-drying it, and reweighing the sample. Indirect methods consist of measuring some soil property affected by soil-water content. Specifically, electrical and thermal conductivity and electrical capacitance of porous materials vary with water content.

Another useful instrument for measuring soil moisture is the neutron probe. When lowered in a small-diameter tube in the ground, determination of soil moisture can be made as a function of depth. The instrument contains a radium-beryllium source of fast neutrons and a detector for slow neutrons. The fast neutrons are slowed by collisions with hydrogen, and because most of the hydrogen in soil is associated with water, the intensity of slow neutrons measured yields, after calibration, the local soil moisture content.[6]

Within the vadose zone a negative-pressure head of water exists, often referred to as *suction,* or *tension* in a positive sense. This tension can be measured by a *tensiometer;* Figure 2.4.4 shows a tensiometer installed in a soil column. The depression Δh in water level measures the local soil tension. Such instruments function in the range from atmospheric pressure (near 1,000 cm of water) to about 200 cm of water (800 cm water tension). Calibration data for soil suction and water content reveal that the relation between the two variables is not single valued; instead, soil structure and compaction, as well as effects of wetting or drying, influence the results.[6]

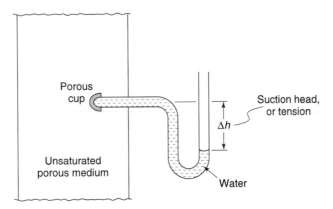

Figure 2.4.4. Illustration of a tensiometer for measuring water tension in unsaturated porous media.

2.4.5   Available Water

Soils absorb and retain water, which may be withdrawn by plants during periods between rain-fall or irrigations. This water-holding capacity is defined by the *available water*, which is the range of plant-available water, the moist end being the field capacity and the dry end the wilting point. *Field capacity* can be defined as the amount of water held in a soil after wetting and after subsequent drainage has become negligibly small. The negligible drainage rate is often assumed after two days; however, different soils possess varying drainage rates so that quantitative values may not be comparable. The *wilting point* defines the water content of soils when plants growing in that soil are reduced to a permanent wilted condition. Because factors such as soil type, volume, plant type, and age influence wilting point, this moisture content can also be variable.

2.5   ZONE OF SATURATION

In the zone of saturation, groundwater fills all of the interstices; hence, the (effective) porosity provides a direct measure of the water contained per unit volume. A portion of the water can be removed from subsurface strata by drainage or by pumping of a well; however, molecular and surface tension forces hold the remainder of the water in place.

2.5.1   Specific Retention

The *specific retention* S_r of a soil or rock is the ratio of the volume of water it will retain after saturation against the force of gravity to its own volume. Thus,

$$S_r = \frac{w_r}{V_t} \qquad (2.5.1)$$

where w_r is the volume occupied by retained water[*] and V_t is the bulk volume of the soil or rock.

2.5.2   Specific Yield

The *specific yield* S_y of a soil or rock is the ratio of the volume of water that, after saturation, can be drained by gravity to its own volume.[17] Therefore,

$$S_y = \frac{w_y}{V_t} \qquad (2.5.2)$$

where w_y is the volume of water drained. Values of S_r and S_y can also be expressed as percentages. Because w_r and w_u constitute the total water volume in a saturated material, it is apparent that

$$V_v = w_r + w_y \qquad (2.5.3)$$

or

$$\alpha = S_r + S_y \qquad (2.5.4)$$

where all pores are interconnecting.

Values of specific yield depend on grain size, shape and distribution of pores, compaction of the stratum, and time of drainage.[40] Representative specific yields for various geologic materials are listed in Table 2.5.1; individual values for a soil or rock can vary considerably from these values. It should be noted that fine-grained materials yield little water, whereas coarse-grained materials permit a substantial release of water and hence serve as aquifers. In

---

[*]It should be noted that the terms *field capacity* and *retained water* refer to the same water content but differ by the zone in which they occur.

Table 2.5.1 Representative Values of Specific Yield (after Johnson[25])

Material	Specific yield (percent)
Gravel, coarse	23
Gravel, medium	24
Gravel, fine	25
Sand, coarse	27
Sand, medium	28
Sand, fine	23
Silt	8
Clay	3
Sandstone, fine grained	21
Sandstone, medium grained	27
Limestone	14
Dune sand	38
Loess	18
Peat	44
Schist	26
Siltstone	12
Till, predominantly silt	6
Till, predominantly sand	16
Till, predominantly gravel	16
Tuff	21

general, specific yields for thick unconsolidated formations tend to fall in the range of 7 to 15 percent, because of the mixture of grain sizes present in the various strata; furthermore, they normally decrease with depth due to compaction.

Specific yield can be measured by a variety of techniques involving laboratory, field, and estimating techniques.[6, 24, 26] Methods based on well-pumping tests, described in Chapter 4, generally give the most reliable results for field measurements.

EXAMPLE 2.5.1

Estimate the average drawdown over an area where 25 million m^3 of water has been pumped through a number of uniformly distributed wells. The area is 150 km^2 and the specific yield of the unconfined aquifer is 25 percent.

SOLUTION

The volume of water drained is $w_y = 25 \times 10^6$ m^3. Eq. 2.5.2 is used to determine the bulk volume, V_t, of the aquifer to extract this volume of water:

$$S_y = \frac{w_y}{V_t}$$

$$0.25 = \frac{25 \times 10^6 \text{ m}^3}{V_t} \rightarrow V_t = 1 \times 10^8 \text{ m}^3$$

Thus, the average water level drop over the area is $\Delta h = \dfrac{V_t}{A} = \dfrac{1 \times 10^8 \text{ m}^3}{150 \times 10^6 \text{ m}^2} = 0.67$ m.

EXAMPLE 2.5.2

The storage coefficient for confined aquifers can be approximated by (Jacob, 1950)

$$S = \gamma \alpha \beta b + \gamma b \eta$$

where γ is the unit weight of water, α is the porosity of the aquifer, β is the reciprocal of the bulk modulus of elasticity of water, η is the reciprocal of the modulus of elasticity of the skeleton of the aquifer, and b is the saturated thickness of the aquifer.

The first and the second terms in the above equation for S represent the fractions of storage attributable to the expansibility of water and the compressibility of the aquifer skeleton, respectively. Assuming reasonable ranges of values, develop an expression to estimate the storage coefficient of a confined aquifer considering the expansion of water only.

SOLUTION   Since we consider only the expansion of water, we must deal with the first term in the given equation. Reasonable ranges of values for the given properties, assuming a temperature range of 5–25°C, are

$$\gamma = 9806 \text{ to } 9779 \text{ N/m}^3$$

$$\beta = 4.85 \times 10^{-10} \text{ to } 4.5 \times 10^{-10} \text{ m}^2/\text{N}$$

$$\alpha = 0.10 \text{ to } 0.50$$

Substituting these values into the expression for the storage coefficient attributable to the expansibility of water yields

$$S = \gamma\alpha\beta b = (9779 \text{ N/m}^3)(0.10)(4.5 \times 10^{-10} \text{ m}^2/\text{N})(b) = (4.4 \times 10^{-7})b$$

or

$$S = \gamma\alpha\beta b = (9806 \text{ N/m}^3)(0.50)(4.85 \times 10^{-10} \text{ m}^2/\text{N})(b) = (2.4 \times 10^{-6})b$$

Thus, in most cases the storage coefficient attributable to the expansion of water would be in the range of $4.4 \times 10^{-7}b$ to $2.4 \times 10^{-6}b$, where b is the saturated thickness of the aquifer.

EXAMPLE 2.5.3   Determine the volume of water released by lowering the piezometric surface of a confined aquifer by 5 m over an area of $A = 1 \text{ km}^2$. The aquifer is 35 m thick and has a storage coefficient of 8.3×10^{-3}.

SOLUTION   The released volume can be determined utilizing the definition of the storage coefficient, $V = (A)(\Delta h)(S) = (1 \times 10^6 \text{ m}^2)(5 \text{ m})(8.3 \times 10^{-3}) = 41,500 \text{ m}^3$.

2.6   GEOLOGIC FORMATIONS AS AQUIFERS

A geologic formation that will yield significant quantities of water has been defined as an aquifer. Many types of formations serve as aquifers.[7, 38] A key requirement is their ability to store water in the rock pores. Porosity may be derived from intergranular spaces or from fractures. Table 2.6.1 summarizes the geologic origin of aquifers in terms of type of porosity and rock type. The roles of various geologic formations as aquifers are briefly described in the following subsections.

2.6.1   Alluvial Deposits

Probably 90 percent of all developed aquifers consist of unconsolidated rocks, chiefly gravel and sand. These aquifers may be divided into four categories, based on manner of occurrence: water courses, abandoned or buried valleys, plains, and intermontane valleys. Water courses consist of the alluvium that forms and underlies stream channels, as well as forming the adjacent floodplains. Wells located in highly permeable strata bordering streams produce large quantities of water, as infiltration from the streams augments groundwater supplies. Abandoned or buried valleys are valleys no longer occupied by the streams that formed them. Although such valleys may resemble water courses in permeability and quantity of groundwater storage, their

Table 2.6.1 Geologic Origin of Aquifers Based on Type of Porosity and Rock Type (after Dept. of Economic and Social Affairs[16])

Type of porosity	Sedimentary		Carbonates	Igneous and metamorphic	Volcanic	
	Consolidated	*Unconsolidated*			*Consolidated*	*Unconsolidated*
Intergranular		Gravelly sand Clayey sand Sandy clay		Weathered zone of granite-gneiss	Weathered zone of basalt	Volcanic ejecta, blocks, and fragments Ash
Intergranular and fracture	Breccia Conglomerate Sandstone Slate		Zoogenic limestone Oolitic limestone Calcareous grit		Volcanic tuff Cinder Volcanic breccia Pumice	
Fracture			Limestone Dolomite Dolomitic limestone	Granite Gneiss Gabbro Quartzite Diorite Schist Mica schist	Basalt Andesite Rhyolite	

recharge and perennial yield are usually less. Extensive plains underlain by unconsolidated sediments exist in the United States. In some places gravel and sandbeds form important aquifers under these plains; in other places they are relatively thin and have limited productivity. These plains flank highlands or other features that served as the source of the sedimentary deposits. The aquifers are recharged chiefly in areas accessible to downward percolation of water from precipitation and from occasional streams. Intermontane valleys are underlain by tremendous volumes of unconsolidated rock materials derived by erosion of bordering mountains. Many of these more or less individual basins, separated by mountain ranges, occur in the western United States. The sand and gravel beds of these aquifers produce large quantities of water, most of which is replenished by seepage from streams into alluvial fans at mouths of mountain canyons.

2.6.2   Limestone

Limestone varies widely in density, porosity, and permeability depending on degree of consolidation and development of permeable zones after deposition. Those most important as aquifers contain sizable proportions of the original rock that have been dissolved and removed.[64] Openings in limestone may range from microscopic original pores to large solution caverns forming subterranean channels sufficiently large to carry the entire flow of a stream.[9, 28] The term *lost river* has been applied to a stream that disappears completely underground in a limestone terrain.* Large springs are frequently found in limestone areas.

The dissolution of calcium carbonate by water causes prevailingly hard groundwater to be found in limestone aquifers; also, by dissolving the rock, water tends to increase the pore space and permeability with time. Solution development of limestone forms a karst terrain,† characterized by solution channels, closed depressions, subterranean drainage through sinkholes, and

---

*The growing interest of venturesome scuba divers in springs, sinkholes, and caves is yielding new information regarding the hydrogeology of limestone aquifers.

†The term *karst* is derived from the German form of the Slavic word *kras* or *krs*, meaning a black waterless place. Also, it is the German name for a district east of Trieste having such a terrain.[44]

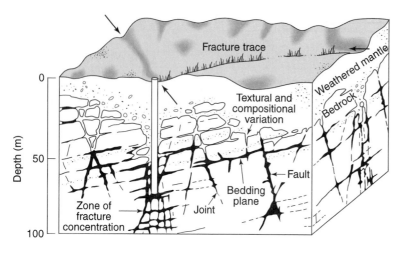

Figure 2.6.1. Diagram showing factors influencing cavity distribution in carbonate rocks (after Lattman and Parizek[32]).

caves (see Figure 2.6.1). Such regions normally contain large quantities of groundwater.[33] Major limestone aquifers occur in the southeastern United States and in the Mediterranean area.[11, 12, 31]* A national karst map is being developed by the U.S. Geological Survey.[18]

The Edwards Aquifer (Figure 2.6.2) in south-central Texas is an example of karst terrain. The limestone and dolomites are exposed at the land surface, (referred to as the outcrop area) where numerous solution openings along vertical joints and sinkholes provide an efficient link between the land surface and water table.[70] Figure 2.6.3 shows a sample of the limestone bedrock from the Edwards Aquifer, illustrating the porousness of the karst formation.

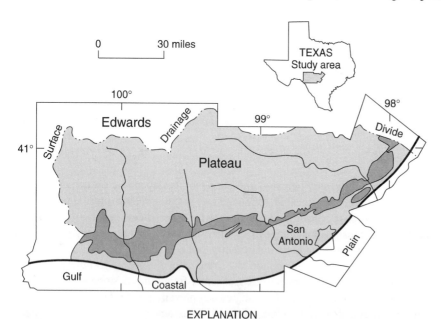

EXPLANATION

▨ Outcrop area of Edwards Aquifer

▢ Artesian area within freshwater zone

── Line separating freshwater zone to the north from saltwater zone to the south

Figure 2.6.2. A large area of karst terrain is associated with the Edwards Aquifer in south-central Texas. Large streams lose a considerable amount of water to ground water as they traverse the outcrop area of the Edwards Aquifer (modified from Brown and Patton[8]).

---

* Awareness of the extensive solution development of limestone along the southern perimeter of Europe continues to increase. Thus, at Gibraltar, St. Michael's Cave was discovered only during fortification work in 1942; it is large enough to serve today as an underground amphitheater for concerts.

Figure 2.6.3. Limestone bedrock from the Edwards Aquifer, a karst formation, is extremely porous and provides for very fast movement of ground water.[21]

2.6.3 Volcanic Rock

Volcanic rock can form highly permeable aquifers; basalt flows in particular often display such characteristics.* The types of openings contributing to the permeability of basalt aquifers include, in order of importance: interstitial spaces in clinkery lava at the tops of flows, cavities between adjacent lava beds, shrinkage cracks, lava tubes, gas vesicles, fissures resulting from faulting and cracking after rocks have cooled, and holes left by the burning of trees overwhelmed by lava.[38] Many of the largest springs in the United States are associated with basalt deposits. Rhyolites are less permeable than basalt, whereas shallow intrusive rocks can be practically impermeable.

2.6.4 Sandstone

Sandstone and conglomerate are cemented forms of sand and gravel. As such, their porosity and yield have been reduced by the cement. The best sandstone aquifers yield water through their joints. Conglomerates have limited distribution and are unimportant as aquifers.

2.6.5 Igneous and Metamorphic Rocks

In solid forms, igneous and metamorphic rocks are relatively impermeable and hence, serve as poor aquifers. Where such rocks occur near the surface under weathered conditions, however, they have been developed by small wells for domestic water supply.

2.6.6 Clay

Clay and coarser materials mixed with clay are generally porous, but their pores are so small that they may be regarded as relatively impermeable. Clayey soils can provide small domestic water supplies from shallow, large-diameter wells.

---

*An excellent example of a highly permeable volcanic aquifer is in Managua, Nicaragua. Here a circular lake contained in an extinct volcanic crater serves as the major municipal water source and yields some 75,000 m³/day. There is no surface inflow, and evaporation exceeds precipitation; hence, the lake, fed entirely by groundwater, acts as a large natural well.

2.7 TYPES OF AQUIFERS

Most aquifers are of large areal extent and may be visualized as underground storage reservoirs, such as the Edwards Aquifer shown in Figure 2.6.2. Water enters a reservoir from natural or artificial recharge; it primarily flows out under the action of gravity or is extracted by wells (see Section 1.6.2). Ordinarily, the annual volume of water removed or replaced represents only a small fraction of the total storage capacity. Aquifers may be classed as unconfined or confined, depending on the presence or absence of a water table, while a leaky aquifer represents a combination of the two types.

2.7.1 Unconfined Aquifer

An *unconfined aquifer* is one in which a water table varies in undulating form and in slope, depending on areas of recharge and discharge, pumpage from wells, and permeability. Rises and falls in the water table correspond to changes in the volume of water in storage within an aquifer. Figure 2.3.1 is an idealized section through an unconfined aquifer; the upper aquifer in Figure 2.7.1 is also unconfined. The outcrop area of the Edwards Aquifer (Figure 2.6.2) is the unconfined portion of an aquifer. Contour maps and profiles of the water table can be prepared from elevations of water in wells that tap the aquifer to determine the quantities of water available and their distribution and movement.

A special case of an unconfined aquifer involves perched water bodies, as illustrated by Figure 2.7.2. This occurs wherever a groundwater body is separated from the main groundwater by a relatively impermeable stratum of small areal extent and by the zone of aeration above the main body of groundwater. Clay lenses in sedimentary deposits often have shallow perched water bodies overlying them. Wells tapping these sources yield only temporary or small quantities of water.

2.7.2 Confined Aquifers

Confined aquifers, also known as *artesian*[*] or *pressure aquifers*, occur where groundwater is confined under pressure greater than atmospheric by overlying relatively impermeable strata.

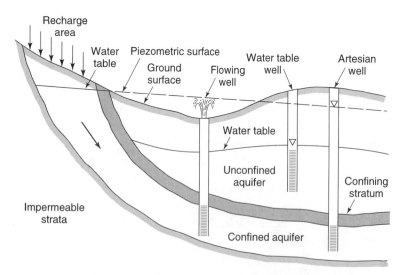

Figure 2.7.1. Schematic cross section illustrating unconfined and confined aquifers.

[*]The word *artesian* has an interesting origin. It is derived from the French *artésien,* meaning "of or pertaining to Artois," the northernmost province of France. Here the first deep wells to tap confined aquifers were drilled and investigated, from about 1750. Originally the term referred to a well with freely flowing water, but at present it is applied to any well penetrating a confined aquifer or simply the aquifer itself.

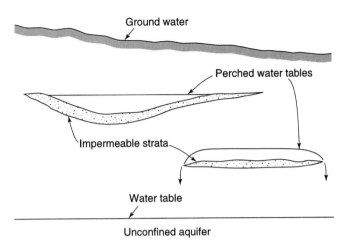

Figure 2.7.2. Sketch of perched water tables.

In a well penetrating such an aquifer, the water level will rise above the bottom of the confining bed, as shown by the artesian and flowing wells of Figure 2.7.1. Water enters a confined aquifer in an area where the confining bed rises to the surface (see Figure 2.6.2); where the confining bed ends underground, the aquifer becomes unconfined. A region supplying water to a confined aquifer is known as a *recharge area;* water may also enter by leakage through a confining bed (see Figure 2.7.1). Rises and falls of water in wells penetrating confined aquifers result primarily from changes in pressure rather than changes in storage volumes. Hence, confined aquifers display only small changes in storage and serve primarily as conduits for conveying water from recharge areas to locations of natural or artificial discharge.

The *piezometric surface*, or *potentiometric surface*, of a confined aquifer is an imaginary surface coinciding with the hydrostatic pressure level of the water in the aquifer (Figure 2.7.1). The water level in a well penetrating a confined aquifer defines the elevation of the piezometric surface at that point. Should the piezometric surface lie above ground surface, a flowing well results. Contour maps and profiles of the piezometric surface can be prepared from well data similar to those for the water table in an unconfined aquifer. It should be noted that a confined aquifer becomes an unconfined aquifer when the piezometric surface falls below the bottom of the upper confining bed. Also, quite commonly an unconfined aquifer exists above a confined one, as shown in Figure 2.7.1.

2.7.3   Leaky Aquifer

Aquifers that are completely confined or unconfined occur less frequently than do *leaky,* or *semiconfined, aquifers.* These are a common feature in alluvial valleys, plains, or former lake basins where a permeable stratum is overlain or underlain by a semipervious aquitard, or semiconfining layer (see Figure 2.7.3). Pumping from a well in a leaky aquifer removes water in two ways: by horizontal flow within the aquifer and by vertical flow through the aquitard into the aquifer.

2.7.4   Idealized Aquifer

For mathematical calculations of the storage and flow of groundwater, aquifers are frequently assumed to be homogeneous and isotropic. A *homogeneous aquifer* possesses hydrologic properties that are everywhere identical. An *isotropic aquifer's* properties are independent of direction. Such idealized aquifers do not exist; however, good quantitative approximations can be obtained by these assumptions, particularly where average aquifer conditions are employed

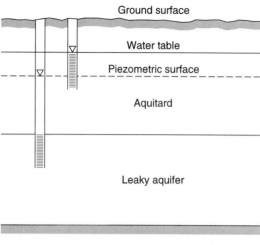

Impermeable strata

Figure 2.7.3. Sketch of a leaky, or semiconfined, aquifer.

on a large scale. Anisotropic aquifers, which possess directional characteristics, are discussed in Chapter 3.

2.8 STORAGE COEFFICIENT

Water recharged to, or discharged from, an aquifer represents a change in the storage volume within the aquifer. For unconfined aquifers this is simply expressed by the product of the volume of aquifer lying between the water table at the beginning and at the end of a period of time and the average specific yield of the formation. In confined aquifers, however, if the aquifer remains saturated, changes in pressure produce only small changes in storage volume. Thus, the hydrostatic pressure within an aquifer partially supports the weight of the overburden while the solid structure of the aquifer provides the remaining support. When the hydrostatic pressure is reduced, such as by pumping water from a well penetrating the aquifer, the aquifer load increases. A compression of the aquifer results, forcing some water from it. In addition, lowering of the pressure causes a small expansion and subsequent release of water. The water-yielding capacity of an aquifer can be expressed in terms of its storage coefficient.

A *storage coefficient* (or *storativity*) is defined as the volume of water that an aquifer releases from or takes into storage per unit surface area of aquifer per unit change in the component of head normal to that surface. For a vertical column of unit area extending through a confined aquifer, as in Figure 2.8.1(a), the storage coefficient S equals the volume of water released from the aquifer when the piezometric surface declines a unit distance. The coefficient is a dimensionless quantity involving a volume of water per volume of aquifer. In most confined aquifers, values fall in the range $0.00005 < S < 0.005$, indicating that large pressure changes over extensive areas are required to produce substantial water yields. Storage coefficients can best be determined from pumping tests of wells (Chapter 4), or from groundwater fluctuations in response to atmospheric pressure or ocean tide variations (see Chapter 6).

The fact that S normally varies directly with aquifer thickness enables the rule-of-thumb relationship[34] for a confined aquifer

$$S = 3 \times 10^{-6}b \qquad (2.8.1)$$

where b is the saturated aquifer thickness in meters to be applied for estimating purposes.

The storage coefficient for an unconfined aquifer corresponds to its specific yield, as shown in Figure 2.8.1(b).

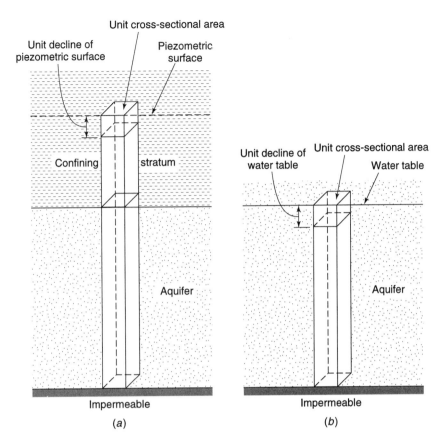

Figure 2.8.1. Illustrative sketches for defining the storage coefficient of (*a*) confined and (*b*) unconfined aquifers.

2.9 GROUNDWATER BASINS/REGIONAL GROUNDWATER FLOW SYSTEMS

A *groundwater basin* may be defined as a hydrogeologic unit containing one large aquifer or several connected and interrelated aquifers.[*] Such a basin may or may not coincide with a physiographic unit. In a valley between mountain ranges, the groundwater basin may occupy only the central portion of the stream drainage basin. In limestone and sandhill areas, drainage and groundwater basins may have entirely different configurations. The concept of a groundwater basin becomes important because of the hydraulic continuity that exists for the contained groundwater resource. In order to ensure continued availability of subsurface water, basin-wide management of groundwater, which is described in Chapter 10, becomes essential.

A regional groundwater flow system can be comprised of subsystems at different scales and a complex hydrogeologic framework, illustrated in Figure 2.9.1. The Edwards Aquifer, illustrated in Figure 2.6.2, is a regional groundwater flow system. Two regional groundwater flow systems, the High Plains Aquifer and the Gulf Coastal Plain Aquifer system, are illustrated in Figure 2.9.2.

2.9.1 High Plains Aquifer

An example of a groundwater basin is the High Plains Aquifer (Figures 2.9.2 and 2.9.3), which underlies an area of about 174,000 square miles extending through parts of Colorado, Kansas, Nebraska, New Mexico, Oklahoma, South Dakota, Texas, and Wyoming. Approximately 20 percent of the irrigated agricultural land in the United States overlies the High Plains aquifer,

---

[*]In practice the term *groundwater basin* is loosely defined; however, it implies an area containing a groundwater reservoir capable of furnishing a substantial water supply.

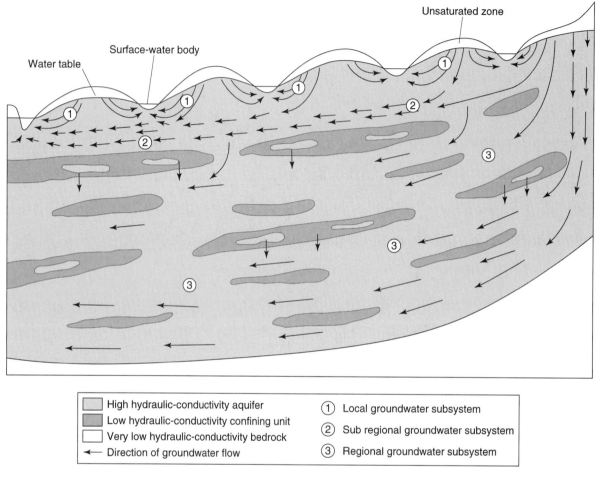

Figure 2.9.1. A regional groundwater flow system that comprises subsystems at different scales and a complex hydrogeologic framework (modified from Sun[53]).

Significant features of this depiction of part of a regional groundwater flow system include (1) local groundwater subsystems in the upper water-table aquifer that discharge to the nearest surface-water bodies (lakes or streams) and are separated by groundwater divides beneath topographically high areas; (2) a subregional groundwater subsystem in the water-table aquifer in which flow paths originating at the water table do not discharge into the nearest surface-water body but into a more distant one; and (3) a deep, regional groundwater flow subsystem that lies beneath the water-table subsystems and is hydraulically connected to them. The hydrogeologic framework of the flow system exhibits a complicated spatial arrangement of high hydraulic-conductivity aquifer units and low hydraulic-conductivity confining units. The horizontal scale of the figure could range from tens to hundreds of miles.[3]

and about 30 percent of the groundwater used for irrigation is withdrawn from the aquifer (U.S. Geological Survey[57]). The aquifer boundary approximates the boundary of the Great Plains Physiographic Province, which is characterized by a flat to gently rolling land surface and moderate precipitation. The region is underlain by sedimentary rocks that dip gently to the east and are upturned with their contact with the Rocky Mountains and other dome mountains, such as the Black Hills. Figure 2.9.3 shows the regional subdivisions of the High Plains Aquifer: the Northern High Plains, the Central High Plains, and the Southern High Plains.

The altitude and configuration of the water table in the High Plains Aquifer is affected mostly by the altitude and configuration of the underlying bedrock surface, the transmissivity of the aquifer, and the rate and distribution of recharge and discharge. Large areas of the aquifer are not continuously saturated or are saturated only in isolated channels in the bedrock surface. The water table in other parts of the aquifer is continuous and slopes eastward at gra-

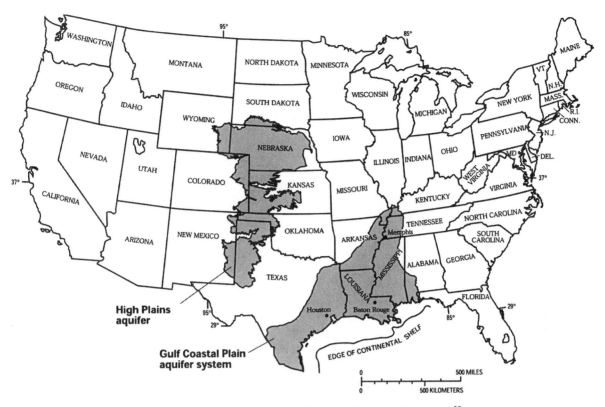

Figure 2.9.2. Location of the High Plains Aquifer and the Gulf Coastal Plain Aquifer system.[55]

dients of about 10 to 40 ft/mi. In most locations, the water levels have declined since irrigation withdrawal became widespread.

Irrigation with groundwater began in the 1800s using windmill-powered pumps, with large-scale withdrawals beginning after the 1930s drought. Because of the large water-level declines in the aquifer, the saturated thickness has substantially decreased, significantly increasing the cost to obtain the water. The result has been to use deeper wells and larger pumps, leading to higher energy costs. Figure 2.9.3 shows the location of selected wells whose hydrographs are plotted in Figure 2.9.4. These hydrographs clearly indicate historical water-level declines. The effects of the groundwater withdrawals on the southern High Plains Aquifer are illustrated in Figure 2.9.5. Figure 2.9.6 shows the changes in groundwater levels and the percent change in saturated thickness of the aquifer from predevelopment to 1997.

2.9.2   Gulf Coastal Plain Aquifer System

The Gulf Coastal Aquifer system (Figure 2.9.2) consists of a large, complex system of aquifers and confining units that underlie about 290,000 square miles.[55] This system extends from Texas to westernmost Florida, including offshore areas to the edge of the Continental Shelf. The aquifer system has been coarsely subdivided into 17 regional aquifers and confining units, most of which are shown in Figure 2.9.7. Groundwater withdrawals have significantly altered the groundwater conditions in the aquifer system, resulting in large-scale regional changes in direction of horizontal flows, changes in vertical direction of flow between aquifers, increases in regional recharge to aquifers, and decreases in regional discharge from aquifers.[55] The simulated widespread reversal of vertical-flow directions from predevelopment to 1987 for the upper part of the Gulf Coastal Plain Aquifer is shown in Figure 2.9.8.

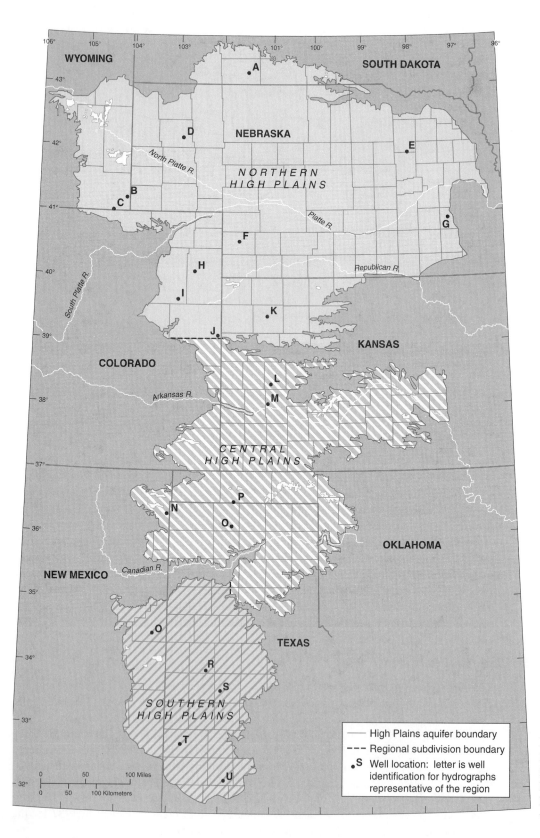

Figure 2.9.3. Regional subdivisions of the High Plains Aquifer, and location of selected wells.[39]

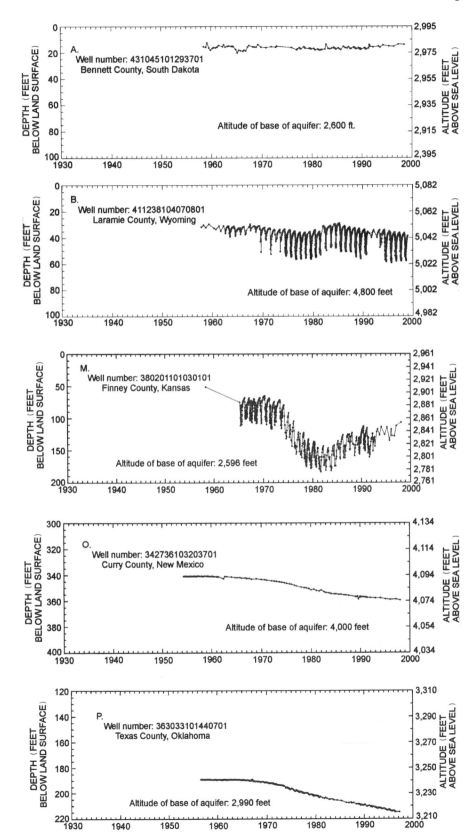

Figure 2.9.4. 1998 Hydrographs for selected wells in the High Plains Aquifer.[39]

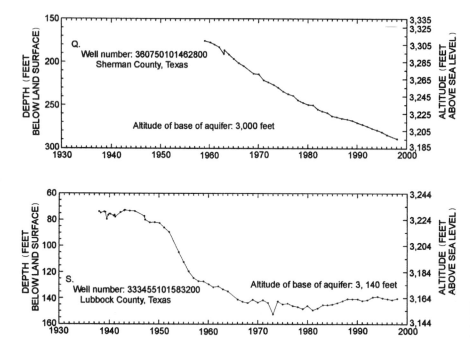

Figure 2.9.4 (continued). 1998 Hydrographs for selected wells is the High Plains Aquifer.[39]

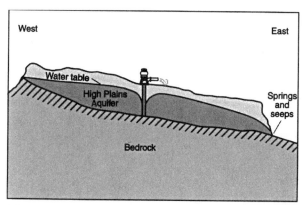

(*a*) vertical scale greatly exaggerated

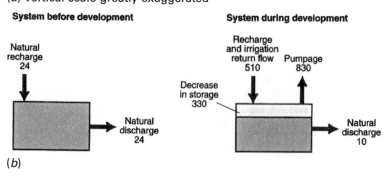

(*b*)

Figure 2.9.5. The effects of groundwater withdrawals on the southern High Plains Aquifer.

(*a*) Schematic cross section of the southern High Plains Aquifer illustrating that groundwater withdrawal in the middle of the southern High Plains Aquifer has a negligible short-term effect on the discharge at the boundaries of the aquifer.[33] (*b*) Water budgets of the southern High Plains Aquifer (all flows in million cubic feet per day) before development and during development.[3, 26, 36]

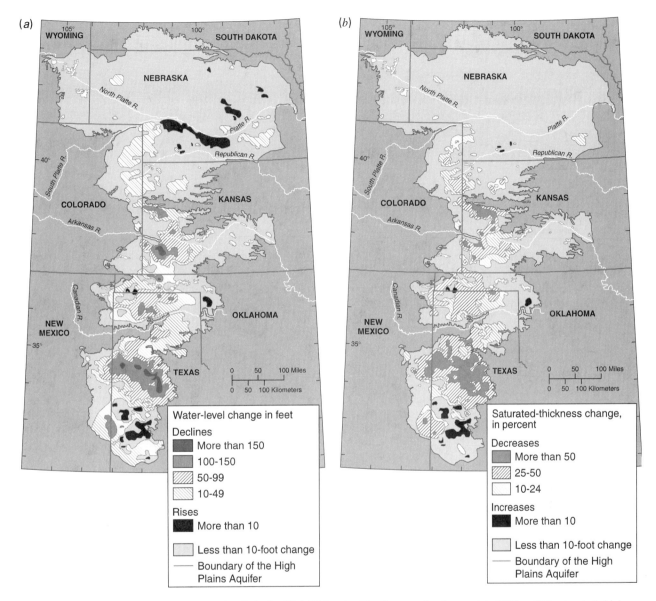

Figure 2.9.6. (*a*) Changes in groundwater levels in the High Plains Aquifer from predevelopment to 1997 and (*b*) saturated thickness in the High Plains Aquifer from predevelopment to 1997. (V. L. McGuire, U.S. Geological Survey, written commun., 1998.)

   Extensive pumping of ground water for irrigation has led to groundwater-level declines in excess of 100 feet in parts of the High Plains Aquifer in Kansas, New Mexico, Oklahoma, and Texas. These large water-level declines have led to reductions in saturated thickness of the aquifer exceeding 50 percent of the predevelopment saturated thickness in some areas. Lower groundwater levels cause increases in pumping lifts. Decreases in saturated thickness result in declining well yields. Surface-water irrigation has resulted in water-level rises in some parts of the aquifer system, such as along the Platte River in Nebraska.[3]

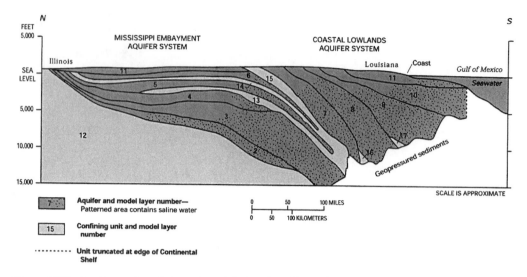

Figure 2.9.7. Aquifers and confining units and designation of layers in a regional model of the Gulf Coastal Plain Aquifer system[69] (as presented in Taylor and Alley[55]).

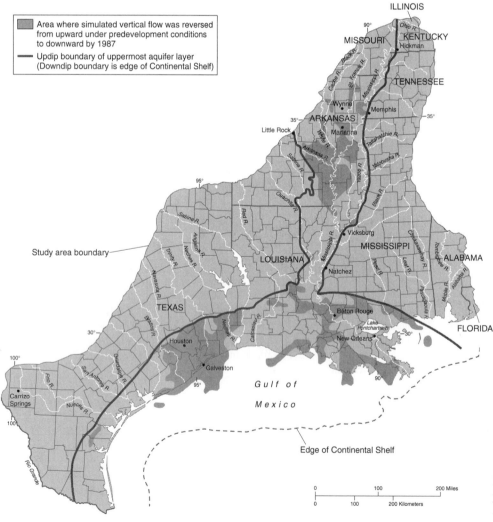

Figure 2.9.8. Areas where vertical flow between uppermost aquifer layers reversed from upward under predevelopment conditions to downward by 1987, as simulated in regional model of Gulf Coastal Plain Aquifer system[22] (as presented in Taylor and Alley[55]).

2.10   SPRINGS

2.10.1   What Are Springs?

A *spring* is a concentrated discharge of groundwater appearing at the ground surface as a current of flowing water. To be distinguished from springs are *seepage areas,** which indicate a slower movement of groundwater to the ground surface. Water in seepage areas may pond and evaporate or flow, depending on the magnitude of the seepage, the climate, and the topography.

Springs occur in many forms and have been classified as to cause, rock structure, discharge, temperature, and variability. Bryan[10] divided all springs into (1) those resulting from nongravitational forces and (2) those resulting from gravitational forces. Under the former category are included *volcanic springs*, associated with volcanic rocks, and *fissure springs*, resulting from fractures extending to great depths in the earth's crust. Such springs are usually thermal (see following section).

Gravity springs result from water flowing under hydrostatic pressure; the following general types are recognized (see Figure 2.10.1):[10]

1. *Depression Springs*—Formed where the ground surface intersects the water table.

2. *Contact Springs*—Created by a permeable water-bearing formation overlying a less permeable formation that intersects the ground surface.

3. *Artesian Springs*—Resulting from releases of water under pressure from confined aquifers either at an outcrop of the aquifer or through an opening in the confining bed.

4. *Impervious Rock Springs*—Occurring in tubular channels or fractures of impervious rock.

5. *Tubular or Fracture Springs*—Issuing from rounded channels, such as lava tubes or solution channels, or fractures in impermeable rock connecting with groundwater.

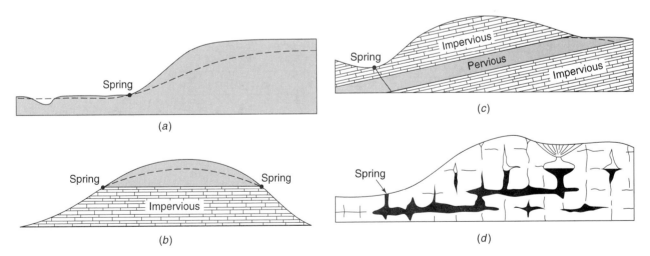

Figure 2.10.1. Diagrams illustrating types of gravity springs. (*a*) Depression spring. (*b*) Contact spring. (*c*) Fracture artesian spring. (*d*) Solution tubular spring (after Bryan,[10] copyright © 1919 by the University of Chicago Press).

---

* *Seepage* is a general term describing the movement of water through the ground or other porous media to the ground surface or surface water bodies. The term is well established in the engineering literature in connection with groundwater movement from and to surface water bodies, particularly where associated with structures such as dams, canals, and levees.

To define magnitudes of springs, Meinzer proposed a classification by discharge shown in Table 2.10.1. Large-magnitude springs occur primarily in volcanic and limestone terrains.[43]

The discharge of a spring depends on the area contributing recharge to the aquifer and the rate of recharge. Adequate spring water to supply the needs of a single family can be obtained from a few hectares, whereas large areas with high rainfalls are necessary to produce a first-magnitude spring. Figure 2.10.2 shows the contributing recharge area ($795 km^2$, 310 square mi) for the Maramec Spring, near Rolla, Missouri. The spring discharges about 1,555 ft^3/sec ($4.4 m^3$/s) into the nearby Maramec River, more than doubling the flow in the river most of the time.[58]

Most springs fluctuate in their rate of discharge. Fluctuations are in response to variations in rate of recharge with periods ranging from minutes to years, depending on geologic and hydrologic conditions. Figure 2.10.3 shows the relationship among precipitation, discharge, and dissolved calcium,[65] indicating the mixing of differing water sources emerging from the spring. The groundwater and surface-water basins coincide with karst features such as caves, sinkholes, losing stream reaches, springs, and conduits in the sedimentary features. Perennial springs drain extensive permeable aquifers and discharge throughout the year, whereas intermittent springs discharge only during portions of the year when sufficient groundwater is recharged to maintain flow. Areas of volcanic rock and sandhills are noted for their perennial springs of nearly constant discharge. Springs that exhibit more or less regular discharge fluctuations not associated with rainfall or seasonal effects are periodic springs. Such fluctuations may be caused by variations in transpiration, by atmospheric pressure changes, by tides affecting confined aquifers, and by natural siphons acting in underground storage basins.

In coastal areas containing limestone or volcanic rock aquifers, large subsurface channels often discharge groundwater through openings to the sea. Such submarine springs can be found along the borders of the Mediterranean Sea and also in Hawaii. Where the discharge is sufficiently large, potable water can be lifted directly from the sea surface.*

Table 2.10.1 Classification of Springs by Discharge (after Meinzer[41])[a]

Magnitude	Mean discharge
First	$>10 m^3$/s
Second	$1-10 m^3$/s
Third	$0.1-1 m^3$/s
Fourth	10–100 l/s
Fifth	1–10 l/s
Sixth	0.1–1 l/s
Seventh	10–100 ml/s
Eighth	<10 ml/s

[a]Another discharge classification of springs, also proposed by Meinzer and based on English units, has been in use for many years in the United States.

*Lucretius, a Roman poet and philosopher of the first century B.C., described a submarine spring in the Mediterranean Sea in his epic poem *De Rerum Natura:* "In the sea at Arados is a fountain of this kind, which wells up with fresh water and keeps off the salt waters all around it . . . a seasonable help in need to thirsting sailors, vomiting forth fresh waters amid the salt."

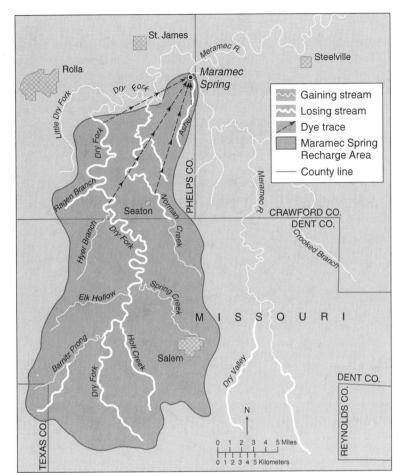

Figure 2.10.2. Gaining and losing streams, dye traces, and the recharge area for Maramec Spring.[58]

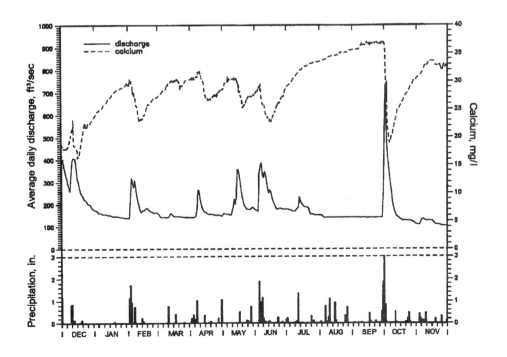

Figure 2.10.3. The relationship between precipitation, discharge, and dissolved calcium, Maramec Spring, 1985–1986.[58]

2.10.2   Edwards Aquifer—Discharge of Springs

The recharge to the Edwards Aquifer (Figure 2.6.2) is derived mainly from seepage into the aquifer from streams that cross the outcrop of the aquifer and direct infiltration of precipitation on the outcrop. The watershed areas providing recharge to the aquifer are shown in Figure 2.10.4. Net recharge into the aquifer from the Guadalupe River Basin is negligible[48]. The estimated annual recharge to the aquifer by basin for 1980 to 2001 is listed in Table 2.10.2. The estimated average annual recharge for the 1980 to 2001 time period is 684,700 ac-ft.

Discharge from the Edwards Aquifer is from wells and springs (refer to Figure 2.10.5). The major discharge from wells is in Bexar, Medina, and Uvalde Counties. The City of San Antonio, Texas, obtains all of its water supply from the aquifer. Table 2.10.3 lists the estimated annual discharge from the Edwards Aquifer by county and lists the portion from wells and from springs. Note the relation of discharge from springs as compared to well discharge as a function of time. Discharge from the Comal Springs (Figure 2.10.6) and San Marcos Springs for 2001 was 414,800 ac-ft, accounting for 78 percent of the discharge for that year.

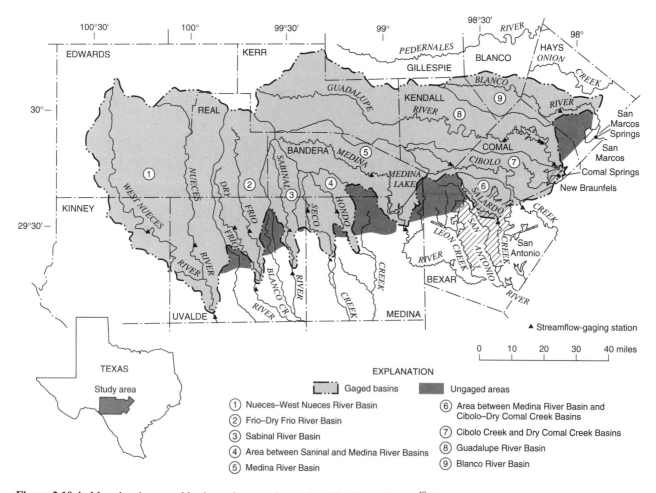

Figure 2.10.4. Map showing gaged basins and ungaged areas (modified from Puente,[48] Figure 1, as presented by Slattery and Thomas[49]).

Table 2.10.2 Estimated Annual Recharge to the Edwards Aquifer by Basin, 1980–2001 (thousands of acre-feet)

Calendar year	Nueces–West Nueces River Basin	Frio-Dry Frio River Basin[1]	Sabinal River Basin[1]	Area between Sabinal and Medina River Basins[1]	Medina River Basin[2]	Area between Medina River Basin and Cibolo–Dry Comal Creek Basin[1]	Cibolo Creek and Dry Comal Creek Basins	Blanco River Basin[1]	Total[3]
1980	58.6	85.6	42.6	25.3	88.3	18.8	55.4	31.8	406.4
1981	205.0	365.2	105.6	252.1	91.3	165.0	196.8	67.3	1,448.4
1982	19.4	123.4	21.0	90.9	76.8	22.6	44.8	23.5	422.4
1983	79.2	85.9	20.1	42.9	74.4	31.9	62.5	23.2	420.1
1984	32.4	40.4	8.8	18.1	43.9	11.3	16.9	25.9	197.9
1985	105.9	186.9	50.7	148.5	64.7	136.7	259.2	50.7	1,003.3
1986	188.4	192.8	42.2	173.6	74.7	170.2	267.4	44.5	1,153.7
1987	308.5	473.3	110.7	405.5	90.4	229.3	270.9	114.9	2,003.6
1988	59.2	117.9	17.0	24.9	69.9	12.6	28.5	25.5	355.5
1989	52.6	52.6	8.4	13.5	46.9	4.6	12.3	23.6	214.4
1990	479.3	255.0	54.6	131.2	54.0	35.9	71.8	41.3	1,123.2
1991	325.2	421.0	103.1	315.2	52.8	84.5	109.7	96.9	1,508.4
1992	234.1	586.9	201.1	566.1	91.4	290.6	286.6	228.9	2,485.7
1993	32.6	78.5	29.6	60.8	78.5	38.9	90.9	37.8	447.6
1994	124.6	151.5	29.5	45.1	61.1	34.1	55.6	36.6	538.1
1995	107.1	147.6	34.7	62.4	61.7	36.2	51.1	30.6	531.3
1996	130.0	92.0	11.4	9.4	42.3	10.6	14.7	13.9	324.3
1997	176.9	209.1	57.0	208.4	63.3	193.4	144.2	82.3	1,134.6
1998	141.5	214.8	72.5	201.4	80.3	86.2	240.9	104.7	1,142.3
1999	101.4	136.8	30.8	57.2	77.1	21.2	27.9	21.0	473.5
2000	238.4	123.0	33.1	55.2	53.4	28.6	48.6	34.1	614.5
2001	297.5	126.7	66.2	124.1	90.0	101.5	173.7	89.7	1,069.4
Average	121.2	134.1	42.2	107.2	61.9	69.9	105.0	43.1	684.7

[1] Includes recharge from ungaged areas (Figure 2.10.4).

[2] Recharge to Edwards aquifer from the Medina River Basin consists entirely of losses from Medina Lake (Puente[48]).

[3] Total might not equal sum of basin values due to rounding.

Source: Slattery and Thomas[49]

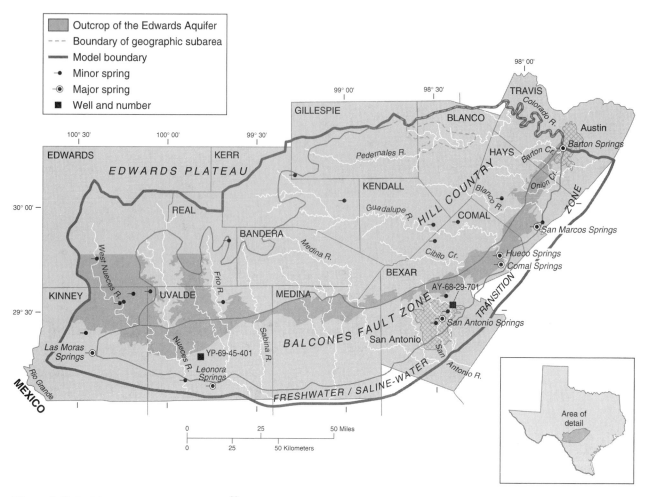

Figure 2.10.5. Edwards Aquifer with springs.[30]

Figure 2.10.6. Comal Springs, Edwards Aquifer.

The highly productive Edwards Aquifer, the first aquifer to be designated as a sole source aquifer under the Safe Drinking Water Act, is the source of water for more than one million people in San Antonio, Texas, some military bases and small towns, and for south-central Texas farmers and ranchers. The aquifer also supplies water to sustain threatened and endangered species habitat associated with natural springs in the region and supplies surface water to users downstream from the major springs. These various uses are in direct competition with groundwater development and have created challenging issues of groundwater management in the region (photograph by Robert Morris, U.S. Geological Survey).

Table 2.10.3 Estimated Annual Discharge from the Edwards Aquifer by County, 1986–2001 (thousands of acre-feet)

Calendar year	Kinney–Uvalde Counties	Medina County	Bexar County	Comal County	Hays County	Total[1]	Well discharge	Spring discharge
1980	151.0	39.9	300.3	220.3	107.9	819.4	491.1	328.3
1981	104.2	26.1	280.7	241.8	141.6	794.4	387.1	407.3
1982	129.2	33.4	305.1	213.2	105.5	786.4	453.1	333.3
1983	107.7	29.7	277.6	186.6	118.5	720.1	418.5	301.6
1984	151.1	46.9	309.7	108.9	85.7	702.3	529.8	172.5
1985	156.9	59.2	295.5	200.0	144.9	856.5	522.5	334.0
1986	[2]91.7	41.9	294.0	229.3	160.4	[2]817.3	429.3	[2]388.1
1987	[2]95.1	15.9	326.6	286.2	198.4	[2]922.0	364.1	[2]558.0
1988	[2]156.7	82.2	317.4	236.5	116.9	[2]909.7	540.0	[2]369.8
1989	156.9	70.5	305.6	147.9	85.6	766.6	542.4	224.1
1990	118.1	69.7	276.8	171.3	94.1	730.0	489.4	240.6
1991	76.6	25.6	315.5	221.9	151.0	790.6	436.3	354.3
1992	76.5	9.3	370.5	412.4	261.3	1,130.2	327.3	802.8
1993	107.5	17.8	371.0	349.5	151.0	996.7	407.3	589.4
1994	95.5	41.1	297.7	269.8	110.6	814.8	424.6	390.2
1995	90.8	35.2	272.1	235.0	127.8	761.0	399.6	361.3
1996	117.6	66.3	286.8	150.2	84.7	705.6	493.6	212.0
1997	[3]29.9	[3]7.0	[3]255.3	243.3	149.2	[3]684.7	[3]300.7	384.0
1998	113.1	51.3	312.8	271.4	169.2	915.9	451.7	464.1
[4]1999	[5]99.8	48.3	298.3	295.2	142.3	884.0	427.8	456.2
2000	89.1	45.1	283.6	226.1	108.4	752.3	414.8	337.5
2001	68.7	33.9	291.6	327.4	175.3	896.9	367.7	529.1

[1] Total might not equal sum of county values due to rounding.

[2] Differs from value in Edwards Underground Water District Bulletins 46–48, table 3, due to correction of an error in the method of computing the Leona Formation underflow.

[3] Does not include irrigation discharge (Bexar, Medina, and Uvalde Counties).

[4] Does not include discharges for domestic supply, stock, and miscellaneous use.

[5] Does not include discharge from Kinney County.

Source: Slattery and Thomas[49]

2.11 HYDROTHERMAL PHENOMENA

2.11.1 Thermal Springs

Thermal springs discharge water having a temperature in excess of the normal local groundwater. The relative terms *warm springs* and *hot springs* are common. Waters of thermal springs are usually highly mineralized and consist for the most part of meteoric water that has been modified in quality by its passage underground.[60]

Hydrothermal phenomena involving the release of water and steam are nearly always associated with volcanic rocks and tend to be concentrated in regions where large geothermal gradients occur. Also, by implication, aquifers must be present that permit water to percolate to great depths—often 1,500 to 3,000 m. This water, heated from below, forms a large convective current that rises to supply hydrothermal areas (see Figure 2.11.1).

A *geyser** is a periodic thermal spring resulting from the expansive force of superheated steam within constricted subsurface channels (see Figure 2.11.2). Water from surface sources

---

*The word *geyser* is derived from the Icelandic word *geysir,* meaning to gush or rage.

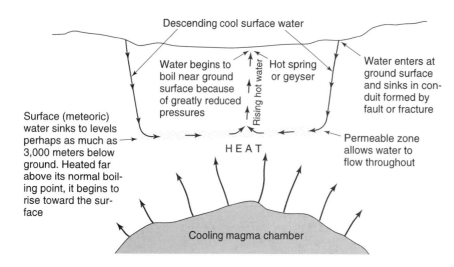

Descending cool surface water

Water begins to boil near ground surface because of greatly reduced pressures

Rising hot water

Hot spring or geyser

Water enters at ground surface and sinks in con-duit formed by fault or fracture

Surface (meteoric) water sinks to levels perhaps as much as 3,000 meters below ground. Heated far above its normal boil-ing point, it begins to rise toward the sur-face

HEAT

Permeable zone allows water to flow throughout

Cooling magma chamber

Figure 2.11.1. Schematic diagram of a hydrothermal system (after Keefer[29]).

Figure 2.11.2. Eruptions of the Midway Geysers in Yellowstone National Park, Wyoming (after Keefer[29]).

and/or shallow aquifers drains downward into a deep vertical tube where it is heated to above the boiling point. With increasing pressure the steam pushes upward; this releases some water at the surface, which reduces the hydrostatic pressure and causes the deeper superheated water to accelerate upward and to flash into steam. The geyser then surges into full eruption for a short interval until the pressure is dissipated; thereafter, the filling begins again and the cycle is repeated.

Another kind of hot spring, known as a *mudpot,* results when only a limited supply of water is available. Here water mixes with clay and undissolved particles brought to the surface, forming a muddy suspension by the small amount of water and steam continuing to bubble to the surface. A *fumarole** is an opening through which only steam and other gases such as car-

---

*The word *fumarole* stems from the Latin *fumus,* meaning smoke.

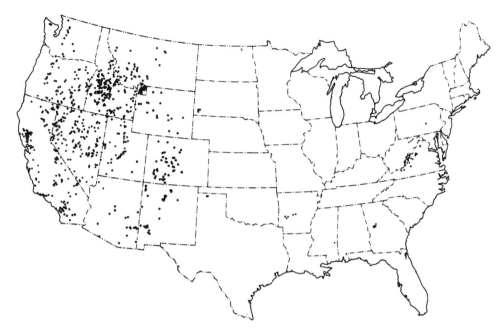

Figure 2.11.3. Thermal springs in the United States (after Waring[60]).

bon dioxide and hydrogen sulfide discharge. These features are normally found on hillsides above the level of flowing thermal springs; water can often be heard boiling underground.[46]

Thermal springs of various kinds are found throughout the world[59] notable areas exist in Iceland, New Zealand, and the Kamchatka Peninsula of Russia. Figure 2.11.3 shows the regional distribution of thermal springs in the United States. Yellowstone National Park in Wyoming, which contains literally thousands of hydrothermal features, possesses the greatest concentration of thermal springs in the world.[29] This area marks the site of an enormous volcanic eruption 600,000 years ago. Today, temperatures of 240°C exist only 300 m below ground surface.

2.11.2   Geothermal Energy Resources

Heat within the earth flows outward at an average rate of 1.5×10^{-6} cal/cm^2/s^{-1} and creates an average geothermal gradient of 1°C/50 m. But in areas of volcanic and tectonic activity, heat flows several orders of magnitude larger than normal have been found. These tremendous reservoirs of heat close to ground surface have been recognized as invaluable sources of energy.[18, 52] But most important from a hydrologic standpoint is the fact that utilization of these geothermal resources invariably involves water as the mechanism for extracting heat. Four types of sources are generally recognized:

1. *Dry Steam Field*—Permeability is so low that groundwater reaching the heat source is limited and is fully vaporized.
2. *Wet Steam Field*—Sufficient groundwater reaches the heat source so that a mixture of water and steam is produced.
3. *Low-Temperature Fields*—Temperatures in the range of 50–80°C are found at shallow depths, enabling the groundwater to be employed directly for heating purposes.

4. *Dry Rock*—In some areas at depths exceeding 3,000 m, temperatures are found in the range of 200–500°C without the presence of groundwater; by injecting water into fractured rock zones through one drill hole, steam can be produced from an adjacent hole.

Geothermal resources exist through the world, but they have been developed in only a few localities.[*] Most homes in Reykjavik, Iceland, are heated by naturally occurring hot water. Notable power plants generating energy from geothermal resources are located in Italy, New Zealand, and the western United States.

2.12   GROUNDWATER IN PERMAFROST REGIONS

Permafrost, or perennially frozen ground, is defined as unconsolidated deposits or bedrock that continuously have had a temperature below 0°C for two years to thousands of years.[67, 68] Figure 2.12.1 illustrates the upper and lower limits of permafrost in terms of the depths at which a 0°C ground temperature occurs. Regions of permafrost in the Northern Hemisphere are shown in Figure 2.12.2. In the continuous-permafrost zone, permafrost is present everywhere to depths of 150–400 m; while in the more southerly discontinuous-permafrost zone, permafrost is perforated by unfrozen zones that depend on local conditions.

Frozen ground creates an impermeable layer that restricts the movement of groundwater, acts as a confining layer, and limits the volume in which liquid water can be stored. In many areas of frozen ground, shallow aquifers are entirely eliminated, thereby requiring that wells be drilled deeper than in similar geologic environments without permafrost. Groundwater can occur above, below, and locally within permafrost.[13] In the continuous-permafrost zone, the

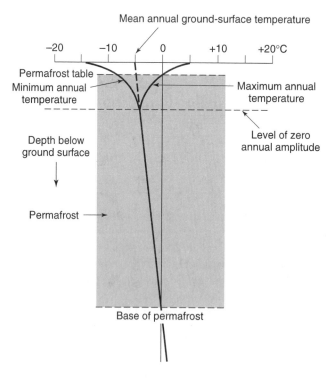

Figure 2.12.1. Location of permafrost below ground surface in relation to ground temperature (after Williams[67]).

---

[*]It has been estimated that the total stored heat in the earth to a depth of 3 km amounts to 2×10^{21} cal and that one percent, or 2×10^{19} cal, of this can be commercially recovered.[63]

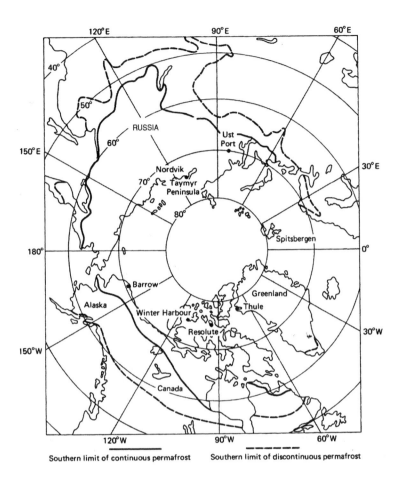

120°E   90°E   60°E

40°

50°

RUSSIA   Ust Port

150°E   60°

Nordvik   30°E
70°   Taymyr
Peninsula
80°

Spitsbergen   0°

180°

Barrow   Greenland
Alaska   Thule
Winter Harbour   30°W
Resolute

150°W

Canada

120°W   90°W   60°W

Southern limit of continuous permafrost   Southern limit of discontinuous permafrost

Figure 2.12.2. Distribution of permafrost in the Northern Hemisphere (after Williams[67]).

best sources of water are in unfrozen alluvium beneath large lakes, in major valleys, and adjacent to riverbeds. In the discontinuous-permafrost zone, groundwater can be produced locally from shallow aquifers; however, because of potential pollution from ground surface, sources beneath the frozen layer are preferable.

2.13   GROUNDWATER IN THE UNITED STATES

Productive aquifers and withdrawals from the wells in the United States are shown in Figures 2.13.1 and 2.13.2. These maps show regions in which moderate to large supplies of water can be obtained from wells. Unconsolidated and consolidated aquifers are indicated in addition to primary geologic formations. Blank areas delineate generally those regions not known to produce yields of more than 3 l/s (48 gpm) to a well.

Areal distribution of groundwater in the United States can best be described by dividing the conterminous United States into the groundwater regions shown in Figures 2.13.1 and 2.13.2. Table 2.13.1 is a summary of the principal physical and hydrologic characteristics of the groundwater regions. Brief geologic and groundwater summaries for each of the 15 regions are provided on the following pages.

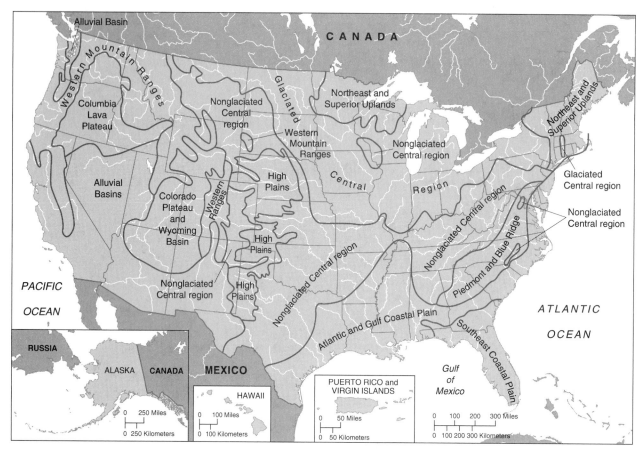

Figure 2.13.1. Groundwater regions in the United States. [The Alluvial Valleys region (region 12) is shown on Figure 2.13.2.][23]

1. Western Mountain Ranges

This mountainous region serves as the principal source of water in the western United States because the bulk of precipitation falls here and thereafter supplies streams and aquifers by its runoff. Rocks are mainly hard and dense; they shed water rather than absorbing it, although weathered surficial rock may locally yield limited groundwater. Some aquifers are to be found in alluvium contained in small intermontane valleys. Because of the thinness and rapid draining of the mantle rock, it is not easy to obtain groundwater from wells. Small springs, wells in valleys, and small surface reservoirs meet most domestic water supply needs.

2. Alluvial Basins

The basins in this region consist of vast depressed areas bounded by adjacent highlands. They are partly filled by erosional debris in the form of alluvium and serve as the storage volumes for water flowing from nearby highlands. The alluvial fill functions as an ideal aquifer and creates the opportunity for development of high-yielding wells. Because of the prevailing arid climate, groundwater development for irrigation is much in demand. Replenishment rates usually are far less than withdrawal rates, so groundwater levels decline as storage is depleted. Locally, artificial recharge (see Chapter 13) has helped to alleviate this problem of overproduction.

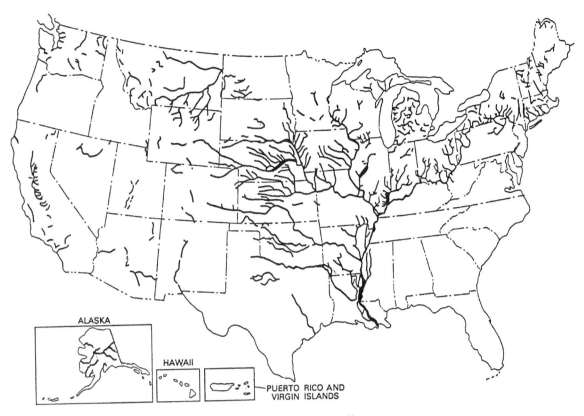

Figure 2.13.2. Alluvial Valleys groundwater region of the United States.[23]

3. Columbia Lava Plateau

This area is formed principally by extrusive volcanic rocks, mainly lava flows, interbedded with or overlain by alluvium and lake sediments. Water originates chiefly from mountains on the perimeter of the region. The lava rocks tend to be highly permeable as a result of tubes and shrinkage cracks and thus form highly productive aquifers. The large volumes of groundwater discharge as major springs or as streams with sustained base flows. Because of the great thickness of the lava flows, groundwater is most readily available in valley bottoms; however, in the higher plateau areas, deep wells are required to extract groundwater for irrigation.

4. Colorado Plateau and Wyoming Basin

This region consists of sedimentary strata, chiefly interbedded sandstone and shale; these are generally horizontal but in places are folded, tilted, or broken by faults. The plateaus are rather high, dry, and deeply dissected by streams. Prospects for large-scale groundwater development are poor; nevertheless, small water supplies for domestic and livestock purposes are widely available. Most aquifers are sandstone beds, although limestone and alluvium yield water in a few places.

5. High Plains

Here alluvium forms a vast plain extending eastward from the Rocky Mountains. The bulk of it is classified as a single stratigraphic unit, the Ogallala Formation, which covers older rocks

Table 2.13.1 Summary of the Principal Physical and Hydrologic Characteristics of the Groundwater Regions of the United States

Region No.	Name	Unconf.: Hydrol. insig.	Unconf.: Minor aquifer / not very productive	Unconf.: Dominant aquifer	Confining beds: Hydrol. insig.	Confining beds: Thin, discontinuous, or very leaky	Confining beds: Interlayered with aquifers	Confined: Hydrol. insig.	Confined: Not highly productive	Confined: Multiple productive aquifers	Confined: The dominant productive aquifer	Arrangement: Single unconfined aquifer	Arrangement: Two interconnected aquifers	Arrangement: Unconfined aquifer, confining bed, confined aquifer	Arrangement: Complex interbedded sequence	Primary: Pores in unconsolidated dep.	Primary: Pores in semiconsolidated rocks	Primary: Tubes and cooling cracks in lava	Secondary: Fractures and faults	Secondary: Solution-enlarged openings	Solubility: Insoluble	Solubility: Mixed soluble and insoluble	Solubility: Soluble	Porosity: Large (>0.2)	Porosity: Moderate (0.01–0.2)	Porosity: Small (<0.01)	Transmissivity: Large (>2,500 m² day⁻¹)	Transmissivity: Moderate (250–2,500 m² day⁻¹)	Transmissivity: Small (25–250 m² day⁻¹)	Transmissivity: Very small (<25 m² day⁻¹)	Recharge: Uplands between streams	Recharge: Losing streams	Recharge: Leakage through confining beds	Discharge: Springs and surface seepage	Discharge: Evaporation and basin sinks	Discharge: Into other aquifers
1	Western Mountain Ranges		×		×											×			×		×				×				×		×	×		×		
2	Alluvial Basins			×			×			×					×	×					×			×			×					×			×	×
3	Columbia Lava Plateau		×			×	×			×					×	×		×	×		×						×					×		×		
4	Colorado Plateau and Wyoming Basin	×					×								×		×		×		×					×	×				×		×	×		
5	High Plains			×	×			×				×				×								×			×							×		
6	Nonglaciated Central Region		×				×			×					×		×		×	×		×			×			×	×		×	×		×		×
7	Glaciated Central Region		×				×			×					×		×		×	×		×				×		×								×
8	Piedmont and Blue Ridge		×		×				×				×						×		×									×	×					
9	Northeast and Superior Uplands		×				×		×				×			×			×		×				×				×		×			×		
10	Atlantic and Gulf Coastal Plain		×				×			×					×	×				×		×				×		×			×		×	×		×
11	Southeast Coastal Plain		×				×				×			×		×				×			×		×			×			×		×	×		
12	Alluvial Valleys			×		×						×				×					×			×		×	×				×			×		
13	Hawaii			×		×						×						×			×			×		×	×				×	×		×		
14	Alaska			×		×									×	×			×			×		×				×				×		×		
15	Puerto Rico and Virgin Islands			×		×				×					×	×				×			×		×			×			×		×	×		×

Source: Heath[23]

80

to thicknesses exceeding 150 m. The sand and the gravel of the formation consitute an aquifer yielding 10 l/s (151 gpm) to more than 60 l/s (451 gpm)of water from individual wells. The region is generally semiarid so that groundwater recharge from precipitation is extremely small. The productiveness of wells has encouraged pumping of groundwater, especially for irrigation in Texas. This water is derived from storage; as a result, water tables have declined substantially for many years.

6. Nonglaciated Central Region

This is a large, complex area characterized by plains and plateaus underlain by consolidated sedimentary rocks. Alluvial deposits of substantial width and thickness form good aquifers but only along major streams. Aquifers in most of the region are dolomitic limestone and sandstone with low to moderate yields. The region includes some of the least productive aquifers in the United States because of low yield, saline water, or both. Wells in some of the karst areas penetrate caverns and exhibit highly variable yields.

7. Glaciated Central Region

Although hydrogeologically similar to the previous region, this area differs fundamentally by the presence of glacial drift deposited by the ice and meltwaters of the continental glaciers. The drift consists mostly of fine-grained rock debris together with beds of water-sorted sand and gravel. In portions of the area, the drift is more than 300 m thick and forms an important aquifer. In this glaciated region, large-diameter wells will yield sufficient water to meet domestic needs of a family. Excellent aquifers can be found along watercourses where rapidly flowing meltwaters removed fine materials and left behind permeable deposits of sand and gravel.

8. Piedmont and Blue Ridge

This mountainous area consists of varying parallel highlands with differing geologic structures, including crystalline rocks, limestone, dolomite, sandstone, and shale. Groundwater productivity ranges from small to moderate or is erratic. Shallow wells can usually obtain small yields for domestic supplies from the weathered rock. Wells of highest average yield occur in the valleys because rocks are more intensely fractured and because of groundwater drainage from the surrounding hills. In the limestone areas, much of the drainage is underground so the chances of obtaining moderately productive wells are good.

9. Northeast and Superior Uplands

Because this region is considerably more rugged than the Glaciated Central Region, there is greater local variation in thickness of the glacial drift. Usually, hilltops are covered by a thin mantle of till; slopes contain even less with bedrock outcrops, and valleys are underlain by thicker drift and till. Near the sea, glacial material is overlain or interbedded with marine deposits. In most of the area, aquifers are not productive. Except for a few areas where bedrock yields moderate quantities of groundwater, the principal groundwater sources are sand and gravel deposits occurring as outwash plains or as channel fillings in the stratified drift.

10. Atlantic and Gulf Coastal Plain

This extensive coastal plain consists of a huge seaward-thickening wedge of generally unconsolidated sedimentary rocks. The sediments are mainly stratified layers of clay, silt, sand, gravel, marl, and limestone. The maximum thickness of deposits beneath the coastline varies from 100 m in the northeast to more than 10,000 m in the south. Essentially all of the area contains extensive and productive aquifers. The principal aquifers are found in sand and gravel beds, while limestone aquifers are most important in the southeastern portion of the region.[51] Problems of seawater intrusion and land subsidence are significant in concentrated pumping localities.

11. Southeast Coastal Plain

The Southeast Coastal Plain region consists of layers of sand clay over semiconsolidated and consolidated carbonate rock. Surface areas are underlain by unconsolidated deposits of clay, sand, gravel, and shell beds, and deeper layers consist of alternating layers of semiconsolidated and consolidated limestones and dolomites. One of the most prolific aquifers in the world is the Floridian Aquifer, which is located in carbonate units. The surficial sand and gravel deposits also have water.

12. Alluvial Valleys Region

The Alluvial Valleys Region (see Figure 2.13.2) consists of thick sand and gravel deposits beneath floodplains and stream terrace deposits. The many river systems in this region have deposited thick sequences of highly porous and permeable sand and gravel—glacial outwash by streams that carried water from melting ice during the Pleistocene era. The resulting long, narrow aquifer systems were the foundations for the modern river systems.

13. Hawaii

Basaltic lava flows make up the bulk of the islands and constitute the most important aquifers. Openings within or between the flows are responsible for the generally high permeability. Near eruptive centers of volcanoes vertical dikes of dense impermeable rock may separate high-level groundwater in inland areas from basal groundwater in the coastal lowlands. The basal water forms a lens of fresh groundwater floating on underlying sea water (see Chapter 14). Groundwater is extensively developed for irrigation and municipal purposes on Maui and Oahu.

14. Alaska

The principal aquifers in this region are bodies of water-sorted sand and gravel within both glacial drift that covers the uplands and glacial outwash and other alluvial deposits extending from the uplands into the lowlands. The most productive aquifers are found in the vast central plateau and on the slopes of the southern mountain system. Permafrost predominates in the permeable deposits of the arctic slope, so groundwater is available only locally near large bodies of surface water. Along the southeastern coastal area, productive alluvial deposits are scarce.

15. Puerto Rico and Virgin Islands

This island region consists of alluvium and limestone overlying and bordering fractured igneous rocks. It is underlain by both limestone and volcanic and intrusive igneous rocks. The islands receive large amounts of rainfall; Puerto Rico has almost 6 ft (2 m) of annual groundwater recharge. The alluvium with sand and gravel in the stream valleys and along the coast is an effective aquifer. There are also limestone areas that are aquifers.

PROBLEMS

2.2.1 A cubical pattern of grain packing is shown below; grain sizes are uniform and each sphere touches all neighboring spheres. Determine the porosity of a sample with this type of grain packing. Does the porosity depend on the grain size?

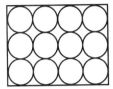

Cubic Packing

2.2.2 Considering the ranges of values for sedimentary materials in Tables 2.2.1 and 2.2.2, comment on a possible relationship between porosity versus grain size. Compare your answer to that of Problem 2.2.1. What conclusion(s) can you draw?

2.2.3 The results of a sieving analysis are tabulated herein. Using these data, prepare a grain size distribution curve for this sample and assess whether the sample is well graded or poorly graded. What would be reasonable porosity values for this sample?

U.S. Standard Sieve Number	Grain size (mm)	Mass retained (g)
4	4.750	11.92
8	2.360	2.66
16	1.180	37.30
20	0.840	87.53
25	0.710	78.07
30	0.600	72.95
40	0.425	98.82
100	0.150	50.85
200	0.075	8.10
Pan	<0.075	1.80
Total sample weight		450.00

2.2.4 The results of a hydrometer test to determine the grain size distribution of a sample of fine sediments are tabulated herein. Plot the grain size distribution and comment on the uniformity of the sample.

Particle grain size (mm)	Percent finer than
1.000	98.900
0.700	95.700
0.500	82.800
0.350	65.600
0.250	32.300
0.175	6.750
0.125	0.600
0.088	0.100
0.0625	0.000

2.5.1 Determine the water level rise in an unconfined aquifer produced by a seasonal precipitation of four inches. The aquifer's porosity is 20 percent and its specific retention is 9 percent.

2.5.2 The leakage from the artificially constructed Tempe Town Lake in Tempe, AZ, can be as low as 0.5 ft/day or as high as 3 ft/day. The lake covers 222 surface acres. If the specific yield of the subsurface formation is 20 percent, estimate the average regional groundwater level rise assuming that the aerial extent of the effect of leakage is

(a) 222 acres

(b) 1 mile^2

(c) 5 mile^2

(d) 25 mile^2

2.5.3 How much water can be produced by lowering the water table of an unconfined aquifer 7 ft over an area of 1 mile^2? The aquifer's porosity and specific retention are 0.38 and 0.15, respectively.

2.8.1 The coefficient of storage of a confined aquifer is found to be 6.8×10^{-4} as a result of a pumping test. The thickness of the aquifer is 50 m and the porosity of the aquifer is 0.25. Determine the fractions of the expansibility of water and compressibility of the aquifer skeleton in making up the storage coefficient of the aquifer. (Refer to Example 2.5.2.)

2.8.2 The specific storage of a 45-m thick confined aquifer is 3.0×10^{-5} m^{-1}. How much water would the aquifer produce if the piezometric surface is lowered by 10 m over an area of 1 km^2?

2.13.1 Prepare a summary (with illustrative figures) addressing the following issues for one or more of the regions listed below:

- Geographical area
- Climate
- Geology
- Surface water (short description of major surface waters)
- Groundwater (types and locations of aquifers and other conditions that have a bearing on the groundwater of the region)
- Groundwater facts (recharge, uses, pollution, salt water intrusion, etc.)

The regions are to be selected from the following:

- Africa
- Artic and Antarctic
- Alaska, Canada, and Greenland
- Australia and Oceana
- China
- Central Europe
- Eastern Europe
- Great Britain
- India

(continued on next page)

- Indochina
- Mexico and Central America
- Middle Eastern Countries

- Russia
- Scandinavian Countries
- South America

REFERENCES

1. Agricultural Research Service, *Field Manual for Research in Agricultural Hydrology*, Agric. Handbk. no. 224, U.S. Dept. of Agric., 215 pp., 1962.

2. Alley, W. M., R. W. Healy, J. W. LaBaugh, and T. E. Reilly, Flow and storage in groundwater systems, *Science,* v. 206, June 14, 2002.

3. Alley, W.M., T. E. Reilly, and O. L. Franke, *Sustainability of Ground-Water Resources, U.S. Geological Survey Circular* 1186, http://water.usgs.gov/pubs/circ/circ1186, U.S. Geological Survey, Denver, CO, 1999.

4. Baver, L. D., et al., *Soil Physics*, 4th ed., John Wiley & Sons, New York, 498 pp., 1972.

5. Bear, J., *Dynamics of Fluids in Porous Media,* Dover Publications, Mineola, NY, 1972.

6. Black, C. A. (ed.), *Methods of Soil Analysis*, Part 1, Agronomy Monograph no. 9, Amer. Soc. Agronomy, Madison, WI, 770 pp., 1965.

7. Blank, N. R., and M. C. Schroeder, Geologic classification of aquifers, *Ground Water,* v. 11, no. 2, pp. 3–5, 1973.

8. Brown, D. S. and J. T. Patton, *Recharge to and from the Edwards Aquifer in the San Antonio Area, Texas, U.S. Geological Survey Open-File Report* 96-181, 1995.

9. Brucker, R. W., et al., Role of vertical shafts in the movement of ground water in carbonate aquifers, *Ground Water,* v. 10, no. 6., pp. 5–13, 1972.

10. Bryan, K., Classification of springs, *Jour. Geology,* v. 27, pp. 522–561, 1919.

11. Burdon, D. J., and N. Papakis, *Handbook of Karst Hydrogeology with Special Reference to the Carbonate Aquifers of the Mediterranean Region,* Inst. Geology and Subsfc. Research, United Nations Spec. Fund, Karst Groundwater Inv., Athens, 276 pp., 1963.

12. Burdon, D. J., and C. Safadi, The karst groundwaters of Syria, *Jour. Hydrology,* v. 2. pp. 324–347, 1964.

13. Cederstrom, D. J., et al., *Occurrence and Development of Ground Water in Permafrost Regions, U.S. Geological Survey Circular* 275, 30 pp., 1953.

14. Charbeneau, R. J., *Groundwater Hydraulics and Pollutant Transport,* Prentice Hall, Upper Saddle River, NJ, 2000.

15. Childs, E. C., *An Introduction to the Physical Basis of Soil Water Phenomena*, John Wiley & Sons, London, 493 pp., 1969.

16. Dept. of Economic and Social Affairs, *Ground-Water Storage and Artificial Recharge*, Natural Resources, Water Ser. no. 2, United Nations, New York, 270 pp., 1975.

17. Dos Santos, A. G., Jr., and E. G. Youngs, A study of the specific yield in land-drainage situations, *Jour. Hydrology,* v. 8, pp. 59–81, 1969.

18. Dutcher, L. C., *Preliminary Appraisal of Ground Water in Storage with Reference to Geothermal Resources in the Imperial Valley Area, California, U.S. Geological Survey Circular* 649, 57 pp., 1972.

19. Epstein, J. B., D. J. Weary, R. C. Orndorff, Z. C. Bailey, and R. C. Kerbo, *National Karst Map Project: An Update*, in U.S. Geological Survey Karst Interest Group Proceedings, Shepherdstown, West Virginia, August 20–22, 2002, ed. E. L. Kuniansky, *U.S. Geological Survey Water-Resources Investigations Report* 02-4174, 2002.

20. Feth, J. H., *Selected References on Saline Ground-Water Resources of the United States, U.S. Geological Survey Circular* 499, 30 pp., 1965.

21. Groundwater Management Districts Association, Newsletter, v. 2, no. 2, http://www.gmdausa.org/newsletter.htm, Spring 2002.

22. Grubb, H. F., *Summary of Hydrology of the Regional Aquifer Systems, Gulf Coastal Plain, South Central United States, U.S. Geological Survey Professional Paper* 1416-A, 1998.

23. Heath, R. C., *Groundwater Regions of the United States, U.S. Geological Survey Water-Supply Paper* 2241, Washington, DC, 1984.

24. Johnson, A. I., *Methods of Measuring Soil Moisture in the Field, U.S. Geological Survey Water-Supply Paper* 1619-U, 25 pp., 1962.

25. Johnson, A. I., *Specific Yield—Compilation of Specific Yields for Various Materials, U.S. Geological Survey Water-Supply Paper* 1662-D, 74 pp., 1967.

26. Johnston, R. H., The hydrologic responses to development in regional sedimentary aquifers, *Ground Water,* v. 27, no. 3, pp. 316–322, 1989.

27. Jones, O. R., and A. D. Schneider, Determining specific yield of the Ogallala aquifer by the neutron method, *Water Resources Research*, v. 5, pp. 1267–1272, 1969.

28. *Karst in China*, Institute of Hydrogeology and Engineering Geology, Chinese Academy of Geological Sciences, Shanghai People's Publishing House, 1976.

29. Keefer, W. R., *The Geologic Story of Yellowstone National Park, U.S. Geological Survey Bull.* 1347, 92 pp., 1971.

30. Kuniansky, E. L., L. Fahlquist, and A. F. Ardis, *Travel Times along Selected Flow Paths of the Edwards Aquifer, Central Texas,* in U.S. Geological Survey Karst Interest Group Proceedings, ed. E. L. Kuniansky, *U.S. Geological Survey Water-Resources Investigations Report* 01-4011, 2001.

31. LaMoreaux, P. E., et al., Hydrology of limestone terranes, annotated bibliography of carbonate rocks, *Geological Survey Alabama Bull.* 94(A), 242 pp., 1970.

32. Lattman, L. H., and R. R. Parizek, Relationship between fracture traces and the occurrence of ground water in carbonate rocks, *Jour. Hydrology,* v. 2, pp. 73–91, 1964.

33. LeGrand, H. E., and V. T. Stringfield, Karst hydrology—A Review, *Jour. Hydrology,* v. 20, pp. 97–120, 1973.

34. Lohman, S. W., *Ground-Water Hydraulics, U.S. Geological Survey Professional Paper* 708, 70 pp., 1972.

35. Lohman, S. W., et al., *Definitions of Selected Ground-Water Terms—Revisions and Conceptual Refinements, U.S. Geological Survey Water-Supply Paper* 1988, 21 pp., 1972.

36. Luckey, R. R., E. D. Gutentag, F. J. Heimes, and J. B. Weeks, *Digital Simulation of the Ground-Water Flow in the High Plains Aquifer in Parts of Colorado, Kansas, Nebraska, New Mexico, Oklahoma, South Dakota, Texas, and Wyoming, U.S. Geological Survey Professional Paper* 1400-D, 1986.

37. Manger, G. E., *Porosity and Bulk Density of Sedimentary Rocks, U.S. Geological Survey Bull.* 1144-E, 55 pp., 1963.

38. Maxey, G. B., and J. E. Hackett, Applications of geohydrologic concepts in geology, *Jour. Hydrology*, v. 1, pp. 35–46, 1963.

39. McGuire, V. L., and B. C. Fisher, *Water-Level Changes in the High Plains Aquifer—1980–1998, U.S. Geological Survey Web Report HP 98*, http://www-ne.cr.usgs.gov/highplains/hp98_web_report/hp98fs.htm, 1998.

40. McQueen, I. S., Evaluating the reliability of specific-yield determinations, *Jour. Research U.S. Geological Survey*, v. 1, pp. 371–376, 1973.

41. Meinzer, O. E., *Outline of Ground-Water Hydrology with Definitions*, U.S. Geological Survey Water-Supply Paper 494, 71 pp., 1923.

42. Meinzer, O.E., *The Occurrence of Ground Water in the United States*, U.S. Geological Survey Water-Supply Paper 489, 321 pp., 1923.

43. Meinzer, O. E., *Large Springs in the United States*, U.S. Geological Survey Water-Supply Paper 557, 94 pp., 1927.

44. Monroe, W. H., *A Glossary of Karst Terminology*, U.S. Geological Survey Water-Supply Paper 1899-K, 26 pp., 1970.

45. Morris, D. A., and A. I. Johnson, *Summary of Hydrologic and Physical Properties of Rock and Soil Materials*, as Analyzed by the Hydrologic Laboratory of the U.S. Geological Survey 1948–1960, *U.S. Geological Survey Water-Supply Paper* 1839-D, 42 pp., 1967.

46. Poland, J. F., et al., *Glossary of Selected Terms Useful in Studies of the Mechanics of Aquifer Systems and Land Subsidence Due to Fluid Withdrawal*, U.S. Geological Survey Water-Supply Paper 2025, 9 pp., 1972.

47. Prill, R. C., *Specific Yield—Laboratory Experiments Showing the Effect of Time on Column Drainage*, U.S. Geological Survey Water-Supply Paper 1662-B, 55 pp., 1965.

48. Puente, C., *Method of Estimating Natural Recharge to the Edwards Aquifer in the San Antonio Area, Texas*, U.S. Geological Survey Water-Investigations Report 78-10, 1978.

49. Slattery, R. N., and D. E. Thomas, *Recharge to and Discharge from the Edwards Aquifer in the San Antonio Area*, 2001, http://tx.usgs.gov/reports/dist/dist-2002–01, San Antonio, TX, October 2002.

50. Soil Survey Staff, *Soil Survey Manual*, U.S. Dept. Agriculture Handbook, no. 18, 503 pp., 1951.

51. Stringfield, V. T., and H. E. LeGrand, Hydrology of carbonate rock terranes—A review with reference to the United States, *Jour. Hydrology*, v. 8, pp. 349–417, 1969.

52. Summers, W. K., *Annotated and Indexed Bibliography of Geothermal phenomena*, New Mexico Bur. Mines and Mineral Resources, Socorro, 665 pp., 1972.

53. Sun, R. J. (ed.), *Regional Aquifer-System Analysis Program of the U.S. Geological Survey—Summary of Projects, 1978–84*, U.S. Geological Survey Circular 1002, 1986.

54. Sun, R. J., and R. H. Johnson, *Regional Aquifer-System Analysis Program of the U.S. Geological Survey, 1978–1992*, U.S. Geological Survey Circular 1099, 1994.

55. Taylor, C. J., and W. M. Alley, *Ground-Water-Level Monitoring and the Importance of Long-Term Water-Level Data*, U.S. Geological Survey Circular 1217, http://water.usgs.gov/pubs/circ/circ1217, Denver, CO, 2001.

56. Thatcher, L., et al., Dating desert ground water, *Science*, v. 134, no. 3472, pp. 105–106, 1961.

57. U.S. Geological Survey, *Ground Water Atlas of the United States*, U.S. Geological Survey Hydrologic Investigations Atlases, http://capp.water.usgs.gov/gwa/gwa.html

  HA 730-B California, Nevada 1995

  HA 730-C Arizona, Colorado, New Mexico, Utah 1995

  HA 730-D Kansas, Missouri, Nebraska 1997

  HA 730-E Oklahoma, Texas 1996

  HA 730-F Arkansas, Louisiana, Mississippi 1998

  HA 730-G Alabama, Florida, Georgia, South Carolina 1990

  HA 730-H Idaho, Oregon, Washington 1994

  HA 730-I Montana, North Dakota, South Dakota, Wyoming 1996

  HA 730-J Iowa, Michigan, Minnesota, Wisconsin 1992

  HA 730-K Illinois, Indiana, Kentucky, Ohio, Tennessee 1995

  HA 730-L Delaware, Maryland, New Jersey, North Carolina, Pennsylvania, Virginia, West Virginia 1997

  HA 730-M Connecticut, Maine, Massachusetts, New Hampshire, New York, Rhode Island, Vermont 1995

  HA 730-N Alaska, Hawaii, Puerto Rico, and U.S. Virgin Islands 1999

58. Vandike, J. E., *The Hydrology of Maramec Spring*, Brochure, Water Resources Report no. 55, Missouri Department of Natural Resources, Division of Geology and Land Survey, Rolla, Missouri, 1996.

59. Vogel, J. C., and D. Ehhalt, The use of carbon isotopes in groundwater studies, *Radioisotopes in Hydrology*, Intl. Atomic Energy Agency, Vienna, pp. 383–395, 1963.

60. Waring, G. A., *Thermal Springs of the United States and Other Countries of the World—A Summary*, U.S. Geological Survey Professional Paper 492, 383 pp., 1965.

61. White, D. E., Thermal waters of volcanic origin, *Bull. Geological Soc. Amer.*, v. 68, pp. 1637–1658, 1957.

62. White, D. E., Magmatic, connate, and metamorphic waters, *Bull. Geological Soc. Amer.*, v. 68, pp. 1659–1682, 1957.

63. White, D. E., *Geothermal Energy*, U.S. Geological Survey Circular. 519, 17 pp., 1965.

64. White, W. B., Conceptual models for carbonate aquifers, *Ground Water*, v. 7, no. 3, pp. 15–21, 1969.

65. Wicks, C. M. and J.A. Hoke, Prediction of quality and quantity of Maramec Spring Water, *Ground Water*, vol. 38, no. 2, pp. 218–225, 2000.

66. Wigley, T. M. L., Carbon 14 dating of groundwater from closed and open systems, *Water Resources Research*, v. 11, pp. 324–328, 1975.

67. Williams, J. R., *Ground Water in Permafrost Regions—An Annotated Bibliography*, U.S. Geological Survey Water-Supply Paper 1792, 294 pp., 1965.

68. Williams, J. R., *Ground Water in the Permafrost Regions of Alaska*, U.S. Geological Survey Professional Paper 696, 83 pp., 1970.

69. Williamson, A. K., and H. F. Grubb, *Groundwater Flow in Gulf Coast Aquifer Systems, South Central United States*, U.S. Geological Survey Professional Paper 1416-F, 2001.

70. Winter, T. C., J. W. Harvey, O. L. Franke, and W. M. Alley, *Groundwater and Surface Water: A Single Source*, U.S. Geological Survey Circular 1139, http://water.usgs.gov/pubs/circ/circ1139, Denver, CO, 1998.

Chapter 3

Groundwater Movement

Groundwater in its natural state is invariably moving. This movement is governed by established hydraulic principles. The flow through aquifers, most of which are natural porous media, can be expressed by what is known as Darcy's law. Hydraulic conductivity, which is a measure of the permeability of the media, is an important constant in the flow equation. Determination of hydraulic conductivity can be made by several laboratory or field techniques. Applications of Darcy's law enable groundwater flow rates and directions to be evaluated. The dispersion, or mixing, resulting from flows through porous media produces irregularities of flow that can be studied by tracers. In the zone of aeration, the presence of air adds a complicating factor to the flow of water.

3.1 DARCY'S LAW

More than a century ago Henry Darcy,[*] a French hydraulic engineer, investigated the flow of water through horizontal beds of sand to be used for water filtration. He reported [24] in 1856:

> I have attempted by precise experiments to determine the law of the flow of water through filters. . . . The experiments demonstrate positively that the volume of water which passes through a bed of sand of a given nature is proportional to the pressure and inversely proportional to the thickness of the bed traversed; thus, in calling s the surface area of a filter, k a coefficient depending on the nature of the sand, e the thickness of the sand bed, $P - H_0$ the pressure below the filtering bed, $P + H$ the atmospheric pressure added to the depth of water on the filter; one has for the flow of this last condition $Q = (ks/e)(H + e + H_0)$, which reduces to $Q = (ks/e)(H + e)$ when $H_0 = 0$, or when the pressure below the filter is equal to the weight of the atmosphere.

This statement, that the flow rate through porous media is proportional to the head loss and inversely proportional to the length of the flow path, is known universally as Darcy's law. It, more than any other contribution, serves as the basis for present-day knowledge of groundwater flow. Analysis and solution of problems relating to groundwater movement and well hydraulics began after Darcy's work.

3.1.1 Experimental Verification

The experimental verification of Darcy's law can be performed with water flowing at a rate Q through a cylinder of cross-sectional area A packed with sand and having piezometers a

---

[*]An interesting summary of the life and accomplishments of Henry Darcy was prepared by Fancher.[31]

distance L apart, as shown in Figure 3.1.1[38, 52] Total energy heads, or fluid potentials, above a datum plane may be expressed by the energy equation

$$\frac{p_1}{\gamma} + \frac{v_1^2}{2g} + z_1 = \frac{p_2}{\gamma} + \frac{v_2^2}{2g} + z_2 + h_L \tag{3.1.1}$$

where p is pressure, γ is the specific weight of water, v is the velocity of flow, g is the acceleration of gravity, z is elevation, and h_L is head loss. Subscripts refer to points of measurement identified in Figure 3.1.1. Because velocities in porous media are usually low, velocity heads may be neglected without appreciable error. Hence, by rewriting, the head loss becomes

$$h_L = \left(\frac{p_1}{\gamma} + z_1 \right) - \left(\frac{p_2}{\gamma} + z_2 \right) \tag{3.1.2}$$

Therefore, the resulting head loss is defined as the potential loss within the sand cylinder, this energy being lost by frictional resistance dissipated as heat energy. It follows that the head loss is independent of the inclination of the cylinder.

Now, Darcy's measurements showed that the proportionalities $Q \sim h_L$ and $Q \sim 1/L$ exist. Introducing a proportionality constant K leads to the equation

$$Q = -KA \frac{h_L}{L} \tag{3.1.3}$$

Expressed in general terms

$$Q = -KA \frac{dh}{dl} \tag{3.1.4}$$

or simply

$$v = \frac{Q}{A} = -K \frac{dh}{dl} \tag{3.1.5}$$

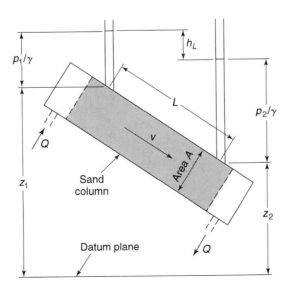

Figure 3.1.1. Pressure distribution and head loss in flow through a sand column.

where v is the *Darcy velocity* or *specific discharge;* K is the *hydraulic conductivity,* a constant that serves as a measure of the permeability of the porous medium; and dh/dl is the hydraulic gradient. The negative sign indicates that the flow of water is in the direction of decreasing head. Equation 3.1.5 states Darcy's law in its simplest form, that the flow velocity v equals the product of the hydraulic conductivity and the hydraulic gradient.

EXAMPLE 3.1.1

A field sample of an unconfined aquifer is packed in a test cylinder (see Figure 3.1.1). The length and the diameter of the cylinder are 50 cm and 6 cm, respectively. The field sample is tested for a period of 3 min under a constant head difference of 16.3 cm. As a result, 45.2 cm^3 of water is collected at the outlet. Determine the hydraulic conductivity of the aquifer sample.

SOLUTION

The cross-sectional area of the sample is

$$A = \frac{\pi D^2}{4} = \frac{\pi (0.06 \text{ m})^2}{4} = 0.00283 \text{ m}^2$$

The hydraulic gradient, dh/dl, is given by

$$\frac{dh}{dl} = \frac{(-16.3 \text{ cm})}{50 \text{ cm}} = -0.326$$

and the average flow rate is

$$Q = \frac{45.2 \text{ cm}^3}{3 \text{ min}} = 15.07 \text{ cm}^3/\text{min} = 0.0217 \text{ m}^3/\text{day}$$

Apply Darcy's law, Equation 3.1.4, to obtain the hydraulic conductivity as

$$Q = -KA\frac{dh}{dl} \rightarrow K = -\frac{Q}{A(dh/dl)} = -\frac{0.0217 \text{ m}^3/\text{day}}{(0.00283 \text{ m}^2)(-0.326)} = 23.5 \text{ m/day}$$
∎

EXAMPLE 3.1.2

A confined aquifer with a horizontal bed has a varying thickness as shown in Figure 3.1.2. The aquifer is inhomogeneous with $K = 12 + 0.006x$, where $x = 0$ at section (1), and the piezometric heads at sections (1) and (2) are 14.2 m and 18.8 m, respectively measured above the upper confining layer. Assuming the flow in the aquifer is essentially horizontal, determine the flow rate per unit width.

SOLUTION

Darcy's law for a constant thickness aquifer is given by Equation 3.1.4,

$$Q = -KA\frac{dh}{dl}$$

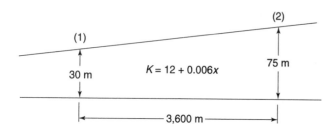

Figure 3.1.2. Aquifer for Example 3.1.2

Since the aquifer thickness is variable in this problem, we must also write the cross-sectional area and the hydraulic gradient as a function of the distance x. Assuming a unit width, $A = b_1 + \dfrac{(b_2 - b_1)x}{L}$, where $b_1 = 30$ m, $b_2 = 75$ m, and $L = 3{,}600$ m, then we have

$$A = 30 + \frac{(75-30)x}{3{,}600} = 30 + 0.0125x$$

Substituting the expressions for A and K into Darcy's equation yields the expression for Q in following form:

$$Q = -(12 + 0.006x)(30 + 0.0125x)\frac{dh}{dx}$$

Rearranging this equation and integrating from section (1) to section (2) yields

$$\int_0^{3600} \frac{1}{(12+0.006x)(30+0.0125x)}\,dx = \int_{14.2}^{18.8} -\frac{1}{Q}\,dh$$

This equation is integrated using partial fraction decomposition to obtain

$$\int_0^{3600}\left[\frac{0.2}{(12+0.006x)} - \frac{0.416}{(30+0.0125x)}\right]dx = \int_{14.2}^{18.8} -\frac{1}{Q}\,dh$$

$$\left[33.333\ln\left(12+0.006x\right) - 33.28\ln\left(30+0.0125x\right)\right]_{x=0}^{x=3{,}600} = -\frac{1}{Q}h\Big]_{h_1=14.2}^{h_2=18.8}$$

$$-26.54 - (-30.36) = -\frac{1}{Q}(18.8 - 14.2)$$

$$Q = -1.20 \ (\text{m}^3/\text{day/m})$$

The minus sign implies that the flow is from section (2) to (1).　■

3.1.2　Darcy Velocity

The velocity v in Equation 3.1.5 is referred to as the *Darcy velocity* because it assumes that flow occurs through the entire cross section of the material without regard to solids and pores. Actually, the flow is limited only to the pore space so that the *average interstitial velocity*

$$v_a = \frac{Q}{\alpha A} \tag{3.1.6}$$

where α is the (effective) porosity. This indicates that for a sand with a porosity of 33 percent, $v_a = 3v$. To define the actual flow velocity, one must consider the microstructure of the rock material. In water flowing through a sand, for example, the pore spaces vary continuously with location within the medium. This means that the actual velocity is nonuniform, involving endless accelerations, decelerations, and changes in direction. Thus, the actual velocity depends on specifying a precise point location within the medium. For naturally occurring geologic materials, the microstructure cannot be specified three-dimensionally; hence, actual velocities can only be quantified statistically.

3.1.3   Validity of Darcy's Law

In applying Darcy's law it is important to know the range of validity within which it is applicable. Because velocity in laminar flow, such as water flowing in a capillary tube, is proportional to the first power of the hydraulic gradient (Poiseuille's law), it seems reasonable to believe that Darcy's law applies to laminar flow in porous media. For flow in pipes and other large sections, the Reynolds number, which expresses the dimensionless ratio of inertial to viscous forces, serves as a criterion to distinguish between laminar and turbulent flow. Hence, by analogy, the Reynolds number has been employed to establish the limit of flows described by Darcy's law, corresponding to the value where the linear relationship is no longer valid.

Reynolds number is expressed as

$$N_R = \frac{\rho v D}{\mu} \tag{3.1.7}$$

where ρ is the fluid density, v the velocity, D the diameter (of a pipe), and μ the (dynamic) viscosity of the fluid. To adapt this criterion to flow in porous media, the Darcy velocity is employed for v and an effective grain size (d_{10}) is substituted for D. Certainly a grain diameter represents only an approximation of the critical flow dimension for which it is intended; however, measuring pore size distribution is a complex research task.

Experiments show that Darcy's law is valid for $N_R < 1$ and does not depart seriously up to $N_R = 10$.[1] This, then, represents an upper limit to the validity of Darcy's law. A range of values rather than a unique limit must be stated because as inertial forces increase, turbulence occurs gradually.[85, 92, 106] The irregular flow paths of eddies and swirls associated with turbulence occur first in the larger pore spaces; with increasing velocity they spread to the smaller pores. For fully developed turbulence the head loss varies approximately with the second power of the velocity rather than linearly.

Fortunately, most natural underground flow occurs with $N_R < 1$, so Darcy's law is applicable. Deviations from Darcy's law can occur where steep hydraulic gradients exist, such as near pumped wells; also, turbulent flow can be found in rocks such as basalt and limestone[17] that contain large underground openings.*

EXAMPLE 3.1.3

The following additional information is given for the aquifer sample in Example 3.1.1. The sample has a median grain size of 0.037 cm and a porosity of 0.30. The test is conducted using pure water at 20°C. Determine the Darcy velocity, average interstitial velocity, and assess the validity of Darcy's law.

SOLUTION

Darcy velocity is computed using Equation 3.1.5:

$$v = -K\frac{dh}{dl} = -(23.54 \text{ m/day})(-0.326) = 7.67 \text{ m/day}$$

The average linear velocity is computed using Equation 3.1.6:

$$v_a = \frac{Q}{\alpha A} = \frac{v}{\alpha} = \frac{7.67 \text{ m/day}}{0.30} = 25.6 \text{ m/day}$$

In order to assess the validity of Darcy's Law we must determine the greatest velocity for which Darcy's law is valid using Equation 3.1.7, $N_R = \frac{\rho v D}{\mu}$, knowing Darcy's law is valid for $N_R < 1$. For water at 20°C, $\mu = 1.005 \times 10^{-3}$ N$_s$/m^2 and $\rho = 998.2$ kg/m^3, so that for $N_R = 1$,

---

*It should also be noted that investigations have shown that Darcy's law may not be valid for very slow water flow through dense clay. Here the effects of electrically charged clay particles on water in the minute pores produce nonlinearities between flow rate and hydraulic gradient.[53]

$$v_{\max} = \frac{\mu}{\rho D} = \frac{1.005 \times 10^{-3} \text{ kg/ms}}{\left(998.2 \text{ kg/m}^3\right)(0.00037 \text{ m})} = 0.00272 \text{ m/s} = 235 \text{ m/day}$$

Then Darcy's law will be valid for Darcy velocities equal to or less than 235 m/day for this sample. Thus, the answer we have found in Example 3.1.1 is valid since $v = 7.67$ m/day < 235 m/day. ■

3.2 PERMEABILITY

3.2.1 Intrinsic Permeability

The permeability of a rock or soil defines its ability to transmit a fluid. This is a property only of the medium and is independent of fluid properties. To avoid confusion with hydraulic conductivity, which includes the properties of groundwater, an *intrinsic permeability k* may be expressed as

$$k = \frac{K\mu}{\rho g} \tag{3.2.1}$$

where K is hydraulic conductivity, μ is dynamic viscosity, ρ is fluid density, and g is acceleration of gravity. Inserting this in Equation 3.1.5 yields

$$k = -\frac{\mu v}{\rho g(dh/dl)} \tag{3.2.2}$$

which has units of

$$k = -\frac{(\text{kg/ms})(\text{m/s})}{\left(\text{kg/m}^3\right)\left(\text{m/s}^2\right)(\text{m/m})} = \text{m}^2 \tag{3.2.3}$$

Thus, intrinsic permeability possesses units of area. Because values of k in Equation 3.2.3 are so small, the U.S. Geological Survey expresses k in square micrometers $(\mu\text{m})^2 = 10^{-12}\text{m}^2$.

In the petroleum industry the value of k is measured by a unit termed the *darcy*, defined as

$$1 \text{ darcy} = \frac{\dfrac{(1 \text{ centipoise})\left(1 \text{ cm}^3/\text{s}\right)}{1 \text{ cm}^2}}{1 \text{ atmosphere/cm}} \tag{3.2.4}$$

By substitution of appropriate units it can be shown that [66]

$$1 \text{ darcy} = 0.987 \ (\mu\text{m})^2 \tag{3.2.5}$$

so the darcy corresponds closely to the intrinsic permeability unit adopted by the U.S. Geological Survey.

3.2.2 Hydraulic Conductivity

For practical work in groundwater hydrology, where water is the prevailing fluid, hydraulic conductivity K is employed. A medium has a unit hydraulic conductivity if it will transmit in unit time a unit volume of groundwater at the prevailing kinematic viscosity* through a cross

---

*Kinematic viscosity equals dynamic viscosity divided by fluid density.

section of unit area, measured at right angles to the direction of flow, under a unit hydraulic gradient. The units are

$$K = -\frac{v}{dh/dL} = -\frac{m/day}{m/m\alpha} = m/day \qquad (3.2.6)$$

indicating that hydraulic conductivity has units of velocity.

3.2.3 Transmissivity

The term *transmissivity T* is widely employed in groundwater hydraulics. It may be defined as the rate at which water of prevailing kinematic viscosity is transmitted through a unit width of aquifer under a unit hydraulic gradient. It follows that

$$T = Kb = (m/day)(m) = m^2/day \qquad (3.2.7)$$

where b is the saturated thickness of the aquifer.

EXAMPLE 3.2.1

A leaky confined aquifer is overlain by an aquitard that is also overlain by an unconfined aquifer. The estimated recharge rate from the unconfined aquifer into the confined aquifer is 0.085 m/year. Piezometric head measurements in the confined aquifer show that the average piezometric head in the confined aquifer is 6.8 m below the water table of the unconfined aquifer. If the average thickness of the aquitard is 4.30 m, find the vertical hydraulic conductivity, K_v, of the aquitard. What type of material could this possibly be?

SOLUTION

Given $v = 0.085$ m/year $= 2.329 \times 10^{-4}$ m/day, Equation 3.2.6 is used to compute the vertical hydraulic conductivity of the aquitard:

$$K = -\frac{v}{dh/dl} = -\frac{2.329 \times 10^{-4} \text{ m/day}}{(6.8 \text{ m}/4.30 \text{ m})} = 1.473 \times 10^{-4} \text{ m/day}$$

From Table 3.2.1, the aquitard is composed of clay.   ∎

3.2.4 Hydraulic Conductivity of Geologic Materials

The hydraulic conductivity of a soil or rock depends on a variety of physical factors, including porosity, particle size and distribution, shape of particles, arrangement of particles, and other factors.[63, 79] In general, for unconsolidated porous media, hydraulic conductivity varies with particle size; clayey materials exhibit low values of hydraulic conductivity, whereas sands and gravels display high values.

An interesting illustration of the variation of hydraulic conductivity with particle size is shown by data in Figure 3.2.1. Here conductivities were measured for two uniform sieved sands. These two sands were then mixed in varying proportions, and the corresponding hydraulic conductivities were again determined. Results show that any mixture of the two sands displays a conductivity less than a linearly interpolated value. The physical explanation lies in the fact that the smaller grains occupy a larger fraction of the space around larger grains than do uniform grains of either size.

Table 3.2.1 contains representative hydraulic conductivities for a variety of geologic materials. It should be noted that these are averages of many measurements; clearly, a range of values exists for each rock type depending on factors such as weathering, fracturing, solution channels, and depth of burial.

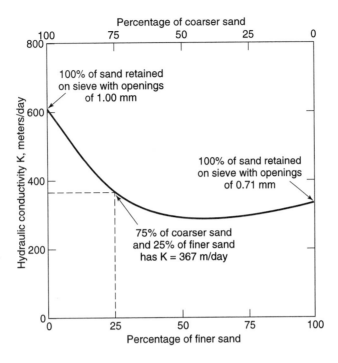

Figure 3.2.1. Hydraulic conductivity of various proportions of two uniform sands (courtesy Illinois State Water Survey).

Table 3.2.1 Representative Values of Hydraulic Conductivity (after Morris and Johnson[75])

Material	Hydraulic conductivity (m/day)	Type of measurement[a]
Gravel, coarse	150	R
Gravel, medium	270	R
Gravel, fine	450	R
Sand, coarse	45	R
Sand, medium	12	R
Sand, fine	2.5	R
Silt	0.08	H
Clay	0.0002	H
Sandstone, fine-grained	0.2	V
Sandstone, medium-grained	3.1	V
Limestone	0.94	V
Dolomite	0.001	V
Dune sand	20	V
Loess	0.08	V
Peat	5.7	V
Schist	0.2	V
Slate	0.00008	V
Till, predominantly sand	0.49	R
Till, predominantly gravel	30	R
Tuff	0.2	V
Basalt	0.01	V
Gabbro, weathered	0.2	V
Granite, weathered	1.4	V

[a]H is horizontal hydraulic conductivity, R is a repacked sample, and V is vertical hydraulic conductivity.

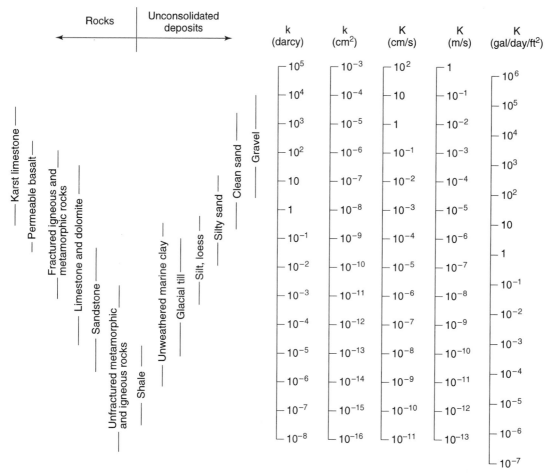

Figure 3.2.2. Range of values of hydraulic conductivity and permeability (Freeze, R. A. and Cherry, J. A., *Groundwater*, Prentice Hall, Englewood Cliffs, NJ, 1979.)

Magnitudes of hydraulic conductivity for various classes of unconsolidated and consolidated rocks are shown in Figure 3.2.2.

3.3   DETERMINATION OF HYDRAULIC CONDUCTIVITY

Hydraulic conductivity in saturated zones can be determined by a variety of techniques, including calculation from formulas, laboratory methods, tracer tests, auger hole tests, and pumping tests of wells.

3.3.1   Formulas

Numerous investigators have studied the relationship of permeability or hydraulic conductivity to the properties of porous media. Several formulas have resulted based on analytic or experimental work. Most permeability formulas have the general form

$$k = cd^2 \qquad\qquad (3.3.1)$$

where c is a dimensionless coefficient, or

$$k = f_s f_\alpha d^2 \tag{3.3.2}$$

where f_s is a grain (or pore) shape factor, f_α is a porosity factor, and d is characteristic grain diameter.[30, 62, 68] Few formulas give reliable estimates of results because of the difficulty of including all possible variables in porous media. For an ideal medium, such as an assemblage of spheres of uniform diameter, hydraulic conductivity can be evaluated accurately from known porosity and packing conditions.

Because of the problems inherent in formulas, other techniques for determining hydraulic conductivity are preferable.

3.3.2   Laboratory Methods

In the laboratory, hydraulic conductivity can be determined by a *permeameter*, in which flow is maintained through a small sample of material while measurements of flow rate and head loss are made.[107] The constant-head and falling-head types of permeameters are simple to operate and widely employed.

The constant-head permeameter shown in Figure 3.3.1a can measure hydraulic conductivities of consolidated or unconsolidated formations under low heads. Water enters the medium cylinder from the bottom and is collected as overflow after passing upward through the material. From Darcy's law it follows that the hydraulic conductivity can be obtained from

$$K = \frac{VL}{Ath} \tag{3.3.3}$$

where V is the flow volume in time t, and the other dimensions, A, L, and h, are shown in Figure 3.3.1a. It is important that the medium be thoroughly saturated to remove entrapped air. Several different heads in a series of tests provide a reliable measurement.

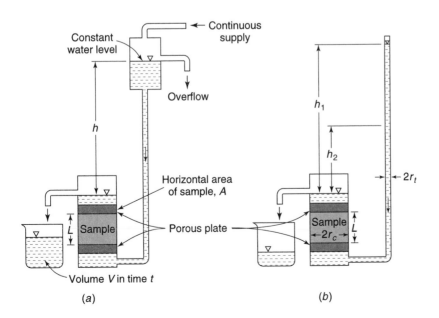

Figure 3.3.1. Permeameters for measuring hydraulic conductivity of geologic samples. (*a*) Constant head. (*b*) Falling head.

A second procedure utilizes the falling-head permeameter illustrated in Figure 3.3.1b. Here water is added to the tall tube; it flows upward through the cylindrical sample and is collected as overflow.

The test consists of measuring the rate of fall of the water level in the tube. The hydraulic conductivity can be obtained by noting that the flow rate Q in the tube

$$Q = \pi r_t^2 dh/dt \qquad (3.3.4)$$

must equal that through the sample, which by Darcy's law is

$$Q = \pi r_c^2 K \, h/L \qquad (3.3.5)$$

After equating and integrating,

$$K = \frac{r_t^2 L}{r_c^2 t} \ln \frac{h_1}{h_2} \qquad (3.3.6)$$

where L, r_t, and r_c are shown in Figure 3.3.1b, and t is the time interval for the water level in the tube to fall from h_1 to h_2.

Permeameter results may bear little relation to actual field hydraulic conductivities. Undisturbed samples of unconsolidated material are difficult to obtain, while disturbed samples experience changes in porosity, packing, and grain orientation, which modify hydraulic conductivities. Then, too, one or even several samples from an aquifer may not represent the overall hydraulic conductivity of an aquifer. Variations of several orders of magnitude frequently occur for different depths and locations in an aquifer. Furthermore, directional properties of hydraulic conductivity may not be recognized.

EXAMPLE 3.3.1

A field sample of medium sand with a median grain size of 0.84 mm will be tested to determine the hydraulic conductivity using a constant-head permeameter. The sample has a length of 30 cm and a diameter of 5 cm. For pure water at 20°C, estimate the range of piezometric head differences to be used in the test.

SOLUTION

The maximum allowable Darcy velocity (assuming $N_R = 1$) for $d = 0.84$ mm is

$$v_{max} = \frac{\mu}{\rho D} = \frac{1.005 \times 10^{-3} \text{ kg/ms}}{\left(998.2 \text{ kg/m}^3\right)(0.00084 \text{ m})} = 0.0012 \text{ m/s} = 103.6 \text{ m/day}$$

Thus, the Darcy velocity in the test must be equal to or less than 103.6 m/day so that Darcy's law will be valid, so that

$$v = -K \frac{dh}{dl} \le 103.6 \text{ m/day} \rightarrow |dh| \le \frac{(103.6 \text{ m/day})(0.30 \text{ m})}{K}$$

For the representative value of hydraulic conductivity for medium sand given in Table 3.2.1,

$$K = 12 \text{ m/day, then } |dh| \le \frac{(103.6 \text{ m/day})(0.30 \text{ m})}{12 \text{ m/day}} \cong 2.6 \text{ m} = 260 \text{ cm}$$

It should be noted that the K value for clean sand ranges approximately from 0.1 m/day to 4,320 m/day. See Figure 3.2.1. Therefore, the early series of tests must be conducted with relatively low piezometric head differences if possible. After analyzing the results of early test data, a better estimate of the maximum allowable piezometric head difference can be made using the above inequality. ∎

EXAMPLE 3.3.2

If the field sample in Example 3.3.1 is tested with a head difference of 5.0 cm and 200 ml of water is collected at the outlet in 15 min, determine the hydraulic conductivity of the sample. What should the maximum allowable piezometric head difference be for a series of tests?

SOLUTION

Equation 3.3.3 is used to compute the hydraulic conductivity in a constant-head permeameter test:

$$K = \frac{VL}{Ath} = \frac{\left(200 \text{ cm}^3\right)\left(30 \text{ cm}\right)}{\left(\dfrac{\pi\left(5 \text{ cm}\right)^2}{4}\right)\left(15 \min \times 60 \dfrac{\text{s}}{\min}\right)\left(5.0 \text{ cm}\right)} = 0.0679 \text{ cm/s} = 58.7 \text{ m/day}$$

Based upon this estimate and referring to Example 3.3.1, the maximum allowable piezometric head difference for tests should be approximately

$$|dh| \le \frac{\left(103.6 \text{ m/day}\right)\left(0.30 \text{ m}\right)}{58.7 \text{ m/day}} \cong 0.53 \text{ m} = 53 \text{ cm}$$ ∎

EXAMPLE 3.3.3

A 20-cm long field sample of silty, fine sand with a diameter of 10 cm is tested using a falling-head permeameter. The falling-head tube has a diameter of 3.0 cm and the initial head is 8.0 cm. Over a period of 8 hr, the head in the tube falls to 1.0 cm. Estimate the hydraulic conductivity of the sample.

SOLUTION

Equation 3.3.6 is used to compute the hydraulic conductivity in a falling-head permeameter test:

$$K = \frac{r_t^2 L}{r_c^2 t} \ln \frac{h_1}{h_2} = \frac{\left(1.5 \text{ cm}\right)^2\left(20 \text{ cm}\right)}{\left(5.0 \text{ cm}\right)^2\left(8 \times 3600 \text{ sec}\right)} \ln \frac{8.0 \text{ cm}}{1.0 \text{ cm}} = 1.3 \times 10^{-4} \text{ cm/s} = 0.112 \text{ m/day}$$ ∎

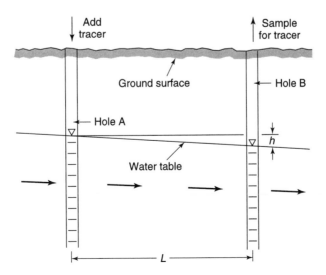

Figure 3.3.2. Cross section of an unconfined aquifer illustrating a tracer test for determining hydraulic conductivity.

3.3.3 Tracer Tests

Field determinations of hydraulic conductivity can be made by measuring the time interval for a water tracer to travel between two observation wells or test holes.[18, 25, 40, 57, 72, 73, 109] For the tracer, a dye, such as sodium fluorescein, or a salt, such as calcium chloride, is convenient, inexpensive, easy to detect, and safe. Figure 3.3.2 shows the cross section of a portion of an unconfined aquifer with groundwater flowing from hole A toward hole B. The tracer is injected as a slug in hole A after which samples of water are taken from hole B to determine the time of passage of the tracer. Because the tracer flows through the aquifer with the average interstitial velocity v_a, then

$$v_a = \frac{K}{\alpha} \frac{h}{L} \tag{3.3.7}$$

where K is hydraulic conductivity, α is porosity, and h and L are shown in Figure 3.3.2. But v_a also is given by

$$v_a = L/t \tag{3.3.8}$$

where t is the travel time interval of the tracer between the holes. Equating these and solving for K yields

$$K = \frac{\alpha L^2}{ht} \tag{3.3.9}$$

Although this procedure is simple in principle, results are only approximations because of serious limitations in the field.

1. The holes need to be close together; otherwise, the travel time interval can be excessively long.

2. Unless the flow direction is accurately known, the tracer may miss the downstream hole entirely. Multiple sampling holes can help, but these add to the cost and complexity of conducting the test.

3. If the aquifer is stratified with layers with differing hydraulic conductivities, the first arrival of the tracer will result in a conductivity considerably larger than the average for the aquifer.

An alternative tracer technique, which has been successfully applied under field conditions, is the *point dilution method*.[28, 45, 53] Here a tracer is introduced into an observation well and thoroughly mixed with the contained water. Thereafter, as water flows into and from the well, repeated measurements of tracer concentration are made.

Analysis of the resulting dilution curve defines the groundwater velocity; this, together with the measured water table gradient and Darcy's law, yields a localized estimate of the hydraulic conductivity and also the direction of groundwater flow.[85] Additional references on tracer tests include Barth et al.[7] on heterogeneous porous media; Mull et al.[76] on carbonate aquifer systems; and Novakowski[77] on divergent radial flow fields. A new Efficient Hydrologic Tracer-test Design (EHTD) has been developed by the U.S. Environmental Protection Agency (EPA).[33–36]

EXAMPLE 3.3.4

A tracer test is conducted to determine the hydraulic conductivity of an unconfined aquifer. The water levels in the two observation wells 20 m apart are 18.4 m and 17.1 m. The tracer injected in the first well arrives at the second observation well in 167 hours. Compute the hydraulic conductivity of the unconfined aquifer given that the porosity of the formation is 0.25.

SOLUTION

Given $\alpha = 0.25$, $L = 20$ m, $h = 18.4$ m $- 17.1$ m $= 1.3$ m, $t = 167$ hours $= 6.96$ days, Equation 3.3.9 is used to compute the hydraulic conductivity of the aquifer:

$$K = \frac{\alpha L^2}{ht} = \frac{(0.25)(20 \text{ m})^2}{(1.3 \text{ m})(6.96 \text{ days})} = 11.1 \text{ m/day}$$

3.3.4 Auger Hole Tests

The auger hole method involves the measurement of the change in water level after the rapid removal of a volume of water from an unlined cylindrical hole. If the soil is loose, a screen may be necessary to maintain the hole. The method is relatively simple and is most adaptable to shallow water table conditions. The value of K obtained is essentially that for a horizontal direction in the immediate vicinity of the hole.

Figure 3.3.3 illustrates an auger hole (also refer to Figure 5.2.4) and the dimensions required for the calculation. It can be shown[10] that hydraulic conductivity is given by

$$K = \frac{C}{864} \frac{dy}{dt} \qquad (3.3.10)$$

where dy/dt is the measured rate of rise in cm/s and the factor 864 yields K values in m/day. The factor C is a dimensionless constant listed in Table 3.3.1 and governed by the variables shown in Figure 3.3.3.

Table 3.3.1 Values of the Factor C for the Auger Hole Test to Determine Hydraulic Conductivity (after Boast and Kirkham[10])

L_w/r_w	y/L_w	$(H-L_w)/L_w$ for impermeable layer								$H-L_w$	$(H-L_w)/L_w$ for infinitely permeable layer			
		0	0.05	0.1	0.2	0.5	1	2	5	∞	5	2	1	0.5
1	1	447	423	404	375	323	286	264	255	254	252	241	213	166
	0.75	469	450	434	408	360	324	303	292	291	289	278	248	198
	0.5	555	537	522	497	449	411	386	380	379	377	359	324	264
2	1	186	176	167	154	134	123	118	116	115	115	113	106	91
	0.75	196	187	180	168	149	138	133	131	131	130	128	121	106
	0.5	234	225	218	207	188	175	169	167	167	166	164	156	139
5	1	51.9	48.6	46.2	42.8	38.7	36.9	36.1		35.8		35.5	34.6	32.4
	0.75	54.8	52.0	49.9	46.8	42.8	41.0	40.2		40.0		39.6	38.6	36.3
	0.5	66.1	63.4	61.3	58.1	53.9	51.9	51.0		50.7		50.3	49.2	46.6
10	1	18.1	16.9	16.1	15.1	14.1	13.6	13.4		13.4		13.3	13.1	12.6
	0.75	19.1	18.1	17.4	16.5	15.5	15.0	14.8		14.8		14.7	14.5	14.0
	0.5	23.3	22.3	21.5	20.6	19.5	19.0	18.8		18.7		18.6	18.4	17.8
20	1	5.91	5.53	5.30	5.06	4.81	4.70	4.66		4.64		4.62	4.58	4.46
	0.75	6.27	5.94	5.73	5.50	5.25	5.15	5.10		5.08		5.07	5.02	4.89
	0.5	7.67	7.34	7.12	6.88	6.60	6.48	6.43		6.41		6.39	6.34	6.19
50	1	1.25	1.18	1.14	1.11	1.07	1.05			1.04			1.03	1.02
	0.75	1.33	1.27	1.23	1.20	1.16	1.14			1.13			1.12	1.11
	0.5	1.64	1.57	1.54	1.50	1.46	1.44			1.43			1.42	1.39
100	1	0.37	0.35	0.34	0.34	0.33	0.32			0.32			0.32	0.31
	0.75	0.40	0.38	0.37	0.36	0.35	0.35			0.35			0.34	0.34
	0.5	0.49	0.47	0.46	0.45	0.44	0.44			0.44			0.43	0.43

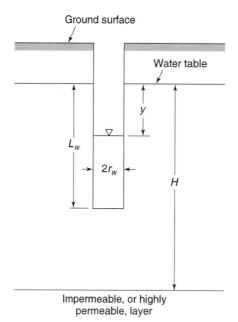

Figure 3.3.3. Diagram of an auger hole and dimensions for determining hydraulic conductivity.

Note that the tabulated values cover the following conditions below the hole: a shallow impermeable layer, an infinite homogeneous stratum, and a shallow, highly permeable (gravel) layer. The value y should correspond to that when dy/dt is measured.

Several other techniques similar to the auger hole test have been developed in which water level changes are measured after an essentially instantaneous removal or addition of a volume of water. With a small-diameter pipe driven into the ground, K can be found by the piezometer, or tube, method.[103] For wells in confined aquifers, the slug method can be employed.[22,66] Here a known volume of water is suddenly injected or removed from a well after which the decline or recovery of the water level is measured in the ensuing minutes. Where a pump is not available to conduct a pumping test on a well, the slug method serves as an alternative approach.

3.3.5   Pumping Tests of Wells

The most reliable method for estimating aquifer hydraulic conductivity is by pumping tests of wells. Based on observations of water levels near pumping wells, an integrated K value over a sizable aquifer section can be obtained. Then, too, because the aquifer is not disturbed, the reliability of such determinations is superior to laboratory methods. Pump test methods and computations are described in Chapter 4.

3.4   ANISOTROPIC AQUIFERS

The discussion of hydraulic conductivity up to this point assumed that the geologic material was homogeneous and isotropic, implying that the value of K was the same in all directions. However, this is rarely the case, particularly for undisturbed unconsolidated alluvial materials.

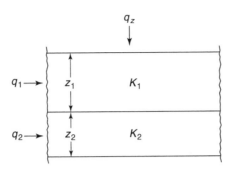

Figure 3.4.1. Diagram of two horizontal strata, each isotropic, with different thicknesses and hydraulic conductivities.

Instead, *anisotropy* is the rule where directional properties of hydraulic conductivity exist. In alluvium this results from two conditions. One is that individual particles are seldom spherical so that when deposited under water they tend to rest with their flat sides down. The second is that alluvium typically consists of layers of different materials, each possessing a unique value of K. If the layers are horizontal, any single layer with a relatively low hydraulic conductivity causes vertical flow to be retarded, but horizontal flow can occur easily through any stratum of relatively high hydraulic conductivity. Thus, the typical field situation in alluvial deposits is to find a hydraulic conductivity K_x in the horizontal direction that will be greater than a value K_z in a vertical direction.

Consider an aquifer consisting of two horizontal layers, each individually isotropic, with different thicknesses and hydraulic conductivities, as shown in Figure 3.4.1. For horizontal flow parallel to the layers, the flow q_1 in the upper layer per unit width is

$$q_1 = K_1 i z_1 \tag{3.4.1}$$

where i is the hydraulic gradient and K_1 and z_1 are as indicated in Figure 3.4.1. Because i must be the same in each layer for horizontal flow, it follows that the total horizontal flow q_x is

$$q_x = q_1 + q_2 = i(K_1 z_1 + K_2 z_2) \tag{3.4.2}$$

For a homogeneous system this would be expressed as

$$q_x = K_x i(z_1 + z_2) \tag{3.4.3}$$

where K_x is the horizontal hydraulic conductivity for the entire system. Equating these and solving for K_x yield

$$K_x = \frac{K_1 z_1 + K_2 z_2}{z_1 + z_2} \tag{3.4.4}$$

which can be generalized for n layers as

$$K_x = \frac{K_1 z_1 + K_2 z_2 + \cdots + K_n z_n}{z_1 + z_2 + \cdots + z_n} \tag{3.4.5}$$

This defines the *equivalent horizontal hydraulic conductivity* for a stratified material.

Now, for vertical flow through the two layers in Figure 3.4.1, the flow q_z per unit horizontal area in the upper layer is

$$q_z = K_1 \frac{dh_1}{z_1} \tag{3.4.6}$$

where dh_1 is the head loss within the first layer. Solving for the head loss, we find

$$dh_1 = \frac{z_1}{K_1} q_z \tag{3.4.7}$$

From continuity, q_z must be the same for the other layer so that the total head loss is

$$dh_1 + dh_2 = \left[\frac{z_1}{K_1} + \frac{z_2}{K_2} \right] q_z \tag{3.4.8}$$

In a homogeneous system

$$q_z = K_z \left[\frac{dh_1 + dh_2}{z_1 + z_2} \right] \tag{3.4.9}$$

where K_z is the vertical hydraulic conductivity for the entire system. Rearranging yields

$$dh_1 + dh_2 = \left[\frac{z_1 + z_2}{K_z} \right] q_z \tag{3.4.10}$$

and equating with Equation 3.4.8, we have

$$K_z = \frac{z_1 + z_2}{\frac{z_1}{K_1} + \frac{z_2}{K_2}} \tag{3.4.11}$$

which can be generalized for n layers as

$$K_z = \frac{z_1 + z_2 + \cdots + z_n}{\frac{z_1}{K_1} + \frac{z_2}{K_2} + \cdots + \frac{z_n}{K_n}} \tag{3.4.12}$$

This defines the *equivalent vertical hydraulic conductivity* for a stratified material.

As mentioned earlier, the horizontal hydraulic conductivity in alluvium is normally greater than that in the vertical direction. This observation also follows from the above derivations; thus, if

$$K_x > K_z \tag{3.4.13}$$

then for the two-layer case from Equations 3.4.4 and 3.4.11

$$\frac{K_1 z_1 + K_2 z_2}{z_1 + z_2} > \frac{z_1 + z_2}{\frac{z_1}{K_1} + \frac{z_2}{K_2}} \tag{3.4.14}$$

which reduces to[67]

$$\frac{z_1}{z_2} (K_1 - K_2)^2 > 0 \tag{3.4.15}$$

Because the left side is always positive, it must be greater than zero, thereby confirming that

$$\frac{K_x}{K_z} \geq 1 \qquad (3.4.16)$$

Ratios of K_x/K_z usually fall in the range of 2 to 10 for alluvium,[75] but values up to 100 or more occur where clay layers are present. For consolidated geologic materials, anisotropic conditions are governed by the orientation of strata, fractures, solution openings, or other structural conditions, which do not necessarily possess a horizontal alignment.

In applying Darcy's law to two-dimensional flow in anisotropic media, the appropriate value of K must be selected for the direction of flow. For directions other than horizontal (K_x) and vertical (K_z), the K value can be obtained from

$$\frac{1}{K_\beta} = \frac{\cos^2 \beta}{K_x} + \frac{\sin^2 \beta}{K_z} \qquad (3.4.17)$$

where K_β is the hydraulic conductivity in the direction making an angle β with the horizontal.

EXAMPLE 3.4.1

An unconfined aquifer consists of three horizontal layers, each individually isotropic. The top layer has a thickness of 10 m and a hydraulic conductivity of 11.6 m/day. The middle layer has a thickness of 4.4 m and a hydraulic conductivity of 4.5 m/day. The bottom layer has a thickness of 6.2 m and a hydraulic conductivity of 2.2 m/day. Compute the equivalent horizontal and vertical hydraulic conductivities.

SOLUTION

Equation 3.4.5 is used to compute the equivalent horizontal hydraulic conductivity:

$$K_x = \frac{K_1 z_1 + K_2 z_2 + K_3 z_3}{z_1 + z_2 + z_3}$$

$$= \frac{(11.6 \text{ m/day})(10 \text{ m}) + (4.5 \text{ m/day})(4.4 \text{ m}) + (2.2 \text{ m/day})(6.2 \text{ m})}{(10 \text{ m} + 4.4 \text{ m} + 6.2 \text{ m})} = 7.25 \text{ m/day}$$

The equivalent vertical hydraulic conductivity is computed using Equation 3.4.12:

$$K_z = \frac{z_1 + z_2 + z_3}{\dfrac{z_1}{K_1} + \dfrac{z_2}{K_2} + \dfrac{z_3}{K_3}}$$

$$= \frac{10 \text{ m} + 4.4 \text{ m} + 6.2 \text{ m}}{\dfrac{10 \text{ m}}{11.6 \text{ m/day}} + \dfrac{4.4 \text{ m}}{4.5 \text{ m/day}} + \dfrac{6.2 \text{ m}}{2.2 \text{ m/day}}} = 4.42 \text{ m/day}$$

Note that the equivalent hydraulic conductivities above are computed based on the assumption that each layer is individually isotropic, that is, $K_x = K_z$ in each layer. ∎

3.5   GROUNDWATER FLOW RATES

From Darcy's law it follows that the rate of groundwater movement is governed by the hydraulic conductivity of an aquifer and the hydraulic gradient. To obtain an idea of the order of magnitude of natural velocities, assume a productive alluvial aquifer with $K = 75$ m/day and a hydraulic gradient $dh/dl = -10$ m/1000 m $= -0.01$. Then from Equation 3.1.5,

$$v = K\frac{dh}{dl} = 75(-0.01) = 0.75 \text{ m/day} \qquad (3.5.1)$$

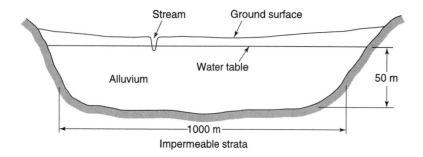

Figure 3.5.1. Cross section of a typical alluvial floodplain containing an unconfined aquifer with groundwater flowing perpendicular to the section (not to scale).

This is approximately equivalent to 0.5 mm/min, which demonstrates the sluggish nature of natural groundwater movement.

If the above flow occurs within and perpendicular to the large alluvial cross section sketched in Figure 3.5.1, then the total flow rate is

$$Q = Av = (50)(1000)(0.75) = 37,500 \text{ m}^3/\text{day} \tag{3.5.2}$$

which, when converted to usual streamflow units, amounts to only 0.43 m³/s. Thus, groundwater typically can be conceived of as a massive, slow-moving body of water.

Groundwater velocities vary widely depending on local hydrogeologic conditions; values from 2 m/year to 2 m/day are normal. Usually, velocities tend to decrease with depth as porosities and permeabilities also decrease. Velocities can range from negligible[*] to those of turbulent streams in underground openings within basalt and limestone. Mechanisms such as wells and drains act to accelerate flows.

An illustration of one-dimensional vertical flow is shown in Figure 3.5.2. Here an aquitard separates an overlying unconfined aquifer from an underlying leaky aquifer. The water table

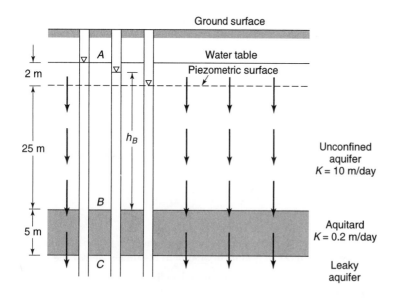

Figure 3.5.2. Diagram illustrating application of Darcy's law for vertically downward flow.

---

[*]All groundwater within the hydrologic cycle should be regarded as in continuous motion, although, it must be granted, some of it flows at extremely small rates.

stands above the piezometric surface so that water moves vertically downward from the unconfined aquifer, through the aquitard, and into the confined aquifer.

EXAMPLE 3.5.1

Determine h_B and the vertical velocity for the situation shown in Figure 3.5.2.

SOLUTION

Assume steady-state conditions. Writing Darcy's law from point A to B with the dimensions indicated in Figure 3.5.2, we have

$$v = K\frac{dh}{dl} = 10\frac{27 - h_B}{27} \tag{3.5.3}$$

and from point B to C,

$$v = K\frac{dh}{dl} = 0.2\frac{h_B + 5 - 30}{5}$$

Solving these yields, $h_B = 26.8$ m* and $v = 0.07$ m/day. ∎

3.6   GROUNDWATER FLOW DIRECTIONS

3.6.1   Flow Nets

For specified boundary conditions flow lines† and equipotential lines can be mapped in two dimensions to form a flow net. The two sets of lines form an orthogonal pattern of small squares. In a few simplified cases, the differential equation governing flow can be solved to obtain the flow net. Flow net analysis techniques have been applied in a number of ways for groundwater studies. Hollett[48] and Fenemore[32] used flow net analysis to provide initial transmissivities prior to trial-and-error numerical model calibration. Rice and Gorelick[82] applied flow net analysis to three field problems and demonstrated the value of the method for the conceptualization of flow systems. Other applications include Scott and Thorpe[87] and vanTonder.[104]

Consider the portion of a flow net shown in Figure 3.6.1. The hydraulic gradient i is given by

$$i = \frac{dh}{ds} \tag{3.6.1}$$

and the constant flow q between two adjacent flow lines by

$$q = K\frac{dh}{ds}dm \tag{3.6.2}$$

for unit thickness. But for the squares of the flow net, the approximation

$$ds \cong dm \tag{3.6.3}$$

can be made so that Equation 3.6.2 reduces to

$$q = K\,dh \tag{3.6.4}$$

---

*It should be noted that the piezometer open at the bottom of the unconfined aquifer (B) displays a water level below the water table (A) because of the head loss associated with vertical flow through the aquifer.

†A flow line is defined here as a line such that the macroscopic velocity vector is everywhere tangent to it.

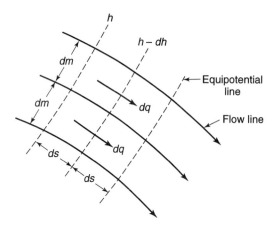

Figure 3.6.1. Portion of an orthogonal flow net formed by flow and equipotential lines.

Applying this to an entire flow net, where the total head loss h is divided into n squares between any two adjacent flow lines, then

$$dh = \frac{h}{n} \qquad (3.6.5)$$

If the flow is divided into m channels by flow lines, then the total flow

$$Q = m \, q = \frac{Kmh}{n} \qquad (3.6.6)$$

Thus, the geometry of the flow net, together with the hydraulic conductivity and head loss, enables the total flow in the section to be computed directly.

In anisotropic media, flow lines and equipotential lines are not orthogonal except when the flow is parallel to one of the principal directions.[8] In order to calculate flows for this situation, the boundaries of a flow section must be transformed so that an isotropic medium is obtained. For the typical alluvial case of $K_x > K_z$, all horizontal dimensions are reduced by the ratio $\sqrt{K_z / K_x}$. This creates a transformed section with an isotropic medium having an equivalent hydraulic conductivity

$$K' = \sqrt{K_x K_z} \qquad (3.6.7)$$

With this transformed section, the flow net can be drawn and flow rate determined.

After the flow net has been defined, it can be converted back to the true anisotropic section by multiplying all horizontal dimensions by $\sqrt{K_x / K_z}$. Figure 3.6.2 illustrates the procedure for an earth dam as well as demonstrates the distortion created by anisotropy in an element of the flow net. The technique can also be extended to anisotropic two-layer systems.[67] Figure 3.6.3 shows contrasting flow nets for channel seepage through layered anisotropic media.

Computer-assisted flow net analysis techniques have been developed and evaluated by Scott.[86]

3.6.2 Flow in Relation to Groundwater Contours

Because no flow crosses an impermeable boundary, flow lines must parallel it. Similarly, if no flow crosses the water table of an unconfined aquifer, it becomes a bounding flow surface. The

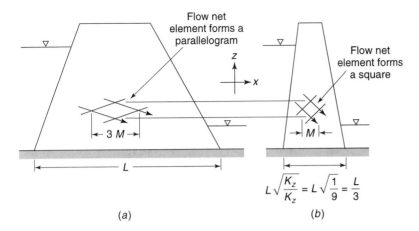

Figure 3.6.2. Illustration of flow net analysis for anisotropic hydraulic conductivity in an earth dam. (*a*) True anisotropic section with $K_x = 9K_z$ (*b*) Transformed isotropic section with $K_x = K_z$

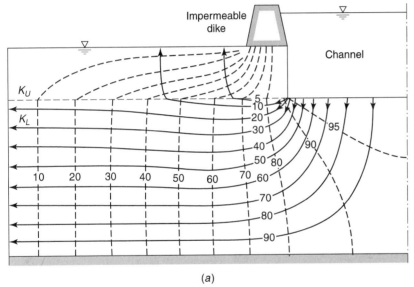

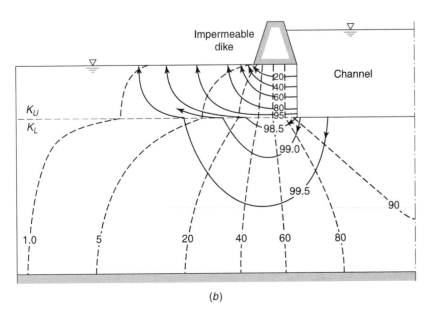

Figure 3.6.3. Flow nets for seepage from one side of a channel through two different anisotropic two-layer systems. (*a*) $K_U/K_L = 1/50$ (*b*) $K_U/K_L = 50$ The anisotropy ratio for all layers is $K_x/K_z = 10$ (after Todd and Bear[95]).

energy head h_E, or fluid potential, from Equation 3.1.2 at any point on the water table can be approximated by

$$h_E = \frac{p}{\gamma} + z \qquad (3.6.8)$$

so that by letting the atmospheric pressure reference be zero, $p = 0$, and $h_E = z$. Therefore, under steady-state conditions, the elevation at any point on the water table equals the energy head and, as a consequence, flow lines lie perpendicular to water table contours. Similarly, flow lines within a confined aquifer are orthogonal to contours of the potentiometric surface.

With only three groundwater elevations known from wells, estimates of local groundwater contours and flow directions can be determined as demonstrated by Figure 3.6.4. From field measurements of static water levels in wells within a basin, a water level contour map can be constructed. Flow lines, sketched perpendicular to contours, show directions of movement. An example appears in Figure 3.6.5.

Contour maps of groundwater levels, together with flow lines, are useful data for locating new wells. Convex contours indicate regions of groundwater recharge, while concave contours are associated with groundwater discharge. Furthermore, areas of favorable hydraulic conductivity can be ascertained from the spacing of contours. The procedure can be illustrated by treating two adjacent flow lines as impermeable boundaries because there can be no flow across a flow line. If the aquifer is uniformly thick, the flow at sections 1 and 2 in Figure 3.6.5 equals

$$q = W_1 v_1 = W_2 v_2 \qquad (3.6.9)$$

where v is velocity and W is the width of the flow section perpendicular to the flow. From Darcy's law

$$W_1 K_1 i_1 = W_2 K_2 i_2 \qquad (3.6.10)$$

which can be rewritten

$$\frac{K_1}{K_2} = \frac{W_2 i_2}{W_1 i_1} \qquad (3.6.11)$$

where K is hydraulic conductivity and i is hydraulic gradient. The ratios W_2/W_1 and i_2/i_1 can be estimated from the water level contour map (see Figure 3.6.5). For the special case of nearly parallel flow lines, Equation 3.6.11 reduces to

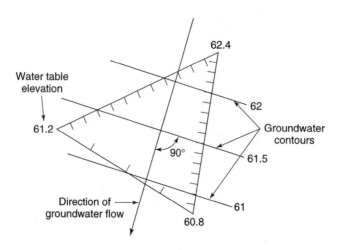

Figure 3.6.4. Estimate of groundwater contours and flow direction from water table elevations in three wells.

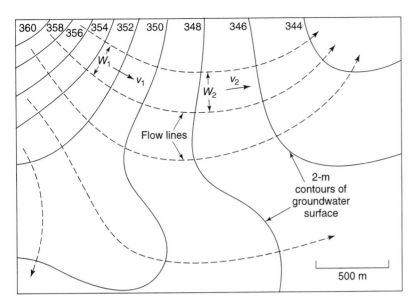

Figure 3.6.5. Contour map of a groundwater surface showing flow lines.

$$\frac{K_1}{K_2} = \frac{i_2}{i_1} \tag{3.6.12}$$

which may be interpreted as indicating that in an area of uniform groundwater flow, areas with wide contour spacings (flat gradients) possess higher hydraulic conductivities than those with narrow spacings (steep gradients). Therefore, in Figure 3.6.5, prospects for a productive well are better near section 2 than 1.

Where a contour map of groundwater levels contains closed contours around a group of wells of known total discharge Q, the transmissivity of the regional aquifer can be calculated. Figure 3.6.6 illustrates such a situation resulting from heavy groundwater pumping in and near Savannah, Georgia in 1957. If a flow net can be constructed, Equation 3.6.6 in the form

$$T = \frac{nQ}{mh} \tag{3.6.13}$$

can be applied where h represents the difference in elevation between any two selected closed contour lines. The typical irregularity of groundwater contours often makes construction of an accurate flow net difficult. As a convenient alternative involving contours but no flow net, Lohman[66] suggested the equation

$$T = \frac{2Q}{\left(L_1 + L_2\right)\Delta h/\Delta r} \tag{3.6.14}$$

where L_1 and L_2 are the lengths of any two concentric closed contours, Δh is the contour interval, and Δr is the average distance between the two closed contours.

Natural permeable boundaries of aquifers include surface water bodies and the ground surface. In a surface water body, the energy head is constant everywhere within the water body and equals the elevation of the water surface; consequently, aquifer flow lines must intersect normal to such a bounding surface. For the ground surface, this does not apply, as only atmospheric pressure exists at the ground surface. Hence, in Equation 3.6.8, by letting $p = 0$, $h_E = z$, which is identical to the case for a water table boundary.

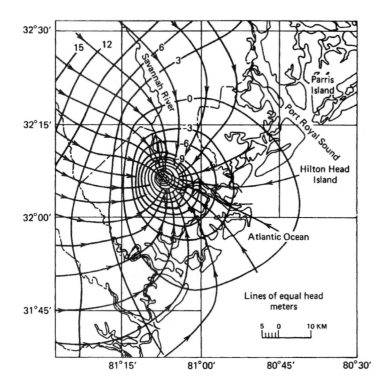

Figure 3.6.6. Contour map of the piezometric surface near Savannah, Georgia, 1957, showing closed contours resulting from heavy local groundwater pumping (after *U.S.G.S. Water-Supply Paper* 1611).

EXAMPLE 3.6.1

Three observation wells are installed to determine the direction of groundwater movement and the hydraulic gradient in a regional aquifer. The distance between the wells and the total head at each well are shown in Figure 3.6.7*a*.

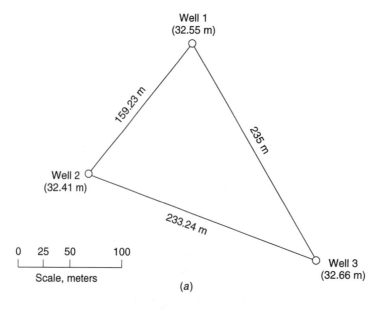

Figure 3.6.7*a.*
Configuration of three observation wells in Example 3.6.1

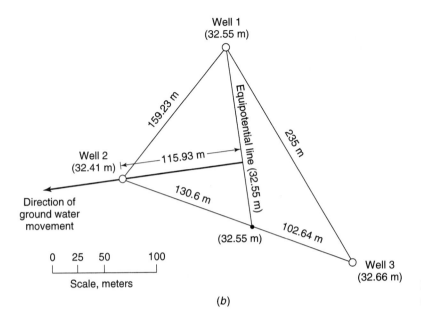

Figure 3.6.7b. Illustration of graphical procedure in Example 3.6.1.

(b)

SOLUTION

Step 1: Identify the well with the intermediate water level—Well 1 in this case.

Step 2: Along the straight line between the wells with the highest head and the lowest head, identify the location of the same head of the well from Step 1. Note that this is accomplished by locating the elevation of 32.55 m between Well 2 and Well 3 in the graphical solution.

Step 3: Draw a straight line between the intermediate well from Step 1 and the point identified in Step 2. This is a segment of the equipotential line along which the total head is the same as that in the intermediate well (i.e., equipotential line of 32.55 m head in this case).

Step 4: Draw a line perpendicular to the equipotential line passing through the well with the lowest head. The hydraulic gradient is the slope of that perpendicular line. Also, the direction of the line indicates the direction of groundwater movement. The graphical procedure above is illustrated in Figure 3.6.7b. The hydraulic gradient is then computed as

$$i = \frac{32.55 \text{ m} - 32.41 \text{ m}}{115.93 \text{ m}} = 0.0012$$

EXAMPLE 3.6.2

The average daily discharge from the Patuxent Formation (see Figure 3.6.8) in the Sparrows Point district of Baltimore, Maryland, in 1945 was estimated as 1×10^6 ft³/day. A flow net of the region is constructed using the available contour lines as shown in Figure 3.6.8. (This example is adapted from Lohman.[66]) Compute the transmissivity of the regional aquifer.

SOLUTION

As shown in the flow net, there are 15 flow channels, hence $m = 15$. There are four equipotential drops from the 60-ft contour line to the 20-ft contour line, so $h = 40$ ft and $n = 4$. Then the overall transmissivity of the district can be computed using Equation 3.6.13:

$$T = \frac{nQ}{mh} = \frac{(4)(1 \times 10^6 \text{ ft}^3/\text{day})}{(15)(40 \text{ ft})} \cong 6700 \text{ ft}^2/\text{day}$$

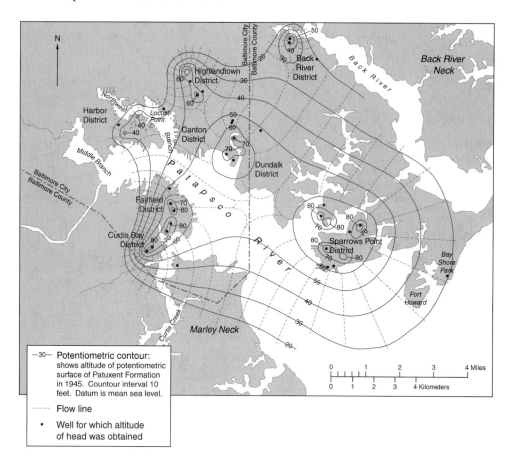

Figure 3.6.8. Map of Baltimore industrial area. Maryland, showing potentiometric surface in 1945 and generalized flow lines in the Patuxent Formation. From Bennett and Meyer[9] (as presented in Lohman[66]).

3.6.3   Flow Across a Water Table

As long as no flow crosses a water table, it serves as a groundwater boundary; however, if flow, such as percolating water, reaches the water table, flow lines no longer parallel the surface as an impermeable boundary.[52] To illustrate this refraction effect, let v_u represent the unsaturated vertical velocity approaching the water table and v_s the saturated velocity below the water table (Figure 3.6.9). The head loss dh for flow along the left flow line below the water table occurs in a distance of $b_s \tan(\delta + \varepsilon)$, as defined in Figure 3.6.9. Thus

$$v_s = Ki = K\frac{dh}{b_s\tan(\delta + \varepsilon)} \qquad (3.6.15)$$

but

$$dh = b_u\tan\delta \qquad (3.6.16)$$

hence

$$v_s = K\frac{b_u\tan\delta}{b_s\tan(\delta + \varepsilon)} \qquad (3.6.17)$$

From continuity

$$\frac{b_u}{b_s} = \frac{v_s}{v_u} \qquad (3.6.18)$$

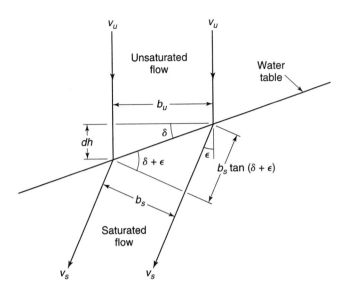

Figure 3.6.9.
Refraction of flow
lines across a water
table (after Jacob[56]).

where b_u and b_s are as shown in Figure 3.6.7 so that

$$v_s = K \frac{v_s \tan \delta}{v_u \tan(\delta + \varepsilon)}$$ (3.6.19)

which, when solved for ε, gives

$$\varepsilon = \tan^{-1}\left(\frac{K}{v_u} \tan \delta\right) - \delta$$ (3.6.20)

This states that for $v_u > 0$, flow lines will form an angle of (90 degrees $- \delta - \varepsilon$) below the water table. For the case of no percolating flow, $v_u = 0$ and $\varepsilon = 90$ degrees $- \delta$, so that v_s parallels the water table.

EXAMPLE 3.6.3

Consider a case where 1.6 m/year of percolating water reaches the water table of an unconfined aquifer. If the hydraulic conductivity and the regional hydraulic gradient of the aquifer are 9.3 m/day and 0.01, respectively, find the deflection angle of the flow lines as they cross the water table of the unconfined aquifer.

SOLUTION

Given $i = 0.01$, $v_u = 1.6$ m/year $= 4.38 \times 10^{-3}$ m/day, $K = 9.3$ m/day, and $\tan \delta = i = 0.01$ so $\delta = 0.573°$. The deflection angle, ε, is computed using Equation 3.6.20:

$$\varepsilon = \tan^{-1}\left(\frac{K}{v_u} \tan \delta\right) - \delta$$

$$= \tan^{-1}\left(\frac{9.3 \text{ m/day}}{4.38 \times 10^{-3} \text{ m/day}} \tan\left(0.573°\right)\right) - 0.573° = 86.7°$$

The angle between the groundwater flow direction and the refracted flow lines is $90° - (\delta + \varepsilon) = 90° - (0.573° + 86.7°) = 2.7°$. ■

3.6.4 Flow Across a Hydraulic Conductivity Boundary

Similar to the above analysis, where flow passes from a region of hydraulic conductivity K_1 to one of K_2, a change in flow direction results. The change of direction can be derived from continuity considerations and expressed in terms of the two K values. Visualizing a flow field as shown in Figure 3.6.10, we see that the normal components of flow approaching and leaving the boundary must be equal; hence, the normal velocities v_n must be such that

$$v_{n_1} = v_{n_2} \tag{3.6.21}$$

or

$$K_1 \frac{dh_1}{dL_1} \cos \theta_1 = K_2 \frac{dh_2}{dL_2} \cos \theta_2 \tag{3.6.22}$$

where θ_1 and θ_2 are angles with the normal shown in Figure 3.6.10. Also, the distance b along the boundary between two adjacent flow lines must be the same on each side of the boundary. From Figure 3.6.10, the distance b can be given as

$$b = \frac{dL_1}{\sin \theta_1} = \frac{dL_2}{\sin \theta_2} \tag{3.6.23}$$

which can be arranged

$$dL_1 \sin \theta_2 = dL_2 \sin \theta_1 \tag{3.6.24}$$

Dividing this equation by Equation 3.6.22 and noting that $dh_1 = dh_2$ between two equipotential lines, gives

$$\frac{K_1}{K_2} = \frac{\tan \theta_1}{\tan \theta_2} \tag{3.6.25}$$

Thus, for saturated flow passing from a medium of one hydraulic conductivity to that of another, a refraction in flow lines occurs such that the ratio of the Ks equals the ratio of the tangents of the angles the flow lines make with the normal to the boundary. Consequences of the relation are illustrated in Figure 3.6.11.

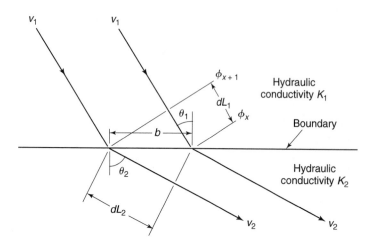

Figure 3.6.10. Refraction of flow lines across a boundary between media of different hydraulic conductivities.

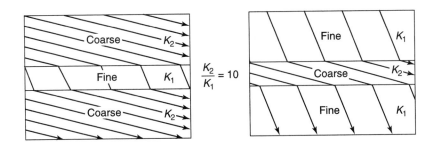

Figure 3.6.11. Refraction across layers of coarse and fine sand with a hydraulic conductivity ratio of 10 (after Hubbert[51] copyright ©1940 by the University of Chicago Press).

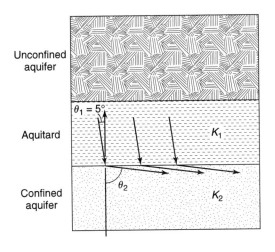

Figure 3.6.12. Example 3.6.4

EXAMPLE 3.6.4

Consider a case where a leaky confined aquifer with 4.5 m/day horizontal hydraulic conductivity is overlain by an aquitard with 0.052 m/day vertical hydraulic conductivity. If the flow in the aquitard is in the downward direction and makes an angle of 5° with the vertical (see Figure 3.6.12), determine θ_2.

SOLUTION

Given $K_1 = 0.052$ m/day, $K_2 = 4.5$ m/day, and $\theta_1 = 5°$, Equation 3.6.25 is used to compute θ_2:

$$\frac{K_1}{K_2} = \frac{\tan\theta_1}{\tan\theta_2} \rightarrow \frac{0.052 \text{ m/day}}{4.5 \text{ m/day}} = \frac{\tan(5^0)}{\tan\theta_2} \rightarrow \theta_2 = 82.5°$$

The flow lines become nearly horizontal as they enter into the confined aquifer. This is a typical case for regional flow systems, as the hydraulic conductivity of a confined aquifer is generally a few orders of magnitude larger than that of the confining layers. ∎

3.6.5   Regional Flow Patterns

Although most groundwater movement in shallow aquifers tends to be nearly horizontal, regional flow patterns can become quite complex. Reasons for this stem from the diversity of field parameters: areas and magnitudes of recharge and discharge, topography, stratigraphy, and anisotropy.[88] Analytic solutions for specified aquifer cross sections by Toth[97, 98] demonstrated that the variability of a water table could produce a variety of flow patterns. Subsequent work by Freeze[37] extended this approach to other subsurface boundary conditions. From these

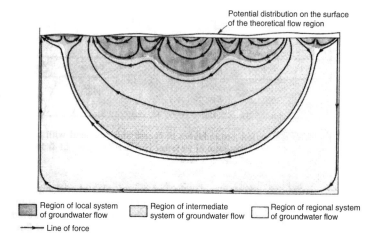

Potential distribution on the surface of the theoretical flow region

Region of local system of groundwater flow

Region of intermediate system of groundwater flow

Region of regional system of groundwater flow

Line of force

Figure 3.6.13 Vertical cross section showing theoretical flow pattern of groundwater through an aquifer with a sloping sinusoidal water table (after Toth[98]).

contributions it is clear that accurate evaluation of groundwater flows is contingent on a detailed knowledge of hydrogeologic conditions.

An illustration of a regional groundwater flow pattern is shown in Figure 3.6.13 for a vertical cross section through an unconfined aquifer. Flow lines were obtained by computer assuming:

1. A homogeneous and isotropic aquifer with impermeable boundaries along the sides and the bottom.

2. A rectangular flow region 6,100 m wide by 3,050 m deep.

3. A sinusoidal potential distribution (equivalent to the water table) with an amplitude of 15 m and a mean slope of 2 percent on the surface of the theoretical flow region.

Local, intermediate, and regional systems of flow are indicated in Figure 3.6.13, as well as flows counter to the mean water table gradient. Although the aquifer portrayed is highly idealized, a similar multiformity in flow pattern can be anticipated for actual aquifers where irregularities of topography, stratigraphy, and anisotropy are introduced.

CASE STUDY *Lake Five-O, Florida*

The purpose of this case study is to examine the head distribution and the groundwater flow near Lake Five-O, which was introduced in Chapter 1. Lake Five-O is a softwater seepage lake in northwestern Florida. The Lake Five-O study area (see Figures 1.6.4 and 1.6.5) has three hydrogeologic units (surficial aquifer, intermediate confining unit, and Upper Floridan aquifer) that influence the hydrology. Figure 3.6.14 illustrates these three units for the hydrogeologic section A–A′, location as shown in Figure 1.6.5.

Grubbs[44] evaluated the temporal and spatial distributions of groundwater flow into and the leakage from Lake Five-O using hydrologic data and simulation models of the shallow groundwater system adjacent to the lake. As shown by the study (hydrologic budget by Grubbs[44]) groundwater inflow to the lake and leakage from the lake to the groundwater system are the dominant components in the total inflow (precipitation plus groundwater inflow) and the total outflow (evaporation plus leakage) budgets of the lake.

As reported by Grubbs,[44] head fluctuations for the two-year (1989–1990) study period were generally consistent with typical seasonal patterns of precipitation. Seasonal precipitation in northwestern Florida causes wet conditions during the summer and dryer conditions during the fall and spring, so that as shown in Figure 3.6.15, heads increased during the wet summer seasons of 1989 and decreased during the dryer spring months of 1989 and 1990 and the fall of 1990. Head data from a monitoring network indicated a consistent pattern of groundwater flow toward the lake, so that water table elevations (a) increased with distance from the lake (Figures 3.6.16 and 3.6.17) and (b) were consistently higher than the stage of the lake.

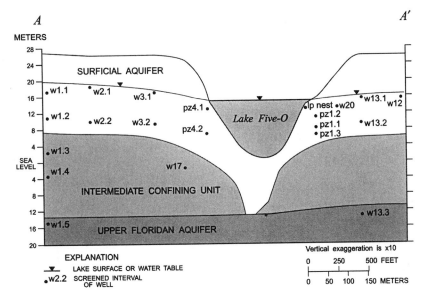

Figure 3.6.14. Hydrogeologic section A–A′ through Lake Five-O showing midpoints of screened intervals of observation wells used to construct the section. (Location of section is shown in figure 1.6.5.) Hydrogeologic section A–A′ revised from Andrews and others[4] (as presented in Grubbs[44]).

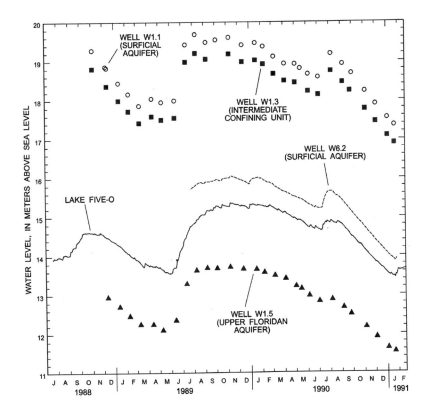

Figure 3.6.15. Water levels of Lake Five-O and selected groundwater monitoring wells, July 1988 to February 1991 (Grubbs[44]).

The head data shown in Figures 3.6.15–3.6.18 indicate a strong potential for downward leakage from Lake Five-O to the Upper Floridan Aquifer. Figure 3.6.18 shows the differences between the stage of the lake and the head in the Upper Floridan aquifer, indicating abrupt declines in head difference during the summer of 1989 and abrupt increases in head differences during the summer of 1990. Leakage from the lake decreased during the summer of 1989 and increased in the summer of 1990. Grubbs[44] notes that the changes in head difference (and the consequent leakage from the lake) were probably due to regional variations in recharge to the Upper Floridan Aquifer during the two-year time period.

(*continues*)

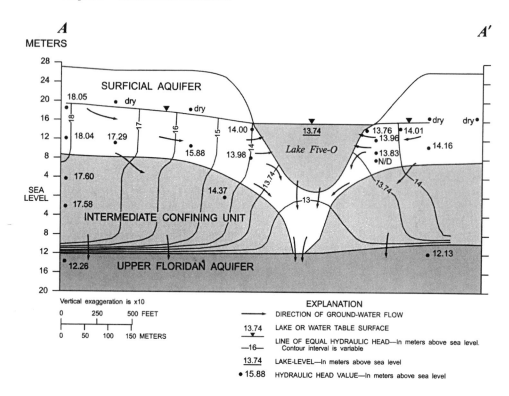

Figure 3.6.16. Hydrogeologic section A–A′ through Lake Five-O, showing vertical head distribution near the lake for April 13, 1989. (Location of section is shown in Figure 1.6.5.) (Grubbs[44])

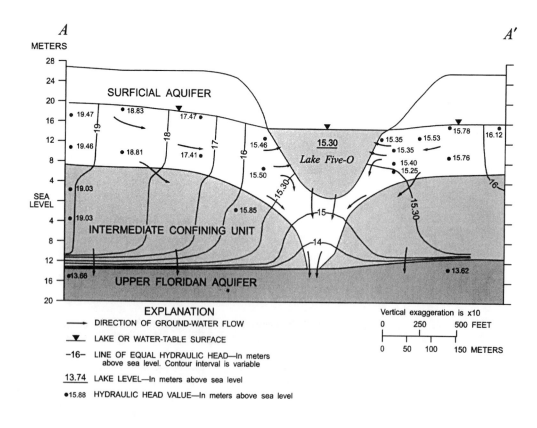

Figure 3.6.17. Hydrogeologic section A–A′ through Lake Five-O, showing vertical head distribution near the lake for January 17, 1990. (Location of section is shown in Figure 1.6.5.) (Grubbs[44])

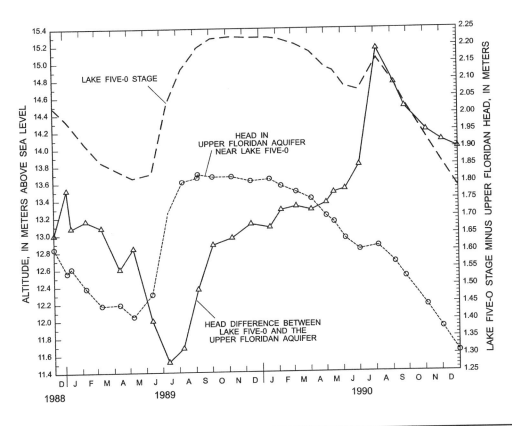

Figure 3.6.18
Comparison of Lake Five-O stage and head in the Upper Floridan Aquifer, December 1988 to January 1991 (Grubbs[44]).

CASE STUDY *Lake Five-O, Florida (continued)*

Head data indicate significant hydraulic conductivity differences within the intermediate confining unit and between the surficial aquifer and intermediate confining unit. Confining properties of the intermediate confining unit are probably not evenly distributed within the unit. A dense, shelly clay layer is located at the base of the unit and is the most effective confining unit in the unit. This is based upon Figures 3.6.16 and 3.6.17, which indicate the large head loss that occurs across the basal clay unit, which is much larger than the head differences between wells 1.3 and 1.4 (refer to Figure 3.6.14 for locations). An abrupt decrease in hydraulic conductivity occurs at contact of the surficial aquifer/intermediate confining units. This is indicated in Figures 3.6.16–3.6.18 by the vertical head losses for the group of wells 1.1 through 1.5. The lower hydraulic conductivity of the intermediate confining unit in relation to the hydraulic conductivity of the surficial aquifer unit limits the downward movement of water.

3.7 DISPERSION

3.7.1 Concept

In saturated flow through porous media, velocities vary widely across any single pore, just as in a capillary tube where the velocity distribution in laminar flow is parabolic. In addition, the pores possess different sizes, shapes, and orientations. As a result, when a labeled miscible liquid, referred to as a *tracer*, is introduced into a flow system, it spreads gradually to occupy an increasing portion of the flow region. This phenomenon is known as *dispersion* and constitutes a nonsteady, irreversible mixing process by which the tracer disperses within the surrounding water.[20, 84]

In a column packed with sand as in Figure 3.7.1a and supplied continuously after a time t_0 with water containing a tracer of concentration c_0, dispersion in the longitudinal direction of

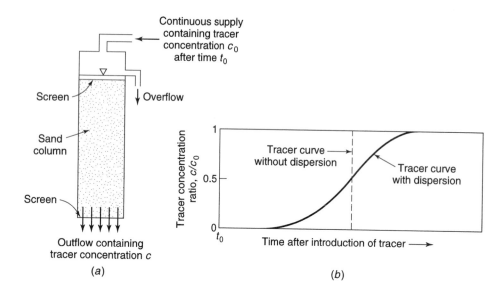

Figure 3.7.1. Longitudinal dispersion of a tracer passing through a sand column. (*a*) Sand column (*b*) Dispersion curve

flow can be measured.[42, 89] From samples of water emerging from the column, a tracer concentration c is found. In Figure 3.7.1b, the solid line shows a typical S-shaped dispersion curve. Besides longitudinal dispersion, lateral dispersion also occurs because water is continually dividing and reuniting as it follows tortuous flow paths around grains of a medium, as illustrated in Figure 3.7.2.[26]

The equation for dispersion in homogeneous and isotropic media for the two-dimensional case has the form

$$\frac{\partial c}{\partial t} = D_L \frac{\partial^2 c}{\partial x^2} + D_T \frac{\partial^2 c}{\partial y^2} - v \frac{\partial c}{\partial x}$$

(3.7.1)

where c is the relative tracer concentration ($0 \le c \le 1$), D_L and D_T are longitudinal and transverse dispersion coefficients, v is fluid velocity, x is the coordinate in the direction of flow, y is the coordinate normal to flow, and t is time. Dimensions of the dispersion coefficients are L^2/T.

Dispersion is essentially a microscopic phenomenon caused by a combination of molecular diffusion and hydrodynamic mixing occurring with laminar flow through porous media. The net result produces a conic form downstream from a continuous point-source tracer (Figure 3.7.3a), and an expanding ellipsoid for a single tracer injection (Figure 3.7.3b).* Many theoretical and experimental studies have provided a better understanding of the phenomenon.[5, 29, 33, 34, 35, 36, 72, 73, 76, 77] Most mathematical descriptions of dispersion are based on statistical concepts because of the difficulties in defining the microstructure of porous media as well as the relative roles of molecular diffusion and mechanical dispersion.[20, 65, 96] Experimental studies have attempted to establish relations between factors such as velocity, media structural properties, and permeability with dispersion coefficients.[35, 47, 49, 76, 92, 99] The pattern of a point tracer as it moves downstream from its source tends to a normal (Gaussian) distribution both longitudinally and transversely, as sketched in Figure 3.7.3b. Furthermore, the longitudinal component is larger than that of the transverse so that the major axis of mixing occurs in the direction of flow.

---

*One may visualize these two forms, by analogy, as a continuous plume and as a single puff of smoke drifting downwind from a smokestack, respectively, even though the flow in porous media is laminar whereas that in the atmosphere is turbulent.

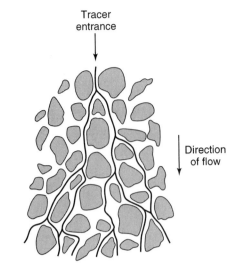

Figure 3.7.2. Lateral dispersion of a tracer originating from a point source in a porous medium.

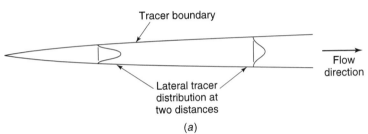

(a)

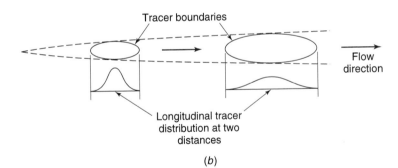

(b)

Figure 3.7.3. Sketches of tracer distribution resulting from dispersion by flow in porous media. (a) Continuous tracer (b) Single slug of tracer

3.7.2 Dispersion and Groundwater Hydrology

In groundwater hydrology, dispersion may be encountered whenever two fluids with different characteristics come into contact. Prime examples of this include tracers for evaluating directions and velocities of groundwater flow, introduction of pollutants into the ground (Chapter 8), artificial recharge of water with one quality into an aquifer containing groundwater of another quality (Chapter 13), and intrusion of saline water into freshwater aquifers (Chapter 14). In general, the magnitude of dispersion for uniform sands can be measured in terms of only a few meters over a travel distance of 10^3 meters.

In addition, the heterogeneity of most geologic materials introduces irregularities of flow with the consequent mixing of a tracer; these effects often far overshadow the effects of microscopic dispersion. The rapid nonsymmetric dispersal of two continuous dye streams in the laboratory demonstration shown in Figure 3.7.4 amply illustrates the important role of

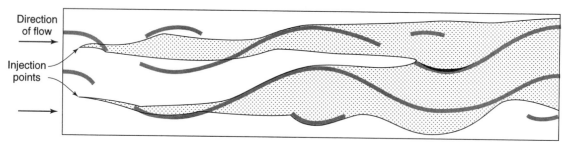

Figure 3.7.4. Dispersion of two dye streams in a heterogeneous porous medium. Dotted areas indicate bands of much higher permeability (after Skibitzke and Robinson[90]).

heterogeneity. In essence, a macroscopic dispersion is superimposed on the flow system. Furthermore, the irregular geometries of groundwater recharge and discharge zones, together with the lack of specific data on aquifer characteristics, preclude quantitative evaluation of dispersion coefficients for most natural groundwater flow situations.

3.8 GROUNDWATER TRACERS

A variety of tracers have been employed for studying dispersion and also for evaluating directions and rates of groundwater flow under field conditions (see Section 3.3.3). An ideal tracer should (1) be susceptible to quantitative determination in minute concentrations, (2) be absent or nearly so from the natural water, (3) not react chemically with the natural water or be absorbed by the porous media, (4) be safe in terms of human health, and (5) be inexpensive and readily available. No tracer completely meets all these requirements, but a reasonably satisfactory tracer can be selected to fit the needs of a particular situation. Possibilities include water-soluble dyes (such as sodium fluorescein), which can be detected by colorimetry; soluble chloride and sulfate salts and sugars, which can be detected chemically; and strong electrolytes, which can be detected by electrical conductivity.

Radioisotopes such as H^3 (tritium), Co^{60}, Rb^{86}, I^{131} have also served as tracers.[41, 54, 58, 59, 94, 108] Naturally occurring radioisotopes, H^3 and C^{14}, not only have enabled residence times of groundwater to be estimated (see Section 2.1) but also have provided a means for evaluating regional groundwater flow.[6, 46, 78, 105] Refer to the work by Field[33–36].

3.9 GENERAL FLOW EQUATIONS

The *control volume* (CV) for a saturated flow is shown in Figure 3.9.1. The sides, which define the *control surface* (CS), have lengths dx, dy, and dz in the coordinate directions. The total volume of the control volume is dxdydz and the volume of water flowing into or out of the control volume is θdxdydz, where θ is the moisture content. The use of the control-volume approach is described in detail in Chow et al.;[21] Crowe et al.;[23] and Mays[69]. The general control volume equation for continuity is applicable:

$$0 = \frac{d}{dt} \int_{CV} \rho \, d\forall + \int_{CS} \rho \mathbf{V} \cdot d\mathbf{A} \qquad (3.9.1a)$$

or

$$\int_{CS} \rho \mathbf{V} \cdot d\mathbf{A} = -\frac{d}{dt} \int_{CV} \rho \cdot d\forall \qquad (3.9.1b)$$

which states that the rate of outflow of groundwater through the control surface (CS) is equal to the rate of change of groundwater stored in the control volume (CV). The elemental volume

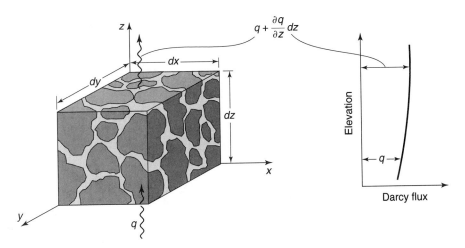

Figure 3.9.1. Control volume for development of the continuity equation in porous medium (from Chow et al[21]).

is $d\forall = dxdydz$, the velocity vector of flow is $\mathbf{V}$, and the volume of flowrate past a given area $d\mathbf{A}$ is $\mathbf{V} \cdot d\mathbf{A}$, where $\mathbf{A}$ is the area vector. Equation 3.9.1b also states that the net rate of outflow of mass (groundwater) from the control volume is equal to the time rate of change of mass (groundwater) stored within the control volume.

The time rate of change of mass stored in the control volume is defined as the time rate of change of fluid mass in storage, expressed as

$$\frac{d}{dt}\int_{CV}\rho\,d\forall = \rho S_S\frac{\partial h}{\partial t}(dxdydz) + \rho W(dxdydz) \tag{3.9.2}$$

where S_s is the specific storage and W is the flow into or out of the control volume, $W = Q/(dxdydz)$. The term $\rho S_S\frac{\partial h}{\partial t}(dxdydz)$ includes two parts: (1) the mass rate of water produced by an expansion of the water under change in density and (2) the mass rate of water produced by the compaction of the porous medium due to change in porosity.

The inflow of water through the control surface at the bottom of the control volume is $qdxdy$ and the outflow at the top is $[q + (\partial q/\partial z)dz]dxdy$ so that the net out flow in the vertical direction z is

$$\int_{CS}\rho\mathbf{V}\cdot d\mathbf{A} = \rho\left(q + \frac{\partial q}{\partial z}dz\right)dxdy - qdxdy$$

$$= \rho dxdydz\frac{\partial q}{\partial z} \tag{3.9.3}$$

where q (in the z direction) in $\frac{\partial q}{\partial z}$ is q_z.

Considering all three directions, we have

$$\int_{CS}\rho\mathbf{V}\cdot d\mathbf{A} = \rho dxdydz\frac{\partial q}{\partial x} + \rho dxdydz\frac{\partial q}{\partial y} + \rho dxdydz\frac{\partial q}{\partial z} \tag{3.9.4}$$

Substituting 3.9.2 and 3.9.4 into 3.9.1 results in

$$0 = \rho S_S\frac{\partial h}{\partial t}dxdydz + \rho W(dxdydz) + \rho dxdydz\frac{\partial q}{\partial x} + \rho dxdydz\frac{\partial q}{\partial y} + \rho dxdydz\frac{\partial q}{\partial z} \tag{3.9.5}$$

Dividing through by $\rho\,dxdydz$ gives

$$S_s\frac{\partial h}{\partial t} + \frac{\partial q}{\partial x} + \frac{\partial q}{\partial y} + \frac{\partial q}{\partial z} + W = 0 \tag{3.9.6}$$

Using Darcy's law, the Darcy flux in each direction is

$$q_x = -K_x \frac{\partial h}{\partial x} \tag{3.9.7a}$$

$$q_y = -K_y \frac{\partial h}{\partial y} \tag{3.9.7b}$$

$$q_z = -K_z \frac{\partial h}{\partial z} \tag{3.9.7c}$$

Subsitituting these definitions of q_x, q_y, and q_z into 3.9.6 and rearranging results in

$$\frac{\partial}{\partial x}\left(K_x \frac{\partial h}{\partial x}\right) + \frac{\partial}{\partial y}\left(K_y \frac{\partial h}{\partial y}\right) + \frac{\partial}{\partial z}\left(K_z \frac{\partial h}{\partial z}\right) = S_s \frac{\partial h}{\partial t} + W \tag{3.9.8}$$

This is the equation for *three-dimensional transient flow through a saturated anisotropic porous medium.* For a *homogenous, isotropic medium* ($K_x = K_y = K_z$) then Equation 3.9.8 becomes

$$\frac{\partial^2 h}{\partial x^2} + \frac{\partial^2 h}{\partial y^2} + \frac{\partial^2 h}{\partial z^2} = \frac{S_s}{K}\frac{\partial h}{\partial t} + \frac{W}{K} \tag{3.9.9}$$

For a steady-state flow, $\frac{\partial h}{\partial t} = 0$, Equation 3.9.9 becomes

$$\frac{\partial^2 h}{\partial x^2} + \frac{\partial^2 h}{\partial y^2} + \frac{\partial^2 h}{\partial z^2} = \frac{W}{K} \tag{3.9.10}$$

For a horizontal confined aquifer of thickness b, $S = S_s b$ and the transmissivity $T = Kb$, so the two-dimensional form of 3.9.9 with $W = 0$ becomes

$$\frac{\partial^2 h}{\partial x^2} + \frac{\partial^2 h}{\partial y^2} = \frac{S}{T}\frac{\partial h}{\partial t} \tag{3.9.11}$$

The governing equation for radial flow can be derived using the control volume approach. Alternatively, Equation 3.9.11 can be converted into radial coordinates using the relation $r = \sqrt{x^2 + y^2}$. This is known as the *diffusion equation*, expressed as

$$\frac{1}{r}\frac{\partial}{\partial r}\left(r\frac{\partial h}{\partial r}\right) = \frac{\partial^2 h}{\partial r^2} + \frac{1}{r}\frac{\partial h}{\partial r} = \frac{S}{T}\frac{\partial h}{\partial t} \tag{3.9.12}$$

where r is the radial distance from a pumped well and t is the time since the beginning of pumping. For steady-state conditions, $\partial h/\partial t = 0$, so Equation 3.9.12 reduces to

$$\frac{1}{r}\frac{\partial}{\partial r}\left(r\frac{\partial h}{\partial r}\right) = 0 \tag{3.9.13}$$

Application to Aquifers. The equations derived above will be applied subsequently to obtain analytic solutions to particular groundwater flow problems. For solution of any problem, idealization of the aquifer and of the boundary conditions of the flow system is necessary. Results may only approximate field conditions; nevertheless, known deviations from assumptions frequently allow analytic solutions to be modified to obtain an answer that otherwise would not have been possible. A common assumption regarding the aquifer is that it is homogeneous and isotropic (see Chapter 2). Often aquifers can be assumed to be infinite in areal extent; if not, boundaries are assumed to be (1) impermeable, such as underlying or overlying rock or clay layers, dikes, faults, or valley walls; or (2) permeable, including surface water bodies in contact with the aquifer, ground surfaces where water emerges from underground, and wells.

3.10   UNSATURATED FLOW

In groundwater hydrology, unsaturated flow is important for downward vertical flow (natural and artificial recharge), upward vertical flow (evaporation and transpiration), movement of pollutants from ground surface, and horizontal flow in the capillary zone above the water table.[91] A large amount of literature exists on unsaturated flow, most of it contributed by soil scientists. Significant summaries of the subject are available in several references.[61, 93, 103]

In order to discuss infiltration, we must first consider the various subsurface flow processes shown in Figure 3.10.1. These processes are infiltration of water to become *soil moisture, subsurface flow* (unsaturated flow) through the soil, and *groundwater flow* (unconfined saturated flow). *Unsaturated flow* refers to flow through porous medium when some of the voids are occupied by air. *Saturated flow* occurs when the voids are filled with water. The *water table* is the interface between the saturated and unsaturated flow where atmospheric pressure prevails. Saturated flow occurs below the water table and unsaturated flow occurs above the water table.

A typical distribution of water content above a water table is sketched in Figure 3.10.2a. If the water table is lowered Δh, then the capillary zone shifts downward. The shaded area serves as a measure of the volume of water drained from above the water table. This water is released by vertical percolation; consequently, specific yield becomes an asymptotic function of time (Figure 3.10.2b).

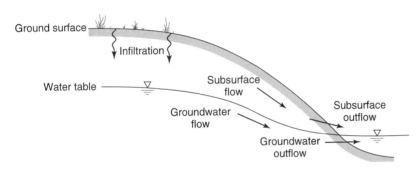

Figure 3.10.1. Subsurface water zones and processes (from Chow et al.[21]).

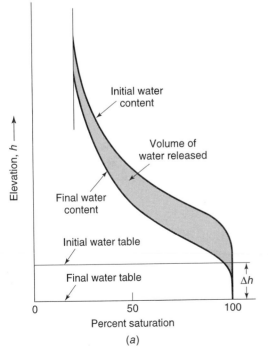

Figure 3.10.2. Movement of water in the zone of aeration by lowering the water table. (*a*) Water content above the water table. (*b*) Specific yield as a function of the time of drainage.

3.10.1 Flow Through Unsaturated Soils

Consider the control volume (element) in Figure (3.10.3b) for an unsaturated soil which has sides of lengths dx, dy, and dz with a volume of $dx\,dy\,dz$. The volume of water contained in the control volume is $\theta\,dx\,dy\,dz$ [where θ is the moisture content]. Flow through the control volume is defined by the *Darcy flux*, $q = Q/A$, which is the volumetric flow rate per unit of soil area. For this derivation, the horizontal fluxes are ignored and only the vertical (z) direction is considered, with z positive upward.

Using the general control volume equation for continuity (3.9.1b)

$$\int_{CS} \rho \mathbf{V} \cdot d\mathbf{A} = \frac{-d}{dt}\int \rho \, d\forall \tag{3.10.1}$$

The time rate of change of mass stored in the element (control volume) is

$$-\frac{d}{dt}\int \rho \, d\forall = \frac{d}{dt}\left(\rho\theta\,dx\,dy\,dz\right) = \rho\theta\,dx\,dy\,dz\,\frac{\partial\theta}{\partial t} \tag{3.10.2}$$

where the density is assumed constant. The net outflow of water is the difference between the volumetric inflow at the bottom ($q\,dx\,dy$) and the volumetric outflow at the top $[q + (\partial q/\partial z)dz]dx\,dy$, so

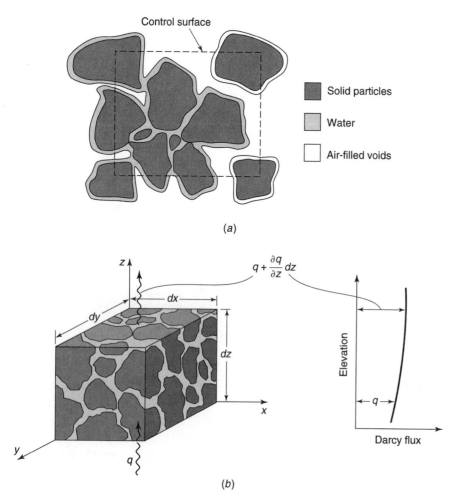

(a)

(b)

Figure 3.10.3. (*a*) Cross-section through an unsaturated porous medium (*b*) Control volume for development of the continuity equation in an unsaturated porous medium (from Chow et al.[21])

$$\int_{cs} \rho \mathbf{V} \cdot d\mathbf{A} = \rho \left(q + \frac{\partial q}{\partial z} dz \right) dx\, dy - \rho q\, dx\, dy = \rho\, dx\, dy\, dz\, \frac{\partial q}{\partial z} \tag{3.10.3}$$

Dividing through by $\rho\, dx\, dy\, dz$ and rearranging results in the following *continuity equation for one-dimensional unsteady unsaturated flow in a porous medium* in the z direction:

$$\frac{\partial \theta}{\partial t} + \frac{\partial q}{\partial z} = 0 \tag{3.10.4}$$

Darcy's law relates the *Darcy flux, q,* to the rate of headloss per unit length of medium. For flow in the vertical direction, the headloss per unit length is the change in total head ∂h over a distance, ∂z, that is, $-\frac{\partial h}{\partial z}$, where the negative sign indicates that total head decreases (as a result of friction) in the direction of flow. Darcy's law for unsaturated flow can be expressed as

$$q = -K_u(\theta) \frac{\partial h}{\partial z} \tag{3.10.5}$$

where $K_u(\theta)$ is the *hydraulic conductivity*, as a function of the moisture content, θ. This law applies to areas that are large compared with the crosssection of individual pores and grains of the medium. Darcy's law describes a steady uniform flow of constant velocity with a net force of zero in a fluid element. In unconfined saturated flow, the forces are gravity and friction. For unsaturated flow, the forces are gravity, friction, and the *suction force* that binds water to soil particles through surface tension.

In unsaturated flow the void spaces are only partially filled with water so that water is attracted to the particle surfaces through electrostatic forces between the water molecule polar bonds and the particle surfaces. This in turn draws water up around the particle surfaces, leaving air in the center of the voids. The energy due to the soil suction forces is referred to as the *suction head (or matric potential)* ψ in unsaturated flow, which varies with moisture content (see subsection 3.11.2). Total head is then the sum of the suction and gravity heads:

$$h = \psi + z \tag{3.10.6}$$

Note that the velocities are so small that there is no term for velocity head in this expression for total head.

Darcy's law can now be expressed as

$$q = -K(\theta) \frac{\partial(\psi + z)}{\partial z} \tag{3.10.7}$$

Darcy's law was originally conceived for saturated flow and was extended by Richards[83] to unsaturated flow with the provision that the hydraulic conductivity is a function of the suction head, that is, $K = K(\psi)$. Also, the hydraulic conductivity can be related more easily to the degree of saturation so that $K = K(\theta)$. Because the soil suction head varies with moisture content and moisture content varies with elevation, the *suction gradient* can be expanded by using the chain rule to obtain

$$\frac{\partial \psi}{\partial z} = \frac{d\psi}{d\theta} \frac{\partial \theta}{\partial z} \tag{3.10.8}$$

in which $\partial \theta / \partial z$ is the *wetness gradient* and the reciprocal of $d\psi/d\theta$, namely $d\theta/d\psi$, is the *specific water capacity*. Now Equation 3.10.7 can be modified to

$$q = -K(\theta) \left(\frac{\partial \psi}{\partial z} + \frac{\partial z}{\partial z} \right) = -K(\theta) \left(\frac{d\psi}{d\theta} \frac{\partial \theta}{\partial z} + 1 \right) = -\left(K(\theta) \frac{d\psi}{d\theta} \frac{\partial \theta}{\partial z} + K(\theta) \right) \tag{3.10.9}$$

The *soil water diffusivity* $D[L^2/T]$ is defined as

$$D = K(\theta)\frac{d\psi}{d\theta} \tag{3.10.10}$$

The diffusivity is considered as the ratio of hydraulic conductivity to the *water capacity of the soil*, $d\theta/d\psi$. Substituting the above expression for D into Equation 3.10.9 results in

$$q = -\left(D\frac{\partial\theta}{\partial z} + K(\theta)\right) \tag{3.10.11}$$

Using the continuity equation (3.10.4) for one-dimensional, unsteady, unsaturated flow in a porous medium, we obtain

$$\frac{\partial\theta}{\partial t} = -\frac{\partial q}{\partial z} = \frac{\partial}{\partial z}\left(D\frac{\partial\theta}{\partial z} + K(\theta)\right) \tag{3.10.12}$$

which is a one-dimensional form of *Richard's equation*. This equation is the governing equation for unsteady unsaturated flow in a porous medium (Richards[83]). For a homogeneous soil, $\partial K/\partial z = 0$, so $\partial\theta/\partial t = \partial/\partial z(D\partial\theta/\partial z)$.

EXAMPLE 3.10.1

Determine the flux for a soil in which the unsaturated hydraulic conductivity is expressed as a function of the suction head as $K = 250(-\psi)^{-2.11}$ in cm/d at depth $z_1 = 80$ cm, $h_1 = -145$ cm, and $\psi_1 = -65$ cm at depth $z_2 = 100$ cm, $h_2 = -160$ cm, and $\psi_2 = -60$ cm.

SOLUTION

The flux is determined using Equation 3.10.5. Hydraulic conductivity is computed using an average value of $\psi = -65 + (60)]/2 = -62.5$ cm. Then $K_u = 250(-\psi)^{-2.11} = 250\,(62.5)^{-2.11} = 0.041$ cm/d. The flux is then

$$q = -K_u\left(\frac{h_1 - h_2}{z_1 - z_2}\right) = -0.041\left[\frac{-145-(-160)}{-80-(-100)}\right] = -0.03 \text{ cm/d}$$

The flux is negative because the moisture is flowing downward in the soil. ∎

EXAMPLE 3.10.2

Determine the soil water diffusivity for a soil in which $\theta = 0.1$ and $K(\theta) = 3 \times 10^{-11}$ mm/s from a relationship of $\psi(\theta)$ at $\theta = 0.1$, $\Delta\psi = 10^7$ mm, and $\Delta\theta = 0.35$.

SOLUTION

Using Equation 3.10.10, we compute the soil water diffusivity as

$$D = K(\theta)\frac{d\psi}{d\theta} = \left(3\times10^{-11} \text{ mm/s}\right)\left(10^7 \text{ mm/0.35}\right) = 8.57\times10^{-4} \text{ mm}^2/\text{s}$$ ∎

3.10.2   Unsaturated Hydraulic Conductivity

Unsaturated flow in the zone of aeration can be analyzed by Darcy's law; however, the *unsaturated hydraulic conductivity* K_u is a function of the water content as well as the negative pressure head (tension).[3, 14, 38, 100] Because part of the pore space is filled with air, the available cross-sectional area available for water flow is reduced; consequently, K_u is always less than the saturated value K.

Although there are hysteresis effects present in the relations of K_u with water content and negative pressure, approximations based on empirical evidence can be stated. Water content data fit the form[31]

$$\frac{K_u}{K} = \left(\frac{S_s - S_o}{1 - S_o}\right)^3 \tag{3.10.13}$$

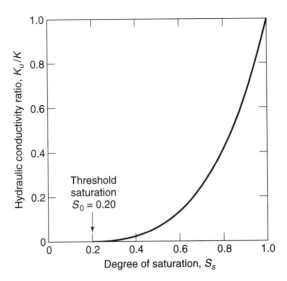

Figure 3.10.4. Ratio of unsaturated to saturated hydraulic conductivity as a function of saturation (after Irmay[55]).

where S_s is the degree of saturation and S_o is the threshold saturation—that part of the voids filled with nonmoving water held primarily by capillary forces. Equation 3.10.13 is plotted in Figure 3.10.4; note that K_u ranges from zero at $S_s = S_o$ to $K_u = K$ at $S_s = 1$, which is saturation.

For hydraulic conductivity and negative pressure, S-shaped relations as indicated in Figure 3.10.5 are generally applicable.[101, 103] These can be approximated by a step function or by

$$\frac{K_u}{K} = \frac{a}{\dfrac{a}{b}(-h)^n + a} \tag{3.10.14}$$

where a, b, and n are constants that vary with particle sizes of unconsolidated material and h is the pressure head measured in centimeters. It can be seen that when $h = 0$, which occurs at atmospheric pressure, $K_u = K$. Orders of magnitude of the constants in Equation 3.10.14 for different soils are as follows:[11]

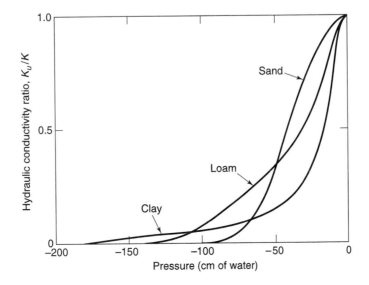

Figure 3.10.5. Median relationships between hydraulic conductivity and soil-water pressure (tension) (after Bouwer[11]).

Material	a	b	n
Medium sands	5×10^9	10^7	5
Fine sands, sandy loams	5×10^6	10^5	3
Loams and clays	5×10^3	5×10^3	2

Hydraulic conductivity is a measure of the ability of soils to transmit water and depends upon both soil properties (total porosity, pore-size distribution, and pore continuity) and fluid properties (viscosity and density).

Table 3.10.1 describes three mathematical relationships for soil water retention and hydraulic conductivity proposed by Brooks and Corey,[13] Campbell,[16] and Van Genuchten.[102] The relationships by Brooks and Corey and Campbell describe only a portion of the soil water-

Table 3.10.1 Soil Water Retention and Hydraulic Conductivity Relationships

Hydraulic soil characteristic	Parameters	Parameter correspondence
Brooks and Corey[13]		
Soil water retention $\dfrac{\theta - \theta_r}{\phi - \theta_r} = \left(\dfrac{h_b}{h}\right)^{\lambda}$	λ = pore-size index h_b = bubbling capillary pressure θ_r = residual water content ϕ = porosity K_s = fully saturated conductivity ($\theta = \phi$) h = matric potential S_e = effective saturation	$\lambda = \lambda$ $h_b = h_b$ $\theta_r = \theta_r$ $\phi = \phi$ $K_s = K_s$ $h = \psi$
Hydraulic conductivity $\dfrac{K(\theta)}{K_s} = \left(\dfrac{\theta - \theta_r}{\phi - \theta_r}\right)^n = (S_e)^n$	$n = 3 + \dfrac{2}{\lambda}$	
Campbell[16]		
Soil water retention $\dfrac{\theta}{\phi} = \left(\dfrac{H_b}{\psi}\right)^{1/b}$	ϕ = porosity H_b = scaling parameter with dimension of length b = constant	$\phi = \phi$ $H_b = h_b$ $b = \dfrac{1}{\lambda}$
Hydraulic conductivity $\dfrac{K(\theta)}{K_s} = \left(\dfrac{\theta}{\phi}\right)^n$	$n = 3 + 2b$	
Van Genuchten[102]		
Soil water retention $\dfrac{\theta - \theta_r}{\phi - \theta_r} = \left[\dfrac{1}{1 + (\alpha h)^n}\right]^m$	ϕ = porosity θ_r = residual water α = constant n = constant m = constant h = absolute value of matric potential	$\phi = \phi$ $\theta_r = \theta_r$ $\alpha = (h_b)^{-1}$ $n = \lambda + 1$ $m = \dfrac{\lambda}{\lambda + 1}$ $h = |\psi|$
Hydraulic conductivity $\dfrac{K(\theta)}{K_s} = \left(\dfrac{\theta - \theta_r}{\phi - \theta_r}\right)^{1/2} \left\{1 - \left[1 - \left(\dfrac{\theta - \theta_r}{\phi - \theta_r}\right)^{1/m}\right]^m\right\}^2$		

θ = water content; ψ = capillary suction, cm; $K(\theta)$ = hydraulic conductivity for given water content, cm/h

Source: Rawls et al.[80]

retention curve for matric potentials less than the *bubbling pressure* (pressure at which air will enter the soil); whereas the Van Genuchten[102] relationship describes the total soil water-retention curve. The water retention relationship for two contrasting soil textures is shown in Figure 3.10.6*a*. It is interesting to note that the sandy loam soil retains less water than the clayey soil. Figure 3.10.6*b* illustrates that the hydraulic conductivity is a nonlinear function of volumetric soil water content, and varies with soil texture. The points for saturated water content—b1 (on the sandy loam topsoil curve) and b2 (on the clayey subsoil curve)—show that the hydraulic conductivity for the sandy loam is higher than for the clay soil, even though the porosity is higher in the clay soil. As shown in the figure, the hydraulic conductivity of both soils decreases rapidly as the soil water content decreases. Figure 3.10.6*c* illustrates that the hydraulic conductivity of the sandy loam decreases more rapidly with a decrease in the matric potential head than that of the clay subsoil. At lower values of the matric potential head, which refers to higher suctions, the hydraulic conductivity of the clay soil is higher. The rate of change of the matric potential head in the sandy soil also decreases much more rapidly than that of the clay soil.

The absorbing (wetting) and desorbing (draining) have differing relationships of the matric potential head as a function of soil water content. This difference is called *hysteresis* and is illustrated in Figure 3.10.7. Hysteresis is caused by the entrapment of air in pockets connecting different size pores during wetting. Cycles of wetting and drying exhibit secondary hysteretic loops (see Figure 3.10.7), which causes considerable difficulty in modeling soil water movement. The $K(\theta)$ has a much smaller hysteresis effect than $\psi(\theta)$ which is why the hydraulic conductivity $K_u(\theta)$ is used in the Darcy equation as opposed to $K_u(\psi)$.

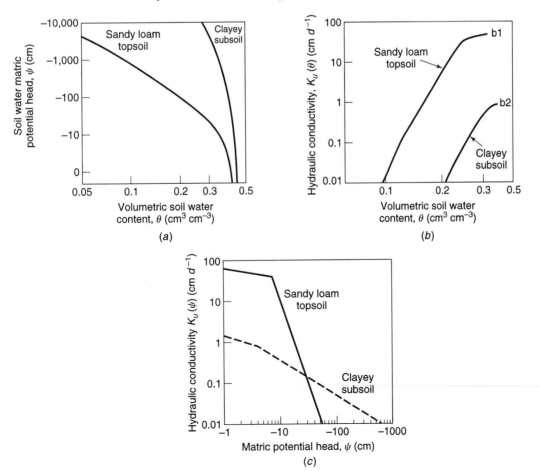

Figure 3.10.6. (*a*) The $\psi(\theta)$, (*b*) $K_u(\theta)$, and (*c*) $K_u(\psi)$ relationships of sandy loam and clayey horizons (Rawls et al[80]).

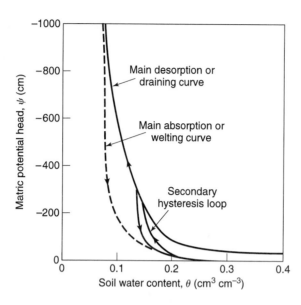

Figure 3.10.7. The effect of hysteresis on the $\psi(\theta)$ function of sandy loam topsoil during wetting and drying (Rawls et al[80]).

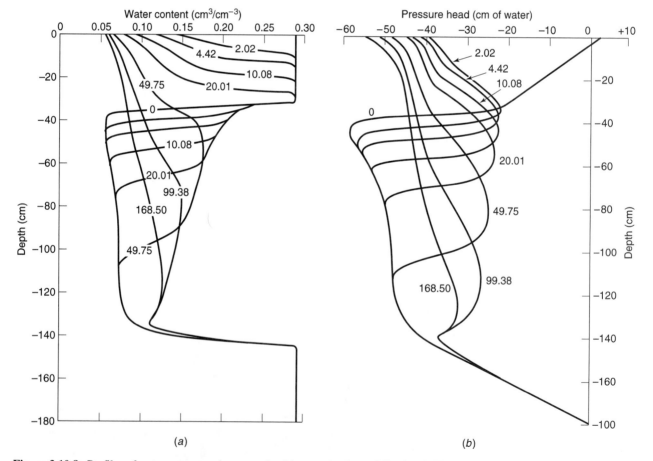

Figure 3.10.8. Profiles of water content and pressure head in a sand column following infiltration of water. Numbers on the curves represent time in minutes after start of infiltration (*a*) Water content (*b*) Pressure head. Data are computer-based numerical results (after Watson in van Schilfgaarde[103]; reproduced from *Drainage for Agriculture*, ASA Monograph No. 17, pp. 368–369, 1974 by permission of the American Society of Agronomy).

3.10.3 Vertical and Horizontal Flows

Illustrative of the vertical drainage of water in unsaturated soils are the data shown in Figure 3.10.8. An initially saturated sand 180 cm in depth was allowed to drain for one hour and was then subjected to infiltration for four min under a surface head of 3 cm. The resulting depth profiles of water content and pressure head are indicated for various time intervals after the end of the infiltration interval. Note that at the start of the subsurface redistribution of water ($t = 0$), saturation (water content = 0.29 cm^3/cm^3) extends to a depth of about 30 cm and corresponds to a linear pressure gradient. For field conditions where stratifications in soil texture can be expected, the complexity of such downward water migration, as from rainfall or irrigation, can be appreciable[39, 103].

Lateral flow occurs above the water table in the capillary zone. The flow rate decreases with the degree of saturation, and hence the hydraulic conductivity (see Figure 3.10.4). The fraction of flow above the water table can be calculated from an equivalent saturated thickness.[71] For aquifers of substantial depth, the flow component above the water table is negligible, but it can be significant in shallow unconfined aquifers.

3.11 KINEMATIC WAVE

Figure 3.11.1 shows an individual soil water wave propagating downward through the soil under gravity drainage. The *kinematic wave model* is a redistribution model based upon gravity drainage (Charbeneau[19, 20] and Morel-Seytoux[74]). Figure 3.11.1a shows the wave immediately after the end of the period of infiltration, consisting of a rectangular portion added to the profile for the antecedent water content. The horizontal line is the *wetting front*, indicating a drop in water content from the wave above to the unwetted soil below. After a short period of drainage, the wave is shown in Figure 3.11.1b, consisting of three parts: (1) a draining part with increasing water content (curved portion of the wave) with depth, (2) a constant or vertical plateau part, and (3) the wetting front. The draining part has not reached the wetting front and is separated from it by the vertical plateau. In Figure 3.11.1c, the drainage profile has caught up with the wetting front and the vertical plateau region no longer exists.

The basic assumption in the kinematic wave theory is that the pressure gradients are negligible. With negligible capillary pressure gradients and with ψ constant within the profile, the soil water diffusivity is zero and Equation 3.10.11 simplifies to

$$q = -(K_u(\theta)) \tag{3.11.1}$$

so that Darcy's law reduces to Equation 3.11.1 for the kinematic wave theory. Equation 3.10.12 then reduces to

$$\frac{\partial \theta}{\partial t} - \frac{\partial}{\partial z}(K_u(\theta)) = 0 \tag{3.11.2}$$

Water content

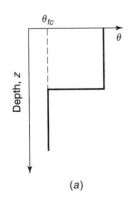

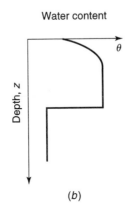

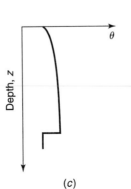

(a) (b) (c)

Figure 3.11.1. Soil moisture profiles for individual kinematic wave (Charbeneau[19]).

so that flow is downward under a unit gradient. Equation 3.11.2 can be expressed as

$$\frac{\partial \theta}{\partial t} + \frac{dK(\theta)}{d\theta}\frac{\partial \theta}{\partial z} = 0 \tag{3.11.3}$$

Charbeneau[19, 20] shows the analytical solution of the kinematic wave model, which can be used to model soil water redistribution and recharge.

3.12 INFILTRATION: THE GREEN–AMPT METHOD

The process of water penetrating into the soil is *infiltration*. The rate of infiltration is influenced by the condition of the soil surface, vegetative cover, and soil properties including porosity, hydraulic conductivity, and moisture constant. Figure 3.12.1 illustrates the distribution of soil moisture within a soil profile during downward movement. These moisture zones are the *saturated zone*, the *transmission zone*, a *wetting zone*, and a *wetting front*. This profile changes as a function of time as shown in Figure 3.12.2.

The *infiltration rate, f,* is the rate at which water enters the soil surface, expressed in in/hr or cm/hr. The *potential infiltration rate* is the rate when water is ponded on the soil surface, so if no ponding occurs, the actual rate is less than the potential rate. Most infiltration equations describe a potential infiltration rate. *Cumulative infiltration, F,* is the accumulated depth of water infiltrated, defined mathematically as

$$F(t) = \int_0^t f(\tau)d\tau \tag{3.12.1}$$

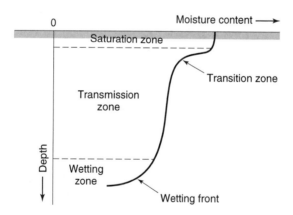

Figure 3.12.1. Moisture zones during infiltration (from Chow et al.[21]).

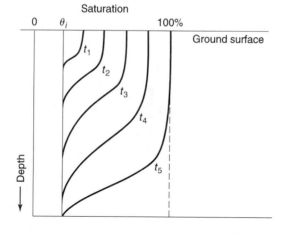

Figure 3.12.2. Moisture profile as a function of time for water added to the soil surface (Chow et al.[21]).

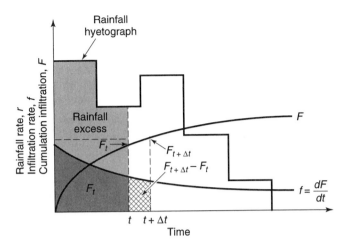

Figure 3.12.3. Rainfall infiltration rate and cumulative infiltration. The rainfall hyetograph illustrates the rainfall pattern as a function of time. The cumulative infiltration at time t is F_t or $F(t)$ and at time $t + \Delta t$ is $F_{t+\Delta t}$ or $F(t + \Delta t)$. The increase in cumulative infiltration from time t to $t + \Delta t$ is $F_{t+\Delta t} - F_t$ or $F(t + \Delta t) - F(t)$ as shown in the figure. *Rainfall excess* is rainfall that is neither retained on the land surface nor infiltrated into the soil (Chow et al.[21]).

and the infiltration rate is the time derivative of the cumulative infiltration given as

$$f(t) = \frac{dF(t)}{dt} \tag{3.12.2}$$

Figure 3.12.3 illustrates a rainfall hyetograph with the infiltration rate and cumulative infiltration curves.

Green and Ampt[43] (1911) proposed the simplified picture of infiltration shown in Figure 3.12.4. The *wetting front* is a sharp boundary dividing initial soil moisture content θ_i below from saturated soil with moisture content η above. The wetting front has penetrated to a depth L in time t since infiltration began. Water is ponded to a small depth h_0 on the soil surface.

Consider a vertical column of soil of unit horizontal cross-sectional area (Figure 3.12.5) and let the control volume be defined around the wet soil between the surface and depth L. If the soil was initially of *moisture content* θ_i throughout its entire depth, the moisture content will increase from θ_i to η (the porosity) as the wetting front passes. The moisture content θ is the ratio of the volume of water to the total volume within the control surface, so the increase

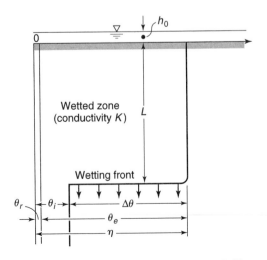

Figure 3.12.4. Variables in the Green-Ampt infiltration model. The vertical axis is the distance from the soil surface, the horizontal axis is the moisture content of the soil (from Chow et al.[21]).

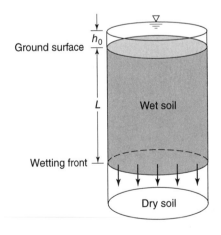

Figure 3.12.5. Infiltration into a column of soil of unit cross-sectional area for the Green–Ampt model (from Chow et al.[21]).

in the water stored within the control volume as a result of infiltration is $L(\eta - \theta_i)$ for a unit cross section. By definition this quantity is equal to F, the cumulative depth of water infiltrated into the soil. Hence

$$F(t) = L(\eta - \theta_i) = L\Delta\theta \tag{3.12.3}$$

where $\Delta\theta = (\eta - \theta_i)$.

Darcy's law may be expressed as

$$q = -K\frac{\partial h}{\partial z} = -K\frac{\Delta h}{\Delta z} \tag{3.12.4}$$

In this case the Darcy flux q is constant throughout the depth and is equal to $-f$, because q is positive upward while f is positive downward. If points 1 and 2 are located respectively at the ground surface and just on the dry side of the wetting front, Equation 3.12.4 can be approximated by

$$f = K\left[\frac{h_1 - h_2}{z_1 - z_2}\right] \tag{3.12.5}$$

The head h_1 at the surface is equal to the ponded depth h_0. The head h_2, in the dry soil below the wetting front, equals $-\psi - L$. Darcy's law for this system is written as

$$f = K\left[\frac{h_0 - (-\psi - L)}{L}\right] \tag{3.12.6a}$$

and if the ponded depth h_0 is negligible compared to ψ and L,

$$f \approx K\left[\frac{\psi + L}{L}\right] \tag{3.12.6b}$$

This assumption ($h_0 = 0$) is usually appropriate for surface water hydrology problems because it is assumed that ponded water becomes surface runoff.

From Equation 3.12.3, the wetting front depth is $L = F/\Delta\theta$, and assuming $h_0 = 0$, substitution into Equation 3.12.6 gives

$$f = K\left[\frac{\psi\,\Delta\theta + F}{F}\right] \tag{3.12.7}$$

Since $f = dF/dt$, Equation 3.12.7 can be expressed as a differential equation in the one unknown F as

$$\frac{dF}{dt} = K\left[\frac{\psi\,\Delta\theta + F}{F}\right]$$

To solve for F, cross-multiply to obtain

$$\left[\frac{F}{F + \psi\,\Delta\theta}\right]dF = K\,dt$$

then divide the left-hand side into two parts

$$\left[\frac{F + \psi\,\Delta\theta}{F + \psi\,\Delta\theta} - \frac{\psi\,\Delta\theta}{F + \psi\,\Delta\theta}\right]dF = K\,dt$$

and integrate

$$\int_0^{F(t)}\left[1-\frac{\psi\,\Delta\theta}{F+\psi\,\Delta\theta}\right]dF = \int_0^t Kdt$$

to obtain

$$F(t) - \psi\Delta\theta\{\ln[F(t)+\psi\Delta\theta] - \ln(\psi\Delta\theta) = Kt \qquad (3.12.8a)$$

or

$$F(t) - \psi\,\Delta\theta\ln\left(1+\frac{F(t)}{\psi\,\Delta\theta}\right) = Kt \qquad (3.12.8b)$$

Equation 3.12.8 is the *Green–Ampt equation* for cumulative infiltration. Once F is computed using Equation 3.12.8, the infiltration rate f can be obtained from Equation 3.12.7 or

$$f(t) = K\left[\frac{\psi\,\Delta\theta}{F(t)}+1\right] \qquad (3.12.9)$$

When the ponded depth h_0 is not negligible, the value of $\psi - h_0$ is substituted for ψ in Equations 3.12.8 and 3.12.9.

Equation 3.12.8 is a nonlinear equation in F, which can be solved by the method of successive substitution by rearranging,

$$F(t) = Kt + \psi\,\Delta\theta\ln\left(1+\frac{F(t)}{\psi\,\Delta\theta}\right) \qquad (3.12.10)$$

Given K, t, ψ and $\Delta\theta$, a trial value F is substituted on the right-hand side (a good trial value is $F = Kt$), and a new value of F calculated on the left-hand side, which is substituted as a trial value on the right-hand side, and so on until the calculated values of F converge to a constant. The final value of cumulative infiltration F is substituted into Equation 3.12.9 to determine the corresponding infiltration rate f.

Equation 3.12.8 can also be solved by Newton's method, which is more complicated than the method of successive substitution but converges in fewer iterations. Referring to Equation 3.12.10, application of the Green–Ampt model requires estimates of the hydraulic conductivity K, the wetting front soil suction head ψ (see Table 13.12.1), and $\Delta\theta$.

The *residual moisture content* of the soil, denoted by θ_r, is the moisture content after it has been thoroughly drained. The *effective saturation* is the ratio of the available moisture $(\theta - \theta_r)$ to the maximum possible available moisture content $(\eta - \theta_r)$, given as

$$s_e = \frac{\theta - \theta_r}{\eta - \theta_r} \qquad (3.12.11)$$

where $\eta - \theta_r$ is called the *effective porosity* θ_e.

The effective saturation has the range $0 \le s_e \le 1.0$, provided $\theta_r \le \theta \le \eta$. For the initial condition, when $\theta = \theta_i$, cross-multiplying Equation 3.12.11 gives $\theta_i - \theta_r = s_e\theta_e$, and the change in the moisture content when the wetting front passes is

$$\Delta\theta = \eta - \theta_i = \eta - (s_e\theta_e + \theta_r);$$
$$\Delta\theta = (1-s_e)\theta_e \qquad (3.12.12)$$

A logarithmic relationship between the effective saturation s_e and the soil suction head ψ can be expressed by the *Brooks–Corey equation* (Brooks and Corey[13])

$$s_e = \left[\frac{\psi_b}{\psi} \right]^{\lambda}$$

(3.12.13)

in which ψ_b and λ are constants obtained by draining a soil in stages, measuring the values of s_e and ψ at each stage, and fitting Equation 3.12.13 to the resulting data.

   Brakensiek et al.[12] presented a method for determining the Green–Ampt parameters using the Brooks–Corey equation. Rawls et al.[81] used this method to analyze approximately 5,000 soil horizons across the United States and determined average values of the Green–Ampt parameters η, θ_e, ψ, and K for different soil classes, as listed in Table 3.12.1. As the soil becomes finer, moving from sand to clay, the wetting front soil suction head increases while the hydraulic conductivity decreases. Table 3.12.1 also lists typical ranges for η, θ_e, and ψ. The ranges are not large for η and θ_e, but ψ can vary over a wide range for a given soil. K varies along with ψ, so the values given in Table 3.12.1 for both ψ and K should be considered typical values that may show a considerable degree of variability in application.[2, 27]

   Table 3.12.2 compares other infiltration methods. Refer to Chow et al.[21] for detailed discussions of each of these methods.

Table 3.12.1 Green–Ampt Infiltration Parameters for Various Soil Classes*

Soil class	Porosity η	Effective porosity θ_e	Wetting front soil suction head ψ (cm)	Hydraulic conductivity K (cm/h)
Sand	0.437 (0.374–0.500)	0.417 (0.354–0.480)	4.95 (0.97–25.36)	11.78
Loamy sand	0.437 (0.363–0.506)	0.401 (0.329–0.473)	6.13 (1.35–27.94)	2.99
Sandy loam	0.453 (0.351–0.555)	0.412 (0.283–0.541)	11.01 (2.67–45.47)	1.09
Loam	0.463 (0.375–0.551)	0.434 (0.334–0.534)	8.89 (1.33–59.38)	0.34
Silt loam	0.501 (0.420–0.582)	0.486 (0.394–0.578)	16.68 (2.92–95.39)	0.65
Sandy clay loam	0.398 (0.332–0.464)	0.330 (0.235–0.425)	21.85 (4.42–108.0)	0.15
Clay loam	0.464 (0.409–0.519)	0.309 (0.279–0.501)	20.88 (4.79–91.10)	0.10
Silty clay loam	0.471 (0.418–0.524)	0.432 (0.347–0.517)	27.30 (5.67–131.50)	0.10
Sandy clay	0.430 (0.370–0.490)	0.321 (0.207–0.435)	23.90 (4.08–140.2)	0.06
Silty clay	0.479 (0.425–0.533)	0.423 (0.334–0.512)	29.22 (6.13–139.4)	0.05
Clay	0.475 (0.427–0.523)	0.385 (0.269–0.501)	31.63 (6.39–156.5)	0.03

*The numbers in parentheses below each parameter are one standard deviation around the parameter value given.

Source: Rawls, Ahuja, Brakensiek, and Shirmohammadi[80] and Rawls, Brakensiek, and Miller[81]

Table 3.12.2 Infiltration Equations

Cumulative infiltration (F_t)	Infiltration rate (f_t)	Comments
Green–Ampt equation $$F_t - \psi\,\Delta\theta \ln\left(1+\frac{F_t}{\psi\,\Delta\theta}\right)=Kt$$	$$f_t = K\left(\frac{\psi\,\Delta\theta}{F_t}+1\right)$$	• Hydraulic conductivity (K) • Wetting front soil suction head (ψ) • Change in moisture content ($\Delta\theta$) $\Delta\theta = \eta - \theta_i$ • Porosity η • Initial moisture content (θ_i)
Horton's equation $F_t = f_c t + (f_0 - f_c)(1 - e^{-kt})/k$	$f_t = f_c + (f_0 - f_c)e^{-kt}$	• f_0 = initial infiltration rate • Constant infiltration rate (f_c) • Decay constant (k)

SCS method $$F_t = \frac{S(P_t - I_a)}{P_t - I_a + S}$$ $p \geq I_a$	$$f_t = \frac{S^2\frac{dP_t}{dt}}{\left(P_t - I_a + S\right)^2}$$	• Potential maximum retention (S)
		$$S = \frac{1000}{CN}-10$$• Dimensionless curve number (CN)
		  $0 \leq CN \leq 100$
		• Initial abstraction I_a
		  $I_a = 0.2S$
		• Total rainfall to time t, P_t

Philip's equation $F_t = St^{1/2} + kt$	$f_t = \tfrac{1}{2}St^{-1/2} + K$	• Sorptivity (S) • Hydraulic conductivity (K)

EXAMPLE 3.12.1

Use the Green–Ampt method to evaluate the infiltration rate and cumulative infiltration depth for a silty clay soil at 0.1-hour increments up to 6 hours from the beginning of infiltration. Assume an initial effective saturation of 20 percent and continuous ponding.

SOLUTION

From Table 3.12.1, for a silty clay soil, $\theta_e = 0.423$, $\psi = 29.22$ cm, and K = 0.05 cm/hr. The initial effective saturation is $s_e = 0.2$, so $\Delta\theta = (1 - s_e)\,\theta_e = (1 - 0.20)0.423 = 0.338$, and $\psi\,\Delta\theta = (29.22)(0.338) = 9.89$ cm. Assuming continuous ponding, the cumulative infiltration F is found by successive substitution in Equation 3.12.10:

$$F = Kt + \psi\,\Delta\theta\,\ln[1 + F/(\psi\,\Delta\theta)] = 0.05\,t + 9.89\,\ln[1 + F/9.89]$$

For example, at time $t = 0.1$ hr, the cumulative infiltration converges to a final value $F = 0.29$ cm. The infiltration rate f is then computed using Equation (3.12.9)

$$f = K(1 + \psi\,\Delta\theta/F) = 0.05(1 + 9.89/F)$$

As an example, at time $t = 0.1$ hr, $f = 0.05(1 + 9.89/0.29) = 1.78$ cm/hr. The infiltration rate and the cumulative infiltration are computed in the same manner between 0 and 6 hours at 0.1 hr or 0.5 hr intervals, as shown in Table 3.12.3.

Table 3.12.3 Infiltration Computations Using the Green–Ampt Method

Time t (hr)	0.0	0.1	0.2	0.3	0.4	0.5	1.0	1.5	2.0
Infiltration rate f (cm/hr)	∞	1.78	1.20	0.97	0.84	0.75	0.54	0.44	0.39
Infiltration depth F (cm)	0.00	0.29	0.43	0.54	0.63	0.71	1.02	1.26	1.47

Time t (hr)	2.5	3.0	3.5	4.0	4.5	5.0	5.5	6.0
Infiltration rate f (cm/hr)	0.35	0.32	0.30	0.28	0.27	0.26	0.25	0.24
Infiltration depth F (cm)	1.65	1.82	1.97	2.12	2.26	2.39	2.51	2.64

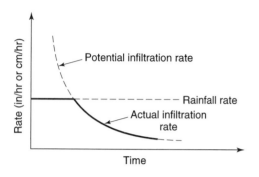

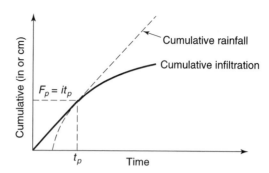

Figure 3.12.6. Ponding time. This figure illustrates the concept of ponding time for a constant intensity rainfall. *Ponding time* is the elapsed time between the time rainfall begins and the time water begins to pond on the soil surface.[69]

EXAMPLE 3.12.2

Ponding time, t_p, is the elapsed time between the time rainfall begins and the time water begins to pond on the soil surface. Develop an equation for ponding time under a constant rainfall intensity, i, using the Green–Ampt infiltration equation (refer to Figure 3.12.6).

SOLUTION

The infiltration rate f and the cumulative infiltration F are related in Equation 3.12.8. The cumulative infiltration at ponding time t_p is $F_p = i\, t_p$, in which i is the constant rainfall intensity (see Figure 3.12.6). Substituting $F_p = i\, t_p$ and the infiltration rate $f = i$ into Equation 3.12.9, we have

$$i = K\left(\frac{\psi\, \Delta\theta}{i t_p} + 1\right)$$

whose solution is

$$t_p = \frac{K\psi\, \Delta\theta}{i(i - K)}$$

which is the ponding time for a constant rainfall intensity. ∎

PROBLEMS

3.1.1 A confined aquifer with a porosity of 0.15 is 30 m thick. The potentiometric surface elevations at two observation wells 1,000 m apart are 52.35 m and 56.90 m. If the horizontal hydraulic conductivity of the aquifer is 25 m/day, determine the flow rate per unit width of the aquifer, specific discharge, and average linear velocity of the flow assuming steady unidirectional flow. How long would it take for a tracer to travel the distance between the observation wells?

3.1.2 A field sample of an aquifer is packed in a test cylinder (see Figure 3.1.1). The cylinder has a length of 120 cm and a diameter of 12 cm. The field sample with a porosity of 0.24 is tested under a constant head difference of 160 cm with water at 10°C. If the estimated hydraulic conductivity of the sample is 30 m/day, calculate (a) the expected total discharge, (b) the specific discharge, (c) the average flow velocity, and (d) the hydraulic gradient along the cylinder.

3.1.3 A confined aquifer with a horizontal bed has a varying thickness as shown in the figure. Assuming the flow in the aquifer is essentially horizontal, determine the flow rate if the piezometric heads at sections (1) and (2) are 23.7 m and 27.1 m, respectively.

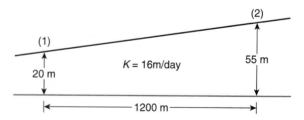

3.1.4 What should be the range of water temperature in Problem 3.1.2 for Darcy's law to be applicable if the median diameter of the field sample is 2.2 mm? Is the water temperature given in the problem appropriate?

3.1.5 Could the same apparatus in Problem 3.1.2 be used to test a field sample having a hydraulic conductivity of 86.4 m/day and a median grain diameter of 5 mm?

3.1.6 If a 50-cm-long field sample with an estimated hydraulic conductivity of 8 m/day is to be tested in the laboratory (see Figure 3.1.1) using water at 20°C, what would be maximum allowable head difference across the column? The sample has a median grain diameter of 0.5 mm.

3.1.7 Rework Problem 3.1.6 with an estimated hydraulic conductivity of 40 m/day and a median grain diameter of 1.5 mm.

3.2.1 The hydraulic conductivity of a medium sand was measured as 11.2 m/day in a laboratory test with water at 25°C. (a) What is the intrinsic permeability of the sample in square micrometers? (b) What is the expected in-situ hydraulic conductivity if the natural groundwater temperature is 10°C?

3.2.2 The intrinsic permeability of a silty sand is 0.36 darcy. Plot the hydraulic conductivity as a function of water temperature from 5°C to 30°C.

3.3.1 A constant head permeameter has a cross-sectional area of 175 cm^2 and is used to measure the hydraulic conductivity of a field sample 40 cm long. If the permeameter discharges 100 ml in 320 seconds under a head difference of 60 cm: (a) what is the hydraulic conductivity as measured in the laboratory? (b) what is the intrinsic permeability of the sample given that water at 20°C is used for the test? (c) is Darcy's law valid under the given conditions ($D = 0.5$ mm)? and (d) what would be the in-situ hydraulic conductivity if the regional groundwater temperature is 12°C?

3.3.2 If the medium sand in Examples 3.3.2 and 3.3.3 were sampled and tested exactly the same way as described in Example 3.3.4, how long would it take for the head to drop from 8 cm to 1 cm?

3.3.3 A 30-cm-long field sample is tested in a falling head permeameter with a 10-cm diameter cylinder. The diameter of the tube is 10 mm. The water level in the tube is 35 cm above the outlet level after one hour of operation, while the level drops to 22 cm after 12 hours of operation. Find the hydraulic conductivity of the sample.

3.3.4 If the silty, fine sand sample in Example 3.3.4 were tested the same way as in Example 3.3.3, how long would it take to collect 200 ml of water at the outlet?

3.3.5 The average flow velocity at a point in an aquifer was found to be 0.68 m/day by means of a point dilution test. If the slope of the piezometric surface at the same point is 0.018 and the aquifer material has a porosity of 0.23, determine the hydraulic conductivity of the aquifer at that point.

3.4.1 The stratification of the same aquifer at three locations is shown below. Calculate the equivalent horizontal and vertical hydraulic conductivities for each location. What happens to the degree of anisotropy?

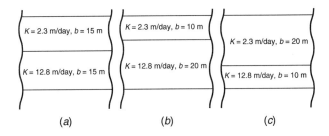

(a)   (b)   (c)

3.4.2 The stratification of a confined aquifer with a horizontal bed varies as follows. Calculate the equivalent horizontal and vertical hydraulic conductivities in each case. What happens to the degree of anisotropy?

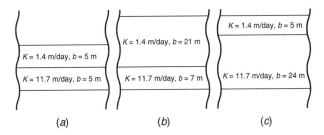

(a)   (b)   (c)

3.4.3 A confined aquifer has a constant thickness of 34 m and is stratified between two observation wells as shown in the following figure. Given the piezometric surface elevations in the figure, determine the flow per unit width of the aquifer. What is the equivalent horizontal hydraulic conductivity of that section of the confined aquifer between the observation wells?

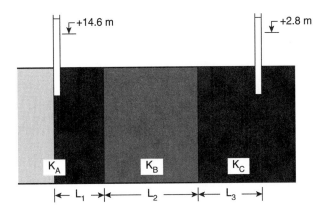

3.5.1 The groundwater temperatures in the United States vary from about 4°C in the northern part to approximately 20°C in the southern part. Assuming a productive alluvial aquifer with an intrinsic permeability of 100 darcys and a hydraulic gradient of 0.01 (see Figure 3.5.1), how much would the production of the same aquifer change between a northern state and a southern state?

3.5.2 A confined aquifer is recharged from an unconfined aquifer through an aquitard as shown in the figure. If the water levels in piezometers A, B, and C are 4.31 m, 4.47 m, and 6.64 m below the ground surface, respectively, determine the permeability of the aquitard.

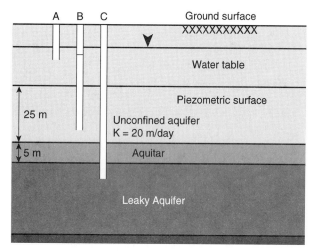

3.6.1 The coordinates and the total heads of three observation wells in a regional aquifer are tabulated below. Determine the hydraulic gradient and direction of groundwater flow.

Well #	x (m)	y (m)	Total head (m)
1	0	0	57.79
2	200	340	55.11
3	190	−150	52.80

3.6.2 Suppose that the average discharge due to the local groundwater pumping near Savannah, Georgia (see Figure 3.6.6), in

1957 was 30,000 m^3 per day. Estimate the transmissivity of the regional aquifer.

3.10.1 Determine the soil water diffusivity for a soil in which $\theta = 0.1$ and $K = 3 \times 10^{-11}$ mm/s from a relationship of $\psi(\theta)$ at $\theta = 0.1$, $\Delta\psi = 10^7$ mm, and $\Delta\theta = 0.35$.

3.12.1 Determine the infiltration rate and cumulative infiltration curves (0 to 5 h) at 1-hr increments for a clay loam soil. Assume an initial effective saturation of 40 percent and continuous ponding.

3.12.2 Rework Problem 3.12.1 using an initial effective saturation of 20 percent.

3.12.3 Rework Example 3.12.1 for a sandy loam soil.

3.12.4 Compute the ponding time and cumulative infiltration at ponding for a sandy clay loam soil with a 30 percent initial effective saturation, subject to a rainfall intensity of 2 cm/h.

3.12.5 Rework Problem 3.12.4 for a silty clay soil.

3.12.6 Determine the cumulative infiltration and the infiltration rate on a sandy clay loam after one hour of rainfall at 2 cm/hr if the initial effective saturation is 25 percent. Assume the ponding depth is negligible in the calculations.

3.12.7 Rework Problem 3.12.6, assuming that any ponded water remains stationary over the soil so that the ponded depth must be accounted for in the calculations.

3.12.8 Use the Green–Ampt method to compute the infiltration rate and cumulative infiltration for a silty clay soil ($\eta = 0.479$, $\psi = 29.22$ cm, $K = 0.05$ cm/hr) at 0.25 hour increments up to hours from the beginning of infiltration. Assume an initial effective saturation of 30 percent and continuous ponding.

3.12.9 Compute the ponding time and cumulative infiltration at ponding for a sandy clay loam soil of 25 percent initial effective saturation for a rainfall intensity of (a) 2 cm/h, (b) 3 cm/h, and (c) 5 cm/h.

3.12.10 Rework Problem 3.12.9 considering a silt loam soil.

REFERENCES

1. Ahmed, N., and D. K. Sunada, Nonlinear flow in porous media, *Jour. Hydraulics Div.*, Amer. Soc. Civil Engrs., v. 95, no. HY6, pp. 1847–1857, 1969.

2. American Society of Agricultural Engineers, Advances in infiltration, Proc. National Conf. on Advances in Infiltration, Chicago, IL, ASAE Publication 11–83, St. Joseph, MI, 1983.

3. Amer. Soc. Testing Matls (ASTM), Permeability and capillarity of soils, *ASTM Spl. Tech. Publ. no. 417*, Philadelphia, 210 pp., 1967.

4. Andrews, W. J., J. P. Oliveros, and J. J. Collins, *Preliminary Report on the Hydrogeology of Lake Five-O and Vicinity, Bay County, Florida, U.S. Geological Survey Water-Resource Investigations Report* 90-4148, 1990.

5. Bachmat, Y., and J. Bear, The general equations of hydrodynamic dispersion in homogeneous, isotropic porous mediums, *Jour. Geophysical Research*, v. 69, pp. 2561–2567, 1964.

6. Back, W., et al., Carbon-14 ages related to ocurrence of salt water, *Jour. Hydraulics Div.*, Amer. Soc. Civil Engrs., v. 96, no. HY11, pp. 2325–2336, 1970.

7. Barth, G. R., T. H. Illangasekare, M. C. Hill, and H. Rajaram, A new tracer-density criterion for heterogeneous porous media, *Water Resources Research,* v. 37, no. 1, pp. 21–31, 2001.

8. Bear, J., and G. Dagan, The relationship between solutions of flow problems in isotropic and anisotropic soils, *Jour. Hydrology*, v. 3, pp. 88–96, 1965.

9. Bennett, R. R. and R. R. Meyer, Geology and ground-water resources of the Baltimore area (Maryland), Maryland Department of Geology, *Mines and Water Resources Bull*. 4, 1952.

10. Boast, C. W., and D. Kirkham, Auger hole seepage theory, *Soil Sci. Soc. Amer. Proc.*, v. 35, pp. 365–373, 1971.

11. Bouwer, H., Unsaturated flow in ground-water hydraulics, *Jour. Hydraulics Div.*, Amer. Soc. Civil Engrs., v. 90, no. HY5, pp. 121–144, 1964.

12. Brakensiek, D. L., R. L. Engleman, and W.J. Rawls, Variation within texture classes of soil water parameters, *Trans. Am Soc. Agric Eng.,* v. 24, no. 2, pp. 335–339, 1981.

13. Brooks, R.H., and A. T. Corey, Hydraulic properties of porous media, *Hydrology Papers,* no. 3, Colorado State University, Fort Collins, CO, 1964.

14. Brooks, R. H., and A. T. Corey, Properties of porous media affecting fluid flow, *Jour. Irrig. Drain. Div.*, Amer. Soc. Civil Engrs., v. 92, no. IR2, pp. 61–88, 1966.

15. Bureau of Reclamation, *Ground Water Manual*, U.S. Dept. Interior, 480 pp., 1977.

16. Campbell, G. S., A simple method for determining unsaturated conductivity from moisture retention data, *Soil Science,* v. 117, pp. 311–314, 1974.

17. Castillo, E., et al., Unconfined flow through jointed rock, *Water Resources Bull.*, v. 8, pp. 266–281, 1972.

18. Cedergren, H. R., *Seepage, Drainage, and Flow Nets*, 2nd ed., John Wiley & Sons, New York, 534 pp., 1977.

19. Charbeneau, R. J., Kinematic models for soil moisture and solute transport, *Water Resources Research*, 17(3), pp. 699–706, 1984.

20. Charbeneau, R. J., *Groundwater Hydraulics and Pollutant Transport*, Prentice Hall, Upper Saddle River, NJ, 2000.

21. Chow, V. T., D. R. Maidment, and L. W. Mays, *Applied Hydrology,* McGraw-Hill, New York, 1988.

22. Cooper, H. H., Jr., et al., Response of a finite-diameter well to an instantaneous charge of water, *Water Resources Research*, v. 3, pp. 263–269, 1967.

23. Crowe, C. T., D. F. Elger, and J. A. Roberson, *Engineering Fluid Mechanics,* 7th edition, John Wiley & Sons, New York, 2001.

24. Darcy, H., *Les fontaines publiques de la ville de Dijon*, V. Dalmont, Paris, 647 pp., 1856.

25. Davis, S. N., D. J. Campbell, H. W., Bentley, and T. J. Flynn, *Ground Water Tracers*, National Ground Water Association, Dublin, OH, 200 pp., 1985.

26. de Josselin de Jong, G., Longitudinal and transverse diffusion in granular deposits, *Trans. Amer. Geophysical Union*, v. 39, pp. 67–74, 1958.

27. Devaurs, M., and G. F. Gifford, Applicability of the Green and Ampt infiltration equation to rangelands, *Water Resource Bulletin,* v. 22, no.1, pp. 19–27, 1986.

28. Drost, W., et al., Point dilution methods of investigating ground water flow by means of radioisotopes, *Water Resources Research*, v. 4, pp. 125–146, 1968.

29. Eldor, M., and G. Dagan, Solutions of hydrodynamic dispersion in porous media, *Water Resources Research*, v. 8, pp. 1316–1331, 1972.

30. Fair, G. M., and L. P. Hatch, Fundamental factors governing the streamline flow of water through sand, *Jour. Amer. Water Works Assoc.*, v. 25, pp. 1551–1565, 1933.

31. Fancher, G., Henry Darcy: Engineer and benefactor of mankind, *Jour. Petr. Tech.*, v. 8, pp. 12–14, Oct. 1956.

32. Fenemor, A. D., A three-dimensional model for management of the Waimea Plains Aquifer, Nelson. Dept. of Scientific and Industrial Research, Hydrology Centre Publication No. 18, 33. Christchurch, New Zealand, 1988.

33. Field, M. S., *Ground-Water Tracing and Drainage Basin Delineation for Risk Assessment Mapping for Spring Protection in Clarke County, Virginia, U.S. Environmental Protection Agency* NCEA-W-0936, 2000.

34. Field, M. S., Efficient hydrologic tracer-test design for tracer-mass estimation and sample-collection frequency, 1. Method development, *Environmental Geology*, v. 42, no. 7, pp. 827–838, 2001.

35. Field, M. S., Efficient hydrologic tracer-test design for tracer-mass estimation and sample-collection frequency, 2. Experimental results, *Environmental Geology*, v. 42, no. 7, pp. 839–850, 2001.

36. Field, M., Tracer-test planning using the efficient hydrologic tracer-test/design program, EPA/600/R-03/034, National Center for Environmental Assessment—Washington Office, U.S. Environmental Protection Agency, Washington, DC, 2003.

37. Freeze, R. A., *Theoretical analysis of regional groundwater flow*, Scientific Ser. no. 3, Inland Waters Branch, Dept. Energy, Mines and Resources, Ottawa, Canada, 202 pp., 1969.

38. Freeze, R. A., Three-dimensional, transient, saturated-unsaturated flow in a groundwater basin, *Water Resources Research*, v. 7, pp. 347–366, 1971.

39. Freeze, R. A., and J. Banner, The mechanism of natural ground-water recharge and discharge, *Water Resources Research*, v. 5, pp. 153–171, 1969; v. 6, pp. 138–155, 1970.

40. Gaspar, E., *Modern Trends in Tracer Hydrology,* CRC Press, Boca Raton, FL, 1987.

41. Gaspar, E., and M. Oncescu, *Radioactive Tracers in Hydrology*, Elsevier, Amsterdam, 342 pp., 1972.

42. Gelhar, L. W., and M. A. Collins, General analysis of longitudinal dispersion in nonuniform flow, *Water Resources Research*, v. 7, pp. 1511–1521, 1971.

43. Green, W.H., and G. A. Ampt, Studies on soil physics, Part I, The flow of air and water through soils, *J. Agric. Sci.,* v. 4, no. 1, pp. 1–24, 1911.

44. Grubbs, J. W., *Evaluation of Ground-Water Flow and Hydrologic Budget for Lake Five-O, A Seepage Lake in Northwestern Florida, U.S. Geological Survey Water-Resources Investigations Report* 94-4145, 1995.

45. Halevy, E., et al., Borehole dilution techniques: A critical review, *Isotopes in Hydrology*, Intl. Atomic Energy Agency, Vienna, pp. 531–564, 1967.

46. Hanshaw, B. B., et al., *Carbonate Equilibria and Radiocarbon Distribution Related to Groundwater Flow in the Floridan Limestone Aquifer, U.S.A., Intl. Assoc. Sci. Hydrology Publ.* 74, pp. 601–614, 1965.

47. Harleman, D. R. F., et al., Dispersion-permeability correlation in porous media, *Jour. Hydraulics Div.*, Amer. Soc. Civil Engrs., v. 89, no. HY2, pp. 67–85, 1963.

48. Hollett, K. J., *Geohydrology and Water Resources of the Papago Farms–Great Plain Area, Papago Indian Reservation, Arizona and the Upper Rio Sonoyta Area, Sonoram Mexico, U.S. Geological Survey Water Supply Paper* 2258, 1985.

49. Hoopes, J. A., and D. R. F. Harleman, Dispersion in radial flow from a recharge well, *Jour. Geophysical Research*, v. 72, pp. 3595–3607, 1967.

50. Horton, R. E., The role of infiltration in the hydrologic cycle, *Trans. Am. Geophysical Union*, v. 14, pp. 446–460, 1933.

51. Hubbert, M. K., The theory of ground-water motion, *Jour. Geol.*, v. 48, pp. 785–944, 1940.

52. Hubbert, M. K., Darcy's law and the field equations of the flow of underground fluids, *Trans. Amer. Inst. Min. and Metal. Engrs.*, v. 207, pp. 222–239, 1956.

53. Intl. Assoc. Hydraulic Research (IAHR), *Fundamentals of Transport Phenomena in Porous Media*, Elsevier, Amsterdam, 392 pp., 1972.

54. Intl. Atomic Energy Agency, Working Group on Nuclear Techniques in Hydrology, *Guidebook on Nuclear Techniques in Hydrology*, Tech. Rept. Ser. no. 91, Vienna, 214 pp., 1968.

55. Irmay, S., On the hydraulic conductivity of unsaturated soils, *Trans. Amer. Geophysical Union*, v. 35, pp. 463–467, 1954.

56. Jacob, C. E., Flow of ground water, *Engineering Hydraulics*, H. Rouse, ed., John Wiley and Sons, New York, pp. 321–386, 1950.

57. Kass, W., *Tracing Technique in Geohydrology*, A. A. Balkema, Rotterdam, Netherlands, 1998.

58. Kaufman, W. J., and D. K. Todd, Application of tritium tracer to canal seepage measurements, *Tritium in the Physical and Biological Sciences*, Intl. Atomic Energy Agency, Vienna, pp. 83–94, 1962.

59. Keeley, J. W., and M. R. Scalf, Aquifer storage determination by radio-tracer techniques, *Ground Water*, v. 7, pp. 17–22, 1969.

60. Kilpatrick, F. A., and E. D. Cobb, *Measurement of discharge using tracers, U.S. Geological Survey Techniques of Water Resources Investigations*, Book 3, Ch. A16, 1985.

61. Kirkham, D., and W. L. Powers, *Advanced Soil Physics*, John Wiley & Sons, New York, 534 pp., 1972.

62. Krumbein, W. C., and G. D. Monk, Permeability as a function of the size parameters of unconsolidated sand, *Trans. Amer. Inst. Min. and Met. Engrs.*, v. 151, p. 153–163, 1943.

63. LeGrand, H. E., and V. T. Stringfield, Development and distribution of permeability in carbonate aquifers, *Water Resources Research*, v. 7, pp. 1284–1294, 1971.

64. Lenhard, R. J., J. C. Parker, and S. Mishra, On the correspondence between Brooks-Corey and Van Genuchten Models, *Journal Irrigation Drainage Engineering*, 15(4), pp. 744–751, 1989.

65. Li, W. H., and G. T. Yeh, Dispersion at the interface of miscible liquids in a soil, *Water Resources Research*, v. 4, pp. 369–378, 1968.

66. Lohman, S. W., *Ground-Water Hydraulics, U.S. Geological Survey Prof. Paper* 708, 70 pp., 1972.

67. Luthin, J. N., ed., *Drainage of agricultural lands*, Agronomy Monograph no. 7, Amer. Soc. Agronomy, Madison, WI, 620 pp., 1957.

68. Masch, F. D., and K. J. Denny, Grain size distribution and its effect on the permeability of unconsolidated sands, *Water Resources Research*, v. 2, pp. 665–677, 1966.

69. Mays, L. W., *Water Resources Engineering*, John Wiley & Sons, New York, 2001.

70. Meigs, L. C., and R. L. Beauheim, Tracer tests in a fractured dolomite, 1. Experimental design and observed tracer recoveries, *Water Resources Research*, v. 37, no. 5, pp. 1113–1128, 2001.

71. Mobasheri, F., and M. Shahbazi, Steady-state lateral movement of water through the unsaturated zone of an unconfined aquifer, *Ground Water*, v. 7, no. 6, pp. 28–34, 1969.

72. Molz, F. J., O. Guven, J. G. Melville, R. D. Crocker, and K. T. Matteson, Performance, analysis, and simulation of a two-well tracer test at the Mobile site, *Water Resources Research*, v. 22, no. 7, pp. 1031–1037, 1986.

73. Molz, F. J., O. Guven, J. G. Melville, R. D. Crocker, and K. T. Matteson, *Performance and Analysis of Aquifer Tracer Tests with Implications for Contaminant Transport Modeling, Technical Report EPA/600/2-86/062*, U.S. Environmental Protection Agency, Washington, DC, 1986.

74. Morel-Seytoux, H. J., Multiphase flow in porous media, in *Developments in Hydraulic Engineering 4*, P. Novak, ed., Ch. 3, Elsevier, New York, 1987.

75. Morris, D. A., and A. I. Johnson, *Summary of Hydrologic and Physical Properties of Rock and Soil Materials, as Analyzed by the Hydrologic Laboratory of the U.S. Geological Survey 1948–60, U.S. Geological Survey Water-Supply Paper 1839-D*, 42 pp., 1967.

76. Mull, D. S., J. L. Smoot, and T. D. Liebermann, *Dye Tracing Techniques Used to Determine Ground-Water Flow in a Carbonate Aquifer System Near Elizabethtown, Kentucky, U.S. Geological Survey Water-Resource Investigations Report WRI* 87-4174, Washington, DC, 1988.

77. Novakowski, K. S., The analysis of tracer experiments conducted in divergent radial flow fields, *Water Resources Research*, v. 28, no. 12, pp. 3215–3225, 1992.

78. Pearson, F. J., Jr., and D. E. White, Carbon 14 ages and flow rates of water in Carrizo Sand, Atascosa County, Texas, *Water Resources Research*, v. 3, pp. 251–261, 1967.

79. Rasmussen, W. C., Permeability and storage of heterogeneous aquifers in the United States, *Intl. Assoc. Sci. Hydrology Publ.* 64, pp. 317–325, 1964.

80. Rawls, W. J., L. R. Ahuja, D. L. Brakensiek, and A. Shirmohammadi, Infiltration and soil water movement, *Handbook of Hydrology*, D. R. Maidment, ed., Ch. 5, McGraw-Hill, New York, 1993.

81. Rawls, W. J., D. L. Brakensiek, and N. Miller, Green-Ampt infiltration parameters from soils data, *J. Hydraulic Div.*, ASCE, v. 109, No.1, pp. 62–70, 1983.

82. Rice, W. A., and S. M. Gorelick, Geological inference from "flow net" transmissivity determination: Three case studies, *Water Resources Bulletin*, v. 21, no. 6, pp. 919–930, 1985.

83. Richards, L. A., Capillary conduction of liquids through porous mediums, *Physics*, v. 1, pp. 318–333, 1931.

84. Rumer, R. R., Longitudinal dispersion in steady and unsteady flow, *Jour. Hydraulics Div.*, Amer. Soc. Civil Engrs., v. 88, no. HY4, pp. 147–172, 1962.

85. Rumer, R. R., and P. A. Drinker, Resistance to laminar flow through porous media, *Jour. Hydraulics Div.*, Amer. Soc. Civil Engrs., v. 92, no. HY5, pp. 155–163, 1966.

86. Scott, D. M., An evaluation of flow net analysis for aquifer identification, *Ground Water*, v. 30, no. 5, pp. 755–764, 1992.

87. Scott, D. M. and H. R. Thorpe, Ground-water resources between the Rakaia and Ashburton Rivers, Ministry of Works and Development for the National Water and Soil Conservation Authority, Hydrology Centre Publication No. 6, Christchurch, New Zealand, 1986.

88. Shahbazi, M., et al., Effect of topography on ground water flow, *Intl. Assoc. Sci. Hydrology Publ.* 77, pp. 314–319, 1968.

89. Shamir, U. Y., and D. R. F. Harleman, Numerical solutions for dispersion in porous mediums, *Water Resources Research*, v. 3, pp. 557–581, 1967.

90. Skibitzke, H. E., and G. M. Robinson, *Dispersion in Ground Water Flowing through Heterogeneous Materials, U.S. Geological Survey Prof. Paper* 386-B, 3 pp., 1963.

91. Smith, W. O., Infiltration in sands and its relation to groundwater recharge, *Water Resources Research*, v. 3, pp. 539–555, 1967.

92. Smith, W. O., and A. N. Sayre, *Turbulence in Ground-Water Flow, U.S. Geological Survey Prof. Paper* 402-E, 9 pp., 1964.

93. Stallman, R. W., Flow in the zone of aeration, *Advances in Hydroscience*, v. 4, in Chow, V. T., ed., pp. 151–195, Academic Press, New York, 1967.

94. Stout, G. E., ed., *Isotope Techniques in the Hydrologic Cycle*, Geophysical Monograph Ser. no. 11, Amer. Geophysical Union, 199 pp., 1967.

95. Todd, D. K., and J. Bear, Seepage through layered anisotropic porous media, *Jour. Hydraulics Div.*, Amer. Soc. Civil Engrs., v. 87, no. HY3, pp. 31–57, 1961.

96. Todorovic, P., A stochastic model of longitudinal diffusion in porous media, *Water Resources Research*, v. 6, pp. 211–222, 1970.

97. Toth, J., A theory of groundwater motion in small drainage basins in Central Alberta, Canada, *Jour. Geophysical Research*, v. 67, pp. 4375–4387, 1962.

98. Toth, J., A theoretical analysis of groundwater flow in small drainage basins, *Jour. Geophysical Research*, v. 68, pp. 4795–4812, 1963.

99. U.S. Geological Survey, *Fluid Movement in Earth Materials, U.S. Geological Survey Professional Paper* 411, Chaps. A to I, 1961–1970.

100. Vachaud, G., Determination of the hydraulic conductivity of unsaturated soils from an analysis of transient flow data, *Water Resources Research*, v. 3, pp. 697–705, 1967.

101. Vachaud, G., and J. L. Thony, Hysteresis during infiltration and redistribution in a soil column at different water contents, *Water Resources Research*, v. 7, pp. 111–127, 1971.

102. Van Genuchten, M. Th., A closed-form equation for predicting the hydraulic conductivity of unsaturated soils, *Soil Science Society Am. J.*, v. 44, pp 892–898, 1980.

103. van Schilfgaarde, J. ed., *Drainage for Agriculture*, Agronomy Monograph no. 17, Amer. Soc. Agronomy, Madison, Wisconsin, 700 pp., 1974.

104. van Tonder, G. J., A computer code for the calculation of the relative transmissivity distribution in an aquifer for steady state ground-water levels, *Water SA*, v. 15, no. 3, pp. 147–152, 1989.

105. Vogel, J. C., Carbon-14 dating of groundwater, *Isotope Hydrology 1970*, Intl. Atomic Energy Agency, Vienna, pp. 225–239, 1970.

106. Ward, J., Turbulent flow in porous media, *Jour. Hydraulics Div.*, Amer. Soc. Civil Engrs., v. 90, no. HY5, pp. 1–12, 1964.

107. Wenzel, L. K., *Methods for Determining Permeability of Water-Bearing Materials with Special Reference to Discharging-Well Methods, U.S. Geological Survey Water-Supply Paper* 887, 192 pp., 1942.

108. Wiebenga, W. A., et al., Radioisotopes as groundwater tracers, *Jour. Geophysical Research*, v. 72, pp. 4081–4091, 1967.

109. Zaluski, M., Tracer Test, *Manual of Applied Field Hydrogeology*, W. D. Weight and J. L. Sonderegger, eds., Ch. 13, McGraw-Hill, New York, 2001.

Chapter 4

Groundwater and Well Hydraulics

$\mathbf{D}$arcy's law and the fundamental equations governing groundwater movement can now be applied to particular situations. Solutions of groundwater flow to wells rank highest in importance. From pumping tests of wells, storage coefficients and transmissivities of aquifers can be determined; furthermore, with these aquifer characteristics known, future declines in groundwater levels associated with pumpage can be calculated. Well flow equations have been developed for steady and unsteady flows, for various types of aquifers, and for several special boundary conditions. For practical application, most solutions have been reduced to convenient graphic or mathematical form, or computer programs.

4.1 STEADY UNIDIRECTIONAL FLOW

Steady flow implies that no change occurs with time. Flow conditions differ for confined and unconfined aquifers and hence need to be considered separately, beginning with flow in one direction.

4.1.1 Confined Aquifer

Let groundwater flow with a velocity v in the x-direction of a confined aquifer of uniform thickness. Then for one-dimensional, steady flow, Equation 3.9.10 reduces to

$$\frac{\partial^2 h}{\partial x^2} = 0 \qquad (4.1.1)$$

which has for its solution

$$h = C_1 x + C_2 \qquad (4.1.2)$$

where h is the head above a given datum and C_1 and C_2 are constants of integration. Assuming $h = 0$ when $x = 0$ and $\partial h/\partial x = -(v/K)$ from Darcy's law, then we have

$$h = -\frac{vx}{K} \qquad (4.1.3)$$

This states that the head decreases linearly, as sketched in Figure 4.1.1, with flow in the x-direction.

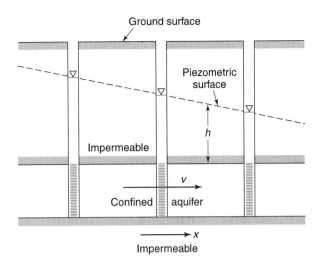

Figure 4.1.1. Steady unidirectional flow in a confined aquifer of uniform thickness.

EXAMPLE 4.1.1

Referring to Figure 4.1.1, if the distance and the observed piezometric surface drop between two adjacent wells are 1,000 m and 3 m, respectively, find an estimate of the time it takes for a molecule of water to move from one well to the other. Assume steady unidirectional flow in a homogeneous silty sand confined aquifer with a hydraulic conductivity $K = 3.5$ m/day and an effective porosity of 0.35.

SOLUTION

First compute the discharge velocity:

$$v = -\frac{hK}{x} = -\frac{(-3\text{m})(3.5 \text{ m/d})}{(1000\text{m})} = 0.0105 \text{ m/d}$$

The pore (seepage) velocity is computed using the velocity:

$$v_p = v/n_e = (0.0105 \text{ m/d})/(0.35) = 0.03 \text{ m/d}$$

It would take 1000 m/(0.03 m/d × 365) ≈ 91.3 years. ∎

4.1.2 Unconfined Aquifer

For the similar flow situation in an unconfined aquifer, direct analytic solution of the Laplace equation is not possible. The difficulty arises from the fact that the water table in the two-dimensional case represents a flow line. The shape of the water table determines the flow distribution, but at the same time the flow distribution governs the water table shape. To obtain a solution, Dupuit[13] assumed (1) the velocity of the flow to be proportional to the tangent of the hydraulic gradient instead of the sine as defined in Darcy's law and (2) the flow to be horizontal and uniform everywhere in a vertical section. These assumptions, although permitting a solution to be obtained, limit the application of the results. For unidirectional flow, as sketched in Figure 4.1.2, the discharge per unit width q at any vertical section can be given as

$$q = -Kh\frac{dh}{dx} \qquad (4.1.4)$$

where K is hydraulic conductivity, h is the height of the water table above an impervious base, and x is the direction of flow. Integrating yields

$$qx = -\frac{K}{2}h^2 + C \qquad (4.1.5)$$

and, if $h = h_0$ where $x = 0$, then the *Dupuit equation*

$$q = \frac{K}{2x}\left(h_0^2 - h^2\right)$$ (4.1.6)

results, which indicates that the water table is parabolic in form.

For flow between two fixed bodies of water of constant heads h_0 and h_1 as in Figure 4.1.2, the water table slope at the upstream boundary of the aquifer (neglecting the capillary zone) is

$$\frac{dh}{dx} = -\frac{q}{Kh_0}$$ (4.1.7)

But the boundary $h = h_0$ is an equipotential line because the fluid potential in a water body is constant; consequently, the water table must be horizontal at this section, which is inconsistent with Equation 4.1.7. In the direction of the flow, the parabolic water table described by Equation 4.1.6 increases in slope. By so doing, the two Dupuit assumptions, previously stated, become increasingly poor approximations to the actual flow; therefore, the actual water table deviates more and more from the computed position in the direction of flow, as indicated in Figure 4.1.2. The fact that the actual water table lies above the computed one can be explained by the fact that the Dupuit flows are all assumed horizontal, whereas the actual velocities of the same magnitude have a downward vertical component so that a greater saturated thickness is required for the same discharge. At the downstream boundary a discontinuity in flow forms because no consistent flow pattern can connect a water table directly to a downstream freewater surface. The water table actually approaches the boundary tangentially above the water body surface and forms a *seepage face*.

The above discrepancies indicate that the water table does not follow the parabolic form of Equation 4.1.6; nevertheless, for flat slopes, where the sine and tangent are nearly equal, it closely predicts the water table position except near the outflow. The equation, however, accurately determines q or K for given boundary heads.[41]

EXAMPLE 4.1.2

A stratum of clean sand and gravel between two channels (see Figure 4.1.2) has a hydraulic conductivity $K = 10^{-1}$ cm/sec, and is supplied with water from a ditch ($h_0 = 6.5$ m deep) that penetrates to the bottom of the stratum. If the water surface in the second channel is 4 m above the bottom of the stratum and its

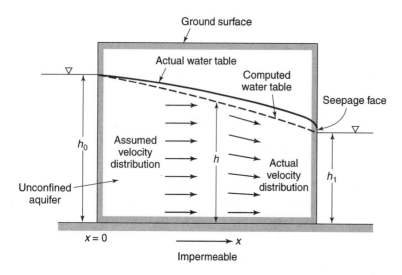

Figure 4.1.2. Steady flow in an unconfined aquifer between two water bodies with vertical boundaries.

distance to the ditch is $x = 150$ m (which is also the thickness of the stratum), estimate the unit flow rate into the gallery.

SOLUTION

The flow is computed using the Dupuit equation (4.1.6) for unit flow, where

$$K = 10^{-1} \text{ cm/sec} = 86.4 \text{ m/day}$$

$$q = \frac{K}{2x}\left(h_0^2 - h^2\right) = \frac{86.4 \text{ m/day}}{2(150 \text{ m})}\left(6.5^2 - 4^2\right)\text{m}^2 = 7.56 \text{ m}^2/\text{day}$$

∎

4.1.3   Base Flow to a Stream

Estimates of the base flow to streams (see Chapter 6) or average groundwater recharge can be computed by applying the above analysis of one-directional flow in an unconfined aquifer. For example, picture the idealized boundaries shown in Figure 4.1.3 of two long parallel streams completely penetrating an unconfined aquifer with a continuous recharge rate W occurring uniformly over the aquifer. With the Dupuit assumptions, the flow per unit thickness is

$$q = -Kh\frac{dh}{dx} \tag{4.1.8}$$

and by continuity

$$q = Wx \tag{4.1.9}$$

Combining these equations and integrating leads to the result

$$h^2 = h_a^2 + \frac{W}{K}\left(a^2 - x^2\right) \tag{4.1.10}$$

where h, h_a, a, and x are as defined in Figure 4.1.3, and K is the hydraulic conductivity. From symmetry and continuity

$$Q_b = 2aW \tag{4.1.11}$$

where Q_b is the base flow entering each stream per unit length of stream channel. If h is known at any point, Q_b or W can be computed provided K is known.

Extensions of this analysis have been applied to design the spacing of parallel drains on agricultural land for specified soil, crop, and irrigation conditions.

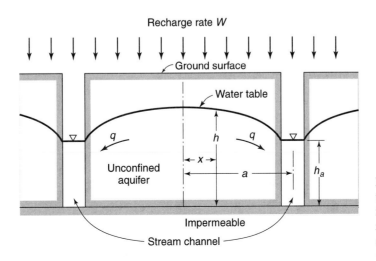

Figure 4.1.3. Steady flow to two parallel streams from a uniformly recharged unconfined aquifer.

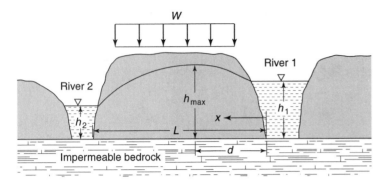

Figure 4.1.4. Unconfined aquifer between two rivers ($x = 0$, $h = h_1$ and $x = L$, $h = h_2$).

Consider the unconfined aquifer between two rivers as shown in Figure 4.1.4 with recharge rate of W. The flow is only in one direction so that the x-axis is aligned parallel to the flow. The flow is then determined by

$$\frac{d}{dx}\left(Kh\frac{dh}{dx}\right) = -W \tag{4.1.12}$$

or

$$\frac{d^2\left(h^2\right)}{dx^2} = -\frac{2W}{K} \tag{4.1.13}$$

Integration of Equation 4.1.13 yields

$$h^2 = \frac{Wx^2}{K} + c_1 x + c_2 \tag{4.1.14}$$

where c_1 and c_2 are constants of integration. Boundary conditions ($h = h_1$ at $x = 0$ and $h = h_2$ at $x = L$) are applied to obtain

$$h^2 = h_1{}^2 - \frac{\left(h_1{}^2 - h_2{}^2\right)x}{L} + \frac{W}{K}(L - x)x \tag{4.1.15}$$

The discharge q_x per unit width at any distance from the origin (see Figure 4.1.4) can be computed using Equation 4.1.8 with dh/dx computed by differentiating Equation 4.1.15:

$$q_x = \frac{K\left(h_1{}^2 - h_2{}^2\right)}{2L} - W\left(\frac{L}{2} - x\right) \tag{4.1.16}$$

where the units are ft²/day or m²/day for q_x, ft/day or m/day for K and W, and ft or m for x, h_1, h_2, and L.

Figure 4.1.4 shows the location where $h = h_{max}$ (a crest in the water table), for the case of infiltration, which is essentially a water divide where $q_x = 0$. The distance d from the origin to the water divide is computed using Equation 4.1.16 with $q_x = 0$ and $x = d$ to obtain

$$d = \frac{L}{2} - \frac{K}{W}\frac{\left(h_1{}^2 - h_2{}^2\right)}{2L} \tag{4.1.17}$$

At $x = d$, $h = h_{max}$ which can then be substituted into Equation 4.1.15 to obtain the following expression for h_{max}:

$$h_{max}^2 = h_1{}^2 - \frac{\left(h_1{}^2 - h_2{}^2\right)d}{L} + \frac{W}{K}(L - d)d \tag{4.1.18}$$

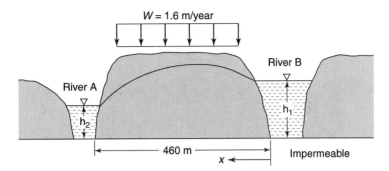

Figure 4.1.5.
Example 4.1.3

EXAMPLE 4.1.3

An unconfined aquifer of clean sand and gravel is located between two fully penetrating rivers (see Figure 4.1.5) and has a hydraulic conductivity of $K = 10^{-2}$ cm/sec. The aquifer is subject to a uniform recharge of 1.6 m/year. The water surface elevations in rivers A and B are 8.5 m and 10 m, respectively, above the bottom. Estimate (a) the maximum elevation of the water table and the location of groundwater divide, (b) the travel times from groundwater divide to both rivers ($n_e = 0.35$), and (c) the daily discharge per kilometer from the aquifer into both rivers.

SOLUTION

(a) The maximum elevation of the water table occurs at the location of the groundwater divide computed using Equation 4.1.17 with $W = 1.6$ m/year $= 0.0044$ m/day and $K = 10^{-2}$ cm/s $= 8.64$ m/day:

$$d = \frac{L}{2} - \frac{K}{W}\frac{\left(h_1^2 - h_2^2\right)}{2L} = \frac{460 \text{ m}}{2} - \frac{8.64 \text{ m/d}}{0.0044 \text{ m/d}}\frac{\left(10^2 - 8.5^2\right)\text{m}^2}{2(460 \text{ m})} = 171 \text{ m from river B}$$

The maximum head at the divide is computed using Equation 4.1.18:

$$h_{\max} = \sqrt{h_1^2 - \frac{\left(h_1^2 - h_2^2\right)d}{L} + \frac{W}{K}(L-d)d}$$

$$= \sqrt{10^2 - \frac{\left(10^2 - 8.5^2\right)(171)}{460} + \frac{0.0044}{8.64}(460-171)(171)} = 10.7\text{m}$$

(b) The average pore velocity is computed using Darcy's law with the Dupuit assumptions:

$$v_A = \left(\frac{K}{n_e}\right)\left(\frac{\Delta h}{\Delta x}\right) = \left(\frac{8.64 \text{ m/d}}{0.35}\right)\left(\frac{10.7 - 8.5}{460 - 171}\frac{\text{m}}{\text{m}}\right) = 0.190 \text{ m/day}$$

So the travel time from the groundwater divide to river A is

$$t = \frac{L_A}{v_A} = \frac{460 \text{ m} - 171 \text{ m}}{0.190 \text{ m/day}} = 1524 \text{ days} = 4.18 \text{ years}$$

Similarly, the travel time from the groundwater divide to river B is computed as

$$v_B = \left(\frac{K}{n}\right)\left(\frac{\Delta h}{\Delta x}\right) = \left(\frac{8.64 \text{ m/d}}{0.35}\right)\left(\frac{10.7 - 10}{171}\right) = 0.101 \text{ m/day}$$

$$t = \frac{L_A}{v_A} = \frac{171\text{m}}{0.104 \text{ m/day}} = 1692 \text{ days} = 4.64 \text{ years}$$

(c) From Equation 4.1.16, for $x = 0$:

$$q_x = \frac{K\left(h_1^2 - h_2^2\right)}{2L} - W\left(\frac{L}{2} - x\right)$$

$$= \frac{(8.64 \text{ m/d})\left(10^2 - 8.5^2\right)\text{m}^2}{2(460)\text{m}} - (0.0044 \text{ m/d})\left(\frac{460}{2} - 0\right)\text{m} = -0.751\left(\text{m}^3/\text{day}\right)/\text{m}$$

The minus sign occurs due to opposite flow direction to the x-axis (see Figure 4.1.5). So, $(0.751 \times 1000 \text{ m}) = 751 \text{ m}^3/\text{day}$ is the daily discharge from the aquifer per kilometer into river B.

Similarly, for $x = 460$ m:

$$q_x = \frac{K\left(h_1^2 - h_2^2\right)}{2L} - W\left(\frac{L}{2} - x\right)$$

$$= \frac{(8.64 \text{ m/d})\left(10^2 - 8.5^2\right)}{2(460)} - (0.0044 \text{ m/d})\left(\frac{460}{2} - 460\right) = 1.27\left(\text{m}^3/\text{day}\right)/\text{m}$$

The daily discharge from the aquifer per kilometer into river A is $(1.27 \times 1000 \text{ m}) = 1270 \text{ m}^3/\text{day}$.   ∎

4.2   STEADY RADIAL FLOW TO A WELL

When a well is pumped, water is removed from the aquifer surrounding the well, and the water table or piezometric surface, depending on the type of aquifer, is lowered. The *drawdown* at a given point is the distance the water level is lowered. A *drawdown curve* (or *cone*) shows the variation of drawdown with distance from the well (see Figure 4.2.1). In three dimensions, the drawdown curve describes a conic shape known as the *cone of depression*, as shown in Figure 4.2.2. Also, the outer limit of the cone of depression (zero drawdown) defines the *area of influence* of the well.

4.2.1   Confined Aquifer

To derive the radial flow equation (which relates the well discharge to drawdown) for a well completely penetrating a confined aquifer, referring to Figure 4.2.1 will prove helpful. The flow is assumed two-dimensional to a well centered on a circular island and penetrating a homogeneous and isotropic aquifer. Because the flow is everywhere horizontal, the Dupuit assumptions apply without error. Using plane polar coordinates with the well as the origin, we obtain the well discharge Q at any distance r as

$$Q = Av = -2\pi r b K \frac{dh}{dr} \qquad (4.2.1)$$

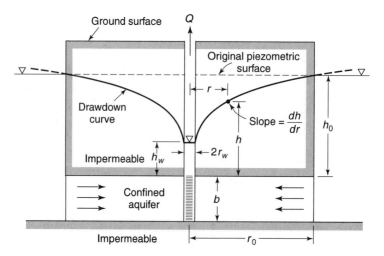

Figure 4.2.1. Steady radial flow to a well penetrating a confined aquifer on an island.

for steady radial flow to the well. Rearranging and integrating 4.2.1 for the boundary conditions at the well, $h = h_w$ and $r = r_w$, and at the edge of the island, $h = h_0$ and $r = r_0$, yield

$$h_0 - h_w = \frac{Q}{2\pi K b} \ln \frac{r_0}{r_w} \tag{4.2.2}$$

or

$$Q = 2\pi K b \frac{h_0 - h_w}{\ln(r_0/r_w)} \tag{4.2.3}$$

with the negative sign neglected.

In the more general case of a well penetrating an extensive confined aquifer, as in Figure 4.2.2, there is no external limit for r. From the above derivation at any given value of r,

$$Q = 2\pi K b \frac{h - h_w}{\ln(r/r_w)} \tag{4.2.4}$$

which shows that h increases indefinitely with increasing r. Yet, the maximum h is the initial uniform head h_0. Thus, from a theoretical aspect, steady radial flow in an extensive aquifer does not exist because the cone of depression must expand indefinitely with time. However, from a practical standpoint, h approaches h_0 with distance from the well, and the drawdown varies with the logarithm of the distance from the well.

The flow net in Figure 4.2.3 illustrates the distribution of flow in a confined aquifer for a fully penetrating well and a 100 percent open hole. Figure 4.2.4 illustrates the flow distribution to a discharging well in a confined aquifer. The well is a fully penetrating, 100 percent open hole. Figure 4.2.5 illustrates the flow net for a well that penetrates 50 percent of the confined aquifer with an open hole. The flow net in Figure 4.2.6 illustrates the distribution of flow in a confined aquifer for a well that penetrates through the upper confining bed but not into the artesian aquifer.

Equation 4.2.4, known as the *equilibrium,* or *Thiem,*[60] *equation,* enables the hydraulic conductivity or the transmissivity of a confined aquifer to be determined from a pumped well

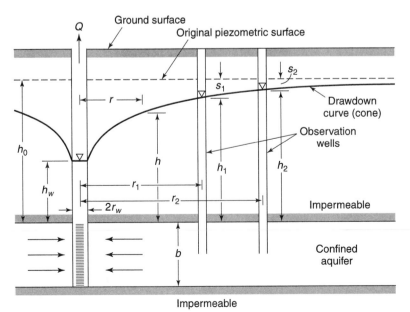

Figure 4.2.2. Radial flow to a well penetrating an extensive confined aquifer.

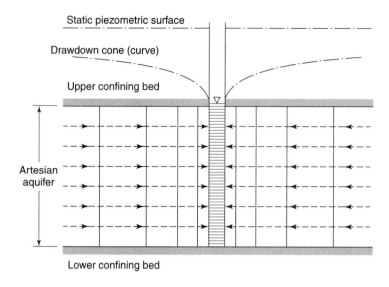

Figure 4.2.3. Distribution of flow to a discharging well in an artesian aquifer—a fully penetrating and 100 percent open hole (from the U.S. Bureau of Reclamation[62]).

that fully penetrates the aquifer. Because any two points define the logarithmic drawdown curve, the method consists of measuring drawdowns in two observation wells at different distances from a well pumped at a constant rate. Theoretically, h_w at the pumped well can serve as one measurement point; however, well losses caused by flow through the well screen and inside the well introduce errors so that h_w should be avoided. The transmissivity is given by

$$T = Kb = \frac{Q}{2\pi(h_2 - h_1)} \ln \frac{r_2}{r_1} \tag{4.2.5}$$

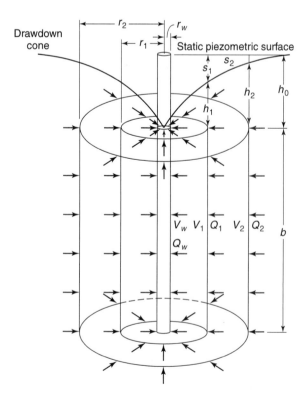

b = Thickness of aquifer
h_0 = Undisturbed artesian head
h_1, h_2 = Undisturbed artesian heads at r_2 respectively when well is discharging
s_1, s_2 = Drawdown at r_1 and r_2 respectively when well is discharging
$Q_w = Q_1 = Q_2$
$A_w = 2\pi r_w b$
$A_1 = 2\pi r_1 b$
$A_2 = 2\pi r_2 b$
$V_w = Q_w/A_w$
$V_1 = Q_1/A_1$
$V_2 = Q_2/A_2$

Figure 4.2.4. Flow distribution to a discharging well in an artesian aquifer—a fully penetrating and 100 percent open hole (from the U.S. Bureau of Reclamation[62]).

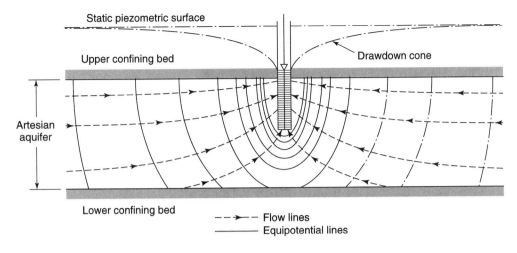

Figure 4.2.5. Distribution of flow to a well in an artesian aquifer—a 50-percent penetrating and open hole (from U.S. Bureau of Reclamation[62]).

where r_1 and r_2 are the distances and h_1 and h_2 are the heads of the respective observation wells.

From a practical standpoint, the drawdown s rather than the head h is measured so that Equation 4.2.5 can be rewritten

$$T = \frac{Q}{2\pi(s_1 - s_2)} \ln \frac{r_2}{r_1}$$

(4.2.6)

where s_1 and s_2 are shown in Figure 4.2.2. In order to apply Equation 4.2.6, pumping must continue at a uniform rate for a sufficient time to approach a steady-state condition—that is, one in which the drawdown changes negligibly with time.[*] The observation wells should be located close enough to the pumping well so that their drawdowns are appreciable and can be readily measured. The derivation assumes that the aquifer is homogeneous and isotropic, is of uniform thickness, and is of infinite areal extent; that the well penetrates the entire aquifer; and that initially the piezometric surface is nearly horizontal.

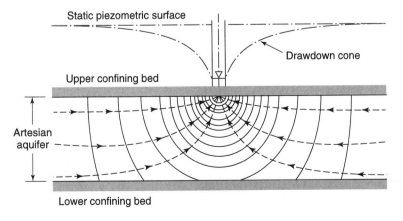

Figure 4.2.6. Distribution of flow to a discharging well—just penetrating to the top of an artesian aquifer. A strong vertical component of flow is established out to a distance approximately equal to the thickness of the aquifer (from U.S. Bureau of Reclamation[62]).

---

[*]In fact, the difference in drawdowns $(s_1 - s_2)$ becomes essentially constant while both values are still increasing so that Equation 4.2.6 generally gives good results after only a few days of pumping.

EXAMPLE 4.2.1

A well fully penetrates a 25-m thick confined aquifer. After a long period of pumping at a constant rate of 0.05 m³/s, the drawdowns at distances of 50 and 150 m from the well were observed to be 3 and 1.2 m, respectively. Determine the hydraulic conductivity and the transmissivity. What type of unconsolidated deposit would you expect this to be?

SOLUTION

Use Equation 4.2.5 to compute the hydraulic conductivity with $Q = 0.05$ m³/s, $r_1 = 50$ m, $r_2 = 150$ m, $s_1 = h_0 - h_1$, and $s_2 = h_0 - h_2$, so $s_1 - s_2 = h_2 - h_1 = 3 - 1.2 = 1.8$ m. $Q = 0.05$ m³/s $= 4320$ m³/day, and

$$K = \frac{Q}{2\pi b(h_2 - h_1)} \ln\left(\frac{r_2}{r_1}\right) = \frac{4320 \text{ m}^3/\text{day}}{2\pi(25 \text{ m})(1.8 \text{ m})} \ln\left(\frac{150}{50}\right) = 16.8 \text{ m/day}$$

The transimissivity is $T = Kb = (16.8$ m/day$)(25$ m$) = 420$ m²/day. Referring to Figure 3.2.2 and Table 3.2.1 with $K = 1.94 \times 10^{-4}$ m/s shows that this aquifer is probably a medium clean sand. ∎

EXAMPLE 4.2.2

A 1-m diameter well penetrates vertically through a confined aquifer 30 m thick. When the well is pumped at 113 m³/hr, the drawdown in a well 15 m away is 1.8 m; in another well 50 m away, it is 0.5 m. What is the approximate head in the pumped well for steady-state conditions and what is the approximate drawdown in the well? Also compute the transmissivity of the aquifer and the radius of influence of the pumping well. Take the initial piezometric level as 40 m above the datum.

SOLUTION

First determine the hydraulic conductivity using Equation 4.2.5: $Q = 113$ m³/hr $= 2712$ m³/day. Then

$$K = \frac{Q}{2\pi b(s_1 - s_2)} \ln\left(\frac{r_2}{r_1}\right) = \frac{2712 \text{ m}^3/\text{day}}{2\pi(30 \text{ m})(1.8 \text{ m} - 0.5 \text{ m})} \ln\left(\frac{50}{15}\right) = 13.3 \text{ m/day}$$

The transmissivity is $T = Kb = 13.3$ m/day $\times 30$ m $= 400$ m²/day.

To compute the approximate head, h_w, in the pumped well, rearrange Equation 4.2.5 and use $h_2 = h_0 - s_2 = 40 - 0.5 = 39.5$ m

$$h_w = h_2 - \frac{Q}{2\pi Kb} \ln\left(\frac{r_2}{r_w}\right) = 39.5 \text{ m} - \frac{2712 \text{ m}^3/\text{day}}{2\pi(13.3 \text{ m/day})(30 \text{ m})} \ln\left(\frac{50 \text{ m}}{0.5 \text{ m}}\right) = 34.5 \text{ m}$$

Drawdown is then

$$s_w = h_0 - h_w = 40 \text{ m} - 34.5 \text{ m} = 5.5 \text{ m}$$

The radius of influence (R) of pumping well can be found by rearranging Equation 4.2.5 and solving for r_0 which is R:

$$R = (r_1)\exp\left[\frac{2\pi Kb(h_0 - h_1)}{Q}\right] = (15 \text{ m})\exp\left[\frac{2\pi(13.3 \text{ m/day})(30 \text{ m})(40 \text{ m} - 38.2 \text{ m})}{2712 \text{ m}^3/\text{day}}\right] = 79 \text{ m}$$ ∎

4.2.2 Unconfined Aquifer

An equation for steady radial flow to a well in an unconfined aquifer also can be derived with the help of the Dupuit assumptions. As shown in Figure 4.2.7, the well completely penetrates the aquifer to the horizontal base and a concentric boundary of constant head surrounds the well. The well discharge is

$$Q = -2\pi r Kh \frac{dh}{dr} \tag{4.2.7}$$

which, when integrated between the limits $h = h_w$ at $r = r_w$ and $h = h_0$ at $r = r_0$, yields

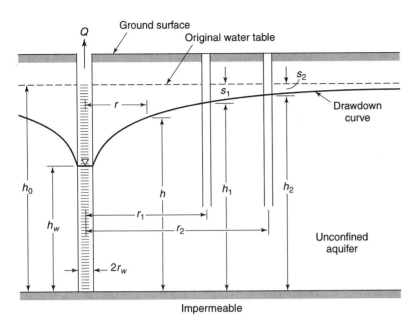

Figure 4.2.7. Radial flow to a well penetrating an unconfined aquifer.

$$Q = \pi K \frac{h_0^2 - h_w^2}{\ln(r_0 / r_w)} \tag{4.2.8}$$

Converting to heads and radii at two observation wells (see Figure 4.2.7),

$$Q = \pi K \frac{h_2^2 - h_1^2}{\ln(r_2/r_1)} \tag{4.2.9}$$

and rearranging to solve for the hydraulic conductivity

$$K = \frac{Q}{\pi\left(h_2^2 - h_1^2\right)} \ln \frac{r_2}{r_1} \tag{4.2.10}$$

This equation fails to accurately describe the drawdown curve near the well because the large vertical flow components contradict the Dupuit assumptions; however, estimates of K for given heads are good. In practice, drawdowns should be small in relation to the saturated thickness of the unconfined aquifer.

The transmissivity can be approximated from Equation 4.2.10 by

$$T \cong K \frac{h_1 + h_2}{2} \tag{4.2.11}$$

Where drawdowns are appreciable, the heads h_1 and h_2 in Equation 4.2.10 can be replaced by $(h_0 - s_1)$ and $(h_0 - s_2)$, respectively, as shown in Figure 4.2.7. Then the transmissivity for the full thickness becomes[3, 37]

$$T = Kh_0 = \frac{Q}{2\pi\left[\left(s_1 - \frac{s_1^2}{2h_0}\right) - \left(s_2 - \frac{s_2^2}{2h_0}\right)\right]} \ln \frac{r_2}{r_1} \tag{4.2.12}$$

Figure 4.2.8 illustrates the flow distribution in an unconfined (free) aquifer, for a fully penetrating well that has the openings in the lower one-third of the aquifer.

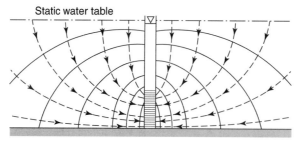

(a) Initial stage in pumping a free aquifer. Most water follows a path with a high vertical component from the water table to the screen.

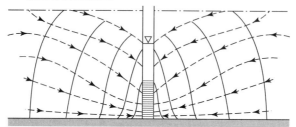

(b) Intermediate stage in pumping a free aquifer. Radial component of flow becomes more pronounced but contribution from drawdown cone in immediate vicinity of well is still important.

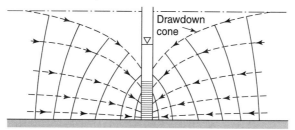

(c) Approximate steady-state stage in pumping a free aquifer. Profile of cone of depression is established. Nearly all water originating near outer edge of area of influence and stable primarily radial flow pattern established.

Figure 4.2.8. Development of flow distribution about a discharging well in a free aquifer—a fully penetrating and 33 percent open hole (from U.S. Bureau of Reclamation[62]).

EXAMPLE 4.2.3

A well penetrates an unconfined aquifer. Prior to pumping the water level (head) is $h_0 = 25$ m. After a long period of pumping at a constant rate of 0.05 m³/s, the drawdowns at distances of 50 and 150 m from the well were observed to be 3 and 1.2 m, respectively. Compute the hydraulic conductivity of the aquifer and the radius of influence of pumping well. What type of deposit is the aquifer material?

SOLUTION

Use Equation 4.2.10 to compute K with $Q = 0.05$ m³/s= 4320 m³/day, $r_1 = 50$ m, $r_2 = 150$ m, $h_1 = 25 - 3 = 22$ m, and $h_2 = 25 - 1.2 = 23.8$ m.

$$K = \frac{Q}{\pi\left(h_2^2 - h_1^2\right)}\ln\left(\frac{r_2}{r_1}\right) = \frac{4320 \text{ m}^3/\text{day}}{\pi\left(23.8^2 - 22^2\right)}\ln\left(\frac{150 \text{ m}}{50 \text{ m}}\right) = 18.3 \text{ m/day}$$

The deposit is probably a medium clean sand. Equation 4.2.10 is used to compute the radius of influence:

$$R = \left(r_1\right)\exp\left[\frac{K\pi\left(h_0^2 - h_1^2\right)}{Q}\right] = \left(50 \text{ m}\right)\exp\left[\frac{\left(18.3 \text{ m/day}\right)\pi\left(25^2 - 22^2\right)}{4320 \text{ m}^3/\text{day}}\right] = 327 \text{ m}$$

EXAMPLE 4.2.4

A well 0.5 m in diameter penetrates 33 m below the static water table. After a long period of pumping at a rate of 80 m³/hr, the drawdowns in wells 18 and 45 m from the pumped well were found to be 1.8 and 1.1 m respectively. (a) What is the transmissivity of the aquifer? (b) What is the approximate drawdown in the pumped well? (c) Determine the radius of influence of the pumping well.

SOLUTION

(a) Use Equation 4.2.10 for steady-state radial flow to a well in an unconfined aquifer to compute the hydraulic conductivity, where $Q = 80$ m³/hr $= 1920$ m³/day; $h_1 = 33 - 1.8 = 31.2$ m.; $h_2 = 33 - 1.1 = 31.9$ m; $r_2 = 45$ m and $r_1 = 18$ m:

$$K = \frac{Q}{\pi\left(h_2^2 - h_1^2\right)}\ln\left(\frac{r_2}{r_1}\right) = \frac{1920 \text{ m}^3/\text{day}}{\pi\left(31.9^2 - 31.2^2\right)}\ln\left(\frac{45}{18}\right) = 12.7 \text{ m/day}$$

The transmissivity is computed as $T = Kb = 12.7$ m/day $\times 33$ m $= 418$ m²/day.

(b) Next compute the head and drawdown at the well. First rearrange Equation 4.2.10 to solve for the head at the well:

$$h_w = \sqrt{h_2^2 - \frac{Q}{\pi K}\ln\left(\frac{r_2}{r_1}\right)} = \sqrt{31.2^2 - \frac{1920 \text{ m}^3/\text{day}}{\pi(12.68 \text{ m/day})}\ln\left(\frac{18 \text{ m}}{0.25 \text{ m}}\right)} = 27.7 \text{ m}$$

The drawdown is computed as $s_w = 33$ m $- 27.7$ m $= 5.3$ m.

(c) The radius of influence of the pumping well is computed by rearranging Equation 4.2.5:

$$R = (r_1)\exp\left(\frac{\pi K\left(h_0^2 - h_1^2\right)}{Q}\right) = (45 \text{ m})\exp\left(\frac{\pi(12.68 \text{ m/day})\left(33^2 - 31.9^2\right)}{1920 \text{ m}^3/\text{day}}\right) = 198 \text{ m}$$

■

4.2.3   Unconfined Aquifer with Uniform Recharge

Figure 4.2.9 shows a well penetrating an unconfined aquifer that is recharged uniformly at rate W from rainfall, excess irrigation water, or other surface-water sources. The flow Q toward the well increases as the well is approached, reaching a maximum of Q_w at the well. The increase

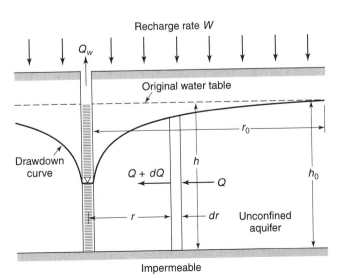

Figure 4.2.9. Steady flow to a well penetrating a uniformly recharged unconfined aquifer.

in flow dQ through a cylinder of thickness dr and radius r comes from the recharged water entering the cylinder from above; hence,

$$dQ = -2\pi r\, dr\, W \tag{4.2.13}$$

Integrating, we obtain

$$Q = -\pi r^2 W + C \tag{4.2.14}$$

but at the well $r \to 0$ and $Q = Q_w$, so that

$$Q = -\pi r^2 W + Q_w \tag{4.2.15}$$

Substituting this flow in the equation for flow to the well (Equation 4.2.7) gives

$$-2\pi r K h \frac{dh}{dr} = -\pi r^2 W + Q_w \tag{4.2.16}$$

Integrating, and noting that $h = h_0$ at $r = r_0$, yield the equation for the drawdown curve:

$$h_0^2 - h^2 = \frac{W}{2K}\left(r^2 - r_0^2\right) + \frac{Q_w}{\pi K}\ln\frac{r_0}{r} \tag{4.2.17}$$

By comparing Equation 4.2.17 with Equation 4.2.8, the effect of the vertical recharge becomes apparent.

It follows that when $r = r_0$, $Q = 0$, so that from Equation 4.2.15

$$Q_w = \pi r_0^2 W \tag{4.2.18}$$

Thus, the total flow of the well equals the recharge within the circle defined by the radius of influence; conversely, the radius of influence is a function of the well pumpage and the recharge rate only. This results in a steady-state drawdown; however, the analysis assumes an idealized circular outer boundary with a constant head and no flow–conditions that rarely occur in the field.

EXAMPLE 4.2.5	A pumping well is to be used to maintain a lowered water table at a construction site. The site is square, 50 m on a side, and the 25-cm diameter well is located at the center of the square, as shown in the figure. The hydraulic conductivity of the unconfined aquifer is estimated to be about 1×10^{-5} m/s or 0.864 m/day. The bottom of the aquifer is approximately horizontal at a depth of 20 m below the ground surface. Under natural conditions, the water table is nearly horizontal at a depth of 1 m below the ground surface and the unconfined aquifer is uniformly recharged at a rate $W = 0.06$ m/day. During the construction period, the water table must be lowered a minimum of 3 m over the site. Assuming steady-state conditions, compute the minimum pumping rate required.
SOLUTION	The given condition is satisfied if the drawdown at any of the corner points is 3 m.

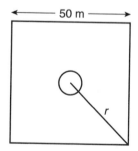

The well discharge is expressed in terms of the radius of influence using Equation 4.2.17 as $Q_w = \pi r_0^2 W =$ (0.06) πr_0^2. Substitute this relationship along with $h_0 = 19$ m, $h = 16$ m, $W = 0.06$ m/day, $K = 0.864$ m/day, and $r = \sqrt{25^2 + 25^2} = 35.35$ m into Equation 4.2.17 to obtain

$$19^2 - 16^2 = \frac{0.06}{2 \times 0.864}\left(35.35^2 - r_0^2\right) + \frac{\left(0.06\pi r_0^2\right)}{\pi \times 0.864}\ln\left(\frac{r_0}{35.35}\right)$$

Solving the above equation using an iterative procedure yields $r_0 \approx 70$ m. The minimum pumping rate is $Q_w = \pi(70^2)\,(0.06) \approx 924$ m³/day or 0.01069 m³/s.   ■

4.3   WELL IN A UNIFORM FLOW

Drawdown curves for well flow presented heretofore have assumed an initially horizontal groundwater surface. A practical situation is that of a well pumping from an aquifer having a uniform flow field, as indicated by a uniformly sloping piezometric surface or water table. Figure 4.3.1 shows sectional and plan views of a well penetrating a confined aquifer with a sloping piezometric surface. It is apparent that the circular area of influence associated with a radial flow pattern becomes distorted; however, for most relatively flat natural slopes the Dupuit radial flow equation can be applied without appreciable error.

For wells pumping from an area with a sloping hydraulic gradient, the hydraulic conductivity can be determined from Equation 4.2.7 by inserting average heads and hydraulic gradients. The resulting expression has the form

$$K = \frac{2Q}{\pi r\left(h_u + h_d\right)\left(i_u + i_d\right)} \tag{4.3.1}$$

for an unconfined aquifer where Q is the pumping rate, h_u and h_d are the saturated thicknesses, and i_u and i_d are the water table slopes at distance r upstream and downstream, respectively, from the well. For a confined aquifer, piezometric slopes replace water table slopes, and $(h_u + h_d)$ is replaced by $2b$ where b is the aquifer thickness.

In Figure 4.3.1, the groundwater divide marking the boundary of the region producing inflow to the well is shown. For a well pumping for an infinite time, the boundary would extend up to the limit of the aquifer. The expression for the boundary of the region producing inflow can be derived by superposition of radial and one-dimensional flow fields to yield

$$-\frac{y}{x} = \tan\left(\frac{2\pi Kbi}{Q}y\right) \tag{4.3.2}$$

where the rectangular coordinates are as shown in Figure 4.3.1 with the origin at the well, b is the aquifer thickness, Q is the discharge rate, i is the natural piezometric slope, and K is hydraulic conductivity. From Equation 4.3.2, the boundary asymptotically approaches the finite limits

$$y_L = \pm\frac{Q}{2Kbi} \tag{4.3.3}$$

as $x \rightarrow \infty$. The boundary of the contributing area extends downstream to a stagnation point where

$$x_L = -\frac{Q}{2\pi Kbi} \tag{4.3.4}$$

It follows that the upstream inflow zone equals $2\pi x_L$.

Equations 4.3.1 to 4.3.3 also apply to unconfined aquifers by replacing b by the uniform saturated aquifer thickness h_0, providing the drawdown is small in relation to the aquifer thickness. An important practical application of these equations concerns determining whether an upstream pollution source will affect a nearby pumping well (see Chapter 8).

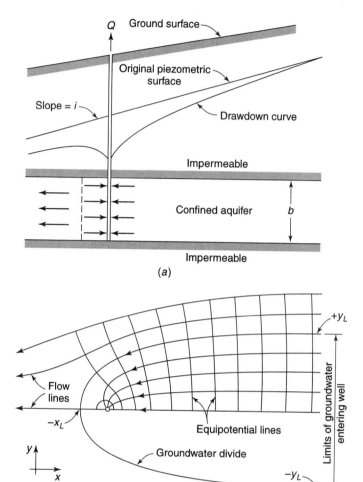

Figure 4.3.1. Flow to a well penetrating a confined aquifer having a sloping plane piezometric surface. (*a*) Vertical section. (*b*) Plan view

EXAMPLE 4.3.1

A fully penetrating production well with a radius of 0.5 m pumps at the rate of 15 L/s from a 35-m thick confined aquifer with a hydraulic conductivity of 20 m/day. If the distance and the observed piezometric head drop between two observation wells were 1000 m and 3 m, respectively, before the production well was installed, determine the longitudinal and transverse limits of groundwater entering the well.

SOLUTION

First determine the slope of the piezometric surface under natural conditions (i.e., before the production well was installed):

$$i = \frac{\Delta h}{\Delta x} = \frac{3 \text{ m}}{1000 \text{ m}} = 0.003$$

It is assumed that the observation wells were aligned with the groundwater flow direction. Then, using Equations 4.3.3 and 4.3.4, compute the limits of groundwater entering the well on a horizontal plane (i.e., plan view) for $Q = 15$ L/s = 1296 m³/day:

$$y_L = \pm \frac{Q}{2Kbi} = \pm \frac{1296 \text{ m}^3/\text{day}}{2(20 \text{ m/day})(35 \text{ m}) \times 0.003} = \pm 308 \text{ m}$$

$$x_L = -\frac{Q}{2\pi Kbi} = -\frac{1296 \text{ m}^3/\text{day}}{2\pi(20 \text{ m/day})(35 \text{ m}) \times 0.003} = -98.2 \text{ m}$$

A practical result is that contaminant sources farther than 98.2 m downstream of the well or ± 308 m in the transverse direction do not impact the well.   ∎

4.4   UNSTEADY RADIAL FLOW IN A CONFINED AQUIFER

4.4.1   Nonequilibrium Well Pumping Equation

When a well penetrating an extensive confined aquifer is pumped at a constant rate, the influence of the discharge extends outward with time. The rate of decline of head times the storage coefficient summed over the area of influence equals the discharge. Because the water must come from a reduction of storage within the aquifer, the head will continue to decline as long as the aquifer is effectively infinite; therefore, unsteady, or transient, flow exists. The rate of decline, however, decreases continuously as the area of influence expands.

The applicable differential equation (see Equation 3.9.12) in plane polar coordinates is

$$\frac{\partial^2 h}{\partial r^2} + \frac{1}{r}\frac{\partial h}{\partial r} = \frac{S}{T}\frac{\partial h}{\partial t} \tag{4.4.1}$$

where h is head, r is radial distance from the pumped well, S is the storage coefficient, T is transmissivity, and t is the time since beginning of pumping. Theis[59] obtained a solution for Equation 4.4.1 based on the analogy between groundwater flow and heat conduction. By assuming that the well is replaced by a mathematical sink of constant strength and imposing the boundary conditions $h = h_0$ for $t = 0$, and $h \rightarrow h_0$ as $r \rightarrow \infty$ for $t \geq 0$, the solution

$$s = \frac{Q}{4\pi T}\int_u^\infty \frac{e^{-u}du}{u} = \frac{Q}{4\pi T}W(u)$$

$$= \frac{Q}{4\pi T}\left[-0.5772 - \ln u + u - \frac{u^2}{2\cdot 2!} + \frac{u^3}{3\cdot 3!} - \frac{u^4}{4\cdot 4!} + \cdots\right] \tag{4.4.2}$$

is obtained, where s is drawdown, Q is the constant well discharge, and

$$u = \frac{r^2 S}{4Tt} \tag{4.4.3}$$

Equation 4.4.2 is known as the *nonequilibrium,* or *Theis, equation.* The integral is a function of the lower limit u and is known as an *exponential integral.* It can be expanded as a convergent series as shown in Equation 4.4.2 and is termed the well function, $W(u)$.

Alternatively, using U.S. customary units (gallon-day-foot system) where s is in ft, Q is in gpm, T is in gpd/ft, u is in ft, and t is in days, we have

$$s = \frac{114.6Q}{T}W(u) \tag{4.4.4a}$$

$$u = \frac{1.87r^2 S}{Tt} \quad (t \text{ in days}) \tag{4.4.4b}$$

$$u = \frac{2693r^2 S}{Tt} \quad (t \text{ in minutes}) \tag{4.4.4c}$$

The nonequilibrium equation permits determination of the formation constants S and T by means of pumping tests of wells. The equation is widely applied in practice and is preferred over the equilibrium equation because (1) a value for S can be determined, (2) only one observation well is required, (3) a shorter period of pumping is generally necessary, and (4) no assumption of steady-state flow conditions is required.

The assumptions inherent in Equation 4.4.2 should be emphasized because they are often overlooked in applying the nonequilibrium equation and thereby can lead to erroneous results. The assumptions include:

1. The aquifer is homogeneous, isotropic, of uniform thickness, and of infinite areal extent.
2. Before pumping, the piezometric surface is horizontal.
3. The well is pumped at a constant discharge rate.
4. The pumped well penetrates the entire aquifer, and flow is everywhere horizontal within the aquifer to the well.
5. The well diameter is infinitesimal so that storage within the well can be neglected.
6. Water removed from storage is discharged instantaneously with decline of head.

Seldom, if ever, are these assumptions strictly satisfied, but recognition of them can create an awareness of the approximations involved for employing the nonequilibrium equation under field conditions. Average values of S and T can be obtained in the vicinity of a pumped well by measuring in one or more observation wells the change in drawdown with time under the influence of a constant pumping rate. Because of the mathematical difficulties encountered in applying Equation 4.4.2, or its equivalent, Equation 4.4.4, several investigators have developed simpler approximate solutions that can be readily applied for field purposes. Three methods, by Theis,[59] Cooper and Jacob,[9] and Chow,[7] are described in the following sections with the necessary tables and/or graphs. An illustrative example accompanies each method.

4.4.2   Theis Method of Solution

Equation 4.4.2 may be simplified to

$$s = \left(\frac{Q}{4\pi T}\right) W(u)$$

(4.4.5)

where $W(u)$, termed the *well function,* is a convenient symbolic form of the exponential integral. Rewriting Equation 4.4.3 as

$$\frac{r^2}{t} = \left(\frac{4T}{S}\right) u$$

(4.4.6)

we can see that the relation between $W(u)$ and u must be similar to that between s and r^2/t because the terms in parentheses in the two equations are constants. Given this similarity, Theis[59] suggested an approximate solution for S and T based on a graphic method of superposition.

A plot on logarithmic paper of $W(u)$ versus u, known as a *type curve,* is prepared. Table 4.4.1 gives values of $W(u)$ for a wide range of u. Values of drawdowns are plotted against values of r^2/t on logarithmic paper of the same size and scale as for the type curve. The observed time–drawdown data are superimposed on the type curve, keeping the coordinate axes of the two curves parallel, and adjusted until a position is found by trial whereby most of the plotted points of the observed data fall on a segment of the type curve. Any convenient point is then selected, and the coordinates of this match point are recorded. With values of $W(u)$, u, s, and r^2/t thus determined, S and T can be obtained from Equations 4.4.5 and 4.4.6.[*]

In areas where several wells exist near a well being test-pumped, simultaneous readings of s in the wells enable distance–drawdown data to be fitted to a type curve in a manner identical to that for time–drawdown data.

---

[*]The computation of r^2/t values can be avoided by plotting s versus t rather than r^2/t. In this case, the type curve must be turned over to obtain coincidence and a match point, but results are identical.

Table 4.4.1   Values of $W(u)$ for Values of u

u	1.0	2.0	3.0	4.0	5.0	6.0	7.0	8.0	9.0
$\times 1$	0.219	0.049	0.013	0.0038	0.0011	0.00036	0.00012	0.000038	0.000012
$\times 10^{-1}$	1.82	1.22	0.91	0.70	0.56	0.45	0.37	0.31	0.26
$\times 10^{-2}$	4.04	3.35	2.96	2.68	2.47	2.30	2.15	2.03	1.92
$\times 10^{-3}$	6.33	5.64	5.23	4.95	4.73	4.54	4.39	4.26	4.14
$\times 10^{-4}$	8.63	7.94	7.53	7.25	7.02	6.84	6.69	6.55	6.44
$\times 10^{-5}$	10.94	10.24	9.84	9.55	9.33	9.14	8.99	8.86	8.74
$\times 10^{-6}$	13.24	12.55	12.14	11.85	11.63	11.45	11.29	11.16	11.04
$\times 10^{-7}$	15.54	14.85	14.44	14.15	13.93	13.75	13.60	13.46	13.34
$\times 10^{-8}$	17.84	17.15	16.74	16.46	16.23	16.05	15.90	15.76	15.65
$\times 10^{-9}$	20.15	19.45	19.05	18.76	18.54	18.35	18.20	18.07	17.95
$\times 10^{-10}$	22.45	21.76	21.35	21.06	20.84	20.66	20.50	20.37	20.25
$\times 10^{-11}$	24.75	24.06	23.65	23.36	23.14	22.96	22.81	22.67	22.55
$\times 10^{-12}$	27.05	26.36	25.96	25.67	25.44	25.26	25.11	24.97	24.86
$\times 10^{-13}$	29.36	28.66	28.26	27.97	27.75	27.56	27.41	27.28	27.16
$\times 10^{-14}$	31.66	30.97	30.56	30.27	30.05	29.87	29.71	29.58	29.46
$\times 10^{-15}$	33.96	33.27	32.86	32.58	32.35	32.17	32.02	31.88	31.76

EXAMPLE 4.4.1

Drawdown was measured during a pumping test at frequent intervals in an observation well 200 ft from a well that was pumped at a constant rate of 500 gpm. The data for this pump test is listed in the table. These measurements show that the water level is still dropping after 4,000 minutes of pumping; therefore, analysis of the test data requires use of the Theis nonequilibrium procedure. Determine T and S for this aquifer.

Pump test data	
Time (min)	Drawdown (ft)
1	0.05
2	0.22
3	0.40
4	0.56
5	0.70
7	0.94
10	1.2
20	1.8
40	2.5
100	3.4
300	4.5
1,000	5.6
4,000	7.0

SOLUTION

Step 1. Plot the time–drawdown data on log–log graph paper. The drawdown is plotted on the vertical axis and the time since pumping started on the horizontal axis (not shown).

Step 2. Superimpose this plot on the type curve sheet of the same size and scale as the time-drawdown plot, so that the plotted points match the type curve. The axes of both graphs must be kept parallel.

Step 3. Select a match point, which can be any point in the overlap area of the curve sheets. It is usually most convenient to select a match point where the coordinates on the type curve are known in advance (e.g., $W(u) = 1$ and $1/u = 1$ or $W(u) = 1$ and $1/u = 10$, etc.). Then determine the value of s and t for this match point:

$$W(u) = 1 \qquad s = 1 \text{ ft} \qquad 1/u = 1 \qquad t = 2 \text{ min}$$

Step 4. Determine T

$$T = \frac{114.6\,Q}{s}\,W(u)$$

$$= \frac{114.6 \times 500}{1} \times 1 = 57300 \text{ gpd / ft}$$

Step 5. Determine S

$$S = \frac{Tt}{\dfrac{1}{u} \times 2693r^2}$$

$$= \frac{57300 \times 2}{1 \times 2693 \times 200^2}$$

$$= 1.06 \times 10^{-3}$$

■

EXAMPLE 4.4.2

A well penetrating a confined aquifer is pumped at a uniform rate of 2,500 m³/day. Drawdowns during the pumping period are measured in an observation well 60 m away; observations of t and s are listed in Table 4.4.2. Using the Theis method determine T and S for this confined aquifer.

SOLUTION

Values of r^2/t in m²/min are computed and appear in the right column of Table 4.4.2. Values of s and r^2/t are plotted on logarithmic paper. Values of $W(u)$ and u from Table 4.4.1 are plotted on another sheet of logarithmic paper of the same size and scale, and a curve is drawn through the points. The two sheets are superposed and shifted with coordinate axes parallel until the observational points coincide with the curve, as shown in Figure 4.4.1. A convenient match point is selected with $W(u) = 1.00$ and $u = 1 \times 10^{-2}$, so that $s = 0.18$ m and $r^2/t = 150$ m²/min $= 216,000$ m²/day. Thus, from Equation 4.4.5,

$$T = \frac{Q}{4\pi s}\,W(u) = \frac{2500(1.00)}{4\pi(0.18)} = 1110 \text{ m}^2 / \text{day}$$

and from Equation 4.4.6,

$$S = \frac{4Tu}{r^2/t} = \frac{4(1110)(1 \times 10^{-2})}{216,000} = 0.000206$$

■

Table 4.4.2 Pumping Test Data

($r = 60$ m)					
t, min	s, m	r^2/t, m²/min	t, min	s, m	r^2/t, m²/min
0	0	∞	18	0.67	200
1	0.20	3,600	24	0.72	150
1.5	0.27	2,400	30	0.76	120
2	0.30	1,800	40	0.81	90
2.5	0.34	1,440	50	0.85	72
3	0.37	1,200	60	0.90	60
4	0.41	900	80	0.93	45
5	0.45	720	100	0.96	36
6	0.48	600	120	1.00	30
8	0.53	450	150	1.04	24
10	0.57	360	180	1.07	20
12	0.60	300	210	1.10	17
14	0.63	257	240	1.12	15

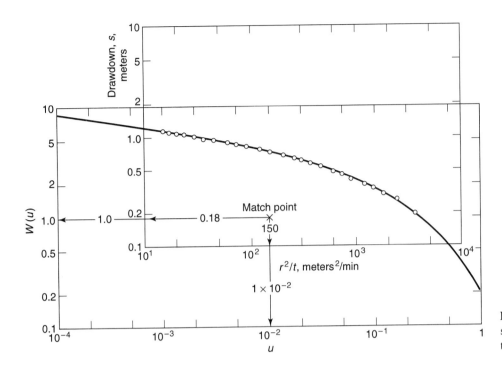

Figure 4.4.1. Theis method of superposition for solution of the nonequilibrium equation.

4.4.3   Cooper–Jacob Method of Solution

It was noted by Cooper and Jacob[9] that for small values of r and large values of t, u is small, so that the series terms in Equation 4.4.4 become negligible after the first two terms. As a result, the drawdown can be expressed by the asymptote

$$s = \frac{Q}{4\pi T}\left(-0.5772 - \ln\frac{r^2 S}{4Tt}\right) \tag{4.4.7}$$

Rewriting and changing to decimal logarithms reduce this to

$$s = \frac{2.30Q}{4\pi T}\log\frac{2.25Tt}{r^2 S} \tag{4.4.8}$$

Therefore, a plot of drawdown s versus the logarithm of t forms a straight line. Projecting this line to $s = 0$, where $t = t_0$ (see Figure 4.4.2), we have

$$0 = \frac{2.30Q}{4\pi T}\log\frac{2.25Tt_0}{r^2 S} \tag{4.4.9}$$

and it follows that for the above to hold true, $\log(1) = 0$ and

$$\frac{2.25Tt_0}{r^2 S} = 1 \tag{4.4.10}$$

resulting in

$$S = \frac{2.25Tt_0}{r^2} \tag{4.4.11}$$

A value for T can be obtained by noting that if $t/t_0 = 10$, then $\log t/t_0 = 1$; therefore, replacing s by Δs, where Δs is the drawdown difference per log cycle of t, Equation 4.4.8 becomes[33]

$$T = \frac{2.30Q}{4\pi\Delta s} \tag{4.4.12}$$

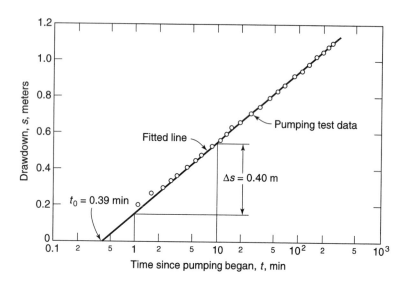

Figure 4.4.2. Cooper–Jacob method for solution of the nonequilibrium equation.

Thus, the procedure is first to solve for T with Equation 4.4.12 and then to solve for S with Equation 4.4.11. The straight-line approximation for this method should be restricted to small values of u ($u < 0.01$) to avoid large errors.

EXAMPLE 4.4.3

Rework Example 4.4.2 using the Cooper–Jacob method.

SOLUTION

From the pumping test data in Table 4.4.2, s and t are plotted on semilogarithmic paper, as shown in Figure 4.4.2. A straight line is fitted through the points, and $\Delta s = 0.40$ m and $t_0 = 0.39$ min $= 2.70 \times 10^{-4}$ day are read. Thus,

$$T = \frac{2.30(2500)}{4\pi(0.40)} = 1144 \ \text{m}^2/\text{day}$$

and

$$S = \frac{2.25 T t_0}{r^2} = \frac{2.25(1144)(2.70 \times 10^{-4})}{(60)^2} = 0.000193$$

∎

EXAMPLE 4.4.4

Using the Cooper–Jacob approximation, compute the rate of piezometric drawdown around a pumping well with respect to time. If the well is pumping at a constant rate of 55 gpm from a sandy confined aquifer with $T = 3{,}600$ ft^2/day and $S = 10^{-4}$, what is the time to reach near–steady-state conditions 200 ft from the pumping well? Assume that near–steady-state conditions are achieved when the drawdown rate falls below 0.5 in/hr (based on accuracy of groundwater level measurements with the available equipment). How does the answer change if the transmissivity of the aquifer is 1,200 ft^2/day?

SOLUTION

First, we must compute the critical time after which the Cooper–Jacob method becomes valid (i.e., $u < 0.01$) at 200 ft:

$$t \geq \frac{r^2 S}{4Tu} = \frac{(200 \ \text{ft})^2 (1 \times 10^{-4})}{4(3600 \ \text{ft}^2/\text{day})(0.01)} \rightarrow t \geq 40 \ \text{min}$$

The drawdown is approximated by

$$s = \frac{Q}{4\pi T}\left(-0.5772 - \ln\frac{r^2 S}{4Tt}\right)$$

which can be rearranged to

$$s = \frac{Q}{4\pi T}\left(-0.5772 - \ln\frac{r^2 S}{4T} + \ln t\right)$$

Taking the derivative of drawdown with respect to time yields

$$\frac{ds}{dt} = \frac{Q}{4\pi T}\frac{1}{t}$$

This relationship implies that according to the Cooper–Jacob approximation, the rate of drawdown is independent of radial distance and is inversely proportional with time. The change in drawdown with respect to time and the time are, respectively,

$$\frac{ds}{dt} = \frac{Q}{4\pi T}\frac{1}{t} = 0.5 \text{ in/hr}$$

$$\frac{10587 \text{ ft}^3/\text{day}}{4\pi\left(3600 \text{ ft}^2/\text{day}\right)}\frac{1}{t} = \frac{0.5}{12} \text{ ft/hr}$$

$$t = 5.6 \text{ hr}$$

Note that the Cooper–Jacob approximation is satisfied so that the near–steady-state conditions at 200 ft will be reached after 5.6 hrs of pumping at this location. If the transmissivity were 1200 ft²/day, the approximation would be valid when $t \geq 120$ min and the drawdown rate at 200 ft would be negligible after 16.8 hours of pumping. Thus, it would take longer to reach steady conditions with a lower transmissivity. ∎

4.4.4 Chow Method of Solution

Chow[7] developed a method of solution with the advantages of avoiding curve fitting and being unrestricted in its application. Again, measurements of drawdown in an observation well near a pumped well are made. The observational data are plotted on semilogarithmic paper in the same manner as for the Cooper–Jacob method. On the plotted curve, choose an arbitrary point and note the coordinates, t and s. Next, draw a tangent to the curve at the chosen point and determine the drawdown difference Δs, in feet, per log cycle of time. Compute $F(u)$ from

$$F(u) = \frac{s}{\Delta s} \tag{4.4.13}$$

and find the corresponding values of $W(u)$ and u from Figure 4.4.3.[*] Finally, compute the formation constant T by Equation 4.4.5 and S by Equation 4.4.6.

EXAMPLE 4.4.5

Repeat Example 4.4.2 using the Chow method.

SOLUTION

In Figure 4.4.4 data are plotted from Table 4.4.2 and point A is selected on the curve where $t = 6$ min $= 4.2 \times 10^{-3}$ day and $s = 0.47$ m. A tangent is constructed as shown; the drawdown difference per log cycle of time is $\Delta s = 0.38$ m. Then $F(u) = 0.47/0.38 = 1.24$, and from Figure 4.4.3, $W(u) = 2.75$ and $u = 0.038$. Hence,

$$T = \frac{Q}{4\pi s}W(u) = \frac{2500}{4\pi(0.47)}2.75 = 1160 \text{ m}^2/\text{day}$$

and

$$S = \frac{4Ttu}{r^2} = \frac{4(1160)\left(4.2\times10^{-3}\right)(0.038)}{(60)^2} = 0.000206$$

∎

---

[*]For $F(u) > 2.0$, $W(u) = 2.30F(u)$, and u is obtained from Table 4.4.1.

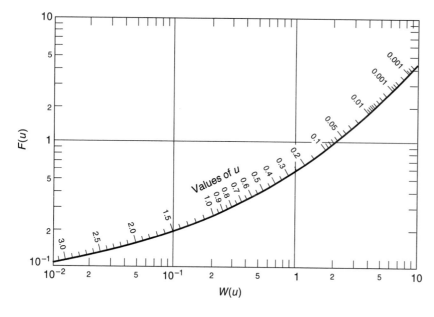

Figure 4.4.3. Relation among $F(u)$, $W(u)$, and u (after Chow[7]).

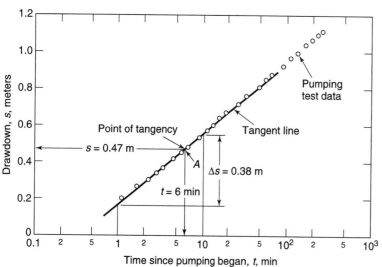

Figure 4.4.4. Chow method for solution of the nonequilibrium equation.

4.4.5   Recovery Test

At the end of a pumping test, when the pump is stopped, the water levels in pumping and observation wells will begin to rise. This is referred to as the *recovery* of groundwater levels, while measurements of drawdown below the original static water level (prior to pumping) during the recovery period are known as *residual drawdowns*. A schematic diagram of change in water level with time during and after pumping is shown in Figure 4.4.5.

It is good practice to measure residual drawdowns because analysis of the data enables transmissivity to be calculated, thereby providing an independent check on pumping test results. Also, costs are nominal in relation to the conduct of a pumping test.* Furthermore, the rate of

---

*In addition, it should be noted that measurement of the recovery within a pumped well will provide an estimate of transmissivity even without an observation well.

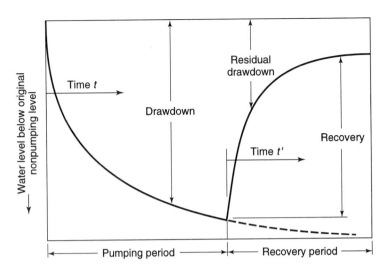

Figure 4.4.5. Drawdown and recovery curves in an observation well near a pumping well.

recharge Q to the well during recovery is assumed constant and equal to the mean pumping rate, whereas pumping rates often vary and are difficult to control accurately in the field.

If a well is pumped for a known period of time and then shut down, the drawdown thereafter will be identically the same as if the discharge had been continued and a hypothetical recharge well with the same flow were superposed on the discharging well at the instant the discharge is shut down. From this principle, Theis[59] showed that the residual drawdown s' can be given as

$$s' = \frac{Q}{4\pi T}\left[W(u) - W(u')\right] \tag{4.4.14}$$

where

$$u = \frac{r^2 S}{4Tt} \quad \text{and} \quad u' = \frac{r^2 S}{4Tt'} \tag{4.4.15}$$

and t and t' are defined in Figure 4.4.5. For r small and t' large, the well functions can be approximated by the first two terms of Equation 4.4.2 so that Equation 4.4.14 can be written as

$$s' = \frac{2.30Q}{4\pi T}\log\frac{t}{t'} \tag{4.4.16}$$

Thus, a plot of residual drawdown s' versus the logarithm of t/t' forms a straight line. The slope of the line equals $2.30Q/4\pi T$ so that for $\Delta s'$, the residual drawdown per log cycle of t/t', the transmissivity becomes

$$T = \frac{2.30Q}{4\pi\Delta s'} \tag{4.4.17}$$

No comparable value of S can be determined by this recovery test method.

EXAMPLE 4.4.6

A well pumping at a uniform rate of 2,500 m³/day was shut down after 240 min; thereafter, measurements of s' and t' tabulated in Table 4.4.3 were made in an observation well. Determine the transmissivity.

SOLUTION

Values of t/t' are computed, as shown in Table 4.4.3, and then plotted versus s' on semilogarithmic paper (see Figure 4.4.6). A straight line is fitted through the points and $\Delta s' = 0.40$ m is determined; then,

$$T = \frac{2.30Q}{4\pi\Delta s'} = \frac{2.30(2500)}{4\pi(0.40)} = 1140 \text{ m}^2/\text{day}$$

Table 4.4.3   Recovery Test Data (pump shut down at $t = 240$ min)

t', min	t, min	t/t'	s', m
1	241	241	0.89
2	242	121	0.81
3	243	81	0.76
5	245	49	0.68
7	247	35	0.64
10	250	25	0.56
15	255	17	0.49
20	260	13	0.55
30	270	9	0.38
40	280	7	0.34
60	300	5	0.28
80	320	4	0.24
100	340	3.4	0.21
140	380	2.7	0.17
180	420	2.3	0.14

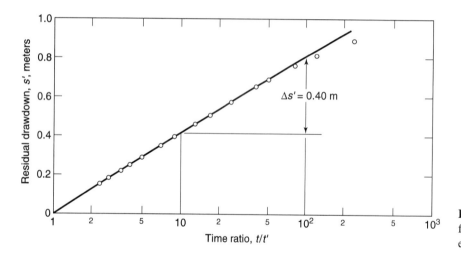

Figure 4.4.6. Recovery test method for solution of the nonequilibrium equation.

4.5   UNSTEADY RADIAL FLOW IN AN UNCONFINED AQUIFER

The previous solution methods for the nonequilibrium equation applied to pumping tests in confined aquifers can also be applied to unconfined aquifers provided that the basic assumptions are satisfied. In general, if the drawdown is small in relation to the saturated thickness, good approximations are possible.[53]

Where drawdowns are significant, the assumption that water released from storage is discharged instantaneously with decline of head is frequently violated in unconfined aquifers. Pumping test data reveal that as a water table is lowered, gravity drainage of water from the unsaturated zone proceeds at a variable rate, known as *delayed yield*.[4, 14, 42] In a series of contributions, Boulton[4–6] developed special type curves for analyzing pumping test data of unconfined aquifers and for taking account of delayed yield.[65] These time–drawdown curves of delayed yield are shown in Figure 4.5.1. The interpretation of any one curve can be considered in three time segments. In the first segment, measured in seconds to a few minutes, water is released essentially instantaneously from storage by compaction of the aquifer and by expan-

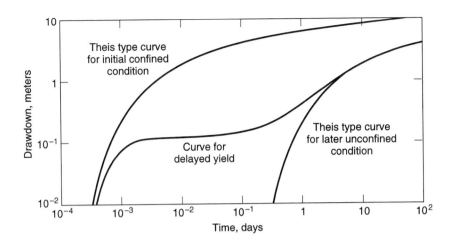

Figure 4.5.1. Type curves of drawdown versus time illustrating the effect of delayed yield for pumping tests in unconfined aquifers (after U.S. Bureau of Reclamation[61]).

sion of entrapped air. This portion of the curve can be fitted by a type curve with a storage coefficient equivalent to that of a confined aquifer. The second segment displays a flattening in slope caused by gravity drainage replenishment from the pore space above the cone of depression. Finally, in the third segment an equilibrium is approached between gravity drainage and the rate of decline of the water table. This condition occurs after several minutes to several days and can be fitted by a type curve with a storage coefficient for an unconfined aquifer.

From a water production standpoint, the storage coefficient obtained from the third segment of the curve in Figure 4.5.1, which is the specific yield, is the most reliable and hence most important. For simplicity a pumping test should be continued sufficiently long enough to define the third segment of the curve; then, by applying one of the solution methods previously described for the nonequilibrium equation, a value for S can be obtained.

The minimum length of pumping test to achieve an accurate estimate of S in an unconfined aquifer depends on the transmissivity of the aquifer. One approach, based on an empirical study of various alluvial aquifer materials, is given by the graphs in Figure 4.5.2. The *delay index* t_d is estimated in Figure 4.5.2a from the composition of aquifer material. Then knowing the distance r between pumping and observation wells, and estimating S and T, an approximation to the minimum pumping time $t_{\min}$ can be calculated from Figure 4.5.2b.

Another approach is simply to ensure that the pumping test duration exceeds the following suggested guidelines[61].

Predominant aquifer material	Minimum pumping time, hours
Silt and clay	170
Fine sand	30
Medium sand and coarser materials	4

Prickett[47] developed a type curve solution for water table conditions based upon Boulton[5]. The following equation for drawdown in an unconfined aquifer with fully penetrating wells and a constant discharge condition was presented later by Neuman:[43]

$$s = \frac{Q}{4\pi T} W\left(u_a, u_y, \eta\right) \tag{4.5.1}$$

where $u_a = \dfrac{r^2 S}{4Tt}$ (applicable for early drawdown data) $\tag{4.5.2}$

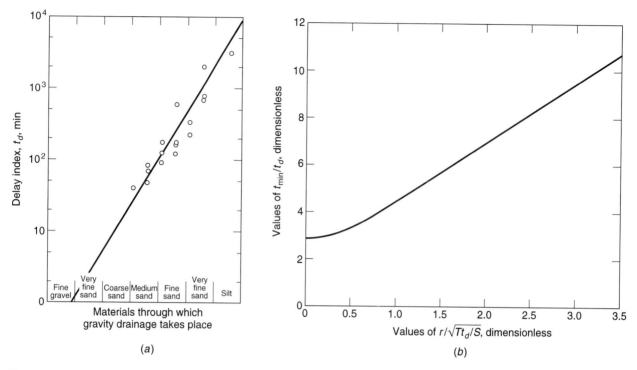

Figure 4.5.2. Empirical method for estimating the minimum length of a pumping test in an unconfined aquifer (after Prickett[47]). *(a)* Empirical relation of delay index to character of materials through which gravity drainage occurs *(b)* Curve for estimating the minimum time t_{min} at which effects of delayed gravity drainage cease to influence drawdown of a pumping well in an unconfined aquifer.

$$u_y = \frac{r^2 S_y}{4Tt} \quad \text{(applicable for later drawdown data)} \quad (4.5.3)$$

$$\eta = \frac{r^2 K_z}{b^2 K_h} \quad (4.5.4)$$

$W(u_a, u_y, \eta)$ is the *unconfined well function* (Figure 4.5.3); K_h and K_v are the horizontal and vertical hydraulic conductivities for an isotropic aquifer $K_v = K_z$ and $\eta = r^2/b^2$; and b is the initial saturated thickness of the unconfined aquifer.

EXAMPLE 4.5.1

(adapted from U.S. Department of the Interior)[62].

A well pumping at 144.4 ft³/min fully penetrates an unconfined aquifer with a saturated thickness of 25 ft. Determine the transmissivity, storativity, specific yield, and horizontal and vertical hydraulic conductivities using the tabulated time–drawdown data in Table 4.5.1 for an observation well located 73 ft away.

SOLUTION

Time–drawdown data (Table 4.5.1) are plotted in Figure 4.5.4, which shows the typical three phases of drawdown for unconfined aquifers. The early drawdown versus time data fit best on the type-a curves for $\eta = 0.06$. The selected match point in Figure 4.5.4 has the following coordinates: ($t = 0.17$ min, $s = 0.57$ ft) and ($1/u_a = 1.0$, $W(u_a, u_y, \eta) = 1.0$). Using Equation 4.5.1 with a discharge of $Q = 144.4$ ft³/min, we find the transmissivity to be

$$T = \frac{Q}{4\pi s} W(u_a, u_y, \eta) = \frac{\left(144.4 \text{ ft}^3/\text{min}\right)}{4\pi(0.57 \text{ ft})}(1.0) = 20.16 \text{ ft}^2/\text{min} \cong 29,900 \text{ ft}^2/\text{day}$$

Next, the storativity value is computed using Equation 4.5.2:

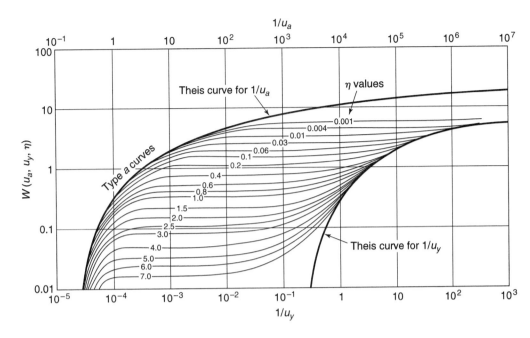

Figure 4.5.3.
Theoretical curves of $W(u_a,u_y,\eta)$ versus $1/u_a$ and $1/u_y$ for an unconfined aquifer (after Neuman[43]).

Table 4.5.1 Time–Drawdown Data for Example 4.5.1.

t (min)	s, feet	t (min)	s, feet	t (min)	s, feet	t (min)	s, feet
0.165	0.12	1.68	0.82	10	1.02	200	1.52
0.25	0.195	1.85	0.84	12	1.03	250	1.59
0.34	0.255	2	0.86	15	1.04	300	1.65
0.42	0.33	2.15	0.87	18	1.05	350	1.7
0.5	0.39	2.35	0.9	20	1.06	400	1.75
0.58	0.43	2.5	0.91	25	1.08	500	1.85
0.66	0.49	2.65	0.92	30	1.13	600	1.95
0.75	0.53	2.8	0.93	35	1.15	700	2.01
0.83	0.57	3	0.94	40	1.17	800	2.09
0.92	0.61	3.5	0.95	50	1.19	900	2.15
1	0.64	4	0.97	60	1.22	1,000	2.2
1.08	0.67	4.5	0.975	70	1.25	1,200	2.27
1.16	0.7	5	0.98	80	1.28	1,500	2.35
1.24	0.72	6	0.99	90	1.29	2,000	2.49
1.33	0.74	7	1	100	1.31	2,500	2.59
1.42	0.76	8	1.01	120	1.36	3,000	2.66
1.5	0.78	9	1.015	150	1.45		

$$S = \frac{4Tu_at}{r^2} = \frac{4(20.16 \text{ ft}^2/\text{min})(1.0)(0.17 \text{ min})}{(73 \text{ ft})^2} = 0.00257$$

Moving the data curve to the right on the type curve to the best late-time match (for $\eta = 0.06$) where $s = 0.57$ ft (see the match point on Figure 4.5.5) yields ($t = 13$ min, $s = 0.57$ ft) and ($1/u_y = 0.1$, $W(u_y, \eta) = 1$). Inserting the appropriate values in Equation 4.5.1 does not change the transmissivity estimate, but using Equation 4.5.3 yields

$$S_y = \frac{4Tu_yt}{r^2} = \frac{4(20.16 \text{ ft}^2/\text{min})(0.1)(13 \text{ min})}{(73 \text{ ft})^2} = 0.02$$

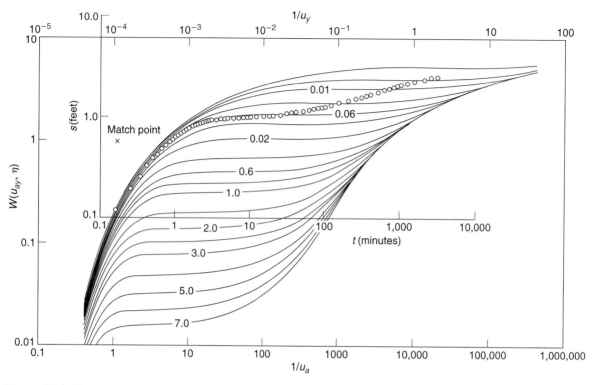

Figure 4.5.4. Type-a curve matching for Example 4.5.1.

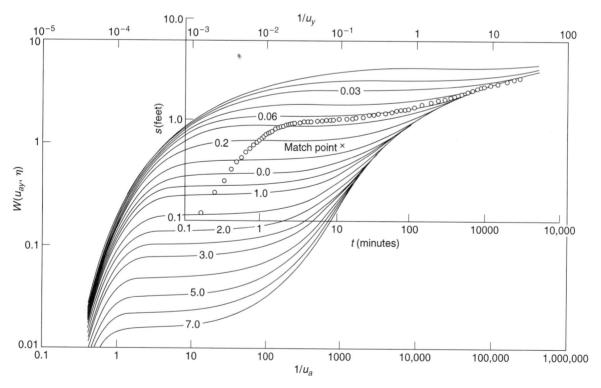

Figure 4.5.5. Type-y curve matching for Example 4.5.1.

The horizontal hydraulic conductivity, K_r or K_h, is computed using

$$K_h = \frac{T}{b} = \frac{20.16 \text{ ft}^2/\text{min}}{25 \text{ ft}^2} = 0.806 \text{ ft/min or } 1160 \text{ ft/day}$$

and the vertical hydraulic conductivity, K_z or K_v, is computed using Equation 4.5.4:

$$K_v = \frac{\eta b^2 K_h}{r^2} = \frac{(0.06)(25 \text{ ft})^2 (1160 \text{ ft/day})}{(73 \text{ ft})^2} = 8.2 \text{ ft/day}$$

∎

4.6   UNSTEADY RADIAL FLOW IN A LEAKY AQUIFER

When a leaky aquifer, as shown in Figure 4.6.1, is pumped, water is withdrawn both from the aquifer and from the saturated portion of the overlying aquitard, or semipervious layer. Lowering the piezometric head in the aquifer by pumping creates a hydraulic gradient within the aquitard; consequently, groundwater migrates vertically downward into the aquifer. The quantity of water moving downward is proportional to the difference between the water table and the piezometric head[8, 29, 58]

Steady-state flow is possible to a well in a leaky aquifer because of the recharge through the semipervious layer. The equilibrium will be established when the discharge rate of the pump equals the recharge rate of vertical flow into the aquifer, provided that the water table remains constant. Solutions for this special steady-state situation are available,[25, 33] but a more general analysis for unsteady flow follows.

When pumping starts from a well in a leaky aquifer, drawdown of the piezometric surface can be given by[19, 21, 25]

$$s = \frac{Q}{4\pi T} W(u, r/B) \tag{4.6.1}$$

where s, Q, and r are defined in Figure 4.6.1, and again

$$u = \frac{r^2 S}{4Tt} \tag{4.6.2}$$

The quantity r/B is given by

$$\frac{r}{B} = \frac{r}{\sqrt{T/(K'/b')}} \tag{4.6.3}$$

where T is the transmissivity of the leaky aquifer, K' is the vertical hydraulic conductivity of the aquitard, and b' is the thickness of the saturated semipervious layer (see Figure 4.6.1). Values of the function $W(u, r/B)$ were tabulated by Hantush.[19] It can be noted that Equation 4.6.1 has the form of the Theis equation (see Equation 4.4.5); in fact, for a confined aquifer, $K' \rightarrow 0$, so that $B \rightarrow \infty$ and $r/B \rightarrow 0$, thereby reducing Equation 4.6.1 to the Theis equation.

Employing this analogy and the Theis method of solution, Walton[64] prepared a family of type curves for $W(u, r/B)$ as presented in Figure 4.6.2. Here values of $W(u, r/B)$ are plotted against $1/u$ for various values of r/B. On another sheet of logarithmic paper of the same scale, s versus t is plotted. Superimposing the two sheets while keeping the coordinate axes parallel, a position is found where most of the data points fall on one of the type curves. Selecting any convenient match point, values of $W(u, r/B)$, $1/u$, s, and t are noted. T is then found from

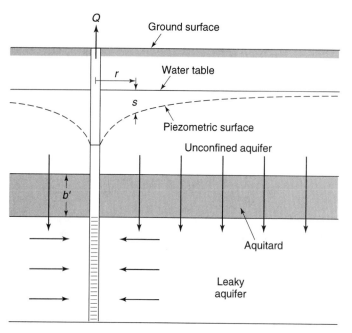

Figure 4.6.1. Well pumping from a leaky aquifer.

Equation 4.6.1 and S from Equation 4.6.2. Finally, from the value of r/B belonging to the type curve of best fit, it is possible to calculate K'/b' from Equation 4.6.3; and if b' is known from field conditions, K' can be evaluated.

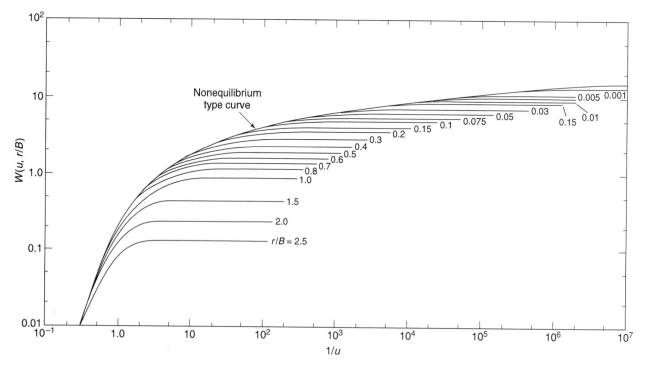

Figure 4.6.2. Type curves for analysis of pumping test data to evaluate storage coefficient and transmissivity of leaky aquifers (after Walton[64]).

EXAMPLE 4.6.1

(adapted from U.S. Department of the Interior)[62].

A well pumping at 600 ft³/min fully penetrates a confined aquifer overlain by a leaky confining layer of 14-ft thickness. Using the tabulated time–drawdown data for an observation well 40 ft away from the pumping well, estimate the transmissivity and storage coefficient of the confined aquifer, and the permeability of the aquitard. Assume that the confining layer does not release water from storage.

Time (min)	Drawdown (ft)	Time (min)	Drawdown (ft)
0	0.00	80	12.02
2	5.65	90	12.26
4	6.96	100	12.33
6	7.72	110	12.37
8	8.00	120	12.41
10	8.71	150	12.69
15	9.47	180	12.85
20	9.99	210	13.09
25	10.35	240	13.13
30	10.70	270	13.25
40	11.14	300	13.33
50	11.46	360	13.37
60	11.62	420	13.41
70	11.86		

SOLUTION

The time–drawdown field data were superimposed on the family type curves for leaky aquifers (Figure 4.6.3). Comparison shows that the best fit occurs for $r/B = 0.03$. The coordinates of the match point selected are

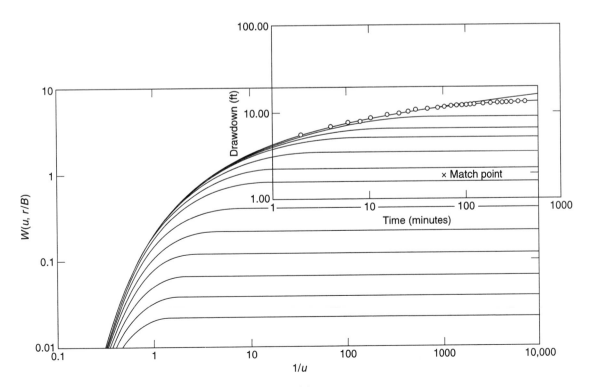

Figure 4.6.3. Leaky type curve matching for Example 4.6.1.

$$\frac{1}{u} = 1000, \qquad W\left(u,\frac{r}{B}\right) = 1.0$$

$$t = 59 \text{ min}, \qquad s = 1.93 \text{ ft}$$

Next we must perform the following unit conversions in order to obtain the transmissivity in units of ft²/day and hydraulic conductivity of the aquitard in units of ft/day for $Q = 600$ ft³/min = 864,000 ft³/day and $t = 59$ min = 0.041 days. The transmissivity and storage coefficient of the confined aquifer are computed using Equations 4.6.1 and 4.6.2 rearranged respectively as

$$T = \frac{Q}{4\pi s} W(u,r/B) = \frac{864,000 \text{ ft}^3/\text{day}}{4\pi(1.93 \text{ ft})}(1.0) = 35,624 \text{ ft}^2/\text{day}$$

$$S = \frac{4Ttu}{r^2} = \frac{4(35,624 \text{ ft}^2/\text{day})(0.041 \text{ days})(0.001)}{(40 \text{ ft})^2} = 0.00365$$

The hydraulic conductivity of the aquitard is computed by rearranging Equation 4.6.3

$$K' = \frac{Tb'(r/B)^2}{r^2} = \frac{(35,624 \text{ ft}^2/\text{day})(14 \text{ ft})(0.03)^2}{(40 \text{ ft})^2} = 0.28 \text{ ft/day}$$ ∎

4.7 WELL FLOW NEAR AQUIFER BOUNDARIES

Where a well is pumped near an aquifer boundary, the assumption that the aquifer is of infinite areal extent no longer holds. Analysis of this situation involves the principle of superposition by which the drawdown of two or more wells is the sum of the drawdowns of each individual well. By introducing imaginary (or *image*) wells, an aquifer of finite extent can be transformed into an infinite aquifer so that the solution methods previously described can be applied.

4.7.1 Well Flow Near a Stream

An example of the usefulness of the method of images is the situation of a well near a perennial stream.[16, 22, 30] It is desired to obtain the head at any point under the influence of pumping at a constant rate Q and to determine what fraction of the pumpage is derived from the stream. Sectional views are shown in Figure 4.7.1 of the real system and an equivalent imaginary system. Note in Figure 4.7.1b that an imaginary recharge well[*] has been placed directly opposite and at the same distance from the stream as the real well. This image well operates simultaneously and at the same rate as the real well so that the buildup (increase of head around a recharge well) and drawdown of head along the line of the stream exactly cancel. This furnishes a constant head along the stream, which is equivalent to the constant elevation of the stream forming the aquifer boundary. Thus, in the plan view of the resulting flow net, illustrated by Figure 4.7.2, a single equipotential line is coincident with the axis of the stream. The resultant asymmetrical drawdown of the real well is given at any point by the algebraic sum of the drawdown of the real well and the buildup of the recharge well, as if these wells were located in an infinite aquifer.

Examples of *hydraulically equivalent aquifer systems* bounded by streams with various configurations are shown in Figure 4.7.3. Note that combinations of both image recharge and pumping wells are required.[33] For the single stream in Figure 4.7.3a, the steady-state drawdown at any point (x,y) is given by

---

[*]A *recharge well* is a well through which water is added to an aquifer; hence, it is the reverse of a pumping well.

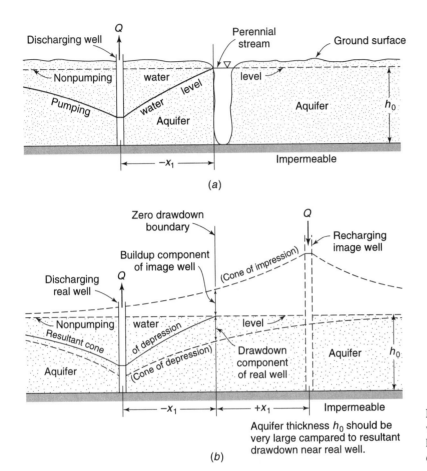

Figure 4.7.1. Sectional views. (*a*) Discharging well near a perennial stream (*b*) Equivalent hydraulic system in an aquifer of infinite areal extent (after Ferris et al.[15])

Aquifer thickness h_0 should be very large campared to resultant drawdown near real well.

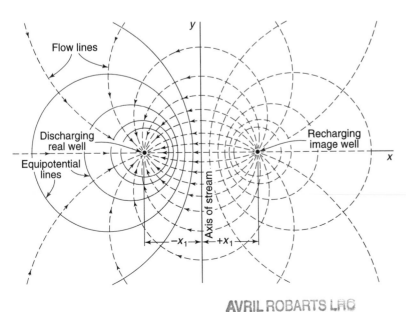

Figure 4.7.2. Flow net for discharging real well and recharging image well (after Ferris, et al.[15]).

$$s = \frac{Q}{4\pi T} \ln \frac{\left(x + x_w\right)^2 + \left(y - y_w\right)^2}{\left(x - x_w\right)^2 + \left(y - y_w\right)^2} \tag{4.7.1}$$

where (x_w, y_w) are the coordinates of the pumped well. Similarly, for the right-angle boundaries of Figure 4.7.3b,

$$s = \frac{Q}{4\pi T} \ln \frac{\left[\left(x - x_w\right)^2 + \left(y + y_w\right)^2\right]\left[\left(x + x_w\right)^2 + \left(y - y_w\right)^2\right]}{\left[\left(x - x_w\right)^2 + \left(y - y_w\right)^2\right]\left[\left(x + x_w\right)^2 + \left(y + y_w\right)^2\right]} \tag{4.7.2}$$

And for the strip aquifer bounded by two straight parallel streams (see Figure 4.7.3c),

$$s = \frac{Q}{4\pi T} \ln \frac{\cosh \dfrac{\pi\left(y - y_w\right)}{2a} + \cosh \dfrac{\pi\left(x + x_w\right)}{2a}}{\cosh \dfrac{\pi\left(y - y_w\right)}{2a} - \cosh \dfrac{\pi\left(x + x_w\right)}{2a}} \tag{4.7.3}$$

and the angles are expressed in radians. Actually, in Figure 4.7.3c, the image wells extend to infinity; however, in practice it is only necessary to include pairs of image wells closest to the real well because others have a negligible influence on the drawdown.

The water level in the wells will draw down initially only under the influence of the pumped well. After a time the effects of the recharge boundary will cause the time rate of drawdown to decrease and eventually reach equilibrium conditions. This occurs when recharge equals the pumping rate, as illustrated in Figure 4.7.4. The total drawdown for equilibrium conditions can be expressed as

$$s_r = s_p - s_i \tag{4.7.4}$$

in which s_r is the drawdown in an observation well near a recharge boundary, s_p is the drawdown due to the pumped well, and s_i is the buildup due to the image well (recharge boundary). The drawdown equation can be written as

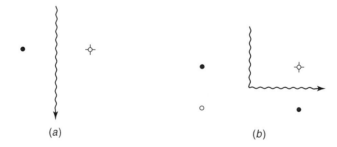

(a)   (b)

+ Real discharging well
o Image discharging well
• Image recharging well

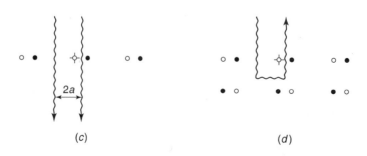

(c)   (d)

Figure 4.7.3. Image well systems for aquifers bounded by streams of various geometries. (a) Unidirectional stream (b) Rectangular stream (c) Two parallel streams (d) U-shaped stream Theoretically, image wells in (c) and (d) extend left and right to infinity.

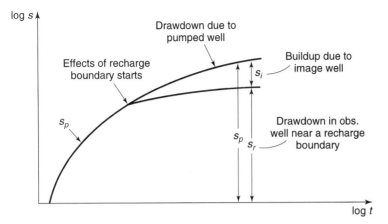

Figure 4.7.4. Recharge boundary effects on time–drawdown curve.

$$s_r = \frac{Q}{4\pi T}\left[W(u_p) - W(u_i)\right]$$
(4.7.5)

where Q is the constant pumping rate [L³/T], T is the transmissivity [L²/T], $W(u_p)$ and $W(u_i)$ are dimensionless, u_p and u_i are

$$u_p = \frac{r_p^2 S}{4Tt_p} \quad \text{and} \quad u_i = \frac{r_i^2 S}{4Tt_i}$$

in which r_i and r_p are in [L] and t is time in [T].

The drawdown in U. S. customary units (the gallon-day-foot system) can be expressed as

$$s_r = \frac{114.6Q}{T}\left[W(u_p) - W(u_i)\right]$$
(4.7.6)

where $u_p = \dfrac{1.87 r_p^2 S}{Tt_p}$ and $u_i = \dfrac{1.87 r_i^2 S}{Tt_i}$.

For large values of time, t, the well functions can be expressed as

$$W(u_p) = -0.5772 - \ln u_p$$
(4.7.7)

and
$$W(u_i) = -0.5772 - \ln u_i$$
(4.7.8)

This allows Equation 4.7.5 to be simplified to

$$s_r = \frac{Q}{4\pi T}\left[-\ln u_p + \ln u_i\right]$$
(4.7.9)

and Equation 4.7.6 is simplified to

$$s_r = \frac{114.6Q}{T}\left[-\ln u_p + \ln u_i\right]$$
(4.7.10)

Now using the gallon-day-foot system with time in minutes

$$u_p = \frac{2693 r_p^2 S}{Tt}$$
(4.7.11)

and

$$u_i = \frac{2693 r_i^2 S}{Tt}$$
(4.7.12)

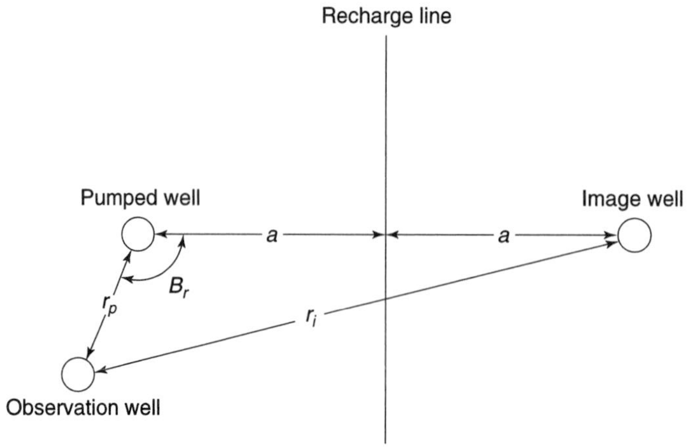

Figure 4.7.5. Definition of terms for Equation 4.7.15.

The drawdown from Equation 4.7.10 is expressed as

$$s_r = \frac{114.6Q}{T}\left[-\ln\left(\frac{2693r_p^2 S}{Tt} \right) + \ln\left(\frac{2693r_i^2 S}{Tt} \right) \right]$$ (4.7.13)

which simplifies to

$$s_r = \frac{528}{T} Q \log\left(\frac{r_i}{r_p} \right)$$ (4.7.14)

Rorabaugh[50] expressed this equation in terms of the distances between the pumped well and the line of recharges as

$$s_r = \frac{528Q\log\sqrt{\left(4a^2 + r_p^2 - 4ar_p \cos B_r \right)/r_p}}{T}$$ (4.7.15)

where a is the distance from the pumped well to the recharge boundary in ft, and B_r is the angle between a line connecting the pumped and image wells and a line connecting the pumped and observation wells. Refer to Figure 4.7.5 for an explanation of terms.

EXAMPLE 4.7.1

A 0.5-m diameter well (200 m from a river) is pumping at an unknown rate from a confined aquifer (see Figure 4.7.6). The aquifer properties are $T = 432$ m²/day and $S = 4.0 \times 10^{-4}$. After eight hours of pumping, the drawdown in the observation well (60 m from the river) is 0.8 m. Compute the rate of pumping and the drawdown in the pumped well. What is the effect of the river on drawdown in the observation well and in the pumped well?

SOLUTION

The following information is given in the above statement: $r_w = 0.25$ m, $T = 432$ m²/day $= 5.0 \times 10^{-3}$ m²/s, $S = 4 \times 10^{-5}$, $t = 8$ hr $= 28{,}800$ s, and $s = 0.8$ m. A recharging image well is placed at the same distance from the river as the pumped well as shown in Figure 4.7.6b.

Equation 4.7.5 is used to compute the discharge from the pumped well knowing the above information:

$$s = \frac{Q}{4\pi T} W(u_p) - \frac{Q}{4\pi T} W(u_i)$$

$$u_p = \frac{r_p^2 S}{4Tt} = \frac{(140)^2 \left(4 \times 10^{-4}\right)}{4\left(5 \times 10^{-3}\right)(28800)} = 1.36 \times 10^{-2}$$

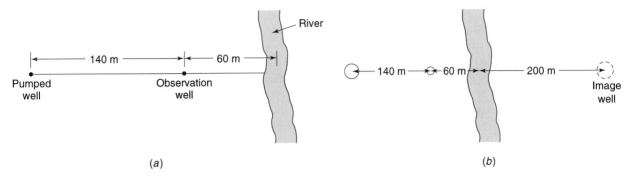

Figure 4.7.6. Example 4.7.1 system. (*a*) Well locations (*b*) Image well location

$$u_i = \frac{r_i^2 S}{4Tt} = \frac{(260)^2(4\times10^{-4})}{4(5\times10^{-3})(28800)} = 4.69\times10^{-2}$$

$$W(u_p) = 3.79 \text{ for } u_p = 1.36\times10^{-2} \text{ and } W(u_i) = 2.54 \text{ for } u_i = 4.69\times10^{-2}$$

Thus the discharge is computed using

$$0.8 = \frac{Q}{4\pi(5\times10^{-3})}(3.79) - \frac{Q}{4\pi(5\times10^{-3})}(2.54)$$

so that $Q = 0.04$ m³/s.
   The drawdown in the pumped well is computed using equation 4.7.5:

$$u_w = \frac{r_w^2 S}{4Tt} = \frac{(0.25)^2(4\times10^{-4})}{4(5\times10^{-3})(28800)} = 4.34\times10^{-8}$$

$$u_i = \frac{(400)^2(4\times10^{-4})}{4(5\times10^{-3})(28800)} = 0.111$$

$$W(u_w) = 16.38 \text{ for } u_w = 4.39\times10^{-8} \text{ and } W(u_i) = 1.75 \text{ for } u_i = 0.111$$

Thus the drawdown is

$$s_w = \frac{0.04}{4\pi(5\times10^{-3})}(16.38) - \frac{0.04}{4\pi(5\times10^{-3})}(1.75) = 9.31 \text{ m}$$

The effect of the river on the wells is to decrease the drawdown, so the reduced drawdown in the observation well is

$$s_{river} = -\frac{Q}{4\pi T}W(u_i) = -\frac{0.04}{4\pi(5\times10^{-3})}(2.54) = -1$$

Similarly, in the pumped well, the reduced drawdown is

$$s_{river} = -\frac{0.04}{4\pi(5\times10^{-3})}(1.75) = -1.11 \text{ m}$$

■

4.7.2 Well Flow Near an Impermeable Boundary

In the case of an impermeable or *barrier boundary,* water cannot flow across the boundary; no water is being contributed to the pumped well from the impervious formation. The cone of depression that would exist for a pumped well in an aquifer of infinite areal extent is shown in Figure 4.7.7*b*. Because of a barrier boundary, the cone of depression shown is no longer valid since there can be no flow across the boundary. Placing an image well, discharging in nature across the barrier boundary, creates the effect of no flow across the boundary. The image well must be placed perpendicular to the barrier boundary and the same distance from the boundary as the real well. The resulting real cone of depression is the summation of the components of both the real and image well depression cones as shown in the Figure 4.7.7*b*. Water levels in wells will decline at an initial rate due only to the influence of the pumped well. As pumping continues the barrier boundary effects will begin as simulated by the image well affecting the real well. When the effects of the barrier boundary are realized, the time rate of drawdown will increase (Figure 4.7.8). When this occurs, the total rate of withdrawal from the aquifer is equal to that of the pumped well plus that of the discharging image well causing the cone of depression of the real well to be deflected downward.

The total drawdown in the real well can be expressed as

$$s_b = s_p + s_i \qquad (4.7.16)$$

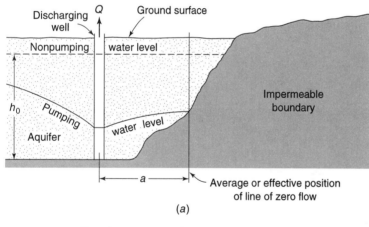

(a)

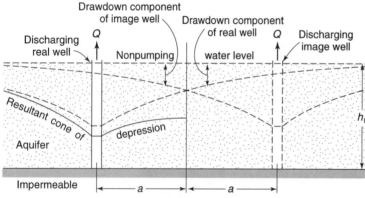

Aquifer thickness h_0 should be very large compared to resultant drawdown near real well

(b)

Figure 4.7.7. Sectional views. (*a*) Discharging well near an impermeable boundary (*b*) Equivalent hydraulic system in an aquifer of infinite areal extent (after Ferris et al.[15])

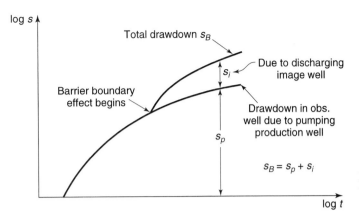

Figure 4.7.8. Barrier boundary effects on time–drawdown curve.

in which s_b is the total drawdown, s_p is the drawdown in an observation well due to pumping of the production well, and s_i is the drawdown due to the discharging image well (barrier boundary). The total drawdown can be expressed as

$$s_b = \frac{Q}{4\pi T} W(u_p) + \frac{Q}{4\pi T} W(u_i) \qquad (4.7.17)$$

where Q is the constant pumping rate [L³/T], T is the transmissivity [L²/T], $W(u_p)$ and $W(u_i)$ are dimensionless, and u_p and u_i are

$$u_p = \frac{r_p^2 S}{4Tt_p} \quad \text{and} \quad u_i = \frac{r_i^2 S}{4Tt_i}$$

in which r_i and r_p are in [L] and t is time in [T].

Drawdown Equation 4.7.16 can also be expressed in U.S. customary units (gal-day-foot) system where S is in ft, Q is in gpm, T is in gpd/ft, r is in ft, and t is in days:

$$\begin{aligned} s_b &= \frac{114.6Q}{T} W(u_p) + \frac{114.6Q}{T} W(u_i) \\ &= \frac{114.6Q}{T} \left[W(u_p) + W(u_i) \right] \end{aligned} \qquad (4.7.18)$$

where $\quad u_p = \dfrac{1.87 r_p^2 S}{Tt_p} \quad$ and $\quad u_i = \dfrac{1.87 r_i^2 S}{Tt_i}$

Now suppose that we choose drawdowns at times t_p and t_i such that $s_p = s_i$, then $W(u_p) = W(u_i)$ and $u_p = u_i$. Then

$$\frac{r_i^2 S}{Tt_i} = \frac{r_p^2 S}{Tt_p}$$

which reduces to

$$\frac{r_i^2}{t_i} = \frac{r_p^2}{t_p} \qquad (4.7.19)$$

Equation 4.7.19 defines the *law of times* which states that for a given aquifer, the times of occurrence of equal drawdown vary directly as the squares of distances from an observation well to a production well and an image well of equal discharge. The law of times can be used to determine the distance from an image well to an observation well, using

$$r_i = r_p \sqrt{\frac{t_i}{t_p}}$$

(4.7.20)

in which r_i is the distance from the image well to the observation well in ft, r_p is the distance from the pumped well to the observation well in ft, t_p is the time after pumping started and before the barrier boundary is effective, and t_i is the time (after pumping started and after the barrier boundary becomes effective) where $s_p = s_i$.

EXAMPLE 4.7.2

A well is pumping near a barrier boundary (see Figure 4.7.9) at a rate of 0.03 m³/s from a confined aquifer 20 m thick. The hydraulic conductivity of the aquifer is 27.65 m/day and its storativity is 3×10^{-5}. Determine the drawdown in the observation well after 10 hours of continuous pumping. What is the fraction of the drawdown attributable to the barrier boundary?

SOLUTION

The following information is given in the above problem statement: $Q = 0.03$ m³/s, $b = 20$ m, $K = 27.65$ m/day $= 3.2 \times 10^{-4}$ m/s, $S = 3 \times 10^{-5}$, $t = 10$ hrs $= 36,000$ s. An image well is placed across the boundary at the same distance from the boundary as the pumped well (as shown in Figure 4.7.9b). The drawdown in the observation well is due to the real well and the imaginary well (which accounts for the barrier boundary). Hence, using Equation 4.7.17

$$s = \frac{Q}{4\pi T} W(u_p) + \frac{Q}{4\pi T} W(u_i)$$

$$u_p = \frac{r_p^2 S}{4Tt} = \frac{(240)^2 (3 \times 10^{-5})}{4(20)(3.2 \times 10^{-4})(36,000)} = 1.88 \times 10^{-3}$$

Next compute the distance from the observation well to the image well: $r_i^2 = 600^2 + 240^2 - 2(600)(300) \cos 30° = 16{,}8185$ m² so $r_i = 410$ m. Using r_i, compute

$$u_i = \frac{168185(3 \times 10^{-5})}{4(20)(3.2 \times 10^{-4})(36,000)} = 5.47 \times 10^{-3}$$

The well functions are now computed or obtained from Table 4.4.1 as $W(u_p) = 5.72$ for $u_p = 1.88 \times 10^{-3}$ and $W(u_i) = 4.64$ for $u_i = 5.47 \times 10^{-3}$.

The drawdown at the observation well is computed as

$$s = \frac{0.03}{4\pi(20)(3.2 \times 10^{-4})} = (5.72 + 4.64) = 3.86 \text{ m.}$$

The drawdown attributable to the barrier boundary is computed as

$$s_i = \frac{Q}{4\pi T} W(u_i) = \frac{0.03}{4\pi(20)(3.2 \times 10^{-4})} (4.64) = 1.73 \text{ m}$$

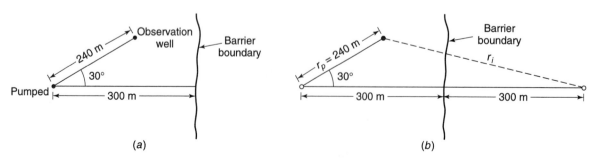

Figure 4.7.9. Example 4.7.2 system. (*a*) Well locations (*b*) Image well location

and the fraction of drawdown attributable to the impermeable boundary is

$$\frac{s_i}{s} = \frac{1.73}{3.86} = 0.45\,(45\%).$$

■

4.7.3 Well Flow Near Other Boundaries

In addition to the previous two examples, the method of images can be applied to a large number of groundwater boundary problems. As before, actual boundaries are replaced by an equivalent hydraulic system, which includes imaginary wells and permits solutions to be obtained from equations applicable only to extensive aquifers. Two boundary conditions to suggest the adaptability of the method are shown in Figure 4.7.10. Figure 4.7.10a shows a discharging well in an aquifer bounded on two sides by impermeable barriers. The image discharge wells I_1 and I_2 provide the required flow but, in addition, a third image well I_3 is necessary to balance drawdowns along the extensions of the boundaries. The resulting system of four discharging wells in an extensive aquifer represents hydraulically the flow system for the physical boundary conditions. Finally, Figure 4.7.10b presents the situation of a well near an impermeable boundary and a perennial stream. The image wells required follow from the previous illustrations.

For a wedge-shaped aquifer, such as a valley bounded by two converging impermeable barriers, the drawdown at any location within the aquifer can be calculated by the same method of images.[54] Consider the aquifer formed by two barriers intersecting at an angle of 45 degrees shown in Figure 4.7.11. Seven image pumping wells plus the single real pumping well form a circle with its center at the wedge apex; the radius equals the distance from the apex to the real pumping well.[15] The drawdown at any point between the two barriers can then be calculated by summing the individual drawdowns. In general, it can be shown that the number of image wells n required for a wedge angle θ is given by

$$n = \frac{360°}{\theta} - 1 \tag{4.7.21}$$

where θ is an aliquot part of 360 degrees.

If the arrangement of two boundaries is such that they are parallel to each other, analysis by the image-well theory requires the use of an image-well system extending to infinity.[15] Each successively added secondary image well produces a residual effect at the opposite boundary.

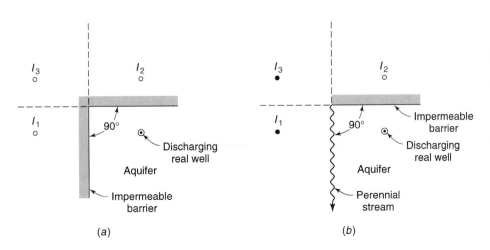

(a)

(b)

Figure 4.7.10. Image well systems for a discharging well near aquifer boundaries. (*a*) Aquifer bounded by two impermeable barriers intersecting at right angles (*b*) Aquifer bounded by an impermeable barrier intersected at right angles by a perennial stream. Open circles are discharging image wells; filled circles are recharging image wells (after Ferris et al.[15]).

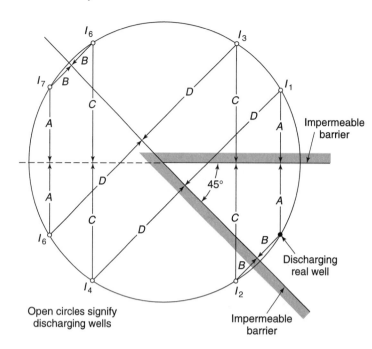

Figure 4.7.11. Image well system for a discharging well in an aquifer bounded by two impermeable barriers intersecting at an angle of 45 degrees (after Ferris et al.[15]).

Open circles signify discharging wells

It is only necessary to add pairs of image wells until the next pair has negligible influence on the sum of all image-well effects out to the point. The use of spreadsheets makes the computation much easier. Image-well systems for several parallel-boundary situations are shown in Figures 4.7.12 and 4.7.13.

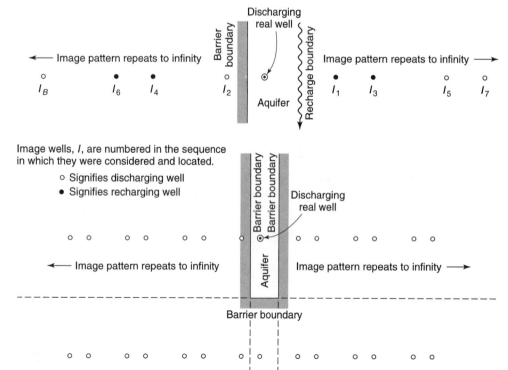

Image wells, *I*, are numbered in the sequence in which they were considered and located.

  o Signifies discharging well
  • Signifies recharging well

Figure 4.7.12. Plans of image-well systems for selected parallel-boundary situations (after Ferris et al.[15]).

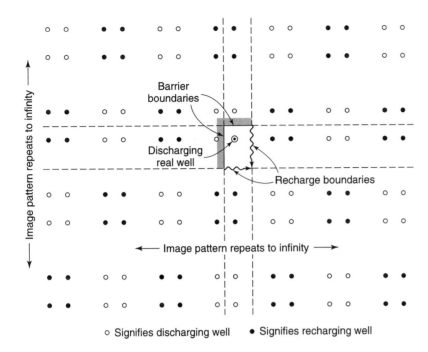

Image pattern repeats to infinity

Barrier boundaries

Discharging real well

Recharge boundaries

← Image pattern repeats to infinity →

○ Signifies discharging well   ● Signifies recharging well

Figure 4.7.13. Plans of image-well systems for a rectangular aquifer (after Ferris et al.[15]).

The above equations involve T for confined aquifers. To adapt them to unconfined aquifers, s should be replaced by $s'' = s - s^2/2h_0$ where h_0 is the initial saturated aquifer thickness. Storage coefficients cannot be calculated from steady-state boundary equations.

Procedures for analyzing unsteady well flows near aquifer boundaries, involving graphic solutions, are available.[15, 21]

4.7.4   Location of Aquifer Boundary

Permeable aquifer boundaries such as streams would normally be visible near a pumping well; however, impermeable subsurface boundaries such as faults or dikes may not be apparent. Where this situation is encountered, the location and orientation of such a barrier can be defined by careful analysis of pumping test data.[15, 61] In the Cooper–Jacob method (see Equation 4.4.8) the slope of the straight line on semilogarithmic paper depends only on the pumping rate and the transmissivity. If an impermeable boundary is present, the rate of drawdown in an observation well will double under the influence of an image pumping well (see Figure 4.7.14a).[*] To determine the location of the image well, straight lines are fitted through the two legs of the data. An arbitrary drawdown s_A is selected and a time t_r for this to occur under the influence of the real well is measured (see Figure 4.7.14a). Similarly, a time t_i for the same drawdown to be produced by the image well is defined. Then, knowing the distance r_r between the real well and the observation well, the distance r_i to the image well (see Figure 4.7.14b) can be found from the law of times (Equation 4.7.20). The distance r_i defines only the radius of a circle on which the image well lies. It requires measurements in two more observation wells in order to define uniquely by intersection of three arcs the location of the image well (see Figure 4.7.14b). The boundary then lies at the midpoint of and perpendicular to a line connecting the real and image wells.

---

[*]It should be noted that if the boundary is a stream recharging the aquifer, an image recharge well is introduced. This produces a slope of equal but opposite sign on the drawdown curve, resulting in a horizontal asymptote.

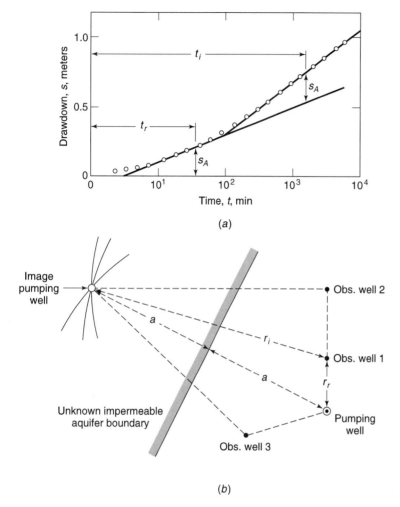

(a)

(b)

Figure 4.7.14. Diagrams illustrating procedure to locate an unknown impermeable aquifer boundary near a pumping well. (*a*) Cooper–Jacob drawdown curve showing effect of an impermeable boundary (*b*) Field situation required to locate an unknown impermeable aquifer boundary

4.8   MULTIPLE WELL SYSTEMS

Where the cones of depression of two nearby pumping wells overlap, one well is said to *interfere* with another because of the increased drawdown and pumping lift created. For a group of wells forming a well field, the drawdown can be determined at any point if the well discharges are known, or vice versa. From the principle of superposition, the drawdown at any point in the area of influence caused by the discharge of several wells is equal to the sum of the drawdowns caused by each well individually. Thus,

$$s_T = s_1 + s_2 + s_3 + \ldots + s_n \qquad (4.8.1)$$

where s_T is the total drawdown at a given point and s_1, s_2, s_3, ..., s_n are the drawdowns at the point caused by the discharge of wells 1, 2, 3, ..., n, respectively. The summation of drawdowns may be illustrated in a simple way by the well line of Figure 4.8.1; the individual and composite drawdown curves are given for $Q_1 = Q_2 = Q_3$. Clearly, the number of wells and the geometry of the well field are important in determining drawdowns. Solutions can be based on the equilibrium or nonequilibrium equation. Equations of well discharge for particular well patterns have been developed.[41, 49]

In general, wells in a well field designed for water supply should be spaced as far apart as possible so their areas of influence will produce a minimum of interference with each other. On

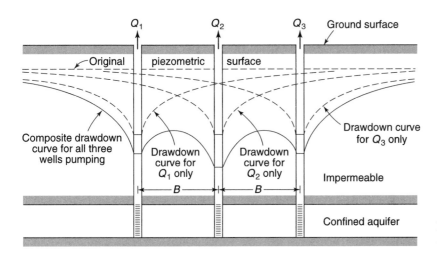

Figure 4.8.1. Individual and composite drawdown curves for three wells in a line.

the other hand, economic factors such as cost of land or pipelines may lead to a least-cost well layout that includes some interference.[21] For drainage wells designed to control water table elevations, it may be desirable to space wells so that interference increases the drainage effect.

EXAMPLE 4.8.1

Three pumping wells located along a straight line are spaced at 200 m apart. What should be the steady-state pumping rate from each well so that the near steady-state drawdown in each well will not exceed 2 m? The transmissivity of the confined aquifer, which all the wells fully penetrate, is 2400 m^2/day and all the wells are 40 cm in diameter. The thickness of the aquifer is 40 m and the radius of influence of each well is 800 m.

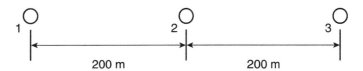

SOLUTION

The following information is given in the above problem statement: $s_1 \leq 2$ m, $s_2 \leq 2$ m, and $s_3 \leq 2$ m, $T = 2{,}400$ m^2/day $= 27.8 \times 10^{-3}$ m^2/s, $r_w = 0.2$ m, $b = 40$ m, $r_0 = 800$ m, and $r = 200$ m. Let Q be the pumping rate from each well and h_0 be the head before pumping started. For well 1, $s_1 = s_{11} + s_{12} + s_{13}$ where s_{ij} is the drawdown in well i due to pumping in well j. Thus, for the other wells, $s_2 = s_{21} + s_{22} + s_{23}$, and $s_3 = s_{31} + s_{32} + s_{33}$. By symmetry, $s_1 = s_3$. The drawdowns in well 1 due to pumping in wells 1, 2, and 3 are respectively

$$s_{11} = \frac{Q \ln\left(\dfrac{r_0}{r_w}\right)}{2\pi T} = \frac{Q \ln\left(\dfrac{800}{0.2}\right)}{2\pi\left(27.8 \times 10^{-3}\right)} = 47.48Q$$

$$s_{12} = \frac{Q \ln\left(\dfrac{r_0}{r_{12}}\right)}{2\pi T} = \frac{Q \ln\left(\dfrac{800}{200}\right)}{2\pi\left(27.8 \times 10^{-3}\right)} = 7.94Q$$

$$s_{13} = \frac{Q \ln\left(\dfrac{r_0}{r_{13}}\right)}{2\pi T} = \frac{Q \ln\left(\dfrac{800}{400}\right)}{2\pi\left(27.8 \times 10^{-3}\right)} = 3.97Q$$

The drawdowns in wells 1 and 3 are identical so total drawdown in the wells is $s_1 = s_3 = 47.48Q + 7.94Q + 3.97Q = 59.39Q$. The drawdowns in well 2 due to pumping in wells 1, 2, and 3 are respectively

$$s_{21} = \frac{Q \ln\left(\dfrac{r_0}{r_{12}}\right)}{2\pi T} = \frac{Q \ln\left(\dfrac{800}{200}\right)}{2\pi\left(27.8 \times 10^{-3}\right)} = 7.94Q$$

$$s_{22} = s_{11} = 47.48Q$$

$$s_{23} = s_{21} = 7.94Q$$

The total drawdown in well 2 is $s_2 = 7.94Q + 47.48Q + 7.94Q = 63.36Q$. The relationships for $s_1 = 59.39Q$ and $s_2 = 63.36Q$ show that for the same discharge from all the wells, more drawdown results at the middle well; therefore, the drawdown in this well governs. So using $s_2 \le 2$ or $63.36Q \le 2$, then the steady-state pumping rate from each well should be $Q \le 3.16 \times 10^{-2}$ m^3/s = 113 m^3/hr. ∎

EXAMPLE 4.8.2

It is required to dewater a construction site 80 m by 80 m. The bottom of the construction will be 1.5 m below the initial water surface elevation of 90 m. Four pumps are to be used in 0.5-m diameter wells at the four corners of the site. The transmissivity and the storage coefficient of the aquifer are 1,600 m^2/day and 0.16, respectively. The site needs to be ready after one month of pumping. Determine the required pumping rate.

SOLUTION

To solve this problem, the least drawdown at the site should be greater than 1.5 m. It can be shown that the potential points of interest that may have the least drawdown are the center of the square (point a) and the midpoint on each side of the square (points b). Approximation is made using the Cooper-Jacob method.

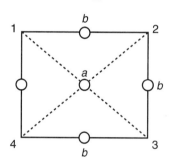

At point a (the center of the square), $r = \sqrt{40^2 + 40^2} = 56.6$ m, and

$$u = \frac{r^2 S}{4Tt} = \frac{(56.6 \text{ m})^2 (0.16)}{4(1600 \text{ m}^2/\text{day})(30 \text{ days})} = 0.00267$$

Since $u < 0.01$, we can use the approximate solution by Cooper-Jacob expressed by Equation 4.4.7:

$$s_a = \frac{Q}{4\pi T}(-0.5772 - \ln(u)) = \frac{Q}{4\pi(1600 \text{ m}^2/\text{day})}(-0.5772 - \ln(0.00267)) = 0.0002661Q$$

Using the principle of superposition and by symmetry, the drawdown caused by the four wells is $s_T = 4 \times s_a = 4 \times 0.0002661Q = 0.0010643Q$ and $s_T = 0.0010643Q = 1.5$ m $\rightarrow Q = 1409$ m^3/day.

At any of the four points represented by b, $r_1 = 40$ m for two of the wells and $r_2 = \sqrt{80^2 + 40^2} = 89.44$ m for the remaining two wells. Then

$$u_1 = \frac{r^2 S}{4Tt} = \frac{(40 \text{ m})^2 (0.16)}{4(1600 \text{ m}^2/\text{day})(30 \text{ days})} = 0.0013333$$

$$u_2 = \frac{r^2 S}{4Tt} = \frac{(89.44 \text{ m})^2 (0.16)}{4(1600 \text{ m}^2/\text{day})(30 \text{ days})} = 0.006666$$

Since $u_1 < 0.01$ and $u_2 < 0.01$, the Cooper–Jacob method of solution can be used again:

$$s_b = 2\left[\frac{Q}{4\pi T}\left(-0.5772 - \ln(u_1)\right)\right] + 2\left[\frac{Q}{4\pi T}\left(-0.5772 - \ln(u_2)\right)\right]$$

$$= 2\left[\frac{Q}{4\pi\left(1600\ \text{m}^2/\text{day}\right)}\left(-0.5772 - \ln(0.0013333)\right)\right] +$$

$$2\left[\frac{Q}{4\pi\left(1600\ \text{m}^2/\text{day}\right)}\left(-0.5772 - \ln(0.006666)\right)\right]$$

$$= 2 \times 0.0003Q + 2 \times 0.0002205Q$$

$$= 1.041 \times 10^{-3}Q = 1.5\ \text{m} \rightarrow Q = 1441\ \text{m}^3/\text{day}$$

Thus the points represented by b are critical and a discharge of 1,441 m³/day from each well is required. ∎

4.9 PARTIALLY PENETRATING WELLS

A well whose length of water entry is less than the aquifer it penetrates is known as a *partially penetrating well*. Figure 4.9.1 illustrates the situation of a partially penetrating well in a confined aquifer. The flow pattern to such wells differs from the radial horizontal flow assumed to exist around fully penetrating wells. The average length of a flow line into a partially penetrating well exceeds that into a fully penetrating well so a greater resistance to flow is thus encountered. For practical purposes this results in the following relationships between two similar wells, one partially and one fully penetrating the same aquifer: if $Q_p = Q$, then $s_p > s$; and if $s_p = s$, then $Q_p < Q$. Here Q is well discharge, s is drawdown at the well, and the subscript p refers to the partially penetrating well. The effect of partial penetration is negligible on the flow pattern and the drawdown beyond a radial distance larger than 0.5 to 2 times the saturated thickness b, depending on the amount of penetration.

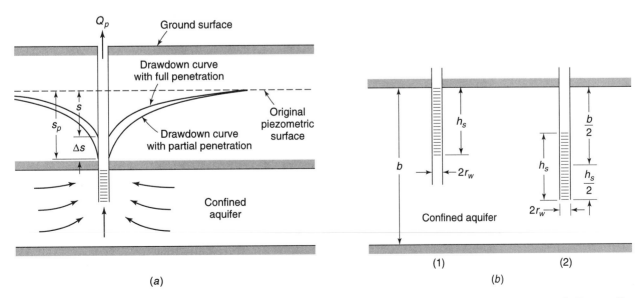

Figure 4.9.1. Partially penetrating wells in a confined aquifer. (*a*) Effect of partially penetrating well on drawdown (*b*) Two configurations of partially penetrating wells

The drawdown s_p at the well face of a partially penetrating well in a confined aquifer (see Figure 4.9.1a) can be expressed as

$$s_p = s + \Delta s \tag{4.9.1}$$

where Δs refers to the additional drawdown resulting from the effect of partial penetration. It can be shown for steady-state conditions and the typical situation[*] in Figure 4.9.1b(1),[28]

$$\Delta s = \frac{Q_p}{2\pi T} \frac{1-p}{p} \ln \frac{(1-p)h_s}{r_w} \tag{4.9.2}$$

where T is transmissivity, p is the penetration fraction ($p = h_s/b$), and h_s and r_w are shown in Figure 4.9.1b(1). Equation 4.9.2 applies where $p > 0.20$.

For the case of a well screen centered in the thickness of the aquifer [see Figure 4.9.1b(2)], the value of Δs is given by

$$\Delta s = \frac{Q_p}{2\pi T} \frac{1-p}{p} \ln \frac{(1-p)h_s}{2r_w} \tag{4.9.3}$$

Equation 4.9.2 can be modified for a well in an unconfined aquifer by defining

$$\Delta s\, 2h_w = \frac{Q_p}{\pi K} \frac{1-p}{p} \ln \frac{(1-p)h_s}{r_w} \tag{4.9.4}$$

where h_w is the saturated thickness at the well with full penetration and the hydraulic conductivity $K = T/h_w$. Then

$$s_p^2 = s^2 + \Delta s\, 2h_w \tag{4.9.5}$$

and similarly for Equation 4.9.3.

Detailed methods for analyzing effects of partial penetration on well flow for steady and unsteady conditions in confined, unconfined, leaky, and anisotropic aquifers have been outlined by Hantush[23, 27] and others.[32, 34, 57] Although evaluating the effects is complicated except for the simplest cases, common field situations often reduce the practical importance of partial penetration.[†] One occurs where an observation well is located more than 1.5 to 2 times the saturated aquifer thickness from a pumping well; in this situation the effect of partial penetration can be neglected for homogeneous and isotropic aquifers. Another applies to many alluvial aquifers with pronounced anisotropy. Here the vertical flow components become small, thereby enabling a pumping well to be approximated as a fully penetrating well in a confined or leaky aquifer with a saturated thickness equal to the length of the well screen.

EXAMPLE 4.9.1

Compare the four cases that are depicted in Figure 4.9.2 where a 1-m diameter well fully/partially penetrates vertically through a confined aquifer whose thickness is 28 m. Comment on their relative efficiencies by evaluating the specific capacity (Q/s) for each case. Take the radius of influence for all cases as 150 m. The pumping rates and the aquifer transmissivity are the same for all cases.

SOLUTION

Case (a): The drawdown in this case where the pumping well fully penetrates is computed by rearranging the Thiem equation (4.2.4):

$$s = \frac{Q}{2\pi T}\left(\ln\left(\frac{R}{r_w}\right)\right) = \frac{Q}{2\pi T}\left(\ln\left(\frac{150}{0.5}\right)\right) = 0.9078\left(\frac{Q}{T}\right)$$

---

[*]The drawdown increment is the same whether partial penetration starts from the top or from the bottom of the aquifer.

[†] It should be noted that any well with 85 percent or more open or screened hole in the saturated thickness may be considered as fully penetrating.

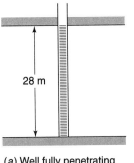 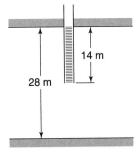 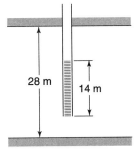 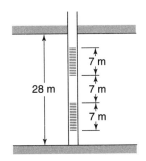

(a) Well fully penetrating

(b) Well penetrating the top 14 m of the confined aquifer

(c) Well penetrating 14 m of the confined aquifer with the well screen centered

(d) Well penetrating 14 m of the confined aquifer with two 7 m screened intervals

Figure 4.9.2. Example 4.9.1 cases

Case (b): The additional drawdown is computed by rearranging Equation 4.9.2:

$$\Delta s = \left[\frac{Q_p}{2\pi T}\right]\left[\frac{1-p}{p}\right]\left[\ln\frac{(1-p)h_s}{r_w}\right] = \left[\frac{Q}{2\pi T}\right]\left[\frac{1-0.5}{0.5}\right]\left[\ln\frac{(1-0.5)(14)}{(0.5)}\right]$$

$$= 0.4200\left(\frac{Q}{T}\right)$$

So the total drawdown becomes s_b $(0.9078 + 0.4200)(Q/T) = 1.3278\ (Q/T)$.

Case (c): The additional drawdown in the case of a well screen centered in the thickness of the aquifer is given by Equation 4.9.3:

$$\Delta s = \left[\frac{Q_p}{2\pi T}\right]\left[\frac{1-p}{p}\right]\left[\ln\frac{(1-p)h_s}{2r_w}\right] = \left[\frac{Q}{2\pi T}\right]\left[\frac{1-0.5}{0.5}\right]\left[\ln\frac{(1-0.5)(14)}{2(0.5)}\right]$$

$$= 0.3097\left(\frac{Q}{T}\right)$$

The total drawdown for this case becomes s_c $(0.9078 + 0.3097)\ (Q/T) = 1.2175\ (Q/T)$

Case (d): This case is equivalent to a screen length of 7 m centered in a 14-m thick aquifer. Again using Equation 4.9.3,

$$\Delta s = \left[\frac{Q_p}{2\pi T}\right]\left[\frac{1-p}{p}\right]\left[\ln\frac{(1-p)h_s}{2r_w}\right] = \left[\frac{Q}{2\pi T}\right]\left[\frac{1-0.5}{0.5}\right]\left[\ln\frac{(1-0.5)(7)}{2(0.5)}\right]$$

$$= 0.1994\left(\frac{Q}{T}\right)$$

The total drawdown for this case becomes s_d $(0.9078 + 0.1994)\ (Q/T) = 1.1072\ (Q/T)$.
Calculate the specific capacity (Q/s) for each case:

$$Q/s_a = 1.1016(T)$$

$$Q/s_b = 0.7531(T)$$

$$Q/s_c = 0.8214(T)$$

$$Q/s_d = 0.9032(T)$$

Or taking case (a), fully penetrating well, as the maximum potential specific capacity, cases (b), (c), and (d) yield 68 percent, 75 percent, and 82 percent of the maximum capacity, respectively. So among the three cases, case (d) is the most effective one. Also, centering the well screen yields higher capacity compared to the case where the same partial penetration starts from the top or from the bottom of the aquifer. ∎

4.10   WELL FLOW FOR SPECIAL CONDITIONS

A variety of solutions to well flow problems have been derived for special aquifer, pumping, and well conditions.[23,38] Inasmuch as these are of less general application than those outlined heretofore and involve more extensive mathematical treatment, they will be omitted here. It is worth noting, however, that solutions have been obtained for the following special conditions:

1. Constant well drawdown[2, 21]
2. Varying, cyclic, and intermittent well discharges[20, 36, 39, 52, 55, 56]
3. Sloping aquifers[21]
4. Aquifers of variable thickness[21]
5. Two-layered aquifers[24, 44]
6. Anisotropic aquifers[10, 23, 27, 43, 66]
7. Aquifer conditions varying with depth[40, 51]
8. Large-diameter wells[46, 67]
9. Collector wells (see Chapter 5)[28, 45]
10. Wells with multiple-sectioned well screens[57]

PROBLEMS

4.1.1 A confined aquifer is 18.5 m thick. The potentiometric surface elevations at two observation wells 822 m apart are 25.96 and 24.62 m. If the horizontal hydraulic conductivity of the aquifer is 25 m/day, determine the flow rate per unit width of the aquifer, specific discharge, and average linear velocity of the flow assuming steady unidirectional flow.

4.1.2 Two confined aquifers are separated by an aquitard as shown below. The piezometric head difference between the upper and lower aquifer measured along a vertical line is 6.5 m. If the vertical hydraulic conductivity of confining unit is 0.046 m/day, determine the direction and magnitude of leakage per km^2 between the upper and lower confined aquifers through the confining unit. Also, estimate the travel time of a water particle through the confining layer between the two aquifers. Estimated thickness of the separation is 4.15 m.

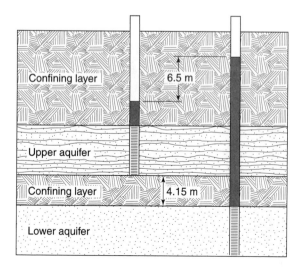

4.1.3 Two observation wells are to be constructed to determine the natural slope of the piezometric surface in a confined aquifer. Assuming a steady unidirectional flow, if the accuracies of piezometric head measurements and horizontal distance measurements are ±3 cm and ±5 cm, respectively, determine the percent error in estimating the slope of the piezometric surface as a function of the distance between the two wells. What would be the accuracy of a hydraulic gradient estimate if a piezometric head drop of 3 m is observed between two wells 100 m apart?

4.1.4 Rework Problem 4.1.1 if a ±10° error is also possible in estimating the groundwater flow direction.

4.1.5 Near steady-state conditions, explain how the hydraulic gradient changes in the flow direction in

(a) a confined aquifer;

(b) an unconfined aquifer.

4.1.6 Consider two strata of the same soil material that lie between two channels. The first stratum is confined and the second one is unconfined, and the water surface elevations in the channels are 24 and 16 m above the bottom of the unconfined aquifer. What should be the thickness of the confined aquifer for which

1) the discharge through both strata are equal,

2) the discharge through the confined aquifer is half of that through the unconfined aquifer?

4.1.7 An unconfined aquifer in a stratum of clean sand and gravel has a hydraulic conductivity of 10^{-2} cm/sec. From two observation wells 200 m apart, the observed water table elevations are 11 and 7 m measured from the bottom of the stratum. Determine the discharge per unit width of the aquifer.

4.1.8 An earthen dam is 200 m across (i.e., the distance from the upstream face to the downstream face) and underlain by impermeable bedrock. The average hydraulic conductivity of the mate-

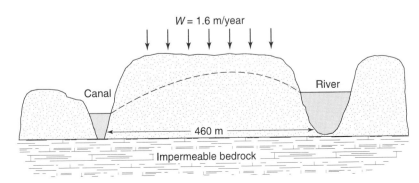

Figure to accompany Problem
4.1.9

rial of which the dam is composed is 0.065 m/day. If the water surface elevations in the reservoir and the tailwaters are 25 and 4.5 m, respectively, estimate the magnitude of leakage from the reservoir to the tailwaters per 100-m width of the dam.

4.1.9 Compute the volume of water that seeps from the channel into the river in the figure above. The water surface elevations in the channel and river with respect to the underlying bedrock are 13 and 10.5 m, respectively. The hydraulic conductivity of formation A is 5.6 m/day and that of formation B is 12.3 m/day.

4.1.10 A canal is constructed parallel to a river 460 m away both fully penetrating an unconfined aquifer of clean sand and gravel as shown in the illustration below. The aquifer has a hydraulic conductivity of $K = 18.5$ m/day and is subject to an average infiltration of 1.6 m/year. The water surface elevation in the canal is 8.5 m and in the river it is 10 m. If the mound between the canal and the river gets contaminated and the river is to remain free of contamination, (a) determine the daily discharge of groundwater into the canal and into the river per kilometer of both; (b) estimate the travel times from the water divide to the canal and to the river ($n_e = 0.35$); (c) assuming that the contaminant travels mainly by advection, propose any operational changes to the given layout to prevent the river from being contaminated.

4.2.1 A well that pumps at a constant rate of 0.5 m³/s fully penetrates a confined aquifer of 34-m thickness. After a long period of pumping, near steady-state conditions, the measured drawdowns at two observation wells 50 and 100 m from the pumping well are 0.9 and 0.4 m, respectively. (a) Calculate the hydraulic conductivity and transmissivity of the aquifer, (b) estimate the radius of influence of the pumping well, and (c) calculate the expected drawdown in the pumping well if the radius of the well is 0.4 m.

4.2.2 A confined aquifer of 10-m thickness and 16.43 m/day hydraulic conductivity is fully penetrated by a pumping well of 0.5 m radius operating at $Q = 425$ m³/day. Determine the drawdown under steady-state conditions in the pumping well and 50 m away from the well. Take the radius of influence of the pumping as 300 m.

4.2.3 What percent increase/decrease would occur in the drawdown of the pumping well if the radius of the well is doubled and the pumping rate is kept the same in Problem 4.2.2? Assume the same radius of influence.

4.2.4 What percent increase/decrease would occur in the well flow if the well diameter is doubled and the drawdown in the well is kept constant in Problem 4.2.2? Assume the same radius of influence.

4.2.5 The initial piezometric surface in a confined aquifer of 20-m thickness is 34 m above the bottom. After a long period of pumping, the piezometric surface stabilizes at 29.3 m above the bottom. The hydraulic conductivity of the aquifer is 12.2 m/day. If the radius of the well is 0.5 m and the radius of influence of the pumping is 500 m, what is the steady-state well discharge?

4.2.6 The initial piezometric head in a confined aquifer that has a thickness of 11.6 m is 85.7 m above sea level. A well with a radius of 0.5 m pumps at a constant rate of 1,240 m³/day. After the cone of depression has achieved equilibrium, the piezometric heads at two observation wells 40 and 95 m from the pumping well are measured as 78.9 and 83.4 m above sea level, respectively. Also, the piezometric head in the pumping well during equilibrium remains at 46.6 m. Determine (a) the aquifer transmissivity, (b) the radius of influence of the pumping, and (c) the total well losses in and around the pumping well.

Figure to accompany Problem
4.1.10

4.2.7 A pumping well of 0.75-m radius fully penetrates an unconfined aquifer of 24-m thickness and produces at a rate of 10 L/s. After a long period of pumping, the drawdown in an observation well 30 m from the pumping well is 1.6 m. The drawdown in another observation well 60 m from the pumping well is 1.1 m. Calculate (a) the hydraulic conductivity of the aquifer, (b) the expected drawdown in the pumping well, and (c) the radius of influence of the pumping well.

4.2.8 After a long period of pumping from an unconfined aquifer at a constant rate of 850 m³/day, the cone of depression reaches equilibrium. The aquifer has an initial saturated thickness of 20 m and a hydraulic conductivity of 8.65 m/day. During the equilibrium, the water levels in an observation well 50 m away and in the pumping well are measured as 18.4 and 9.9 m. Determine (a) the radius of influence of the pumping, (b) the radial distance where the steady state drawdown is 5 cm, (c) the expected drawdown in the pumping well ($r_w = 0.4$ m), and (d) the total well head losses.

4.2.9 Water is pumped at a constant rate of 500 m³/day from an unconfined aquifer whose thickness is 15 m and hydraulic conductivity is 5.5 m/day. Given that the steady-state drawdown 50 m from the pumping well is 2.5 m, plot the water table profile under steady-state conditions for $r > 1.5H$, where H is the initial saturated thickness of the aquifer.

4.2.10 Rework Problem 4.2.5 in the absence of the recharge. (*Hint:* You may have to guess the radius of influence for this problem.)

4.2.11 Rework Problem 4.2.5 with the recharge rate doubled (i.e., $W = 0.12$ m/day). Approximately what percentage of the extracted water comes from the aquifer itself?

4.2.12 A well with a radius of 0.5 m pumps at the rate of 15 L/s from an unconfined aquifer that is uniformly recharged at a rate of 0.6 m/day. Without pumping from the well, the water table is nearly horizontal and the aquifer thickness is 30 m. The hydraulic conductivity of the aquifer is 20 m/day. Determine the radius of influence of the well and the approximate drawdown at the well location near steady-state conditions.

4.3.1 A confined aquifer of 35-m thickness is contaminated as shown in the plan view below. The regional groundwater gradient is $i = 0.003$ and the confined aquifer has a hydraulic gradient of 20

m/day. A capture well is proposed to clean up the contamination. The coordinates of the limits of contamination with respect to the capture well are tabulated below. Determine the minimum required pumping rate for the cleanup and delineate the capture zone. Assume that the plume remains relatively still before or during the operation of the capture well. Note that the capture well is located at (0 m, 0 m).

x (m)	y (m)
−15	35
−29	9.5
−20	−28.8
5	−26
38	−29
62	−1
14.2	33.4
−15	35

4.4.1 A fully penetrating production well pumps from a confined aquifer at a constant rate of 64 L/s. If the coefficients of transmissivity and storage of the aquifer are 1,240 m²/day and 4×10^{-4}, respectively, estimate the drawdowns at a distance of 200 m from the pumping well for pumping periods of 8 hours, 30 days, and 6 months using the Theis equation. Also, estimate the radius of influence of the pumping well after 1 hour of pumping.

4.4.2 Rework Problem 4.4.1 using the Cooper–Jacob method of solution.

4.4.3 Given that $T = 125$ m²/day, $S = 10^{-4}$, $t = 2,693$ min, $Q = 5,500$ m³/day, and $r = 305$ m for a confined aquifer, compute the drawdown.

4.4.4 Using the Cooper–Jacob method of solution, plot drawdown versus time for four observation wells 5, 50, 100, and 500 m from a well pumping at a constant rate of 300 m³/day in a sandy confined aquifer with $b = 12$ m, $K = 25$ m/day, and $S = 10^{-4}$. Given that depths to water in wells and piezometers can be measured with an accuracy of about 0.5 cm, how long does it take to reach near-steady conditions? Also, calculate the volume of water accumulated above ground until the near-steady conditions are achieved at those distances.

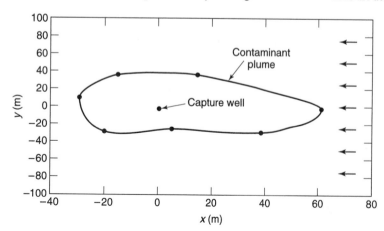

4.4.5 Drawdown was measured during a pumping test in a confined aquifer at frequent intervals in an observation well 200 ft from a well that was pumped at a constant rate of 500 gpm. The data for this pump test is listed below. Determine T and S for this aquifer.

Pump test data	
Time (min)	Drawdown (ft)
1	0.05
2	0.22
3	0.40
4	0.56
5	0.70
7	0.94
10	1.2
20	1.8
40	2.5
100	3.4
300	4.5
1,000	5.6
4,000	7

4.4.6 For the time–drawdown data listed below for a confined aquifer, calculate T and S using Jacob's approximation. After computing T and S, check to see that the basic assumption of this approximation is satisfied. For the values of T and S that you computed, after how many minutes of pumping would Jacob's approximation be valid? The discharge is $Q = 1,500$ gpm and the radius $r = 300$ ft.

Time after pumping started (min)	Drawdown (ft)
1	0.45
2	0.74
3	0.91
4	1.04
6	1.21
8	1.32
10	1.45
30	2.02
40	2.17
50	2.30
60	2.34
80	2.50
100	2.67
200	2.96
400	3.25
600	3.41
800	3.50
1,000	3.60
1,440	3.81

4.4.7 For the time–drawdown data recorded during the recovery phase of a pump test, determine the transmissivity and storativity of the confined aquifer. The pumping rate is 162.9 ft^3/min and the pump is shut off at 800 min.

Recovery of test well		Recovery of observation well	
Time (min)	s' (ft)	Time (min)	s' (ft)
800	12.5	800	1.86
803	20	805	1.78
808	5	810	1.64
813	0.5	815	1.53
820	1.5	820	1.45
880	1	825	1.37
940	0.80	830	1.32
995	0.69	840	1.22
1,055	0.59	850	1.15
1,115	0.51	860	1.09
1,175	0.49	870	1.03
1,235	0.46	880	0.97
1,295	0.38	890	0.94
1,360	0.34	900	0.90
1,416	0.33	910	0.87
1,418	0.33	920	0.85
1,527	0.22	980	0.70
1,600	0.22	1,040	0.61
		1,100	0.54
		1,160	0.49
		1,220	0.46
		1,280	0.40
		1,340	0.36
		1,400	0.36
		1,460	0.34
		1,520	0.31
		1,600	0.29

4.4.8 (The data given in this problem also refer to Problems 4.4.9 and 4.7.1 below.) A fully penetrating pumping well in a confined aquifer is located between a suspected barrier boundary and a known recharge boundary as shown in the figure on the next page. Drawdown data from the observation well in the figure during a typical pumping test of the well in question are given in the following table. If the pumping is constant at a rate of 100 gpm and the observation well is 200 ft away from the pumping well, determine the transmissivity and the storage coefficient of the aquifer using the given aquifer test data.

Data from observation well	
Time (min)	Drawdown (ft)
2	0.025
3	0.04

(*continues*)

Data from observation well	
Time (min)	Drawdown (ft)
4	0.06
5	0.07
6	0.085
7	0.095
8	0.105
9	0.115
10	0.12
15	0.16
20	0.18
30	0.22
40	0.25
50	0.27
60	0.285
70	0.3
80	0.33
90	0.35
100	0.37
150	0.45
200	0.54
300	0.58
400	0.64
500	0.69
600	0.74

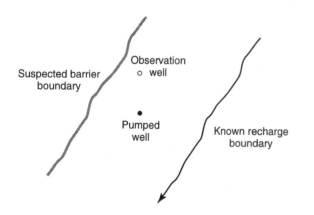

4.4.9 Determine the aquifer properties in Problem 4.4.8 using the Cooper–Jacob method of solution.

4.5.1 (This problem is adapted from *U.S. Geological Survey Water-Resources Investigations Report* 89-4081 (1989).) The Galena-Platteville Aquifer Test was performed on June 8–15, 1987, with a 6-inch-diameter fully penetrating well pumping under a constant rate (Q) of 19.4 gal/min for 54 hr. Eight partially penetrating observation wells in the Galena-Platteville Aquifer were monitored during the pumping test. The time–drawdown data for observation well PZ-1, which is 108 ft from the pumping

well, is given in the following table. The average thickness of the aquifer near this observation well is 84 ft. The test also reported that the Galena-Platteville Aquifer has a heterogeneous and anisotropic nature and consists of two media, fractures and porous rock. The estimated hydraulic conductivity and specific yield of the aquifer are respectively 8 ft/day and 0.049 from the PZ-1 observation well data. Using the given time–drawdown data and the Neuman Type A and Type Y curves for unconfined aquifers, make your own estimate of the aquifer hydraulic conductivity and specific yield.

Time (min)	Drawdown (ft)	Time (min)	Drawdown (ft)
0.9	0.012	100	0.805
1	0.014	230	0.842
1.5	0.032	320	0.86
2	0.052	410	0.877
2.5	0.072	450	0.883
3	0.093	530	0.891
3.5	0.111	630	0.903
4	0.13	810	0.946
4.5	0.148	890	0.969
5	0.165	980	0.98
6	0.198	1070	0.976
7	0.254	1190	0.991
8	0.277	1350	1.006
9	0.299	1525	1.031
10	0.319	1720	1.065
20	0.509	2130	1.085
30	0.687	2250	1.114
40	0.754	2400	1.136
50	0.774	2500	1.149
60	0.783	2600	1.159
70	0.788	2810	1.175
80	0.791	3140	1.204
90	0.8		

4.6.1 (This problem is adapted from the *Ground Water Manual*.[62]) The drawdown versus time data for an observation well 160 feet from the pumping well of the same aquifer as in Example 4.6.1 is given below. Estimate the transmissivity and storage coefficient of the confined aquifer and the permeability of the aquitard. Compare the results with the answers to Example 4.6.1.

Time (min)	Drawdown (ft)
4	2.15
6	2.86
8	3.46
10	3.78
15	4.58
20	5.09

(*continues*)

Time (min)	Drawdown (ft)
25	5.49
30	5.85
40	6.37
50	6.64
60	6.8
70	6.96
80	7.16
90	7.36
100	7.44
110	7.52
120	7.56
150	7.64
180	7.88
210	7.92
240	7.96
270	7.96
300	7.96
360	7.95
420	7.96

4.7.1 What would the drawdown in the pumping well be in Problem 4.4.8 at the end of pumping at a constant rate of 100 gpm for a continuous period of 180 days? The pumping well with a radius of 1 ft is located 500 ft away from the barrier boundary and 1,000 ft away from the recharge boundary.

4.7.2 (This problem is adapted from the *Ground Water Manual*.[62]) Drawdown versus time data for an observation well 100 ft from the pumping well in a pump test are tabulated below. Identify the type of boundary and determine the radius of the image well from the observation well. What additional information would you need to locate the boundary? (*Hint:* Use the method described at the end of Section 4.7)

Time (min)	Drawdown (ft)
5	0.08
10	0.22
15	0.32
20	0.41
25	0.49
30	0.56
40	0.67
50	0.77
60	0.85
70	0.95
80	1.01
90	1.08
100	1.14
110	1.20

(continues)

120	1.25
180	1.51
240	1.70
300	1.87
360	1.99
420	2.10
480	2.20
540	2.28
600	2.36
660	2.46
720	2.50
840	2.63
960	2.77

4.7.3 A production well fully penetrating a nonleaky isotropic artesian aquifer delimited by two barrier boundaries (perpendicular to each other) was continuously pumped at a constant rate of 1,485 gpm for a period of four hours. The drawdowns in the following table were observed at a distance of 300 ft in a fully penetrating observation well. Compute the coefficients of transmissivity and storage of the aquifer and the distances to each image well from the observation well.

t (min)	s (ft)
2	0.80
3	0.92
4	1.06
5	1.17
6	1.23
7	1.32
8	1.37
9	1.43
10	1.48
20	1.88
30	2.11
40	2.34
50	2.52
60	2.70
70	2.83
80	3.00
90	3.17
100	3.30
200	4.21
300	4.43

4.8.1 It is required to dewater a construction site 80 m by 80 m. The bottom of the construction will be 1.5 m below the initial water surface elevation of 90 m. Four pumps are to be used in 0.5-m diameter wells at the four corners of the site. Determine the required pumping rate. The aquifer has $T = 1600$ m^2/day and the wells each have a radius of influence of 600 m.

4.8.2 Reposition the wells in Example 4.8.1 such that they form an equilateral triangle (same spacings). For the same restrictions on the drawdown, will the discharge decrease or increase? If so, by what percent? If not, what difference do you perceive between the two problems?

4.8.3 Rework Problem 4.8.1 as if the site is to be ready after one month of pumping. Assume the storage coefficient of the aquifer is $S = 0.16$.

4.8.4 Two pumping wells 1,000 m away fully penetrate the same confined aquifer. One of the wells pumps at a rate of 1,240 m^3/day. The second well pumps at 850 m^3/day. If the aquifer has a transmissivity of 2,000 m^2/day and a storage coefficient of 4×10^{-4}, when would the wells start interfering with each other?

REFERENCES

1. Barlow, P. M. and A. F. Moench, *WTAQ—A Computer Program for Calculating Drawdowns and Estimating Hydraulic Properties for Confined and Water-Table Aquifers, U.S. Geological Survey Water–Resources Investigations Report* 99-4225, 1999.

2. Bennett, G. D. and E. P. Patten, Jr., *Constant-Head Pumping Test of a Multiaquifer Well to Determine Characteristics of Individual Aquifers, U.S. Geological Survey Water-Supply Paper* 1536-G, pp. 181–203, 1962.

3. Bentall, R., *Methods for Determining Permeability, Transmissibility, and Drawdown, U.S. Geological Survey Water-Supply Paper* 1536-I, pp. 243–341, 1963.

4. Boulton, N. S., The drawdown of the water-table under non-steady conditions near a pumped well in an unconfined formation, *Proc. Inst. Civil Engrs.*, v. 3, pt. III, pp. 564–579, 1954.

5. Boulton, N. S., Analysis of data from non-equilibrium pumping tests allowing for delayed yield from storage, *Proc. Inst. Civil Engrs.*, v. 26, pp. 469–482, 1963.

6. Boulton, N. S. and T. D. Streltsova, New equations for determining the formation constants of an aquifer from pumping test data, *Water Resources Research*, v. 11, pp. 148–153, 1975.

7. Chow, V. T., On the determination of transmissibility and storage coefficients from pumping test data, *Trans. Amer. Geophysical Union*, v. 33, pp. 397–404, 1952.

8. Cooley, R. L. and C. M. Case, Effect of a water table aquitard on drawdown in an underlying pumped aquifer, *Water Resources Research*, v. 9, pp. 434–447, 1973.

9. Cooper, H. H., Jr., and C. E. Jacob, A generalized graphical method for evaluating formation constants and summarizing well-field history, *Trans. Amer. Geophysical Union*, v. 27, pp. 526–534, 1946.

10. Dagan, G., A method of determining the permeability and effective porosity of unconfined anisotropic aquifers, *Water Resources Research*, v. 3, pp. 1059–1071, 1967.

11. Dawson, K. and J. Istok, *Aquifer Testing: Design and Analysis of Pumping and Slug Tests,* Lewis Publishers, Boca Raton, FL, 1991.

12. Driscoll, F. G, *Groundwater and Wells*, 2nd ed., Johnson Division, St. Paul, MN, 1986.

13. Dupuit, J., *Études Théoriques et Pratiques sur La Mouvement des Eaux dans Les Canaux Découverts et à Travers les Terrains Perméables*, 2nd ed., Dunod, Paris, 304 pp., 1863.

14. Ehlig, C. and J. C. Halepaska, A numerical study of confined-unconfined aquifers including effects of delayed yield and leakage, *Water Resources Research*, v. 12, pp. 1175–1183, 1976.

15. Ferris, J. G. et al., Theory of aquifer tests, *U.S. Geological Survey Water-Supply Paper* 1536-E, pp. 69–174, 1962.

16. Glover, R. E. and G. G. Balmer, River depletion resulting from pumping a well near a river, *Trans. Amer. Geophysical Union*, v. 35, pp. 468–470, 1954.

17. Halford, K. J. and E. L. Kuniansky, *Spreadsheets for the Analysis of Aquifer-Test and Slug-Test Data, U.S. Geological Survey Open-File Report* 02-197, http://water.usgs.gov/pubs/of/ofr02197, 2003.

18. Hall, P. and J. Chen, *Water Well and Aquifer Test Analysis*, Water Resources Publications, Littleton, CO, 1996.

19. Hantush, M. S., Analysis of data from pumping tests in leaky aquifers, *Trans. Amer. Geophysical Union*, v. 37, pp. 702–714, 1956.

20. Hantush, M. S., Drawdown around wells of variable discharge, *Jour. Geophysical Research*, v. 69, pp. 4221–4235, 1964.

21. Hantush, M. S., Hydraulics of wells, in *Advances in Hydroscience*, V. T. Chow, ed., v. 1, Academic Press, New York, pp. 281–432, 1964.

22. Hantush, M. S., Wells near streams with semipervious beds, *Jour. Geophysical Research*, v. 70, pp. 2829–2838, 1965.

23. Hantush, M. S., Wells in homogeneous anisotropic aquifers, *Water Resources Research*, v. 2, pp. 273–279, 1966.

24. Hantush, M. S., Flow to wells in aquifers separated by a semipervious layer, *Jour. Geophysical Research*, v. 72, pp. 1709–1720, 1967.

25. Hantush, M. S. and C. E. Jacob, Non-steady radial flow in an infinite leaky aquifer, *Trans. Amer. Geophysical Union*, v. 36, pp. 95–112, 1955.

26. Hantush, M. S. and I. S. Papadopulos, Flow of ground water to collector wells, *Jour. Hydraulics Div.*, Amer. Soc. Civil Engrs., v. 88. no. HY5, pp. 221–244, 1962.

27. Hantush, M. S. and R. G. Thomas, A method for analyzing a drawdown test in anisotropic aquifers, *Water Resources Research*, v. 2, pp. 281–285, 1966.

28. Huisman, L., *Groundwater Recovery*, Winchester, NY, 336 pp., 1972.

29. Jacob, C. E., Radial flow in a leaky artesian aquifer, *Trans. Amer. Geophysical Union*, v. 27, pp. 198–208, 1946.

30. Jenkins, C. T., Techniques for computing rate and volume of stream depletion by wells, *Ground Water*, v. 6, no. 2, pp. 37–46, 1968.

31. Kasenow, M., *Introduction to Aquifer Analysis,* Water Resources Publications, Littleton, CO, 1997.

32. Kipp, K. L., Jr., Unsteady flow to a partially penetrating, finite radius well in an unconfined aquifer, *Water Resources Research*, v. 9, pp. 448–462, 1973.

33. Kruseman, G. P. and N. A. de Ridder, Analysis and evaluation of pumping test data, International Institute for Land Reclamation and Improvement, The Netherlands, 1994.

34. Lakshminarayana, V. and S. P. Rajagopalan, Type-curve analysis of time-drawdown data for partially penetrating wells in unconfined anisotropic aquifers, *Ground Water*, v. 16, pp. 328–333, 1978.

35. Lee, J., *Well Testing,* Society of Petroleum Engineers of AIME, New York, 1982.

36. Lennox, D. H. and A. Vanden Berg, Drawdowns due to cyclic pumping, *Jour. Hydraulics Div.*, Amer. Soc. Civil Engrs., v. 93, no. HY6, pp. 35–51, 1967.

37. Lohman, S. W., *Ground-Water Hydraulics, U.S. Geological Survey Professional Paper* 708, 70 pp., 1972.

38. Milojevic, M., Radial collector wells adjacent to the river bank, *Jour. Hydraulics Div.*, Amer. Soc. Civil Engrs., v. 89, no. HY6, pp. 133–151, 1963.

39. Moench, A. Ground-water fluctuations in response to arbitrary pumpage, *Ground Water*, v. 9, no. 2, pp. 4–8, 1971.

40. Moench, A. F. and T. A. Prickett, Radial flow in an infinite aquifer undergoing conversion from artesian to water table conditions, *Water Resources Research*, v. 8, pp. 494–499, 1972.

41. Muskat, M., *The Flow of Homogeneous Fluids through Porous Media*, McGraw-Hill, New York, 763 pp., 1937.

42. Neuman, S. P., Theory of flow in unconfined aquifers considering delayed response of the water table, *Water Resources Research*, v. 8, pp. 1031–1045, 1972.

43 Neuman, S. P., Analysis of pumping test data from anisotropic unconfined aquifers considering delayed gravity response, *Water Resources Research*, v. 11, pp. 329–342, 1975.

44. Neuman, S. P. and P. A. Witherspoon, Field determination of the hydraulic properties of leaky multiple aquifer systems, *Water Resources Research*, v. 8, pp. 1284–1298, 1972.

45. Newcome, R., Jr., *Pumping Tests of the Coastal Plain Aquifers in South Carolina, State of South Carolina Water Resources Commission Report 174*, 1993.

46. Papadopulos, I. S. and H. H. Cooper, Jr., Drawdown in a well of large diameter, *Water Resources Research*, v. 3, pp. 241–244, 1967.

47. Prickett, T. A., Type-curve solution to aquifer tests under water-table conditions, *Ground Water*, v. 3, no. 3, pp. 5–14, 1965.

48. Prudic, D. E., *Estimates of Hydraulic Conductivity from Aquifer-Test Data and Specific-Capacity Data, Gulf Coast Regional Aquifer Systems, South-Central United States, U. S. Geological Survey Water-Resources Investigations Report* 90-4121, 1991.

49. Rao, D. B. et al., Drawdown in a well group along a straight line, *Ground Water*, v. 9, no. 4, pp. 12–18, 1971.

50. Rorabaugh, M. I., Graphical and theoretical analysis of step-drawdown test of artesian well, *Proc. Amer. Soc. Civil Engrs.*, v. 79, sep. 362, 23 pp., 1953.

51. Rushton, K. R. and Y. K. Chan, Pumping test analysis when parameters vary with depth, *Ground Water*, v. 14, pp. 82–87, 1976.

52. Sheahan, N. T., Determining transmissibility from cyclic discharge, *Ground Water*, v. 4, no. 3, pp. 33–34, 1966.

53. Stallman, R. W., Effects of water table conditions on water level changes near pumping wells, *Water Resources Research*, v. 1, pp. 295–312, 1965.

54. Stallman, R. W. and I. S. Papadopulos, *Measurement of Hydraulic Diffusivity of Wedge-Shaped Aquifers Drained by Streams, U.S. Geological Survey Professional Paper* 514, 50 pp., 1966.

55. Sternberg, Y. M., Transmissibility determination from variable discharge pumping tests, *Ground Water*, v. 5, no. 4, pp. 27–29, 1967.

56. Sternberg, Y. M., Simplified solution for variable rate pumping test, *Jour. Hydraulics Div.*, Amer. Soc. Civil Engrs., v. 94, no. HY1, pp. 177–180, 1968.

57. Sternberg, Y. M., Efficiency of partially penetrating wells, *Ground Water*, v. 11, no. 3, pp. 5–8, 1973.

58. Streltsova, T. D., On the leakage assumption applied to equations of groundwater flow, *Jour. Hydrology*, v. 20, pp. 237–253, 1973.

59. Theis, C. V., The relation between the lowering of the piezometric surface and the rate and duration of discharge of a well using ground-water storage, *Trans. Amer. Geophysical Union*, v. 16, pp. 519–524, 1935.

60. Thiem, G. *Hydrologische Methoden*, Gebhardt, Leipzig, 56 pp., 1906.

61. U.S. Bureau of Reclamation, *Ground Water Manual*, U.S. Dept. Interior, 480 pp., 1977.

62. U.S. Bureau of Reclamation, *Ground Water Manual,* U.S. Government Printing Office, Denver, CO, 1981.

63. Walton, W., *Aquifer Test Analysis with Windows*[TM], CRC Press, Boca Raton, FL, 1996.

64. Walton, W. C., *Leaky Artesian Aquifer Conditions in Illinois, Illinois State Water Survey Rept. Invest.* 39, Urbana, 27 pp., 1960.

65. Walton, W. C., Comprehensive analysis of water-table aquifer test data, *Ground Water*, v. 16, pp. 311–317, 1978.

66. Weeks, E. P., Determining the ratio of horizontal to vertical permeability by aquifer-test analysis, *Water Resources Research*, v. 5, pp. 196–214, 1969.

67. Wigley, T. M. L., Flow into a finite well with arbitrary discharge, *Jour. Hydrology*, v. 6, pp. 209–213, 1968.

Chapter 5

Water Wells

A *water well* is a hole or shaft, usually vertical, excavated in the earth for bringing groundwater to the surface. Occasionally wells serve other purposes, such as for subsurface exploration and observation, artificial recharge, and disposal of wastewaters. Many methods exist for constructing wells; selection of a particular method depends on the purpose of the well, the quantity of water required, depth to groundwater, geologic conditions, and economic factors. Shallow wells are dug, bored, driven, or jetted; deep wells are drilled by cable tool or rotary methods. Attention to proper design can ensure efficient and long-lived wells. After a well has been drilled, it should be completed, developed for optimum yield, and tested. Wells should be sealed against entrance of surface pollution and given periodic maintenance. Wells of horizontal extent are constructed where warranted by special groundwater situations.

5.1 TEST HOLES AND WELL LOGS

Before drilling a well in a new area, it is common practice to drill a test hole. The purpose of a test hole is to determine depths to groundwater, quality of water, and physical character and thickness of aquifers without the expense of a regular well, which might prove to be unsuccessful. Diameters seldom exceed 20 cm. Test holes may be put down by any method for well construction; however, cable tool, rotary, and jetting methods are commonly employed. If the test hole appears suitable as a site for a finished well, it can be reamed with hydraulic rotary equipment to convert it to a larger permanent well.

During drilling of a test hole or well, a careful record, or log, is kept of the various geologic formations and the depths at which they are encountered (see Chapter 11). A helpful method is to collect samples of cuttings in containers, labeling each with the depth where obtained. Later these can be studied and analyzed for grain size distribution. Most states require licensed well drillers to submit logs—recording depth, color, character, size of material, and structure of the strata penetrated—for wells they drill. Proper identification of strata in the hydraulic rotary method requires careful analysis because drilling mud is mixed with each sample. A drilling-time log (again see Chapter 11) is sometimes helpful in this respect.

5.2 METHODS FOR CONSTRUCTING SHALLOW WELLS

Shallow wells, generally less than 15 m in depth, are constructed by digging, boring, driving, or jetting.[38, 83, 96] The methods are briefly described in the following paragraphs; Table 5.2.1 lists their applications.

Table 5.2.1 Water Well Construction Methods and Applications (after U.S. Soil Conservation Service[97])

Method	Materials for which best suited	Water table depth for which best suited (m)	Usual maximum depth (m)	Usual diameter range (cm)	Usual casing material	Customary use	Yield (m³/day)[a]	Remarks
Augering Hand auger	Clay, silt, sand, gravel less than 2 cm	2–9	10	5–20	Sheet metal	Domestic, drainage	15–250	Most effective for penetrating and removing clay. Limited by gravel over 2 cm. Casing required if material is loose.
Power auger	Clay, silt, sand, gravel less than 5 cm	2–15	25	15–90	Concrete, steel or wrought iron pipe	Domestic, irrigation, drainage	15–500	Limited by gravel over 5 cm, otherwise same as for hand auger.
Driven wells Hand, air hammer	Silt, sand, gravel less than 5 cm	2–5	15	3–10	Standard weight pipe	Domestic, drainage	15–200	Limited to shallow water table, no large gravel.
Jetted wells Light, portable rig	Silt, sand, gravel less than 2 cm	2–5	15	4–8	Standard weight pipe	Domestic, drainage	15–150	Limited to shallow water table, no large gravel.
Drilled wells Cable tool	Unconsolidated and consolidated medium hard and hard rock	Any depth	450[b]	8–60	Steel or wrought-iron pipe	All uses	15–15,000	Effective for water exploration. Requires casing in loose materials. Mudscow and hollow rod bitsdeveloped for drilling unconsolidated fine to medium sediments.
Rotary	Silt, sand, gravel less than 2 cm; soft to hard consolidated rock	Any depth	450[b]	8–45	Steel or wrought-iron pipe	All uses	15–15,000	Fastest method for all except hardest rock. Casing usually not required during drilling. Effective for gravel envelope wells.
Reverse-circulation rotary	Silt, sand, gravel, cobble	2–30	60	40–120	Steel or wrought-iron pipe	Irrigation, industrial, municipal	2,500–20,000	Effective for large-diameter holes in unconsolidated and partially consolidated deposits. Requires large volume of water for drilling. Effective for gravel envelope wells.
Rotary-percussion	Silt, sand, gravel less than 5 cm; soft to hard consolidated rock	Any depth	600[b]	30–50	Steel or wrought-iron pipe	Irrigation, industrial, municipal	2,500–15,000	Now used in oil exploration. Very fast drilling. Combines rotary and percussion methods (air drilling) cuttings removed by air. Would be economical for deep water wells.

[a] Yield influenced primarily by geology and availability of groundwater.

[b] Greater depths reached with heavier equipment.

5.2.1   Dug Wells

Dating from Biblical times, dug wells have furnished countless water supplies throughout the world.[*] Depths range up to 20 m or more, depending on the position of the water table, while diameters are usually 1 to 10 m. Dug wells can yield relatively large quantities of water from shallow sources and are most extensively employed for individual water supplies in areas containing unconsolidated glacial and alluvial deposits.[106] Their large diameters permit storage of considerable quantities of water if the wells extend some distance below the water table.

In the past, all dug wells were excavated by hand, and even today the same method is widely employed. A typical dug well in under-developed portions of the world is often no more than an irregular hole in the ground that intersects the water table (see Fig. 5.2.1). A pick and shovel are the basic implements. Loose material is hauled to the surface in a container by means of suitable pulleys and lines. Large dug wells can be constructed rapidly with portable excavating equipment such as clamshell and orange-peel buckets. For safety and to prevent caving, lining (or cribbing) of wood or sheet piling should be placed in the hole to brace the walls.

A modern dug well is permanently lined with a casing (often referred to as a *curb*) of wood staves, brick, rock (Fig. 5.2.2), concrete, or metal. Curbs should be perforated or contain openings for entry of water and must be firmly seated at the bottom. Dug wells should be deep enough to extend a few meters below the water table. Gravel should be backfilled around the curb and at the bottom of the well to control sand entry and possible caving. A properly

Figure 5.2.1. Women gathering water from a crude dug well in the Shinyanga Region of Tanzania, East Africa (courtesy DHV Consulting Engineers, Amersfoort, The Netherlands).

---

[*]See also the description of qanats in Chapter 1.

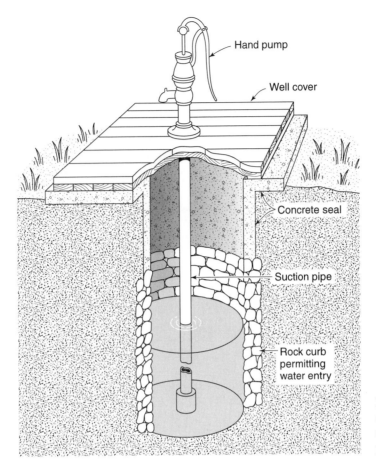

Figure 5.2.2. A domestic dug well with a rock curb, concrete seal, and hand pump.

constructed dug well penetrating a permeable aquifer can yield 2,500 to 7,500 m^3/day, although most domestic dug wells yield less than 500 m^3/day.

A serious limitation of large open dug wells involves the ease of their pollution by surface water, airborne material, and objects falling or finding entrance into the wells.

5.2.2 Bored Wells

Where a water table exists at a shallow depth in an unconsolidated aquifer, bored wells can furnish small quantities of water at minimum cost. *Bored wells* are constructed with hand-operated or power-driven earth augers. Hand augers are available in several shapes and sizes, all operating with cutting blades at the bottom that bore into the ground with a rotary motion. When the blades are full of loose earth, the auger is removed from the hole and emptied; the operation is repeated until the desired hole depth is reached. Hand-bored wells seldom exceed 20 cm in diameter and 15 m in depth. Power-driven augers will bore holes up to 1 m in diameter and, under favorable conditions, to depths exceeding 30 m. The auger consists of a cylindrical steel bucket (Figure 5.2.3) with a cutting edge projecting from a slot in the bottom. The bucket is filled by rotating it in the hole by a drive shaft of adjustable length. When full, the auger is hoisted to the surface and the excavated material is removed through hinged openings on the side or bottom of the bucket. Reamers, attached to the top of the bucket, can enlarge holes to diameters exceeding the auger size.

A continuous-flight power auger (Figure 5.2.4) has a spiral extending from the bottom of the hole to the surface. Cuttings are carried to the surface as on a screw conveyor, while sec-

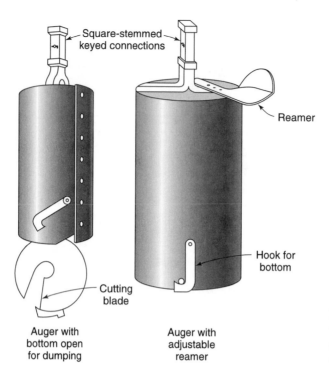

Square-stemmed keyed connections

Reamer

Cutting blade

Hook for bottom

Auger with bottom open for dumping

Auger with adjustable reamer

Figure 5.2.3. Augers for boring wells. Spiral flight augers are also widely employed for small-diameter holes.

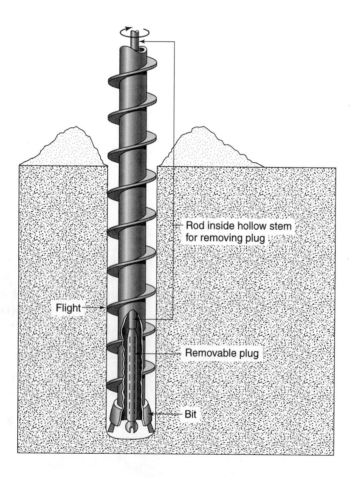

Rod inside hollow stem for removing plug

Flight

Removable plug

Bit

Figure 5.2.4. Hollow-stem auger drilling. The hollow-stem, continuous-flight auger bores into soft soils, carrying the cuttings upward along the flights. When the desired depth is reached, the plug is removed from the bit and withdrawn from inside the hollow stem. A well point on a casing ($1\frac{1}{4}$ in or 2 in) can then be inserted to the bottom of the hollow stem and the auger pulled out, leaving the small-diameter monitoring well in place (M. L. Scalf[83]).

tions of the auger may be added as depth increases. The simple equipment, usually truck-mounted, can be operated rapidly by one person and functions to depths exceeding 50 m in unconsolidated formations that do not contain large boulders.

Augers work best in formations that do not cave. Where loose sand and gravel are encountered in a large-diameter hole, or the boring reaches the water table, it may be necessary to lower a concrete or metal casing to the bottom of the hole and continue boring inside. After the desired depth is reached, the permanent well casing and screen are centered in the hole, the outer casing is removed, and the peripheral space is backfilled with gravel. Augers sometimes supplement other well-drilling methods where sticky clay formations are encountered; here augers are more effective than any other penetrating device.

5.2.3   Driven Wells

A *driven well* consists of a series of connected lengths of pipe driven by repeated impacts into the ground to below the water table. Water enters the well through a *drive* (or *sand*) *point* at the lower end of the well (Fig. 5.2.5). This consists of a screened cylindrical section protected during driving by a steel cone at the bottom. Diameters of driven wells are small, most falling in the range of 3 to 10 cm. Standard-weight water pipe having threaded couplings serves for casing. Most depths are less than 15 m, although a few exceed 20 m. As suction-type pumps extract water from driven wells, the water table must be near the ground surface if a continuous

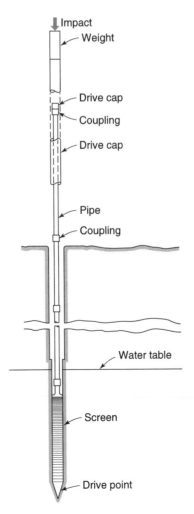

Figure 5.2.5. A driven well with driving mechanism.

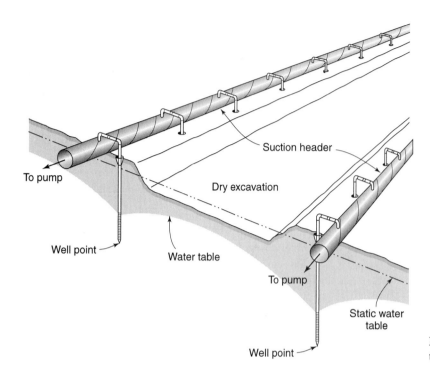

To pump

Suction header

Dry excavation

Well point

Water table

To pump

Static water table

Well point

Figure 5.2.6. A well-point system to dewater an excavation site.

water supply is to be obtained. For best results the water table should be within 3 to 5 m of ground surface in order to provide adequate drawdown without exceeding the suction limit. Yields from driven wells are small, with discharges of about 100–250 m^3/day.

Driven wells are best suited for domestic supplies, for temporary water supplies, and for exploration and observation. Batteries of driven wells connected by a suction header to a single pump are effective for localized lowering of the water table. Such installations, known as *well-point systems*, are particularly advantageous for dewatering excavations for foundations and other subsurface construction operations.[74] Figure 5.2.6 illustrates how a well-point installation reduces the groundwater level to furnish a dry excavation.

Driven wells are limited to unconsolidated formations containing no large gravel or rocks that might damage the drive point. To drive a well, the pipe casing and threads should be protected at the top with a drive cap (see Figure 5.2.5). Driving can be done with a maul, sledge, drop hammer, or air hammer. It is usually good practice to place (by boring or driving) an outer protective casing to at least 3 m below ground surface. Screens are available in a variety of opening sizes, the choice depending on the size of particles in the water-bearing stratum.

Important advantages of driven wells are that they can be constructed in a short time, at minimum cost, and even by one person.

5.2.4   Jetted Wells

Jetted wells are constructed by the cutting action of a downward-directed stream of water. The high-velocity stream washes the earth away, while the casing, which is lowered into the deepening hole, conducts the water and cuttings up and out of the well. Small-diameter holes of 3 to 10 cm are formed in this manner (although the method is capable of producing diameters up to 30 cm or more) to depths greater than 15 m. Jetted wells have only small yields and are best adapted to unconsolidated formations. Because of the speed of jetting a well and the portability of the equipment, jetted wells are useful for exploratory test holes, observation wells, and well-point systems.[11, 66]

Figure 5.2.7. Various designs of jetting drill bits.

Various types of jetting drill bits are shown in Figure 5.2.7. In penetrating clays and hardpans, the drill pipe is raised and lowered sharply, causing the bit to shatter the formation. During the jetting operation, the drill pipe is turned slowly to ensure a straight hole. To complete a shallow jetted well after the casing extends to below the water table, the well pipe with screen attached is lowered to the bottom of the hole inside the casing. The outer casing is then pulled, gravel is inserted in the outer space, and the well is ready for pumping.

A simplification of the above procedure can be obtained by using a self-jetting well point. This consists of a tube of brass screen ending in a jetting nozzle, which is screwed to the well

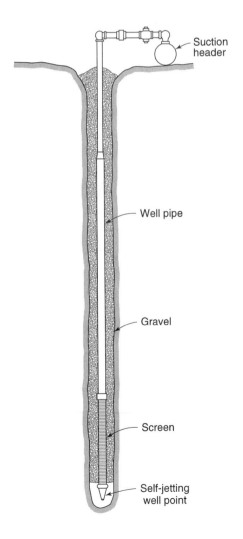

Figure 5.2.8. Jetted well with self-jetting well point.

pipe (Figure 5.2.8). As soon as the well point has been jetted to the required depth, the well is completed and ready for pumping. Gravel should be added around the drill pipe for permanent installations.

5.3   METHODS FOR DRILLING DEEP WELLS

Most large, deep, high-capacity wells are constructed by drilling. Construction can be accomplished by the cable tool method or by one of several rotary methods.[21, 47, 53] Each method has particular advantages, so experienced drillers endeavor to have equipment available for a diversity of drilling approaches.[1, 2, 47] Applications of drilling methods are listed in Table 5.3.1, while Table 5.3.2 indicates the performance of the methods in various geologic formations.

Examples of the construction of deep wells in unconsolidated and consolidated formations are shown in Figures 5.3.1(*a*)–5.3.1(*l*), respectively, taken from standard specifications for wells prepared by the American Water Works Association. The construction procedure of a successful well is dependent on local conditions encountered in drilling; hence, each well should be treated as an individual project. Figure 5.3.2 illustrates modern truck-mounted drill rigs. Construction methods differ regionally within the United States and also from one driller to another. General construction methods are described in the following sections.

Table 5.3.2 Performance of Drilling Methods in Various Types of Geologic Formations (after Speedstar Div.[88])

Type of formation Rotary percussion[a]	Cable tool	Rotary	
Dune sand	Difficult	Rapid	NR
Loose sand and gravel	Difficult	Rapid	NR
Quicksand	Difficult, except in thin streaks Requires a string of drive pipe	Rapid	NR
Loose boulders in alluvial fans or glacial drift	Difficult; slow but generally can be handled by driving pipe	Difficult, frequently impossible	NR
Clay and silt	Slow	Rapid	NR
Firm shale	Rapid	Rapid	NR
Sticky Shale	Slow	Rapid	NR
Brittle shale	Rapid	Rapid	NR
Sandstone, poorly cemented	Slow	Slow	NR
Chert nodules	Rapid	Slow	NR
Limestone	Rapid	Rapid	Very rapid
Limestone with chert nodules	Rapid	Slow	Very rapid
Limestone with small cracks or fractures			Very rapid
Limestone, cavernous	Rapid	Slow to impossible	Difficult
Dolmite	Rapid	Rapid	Very rapid
Basalts, thin layers in sedimentary rocks	Rapid	Slow	Very rapid
Basalts, thick layers	Slow	Slow	Rapid
Metamorphic rocks	Slow	Slow	Rapid
Granite	Slow	Slow	Rapid

[a]NR: not recommended.

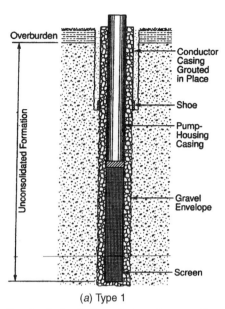

(a) Type 1

Gravel-packed well with conductor casing grouted in
place and gravel envelope extending to surface.

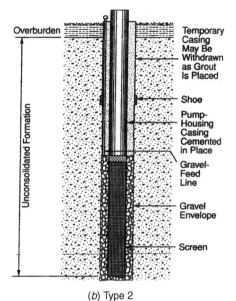

(b) Type 2

Gravel-packed well with well casing cemented in
place and gravel envelope terminated above the top
of the screen with gravel feed line.

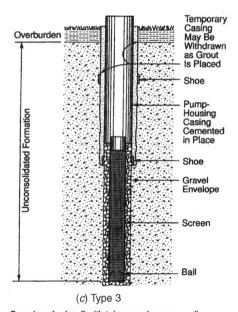

(c) Type 3

Gravel-packed well with telescoped screen, well
casing cemented in place, and gravel envelope
terminated above top of screen.

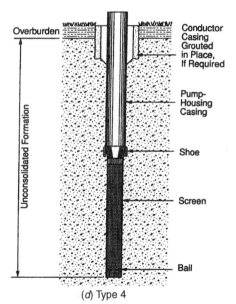

(d) Type 4

Naturally developed well with telescoped screen,
well pump-housing casing driven or jacked into
place, and the conductor sealed as locally required.

Figure 5.3.1. Examples of well construction in unconsolidated and consolidated formations (reprinted from American Water Works Association *AWWA Standard for Deep Wells*,[3] by permission of the American Water Works Association, 6666 West Quincy Avenue, Denver, CO 80235).

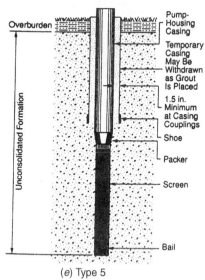

(e) Type 5

Naturally developed well with telescoped screen, temporary casing driven or jacked into place, and the pump-housing casing sealed in to prevent contamination.

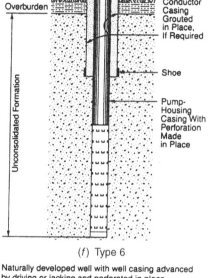

(f) Type 6

Naturally developed well with well casing advanced by driving or jacking and perforated in place.

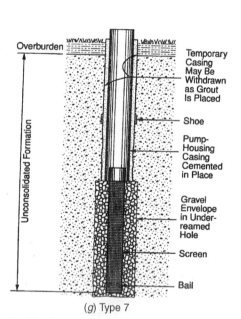

(g) Type 7

Gravel-packed well with underreamed borehole for screen and pump-housing casing cemented in place.

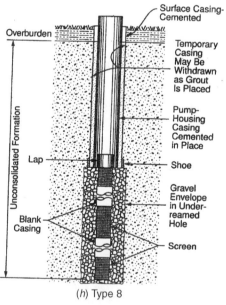

(h) Type 8

Gravel-packed well with underreamed borehole for screens in multiple unconsolidated aquifers.

Figure 5.3.1 (continued)

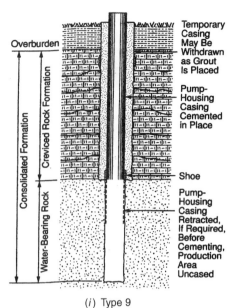

(*i*) Type 9

Well with open hole completion in consolidated rock and well casing cemented in place.

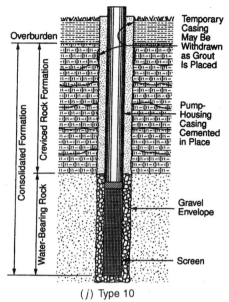

(*j*) Type 10

Gravel-packed well completed in consolidated rock with well casing cemented in place.

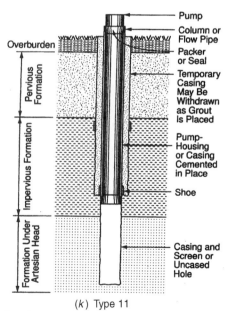

(*k*) Type 11

Open hole or screened well completion in an artesian aquifer where the piezometric level is above the ground elevation.

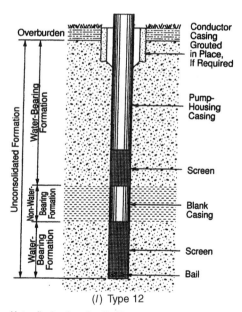

(*l*) Type 12

Naturally developed well with screen and casing installed in place in an open hole. Blank casing in non-water-bearing formation is optional.

Figure 5.3.1 (*continued*)

Figure 5.3.2. Rotary drill rig designed for air/mud applications. This rig is truck-engine powered and the single-tubed derrick accommodates a casing hammer, dual wall systems, or drill and drive systems (courtesy of Ingersoll Rand).

5.3.1   Cable Tool Method

Wells drilled by the cable tool (also *percussion or standard*) method are constructed with a standard well-drilling rig, percussion tools, and a bailer.[38, 40, 95] The method is capable of drilling holes of 8 to 60 cm in diameter through consolidated rock materials to depths of 600 m. In unconsolidated sand and gravel, especially quicksand, it is least effective because the loose material slumps and caves around the bit. Drilling is accomplished by regular lifting and dropping of a string of tools. On the lower end, a bit with a relatively sharp chisel edge breaks the rock by impact.

From top to bottom, a string of tools consists of a *swivel socket, a set of jars, a drill stem*, and *a drilling bit* (Figure 5.3.3). The total weight may amount to several thousand kilograms.

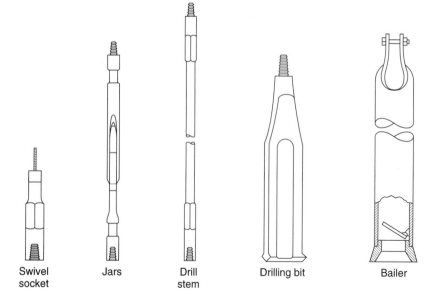

Swivel socket   Jars   Drill stem   Drilling bit   Bailer

Figure 5.3.3. Basic well-drilling tools for the cable tool method.

Tools are made of steel and are joined with tapered box-and-pin screw joints. The most important part of the string of tools is the bit, which does the actual drilling. Bits are manufactured in lengths of 1 to 3 m and weigh up to 1,500 kg. Variously shaped bits are made for drilling in different rock formations. The drill stem is a long steel bar that adds weight and length to the drill so that it will cut rapidly and vertically. A set of jars consists of a pair of narrow connecting links. They have no direct effect on the drilling; their purpose is only to loosen the tools should they stick in the hole. Under normal tension on the drilling line the jars remain fully extended. When tools become stuck, the line is slackened to allow the links to open to their full length, whereupon an upstroke of the line will cause the upper section of the jars to impart an upward blow to the tools. The swivel socket attaches the drilling cable to the string of tools.

Drill cuttings are removed from the well by a *bailer* or *sand bucket* (Figure 5.3.3). Although several models are manufactured, a bailer consists essentially of a section of pipe with a valve at the bottom and a ring at the top for attachment to the bailer line. When lowered into the well, the valve permits cuttings to enter the bailer but prevents them from escaping. After filling, the bailer is hoisted to the surface and emptied. Bailers are available in a range of diameters, lengths of 3 to 8 m, and capacities up to 0.25 m^3.

The drilling rig for the cable tool method consists of a mast, a multiline hoist, a walking beam, and an engine. In most present-day designs the entire assembly is truck-mounted (again see Figure 5.3.2) for ready portability.

During drilling, the tools make 20–40 strokes per minute, ranging from 40 to 100 cm in length. The drilling line is rotated so that the bit forms a round hole, and additional line is let out as needed so that the bit will always strike the bottom of the hole. Water should be added to the hole if none is encountered to form a paste with the cuttings, thereby reducing friction on the falling bit. After the bit has cut 1 or 2 m through a formation, the string of tools is lifted to the surface and the hole is bailed. In unconsolidated formations, casing should be maintained to near the bottom of the hole to avoid caving. Casing is driven down by means of drive clamps fastened to the drill stem; the up and down motion of the tools striking the top of the casing, protected by a *drive head*, sinks the casing. On the bottom of the first section of casing, a *drive shoe* (see Figure 5.3.1) with a beveled cutting edge is fastened to protect the casing as it is being driven.

As casing is driven deeper in unconsolidated formations, the vibration causes the sides of the hole to collapse against it. Frictional forces increase until further driving becomes impossible. When this occurs, a smaller diameter casing is telescoped inside of the casing already in the hole; thereafter, drilling is continued using a smaller diameter bit. In deep holes, several such reductions in casing may be required.

In drilling any deep well it is important that proper alignment be maintained so as not to interfere with pump installation and operation. The greatest problem occurs in drilling through rock formations. Some drillers have found that holes tending to bend can be corrected by detonating explosives at the bottom. This shatters the surrounding rock and permits drilling to progress vertically.

The cable tool rig is highly versatile in its ability to drill satisfactorily over a wide range of geologic conditions.* Its major drawbacks are its slower drilling rate, its depth limitation, the necessity of driving casing coincidentally with drilling in unconsolidated materials, and the difficulty of pulling casing from deep holes. The simplicity of design, ruggedness, and ease of maintenance and repair of the rigs and tools are important advantages in isolated areas.[94] Also, less water is required for drilling than with other methods, a matter of concern in arid and semiarid regions. Furthermore, sampling and formation logging are simpler and more accurate with a cable tool rig.

---

*In particular, cable tool rigs can drill through boulders and fractured, fissured, broken, or cavernous rocks, which often are beyond the capabilities of other types of equipment (see Table 5.3.1).

5.3.2 Rotary Method

A rapid method for drilling in unconsolidated strata is the rotary method.[73] Deep wells up to 45 cm in diameter, and even larger with a reamer, can be constructed. The method operates continuously with a hollow rotating bit through which a mixture of clay and water, or *drilling mud*, is forced. Material loosened by the bit is carried upward in the hole by the rising mud. No casing is ordinarily required during drilling because the mud forms a clay lining, or *mud cake*, on the wall of the well by filtration. This seals the walls, thereby preventing caving, entry of groundwater, and loss of drilling mud.

Drill bits are available in various forms; a group of conical roller gears with teeth that scrape, grind, and fracture the rock is a common design (see Figure 5.3.4). The typical string of tools consists of a bit, a drill collar (which adds weight to the bit and aids in maintaining hole alignment), and a drill pipe that extends to the ground surface. The upper end of the drill pipe is attached to the *kelly*—a square section of drill rod. The drill is turned by a rotating table that fits closely around the kelly and allows the drill rod to slide downward as the hole deepens. The drilling rig for a rotary outfit consists of a derrick, or mast, a rotating table, a pump for the drilling mud, a hoist, and the engine.

Drilling mud consists of a suspension of water, bentonite, clay, and various organic additives. Maintenance of the correct mud in terms of weight, viscosity, jelling strength, and low percentage of suspended solids is important for trouble-free drilling.[94] Organic additives that degrade with time and thereby cause the mud cake to break down within a few days are a recent innovation. The drilling mud leaves the drill pipe through the bit, where it cools and lubricates the cutting surface, entrains drill cuttings, and carries the drill cuttings upward within the annular space between drill pipe and hole wall as the fluid returns to ground surface (see Figure 5.3.5). The drilling mud then overflows into a ditch and passes into a settling pit. Here the cuttings settle out; thereafter, the mud is picked up by the pump for recirculation in the hole.

Rotary drilling is employed for oil wells and its application to water-well drilling is steadily increasing. Advantages are the rapid drilling rate, the avoidance of placement of a casing during drilling, and the convenience for electric logging (see Chapter 12). Disadvantages include high equipment cost, more complex operation, the need to remove the mud cake during well development, and the problem of lost circulation in highly permeable or cavernous geologic formations.

5.3.3 Air Rotary Method

Rotary drilling can also be accomplished with compressed air in place of drilling mud. The technique is rapid and convenient for small-diameter holes in consolidated formations where a clay lining is unnecessary to support the walls against caving. Larger diameter holes can be drilled by employing foams and other air additives.[21] Drilling depths can exceed 150 m under

(a)

(b)

(c)

Figure 5.3.4. Examples of rotary drill bits. (*a*) Fishtail bit (*b*) Cone-type rock bit (*c*) Carbide button bit (after Speedstar Div.[88])

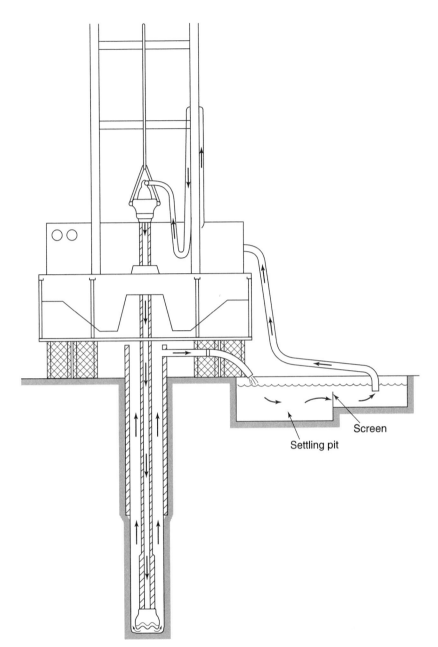

Screen

Settling pit

Figure 5.3.5. Drilling mud circulation system for the rotary method (after Speedstar Div.[88]).

favorable circumstances. An important advantage of the air rotary method is its ability to drill through fissured rock formations with little or no water required.

5.3.4   Rotary-Percussion Method

A recently developed rotary-percussion procedure using air as the drilling fluid provides the fastest method for drilling in hard-rock formations. A rotating bit, with the action of a pneumatic hammer, delivers 10 to 15 impacts per second to the bottom of the hole. Penetration rates of as much as 0.3 m/min have been achieved. Where caving formations or large quantities of water are encountered, a change to conventional rotary drilling with mud usually becomes necessary.

5.3.5   Reverse-Circulation Rotary Method

The reverse-circulation rotary method has become increasingly popular as a means for drilling large-diameter holes in unconsolidated formations. Water is pumped up through the drill pipe employing a large-capacity centrifugal or jet pump similar to those for gravel dredges. Discharge from the hole flows into a large pit where cuttings settle out. Thereafter the water runs through a ditch and back into the hole so that the water level in the hole is maintained at ground surface.

To avoid erosion of the sides of the hole, downward velocities must be restricted; therefore, the minimum hole diameter is about 40 cm. Drilling bits range in diameter from 0.4 to 1.8 m. The velocity of water up the drill pipe usually exceeds 2 m/s.

The water table should be 3 to 4 m below ground surface in order to obtain an effective head differential between well and aquifer. With this difference, fine-grained particles suspended in the water aid in stabilizing the walls. Where the water table is closer to the surface, casing can be extended above ground surface to increase the head. On the other hand, where the water table is deep, it may be necessary to place surface casing to minimize water loss.

For unconsolidated formations the reverse-circulation rig is probably the most rapid drilling equipment available. It requires a large volume of readily available water. Such rigs normally can drill to depths of 125 m; modifications with airlift pumping can extend this depth range.* Because reverse-circulation holes have large diameters, completed wells are usually gravel packed.

5.4   WELL COMPLETION

After a well has been drilled, it must be completed. This can involve placement of casing, cementing of casing, placement of well screens, and gravel packing; however, wells in hard-rock formations can be left as open holes so that these components may not be required.

5.4.1   Well Casings

Well casing serves as a lining to maintain an open hole from ground surface to the aquifer. It seals out surface water and any undesirable groundwater and also provides structural support against caving materials outside the well. Materials commonly employed for well casings are wrought iron, alloyed or unalloyed steel, and ingot iron.[5] Joints normally consist of threaded couplings or are welded, the object being to secure watertightness. In cable-tool drilling, the casing is driven into place; in rotary methods, the casing is smaller than the drilled hole and hence can be lowered into place. Polyvinyl chloride pipe is widely employed as casing for shallow, small-diameter observation wells.

Surface casing is installed from ground surface through upper strata of unstable or fractured materials into a stable and, if possible, relatively impermeable material.[94] Such surface casing serves several purposes, including: (1) supporting unstable materials during drilling, (2) reducing loss of drilling fluids, (3) facilitating installation or removal of other casing, (4) aiding in placing a sanitary seal, and (5) serving as a reservoir for a gravel pack. This casing may be temporary during drilling or it may be permanent. Recommended minimum diameters of surface casing are given in Table 5.4.1.

---

*A versatile modification of reverse rotary drilling is the *dual-tube* method. This involves a rotating outer casing connected to an inner concentric pipe with air serving as the circulating fluid. Rapid drilling rates are possible with diameters up to 40 cm and depths to 400 m. A particular advantage of the method is that continuous and accurate geologic and groundwater quality samples can be obtained as a function of depth.

Table 5.4.1 Recommended Minimum Diameters for Well Casings and Screens
(after U.S. Bureau of Reclamation[94])

| Well yield (m³/day) | Nominal pump chamber casing diameter (cm) | Surface Casing Diameter (cm) | | Nominal screen diameter (cm) |
		Naturally developed wells	Gravel-packed wells	
< 270	15	25	45	5
270–680	20	30	50	10
680–1,900	25	35	55	15
1,900–4,400	30	40	60	20
4,400–7,600	35	45	65	25
7,600–14,000	40	50	70	30
14,000–19,000	50	60	80	35
19,000–27,000	60	70	90	40

Pump chamber casing comprises all casing above the screen in wells of uniform diameter. For telescoping wells it is the casing within which the pump bowls are set. Recommended minimum diameters are listed in Table 5.4.1. Normally, the pump chamber casing should have a nominal diameter at least 5 cm larger than the nominal diameter of the pump bowls. Nonmetallic pipes are sometimes employed where corrosion or incrustation is a problem. Possibilities include ceramic-clay, concrete, asbestos-cement, plastic, and fiberglass-reinforced plastic pipe; strengths of these materials are not comparable to steel pipe, however.

5.4.2   Cementing

Wells are cemented in the annular space surrounding the casing to prevent entrance of water of unsatisfactory quality, to protect the casing against exterior corrosion, and/or to stabilize caving rock formations. Cement grout, consisting of a mixture of cement and water and sometimes various additives, can be placed by a dump bailer, by a tremie pipe, or by pumping.[5, 21] It is important that the grout be introduced at the bottom of the space to be grouted to ensure that the zone is properly sealed.

5.4.3   Screens

In consolidated formations, where the material surrounding the well is stable, groundwater can enter directly into an uncased well. In unconsolidated formations, however, wells are equipped with screens. These stabilize the sides of the hole, prevent sand movement into the well, and allow a maximum amount of water to enter the well with a minimum of hydraulic resistance.

In the cable-tool method of drilling, screens are normally placed by the *pullback method.* After casing is in place, the screen is lowered inside, and the casing is pulled up to near the top of the screen. A lead packer ring on the top of the screen is flared outward to form a seal between the inside of the casing and the screen. For the rotary method of drilling without casing, screens are lowered into place as drilling mud is diluted and again are sealed by a lead packer to an upper permanent casing. Screens are also sometimes placed by the *bail-down method,* involving bailing out material below the screen until the screen section is lowered to the desired aquifer depth.

The placement of the well screen or open interval of a discharging well can greatly influence areas contributing recharge to the well and vulnerability of the well to contamination. The "degree of separation" refers to a combination of two factors. The first is the distance between the water table and the top of the well screen or open interval of the well (greater distance

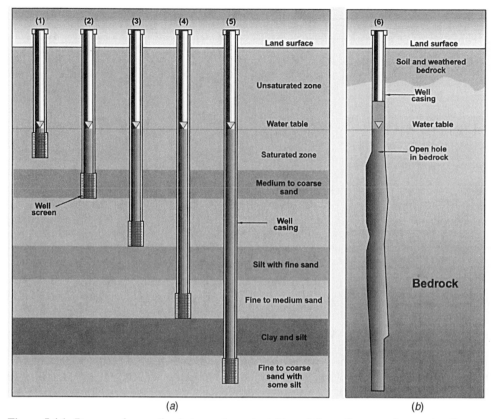

Figure 5.4.1. Degrees of separation between the water table and the well screen of a cased well or the open hole of an uncased well.[34]

implies greater separation). The second is the thickness and vertical and horizontal conductivities of confining units and aquifers between the water table and open interval of the well (in particular, greater thickness and lower hydraulic conductivity of low-conductivity confining units imply greater separation). The wells in Figure 5.4.1 part a illustrate various degrees of separation in a layered aquifer. In general, greater degrees of separation lead to a more complex three-dimensional configuration of groundwater flowpaths, longer travel times from the water table or surface-water body to a discharging well, extension of the contributing area farther from the discharging well, and possibly a more discontinuous and complex shape of the contributing area. A separate sketch for bedrock wells is provided in part b (well 6) to illustrate a practice in some areas to install well casing only to the top (surface) of consolidated rock. As a result, the bottom of the well casing can be above or only a few feet below the water table. In these situations, the well can be particularly vulnerable to possible contamination from groundwater near the water table; that is, no separation exists between the water table and the top of the operation interval of the bedrock well. Homeowner wells and virtually all public-supply wells tapping unconsolidated deposits (wells 1–5) usually are screened some distance below the water table.[34]

In the past, well casings were often perforated in place by a special cutting knife. This practice is now generally discontinued because of the large irregular openings created, the small percentage of open area obtained, and the difficulty of controlling entry of sand with water during pumping. More common is the use of preperforated casing, constructed by sawing, machining, or torch-cutting slots in the casing. Slot openings range from less than 1 to 6 mm; with larger slots the maximum percentage of open area is about 12 percent.[94] Openings by sawing or machining can be properly sized, whereas torch-cut slots tend to be large, irregular, and conducive to sand entry.

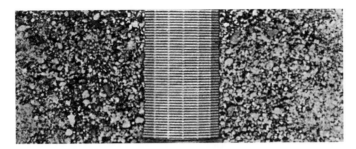

Figure 5.4.2. Continuous slot wire-wound well screen in an unconsolidated formation. The grain size distribution around the screen illustrates a properly developed well (courtesy Johnson Div., UOP Inc.[53]).

A major factor in controlling head loss through a perforated well section is the percentage of open area. For practical purposes a minimum open area of 15 percent is desirable; this value is readily obtained with many commercial screens but not with preperforated casing.[92, 94]

Manufactured screens are preferred to preperforated casing because of the ability to tailor opening sizes to aquifer conditions and because of larger percentages of open area that can be achieved. Several types are available: punched, stamped, louvred, wire-wound perforated pipe, and continuous-slot wire-wound screens. The latter type, consisting of a continuous winding of round or specially shaped wire on a cage of vertical rods, is the most efficient, possesses the largest open area, and can be closely matched to aquifer gradations (see Figure 5.4.2). Although such screens are more expensive, they may prove to be more economical, especially for thin but highly productive aquifers.

Screens are available in a range of diameters; selection of screen diameter should be made on the basis of the desired well yield and aquifer thickness. Recommended minimum screen diameters are included in Table 5.4.1. To minimize well losses and screen clogging, entrance velocities should be kept within specified limits. Because aquifers composed of finer grained materials tend to clog more easily than do those of coarser materials, field experience has indicated that there is a relationship between hydraulic conductivity of an aquifer and screen entrance velocity, as shown in Table 5.4.2.

To express the velocities in Table 5.4.2 in terms of screen size, the following equation can be applied:

$$v_s = \frac{Q}{c\pi d_s L_s P} \tag{5.4.1}$$

where v_s is the optimum screen entrance velocity, Q is well discharge, c is a clogging coefficient (estimated at 0.5 on the basis that approximately 50 percent of the open area of a screen

Table 5.4.2 Optimum Entrance Velocity of Water Through a Well Screen (after Walton[101])

Hydraulic conductivity of aquifer (m/day)	Optimum screen entrance velocity (m/min)
> 250	3.7
250	3.4
200	3.0
160	2.7
120	2.4
100	2.1
80	1.8
60	1.5
40	1.2
20	0.9
< 20	0.6

will be blocked by aquifer material), d_s is the screen diameter, L_s is the screen length, and P is the percentage of open area in the screen (available from manufacturer's specifications). Thus, for a given aquifer material, aquifer thickness, well yield, and type of screen, the appropriate diameter and length of well screen can be selected.

Screens are made of a variety of metals and metal alloys, plastics, concrete, asbestos-cement, fiberglass-reinforced epoxy, coated base metals, and wood.[94] Because a well screen is particularly susceptible to corrosion and incrustation, nonferrous metals, alloys, and plastics are often selected to prolong well life and efficient operation. Table 5.4.3 lists the more common metallic screen materials in the order of increasing cost.

A significant characteristic of a well screen is its slot size, which should be determined from mechanical analyses of formation samples obtained during drilling of the well or a pilot hole.* If the uniformity coefficient of an aquifer sample for a naturally developed well (without a gravel pack) is 5 or less, the selected slot size should retain 40 to 50 percent of the aquifer material. For a uniformity coefficient greater than 5, the slot size should retain 30 to 50 percent of the aquifer material.[94] In essence, the screen permits finer material to enter the well and to be removed by bailing during development of the well.[36, 48, 87, 98] But coarser material is retained outside, forming a permeable envelope around the well (Figure 5.4.2). An illustration of how a well screen can form a natural gravel pack around a well is shown in Figure 5.4.3. Here, with 40 percent of the aquifer material retained, the resulting gravel envelope has a permeability more than 30 times greater than that of the aquifer.

Where a well screen is to be surrounded by an artificial gravel pack, the size of the screen openings is governed by the size of the gravel (see next section).

5.4.4 Gravel Packs

A gravel-packed well is one containing an artificially placed gravel screen or envelope surrounding the well screen (see Figures 5.3.1(a–c)). A gravel pack (1) stabilizes the aquifer, (2) minimizes sand pumping, (3) permits use of a large screen slot with a maximum open area,

Table 5.4.3 Metallic Well Screen Materials and Their Resistance to Acid and Corrosion (after U.S. Bureau of Reclamation[94])

Material[a]	Acid resistance	Corrosion resistance in normal groundwater
Low-carbon steel	Poor	Poor[b,c]
Toncan and Armco iron	Poor	Fair[b,c]
Admiralty red brass	Good	Good[c]
Silicon red brass	Good	Good[c]
304 stainless steel	Good	Very good
Everdure bronze	Very good	Very good[d]
Monel metal	Very good	Very good[d]
Super nickel	Very good	Very good[d]

[a]Materials are listed in order of increasing cost.

[b]Not recommended for permanent installations where incrustation is a serious problem.

[c]Not recommended for permanent installations where sulfate-reducing or similar bacteria are present or where water contains more than 60 mg/l SO_4.

[d]Recommended only in areas where corrosion is very aggressive.

---

*Screen manufacturers will often recommend the most satisfactory slot size based on the grain size analysis of a given aquifer.

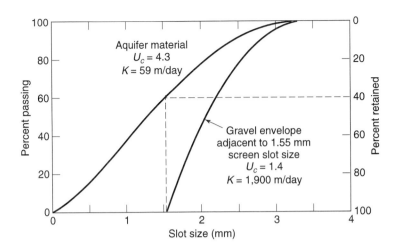

Figure 5.4.3. Development of a natural gravel pack around a well screen from a well-graded unconsolidated aquifer (after Mogg[71]).

and (4) provides an annular zone of high permeability, which increases the effective radius and yield of the well. Maximum grain size of a pack should be near 1 cm, while the thickness should be in the range of 8 to 15 cm.

Various formulas for relating gravel pack grain-size gradations to aquifer grain-size gradations have been developed.[52, 60, 71, 84] Criteria conforming to U.S. Bureau of Reclamation field experience are summarized in Table 5.4.4. The selected gravel should be washed and screened siliceous material that is rounded, abrasive-resistant, and dense. Gravel should be placed in such a manner as to ensure complete filling of the annular space and to minimize segregation. A common procedure is to extend two tremie pipes to the bottom of the well on opposite sides of the screen. Gravel is poured, washed, or pumped into the tremie pipes; these are then withdrawn in stages as the pack is placed. In cable-tool holes, the inner casing and screen are set inside the blank outer casing, the annular space is filled with gravel, and thereafter the outer casing is pulled out of the well. In sandy aquifers, where a gravel pack is most essential, deep wells should be constructed by the rotary or reverse-circulation rotary method. The drilling fluid should be circulated and diluted with water before the gravel is introduced.

Table 5.4.4 Criteria for Selection of Gravel Pack Material (after U.S. Bureau of Reclamation[94])

Uniformity coefficient (U_c) of aquifer	Gravel pack criteria	Screen slot size
< 2.5	(a) U_c between 1 and 2.5 with the 50% size not greater than six times the 50% size of the aquifer (b) If (a) is not available, U_c between 2.5 and 5 with 50% size not greater than nine times the 50% size of the aquifer	≤ 10% passing size of the gravel pack
2.5–5	(a) U_c between 1 and 2.5 with the 50% size not greater than nine times the 50% size of the formation (b) If (a) is not available, U_c between 2.5 and 5 with 50% size not greater than 12 times the 50% size of the aquifer	≤ 10% passing size of the gravel pack
> 5	(a) Multiply the 30% passing size of the aquifer by 6 and 9 and locate the points on the grain-size distribution graph on the same horizontal line. (b) Through these points draw two parallel lines representing materials with U_c ≤ 2.5. (c) Select gravel pack material that falls between the two lines.	≤ 10% passing size of the gravel pack

5.5 WELL DEVELOPMENT

Following completion, a new well is developed to increase its specific capacity, prevent sanding, and obtain maximum economic well life. These results are accomplished by removing the finer material from the natural formations surrounding the perforated sections of the casing. Where a well has been gravel packed, much of the same purpose has been accomplished, although development is still beneficial. The importance of developing wells cannot be underestimated; all too often development is not carried out adequately to produce full potential yields.

Development procedures are varied and include pumping, surging, use of compressed air, hydraulic jetting, addition of chemicals, hydraulic fracturing, and use of explosives.[53, 94] These are briefly described in subsequent paragraphs.

5.5.1 Pumping

This procedure involves pumping a well in a series of steps from a low discharge to one exceeding the design capacity.[*] To be most effective, the intake area of the pump should extend to near the center of the screened section. At each step the well is pumped until the water clears, after which the power is shut off and water in the pump column surges back into the well. The step is repeated until only clear water appears. The discharge rate is then increased and the procedure repeated until the final rate is the maximum capacity of the pump or well. This irregular and noncontinuous pumping agitates the fine material surrounding the well so that it can be carried into the well and pumped out. The coarser fraction entering the well is removed by a bailer or sand pump from the bottom. This development method by pumping is recommended as a finishing procedure after any of the development techniques described subsequently.

5.5.2 Surging

Another method for developing a well is by the up-and-down motion of a surge block attached to the bottom of a drill stem. Such blocks are particularly applicable with a cable tool rig.[94] Solid, vented, and spring-loaded surge blocks, often constructed by well-drilling contractors, are employed.[†] The cylindrical block is 2 to 5 cm smaller than the well screen and fitted with belting, rubber, or leather that will not damage the screen. As the block is moved up and down in the screen, a surging action is imparted to the water. The downstroke causes backwash to break up any bridging that may occur, while the upstroke pulls dislodged sand grains into the well.

Initially, surging should begin with a slow stroke at the bottom of the screen and progress to the top of the screen. This should then be repeated with increasingly faster strokes. The procedure is completed when material accumulating in the bottom of the well becomes negligible. For wells in rock aquifers, surging can be accomplished in the casing above open holes.

5.5.3 Surging with Air

To develop wells with compressed air, an air compressor is connected to an air pipe into the well. Around the air pipe a discharge pipe is fitted, as shown in Figure 5.5.1. (Both pipes should be capable of being shifted vertically by clamps. Initially, the pipes extend to near the bottom of the screened section; for efficient operation, the water depth in the discharge pipe

---

[*]In the field this technique is sometimes referred to as "rawhiding" a well. If a high discharge occurs initially, "bridging" (wedging sand grains around individual perforations formed by the sudden pull on the sand toward the well) can prevent fine material from being removed and reduce the effectiveness of the development process.

[†]Surging can also be accomplished with a flap-valve bailer, if a close fit exists within the well screen.

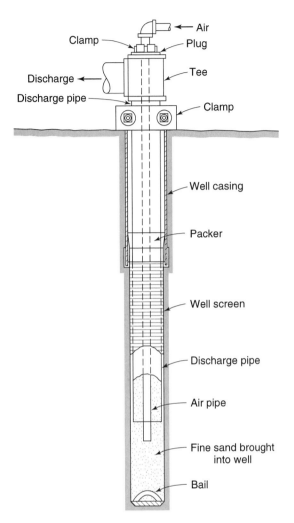

Clamp

Discharge ←

Discharge pipe

Air

Plug

Tee

Clamp

Well casing

Packer

Well screen

Discharge pipe

Air pipe

Fine sand brought
into well

Bail

Figure 5.5.1. Installation for well development with compressed air.

should exceed two-thirds the length of the pipe. To begin the development, the air pipe is closed and the air pressure is allowed to build up to 0.7 to 1.0×10^6 Pa, whereupon it is released suddenly into the well by means of a quick-opening valve. The inrush of air creates a powerful surge within the well, first increasing then decreasing the pressure as water is forced up the discharge pipe. The process loosens the fine material surrounding the perforations; the material may then be brought into the well by continuous air injection, creating an airlift pump. The operation is repeated at intervals along the screened section until sand accretion becomes negligible.

5.5.4 Backwashing with Air

In the backwashing method, the top of the well is fitted with an airtight cover. Discharge and air pipes are installed similar to the previous method, together with a separate short air pipe and a three-way valve, as shown in Figure 5.5.2. Compressed air is released through the long air pipe, forcing air and water out of the well through the discharge pipe. After the water clears, the air supply is shut off and the water is allowed to return to its static level. The three-way

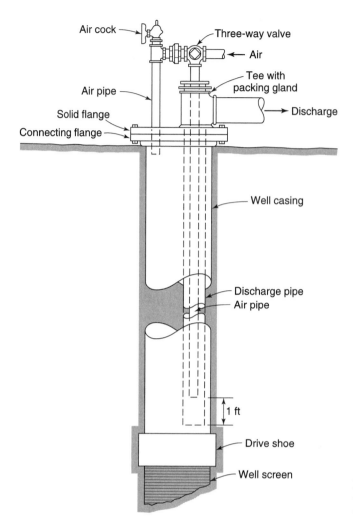

Figure 5.5.2. Installation for well development by backwashing with air.

valve is then turned to admit air into the top of the well through the short air pipe. This backwashes the water from the well through the discharge pipe and at the same time agitates the sand grains surrounding the well. Air is forced into the well until it begins escaping from the discharge pipe, after which the three-way valve is turned and the air supply is again directed down the long air pipe to pump the well. Backwashing is repeated until the well is fully developed.

5.5.5 Hydraulic Jetting

Jetting with a high-velocity stream of water is an effective development technique in open rock holes and in wells containing screens with large percentage openings (see Figure 5.5.3). The jet nozzle, mounted horizontally, is attached to a string of pipe, which is connected through a swivel and hose to a high-pressure, high-capacity pump. The jet head is slowly rotated to successively higher levels. Fine-grained material from unconsolidated aquifers is carried into the well by the turbulent flow; in addition, the method is particularly effective in developing gravel-packed wells.

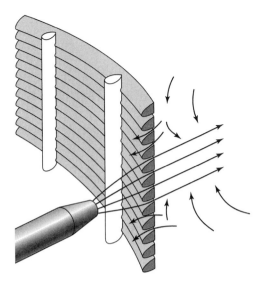

Figure 5.5.3. High-velocity hydraulic jetting through a continuous slot wire-wound well screen for well development (after Johnson Div., UOP Inc.[53]).

5.5.6 Chemicals

Open-hole wells in limestone or dolomite formations can be developed by adding hydrochloric acid to water in the well. The solvent action removes fine particles and tends to widen fractures leading into the well bore. Normally this procedure would be followed by one of the previously described development methods. Hydrofluoric acid can be similarly employed for rocks containing silicates.

For most development methods, adding one of the polyphosphates* to water in the well will aid the development process.[94] These compounds act as deflocculants and dispersants of clays and other fine-grained materials, thereby enabling the mud cake on the wall of a hole and the clay fractions in an aquifer to be more readily removed by the development.

Blocks of solid carbon dioxide (dry ice) are sometimes added to a well after acidizing and surging with compressed air to complete well development. The accumulation of gaseous carbon dioxide released by sublimation builds up a pressure within the well; upon release this causes a burst of muddy water from the well.

5.5.7 Hydraulic Fracturing

Hydraulic fracturing, a technique borrowed from the petroleum industry, is occasionally employed to enhance the yield of open-hole rock wells. Inflatable packers on a pipe extending to ground surface isolate a section of aquifer. After filling the pipe and isolated section with water, pump pressure is applied to fracture the rock. Sand is sometimes pumped into the section to force the grains into the rock fractures, to maintain the openings.

5.5.8 Explosives

Detonation of explosives in rock wells often increases yields by enlarging the hole, increasing rock fractures, and removing fine-grained deposits on the face of the well bore.

---

*Common household phosphate-based detergents can serve as a substitute; excessive foaming when the well is pumped may result.

5.6   TESTING WELLS FOR YIELD

Following development of a new well, it should be tested to determine its yield and drawdown. This information provides a basis for determining the water supply available from the well, for selecting the type of pump, and for estimating the cost of pumping. A test is accomplished by measuring the static water level, after which the well is pumped at a maximum rate until the water level in the well stabilizes. The depth to water is then noted (techniques for measuring water well depths are described in Chapter 12). The difference in depths is the drawdown, and the discharge–drawdown ratio is an estimate of the specific capacity of the well. The discharge can be determined by any of several measuring devices connected to the discharge pipe.

5.7   PUMPING EQUIPMENT

Well pumps produce flow by transforming mechanical energy to hydraulic energy. A wide variety of pumps are produced annually by many manufacturers. The selection of a particular size and type of pump depends on several factors, including: (1) pumping capacity, (2) well diameter and depth, (3) depth and variability of pumping level, (4) straightness of the well, (5) sand pumping, (6) total pumping head, (7) duration of pumping, (8) type of power available, and (9) costs.[94] Table 5.7.1 summarizes characteristics of pumps most frequently employed in wells.

5.7.1   Total Pumping Head

The total pumping head, or total dynamic head, of a pump represents the total vertical lift of the water from the well. As shown in Figure 5.7.1, this total head consists of three components: (1) the drawdown inside the well (including aquifer and well losses); (2) the static head, being the difference between the static groundwater level and the static discharge elevation; and (3) friction losses due to flow through the intake and discharge pipes. As indicated in the figure, the total pumping head increases with discharge.

5.7.2   Pumps for Shallow Wells

For shallow wells where only small discharges are needed, hand-operated pitcher pumps, turbine pumps, and gear pumps may be installed. Discharges range up to 500 m^3/day. Suction lifts should not exceed about 7 m for efficient and continuous service.

Where a larger discharge is required from a shallow well, a centrifugal pump is commonly employed. The assembly may be mounted with a horizontal or a vertical shaft. The horizontal design is efficient, easy to install and maintain, and usually connected directly to an electric motor. Because of the low suction head, the pump is often placed a short distance above the water level in large diameter wells.

5.7.3   Pumps for Deep Wells

In deep wells requiring high lifts, large-capacity pumps serving irrigation, municipal, or industrial water requirements are installed. Several types of pumps are suitable for deep-well operation: plunger, displacement, airlift, jet, and, most importantly, deep-well turbine and submersible.

The deep-well turbine pump has been widely adopted for large, deep, high-capacity wells (see Figure 5.7.2a). This type of pump has its impeller suspended vertically on a long drive shaft within a discharge pipe. The bowl of the pump contains the impeller and guide vanes;

Table 5.7.1 Characteristics of Pumps Frequently Employed in Wells (after U.S. Public Health Service[96])

Type of pump	Practical suction lift[a]	Usual well-pumping depth	Usual pressure heads	Advantages	Disadvantages
Reciprocating:					
Shallow well	6–7 m	6–7 m	30–60 m	Positive action; discharge against variable heads; pumps water containing sand and silt; especially adapted to low capacity and high lifts.	Pulsating discharge; subject to vibration and noise; maintenance cost may be high; may cause destructive pressure if operated against closed valve
Deep well	6–7 m	Up to 180 m	Up to 180 m above cylinder		
Centrifugal:					
Shallow well					
Straight centrifugal (single stage)	6 m max.	3–6 m	30–45 m	Smooth, even flow; pumps water containing sand and silt; pressure on system is even and free from shock; low starting torque; usually reliable and good service life	Loses prime easily; efficiency depends on operating under design heads and speed
Regenerative vane turbine type (single impeller)	8 m max.	8 m	30–60 m	Same as straight centrifugal except not suitable for pumping water containing sand or silt; self-priming	Same as straight centrifugal except maintains priming easily
Deep well					
Vertical line shaft turbine (multistage)	Impellers submerged	15–90 m	30–250 m	Same as shallow well turbine	Efficiency depends on operating under design head and speed; requires straight well large enough for turbine bowls and housing; lubrication and alignment of shaft critical; abrasion from sand
Submersible turbine (multistage)	Pump and motor submerged	15–120 m	15–120 m	Same as shallow well turbine; easy to frost-proof installation; short pump shaft to motor	Repair to motor or pump requires pulling from well; sealing of electrical equipment from water vapor critical; abrasion from sand
Jet:					
Shallow well	4–6 m below ejector	Up to 4–6 m below ejector	25–45 m	High capacity at low heads; simple in operation; does not have to be installed over well; no moving parts in well	Capacity reduces as lift increases; air in suction or return line will stop pumping
Deep well	4–6 m below ejector	7–35 m 60 m max.	25–45 m	Same as shallow well jet	Same as shallow well jet
Rotary:					
Shallow well (gear type)	7 m	7 m	15–75 m	Positive action; discharge constant under variable heads; efficient operation	Subject to rapid wear if water contains sand or silt; wear of gears reduces efficiency
Deep well (helical rotary type)	Usually submerged	15–150 m	30–150 m	Same as shallow well rotary; only one moving pump device in well	Same as shallow well rotary except no gear wear

[a]Practical suction lift at sea level. Reduce lift 0.3 m for each 300 m above sea level.

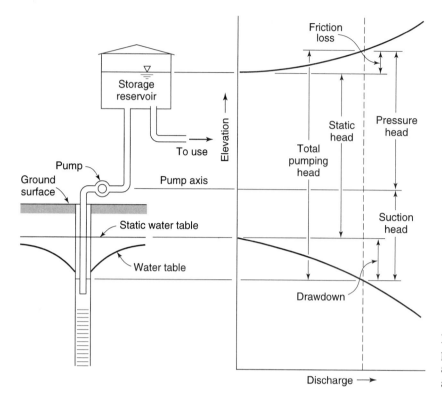

Figure 5.7.1. Diagram illustrating total pumping head for a well supplying a storage reservoir. Note the increase in head as a function of well discharge.

several bowls connected in series for higher heads form a multiple-stage pump. The pump is usually driven by an electric motor at the ground surface and connected by a long vertical shaft positioned by bearings within the discharge pipe. Deep-well pumps, being submerged, require no priming and are capable of operating under a wide range of water levels without having to be reset.

The submersible pump is simply a deep-well turbine pump close-coupled to a small-diameter submersible electric motor, as shown in Figure 5.7.2b. The efficiency of the pump is increased by direct coupling, while effective cooling results from complete immersion. Pump sizes range from small units that fit inside an 8-cm casing to large-capacity units involving numerous stages. An advantage of submersible pumps is that they can lift water from deep wells where long shafts in crooked casings might prohibit installation of a deep-well turbine pump. Other advantages, which account for increasing installations of this type of pump, include ease of maintenance, freedom from noise, protection from weathering and flooding, and avoidance of a large above ground installation.

5.8   PROTECTION OF WELLS

5.8.1   Sanitary Protection

Wherever groundwater pumped from a well is intended for human consumption, proper sanitary precautions must be taken to protect the water quality. Pollution sources may exist either above or below ground surface (see Chapter 8). Precautions apply equally to springs; Figure 5.8.1 shows, for example, a typical method for protecting a spring water supply.

Surface pollution can enter wells either through the annular space outside of the casing or through the top of the well itself. To close avenues of access for undesirable water outside of the casing, the annular space should be filled with cement grout as shown for deep wells in Fig-

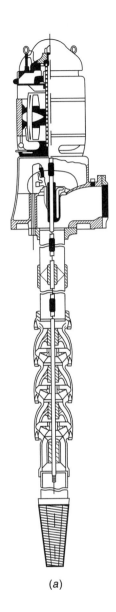

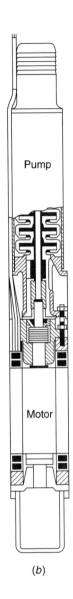

Pump

Motor

Figure 5.7.2. Pumps for deep wells. (*a*) Turbine (*b*) Submersible (after Anderson[6])

(*a*)   (*b*)

ure 5.3.1. Entry through the top of the well can be avoided by providing a watertight cover to seal the top of the casing.[12] Some pumps are available with closed metal bases that provide the necessary closure. For pumps with an open-type base, or where the pump is not placed directly over the well, a seal is required for the annular opening between the discharge pipe and casing.[*] Seals may be made of metal or lead packing; asphaltic and mastic compounds are also satisfactory. Covers around the well should be made of concrete, should be elevated above the adjacent land level, and should slope away from the well (see Figure 5.8.2).

Whenever a new well is completed or an old well repaired, contamination from equipment, well materials, or surface water may be introduced into the well. Addition and agitation of a chlorine compound will disinfect the well. Following disinfection, the well should be pumped

---

[*]It is desirable to provide a small opening in or below the pump base to allow for periodic water level measurements.

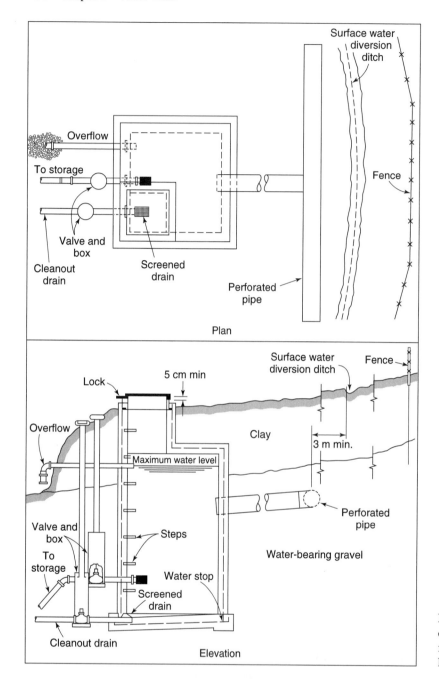

Figure 5.8.1. Plan and elevation views of a developed spring showing a typical method for providing sanitary protection (after U.S. Public Health Service [96]).

to waste until all traces of chlorine are removed. As a final check on the potability of the water, a sample should be collected and sent to a laboratory for bacteriological examination.

5.8.2 Frost Protection

In regions where winter frost occurs, it is important to protect pumps and water lines from freezing. A typical method for frostproofing a domestic well is shown in Figure 5.8.3. The pitless adapter, attached to the well casing, provides access to the well, while the discharge pipe runs about 2 m underground to the basement of the house.

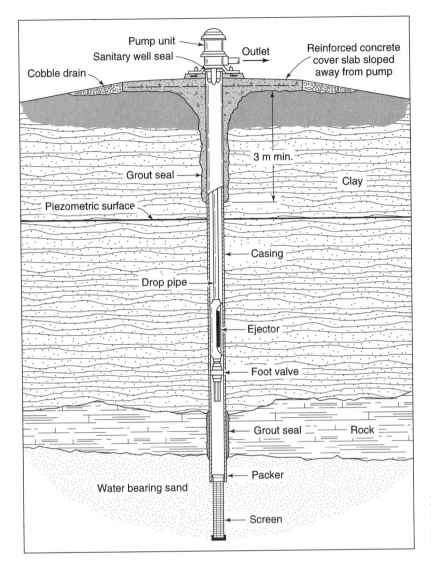

Figure 5.8.2. A drilled well showing grout seal, concrete slab, and well seal for sanitary protection (after U.S. Public Health Service[96]).

5.8.3 Abandonment of Wells

Whenever a well is abandoned, for whatever reason, it should be sealed by filling it with clay, concrete, or earth. Not only is surface contamination then unable to enter the well, but sealing serves other useful purposes: prevents accidents, avoids possible movement of inferior water from one aquifer to another, and conserves water in flowing wells.[20]

5.9 WELL REHABILITATION

A new well, properly drilled, cased, and developed, will give years of satisfactory service with little attention. Many wells fail, however; that is, they yield decreasing quantities of water with time.* *Well rehabilitation* refers to the treatment of a production well by mechanical, chemical, or other means to recover as much as possible of the lost production capacity.[58, 59] Table 5.9.1 lists well rehabilitation methods and their applications to various types of aquifers.

---

*Frequently the pump rather than the well is at fault; hence, it should be checked before beginning any extensive well repair.

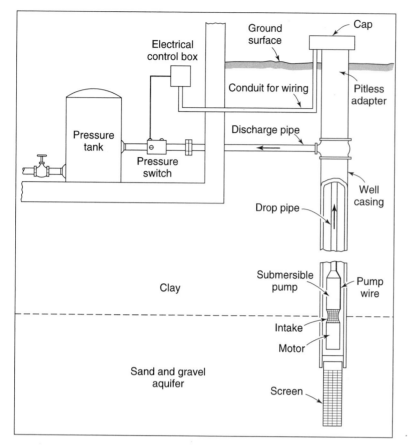

Figure 5.8.3. Diagram of a domestic well installation with a pitless adapter to protect the well from frost (after Gibb[37]).

One cause of failure is depletion of the groundwater supply. Not a fault of the well, this trouble can sometimes be remedied by decreasing pumping drafts, resetting the pump, or deepening the well. A second cause of well trouble results from faulty well construction. Such items as poor casing connections, improper perforations or screens, incomplete placement of gravel packs, and poorly seated wells are typical of difficulties encountered. Depending on the particular situation as determined from a television or photographic survey of the well (see Chapter 12), it may be possible to repair the well, but sudden failures involving entrance of sand or collapse of a casing often require replacement of the entire well.

The third and most prevalent cause of well failure results from corrosion or incrustation of well screens.[8, 62, 81] Corrosion may result from direct chemical action of the groundwater or from electrolytic action caused by the presence of two different metals in the well. The effects of corrosion can be minimized by selecting nonmetallic well screens or ones of corrosion-resistant metal (such as nickel, copper, or stainless steel), and by providing cathodic protection.[*] If the damage is localized, it may be possible to insert a liner inside the screen to prevent excessive sand pumping.

Incrustation is caused by precipitation on or near well screens of materials carried in solution by groundwater.[72] The sudden pressure drop associated with water entering a well under heavy pumping can release carbon dioxide and cause precipitation of calcium carbonate. Another cause of incrustation stems from the presence of oxygen in a well; this can change soluble ferrous iron to insoluble ferric hydroxide. Screens can be cleaned by shooting a string

---

[*]One method of providing cathodic protection for a well is to introduce a metal low on the electrochemical scale that will be corroded instead of the well casing. Rods of magnesium suspended in the well water serve this purpose.

Table 5.9.1 Rehabilitation Methods and Their Applications to Various Types of Aquifers (after Erickson[30])

Method	Unconsolidated aquifers	Consolidated sandstone	Consolidated limestone
Muriatic acid[a] followed by chlorine	Removes iron, sulfur, and carbonate deposits	Not usually effective	Sometimes beneficial; best results obtained by pressure acidizing
Polyphosphate followed by chlorine	Removes fine silt, clay, colloids, disseminated shale, and soft iron deposits	Not usually effective	Not usually effective
Dynamiting	Not recommended	Effective for all types of well-screen deposits	Effective when very large charges are used
Compressed air	Removes plugging deposits of silt and fine sand in areas adjacent to screens	Not used	Not used
Dry ice	Same as compressed air	Used only rarely, to remove cuttings from the face of a new production well	Not usually effective
Surging	Same as compressed air	Rarely used	Rarely used
Chlorine[b]	Removes iron and slime-forming bacteria	Removes iron and slime-forming bacteria	Removes iron and slime-forming bacteria
Caustic soda	Removes oil scum left by oil-lubricating pumps	Removes oil scum left by oil-lubricating pumps	Removes oil scum left by oil-lubricating pumps

[a]Not to be used with concrete screens.

[b]Usually used in a concentration of 500 mg/1.

of vibratory explosives in the well or by adding hydrochloric acid (HCl) or sulfamic acid (H_2NSO_3H) to the well, followed by agitation and surging. Where slime-forming organisms block screens, particularly in recharge wells, treatment with chlorine gas or hypochlorite solutions[94] can remedy the problem. For improving yields of rock wells, acidizing or shooting with explosives is generally effective.[102]

5.10   HORIZONTAL WELLS

Subsurface conditions often preclude groundwater development by normal vertical wells. Such conditions may involve aquifers that are thin, poorly permeable, or underlain by permafrost or saline water. In other circumstances, where groundwater is to be derived primarily from infiltration of streamflow, a horizontal well system may be advantageous. Also, in developing areas of the world, a horizontal well may be far less costly than a vertical well.

5.10.1   Infiltration Galleries

An *infiltration gallery* is a horizontal conduit for intercepting and collecting groundwater by gravity flow.[9] Qanats, described in Chapter 1, illustrate one type of gallery. Galleries, normally constructed at the water table elevation, discharge into a sump where a pump lifts the water to ground surface for use. In Europe and the United States, many infiltration galleries are laid parallel to riverbeds, where with induced infiltration an adequate perennial water supply can be obtained (see Chapter 13). Depending on the type of aquifer penetrated, galleries may be unlined or lined with vitrified clay, brick, concrete, or cast iron.*

---

*Deep infiltration galleries exist in the chalk aquifers of southeastern England. Here the permeability of solid chalk is quite low, and unlined horizontal tunnels (or adits), roughly 2 m in diameter and extending for distances up to 2 km, intersect a maximum number of fissures from which most of the water is obtained.

In Alaska, galleries have been widely employed to obtain water supplies where underlying permafrost (see Chapter 2) would not contribute groundwater.[33] On oceanic islands, galleries have the particular advantage of enabling fresh water to be collected with little disturbance of underlying saline water (see Chapter 14). Such installations are well known in Hawaii (see Figure 5.10.1), for example, where they are unlined in basalt,[78, 103] and in Barbados, where they are unlined in coral limestone.

5.10.2   Horizontal Pipes

On sloping ground surfaces, small-diameter horizontal holes can be drilled by the rotary method.[93, 107] Perforated pipes placed in these holes tap groundwater that would otherwise be discharged by seepage or from small springs. Two examples are shown in Figure 5.10.2. Such horizontal pipes provide sanitary and lowcost water, and, in addition, enable flow to be controlled by valves at the discharge ends. The same technique has also been widely employed to drain side slopes, such as in highway cuts, in order to reduce the possibility of landslides.

5.10.3   Collector Wells

For cities and industries located near rivers, the problem of obtaining high-quality, low-temperature water at reasonable cost has become increasingly difficult. In Europe and the United States, groundwater pumped from collector wells tapping permeable alluvial aquifers has often proved to be a successful solution.[39, 89] If located adjacent to a surface water source, a collector well lowers the water table and thereby induces infiltration of surface water through the bed of the water body to the well (see Chapter 13). In this manner greater supplies of water

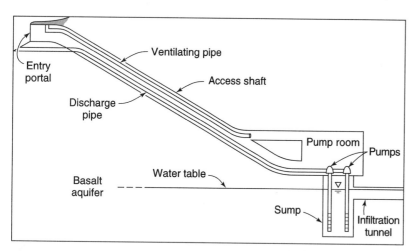

Figure 5.10.1. Cross section of an infiltration gallery in Hawaii. These are locally known as *Maui-type wells* because they were first constructed on the island of Maui to provide water for irrigation of sugar cane (after Watson[103]).

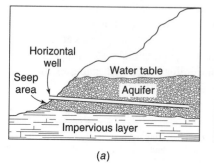

(a)

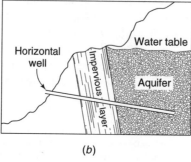

(b)

Figure 5.10.2. Examples of horizontal wells consisting of small-diameter perforated pipes drilled into hillsides. These wells should have a downward slope into the aquifer to prevent formation of a vacuum inside the pipes (after Welchert and Freeman[107]). (*a*) Contact spring formation (*b*) Dike spring formation

can be obtained than would be available from groundwater alone. Analytic solutions of the flow to a collector well have been developed (see Chapter 4).

Plan and elevation views of a collector well are shown in Figure 5.10.3. The central cylinder, consisting of a monolithic concrete caisson about 5 m in diameter, is sunk into the aquifer by excavating the inside earth material. After the requisite depth is reached, a thick concrete plug is poured to seal the bottom. Perforated pipes, 15 to 20 cm in diameter, are jacked hydraulically into the water-bearing formation through precast portholes in the caisson to form a radial pattern of horizontal pipes. During construction, fine-grained material is washed into the cais-

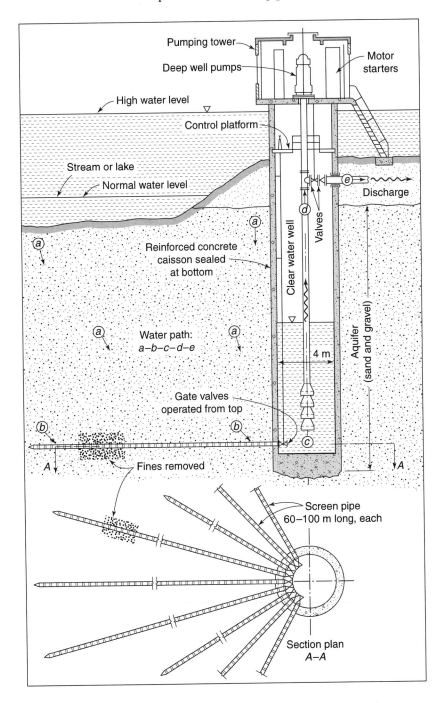

Figure 5.10.3. A collector well located near a surface water body (courtesy Ranney Method Water Supplies, Inc.).

son so that natural gravel packs form around the perforations (again see Figure 5.10.3). The number, length, and radial pattern of the collector pipes can be varied to obtain the maximum capacity; usually more pipes are extended toward than away from the surface water source.

The large area of exposed perforations in a collector well causes low inflow velocities, which minimize incrustation, clogging, and sand transport. Polluted river water is filtered by its passage through the unconsolidated aquifer to the well. The initial cost of a collector well exceeds that of a vertical well; however, advantages of large yields, reduced pumping heads, and low maintenance costs are factors to be considered. Yields vary with local conditions; the average for a large number of such wells approximated 27,000 m^3/day. Installations bordering streams with regulated stages may encounter decreases in yield with time because of sediment deposition on streambeds.[39]

Collector wells can also function in permeable aquifers removed from surface water. Several such installations gave an average yield of about 15,000 m^3/day.

5.11   CHARACTERISTIC WELL LOSSES

5.11.1   Well Losses

The drawdown at a well includes not only that of the logarithmic drawdown curve at the well face, but also a *well loss* caused by flow through the well screen and flow inside of the well to the pump intake. Because the well loss is associated with turbulent flow, it may be indicated as being proportional to an nth power of the discharge, as Q^n, where n is a constant greater than one. Jacob[51] suggested that a value $n = 2$ might be reasonably assumed, but Rorabaugh[82] pointed out that n can deviate significantly from 2. An exact value for n cannot be stated because of differences of individual wells; detailed investigations of flows inside and outside of wells show that considerable variations occur from assumed flow distributions.

Taking account of the well loss, the total drawdown s_w at the well may be written for the steady-state confined case:

$$s_w = \frac{Q}{2\pi T} \ln \frac{r_0}{r_w} + CQ^n \tag{5.11.1}$$

where C is a constant governed by the radius, construction, and condition of the well. For simplicity let

$$B = \frac{\ln(r_0/r_w)}{2\pi T} \tag{5.11.2}$$

so that

$$s_w = BQ + CQ^n \tag{5.11.3}$$

Therefore, as shown in Figure 5.11.1, the total drawdown s_w consists of the formation loss BQ and the well loss CQ^n.

Consideration of Equation 5.11.3 provides a useful insight to the relation between well discharge and well radius. From Equations 4.2.3 and 4.2.7, it can be seen that Q varies inversely with $\ln(r_0/r_w)$, if all other variables are held constant. This shows that discharge varies only a small amount with well radius. For example, doubling a well radius increases the discharge only 10 percent. When the comparison is extended to include well loss, however, the effect is significant. Doubling the well radius doubles the intake area, reduces entrance velocities to almost half, and (if $n = 2$) cuts the frictional loss to less than a third. For axial flow within the well, the area increases four times, reducing this loss an even greater extent.

It is apparent that the well loss can be a substantial fraction of total drawdown when pumping rates are large, as illustrated by Figure 5.11.2. With proper design and development

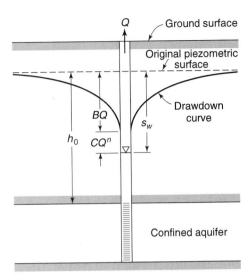

Figure 5.11.1. Relation of well loss CQ^n to drawdown for a well penetrating a confined aquifer.

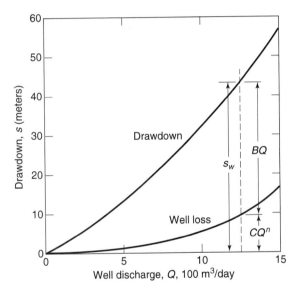

Figure 5.11.2. Variation of total drawdown s_w, aquifer loss BQ, and well loss CQ^n with well discharge (after Rorabaugh[82]).

of new wells, well losses can be minimized. Clogging or deterioration of well screens can increase well losses in old wells.[54] Based on field experience, Walton[101] suggested criteria for the well loss coefficient C in Equation 5.11.3. These are presented in Table 5.11.1 to aid in evaluating the condition of a well.

5.11.2 Evaluation of Well Loss

To evaluate well loss, a *step-drawdown pumping test* is required. This consists of pumping a well initially at a low rate until the drawdown within the well essentially stabilizes.[51, 64, 86] The discharge is then increased through a successive series of steps as shown by the time–drawdown data in Figure 5.11.3a. Incremental drawdowns Δs for each step are determined from approximately equal time intervals. The individual drawdown curves should be extrapolated with a slope proportional to the discharge in order to measure the incremental drawdowns.

From Equation 5.11.3 and letting $n = 2$, we have

$$\frac{s_w}{Q} = B + CQ \qquad (5.11.4)$$

Table 5.11.1 Relation of Well Loss Coefficient to Well Condition (after Walton[101])

Well loss coefficient C (min^2/m^5)	Well condition
< 0.5	Properly designed and developed
0.5 to 1.0	Mild deterioration or clogging
1.0 to 4.0	Severe deterioration or clogging
> 4.0	Difficult to restore well to original capacity

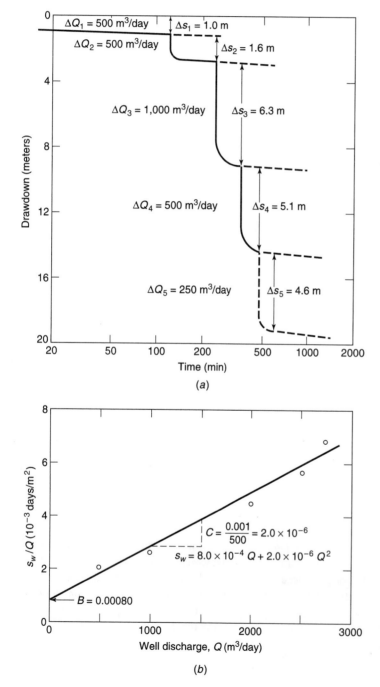

(a)

(b)

Figure 5.11.3. Step-drawdown pumping test analysis to evaluate well loss (after Bierschenk[13]). (a) Time–drawdown data from step-drawdown pumping test (b) Determination of B and C from graph of S_w/Q versus Q

Therefore, by plotting s_w/Q versus Q (see Figure 5.11.3) and fitting a straight line through the points, the well loss coefficient C is given by the slope of the line and the formation loss coefficient B by the intercept $Q = 0$.

Rorabaugh[82] presented a modification of this graphic analysis to determine n in cases where it deviates significantly from 2.

EXAMPLE 5.11.1

The time–drawdown data from a step-drawdown pumping test in an existing well is given below. Determine the well loss coefficient C and the formation loss coefficient B assuming $n = 2$, and find the corresponding contributions of the formation loss and well loss for the given drawdown data.

Q (m³/day)	s_w (m)
500	1.14
1,000	2.66
1,500	5.57
2,000	8.82
2,500	13.54
3,000	18.79
3,500	23.67

SOLUTION

Using the given time–drawdown data, we plot values of s_w/Q versus Q in Figure 5.11.4. Then, a straight line is fitted through the data points.

The equation for the best-fit straight line is given by

$$y = 1.597 \times 10^{-6}x + 0.001307$$

or

$$s_w/Q = 1.597 \times 10^{-6}Q + 0.001307$$

Thus, from Equation 5.11.4, the formation loss coefficient is $B = 0.001307$ days/m² and the well loss coefficient is $C = 1.597 \times 10^{-6}$ day²/m⁵ or $C = 3.31$ min²/m⁵. Note that this coefficient indicates severe deterioration or clogging in the well according to Walton criteria (see Table 5.11.1).[101]

Using the coefficients computed above, we can calculate the drawdown due to formation loss and the additional drawdown due to well loss for any discharge. The theoretical drawdown data based on the estimated coefficients are reproduced for the given discharges in Table 5.11.2. As seen in the table, the theoretical drawdowns do not match the measured data exactly because of the $n = 2$ assumption. On the other hand, they are close enough for practical purposes.

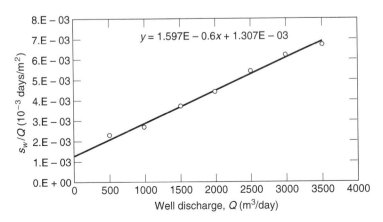

Figure 5.11.4. Drawdown data for Example 5.11.1.

Table 5.11.2 Calculated Well Loss Parameters for Example 5.11.1

Q (m³/day)	BQ (m)	CQ^2 (m)	s_w (m)
500	0.65	0.40	1.05
1,000	1.31	1.60	2.90
1,500	1.96	3.59	5.55
2,000	2.61	6.39	9.00
2,500	3.27	9.98	13.25
3,000	3.92	14.37	18.29
3,500	4.57	19.56	24.14

Finally, the theoretical drawdown curves relating the formation and well losses to the well discharge are shown in Figure 5.11.5. As can be seen, the well losses become more significant as the well discharge increases and even larger than the formation loss beyond approximately $Q = 800$ m³/day. ∎

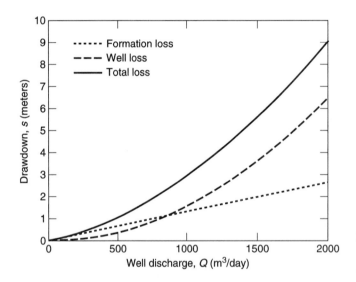

Figure 5.11.5 Example 5.11.1: Comparison of formation and well losses.

EXAMPLE 5.11.2

The purpose of this example is to illustrate the Rorabaugh graphical procedure[82] to determine the value of n, the formation loss coefficient B, and the well loss coefficient C in Equation 5.11.3 using the time-drawdown data from a step-drawdown pumping test (listed below)

Q (m³/day)	s_w (m)
500	2.40
1,000	5.38
1,500	9.28
2,000	14.36
2,500	20.82
3,000	28.87
3,500	38.70

SOLUTION

Step 1 Rearrange Equation 5.11.3 as follows:

$$s_w = BQ + CQ^n$$

$$\frac{s_w}{Q} - B = CQ^{n-1}$$

$$\log\left(\frac{s_w}{Q} - B\right) = \log(C) + (n-1)\log(Q)$$

This represents a straight-line plot of $(s_w/Q - B)$ versus Q on log–log scale with slope $(n-1)$ and intercept C. However, such a plot cannot be made since the formation loss coefficient B is not known.

Step 2 Rorabaugh's procedure requires a family of plots for incremental values of the formation loss coefficient B until a straight line plot is obtained. This is done for the given time–drawdown data and the resulting plots are shown in Figure 5.11.6.

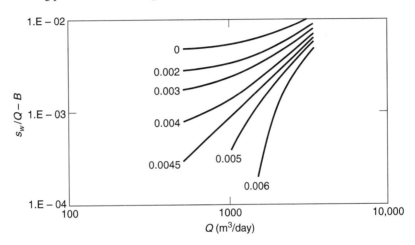

Figure 5.11.6. Example 5.11.2: Plots for Rorabaugh procedure.

Step 3 Referring to Figure 5.11.6, we observe a straight line for $B = 0.0045$ days/m^2. The slope of this line is 1.6; $n - 1 = 1.6$ and $n = 2.6$ is obtained.

Step 4 The value of C is obtained by taking an arbitrary combination of Q and s_w and applying Equation 5.11.3: $Q = 2000$ m^3/day and $s_w = 14.36$ m,

$$s_w = BQ + CQ^n$$
$$14.36 \text{ m} = (0.0045 \text{ days/m}^2)(2000 \text{ m}^3/\text{day}) + C(2000 \text{ m}^3/\text{day})^{2.6}$$
$$C = 1.4 \times 10^{-8}$$

Step 5 Using the values of B, C, and n, the formation and well losses for the given discharges are calculated and tabulated in Table 5.11.3.

Table 5.11.3 Calculated Drawdown Parameters for Example 5.11.2

Q (m^3/day)	BQ (m)	CQ^2 (m)	s_w (m)
500	2.25	0.15	2.40
1,000	4.50	0.88	5.38
1,500	6.75	2.53	9.28
2,000	9.00	5.36	14.36
2,500	11.25	9.57	20.82
3,000	13.50	15.37	28.87
3,500	15.75	22.95	38.70

■

5.12   SPECIFIC CAPACITY AND WELL EFFICIENCY

If discharge is divided by drawdown in a pumping well, the *specific capacity* of the well is obtained. This is a measure of the productivity of a well; clearly, the larger the specific capacity, the better the well. Starting from the Cooper-Jacob approximation of the nonequilibrium equation (Equation 4.4.8) and including the well loss,

$$s_w = \frac{2.30Q}{4\pi T}\log\frac{2.25Tt}{r_w^2 S} + CQ^n \tag{5.12.1}$$

so that the specific capacity is

$$\frac{Q}{s_w} = \frac{1}{(2.30/4\pi T)\log\left(2.25Tt/r_w^2 S\right) + CQ^{n-1}} \tag{5.12.2}$$

This indicates that the specific capacity decreases with Q and t; the well data plotted in Figure 5.12.1 demonstrate this effect. For a given discharge a well is often assumed to have a constant specific capacity. When using this approximation for $u < 0.01$, it can be seen that the change with time is minor.

Any significant decline in the specific capacity of a well can be attributed either to a reduction in transmissivity due to a lowering of the groundwater level in an unconfined aquifer or to an increase in well loss associated with clogging or deterioration of the well screen.

If a pumping well is assumed to be 100 percent efficient ($CQ^n = 0$), then the specific capacity from Equation 5.12.2 can be presented in the graphic form of Figure 5.12.2. Here specific capacity at the end of one day of pumping is plotted as a function of S, T, and a well diameter of 30 cm.[65] This graph provides a convenient means for estimating T from existing pumping wells; any error in S has a small effect on T.[35, 49] Refer to Prudic[79] for application of estimating hydraulic conductivity from aquifer-test data and specific-capacity data. The factors that affect estimates of transmissivity based on specific capacity are shown in Figure 5.12.3.

Figure 5.12.2 yields a theoretical specific capacity (Q/BQ) for known values of S and T in an aquifer. This computed specific capacity, when compared with one measured in the field (Q/s_w), defines the approximate efficiency of a well.[13] Thus, for a specified duration of pumping, the well efficiency E_w is given as a percentage by

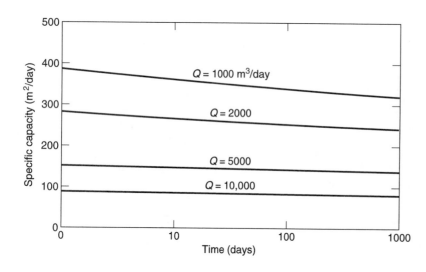

Figure 5.12.1. Variation in specific capacity of a pumping well with discharge and time (after Jacob[51]).

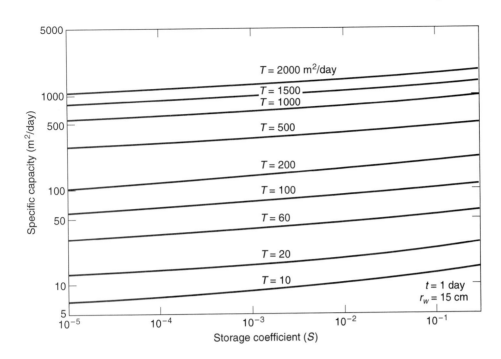

Figure 5.12.2. Graph relating specific capacity to transmissivity and storage coefficient from the Cooper-Jacob nonequilibrium equation for a confined aquifer (after Bentall[10]).

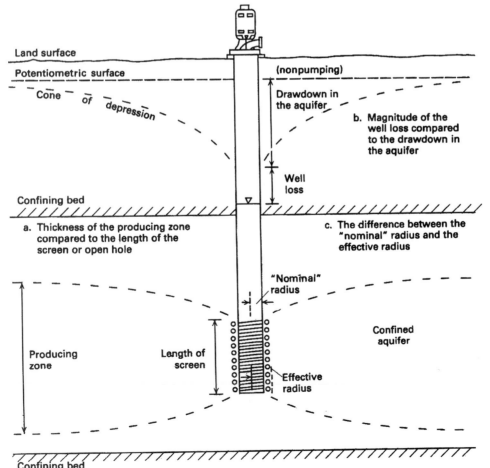

Figure 5.12.3. Factors affecting estimates of transmissivity based on specific capacity (Heath[45]).

$$E_w = 100\,\frac{Q/s_w}{Q/BQ} = 100\,\frac{BQ}{s_w} \text{ or } E_w = 100\left(\frac{BQ}{BQ+CQ^2}\right) \tag{5.12.4}$$

Another method for recognizing an inefficient well is to note its initial recovery rate when pumping is stopped. Where the well loss is large, this drawdown component recovers rapidly by drainage into the well from the surrounding aquifer. A rough rule of thumb for this purpose is if a pump is shut off after one hour of pumping and 90 percent or more of the drawdown is recovered after five minutes, it can be concluded that the well is unacceptably inefficient.

5.13   SLUG TESTS

5.13.1   Definition

Pump tests are typically expensive to conduct because of the installation costs of wells. Slug testing has been used for several years as a cost-effective and quick method of estimating the hydraulic properties of aquifers. More recently (since the 1980s) it has gained even more popularity (a) in obtaining estimates of hydraulic properties of contaminated aquifers where treating the pumped water is not desirable or feasible and (b) in field investigations of low-permeability materials, particularly for studies of potential waste storage or disposal sites. The materials at these sites may have a hydraulic conductivity that is too low for pump tests.

Slug tests consist of measuring the recovery of head in a well after near instantaneous change in the head at that well. A solid object (slug) is rapidly introduced into or removed from the well, causing a sudden change (increase or decrease) in the water level in the well. Tests can also be performed by introducing an equivalent volume of water into the well; or, an equivalent volume of water can be removed from the well, causing a sudden decrease in the water level. Following the sudden change in head, the water level returns to the static water level. While the water level is returning to the static level, the head is measured as a function of time (referred to as the *response data*). This response data is used to determine the hydraulic properties of the aquifer using one of several methods of analyses. Butler[17] presents a comprehensive summary of slug-test methodology and analysis.

5.13.2   Design Guidelines

Butler[17] presented the following guidelines for slug tests (also see Butler et al.,[19] Dawson and Istok,[28] Driscoll,[29] and Weight and Sonderegger[105] for further discussions):

1. Well-drilling methods, such as the driving based cable-tool, pneumatic/hydraulic hammering, or rotasonic methods that minimize the generation of drilling debris should be employed.
2. Well development efforts should focus on developing discrete intervals along the well screen. Special care should be given to prevent the vertical flow within the filter pack from weakening the effectiveness of development activities. Among the measures for doing so are developing before the filter pack is installed, using the filter packs that minimize the vertical flow, or using natural filter packs in unstable formations.
3. Further development activities may be needed in some cases. The identification of a low permeability skin is the clearest indication of such a need. A preliminary analysis of the response data using a theoretical model and a resulting implausible specific storage estimate is the strongest evidence of a low-*K* skin.
4. The nominal screen length should be used as the effective screen length in the analyses.

5. The effective screen radius should be taken as the radius of the filter pack in wells with artificial filter packs. On the other hand, the nominal screen radius may be more appropriate to use in the analyses for wells with natural filter packs where development has been limited.

6. The effective radius of the well casing should be taken as the nominal radius of the well casing. However, a comparison of the theoretical and actual rise/drop of water level in the well following the injection of the slug will indicate the reasonableness of this replacement.

7. A minimum number of three slug tests with initial displacement values varying by at least a factor of two should be performed at each well. Both the rising head and falling head tests should be employed to reveal a possible skin-related directional dependence. The possible existence of a dynamic skin can be identified if the first and last tests of the series at the same well employ the same initial displacement. All these variations and combinations should rate the effectiveness of well-development efforts and the validity of conventional slug test theories.

8. A series of tests consisting of a combination of rising and falling head tests should be planned and employed.

5.13.3 Performance of Slug Tests

The equipment (Figure 5.13.1) used in slug tests include devices for initiating the test, devices for measuring changes in head during the test, and devices for storing the head measurements. The most common method for initiating the test is to use a solid object (or slug) and introduce (or remove) it into (from) the well. A conventional slug would be a piece of stainless steel or PVC pipe filled with sand or similar material and capped at both ends. Both conventional and streamlined solid slugs have been used. The majority of slug tests are performed with sensors (pressure transducers) that measure the pressure exerted by the overlying column of water. The pressure transducer most commonly used is the semiconductor strain-gauge transducer. Electric tapes are also commonly used to measure water levels in slug tests of moderate to low hydraulic conductivity.

Slug tests may also be performed pneumatically. This method involves pressurizing the air column in a sealed well by injecting compressed air or nitrogen gas to depress the water level

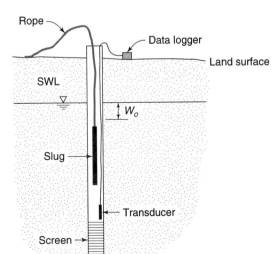

Figure 5.13.1. Slug testing schematic (Weight and Wittman;[106] used with permission of Groundwater Publishing Co.).

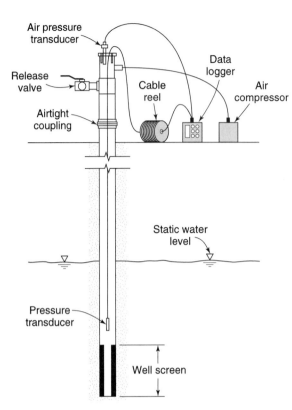

Air pressure
transducer

Release
valve

Airtight
coupling

Data
logger

Cable
reel

Air
compressor

Static water
level

Pressure
transducer

Well screen

Figure 5.13.2. Hypothetical cross section displaying a well at which the pneumatic method is being used for test initiation. Note that there must be an airtight seal at the point at which the cable for the submerged transducer passes through the top of the well head (figure not to scale; after McLane et al.[69] as presented in Butler[17]).

by driving water out of the well into the aquifer (see Figure 5.13.2). Water levels continue to drop until the decrease in pressure head of the water is equal to the magnitude of the increase of pressure head of the air column. Then the test is initiated by rapidly depressurizing the air column. A variation of this procedure would be to apply a vacuum to the air column followed by a sudden release of the vacuum to initiate the test.

Advantages of the pneumatic approach include the following: no water is handled during the test, test initiation is very rapid, and only the transducer and cable have to be cleaned prior to moving to a new well. Another advantage is that if both the pressurization and depressurization are done rapidly with respect to the formation response, then a pair of falling-head and rising-head tests, respectively, can be performed. The pneumatic approach requires a special airtight well-head apparatus, which can be a major disadvantage for the method. Greene and Shapiro[41, 42] describe methods for conducting pneumatic slug tests and computation of type curves for estimating transmissivity and storativity. Butler[17] provides a more detailed discussion of the various devices used for different slug tests.

Butler[17] provides the following four performance guidelines for slug tests:

1. The measurement of the response of a formation to a slug test must be done with appropriate equipment, usually selected based on the expected hydraulic conductivity of the formation. A pressure transducer connected to a data-acquisition device is most appropriate for obtaining and storing head measurements in a slug test where the formation has a moderate to high hydraulic conductivity. For formations that are less permeable, use of an electric tape is also acceptable. The equipment used for head measurements should always be checked before and after the slug test at a particular well.

2. The initial displacement of water in the well and the resulting rise or drop of the water level must be instantaneous relative to the response of the formation. Pneumatic- or packer-based systems should be used for test initiation in formations of very high permeability. In less

permeable formations, use of a solid slug is also acceptable. Techniques such as the addition or removal of water that may threaten the instantaneous test initiation should be avoided.

3. The initial head following the instantaneous displacement of water in the well must be measured accurately. A comparison of the expected initial head based on the known volume of displacement (i.e., slug volume) and the measured initial head should reveal the appropriateness of using the nominal casing diameter as the effective casing diameter. If the nominal casing diameter together with the known volume of displacement does not provide a reasonable estimate of the initial head, the effective radius can be adjusted based on a mass balance.

4. There should be enough time before a slug test is repeated in the same well that the residual deviation from the static level is less than 5% of the initial head change. If it is suspected that the formation response is affected by an incomplete recovery of the water level from the previous test, the test data can be plotted as in the Hvorslev method (mentioned in the next section). A near-linear plot of the log of the deviation of head from the static water level versus time would deny an incomplete recovery from the previous test.

5.13.4   Methods for Analyzing Slug Test Data

Slug-test analysis methods were first developed during the 1950s by Hvorslev[46] and Ferris and Knowles.[31] Hvorslev's simple method led to its use for both confined and unconfined aquifers as an estimate of the hydraulic conductivity within the screened interval. Cooper et al.[26] and Papadopulos et al.[76] made improvements of the methods by Ferris and Knowles. Later, Bouwer and Rice[15] and Bouwer[14] developed a method for analysis in confined and unconfined aquifers that takes into account aquifer geometry and partial penetration effects. Campbell et al.[22] summarized various methods with their relative merits and an extensive list of references.

Methods for analyzing slug test data have been reported in various reports and books, including references 17, 28, 42, 43, 85, and 100. Other references on aquifer tests include Hall and Chen[44], Kasenow[55], Kruseman and de Ridder[61], and Lee[63].

Many government computer software packages and spreadsheets have been developed, including those by Greene and Shapiro[42], Halford and Kuniansky[43], and Sepulveda.[85] The private sector has also developed many computer software packages and spreadsheets with graphic user interfaces. Figure 5.13.3 presents eight conditions for slug tests. Various methods for the analysis of slug tests in these eight cases (discussed in the following sections of this chapter) may be outlined as follows.

Case 1: Unconfined aquifer, partially/fully penetrating well, water level variations above the well screen (rising head test).

Available methods:

  (i) Bouwer and Rice method[14, 15]

 (ii) Dagan method[17]

(iii) KGS (Kansas Geological Survey) model[17]

Case 2: Unconfined aquifer, partially/fully penetrating well, water level drops below the top of well screen (rising head test). (*Note:* A double straight line effect or a concave-upward curvature may be observed when the response data is plotted as the logarithm of head vs. time. This is more likely if there is an artificial filter pack in a well screened across a low hydraulic conductivity unit.)

Available methods:

  (i) Bouwer and Rice method (an effective casing radius may be needed)

 (ii) Dagan method

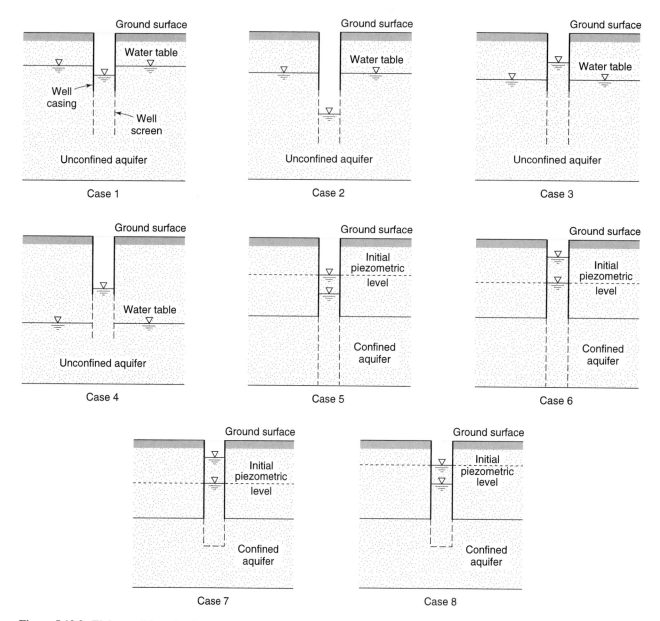

Figure 5.13.3. Eight conditions for slug tests.

Case 3: Unconfined aquifer, partially/fully penetrating well, water level variations above the well screen (falling head test). The conditions for Case 1 also apply to this case.

Case 4: Unconfined aquifer, partially/fully penetrating well, well screen intersects the water table (falling head test). The analysis methods would overestimate the hydraulic conductivity of the formation in this case because the water from the well would drain both through the saturated portion of the well screen and into the vadose zone. The drainage into the vadose zone increases the rate of fall of the water level beyond that caused by the drainage into the saturated aquifer, resulting in an overestimate of the hydraulic conductivity.

Case 5: Confined aquifer, fully penetrating well, rising head test.

Applicable methods:

(i) Cooper-Bredehoeft-Papadopulos method[26]

(ii) Hvorslev method[46]

Case 6: Confined aquifer, fully penetrating well, falling head test. The methods for Case 5 also apply to this case.

Case 7 and Case 8: Confined aquifer, partially penetrating well, falling head test.

Applicable methods:

(i) Hvorslev method (different shape factors are used than that for fully penetrating wells)

(ii) Confined extensions of the Dagan method[17]

(iii) KGS model (developed by extending the Cooper et al. method to the case of a partially penetrating well[17])

(iv) Bouwer and Rice method (while, theoretically, the accuracy of the hydraulic conductivity estimates increase as the distance between the top of well screen and the upper confining layer increases)

5.14   SLUG TESTS FOR CONFINED FORMATIONS

Butler[17] points out that the vast majority of slug tests performed on fully penetrating wells are analyzed using the Cooper, Bredehoeft, and Papadopulos[26] or Hvorslev[46] methods, which can be used for fully or partially penetrating wells. Both these methods are discussed in the sections below. A newer method by Peres et al.,[77] based upon a deconvolution approach, also has potential, as discussed by Butler,[17] for both partially and fully penetrating wells. Also, methods by Dagan[27] and the KGS model (Hyder et al.[50]), described later, can be used for partially penetrating wells in confined aquifers.

5.14.1   Cooper, Bredehoeft, and Papadopulos Method

This method (Cooper et al.[26]) is based upon a fully penetrating well or open borehole, as illustrated in Figure 5.14.1. A slug or a known volume of water is introduced into the well over a short time period of a few seconds, causing the water to rise to a maximum height H_0 above the initial water level. As the water level declines, the head above the original level, $H(t)$, is measured as a function of time. A plot of the ratio $H(t)/H_0$ as function of time can be developed with H/H_0 on an arithmetic scale and time with a log scale.

This methodology is based upon assumptions that include a homogeneous formation and Darcian flow, instantaneous introduction of the slug, hydrologic boundaries in the plane of the formation a large distance from the test well, and negligible well losses. Cooper et al.[26] developed the following analytical solution to the mathematical model:

$$\frac{H(t)}{H_0} = f(\beta, \alpha) \qquad (5.14.1)$$

β is a dimensionless time parameter defined as

$$\beta = \frac{K_r B t}{r_c^2} = \frac{T t}{r_c^2} \qquad (5.14.2)$$

where K_r is the radial component of hydraulic conductivity, B is the formation thickness, and t is the time since the initial displacement of water in the well. α is the dimensionless storage parameter defined as

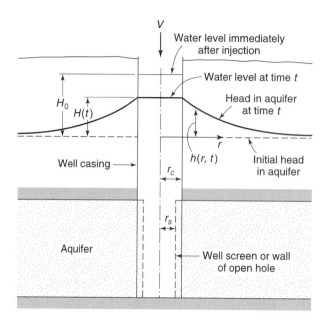

Figure 5.14.1. Well into which a volume, V, of water is suddenly injected for a slug test of a confined aquifer (Cooper et al. [26]).

$$\alpha = \frac{r_s^2 S_s}{r_c^2} \tag{5.14.3}$$

where r_s is the effective radius of well screen, r_c is the effective radius of well casing, and S_s is the specific storage. If time is measured in seconds and all dimensions are in centimeters, then the value of T is cm^2/s. If time is in seconds and all dimensions are in feet, then transmissivity T is in ft^2/s.

Tables 5.14.1a and b provide values of $H(t)/H_0$ for Equation 5.14.1. Figure 5.14.2 presents the type curves developed by Papadopulos et al.[76] for slug tests in wells of finite diameter.

Steps in the Cooper, Bredehoeft, Papadopulos method are as follows:

1. Plot the normalized response (field) data $(H(t)/H_0)$ versus the logarithm of the time since the test began on the same scale and size as the type curves.

2. Overlay the data plot on the type-curves moving parallel along the log-time axis of the response data until one of the α curves approximately matches the plot of the response (field) data. Record the value of α.

3. Select match points from each plot. As an example, set $\beta = 1$ and read the corresponding real time $t_{1.0}$.

4. Use Equation 5.14.2 to solve for the hydraulic conductivity and then solve for the transmissivity.

5. Use Equation 5.14.3 to solve for the specific storage (storage coefficient).

EXAMPLE 5.14.1

A slug test is performed in a confined aquifer of 5 m thickness with a casing radius of 8.8 cm and a screen radius of 5.6 cm. After the slug of water is injected in the well, the water level rises by 0.52 m following which the water level drops with time as shown in Table 5.14.2. Estimate the transmissivity, hydraulic conductivity, and storage coefficient using the Cooper-Bredehoeft-Papadopulos method.

SOLUTION

Using a plot of H/H_0 versus time on a semilogarithmic paper, this field curve is superimposed on the type curves (Figure 5.14.2) with the arithmetic axis coincident and shifted horizontally until a best fit is obtained on one of the type curves. In this case the field data fits best to the type curve for $\alpha = 10^{-4}$. The

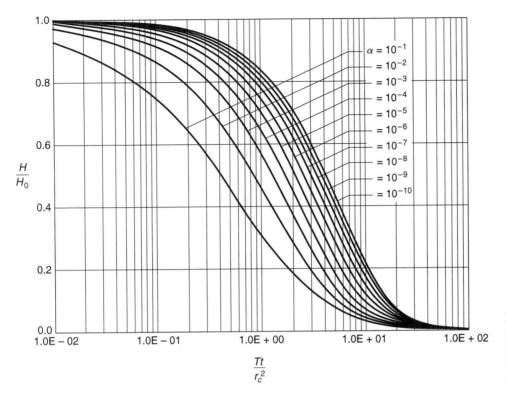

$\dfrac{H}{H_0}$

$\dfrac{Tt}{r_c^2}$

Figure 5.14.2. Selected type curves for the Cooper, Bredehoeft, and Papadopulos method (Papadopulos et al.[76]).

Table 5.14.1a Values of $H(t)/H_0$

	H/H_0				
Tt/r_c^2	$\alpha = 10^{-1}$	$\alpha = 10^{-2}$	$\alpha = 10^{-3}$	$\alpha = 10^{-4}$	$\alpha = 10^{-5}$
1.00×10^{-3}	0.9771	0.9920	0.9969	0.9985	0.9992
2.15×10^{-3}	0.9658	0.9876	0.9949	0.9974	0.9985
4.64×10^{-3}	0.9490	0.9807	0.9914	0.9954	0.9970
1.00×10^{-2}	0.9238	0.9693	0.9853	0.9915	0.9942
2.15×10^{-2}	0.8860	0.9505	0.9744	0.9841	0.9888
4.64×10^{-2}	0.8293	0.9187	0.9545	0.9701	0.9781
1.00×10^{-1}	0.7460	0.8655	0.9183	0.9434	0.9572
2.15×10^{-1}	0.6289	0.7782	0.8538	0.8935	0.9167
4.64×10^{-1}	0.4782	0.6436	0.7436	0.8031	0.8410
1.00×10^{0}	0.3117	0.4598	0.5729	0.6520	0.7080
2.15×10^{0}	0.1665	0.2597	0.3543	0.4364	0.5038
4.64×10^{0}	0.07415	0.1086	0.1554	0.2082	0.2620
7.00×10^{0}	0.04625	0.06204	0.08519	0.1161	0.1521
1.00×10^{1}	0.03065	0.03780	0.04821	0.06355	0.08378
1.40×10^{1}	0.02092	0.02414	0.02844	0.03492	0.04426
2.15×10^{1}	0.01297	0.01414	0.01545	0.01723	0.01999
3.00×10^{1}	0.009070	0.009615	0.01016	0.01083	0.01169
4.64×10^{1}	0.005711	0.005919	0.006111	0.006319	0.006554
7.00×10^{1}	0.003722	0.003809	0.003884	0.003962	0.004046
1.00×10^{2}	0.002577	0.002618	0.002653	0.002688	0.002725
2.15×10^{2}	0.001179	0.001187	0.001194	0.001201	0.001208

Source: From Cooper et al.[26]

Table 5.14.1b Values of $H(t)/H_0$

Tt/r_c^2	$\alpha = 10^{-6}$	$\alpha = 10^{-7}$	$\alpha = 10^{-8}$	$\alpha = 10^{-9}$	$\alpha = 10^{-10}$
0.001	0.9994	0.9996	0.9996	0.9997	0.9997
0.002	0.9989	0.9992	0.9993	0.9994	0.9995
0.004	0.9980	0.9985	0.9987	0.9989	0.9991
0.006	0.9972	0.9978	0.9982	0.9984	0.9986
0.008	0.9964	0.9971	0.9976	0.9980	0.9982
0.01	0.9956	0.9965	0.9971	0.9975	0.9978
0.02	0.9919	0.9934	0.9944	0.9952	0.9958
0.04	0.9848	0.9875	0.9894	0.9908	0.9919
0.06	0.9782	0.9819	0.9846	0.9866	0.9881
0.08	0.9718	0.9765	0.9799	0.9824	0.9844
0.1	0.9655	0.9712	0.9753	0.9784	0.9807
0.2	0.9361	0.9459	0.9532	0.9587	0.9631
0.4	0.8828	0.8995	0.9122	0.9220	0.9298
0.6	0.8345	0.8569	0.8741	0.8875	0.8984
0.8	0.7901	0.8173	0.8383	0.8550	0.8686
1	0.7489	0.7801	0.8045	0.8240	0.8401
2	0.5800	0.6235	0.6591	0.6889	0.7139
3	0.4554	0.5033	0.5442	0.5792	0.6096
4	0.3613	0.4093	0.4517	0.4891	0.5222
5	0.2893	0.3351	0.3768	0.4146	0.4487
6	0.2337	0.2759	0.3157	0.3525	0.3865
7	0.1903	0.2285	0.2655	0.3007	0.3337
8	0.1562	0.1903	0.2243	0.2573	0.2888
9	0.1292	0.1594	0.1902	0.2208	0.2505
10	0.1078	0.1343	0.1620	0.1900	0.2178
20	0.02720	0.03343	0.04129	0.05071	0.06149
30	0.01286	0.01448	0.01667	0.01956	0.02320
40	0.008337	0.008898	0.009637	0.01062	0.01190
50	0.006209	0.006470	0.006789	0.007192	0.007709
60	0.004961	0.005111	0.005283	0.005487	0.005735
80	0.003547	0.003617	0.003691	0.003773	0.003863
100	0.002763	0.002803	0.002845	0.002890	0.002938
200	0.001313	0.001322	0.001330	0.001339	0.001348

Source: From Papadopulos et al.[76]

time that corresponds to the vertical axis is for $\beta = 1.0$. At the axis for $\beta = 1.0$, the $t = 133$ seconds. Substituting $t = 133s$ and $r_c = 8.8$ cm into Equation 5.14.2 yields the hydraulic conductivity

$$K_r = \frac{\beta r_c^2}{Bt_{1.0}} = \frac{(1.0)(8.8 \text{ cm})^2}{(500 \text{ cm})(133 \text{ s})} = 1.1645 \times 10^{-3} \text{ cm/s} = 1.006 \text{ m/day}$$

and the transmissivity is given by

$$T = K_r B = (1.006 \text{ m/day})(5 \text{ m}) = 5.03 \text{ m}^2/\text{day}$$

Substituting $r_c = 8.8$ cm and $r_s = 5.6$ cm, and $\alpha = 10^{-4}$ into Equation 5.14.3 and solving for S yields

$$S = \frac{r_c^2 \alpha}{r_s^2} = \frac{(8.8 \text{ cm})^2 (10^{-4})}{(5.6 \text{ cm})^2} = 2.47 \times 10^{-4}$$

■

Table 5.14.2 Water Level after Slug Test for Example 5.14.1

Time (s)	H (m)	H/H₀
2	0.51	0.99
4	0.51	0.98
6	0.50	0.97
8	0.50	0.96
10	0.49	0.95
20	0.48	0.92
30	0.46	0.89
50	0.43	0.83
80	0.40	0.76
100	0.37	0.72
120	0.35	0.68
150	0.32	0.62
200	0.29	0.55
250	0.25	0.48
300	0.22	0.42
400	0.17	0.33
500	0.14	0.27
800	0.07	0.13
1,000	0.06	0.11

5.14.2   Hvorslev Method

This method differs from the Cooper, Bredehoeft, and Papadopulos method in the following three aspects:

1. Although the hydraulic head in the formation still varies with time, the specific storage is assumed to be so small that its effects can be neglected, so that any change in head in the well is instantly propagated throughout the flow system (a quasi–steady-state representation of the slug-induced flow).

2. The slug test need not be introduced in an instantaneous fashion.

3. Lateral constant-head boundaries are at a finite distance from the test well.

The analytical solution for this methodology is expressed as[23]

$$\ln\left[\frac{H(t)}{H_0}\right] = -\frac{2K_r L_e t}{r_c^2 \ln(R_e/r_s)} \tag{5.14.4}$$

where R_e is the effective radius of the slug test and L_e is the effective length of the well screen.

A semilog plot of the normalized heads $H(t)/H_0$, versus time, is a straight line. This method calculates the slope of that line, which in turn is used to estimate the radial component of the hydraulic conductivity. The steps in the Hvorslev method are as follows:

1. Plot the logarithm of the normalized response data versus the time since the test began. A straight line is fit to the plot.

2. Calculate the slope of the straight line. One method is to estimate the time (referred to as T_0) at which the normalized head equals 0.368 (natural log of 0.368 is –1), which is defined as the basic time lag according to Hvorslev.[46] At the start of the test, the logarithmic head term is one and the time term is zero, so the slope is simply the $\log_{10}$ of 0.368 over T_0, which is $-1/T_0$ in terms of the natural logarithm.

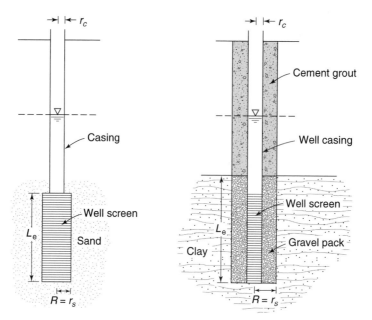

Figure 5.14.3. Piezometer geometry for Hvorslev method. Note that for a piezometer installed in a low-permeability unit the value R is the radius of the highest permeable zone that includes the gravel pack zone and L_e is the length of the gravel pack zone.[32]

3. Calculate the radial component of the hydraulic conductivity using Equation 5.14.4 written in terms of the slope calculated using a normalized head of 0.368,

$$K_r = \frac{r_c^2 \ln(R_e/r_s)}{2L_e T_0} \qquad (5.14.5)$$

in which T_0 is the time at which a normalized head of 0.368 is obtained. If a gravel pack zone exists, which is usually installed while testing a low permeability unit, the radius of the well screen should include the gravel pack zone. Similarly, L_e is the length of the gravel pack zone in such a case (see Figure 5.14.3).

Equation 5.14.5 requires an estimate of the effective radius of the slug test, R_e, an empirical parameter that is a function of α. Typically values of either the well screen length or 200 times the effective radius of the well screen are used for R_e (Butler[17]). The assumption of a straight-line fit of the semilog plot of the test data is appropriate as long as the effect of the elastic storage mechanisms can be neglected. When the storage effects have an effect on the response data, the data will display a distinct concave-upward curvature, which is primarily a function of the dimensionless storage parameter, α. Chirlin[23] provides a theoretical explanation for the behavior. The Hvorslev method does not require that the slug be introduced nearly instantaneously as in the Cooper, Bredehoeft, and Papadopulos method. As previously mentioned this is a quasi–steady-state representation of the slug-induced flow, eliminating the need for a nearly instantaneous initiation, with the only assumption that the slug introduction is completed prior to the collection of the response data. The Hvorslev method can therefore be used to analyze response data that have been affected by noninstantaneous slug introduction.

EXAMPLE 5.14.2

A slug test is performed by adding a slug of 3,000 cm³ of water into a piezometer that is screened 12 m vertically in medium sand. The injection of the slug causes the water level in the well to rise 0.56 m. The radius of the well screen is 5.8 cm while the radius of the well casing is 4.1 cm. The water level recovery in the well is recorded using a pressure transducer as tabulated on the next page. Determine the hydraulic conductivity in the vicinity of the screened portion of the formation using the Hvorslev method.

Elapsed Time (s)	Change in water level (m)	$-H/H_0$
0	0.560	1.000
1	0.387	0.691
2	0.198	0.354
3	0.141	0.252
4	0.087	0.155
5	0.050	0.089
6	0.035	0.062
7	0.017	0.030
8	0.010	0.018
9	0.006	0.011
10	0.000	0.000

From the plot (Figure 5.14.4) of the data (i.e., normalized deviation of head from static water level versus time), the time for the head to drop to 37 percent of the initial head change is $T_0 = 2$ seconds. Also given in the problem statement is the well construction information:

$$r_c = 4.1 \text{ cm}, \quad r_s = 5.8 \text{ cm}, \quad L_e = 12 \text{ m}$$

Finally, taking the effective radius of the slug test, R_e, as the well screen length, L_e, and substituting into Equation 5.14.5 yields

$$K_r = \frac{r_c^2 \ln(R_e/r_s)}{2L_e T_0} = \frac{(4.1 \text{ cm})^2 \ln(1200 \text{ cm}/5.8 \text{ cm})}{(2)(1200 \text{ cm})(2.0 \text{ s})} = 0.01867 \text{ cm/s} = 16.13 \text{ m/day}$$

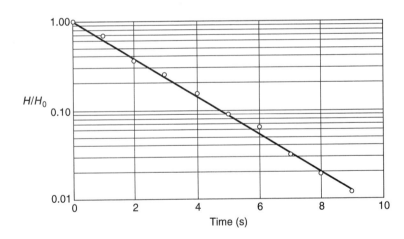

Figure 5.14.4. Dataplot for Example 5.14.2. ∎

5.15   SLUG TESTS FOR UNCONFINED FORMATIONS

Methods for unconfined formations are primarily classified on the basis of whether or not the well screen intersects the water table (see Figure 5.15.1a and b). Butler[17] classifies these as: a) well screen below the water table, b) well screen intersects the water table, and c) well screen below the water table, in which case the methods can be subdivided according to their handling of the storage mechanism. When the well screen is below the water table, the change in saturated thickness during a test is typically quite small, so a linear model can be used. When a well screen intersects the water table, assumptions of a constant saturated thickness or a constant effective screen length may not be correct, resulting in the need for a nonlinear model.

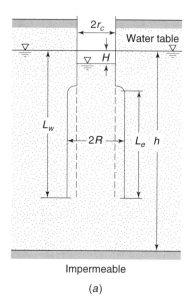

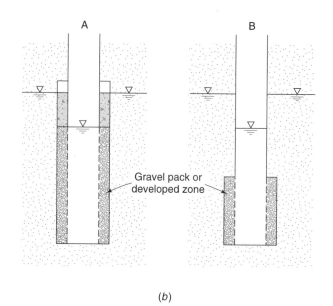

(a)

(b)

Figure 5.15.1. (*a*) Geometry and symbols for a slug test on a partially penetrating screened well in an unconfined aquifer with a gravel pack around the screen[14] (*b*) Slug test for borehole with groundwater level below A and above B top of screen or perforated section[14]

According to Butler,[17] the vast majority of slug tests performed in unconfined formations in wells screened below the water table are analyzed using the Bouwer and Rice method[15], the Dagan method[27], or the KGS model with the unconfined variant (Hyder et al.[50]).

5.15.1 Bouwer and Rice Method

The two key assumptions of this method are that the effects of elastic storage mechanisms can be neglected and the water table level does not change; consequently the saturated thickness of the formation does not change during the test. The original model of Bouwer and Rice was defined for isotropic conditions. Zlotnik[110] extended the method to the general anisotropic case.

As pointed out by Bouwer and Rice[15], the slug test applies theoretically to aquifers where the upper boundary is a plane source (rising water level test) or sink (falling water-level test), as in an unconfined aquifer. Because most of the head difference, H, between the static water table and the water level in the well is dissipated in the vicinity around the screen of the perforated section, the method should also be applicable to situations where the upper boundary of the aquifer is an impermeable or semi-impermeable upper confining layer. Thus the method should give reasonable results for the hydraulic conductivity in confined, semiconfined, or stratified aquifers. Theoretically, the greater the distance between the top of the screened or open section of the well and the upper confining layer, the more accurate the estimate of the hydraulic conductivity. However, as Bouwer[14] points out, source boundaries of groundwater flowing into the well in response to lowering the water level are difficult to define because of the elastic deformation of the aquifer material and confining and interbedded fine-textured layers, and because of leakage through semiconfining layers.

The analytical solution of the method is (Bouwer and Rice[15]; Bouwer[14]; and Zlotnik[110]):

$$K_r = \frac{r_c^2 \ln(R_e/r_w)}{2L_e} \frac{1}{t} \ln\left[\frac{H_0}{H_t}\right] \tag{5.15.1}$$

where the symbols are defined as follows:

K_r hydraulic conductivity of the formation [L/T]

r_c radius of the well casing [L]

r_w radius of the gravel envelope [L]

R_e the effective radial distance over which head is dissipated [L]

L_e the length of the screen or open section of the well through which water can enter

H_0 the drawdown at time $t = 0$ [L]

H_t the drawdown at time $t = t$ [L]

t the time since $H = H_0$ [T]

This method is based upon the plot of the normalized head (H_t/H_0) versus time as a straight line and involves calculating the slope of the straight-line fit to the response data. Then the slope is used to estimate the hydraulic conductivity. The steps in the method are as follows (Butler[17]):

1. Plot the logarithm of the normalized response data (H_t/H_0) versus the time since the test began and fit a straight line through the data.

2. Calculate the slope of the straight-line fit. Using the time lag T_0, write the slope in terms of the natural logarithm as $-1/T_0$.

3. Assume the anisotropy ratio to be 1. Estimate the effective radius parameter (Butler[17]).

4. Compute the radial component of the hydraulic conductivity using Equation 5.15.1.

As with the Hvorslev method, the effective radius, R_e, is considered as a parameter. Bouwer and Rice[15] and Bouwer[14] presented a method for estimating the dimensionless ratio, $\ln(R_e/r_w)$. If L_w is less than h, the saturated thickness of the aquifer, then the well is partially penetrating and

$$\ln \frac{R_e}{r_w} = \left[\frac{1.1}{\ln(L_w/r_w)} + \frac{A + B\ln\left[(H - L_w)/r_w\right]}{L_e/r_w} \right]^{-1} \qquad (5.15.2)$$

If L_w is equal to h, then the well is fully penetrating and

$$\ln \frac{R_e}{r_w} = \left[\frac{1.1}{\ln(L_w/r_w)} + \frac{C}{L_e/r_w} \right]^{-1} \qquad (5.15.3)$$

where A, B, and C are dimensionless numbers found from Figure 5.15.2.

Theoretically, this slug test method applies to any diameter of the well, and from a practical viewpoint the well diameter should be selected so that the geometric parameters apply to Figure 5.15.2. The larger r_w and L_e, the larger the portion of the aquifer on which the hydraulic conductivity is determined.

Bouwer[14] discussed the fact that users of the method have observed that when plotting the log $H(t)$ versus t as in Figure 5.15.3, sometimes a double straight-line effect results, as illustrated in Figure 5.15.4. This discussion assumes that a bail-down test (water level is lowered by bailing or pumping so that water flows from the aquifer into the well) was performed. The steep straight-line segment A–B is most likely due to a highly permeable zone around the well (gravel pack or developed zone), which sends water into the well immediately after the water level has been lowered (A in Figure 5.15.1b). When the water level in the permeable zone around the well has been drained to the water level in the well, the flow into the well slows down and forms the less steep segment B–C shown in Figure 5.15.4. This less steep

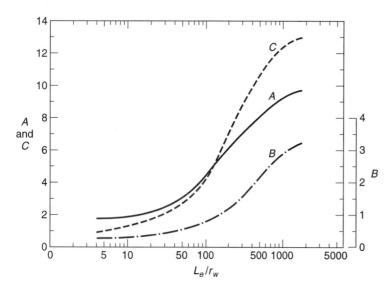

Figure 5.15.2. Dimensionless parameters A, B, and C plotted as a function of L_e/r_w. These parameters are used in the determination of $\ln(R_e/r_w)$.[14]

segment is indicative of the flow from the undisturbed aquifer into the well; therefore segment B–C should be used in calculating the hydraulic conductivity.

Next consider the situation when the groundwater level is above the top of the screen or perforated section (B in Figure 5.15.1b) and during the test the water level in the well is not lowered enough to drop below the top of the open section. In this situation the gravel envelope or developed zone around the open section cannot drain and the inflow into the hole is then immediately controlled by the aquifer, such that the double straight-line effect does not occur. In the situation that the double straight line still occurs, this may indicate leakage around the casing or grouting above the gravel pack.

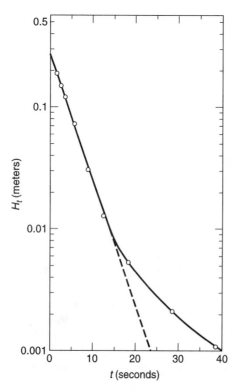

Figure 5.15.3. Head in a borehole as a function of time for a well in Salt River Bed, 27th Avenue, Phoenix, AZ. Note that the data form a straight line during the early part of the test but eventually deviate from the straight line.[14]

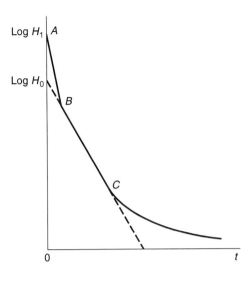

Figure 5.15.4. Head in a borehole as a function of time forms two straight-line segments during the early part of the test but eventually deviates from the straight line (Bouwer[14]).

Bouwer[14] points out that when the double straight line is caused by a gravel pack around the well, the effective well radius should be taken as the radial distance from the center of the well to the outer surface of the gravel pack. When the double-straight line is caused by a naturally developed zone around the well, r_w is more difficult to evaluate, resulting in the need for experience and good judgment. It may be possible to estimate r_w from the value of H at point B in Figure 5.15.4.

EXAMPLE 5.15.1

A slug test is performed by bailing 1,230 cm^3 of water from the test well. The bailing causes a near instantaneous drop of 0.61 m of the water level in the well. Head measurements during the recovery phase of the water level using an electric tape are presented in Table 5.15.1. Using the given well construction information, determine the hydraulic conductivity of the formation using the Bouwer and Rice method.

SOLUTION

First, determine if the measured drop of water level agrees with removed volume of water.

A conceptual picture of the problem shows that the water level is still above the screened interval following the bailing (Figure 5.15.5). Thus, the expected drop of water level corresponding to a removal of 1,230 cm^3 water can be calculated utilizing the casing radius:

$$H_0 = \frac{V}{\pi r_c^2} = \frac{1230 \text{ cm}^3}{\pi (2.55 \text{ cm})^2} = 60.2 \text{ cm}$$

This result is consistent with the measured drop of water level, 61 cm, given in the problem statement.

The logarithm of the normalized response data versus time is plotted in Figure 5.15.6.

Next, we must rearrange Equation 5.15.1:

$$K = \frac{r_c^2 \ln(R_e/r_w)}{2L_e} \frac{1}{t} \ln\left[\frac{H_0}{H_t}\right]$$

Taking $H_t = 0.368 H_0$ yields

$$K = \frac{r_c^2 \ln(R_e/r_w)}{2L_e} \frac{1}{T_0} \ln\left[\frac{H_0}{0.368 H_0}\right] = \frac{r_c^2 \ln(R_e/r_w)}{2L_e} \frac{1}{T_0}$$

From the given geometry,

$$r_c = 2.55 \text{ cm}, \quad r_w = 10.15 \text{ cm}, \quad L_e = 3 \text{ m}, \quad L_w = 4.4 \text{ m}$$

Table 5.15.1 Recovery from Bailing Test (Example 5.15.1)

Elapsed Time (s)	Change in water level (m)	H/H_0
0	0.610	1.000
2	0.583	0.956
6	0.551	0.903
10	0.526	0.862
15	0.508	0.833
26	0.482	0.790
42	0.450	0.738
61	0.418	0.685
77	0.396	0.649
92	0.377	0.618
115	0.350	0.574
135	0.329	0.540
147	0.318	0.521
168	0.300	0.492
189	0.284	0.466
207	0.271	0.445
227	0.258	0.423
257	0.240	0.394
286	0.224	0.368
315	0.209	0.343
346	0.196	0.321
386	0.180	0.295
430	0.164	0.269
574	0.121	0.198
668	0.099	0.162
731	0.092	0.150
938	0.077	0.126
1,115	0.066	0.109

Boring diameter = 20.3 cm

Casing diameter = 5.1 cm

Depth to bottom of well = 8.2 m BGS (below ground surface)

Depth to top of casing = 5.2 m BGS

Depth to water = 3.8 m BGS

Saturated thickness = 4.4 m

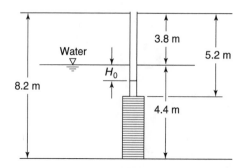

Figure 5.15.5. Conditions for slug test in Example 5.15.1.

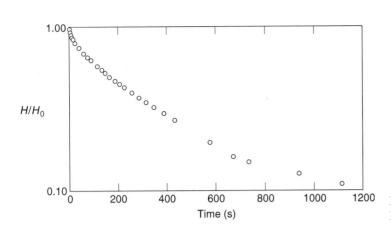

Figure 5.15.6. Recovery from bailing test for Example 5.15.1.

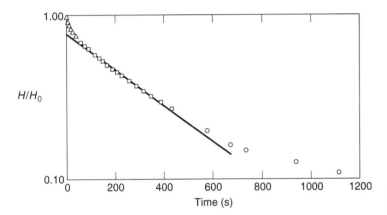

Figure 5.15.7. Response to bailing test for Example 5.15.1.

Since L_w is equal to h, the saturated thickness of the aquifer, Equation 5.15.3 must be used:

$$\ln\frac{R_e}{r_w} = \left[\frac{1.1}{\ln(L_w/r_w)} + \frac{C}{L_e/r_w}\right]^{-1}$$

$L_e/r_w = 300\ \text{cm}/10.15\ \text{cm} \cong 30 \quad \rightarrow \quad \text{Figure 5.15.2} \quad \rightarrow \quad C = 2.2$

$$\ln\frac{R_e}{r_w} = \left[\frac{1.1}{\ln(440\ \text{cm}/10.15\ \text{cm})} + \frac{2.2}{300\ \text{cm}/10.15\ \text{cm}}\right]^{-1} = 2.73$$

Note that the points of the response data do not follow a straight line exactly in Figure 5.15.7. Instead, an early quick response is followed by a relatively straight line segment. Finally, the data points start deviating from this straight line for large t values. Overall, the response data indicate a double straight-line effect. The relatively quick early time response may be attributed to the effect of the gravel pack zone, if there is any. However, because the water level changes occur above the screened portion of the well in this case, it is more likely that there is leakage around the casing or grouting above the gravel pack. A sensitivity analysis could be made using different parts of the data to find the hydraulic conductivity. However, the response data will be separated into three segments for time intervals $t = 0$–26 s, 42–430 s, and 574–$1{,}115$ s in the current solution.

Note that a new h_o value emerges when the straight line is extrapolated backward because of the double straight-line effect, such that $(H_0)_{new} \cong 0.778 H_0$.

$T_0 = 388.5$ s for the first straight line segment based on the new h_0 value. Thus,

$$K = \frac{r_c^2 \ln(R_e/r_w)}{2L_e} \frac{1}{T_0} = \frac{(2.55 \text{ cm})^2 (2.73)}{(2)(300 \text{ cm})} \frac{1}{388.5 \text{ s}} = 7.6 \times 10^{-5} \text{ cm/s} = 0.0658 \text{ m/day}$$ ∎

5.15.2 Dagan Method

Dagan[27] presented a method based on a mathematical model similar to that for the Bouwer and Rice method with the only difference that a constant-head boundary is not assumed at the finite radial distance from the test well, so that the hydrologic boundaries in the lateral plane are assumed to be at an infinite distance from the well.

The steps in applying this method are as follows (Butler[17]):

1. Plot the log of the normalized response data versus the time since the test began and fit a straight line to the plot. Compute the slope of the straight line.

2. Estimate the parameter ψ for the particular well formation configuration using

$$\Psi = \frac{\sqrt{K_z/K_r}}{L_e/r_w}$$

where K_z/K_r is the anisotropy ratio. In most cases, 1 is used for the anisotropy ratio.

3. Using ψ, the normalized distance from the water table $(L_w + L_e)/L_e$, and the normalized length of the well screen L_e/h, select a value of P, the dimensionless flow parameter, from Table 5.15.2. Table 5.15.3 is used to determine values of P for confined aquifer conditions.

4. Estimate the hydraulic conductivity (radial component) using

$$K_r = \frac{r_c^2 (1/P)}{2L_e T_0} \qquad (5.15.4)$$

Table 15.15.2a Tabulated Values of the Dimensionless Flow Parameter, P, Used in Dagan Method for Wells Screened Below the Water Table

ψ	\multicolumn{5}{c}{$(L_w + L_e)/L_e$}
	8.0	4.0	2.0	1.5	1.05
0.20	0.646	0.663	0.705	0.756	1.045
0.10	0.477	0.487	0.505	0.531	0.687
0.067	0.409	0.416	0.429	0.446	0.562
0.050	0.367	0.373	0.385	0.397	0.491
0.033	0.322	0.325	0.335	0.352	0.414
0.025	0.294	0.297	0.305	0.322	0.370
0.020	0.276	0.278	0.287	0.301	0.342
0.013	0.247	0.249	0.255	0.269	0.300
0.010	0.230	0.231	0.238	0.250	0.276
0.0067	0.211	0.210	0.213	0.227	0.248
0.0050	0.198	0.199	0.201	0.213	0.230

Note: Values generated with the semianalytical solution of Cole and Zlotnik[25], courtesy of K. D. Cole (as presented in Butler[17]). Values for $L_e/h \le 0.05$.

Table 15.15.2b Tabulated Values of the Dimensionless Flow Parameter, P, Used in Dagan Method for Wells Screened Below the Water Table

ψ	\multicolumn{6}{c}{L_e/h}
	1.0	0.83	0.67	0.50	0.20	0.10
0.20	1.289	0.723	0.631	0.576	0.510	0.492
0.10	0.800	0.510	0.460	0.428	0.390	0.380
0.050	0.536	0.384	0.354	0.335	0.312	0.306
0.025	0.388	0.305	0.286	0.273	0.258	0.254
0.010	0.279	0.238	0.227	0.219	0.209	0.206

Note: Values generated with the semianalytical solution of Cole and Zlotnik[25], courtesy of K. D. Cole (as presented in Butler[17]). Values for $(L_w + L_e) = h$.

Table 5.15.3 Tabulated Values of the Dimensionless Flow
Parameter, P, Used in Confined Formation for the Dagan Method

ψ	$(L_w+L_e)/L_e$				
	8.0	4.0	2.0	1.5	1.05
0.20	0.741	0.727	0.681	0.640	0.561
0.10	0.539	0.533	0.505	0.483	0.432
0.067	0.458	0.455	0.432	0.416	0.377
0.050	0.412	0.408	0.390	0.378	0.345
0.033	0.359	0.357	0.343	0.331	0.307
0.025	0.328	0.325	0.314	0.305	0.285
0.020	0.307	0.305	0.295	0.288	0.270
0.013	0.275	0.273	0.263	0.259	0.245
0.010	0.254	0.254	0.246	0.240	0.230
0.0067	0.232	0.230	0.224	0.218	0.211
0.0050	0.218	0.216	0.210	0.205	0.199

Note: Values generated with the semianalytical solution of Cole and
Zlotnik,[25] courtesy of K. D. Cole (as presented in Butler[17]). Values for
$L_e/h \le 0.05$.

$$\Psi = \frac{\sqrt{K_z/K_r}}{L_e/r_w}$$

Practical issues of concern discussed for the Bouwer and Rice method are also applicable to this method. These include the uncertainty about the anisotropy, which interval to fit with a straight line, and the impact of the noninstantaneous slug.

EXAMPLE 5.15.2

Using the slug test data in Table 5.15.4, determine the hydraulic conductivity of the formation by the Dagan method.

SOLUTION

A conceptual picture of the slug test is shown in Figure 5.15.8.

From this picture,

$$r_c = 2.55 \text{ cm}, \quad r_w = 10 \text{ cm}, \quad L_e = 3 \text{ m}, \quad L_w = 10.3 \text{ m}, \quad h = 15 \text{ m}$$

The response data on semilogarithmic scale is plotted in Figure 5.15.9. From this plot, for $H_t = 0.368 H_0$, $T_0 = 7.9$ seconds. Next, determine the ψ parameter for the given well geometry. Assuming the anisotropy ratio is one,

$$\Psi = \frac{\sqrt{K_z/K_r}}{L_e/r_w} = \frac{1}{300 \text{ cm}/10 \text{ cm}} = 0.0333$$

$$\frac{L_w + L_e}{L_e} = \frac{10.3 \text{ m} + 3 \text{ m}}{3 \text{ m}} = 4.43$$

Using the above parameters, the dimensionless flow parameter from Table 5.15.2 can be determined: $P = 0.325$.

Finally, the hydraulic conductivity estimate can be made using Equation 5.15.4:

$$K_r = \frac{r_c^2 (1/P)}{2 L_e T_0} = \frac{(2.55 \text{ cm})^2 (1/0.325)}{(2)(300 \text{ cm})(7.9 \text{ s})} = 0.004221 \text{ cm/s} = 3.65 \text{ m/day}$$

∎

Table 5.15.4 Slug Test Data for Example 5.15.2

Elapsed Time (s)	Change in water level (m)
0	0.640
1	0.617
2	0.491
3	0.433
4	0.385
5	0.340
6	0.305
7	0.268
8	0.241
9	0.211
10	0.189
11	0.169
12	0.147
13	0.130
14	0.115
15	0.101
16	0.089
17	0.078
18	0.068
19	0.058
20	0.051
21	0.044
22	0.038
23	0.032

Boring diameter = 20 cm

Casing diameter = 5.1 cm

Depth to bottom of well = 17 m BGS (below ground surface)

Depth to top of casing = 14 m BGS

Depth to water = 6.7 m BGS

Saturated thickness = 15 m

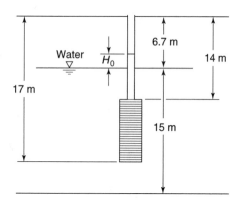

Figure 5.15.8. Slug test for Example 5.15.2.

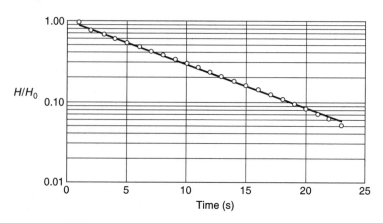

Figure 5.15.9. Normalized response data for Example 5.15.2.

5.15.3 KGS Model

Unlike the Bouwer and Rice method or the Dagan method, the KGS (Kansas Geological Survey) method incorporates the storage properties of the media into the analysis. Hyder et al.[50] presented the analytical solution to this model as

$$\frac{H(t)}{H_0} = f\left(\beta,\ \alpha,\ \psi,\ L_e/L_w,\ L_e/h\right) \tag{5.15.5}$$

Plotting the normalized head versus the logarithm of β for a particular well-formation configuration defined as $(\psi,\ L_e/L_w,\ L_e/h)$ develops a series of type curves (Figure 5.15.10), each referring to a different α. The analysis procedure consists of fitting one of the α curves to the field data (normalized response data versus logarithm of the time since the test began) and

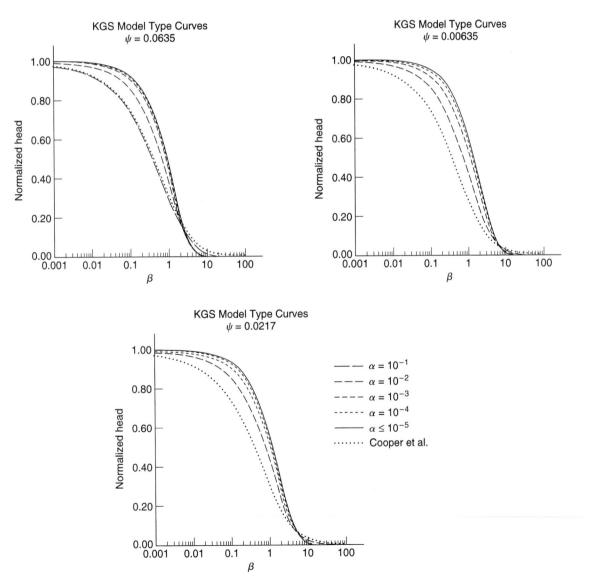

Figure 5.15.10. Normalized head $(H(t)/H_0)$ vs. logarithm of β type curves generated using the KGS model for different values of ψ.[17]

select match points (e.g., 1.0) and read the value of $t_{1.0}$. Using the definition of the radial component of the hydraulic conductivity is

$$K_r = \frac{r_c^2}{L_e t_{1.0}}$$ (5.15.6)

And the specific storage is

$$S_s = \frac{r_c^2}{r_w^2 L_e}$$ (5.15.7)

5.16   SLUG TESTS FOR HIGH CONDUCTIVITY FORMATIONS

Slug tests performed in high transmissivity (high conductivity) aquifers may produce oscillatory data. In fact, the data may exhibit an underdamped oscillatory behavior, where the oscillations occur as a damped sine wave, indicating that inertial forces are significant. Figure 5.16.1 illustrates the slug test results with underdamped oscillatory data. Methods have been developed for the analysis of underdamped oscillatory slug-test response data.[57, 67, 70, 90, 99]

These methods are based upon solutions of the conservation of momentum written in terms of the dimensionless deviation of the water level ($w_d = w/H_0$) from the static level in the test well, where w is the actual deviation, and dimensionless time, $t_d = (g/L_e)^{0.5}t$. The conservation of momentum is expressed as

$$\frac{d^2 w_d}{dt_d^2} + C_d \frac{dw_d}{dt_d} + w_d = 0$$ (5.16.1)

where C_d is the dimensionless dampening parameter.

For the van der Kamp model[99], the dimensionless dampening parameter, $C_d \ll 2$ (for response data that are well within the underdamped region), is expressed as

$$C_d = \sqrt{\frac{g}{L_e}} \frac{r_c^2 \ln\left[\left(1.27/r_s^2\right)\left(L_e/g\right)^{0.5}\left(K_r/S_s\right)\right]}{8 K_r B}$$ (5.16.2)

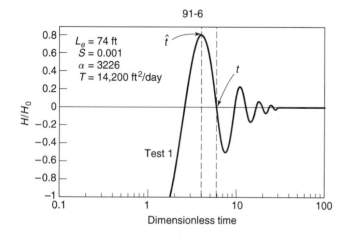

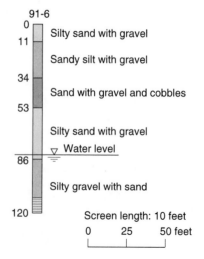

Figure 5.16.1. Slug test results and drillhole schematic for well 91-6.[106]

For the Kipp model[57]

$$C_d = \sqrt{\frac{g}{L_e}} \, \frac{r_c^2 \ln\left[(L_e/g)^{0.5} \left(K_r/S_s r_s^2 \right) \right]}{4 K_r B} \tag{5.16.3}$$

For the Springer and Gelhar model[90]

$$C_d = \sqrt{\frac{g}{L_e}} \, \frac{r_c^2 \ln\left[R_e/r_s \right]}{2 K_r B} \tag{5.16.4}$$

For the McElwee et al. model[67]

$$C_d = \sqrt{\frac{g}{L_e}} \, \frac{r_c^2 \ln\left[1/(2\varphi) + \left(1 + (1/(2\psi))^2 \right)^{0.5} \right]}{2 K_r B} \tag{5.16.5}$$

The Kipp method accounts for both well-bore storage and inertial effects, producing a series of type curves that allow for oscillatory responses.

5.17 WELL-SKIN EFFECT

The process of drilling, installing, and developing a well often causes the material in the immediate vicinity of the well to have altered characteristics from that of the original formation. These altered characteristics cause the well-skin effect, which may have a large impact on the estimate of the hydraulic conductivity from a slug test. The method by Ramey et al.[80] has been the most common approach for incorporating the well-skin effect in fully penetrating wells. In partially penetrating wells, the method by Ramey et al.[80] and the KGS model can be used. See Butler[17] for more detail on the well-skin effect.

PROBLEMS

5.11.1 The drawdown data from a step-drawdown pumping test are tabulated below. Using the given data and assuming $n = 2$, (a) calculate the formation and well loss coefficients, (b) comment on the validity of the $n = 2$ assumption, (c) assess the well condition, and (d) calculate the specific capacity and efficiency of the well for a pumping rate of 1,250 m³/day.

Q (m³/day)	s_w (m)
500	1.14
1,000	2.5
1,500	4.22
2,000	6.43
2,500	9.12
3,000	11.97
3,500	14.87

5.11.2 Using the Rorabaugh procedure, determine the formation loss coefficient, well loss coefficient, and the power n of discharge in Equation 4.67 for the drawdown data from a step-drawdown pumping test below.

Q (m³/day)	S_w (m)
250	0.32
500	0.87
750	1.78
1,000	3.12
1,250	4.95
1,500	7.34
1,750	10.33

5.11.3 A new pumping well with a radius of 0.5 m is constructed and the drawdown data during the first six months are given on page 274. The well pumps at a constant rate of 1,500 m³/day from a confined aquifer with a transmissivity of 640 m²/day and a storage coefficient of 4.5×10^{-4}. If the aquifer is isotropic, homogeneous, and infinite in areal extent, assess the well condition for the first six months following its construction.

t (days)	s_w (m)
2	3.34
4	3.47
6	3.55
8	3.60
10	3.64
15	3.72
20	3.79
30	3.88
60	4.13
90	4.36
120	4.59
150	4.87
180	5.34

5.11.4 The drawdown in an existing pumping well with a radius of 0.5 m is measured as 3.88 m. If the well has been pumping at a constant rate of 750 m³/day for a long period of time so that near steady-state conditions are established (the water level in the pumping well is essentially stabilized), estimate the transmissivity of the confined aquifer. Assume the well is 100 percent efficient. Would the actual transmissivity be larger or smaller than your estimate considering the well losses?

5.11.5 An existing well has been pumping from a confined aquifer at a constant rate of 1,500 m³/day for three months. The drawdown in the pumping well with a radius of 0.5 m is measured as 4.42 m at the end of three months. Estimate the aquifer transmissivity assuming the well losses are negligible. Perform a sensitivity analysis considering the well losses.

5.11.6 Resolve Problem 4.7.1 including well losses. Use $C = 0.8 \times 10^{-4}$ ft/gpm² and $n = 2$.

5.14.1 The water level data from a slug test are tabulated below. The initial water level following the injection of the slug is 0.46 m. The casing radius of the well is 7.8 cm and the radius of the screen is 5.4 cm. If the test is performed in a confined aquifer with a thickness of 12 m, determine the transmissivity, hydraulic conductivity, and storage coefficient using the Cooper-Bredehoeft-Papadopulos method.

Time (s)	H (m)	H/H_0
2	0.41	0.89
6	0.36	0.78
10	0.31	0.68
20	0.25	0.54
45	0.15	0.32
80	0.09	0.20
100	0.06	0.14
120	0.05	0.11

5.14.2 A casing with a radius of 6.8 cm is installed through a confined aquifer. A screen with a radius of 5.6 cm penetrates the full thickness of the formation, 8 m. A slug of water is injected instantaneously, raising the water level in the well 0.51 m. The water level recovery in the well is tabulated below. Estimate the transmissivity, hydraulic conductivity, and storage coefficient using the Cooper-Bredehoeft-Papadopulos method.

Time (s)	H (m)	H/H_0
3	0.46	0.910
9	0.39	0.765
15	0.34	0.660
30	0.23	0.450
45	0.16	0.310
60	0.11	0.220
90	0.06	0.125

5.14.3 The results of a slug test are given in Table P5.14.3. Well construction information is as follows:

Boring diameter = 20 cm
Casing diameter = 5.1 cm
Depth to bottom of well = 17 m BGS
Depth to top of casing = 14 m BGS
Depth to water = 6.7 m BGS
Saturated thickness = 15 m

If the initial water level rise following the injection of the slug is 0.64 m, determine the hydraulic conductivity of the formation using the Hvorslev method.

Table P5.14.3 Slug Test Results for Problem 5.14.3

Elapsed time (s)	Change in water level (m)
0	0.640
1	0.617
2	0.491
3	0.433
4	0.385
5	0.340
6	0.305
7	0.268
8	0.241
9	0.211
10	0.189
11	0.169
12	0.147
13	0.130
14	0.115
15	0.101
16	0.089
17	0.078
18	0.068
19	0.058
20	0.051
21	0.044
22	0.038
23	0.032

5.14.4 A slug test is performed by lowering the static water level in the well by 0.60 m at time $t = 0$ sec. Following this, the water level recovery in the well occurs as tabulated below. Given the well construction information, determine the hydraulic conductivity of the formation using the Hvorslev method.

Elapsed time (s)	Change in water level (m)
0	0.600
22	0.550
43	0.510
65	0.475
92	0.436
134	0.387
167	0.344
225	0.298
263	0.258
298	0.228
350	0.192
415	0.156
472	0.134
542	0.104
628	0.082

Boring diameter = 20 cm
Casing diameter = 5.2 cm
Depth to bottom of well = 8.2 m BGS
Depth to top of casing = 5.2 m BGS
Depth to water = 3.8 m BGS
Saturated thickness = 4.4 m

5.15.1 Rework Problem 5.14.3 using the Bouwer and Rice method. Compare your results to the results of Problem 5.14.3. Are they consistent?

5.15.2 The response data from a slug (bail-down) test are tabulated in Table P5.15.2. The well construction details are shown in Figure P5.15.2. The initial head drop following the bailing is 0.99 m. Assess the feasibility of the following methods for the given well geometry and response data.

1. Cooper-Bredehoeft-Papadopulos method
2. Hvorslev method
3. Bouwer and Rice method
4. Dagan method

Determine the hydraulic conductivity of the formation using the feasible methods and compare the results.

Table P5.15.2 Slug Test Data for Problem 5.15.2

Time (s)	H/H_0
0	1.000
2	0.960
4	0.930
5	0.918
10	0.867

Table P5.15.2 (*continued*) Slug Test Data for Problem 5.15.2

Time (s)	H/H_0
15	0.829
20	0.792
29	0.732
39	0.675
48	0.629
57	0.588
67	0.548
78	0.509
87	0.480
97	0.451
115	0.407
133	0.369
156	0.332
179	0.302
219	0.263
255	0.239
292	0.220
355	0.197
449	0.176
563	0.160
626	0.154
738	0.145
795	0.140
879	0.135

Boring diameter = 23 cm
Casing diameter = 10.2 cm
Depth to bottom of well = 36.6 m BGS
Depth to top of casing = 24.4 m BGS
Depth to water = 24.4 m BGS
Saturated thickness = 15.2 m

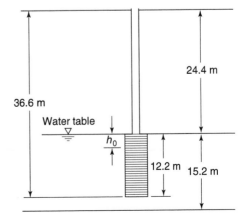

Figure P5.15.2. Well construction details for Problem P5.15.2

5.16.1 Perform a literature search in order to describe the van der Kamp method[99] for performing slug tests in high-conductivity formations. Start with the reference Wylie and Magnuson[108].

5.16.2 Perform a literature search in order to describe the Kipp methodology[57] for slug tests in high conductivity formations. Start with the reference by Weight and Wittman.[106]

5.16.3 Develop a write-up on the well-skin effect.

5.16.4 Describe the mathematical formulation of the Cooper et al. method and describe the assumptions.

5.16.5 Describe the mathematical formulation of the Hvorslev method and describe the differences in the mathematical formulations between this method and the Cooper et al. method.

5.16.6 Describe the mathematical formulation of the Bouwer and Rice method. Describe how the two key assumptions of this method are reflected in the mathematical formulation.

5.16.7 Describe the mathematical formulation of the Dagan method. Describe the differences in assumptions with the Bouwer and Rice method.

REFERENCES

1. Ahmad, N., *Tubewells, Construction and Maintenance,* Scientific Research Stores, Lahore, Pakistan, 250 pp., 1969.

2. Ahrens, T. P., Basic considerations of well design, *Water Well Jour.,* v. 24, no. 4, pp. 45–50; no. 5, pp. 49–51; no. 6, pp. 47–51; no. 8, pp. 35–37, 1970.

3. Amer. Water Works Assoc., *AWWA Standard for Water Wells,* ANSI/AWWA A100–97, Denver, CO, Feb. 1998.

4. Amer. Water Works Assoc., *Getting the Most from Your Well Supply,* Proceedings AWWA Seminar, New York, 62 pp., 1972.

5. Amer. Water Works Assoc., *Ground water,* AWWA Manual M21, 130 pp., 1973.

6. Anderson, K. E., *Water Well Handbook,* 2nd ed., Missouri Water Well & Pump Contractors Assoc., Rolla, 281 pp., 1967.

7. Barlow, P. M. and A. F. Moench, *WTAQ—Computer Program for Calculating Drawdowns and Estimating Hydraulic Properties for Confined and Water-Table Aquifers, U.S. Geological Survey Water-Resources Investigations Report* 99-4425, 1999.

8. Barnes, I. and F. E. Clarke, *Chemical Properties of Ground Water and Their Corrosion and Encrustation Effects on Wells, U.S. Geological Survey Professional Paper* 498-D, 58 pp., 1969.

9. Bennett, T. W., On the design and construction of infiltration galleries, *Ground Water,* v. 8, no. 3, pp. 16–24, 1970.

10. Bentall, R., *Methods for Determing Permeability, Transmissibility, and Drawdown. U.S. Geological Survey Water-Supply Paper* 1536-I, pp. 243–341, 1963.

11. Bentall, R., *Methods of Collecting and Interpreting Ground-Water Data. U.S. Geological Survey Water-Supply Paper* 1544-H, 97 pp., 1963.

12. Bernhard, A. P., Protection of water-supply wells from contamination by wastewater, *Ground Water,* v. 11, no. 3, pp. 9–15, 1973.

13. Bierschenk, W. H., Determining well efficiency by multiple step–drawdown tests, *Intl. Assoc. Sci. Hydrology Publ. 64,* pp. 493–507, 1964.

14. Bouwer, H., The Bouwer and Rice Slug Test—An Update, *Ground Water,* v. 27, no. 3, pp. 304–309, 1989.

15. Bouwer, H., and R. C. Rice, A slug test for determining hydraulic conductivity of unconfined aquifers with completely or partially penetrating wells, *Water Resources Research,* v. 12, pp. 423–428, 1976.

16. Butler, J. J., Jr., The role of pumping tests in site characterization: Some theoretical considerations, *Ground Water,* v. 28, no. 3, pp. 394–402, 1990.

17. Butler, J. J. Jr., *The Design, Performance, and Analysis of Slug Tests,* Lewis Publishers, CRC Press LLC, Boca Raton, FL, 1998.

18. Butler, J. J., Jr., and J. M. Healey, Relationship between pumping-test and slug-test parameters: Scale effect or artifact, *Ground Water,* v. 36, no. 2, pp. 305–313, 1998.

19. Butler, J. J., Jr., C. D. McElwee, and W. Liu, Improving the quality of parameter estimates obtained from slug tests, *Ground Water,* v. 34, no. 3, pp. 480–490, 1996.

20. California Dept. of Water Resources, *Water well standards: State of California,* Bull. 74, Sacramento, 205 pp., 1968.

21. Campbell, M. D., and J. H. Lehr, *Water Well Technology,* McGraw-Hill, New York, 681 pp., 1973.

22. Campbell, M. D., M. S. Starrett, J. D. Fowler, and J. J. Klein, Slug tests and hydraulic conductivity, *Ground Water Management,* v. 2, 1990.

23. Chirlin, G. R., A critique of the Hvorslev method for slug test analysis: The fully penetrating well, *Ground Water Monitor Rev.,* v. 9, no. 2, 1989.

24. Chirlin, G. R., The slug test: The first four decades, *Ground Water Management,* v. 1, 1990.

25. Cole, K. D., and V. A. Zlotnik, Modification of Dagan's numerical method for slug and packer test interpretation, in *Computational Methods in Water Resources* X. Peters et al., ed., Kluwer Academic, Dordrecht, 1994.

26. Cooper, H. H., Jr., J. D. Bredehoeft, and I. S. Papadopulos, Response of a finite-diameter well to an instantaneous charge of water, *Water Resources Research,* v. 3, no.1, pp. 11–21, 1967.

27. Dagan, G., A note on packer, slug, and recovery tests in unconfined aquifers, *Water Resources Research,* v. 14, no. 5, 1978.

28. Dawson, K., and J. Istok, *Aquifer Testing: Design and Analysis of Pumping and Slug Tests,* Lewis Publishers, Boca Raton, FL, 1991.

29. Driscoll, F. G., *Groundwater and Wells,* 2nd ed., Johnson Division, St. Paul, MN, 1986.

30. Erickson, C. R., Cleaning methods for deep wells and pumps, *Jour. Amer. Water Works Assoc.,* v. 53, pp. 155–162, 1961.

31. Ferris, J. G., and D. B. Knowles, *Slug Test for Estimating Transmissibility,* U.S. Geological Survey Note 26, 1954.

32. Fetter, C. W., *Applied Hydrogeology,* Prentice Hall, Upper Saddle River, NJ, 2001.

33. Feulner, A. J., *Galleries and Their Use for Development of Shallow Groundwater Supplies, with Special Reference to Alaska, U.S. Geological Survey Water-Supply Paper* 1809-E, 16 pp., 1964.

34. Franke, O. L., *Estimating Areas Contributing Recharge to Wells: Lessons from Previous Studies, U.S. Geological Survey Circular* 1174, 1998.

35. Gabrysch, R. D., The relation between specific capacity and aquifer transmissibility in the Houston area, Texas, *Ground Water,* v. 5, no. 4., pp. 9–14, 1968.

36. Garg, S. P., and J. Lal, Rational design of well screens, *Jour. Irrig. Drainage Div.,* Amer. Soc. Civil Engrs., v. 97, no. IR1, pp. 131–147, 1971.

37. Gibb, *J. P., Wells and pumping systems for domestic water supplies,* Illinois State Water Survey Circular 117, Urbana, 17 pp., 1973.

38. Gibson, U. P., and R. D. Singer, *Water Well Manual,* Premier Press, Berkeley, CA, 156 pp., 1971.

39. Gidley, H. K., and J. H. Millar, Performance records of radial collector wells in Ohio River Valley, *Jour. Amer. Water Works Assoc.,* v. 52, pp. 1206–1210, 1960.

40. Gordon, R. W., *Water well drilling with cable tools,* Bucyrus-Erie Co., South Milwaukee, WI, 230 pp., 1958.

41. Greene, E. A., and A. M. Shapiro, *Methods of Conducting Air-Pressurized Slug Tests and Computation of Type Curves for Estimating Transmissivity and Storativity, U.S. Geological Survey Open-File Report* 95-424, 1995.

42. Greene, E. A., and A. M. Shapiro, AIRSLUG: A Fortran program for the computation of type curves to estimate transmissivity and storativity of prematurely terminated air-pressurized slug tests, *Ground Water,* v. 36, no. 2, pp. 373–376, 1998.

43. Halford, K. J., and E. L. Kuniansky, *Spreadsheets for the Analysis of Aquifer-Test and Slug-Test Data, U.S. Geological Survey Open-File Report* 02-197, http://pubs.usgs.gov/of/ofr02197, 2002.

44. Hall, P., and J. Chen, *Water Well and Aquifer Test Analysis,* Water Resources Publications, Littleton, CO, 1996.

45. Heath, R. C., *Basic Ground-Water Hydrology, U.S. Geological Water Survey Water-Supply Paper* 2220, 1983.

46. Hrovslev, M. J., Time lag and soil permeability in groundwater observations, U.S. Army Corps of Engineers Waterway Experimentation Station, Bulletin 36, Vicksburg, MS 1951.

47. Huisman, L., *Groundwater Recovery,* Winchester Press, New York, 336 pp., 1972.

48. Hunter Blair, A., Well screens and gravel packs, *Ground Water,* v. 8, no. 1, pp. 10–21, 1970.

49. Hurr, R. T., A new approach for estimating transmissibility from specific capacity, *Water Resources Research,* v. 2, pp. 657–664, 1966.

50. Hyder, Z., J. J. Butler, Jr., C. D. McElwee, and W. Z. Liu, Slug tests in partially penetrating wells, *Water Resources Research,* v. 30, no. 11, 1994.

51. Jacob, C. E., Drawdown test to determine effective radius of artesian well, *Trans. Amer. Soc. Civil Engrs.,* v. 112, pp. 1047–1070, 1947.

52. Johnson, A. I., et al., *Laboratory Study of Aquifer Properties and Well Design for an Artificial-Recharge Site, U.S. Geological Survey Water-Supply Paper* 1615-H, 42 pp., 1966.

53. Johnson Div., UOP Inc., *Ground Water and Wells,* 2nd ed., Edward E. Johnson, St. Paul, MN, 440 pp., 1972.

54. Karanjac, J., Well losses due to reduced formation permeability, *Ground Water,* v. 10, no. 4, pp. 42–49, 1972.

55. Kasenow, M. *Introduction to Aquifer Analysis,* Water Resources Publications, Littleton, CO, 1997.

56. Kemblowski, M. W., and C. L. Klein, An automated numerical evaluation of slug test data, *Ground Water,* v. 26, pp. 435–438, 1988.

57. Kipp, K. L., Jr., Type curve analysis of inertial effects in the response of a well to a slug test, *Water Resources Research,* v. 21, no. 9, pp. 1397–1408, 1985.

58. Koenig, L., Survey and analysis of well stimulation performance, *Jour. Amer. Water Works Assoc.,* v. 52, pp. 333–350, 1960.

59. Koenig, L., Relation between aquifer permeability and improvement achieved by well stimulation, *Jour. Amer. Water Works Assoc.,* v. 53, pp. 652–670, 1961.

60. Kruse, G., *Selection of gravel packs for wells in unconsolidated aquifers,* Colorado State Univ. Exp. Sta. Tech. Bull. 66, Fort Collins, 22 pp., 1960.

61. Kruseman, G. P., and N. A. deRider, *Analysis and Evaluation of Pumping Test Data,* International Insttiue for Land Reclamation and Improvement, The Netherlands, 1994.

62. Larson, T. E., *Corrosion by domestic waters,* Bull. 59, Illinois State Water Survey, Urbana, 48 pp., 1975.

63. Lee, J., *Well Testing,* Society of Petroleum Engineers of AIME, New York, 1982.

64. Lennox, D. H., Analysis and application of step–drawdown test, *Jour. Hydraulics Div., Amer. Soc. Civil Engrs.,* v. 92, no. HY6, pp. 25–48, 1996.

65. Logan, J., Estimating transmissibility from routine production tests of water well.s, *Ground Water,* v. 2, no. 1, pp. 36–37, 1964.

66. Matlock, W. G., Small diameter wells drilled by jet-percussion method, *Ground Water,* v. 8, no. 1, pp. 6–9, 1970.

67. McElwee, C. D., J. J. Butler, Jr., G. C. Bohling, *Nonlinear Analysis of Slug Tests in Highly Permeable Aquifers Using a Hvorslev-type approach, Kansas Geological Survey, Open-File Report* 92-39, Lawrence, KS, 1992.

68. McElwee, C. D., J. J. Butler, Jr., and G. C. Bohling, Sensitivity Analysis of slug tests: Part I, The slugged well, *Journal of Hydrology,* v. 164, pp. 53–67, 1995.

69. McLane, G. A., D. A. Harrity, and K. O. Thomsen, A pneumatic method for conducting rising and falling head tests in highly permeable aquifers, Proc. 1990 NWWA Outdoor Action Conf., National Water Well Association, 1990.

70. McElwee, C. D., and M. Zenner, Unified analysis of slug tests including nonlinearities, inertial effects, and turbulence, *Kansas Geological Survey Open-File Report* 93–45, 1993.

71. Mogg, J. L., The technical aspects of gravel well construction, *Jour. New England Water Works Assoc.,* v. 77, pp. 155–164, 1963.

72. Mogg, J. L., Practical construction and incrustation guide lines for water wells, *Ground Water,* v. 10, no. 2, pp. 6–11, 1972.

73. Moore, P. L., *Drilling Practices Manual,* Petroleum Publishing, Tulsa, OK, 448 pp., 1974.

74. Noble, D. G., Well points for dewatering, *Ground Water,* v. 1, no. 3, pp. 21–26, 1963.

75. Pandit, N. S., and R. F. Miner, Interpretation of slug test data, *Ground Water,* v. 24, pp. 743–749, 1986.

76. Papadopulos, S. S., J. D. Bredehoeft, and H. H. Cooper, Jr., On the analysis of "slug test" data, *Water Resources Research,* v. 9, no. 4, pp. 19–21, 1973.

77. Peres, A. M., M. Onur, and A. C. Reynolds, A new analysis procedure for determining aquifer properties from slug test data, *Water Resources Research,* v. 25, no. 7, 1989.

78. Peterson, F. L., Water development on tropic volcanic islands—Type example: Hawaii, *Ground Water,* v. 10, no. 5, pp. 18–23, 1972.

79. Prudic, D. E., *Estimates of Hydraulic Conductivity from Aquifer-Test Data and Specific-Capacity Data, Gulf Coast Regional Aquifer Systems, South-Central United Sates, U.S. Geological Survey Water-Resources Investigations Report* 90-4121, 1991.

80. Ramey, H. J., Jr., R.G. Agarwal, and I. Martin, Analysis of "slug test" or DST flow period data, *Journal Canadian Petroleum Technology,* vol. 24, 1975.

81. Ritchie, E. A., Cathodic protection wells and ground-water pollution, *Ground Water,* v. 14, pp. 146–149, 1976.

82. Rorabaugh, M. I., Graphical and theoretical analysis of step–drawdown test of artesian well, *Proc. Amer. Soc. Civil Engrg.,* v. 79, Sept 352, 23 pp., 1953.

83. Scalf, M. L. et al., *Manual of Ground Water Sampling Procedures,* National Water Well Association, 1981.

84. Schwartz, D. H. Successful sand control design for high rate oil and water wells, *Jour. Petr. Tech.,* v. 21, pp. 1193–1198, 1969.

85. Sepulveda, N., *Computer Algorithm for the Analysis of Underdamped and Overdamped Water-Level Responses in Slug Tests, U.S. Geological Survey Water-Resources Investigations Report* 91-4162, 1992.

86. Sheahan, N. T., Type-curve solution of step–drawdown test, *Ground Water,* v. 9, no. 1, pp. 25–29, 1971.

87. Soliman, M. M., Boundary flow considerations in the design of wells, *Jour. Irrig. Drain. Div.,* Amer. Soc. Civil Engrs., v. 91, no. IR1, pp. 159–177, 1965

88. Speedstar Div., *Well Drilling Manual,* Koehring Co., Enid, OK, 72 pp. (no date).

89. Spiridonoff, S. V., Design and use of radial collector wells, *Jour. Amer. Water Works Assoc.,* v. 56, pp. 689–698, 1964.

90. Springer, R. K., and L. W. Gelhar, *Characterization of Large-Scale Aquifer Heterogeneity in Glacial Outwash by Analysis of Slug Tests with Oscillatory Responses, Cape Cod, Massachusetts, U.S. Geological Survey Water Resources Investigations Report* 91-4034, 1991.

91. Stow, G.R.S., Modern water-well drilling techniques in use in the United Kingdom, *Ground Water,* v. 1, no. 3, pp. 3–12, 1963.

92. Stramel, G. J., Maintenance of well efficiency, *Jour. Amer. Water Works Assoc.,* v. 57, pp. 996-1010, 1965.

93. U.S. Bureau of Mines, *Horizontal Boring Technology: A State of the Art Study,* Info. Circular 8392, 86 pp., 1968.

94. U.S. Bureau of Reclamation, *Ground Water Manual,* U.S. Dept. of the Interior, 480 pp., 1977.

95. U.S. Environmental Protection Agency, *Manual of Water Well Construction Practices, Rept. Environmental Protection Agency-570/9-75-001,* Washington, DC, 156 pp., 1976.

96. U.S. Public Health Service, *Manual of Individual Water Supply Systems,* Publ. no. 24, 121 pp., 1962.

97. U.S. Soil Conservation Service, *Engineering Field Manual for Conservation Practices,* 995 pp., 1969.

98. Vaadia, Y., and V. H. Scott, Hydraulic properties of perforated well casings, *Jour. Irrig. Drainage Div.,* Amer. Soc. Civil Engrs., v. 84, no. IR1, Paper 1505, 26 pp., 1958.

99. van der Kamp, G., Determining aquifer transmissivity by means of well response tests: The underdamped case, *Water Resources Research,* v. 12, no. 1, pp. 71–77, 1976.

100. Walton, W., *Aquifer Test Analysis with Windows*TM, CRC Press, Boca Raton, FL, 1996.

101. Walton, W. C., *Selected analytical methods for well and aquifer evaluation,* Bull. 49, Illinois State Water Survey, Urbana, 81 pp., 1962.

102. Walton, W. C., and S. Csallany, *Yields of deep sandstone wells in Northern Illinois,* Rept. of Inv. 43, Illinois State Water Survey, Urbana, 47 pp., 1962.

103. Watson, L. J., Development of ground water in Hawaii, *Jour. Hydraulics Div.,* Amer. Soc. Civil Engrs., v. 90, no. HY6, pp. 185–202, 1964.

104. Watt, S. B., and W. E. Wood, *Hand Dug Wells and Their Construction,* Intermediate Technology, London, 234 pp., 1976.

105. Weight, W. D., and J. L. Sonderegger, *Manual of Applied Field Hydrogeology,* McGraw-Hill, New York, 2001.

106. Weight, W. D., and G. P. Wittman, Oscillatory slug-test data sets: A comparison of two methods, *Ground Water,* v. 37, no. 6, pp. 827–835, 1999.

107. Welchert, W. T., and B. N. Freeman, Horizontal wells, *Jour. Range Management,* v. 26, pp. 253–256, 1973.

108. Wylie, A. and S. Magnuson, Spreadsheet modeling of slug tests using the van der Kamp method, *Ground Water,* v. 33, no. 2, pp. 326–329, 1995.

109. Yang, Y. J., and T. M. Gates, Wellbore skin effect in slug-test data analysis for low-permeability geologic materials, *Ground Water,* v. 35, no. 6, pp. 931–937, 1997.

110. Zlotnik, V., Interpretation of slug and packer tests in anisotropic aquifers, *Ground Water,* v. 32, no. 5, 1994.

Chapter 6

Groundwater Levels and Environmental Influences

A groundwater level, whether it be the water table of an unconfined aquifer or the piezometric surface of a confined aquifer, indicates the elevation of atmospheric pressure of the aquifer. Any phenomenon that produces a change in pressure on groundwater will cause the groundwater level to vary. Differences between supply and withdrawal of groundwater cause levels to fluctuate. Streamflow variations are closely related to groundwater levels. Other diverse influences on groundwater levels include meteorological and tidal phenomena, urbanization, earthquakes, and external loads. Finally, subsidence of the land surface can occur due to changes in underlying groundwater conditions.

6.1 TIME VARIATIONS OF LEVELS

6.1.1 Secular Variations

Secular variations of groundwater levels are those extending over periods of several years or more. Alternating series of wet and dry years, in which the rainfall is above or below the mean, will produce long-period fluctuations of levels. The long records of rainfall and groundwater levels shown in Figure 6.1.1 illustrate this point. Rainfall is not an accurate indicator of groundwater level changes. Recharge is the governing factor (assuming annual withdrawals are constant); it depends on rainfall intensity and distribution and amount of surface runoff. Figure 6.1.2 illustrates the recharge occurring during and immediately following periods of precipitation.

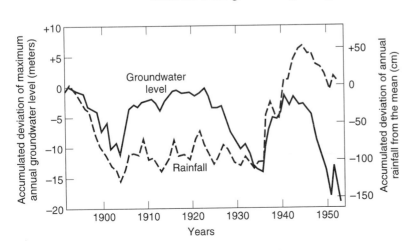

Figure 6.1.1. Secular variations of maximum annual groundwater level and annual rainfall in San Bernardino Valley, California.

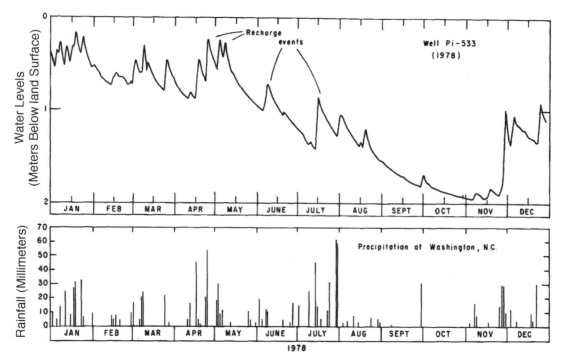

Figure 6.1.2. Fluctuation of the water table in the coastal plain of North Carolina (Heath[34]).

In other instances, pronounced trends may be noted. Thus, in overdeveloped basins where draft exceeds recharge, a downward trend of groundwater levels may continue for many years. Figure 6.1.3a dramatically illustrates the decline in piezometric surface of a deep sandstone aquifer as a result of nearly a century of intensive pumping in the Chicago metropolitan area. Maximum declines in heads occurred around 1980, after which public water suppliers shifted their source of water from groundwater to additional withdrawals from Lake Michigan. This shift resulted in a significant decrease in aquifer withdrawals and a recovery of heads as shown in Figure 6.1.3b.

Figures 6.1.4–6.1.6 show, respectively, the inferred potentiometric surface of the Memphis (Tennessee) aquifer prior to groundwater development, the potentiometric surface in 1995 with cones of depression, and the declining water levels in two observation wells.

6.1.2 Seasonal Variations

Many groundwater levels show a seasonal pattern of fluctuation. This results from influences such as rainfall and irrigation pumping that follow well-defined seasonal cycles. The variations shown in Figure 6.1.7 are typical for areas subject to frozen ground in winter. Highest levels occur in late spring and are lowest in winter. In irrigated areas where frozen ground is not a factor, lowest levels normally occur during fall at the end of the irrigation season. The amplitude depends on recharge, pumpage, and the type of aquifer; confined aquifers normally display a greater range in levels than do unconfined aquifers.

6.1.3 Short-Term Variations

Groundwater levels often display characteristic short-term fluctuations governed by the primary use of groundwater in a locality. Clearly defined diurnal variations may be associated with municipal water-supply wells. Similarly, weekly patterns occur with pumping for industrial and municipal purposes.

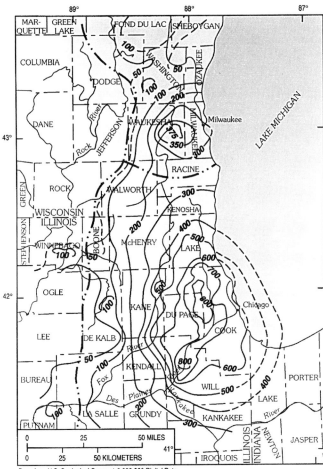

Explanation

— *700* – –   **Line of equal water-level decline, 1864–1980**—Dashed
where approximate. Interval, in feet, is variable

— — · · —   **Major ground-water divide**

Figure 6.1.3a. Decline in heads (water levels) in the
Cambrian–Ordovician confined aquifer, Chicago and
Milwaukee areas, 1864–1980 (modified from Avery[3] as
presented in Alley et al.[1]).

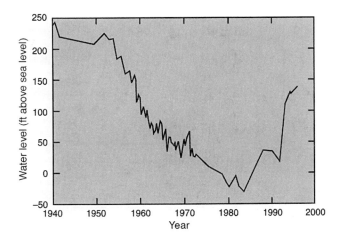

Figure 6.1.3b. Representative
trend of water levels for a deep
well in Cook County, Chicago
area, since 1940 (from
Visocky[89]).

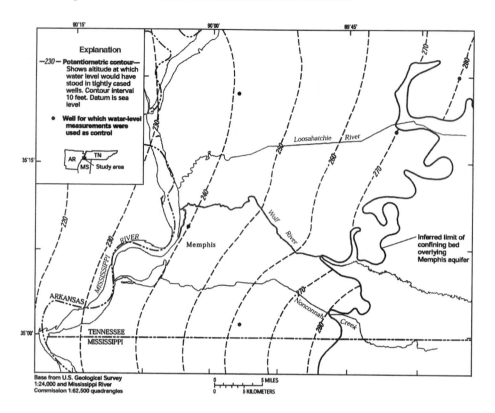

Figure 6.1.4. Inferred potentiometric surface of the Memphis aquifer prior to groundwater development. The observation wells shown were selected for their early records away from initial pumping centers (modified from Criner and Parks, 1976, as presented in Taylor and Alley[80]).

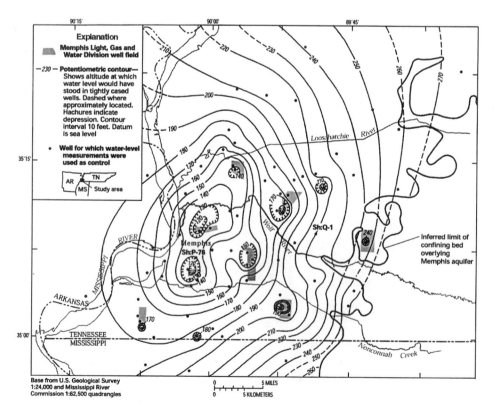

Figure 6.1.5. Potentiometric surface of the Memphis aquifer in 1995 showing cones of depression and location of observation wells Sh:P–76 and Sh:Q–1 (modified from Kingsbury, 1996, as presented in Taylor and Alley[80]).

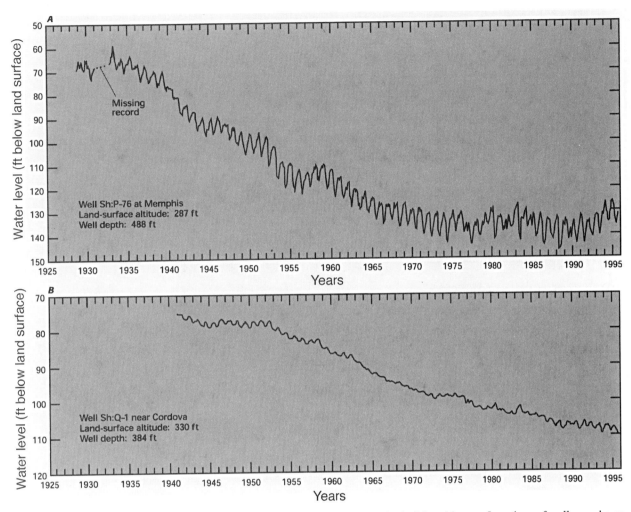

Figure 6.1.6. Declining water-level trends in two long-term observation wells in the Memphis area. Locations of wells are shown in Figure 6.1.5 (Taylor and Alley[80]).

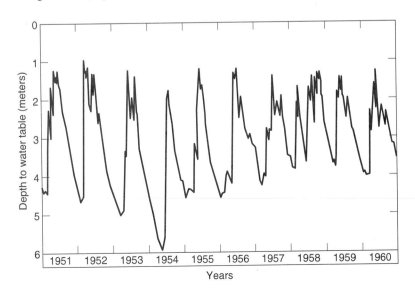

Figure 6.1.7. Seasonal fluctuations of the water table in a glacial till aquifer in Ohio. Well depth is 9 m (after Klein and Kaser[51]).

6.2 STREAMFLOW AND GROUNDWATER LEVELS

Where a stream channel is in direct contact with an unconfined aquifer, the stream may recharge the groundwater or receive discharge from the groundwater, depending on the relative levels. A *gaining stream* is one receiving groundwater discharge; a *losing stream* is one recharging groundwater (see Figure 6.2.1). Often a gaining stream may become a losing one, and conversely, as the stream stage changes.[50, 59] Figure 6.2.2 illustrates water table contours and groundwater flow direction in relation to stream stages.

The term *rising water* is applied to marked increases in streamflow in reaches where a subsurface restriction forces groundwater to the surface. Figure 6.2.3 illustrates the phenome-

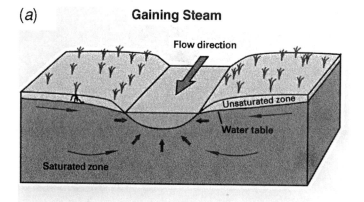

(a) **Gaining Steam**

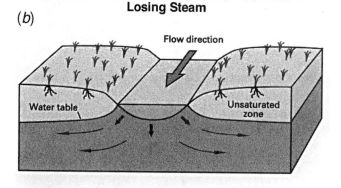

(b) **Losing Steam**

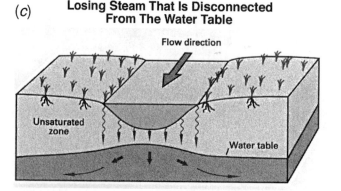

(c) **Losing Steam That Is Disconnected From The Water Table**

Figure 6.2.1. Interaction of streams and ground water (modified from Winter et al.[96], as presented in Alley et al.[1]). Gaining streams (*a*) receive water from the groundwater system, whereas losing streams (*b*) lose water to the groundwater system. For groundwater to discharge to a stream channel, the altitude of the water table in the vicinity of the stream must be higher than the altitude of the stream-water surface. Conversely, for surface water to seep to ground water, the altitude of the water table in the vicinity of the stream must be lower than the altitude of the stream surface. Some losing streams (*c*) are separated from the saturated groundwater system by an unsaturated zone.

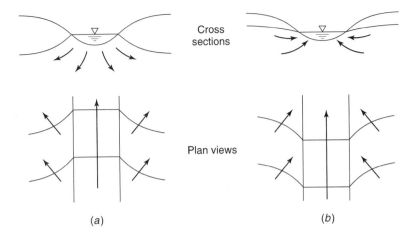

Figure 6.2.2. Water table contours and ground-water flow directions in relation to stream stages. (*a*) Losing stream (*b*) Gaining stream

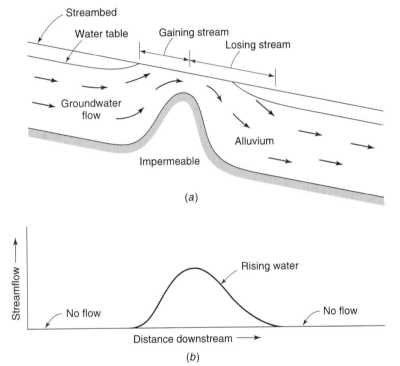

Figure 6.2.3. Illustration of rising water in a stream channel from emerging groundwater flow. (*a*) Cross section along stream channel in an alluvial valley (*b*) Streamflow as a function of distance along the stream

non for a situation where a dry stream channel exists above and below the convergent section. Figure 6.2.4 illustrates the dynamic interface between groundwater and streams.

6.2.1 Bank Storage

During a flood period of a stream, groundwater levels are temporarily raised near the channel by inflow from the stream. The volume of water stored and released after the flood is referred to as the *bank storage*. Field data are rarely adequate to evaluate bank storage and its rate of inflow and outflow; therefore, analytic or model approaches are necessary to obtain quantitative estimates for specified boundary conditions.

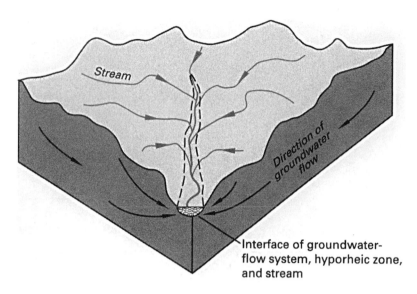

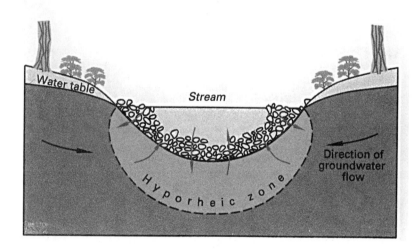

Figure 6.2.4. The dynamic interface between groundwater and streams (modified from Winter et al.,[96] as presented in Alley et al.[1]). Streambeds are unique environments where groundwater that drains much of the subsurface of landscapes interacts with surface water that drains much of the surface of landscapes. Mixing of surface water and groundwater takes place in the hyporheic zone where microbial activity and chemical transformations commonly are enhanced.

Figure 6.2.5 illustrates idealized groundwater conditions adjacent to a flooding stream. A flood hydrograph of sinusoidal form (Figure 6.2.5a) was superimposed on an aquifer and stream situation sketched in Figure 6.2.5b. As a result of the flood, the bank storage increased and then decreased; the variation of the volume of water in storage is depicted in Figure 6.2.5c. The derivative of the volume curve yields the groundwater flow curve (Figure 6.2.5d). From this it can be seen that a stream fluctuation produces large variations in magnitude and direction of local groundwater flow.

Cooper and Rorabaugh[13] derived solutions for changes in groundwater head near the stream, groundwater flow to the stream, and bank storage. Their comprehensive analysis also included a family of asymmetric flood-wave stage hydrographs, which facilitate study of the effects of a wide variety of flood shapes on groundwater.

6.2.2 Base Flow

Streamflow originating from groundwater discharge is referred to as *groundwater runoff* or *base flow*. During periods of precipitation, streamflow is derived primarily from surface runoff;

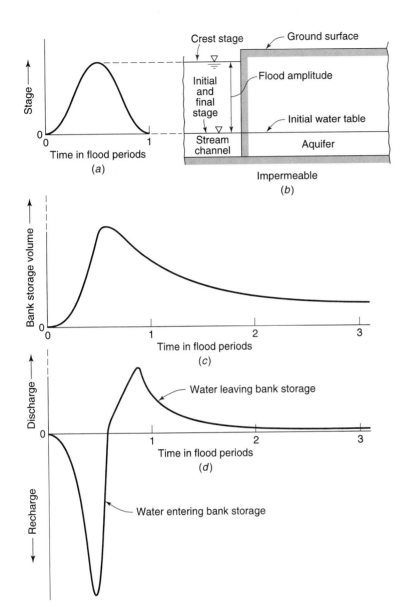

Figure 6.2.5. Groundwater in relation to a flooding stream. (*a*) Flood hydrograph (*b*) Vertical cross section of field conditions (*c*) Volume of bank storage as a function of time (*d*) Groundwater flow to and from bank storage. Results are from a laboratory model (after Todd[81]).

whereas during extended dry periods, all streamflow may be contributed by base flow.[32, 60] Typically, base flow is not subject to wide fluctuations and is indicative of aquifer characteristics within a basin.[49, 57, 58]

To estimate base flow, a rating curve of groundwater runoff can be prepared by plotting mean groundwater stage (water table level) within the basin against streamflow during periods when all flow is contributed by groundwater.[74] Figure 6.2.6*a* shows rating curves for a small drainage basin in Illinois. Data were fitted by two curves: one for the April–October period when evapotranspiration from groundwater is significant, and the other for the November–March period when evapotranspiration is minimal. With these rating curves and the mean groundwater stages for one year (see Figure 6.2.6*b*), the separation of surface runoff and base flow hydrographs shown in Figure 6.2.6*c* could be achieved. It can be noted that frozen ground impeded groundwater recharge during February and March and that base flows were largest during the spring and summer months. Groundwater contributed 33 percent of total streamflow for the year.

Streamflow at any instant contains groundwater contributed at previous times and different locations within the drainage area. During and after a storm period in a small drainage basin, the water table will rise, causing the base flow to increase also (Figure 6.2.6a). But superimposed on this will be the bank storage fluctuation (Figure 6.2.5d). The effects of these two variations are shown schematically in Figure 6.2.7.

An alternative approach to determining the separation of total streamflow into surface-runoff and groundwater components during flood periods can be accomplished from measurements of chemical concentrations.[53, 88] Total dissolved solids or any major ion will serve the purpose with the equation

$$C_{TR}Q_{TR} = C_{GW}Q_{GW} + C_{SR}Q_{SR} \qquad (6.2.1)$$

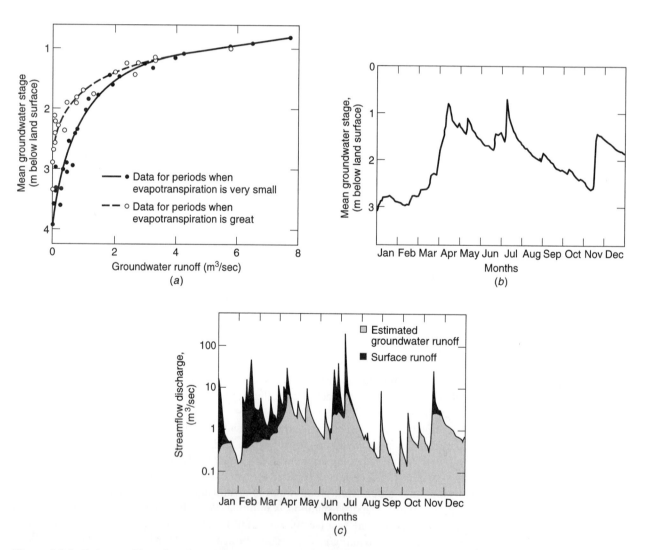

Figure 6.2.6. Estimate of base flow for Panther Creek near Peoria, Illinois (drainage area: 246 km^2). (*a*) Rating curves of mean groundwater stage versus groundwater runoff (base flow) (*b*) Mean groundwater stage for 1951 (*c*) Streamflow hydrograph for 1951 showing surface runoff and base flow components (after Schicht and Walton[73])

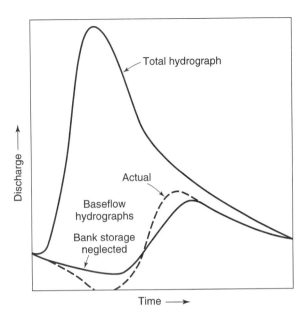

Figure 6.2.7. Schematic diagram of the variation of base flow during a flood hydrograph with and without effects of bank storage (after Singh[75]).

where C is ionic concentration, Q is streamflow, TR is total runoff, GW is groundwater contribution (base flow), and SR is surface runoff. Solving for the base flow,

$$Q_{GW} = [(C_{TR} - C_{SR})/(C_{GW} - C_{SR})]Q_{TR} \qquad (6.2.2)$$

where

$$Q_{TR} = Q_{GW} + Q_{SR} \qquad (6.2.3)$$

Values of C_{GW} are measured during rainless periods, C_{SR} is measured in small tributary streams during storm events, and C_{TR} is measured during the peak flow period in the main stream. Measurements by this method for three small basins (6 to 13 km^2) showed that groundwater contributed from 32 to 42 percent of the total flow at peak discharge.[64]

6.2.3 Base Flow Recession Curve

A *recession curve* shows the variation of base flow with time during periods of little or no rainfall over a drainage basin (see Chow et al.[9]). In essence, it is a measure of the drainage rate of groundwater storage from the basin.[21, 94] If large, highly permeable aquifers are contained within a drainage area, the base flow will be sustained even through prolonged droughts; if the aquifers are small and of low permeability, the base flow will decrease relatively rapidly and may even cease.[17, 83]

Analyses of streamflow hydrographs show that the recession curve can often be fitted by the equation

$$Q = Q_0 K^t \qquad (6.2.4)$$

where Q is streamflow at time t in days after a given discharge Q_0, and K is a recession constant governed by the hydrogeologic characteristics of the basin.[4, 9, 52] The value of K can be empirically determined from the slope of a straight line fitted to a series of consecutive discharges plotted on semilogarithmic paper, as shown in Figure 6.2.8. Typical values lie in the range of 0.89 and 0.95. Thus, prior knowledge of the shape of the recession curve enables future estimates to be made of streamflow during rainless periods.

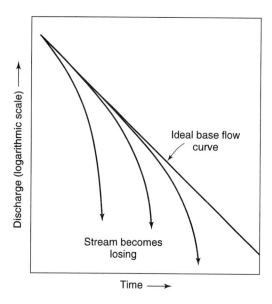

Figure 6.2.8. Base flow recession curves of streamflow for varying magnitudes of evapotranspiration losses from groundwater.

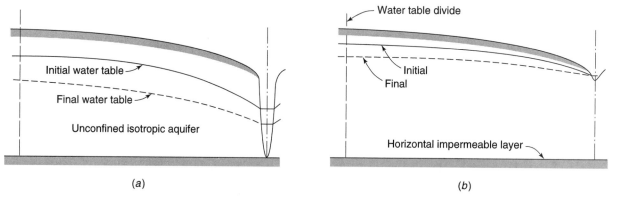

(a)   *(b)*

Figure 6.2.9. Water table and stream channel conditions affecting base flow. *(a)* Fully penetrating stream *(b)* Partially penetrating stream (after Singh[75])

An analytic study of base flow by Singh[75] demonstrated that base flow recession curves depend on the degree to which a stream channel is entrenched in an aquifer. For a fully penetrating stream (see Figure 6.2.9a), recession curves do not plot as straight lines on semilogarithmic paper; instead, the recession rate continuously decreases with time, forming a concave curve. But for deep aquifers and partially penetrating streams (see Figure 6.2.9b), the straight-line approximation is generally applicable. The value of K in Equation 6.2.4 varies directly with the degree of stream entrenchment.

These approaches to base flow assume that groundwater drains only toward the stream channel. Groundwater also can flow downward to an underlying leaky aquifer and can be lost by evapotranspiration to the atmosphere.[75] Where these diversions are significant, the recession curve will be deflected downward. In semiarid regions where streamflow is intermittent, evapotranspiration losses become significant; this causes the recession curve to steepen (see Figure 6.2.8) until streamflow finally ceases.

6.3   FLUCTUATIONS DUE TO EVAPOTRANSPIRATION

Unconfined aquifers with water tables near ground surface frequently exhibit diurnal fluctuations that can be ascribed to evaporation and/or transpiration. Both processes cause a discharge of groundwater into the atmosphere and have nearly the same diurnal variation because of their high correlation with temperature.

6.3.1   Evaporation Effects

Evaporation from groundwater increases as the water table approaches ground surface. The rate also depends on the soil structure, which controls the capillary tension above the water table and hence its hydraulic conductivity (see Chapter 3). Computation of actual evaporation from bare soil is complicated by variations in external evaporative conditions at the soil surface.[26, 35, 87] For isothermal conditions, upward movement is essentially all in the liquid phase, but a soil may have a high surface temperature, causing it to dry out, establishing upward vapor movement in response to a vapor pressure gradient.[69]

Field measurements of groundwater evaporation from tanks filled with soil (lysimeters) have been made. Water tables were maintained at prescribed depths below ground surface. Results, expressed as a percentage of pan evaporation at ground surface, are shown in Figure 6.3.1. For water tables within one meter of ground surface, evaporation is largely controlled by atmospheric conditions, but below this soil properties become limiting and the rate decreases markedly with depth.

6.3.2   Transpiration Effects

Where the root zone of vegetation reaches the saturated stratum, the uptake of water by roots equals (for practical purposes) the transpiration rate. Figure 6.3.2 shows water level variations measured in a well in a thicket of willows. Rapid foliage growth during August (Figure 6.3.2*a*) caused daily fluctuations averaging about 10 cm with the water table between 1.5 and 1.8 m below ground surface. Heavy frosts occurred in early October and most leaves had fallen

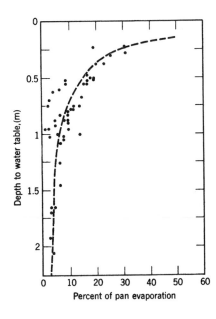

Figure 6.3.1. Groundwater evaporation, expressed as a percentage of pan evaporation, as a function of depth to water table (after White[95]).

by mid-October; thereafter, diurnal fluctuations were negligible (Figure 6.3.2*b*), with the vegetation dormant.

Magnitudes of transpiration fluctuations depend on the type of vegetation, season, and weather. Hot, windy days produce maximum drawdowns, whereas cool, cloudy days show only small variations. Fluctuations begin with the appearance of foliage and cease after killing frosts. Cutting of plants eliminates or materially reduces amplitudes. Transpiration discharge does not occur in nonvegetated areas, such as plowed fields, or in areas where the water table is far below ground surface. After rain on high water table vegetated land, the water table rises sharply as the increased soil moisture meets the transpiration demand and reduces the groundwater discharge; but on cleared land or when vegetation is dormant, little or no rise is evident.

The effect of transpiration on the interaction of groundwater with prairie pothole wetlands is illustrated in Figure 6.3.3. The wetland, near Jamestown, North Dakota, receives groundwater discharge most of the time; however, transpiration of groundwater by plants around the perimeter of the wetland can cause water to seep from the wetland, creating cones of depression. Seepage from wetlands commonly is assumed to be groundwater recharge, but in this case the water is actually lost to transpiration. This process results in depressions of the water table. The transpiration-induced depressions in the water table commonly are filled by recharge during the following spring and then form again by late summer almost every year.[96]

6.3.3 Evapotranspiration Effects

From a practical standpoint it is often difficult to segregate evaporation and transpiration losses from groundwater; therefore, the combined loss, referred to as *evapotranspiration*

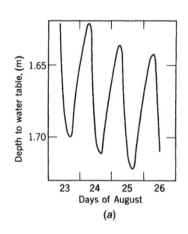

(a)

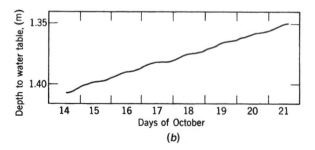

(b)

Figure 6.3.2. Effect of transpiration discharge on groundwater levels near Milford, Utah. (*a*) In summer (*b*) After frost (after White[95])

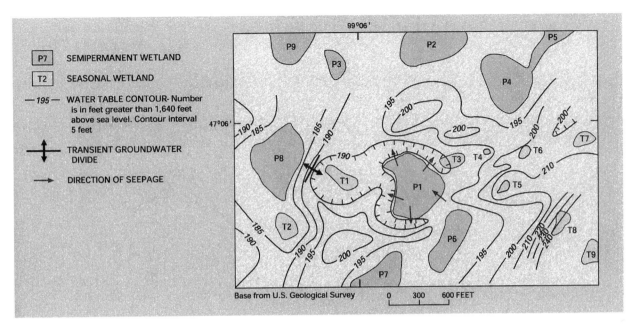

Figure 6.3.3. Transpiration directly from groundwater causes cones of depression to form by late summer around the perimeter of prairie pothole Wetland P1 in the Cottonwood Lake area in North Dakota (modified from T. C. Winter and D. O. Rosenberry, 1995). The interaction of ground water with prairie pothole wetlands in the Cottonwood Lake area, east-central North Dakota, 1979–1990: (*Wetlands*, v. 15, no. 3, pp. 193–211, as presented in Winter et al.[96])

(or *consumptive use*) is typically the quantity normally measured or calculated. The variation of evapotranspiration with water table depth is sketched in Figure 6.3.4 for three groundcover conditions. It is apparent that the deeper the roots, the greater the depth at which water losses occur. Even with relatively deep water tables, evapotranspiration does not necessarily become zero because upward transport can still occur, albeit minimally, in the vapor phase.

The pattern of diurnal fluctuation resulting from discharge of groundwater is nearly identical for evaporation and transpiration. The maximum water table level occurs in midmorning (see Figure 6.3.5) and represents a temporary equilibrium between discharge and recharge from surrounding groundwater. From midmorning until early evening, losses exceed recharge and the level falls. The steep slope near midday indicates maximum discharge associated with highest temperatures. The evening minimum again represents an equilibrium point, while the rise during the night hours is recharge in excess of discharge.

White[95] suggested a method for computing the total quantity of groundwater withdrawn by evapotranspiration during a day. If it is assumed that evapotranspiration is negligible from midnight to 4 A.M. and, further, that the water table level during this interval approximates the mean for the day, then the hourly recharge from midnight to 4 A.M. may be taken as the average rate for the day. If we let *h* equal the hourly rate of rise of the water table from midnight to 4 A.M., as shown by the upper curve in Figure 6.3.5, and *s* the net fall or rise of the water table during the 24-hour period, then as a good approximation the diurnal volume of groundwater discharge per unit area becomes

$$V_{ET} = S_y(24h \pm s) \tag{6.3.1}$$

where S_y is the specific yield near the water table. Actually, as pointed out by Troxell,[82] the rate of groundwater recharge to the vegetated area varies inversely with the water table level. Thus,

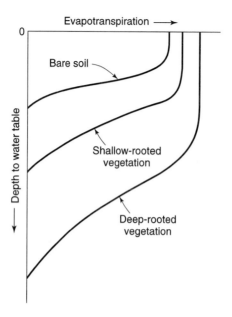

Figure 6.3.4. Generalized variation of evapotranspiration from groundwater with water table depth for three groundcover conditions (after Bouwer[5]).

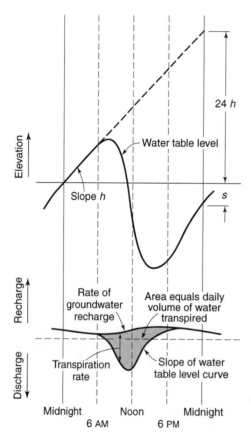

Figure 6.3.5. Interrelations of water table level, recharge, and evapotranspiration fluctuations (after Troxell[82]).

the difference between the recharge rate and the slope of the groundwater level curve gives the evapotranspiration rate. The lower portion of Figure 6.3.5 illustrates this; the area between the two curves is a measure of the daily volume of water released to the atmosphere.

6.4 FLUCTUATIONS DUE TO METEOROLOGICAL PHENOMENA

6.4.1 Atmospheric Pressure

Changes in atmospheric pressure (barometric tides) produce sizable fluctuations in wells penetrating confined aquifers.[10, 28] The relationship is inverse; that is, increases in atmospheric pressure produce decreases in water levels, and conversely. When atmospheric pressure changes are expressed in terms of a column of water, the ratio of water level change to pressure change expresses the *barometric efficiency* of an aquifer. Thus,

$$B = \frac{\gamma \Delta h}{\Delta p_a} \tag{6.4.1}$$

where B is barometric efficiency, γ is the specific weight of water, Δh is the change in piezometric level, and Δp_a is the change in atmospheric pressure. Most observations yield values in the range of 20 to 70 percent.

The effect is apparent in data shown in Figure 6.4.1. The upper curve indicates observed water levels in a well in Iowa City, Iowa, penetrating a confined aquifer. The lower curve shows atmospheric pressure inverted, expressed in meters of water, and multiplied by 0.75. A close correspondence of major fluctuations exists in the two curves; the equality of amplitudes indicates that the barometric efficiency of the aquifer is about 75 percent.

The phenomenon can be explained by recognizing that aquifers are elastic bodies.[70, 71] If Δp_a is the change in atmospheric pressure and Δp_w is the resulting change in hydrostatic pressure at the top of a confined aquifer, then

$$\Delta p_a = \Delta p_w + \Delta s_c \tag{6.4.2}$$

where Δs_c is the increased compressive stress on the aquifer (Figure 6.4.2). At a well penetrating the confined aquifer, the relation

$$p_w = p_a + \gamma h \tag{6.4.3}$$

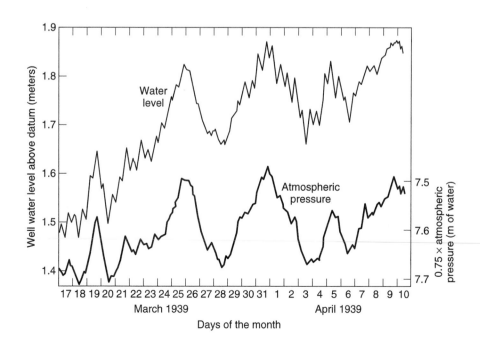

Figure 6.4.1. Response of water level in a well penetrating a confined aquifer to atmospheric pressure changes, showing a barometric efficiency of 75 percent (after Robinson[71]).

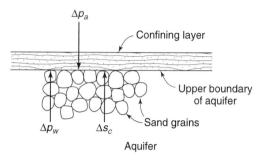

Figure 6.4.2. Idealized distribution of forces at the upper boundary of a confined aquifer resulting from a change in atmospheric pressure.

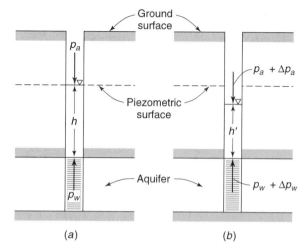

(a)   (b)

Figure 6.4.3. Effect of an increase in atmospheric pressure on the water level of a well penetrating a confined aquifer.

exists as shown in Figure 6.4.3a, where γ is the specific weight of water. Let the atmospheric pressure increase by Δp_a, then

$$p_w + \Delta p_w = p_a + \Delta p_a + \gamma h' \tag{6.4.4}$$

as shown in Figure 6.4.3b. Substituting for p_w from Equation 6.4.3 yields

$$\Delta p_w = \Delta p_a + \gamma(h' - h) \tag{6.4.5}$$

But from Equation 6.4.2 it is apparent that $\Delta p_w < \Delta p_a$, indicating that $h' < h$. Generally, therefore, the water level in a well falls with an increase in atmospheric pressure. It follows that the converse is also true.*

Jacob[48] developed expressions relating barometric efficiency of a confined aquifer to aquifer and water properties, including the storage coefficient. Gilliland[28] showed that changes in soil moisture from infiltrating precipitation can affect the magnitude of barometric efficiency.

For an unconfined aquifer, atmospheric pressure changes are transmitted directly to the water table, both in the aquifer and in a well; hence, no pressure difference occurs. Air entrapped in pores below the water table is affected by pressure changes, however, causing fluctuations similar to but smaller than that observed in confined aquifers.[59, 63] Temperature fluctuations in the capillary zone will also induce water table fluctuations where entrapped air is present.[78, 84]

Atmospheric pressure fluctuations do affect water tables substantially on small, permeable oceanic islands. The response of sea-level changes to atmospheric pressure is essentially *isostatic;* that is, sea level adjusts to a constant mass of the ocean–atmosphere column. This causes the ocean to act as an inverted barometer with sea level rising about 1 cm to compensate for a drop in atmospheric pressure of 1 mb. These fluctuations amount to about 20 cm in the open ocean and are transmitted as long-term tides to the water table (see the subsequent section on ocean tides). Data from Bermuda by Vacher[86] shown in Figure 6.4.4 illustrate the fluctuations.

---

*Atmospheric pressure waves created by nuclear explosions in the Soviet Union have caused fluctuations of the piezometric surface in limestone aquifers in England.[38] One fluctuation displayed an amplitude of 0.46 cm in response to a pressure wave of 900–1,000 microbars.

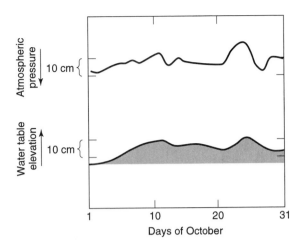

Figure 6.4.4. Variations in atmospheric pressure and water table elevation during October 1973 at Devonshire Post Office, Bermuda. Note that the variation is inverse with a rise in pressure associated with a decline in water level (after Vacher[86]).

6.4.2 Rainfall

As described previously, rainfall is not an accurate indicator of groundwater recharge because of surface and subsurface losses as well as travel time for vertical percolation. The travel time may vary from a few minutes for shallow water tables in permeable formations to several months or years for deep water tables underlying sediments with low vertical permeabilities. Furthermore, in arid and semiarid regions, recharge from rainfall may be essentially zero. Shallow water tables show definite responses to rainfall, as Figure 6.4.5 indicates. Water levels shown are the average for 25 observation wells; greatest fluctuations occurred in the upper portions of the basin and smallest near the basin outlet.

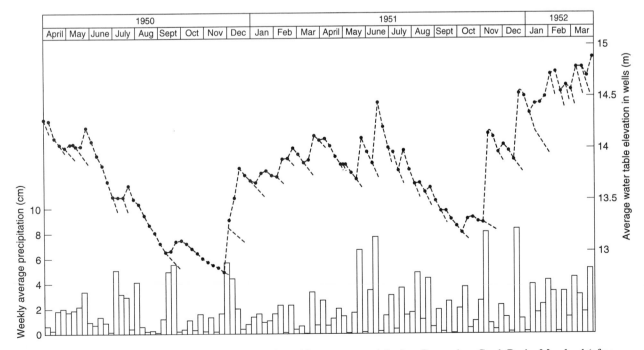

Figure 6.4.5. Variation in average water table level and weekly average precipitation, Beaverdam Creek Basin, Maryland (after Rasmussen and Andreasen[68]).

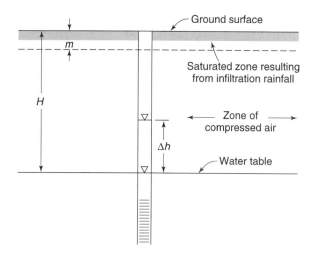

Figure 6.4.6. Water table rise in an observation well resulting from infiltrating rainfall sealing the ground surface and compressing air above the water table.

Groundwater levels may show seasonal variations due to rainfall, but often these include natural discharge and pumping effects as well. Droughts extending over a period of several years contribute to declining water levels.

Where the unsaturated zone above a water table has a moisture content less than that of specific retention (see Chapter 2), the water table will not respond to recharge from rainfall until this deficiency has been satisfied. Thereafter, the rise Δh will amount to

$$\Delta h = P_i/S_y \qquad (6.4.6)$$

where P_i is that portion of precipitation that percolates to the water table and S_y is specific yield.

An interesting phenomenon occasionally noted in observation wells is a nearly instantaneous response of shallow water tables to rainfall. This may be explained by the pressure increase of air trapped in the zone of aeration when rainfall seals surface pores and infiltrating water compresses the underlying air. If the zone containing interconnected air-filled pores (H in Figure 6.4.6) is compressed to a thickness $H - m$, then the pressure above the water table is increased by $m/(H - m)$ of an atmosphere, causing the water level in an observation well to rise

$$\Delta h = \frac{m}{H - m}(10)\text{m} \qquad (6.4.7)$$

For shallow water tables the rise (which occurs only in the well) can be an order of magnitude larger than the depth of infiltrating rainfall; however, escaping air soon dissipates the effect. Similarly, when water is applied uniformly to the top of a dry column of sand in the laboratory, the air is compressed until released by a spontaneous upward eruption.

6.4.3 Wind

Minor fluctuations of water levels are caused by wind blowing over the tops of wells. The effect is identical to the action of a vacuum pump. As a gust of wind blows across the top of a casing, the air pressure within the well is suddenly lowered and, as a consequence, the water level quickly rises. After the gust passes, the air pressure in the well rises and the water level falls.

6.4.4 Frost

In regions of heavy frost, it has been observed that shallow water tables decline gradually during the winter and rise sharply in early spring before recharge from the ground surface could

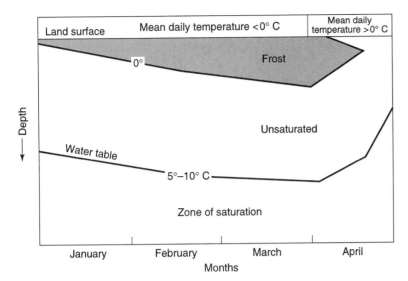

Figure 6.4.7. Sketch illustrating the variation in depth to water table in response to winter frost conditions (after Schneider[74]).

occur (see Figure 6.4.7).[74] This fluctuation can be attributed to the presence of a frost layer above the water table. During winter, water moves upward from the water table by capillary movement and by vapor transfer to the frost layer, where it freezes. Vapor migration occurs in response to the thermal gradient and to the fact that vapor pressure over ice is less than that over liquid water at 0°C. In early spring, approximately when the mean air temperature reaches 0°C, the frost layer begins thawing from the bottom; consequently, meltwater percolates downward to rejoin the water table.

6.5 FLUCTUATIONS DUE TO TIDES

6.5.1 Ocean Tides

In coastal aquifers in contact with the ocean, sinusoidal fluctuations of groundwater levels occur in response to tides. If the sea level varies with a simple harmonic motion, a train of sinusoidal waves is propagated inland from the submarine outcrop of the aquifer. With distance, inland amplitudes of the waves decrease and the time lag of a given maximum increases. The problem has been solved by analogy to heat conduction in a semi-infinite solid subject to periodic temperature variations normal to the infinite dimension.[31, 93]

For simplicity, consider the one-directional flow in a confined aquifer as shown in Figure 6.5.1a. From Equation 3.9.11, the applicable differential equation governing the flow is

$$\frac{\partial^2 h}{\partial x^2} = \frac{S}{T} \frac{\partial h}{\partial t} \tag{6.5.1}$$

where h is the net rise or fall of the piezometric surface with reference to the mean level, x is the distance inland from the outcrop, S is the storage coefficient of the aquifer, T is transmissivity, and t is time. If we let the amplitude, or half-range, of the tide be h_0 (see Figure 6.5.1a), the applicable boundary conditions include $h = h_0 \sin \omega t$ at $x = 0$ and $h = 0$ at $x = \infty$. The angular velocity is ω; for a tidal period t_0,

$$\omega = \frac{2\pi}{t_0} \tag{6.5.2}$$

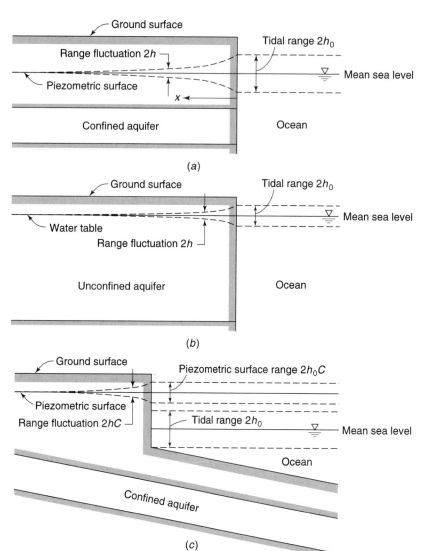

Figure 6.5.1. Groundwater level fluctuations produced by ocean tides. (*a*) Confined aquifer (*b*) Unconfined aquifer (*c*) Loading of a confined aquifer

The solution of Equation 6.5.1 with these boundary conditions is

$$h = h_0 e^{-x\sqrt{\pi S/t_0 T}} \sin\left(\frac{2\pi t}{t_0} - x\sqrt{\pi S / t_0 T}\right) \tag{6.5.3}$$

From this it follows that amplitude h_x of groundwater fluctuations at a distance x from the shore equals

$$h_x = h_0 e^{-x\sqrt{\pi S/t_0 T}} \tag{6.5.4}$$

The time lag t_L of a given maximum or minimum after it occurs in the ocean can be obtained by solving the quantity within the parentheses of Equation 6.5.3 for t, so that

$$t_L = x\sqrt{t_0 S / 4\pi T} \tag{6.5.5}$$

The waves travel with a velocity

$$v_w = \frac{x}{t_L} = \sqrt{4\pi T / t_0 S} \tag{6.5.6}$$

and the wavelength is given by

$$L_w = v_w t_0 = \sqrt{4\pi t_0 T/S} \qquad (6.5.7)$$

Substituting the wavelength for x in Equation 6.5.4 shows that the amplitude decreases by a factor $e^{-2\pi}$, or 1/535, for each wavelength. Water flows into the aquifer during half of each cycle and out during the other half. By Darcy's law, the quantity of flow V per half-cycle per foot of coast is

$$V = \int_{-t_0/8}^{3t_0/8} q\, dt = T\int_{-t_0/8}^{3t_0/8} \left(\frac{\partial h}{\partial x}\right)_{x=0} dt \qquad (6.5.8)$$

where q is the flow per foot of coast. Differentiating Equation 6.5.3 to obtain $\partial h/\partial x$ and integrating yield

$$V = h_0\sqrt{2t_0 ST/\pi} \qquad (6.5.9)$$

The above analysis is also applicable as a good approximation to water table fluctuations of an unconfined aquifer if the range of fluctuation is small in comparison to the saturated thickness (Figure 6.5.1b). Figure 6.5.2 shows fluctuations in wells penetrating an unconfined aquifer at various distances from a surface water level varying approximately with a sinusoidal pattern.

Just as atmospheric pressure changes produce variations of piezometric levels, so do tidal fluctuations vary the load on confined aquifers extending under the ocean floor (Figure 6.5.1c). Contrary to the atmospheric pressure effect, tidal fluctuations are direct; that is, as the sea level increases, the groundwater level does also. Figure 6.5.3 illustrates the effect for a well only 30 m from shore. The ratio of piezometric level amplitude to tidal amplitude is known as the *tidal*

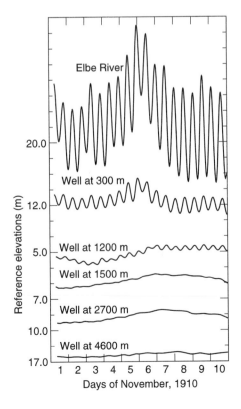

Figure 6.5.2. Fluctuations of the Elbe River and water table levels in wells at various distances from the river (after Werner and Noren[93]).

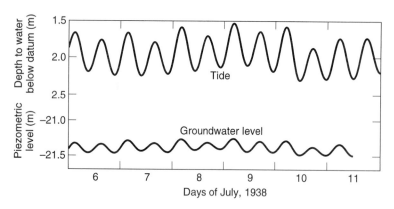

Figure 6.5.3. Tidal fluctuations and induced piezometric surface fluctuations observed in a well 30 m from shore at Mattawoman Creek, Maryland (after Meinzer[56]).

efficiency of the aquifer. Jacob[48] showed that tidal efficiency C is related to barometric efficiency B by

$$C = 1 - B \qquad (6.5.10)$$

Thus, tidal efficiency is a measure of the incompetence of overlying confining beds to resist pressure changes. Aquifer response to loading rather than head change at the outcrop requires that the amplitude given by Equation 6.5.3 be multiplied by C.

6.5.2 Earth Tides

Regular semidiurnal fluctuations of small magnitude have been observed in piezometric surfaces of confined aquifers located at great distances from the ocean.[70] After correcting well levels for atmospheric pressure changes, these fluctuations appear quite distinctly in certain wells where the phenomenon has been investigated. Figure 6.5.4 shows fluctuations over a lunar cycle from a 250-m well tapping a confined aquifer in Iowa City.

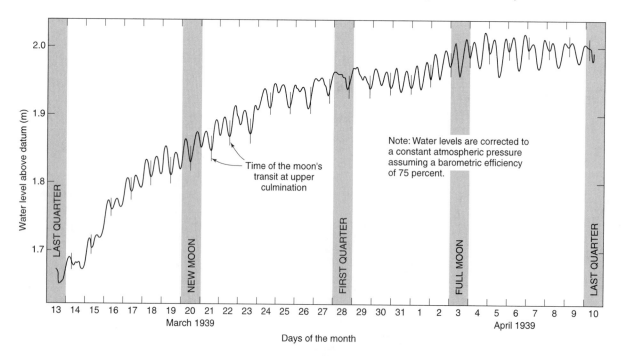

Figure 6.5.4. Water level fluctuations in a confined aquifer produced by earth tides (after Robinson[71]).

These fluctuations result from earth tides, produced by the attraction exerted on the earth's crust by the moon and, to a lesser extent, the sun. Robinson's observations,[71] based on analyses of well records, make convincing evidence: (1) two daily cycles of fluctuations occur about 50 min later each day, as does the moon; (2) the average daily retardation of cycles agree closely with that of the moon's transit; (3) the daily troughs of the water level coincide with the transits of the moon at upper and lower culmination; and (4) periods of large regular fluctuations coincide with periods of new and full moon, whereas periods of small irregular fluctuations coincide with periods of first and third quarters of the moon. All of these facts may be noted in the data of Figure 6.5.4. Bredehoeft[6] has pointed out that wells serve as sensitive indicators of this dilatation of the earth's crust.

At times of new and full moon, the tide-producing forces of the moon and sun act in the same direction; then ocean tides display a greater than average range. But when the moon is in the first or third quarter, tide-producing forces of the sun and moon act perpendicular to each other, causing ocean tides of smaller than average range. The coincidence of the time of low water with that of the moon's transit can be explained by reasoning that at this time tidal attraction is maximum; therefore, the overburden load on the aquifer is reduced, allowing the aquifer to expand slightly.

6.6 URBANIZATION

The process of urbanization often causes changes in groundwater levels as a result of decreased recharge and increased withdrawal. In rural areas, water supplies are usually obtained from shallow wells, while most of the domestic wastewater is returned to the ground through cesspools or septic tanks. Thus, a quantitative balance in the hydrologic system remains. As population increases, many individual wells are abandoned in favor of deeper public wells. Later, with the introduction of sewer systems, storm water and wastewater typically discharge to a nearby surface water body (see Figure 6.6.1). Thus, three conditions disrupt the subsurface hydrologic balance and produce declines in groundwater levels:

1. Reduced groundwater recharge due to paved surface areas and storm sewers

2. Increased groundwater discharge by pumping wells

3. Decreased groundwater recharge due to export of wastewater collected by sanitary sewers

Effects of the urbanization trend are well illustrated on Long Island, New York, in Figure 6.6.2.[12, 20] Here the above conditions have all been present, leading not only to a decline in water tables but also to groundwater pollution (see Chapter 8), seawater intrusion (see Chapter 14), and reduced streamflow. Artificial recharge efforts (see Chapter 13) are underway to counteract these undesirable results of urbanization.

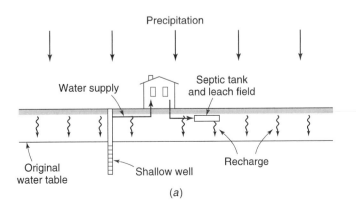

Figure 6.6.1. Schematic diagram illustrating how urbanization can cause lowering of water table elevation. (*a*) Rural situation

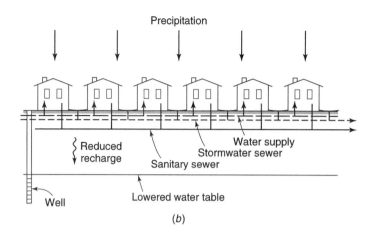

Figure 6.6.1 (*continued*). Schematic diagram illustrating how urbanization can cause lowering of water table elevation. (*b*) Urban development

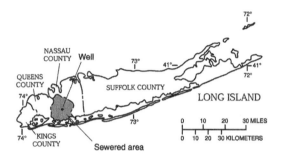

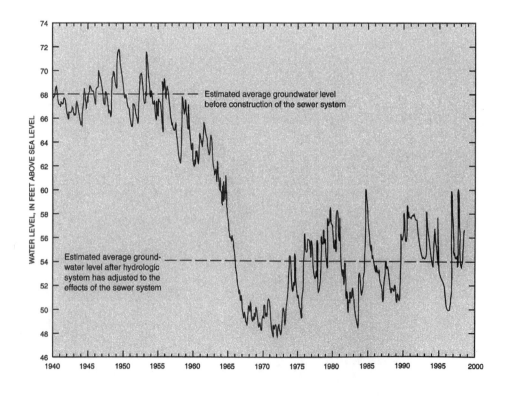

Figure 6.6.2. Water-level record for a well completed in the upper glacial aquifer in west–central Nassau County, Long Island, New York (as presented in Alley et al.[1]).

Figure 6.6.3 illustrates the effects of urbanization on Las Vegas, Nevada. Figure 6.6.4 illustrates the differences in water budget components for predevelopment times and the present. In the present water budget, even though the usage is greatly increased, only a small fraction of the water in the Las Vegas Valley is consumed. Most of the water is either returned to the aquifer system or evaporated or discharged into the Colorado River system (Galloway et al.[23]).

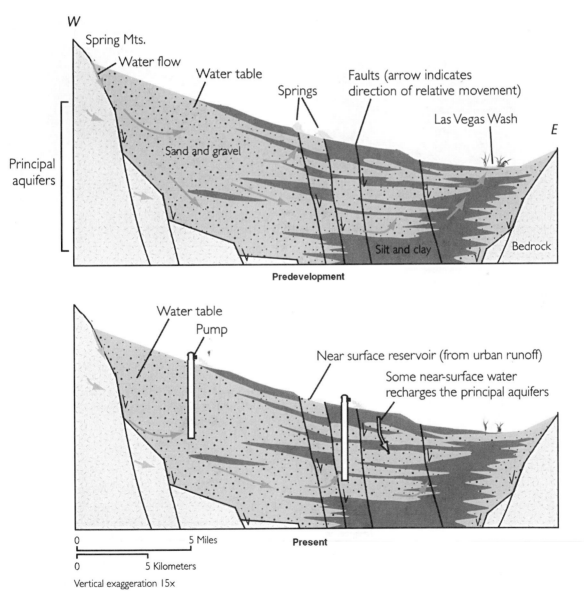

Figure 6.6.3. Effect of urbanization on Las Vegas, Nevada (Galloway et al.[23]).

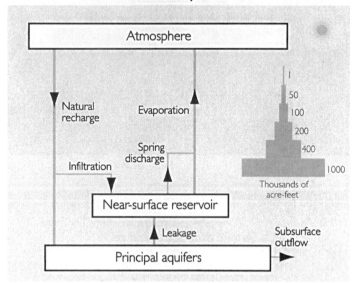

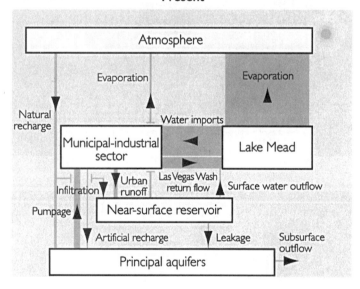

Figure 6.6.4. Comparison of water budgets for predevelopment period and the present for Las Vegas, Nevada (Galloway et al.[23]).

6.7 EARTHQUAKES

Observations reveal that earthquakes have a variety of effects on groundwater.[90, 91] Most spectacular are sudden rises or falls of water levels in wells, changes in discharge of springs, appearance of new springs, and eruptions of water and mud out of the ground. More commonly, however, earthquake shocks produce small fluctuations (*hydroseisms*) in wells penetrating confined aquifers. A good example is furnished by the water level record on an expanded time scale shown in Figure 6.7.1. This earthquake was centered on the Argentina–Chile border, nearly 8,000 km from the recording well in Milwaukee. Although little is known of the quantitative effects of earthquakes on groundwater, these fluctuations result from compression and expansion (dilatation) of elastic confined aquifers by the passage of earthquake (Rayleigh) waves.[14]

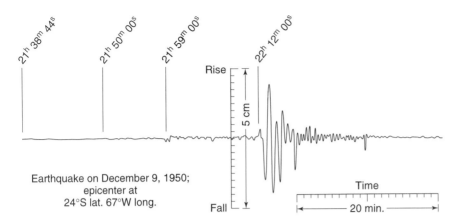

Figure 6.7.1. Water level fluctuations in a well at Milwaukee, Wisconsin, resulting from an earthquake centered on the Argentina–Chile border (after Vorhis[91])

These waves travel at speeds of approximately 200 km/min so that fluctuations appear after little more than one hour even from the most distant earthquake centers.

Looking at the converse situation, field studies have revealed that injection of wastewater into a deep well can trigger earthquakes.[33] Evidence stems from injection of chemical-manufacturing waste fluids near Denver, Colorado, into a well 3,671 m deep, penetrating sedimentary rocks into Precambrian crystalline rocks. Figure 6.7.2 shows the time variation of fluid injection and earthquake frequency. Most earthquake magnitudes were small—within the range of 1.5 to 4.4 on the Richter scale. It is believed that the mechanism by which fluid injection triggered the earthquakes stems from a reduction of frictional resistance to faulting, a reduction that occurs with increase in pore pressure. Knowledge of this phenomenon has stimulated research into the possibility of injecting water into potentially dangerous fault zones.

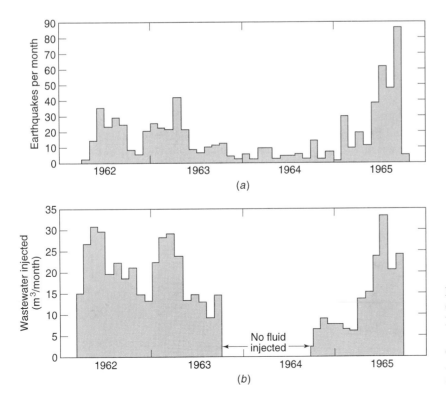

Figure 6.7.2. (*a*) Earthquake frequency (*b*) Volume of injected wastewater. Near Denver, Colorado, during the period 1962–1965 (after Healy et al.;[33] copyright © 1968 by American Association for the Advancement of Science).

This might trigger minor earthquakes, ease stresses along a fault, and hence prevent the sudden release of accumulated energy that results in disastrous earthquakes.

The Hegben Lake, Montana, earthquake of August 1959 caused fluctuations in wells throughout the United States and even in Hawaii and Puerto Rico.[15] The main shock was recorded by water level fluctuations that ranged to more than 3.0 m; values exceeded 0.3 m in nine states. Even more dramatic were the hydroseismic responses to the Anchorage, Alaska, earthquake of March 27, 1964. This large-magnitude earthquake (8.4 to 8.8 on the Richter scale) affected groundwater levels throughout the United States, with the largest fluctuation exceeding 7.0 m at a well in South Dakota.[92] Hydroseisms were also recorded at such distant locations as Denmark, Egypt, South Africa, the Philippines, and Australia.

6.8 EXTERNAL LOADS

The elastic property of confined aquifers results in changes in hydrostatic pressure when changes in loading occur. Some of the best examples are exhibited by wells located near railroads where passing trains produce measurable fluctuations of the piezometric surface. Figure 6.8.1 illustrates changes in water level produced by a train stopping and starting near a well at Smithtown, New York.

The application of a load compresses the aquifer and increases the hydrostatic pressure. Thereafter the pressure decreases and approaches its original value asymptotically as water flows radially away from the point where the load is applied. Thus, initially the load is shared by the confined water and the solid material of the aquifer; however, as the water flows radially outward, an increasing proportion of the load is borne by the structure of the aquifer. The schematic diagram after Jacob[47] in Figure 6.8.2 shows this effect. Here a point load is instantaneously applied. The lower surface of the aquifer is assumed fixed; lengths of arrows indicate the relative magnitudes of flow velocities at various distances from the load. During the interval from A to B, the hydrostatic pressure decreases and the deflection of the upper surface of the aquifer increases. Subsequently, when the load is removed, the pressure drops to a minimum and then recovers toward its initial value as shown by times C and D.

6.9 LAND SUBSIDENCE AND GROUNDWATER

Land subsidence is a gradual settling or sudden sinking of the earth's surface caused by the subsurface movement of earth materials. Land subsidence is a global problem; in the United States more than 17,000 square miles in 45 states have been directly affected by subsidence. Principal causes are aquifer-system compaction, mining of groundwater, drainage of organic soils, underground mining, hydrocompaction, natural compaction, sinkholes, and thawing per-

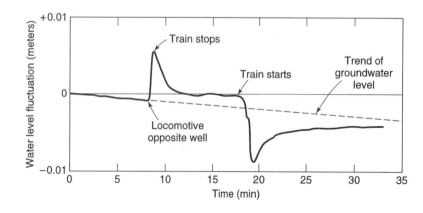

Figure 6.8.1. Water level fluctuations in a confined aquifer produced by a train stopping and starting near an observation well (after Jacob[47]).

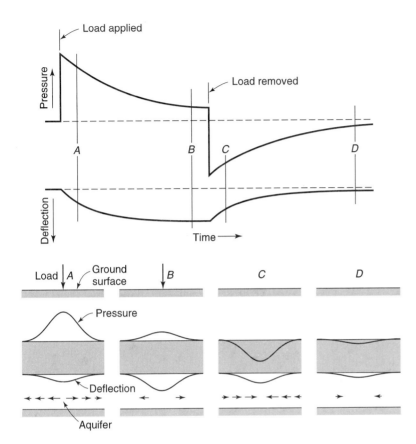

Figure 6.8.2. Hydrostatic pressure variations and aquifer deflections resulting from a point load applied and later removed from the ground surface above a confined aquifer (after Jacob[47]).

mafrost. More than 80 percent of the identified subsidence in the United States is a consequence of our exploitation of groundwater. Figure 6.9.1 shows the locations in the United States where subsidence has been attributed to groundwater pumpage. The continuing and increasing development of land and water resources threatens to worsen existing land-subsidence problems and initiate new problems. Galloway et al.[23] reported on the land subsidence due to mining groundwater in several areas of the United States, including Santa Clara Valley, California; San Joaquin Valley, California; Houston–Galveston, Texas; Las Vegas, Nevada; and south–central Arizona.

The relation between changes in groundwater levels and compression of the aquifer system is based upon the principle of effective stress. When groundwater levels are lowered, the support provided by the pore-fluid pressure is transferred to the skeleton of the aquifer system, which compresses as shown in Figure 6.9.2. When the pore-fluid pressure is increased, such as when groundwater recharges the aquifer system, support previously provided by the skeleton is transferred to the fluid and the skeleton expands. The skeleton alternately undergoes compression and expansion as the pore-fluid pressure fluctuates with aquifer-system discharge and recharge. This fully recoverable deformation occurs in all aquifer systems, commonly resulting in seasonal, reversible displacements in land surface of up to one inch or more in response to seasonal changes in pumpage[23].

Preconsolidation stress is the maximum level of past stressing of a skeletal element. When the load on an aquitard skeleton exceeds the preconsolidation stress, the aquitard skeleton can undergo irreversible compaction, resulting in a permanent reduction of pore volume as the pore fluid is forced out of the aquitards into the aquifers. This process is illustrated in Figure 6.9.3. In confined aquifer systems subject to large-scale overdraft, the volume of water resulting from

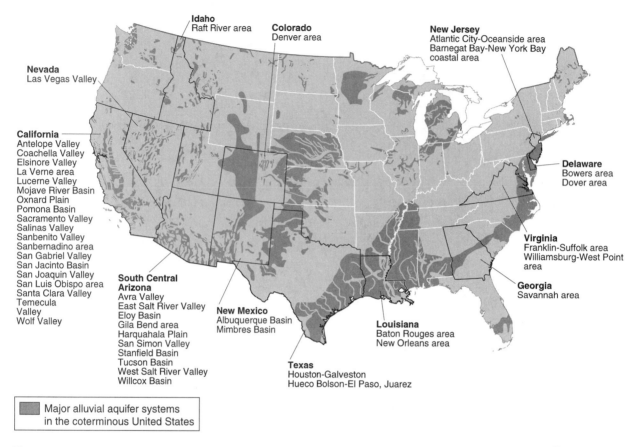

Figure 6.9.1. Areas in the United States where subsidence has been attributed to groundwater pumpage (Galloway et al.[23]).

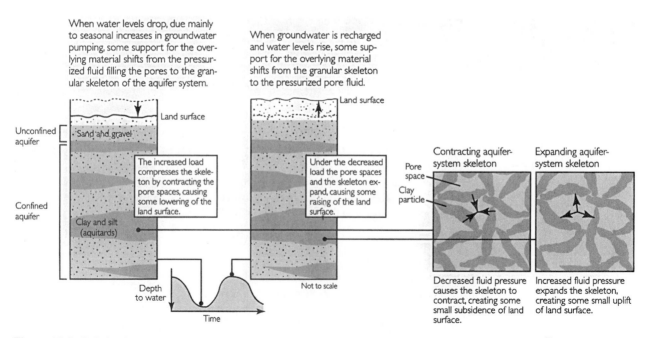

Figure 6.9.2. Relation between changes in groundwater level and compression of aquifer system (Galloway et al[23]).

irreversible aquitard compaction is essentially equal to the volume of subsidence. This volume typically can range from 10 to 30 percent of the total volume of water pumped, representing a one-time mining of the stored groundwater and resulting in a small permanent reduction in storage capacity[23]. The concepts of aquitard drainage and aquifer-system compaction are further illustrated in Figure 6.9.4.

Figure 6.9.5 illustrates the predeveloped conditions and postdeveloped conditions (with land subsidence) of the San Joaquin Valley in California. At least four distinct phenomena have been identified.[2]

6.9.1 Lowering of Piezometric Surface

Land subsidence has been observed to accompany extensive lowering of the piezometric surface in regions of heavy pumping from confined aquifers. Figure 6.9.6 and 6.9.7, respectively, illustrate the trend in subsidence in the Houston area due to groundwater pumpage and the arrested subsidence by substituting imported water. Figure 6.9.8 shows land subsidence in south–central Arizona, illustrating another regional subsidence.

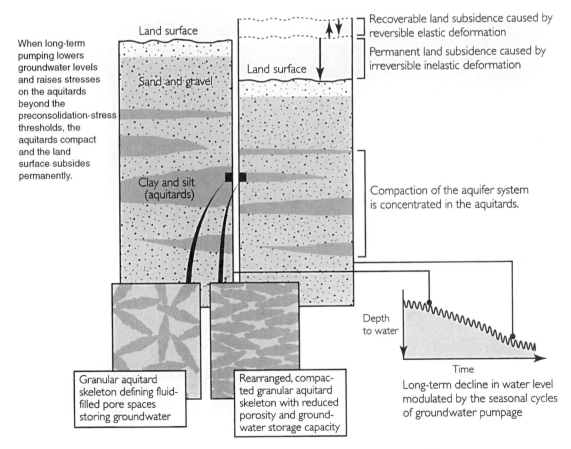

Figure 6.9.3. Inelastic compaction irreversibly altering aquifer system (Galloway et al.[23]).

This principle describes the relation between changes in water levels and deformation of the aquifer system.

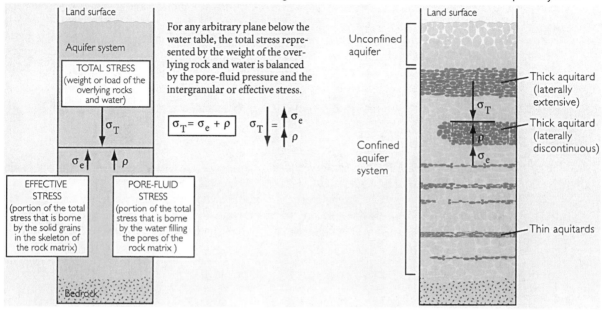

Figure 6.9.4a. Aquitard drainage and aquifer-system compaction; the principle of effective stress (Galloway et al.[23]).

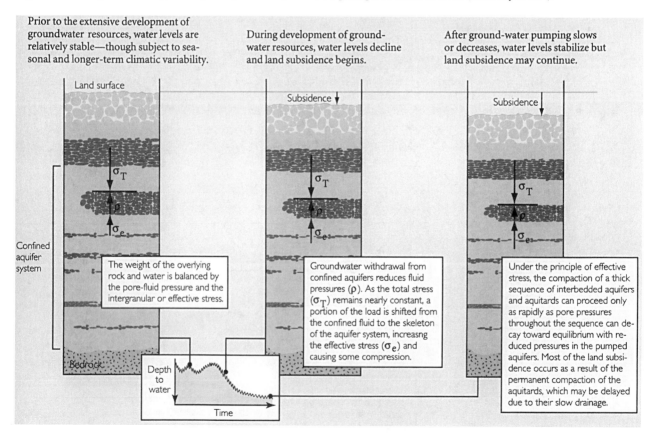

Figure 6.9.4b. Aquitard drainage and aquifer-system compaction; prolonged changes in groundwater levels induce subsidence (Galloway et al.[23]).

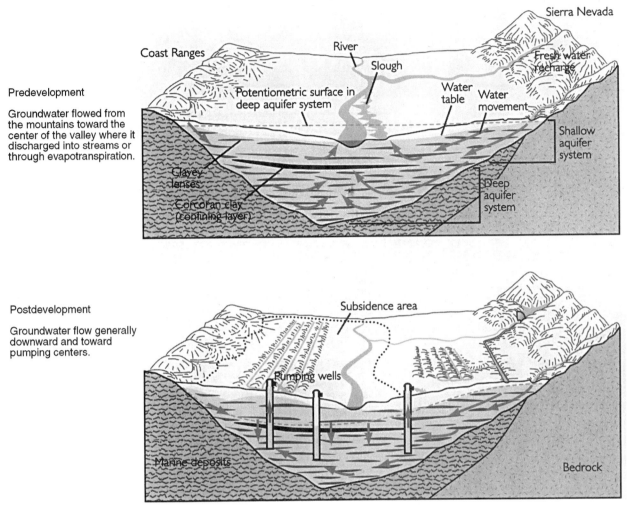

Figure 6.9.5. Comparison of pre- and postdevelopment concepts for the San Joaquin Valley in California (Galloway et al.[23]).

The explanation for this subsidence is based on fundamentals of soil mechanics.[16, 54] Consider the pressure diagram for a confined aquifer overlain by an unconfined aquifer shown in Figure 6.9.9. Initially, the total (geostatic) pressure p_t at any depth (see Figure 6.9.9a) is

$$p_t = p_h + p_i \tag{6.9.1}$$

where p_h is the hydraulic pressure and p_i is the intergranular pressure. If pumping in the confined aquifer lowers the piezometric surface while the water table remains unchanged due to an impermeable clay layer separating the aquifers (Figure 6.9.9b) then Equation 6.9.1 becomes

$$p_t = p_h' + p_i' \tag{6.9.2}$$

Note that $p_h' < p_h$ and $p_i' > p_i$ for both the confined aquifer and the clay layer. Adjustments to these new pressure distributions will take place essentially instantaneously in the permeable, coarse-grained aquifer. But in the relatively impermeable, fine-grained clay, this adjustment may take months to years. Because clayey materials are highly compressible, the increased

Subsidence trends reflect patterns of resource
development that shifted inland from coastal
oil and gas extraction to groundwater extraction for muicipal and industrial supplies.

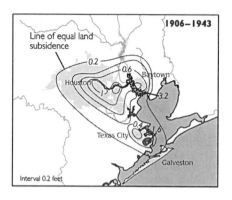

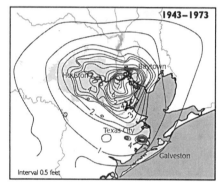

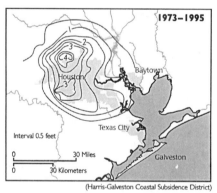

Figure 6.9.6. Subsidence trends in
Houston, Texas area (Galloway et al.[23]).

intergranular pressure $(p_i' - p_i)$ causes the clay layer to be compacted. This reduces its porosity, while water contained in the clay pores is squeezed downward into the confined aquifer.

The volume of water displaced from the clay equals the reduction in clay volume and also
the volumetric land surface subsidence. Similarly, the reduction in thickness of the clay layer
equals the vertical land subsidence. The amount of compaction is a function of the thickness
and vertical permeability of the clay, of the time and magnitude of piezometric surface decline,
and of the microstructure of the clay.[66] Because sand and gravel deposits are relatively incom-
pressible, the increased intergranular pressure has a negligible effect on the aquifer itself.

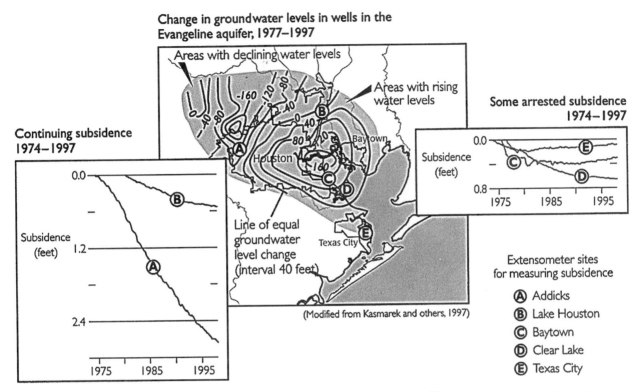

Figure 6.9.7. Houston, Texas, area subsidence and arrested subsidence (Galloway et al.[23]). The Harris–Galveston Coastal Subsidence District has arrested subsidence along the western margins of Galveston Bay by substituting imported water for groundwater. A new challenge is to manage groundwater use north and west of Houston where water levels are declining and subsidence is increasing.

Land subsidence resulting from compaction of fine-grained sediments occurs at several locations in the United States[22] and throughout the world.[19, 25] The problem has been extensively studied by Poland and others.[46, 65, 66, 67*] Compaction of clay is largely inelastic and permanent; consequently, the only effective control measure for this type of subsidence is increasing piezometric levels by reducing pumping and by recharge of water through injection wells (see Chapter 13). Withdrawal of oil and gas produces the same problem of land subsidence; major areas affected exist in the United States, Italy, Japan, and Venezuela.[65]

6.9.2 Hydrocompaction

Collapse of the ground surface has been observed to occur when water is applied to certain types of soils. Particularly susceptible are (1) loose, moisture-deficient alluvial deposits, including mud flows; and (2) moisture-deficient loess deposits.[55, 77] An example of this type of shallow subsidence is found on the arid west side of the San Joaquin Valley, California. Soils characteristically are desiccated with a high void content and low density (1.1 to 1.4 g/cm^3). Most of these soils have never been saturated since deposition, but when irrigation water, for

---

*The periodic flooding of Piazza San Marco in Venice from the Grand Canal provides dramatic evidence that subsidence from groundwater pumpage threatens this beautiful city. A decline of 20 cm over the last 50 years has been reported. Recent restrictions on pumpage in the industrial suburb of Porto Marghera are expected to halt this subsidence rate.[25]

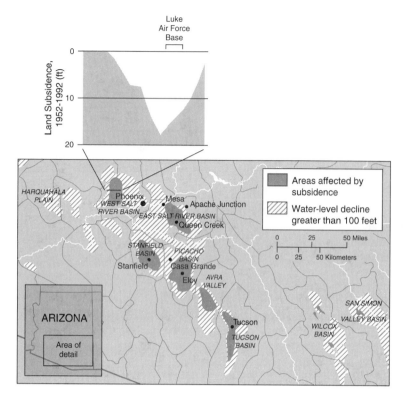

Figure 6.9.8. Land subsidence in south–central Arizona (modified from Carpenter[7], as presented in Alley et al.[1]). Groundwater development for agriculture in the basin-fill aquifers of south–central Arizona began in the late 1800s, and by the 1940s many of the basins had undergone intensive ground-water development. Groundwater depletion has been widespread over these basins, and locally, water-level declines have exceeded 300 feet. These water-level declines have resulted in regional subsidence, exceeding 10 feet in some areas. A profile near Luke Air Force Base illustrates that subsidence is greater near the center of basins, where the aggregate thickness of the fine-grained sediments is generally greater. In conjunction with widespread subsidence, numerous earth fissures have formed at and near the margins of subsiding basins or near exposed or shallow buried bedrock.

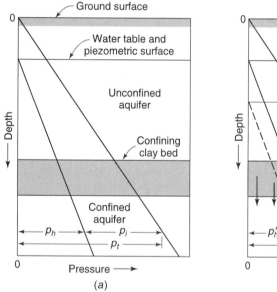

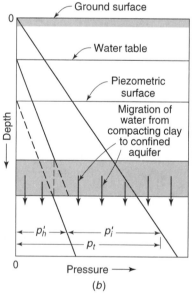

Figure 6.9.9. Graph of hydraulic and intergranular pressures as a function of depth for an unconfined aquifer overlying a confined aquifer (after Poland et al.[67]). (*a*) Initial condition with water table and piezometric surface at same elevation (*b*) Subsequent condition with piezometric surface lowered

example, is applied, their internal high void structure collapses, resulting in an erratic subsidence of the land surface.

To define the magnitude of this subsidence, a test pond 30 m by 30 m was constructed on flat land, and benchmarks anchored at various depths were installed. Water to a depth of 0. m was admitted in early October 1956. Subsidence of the various benchmarks appears in Figure 6.9.10. As the wetting front moved downward, the bench marks progressively subsided. At

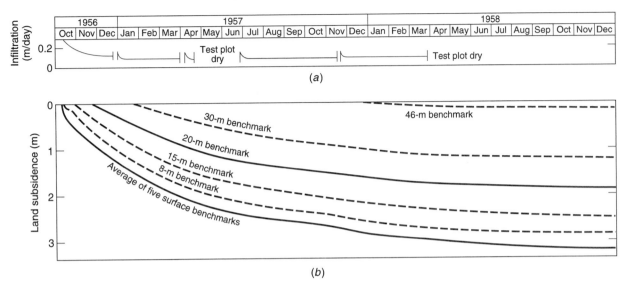

Figure 6.9.10. Hydrocompaction and land subsidence resulting from water application on a test plot containing low-density soil, San Joaquin Valley, California (after Lofgren[55]; courtesy The Geological Society of America, 1969). (*a*) Infiltration from test plot (*b*) Subsidence of benchmarks anchored at ground surface and at various depth intervals

ground surface, the change in level amounted to more than 3 m, while at the 45 m depth no effect was observed until after 16 months.

Shallow subsidence can influence irrigation, drainage, sewerage, and transportation systems.* Sprinkler irrigation and pipelines for water conveyance are best suited in these terrains.

6.9.3 Dewatering of Organic Soils

In flat peat or muck land with a shallow water table, lowering of the water table, such as by drainage, produces land subsidence. Causes include (1) shrinkage due to desiccation, (2) consolidation by loss of the buoyant force of groundwater, (3) compaction with tillage, (4) wind erosion, (5) burning, and (6) biochemical oxidation.[46] Investigations have shown that the rate of subsidence is proportional to the depth to the water table. To conserve the life of organic soils, the water table should be maintained as high as crop requirements and field conditions will permit.

Subsidence of organic soils has been noted in the Netherlands, the Soviet Union, and at various locations in the United States. During the last 70 years extensive drainage for agricultural purposes of islands in the Sacramento–San Joaquin Delta, California, has lowered the land surface greater than 15 ft over much of the area, as shown in Figure 6.9.11. This has necessitated construction of a vast network of perimeter levees to prevent inundation of the depressed islands by floods or high tides.

6.9.4 Sinkhole Formation

Catastrophic land subsidence leading to the formation of sinkholes can also be associated with declines in groundwater levels. Soluble rocks such as dolomite and limestone are slowly

---

*More than 100 km along the California Aqueduct, a concrete-lined canal carrying some 25×10^6 m³/day of water to Southern California, cross formations susceptible to shallow land subsidence. To avoid the danger of subsidence from canal leakage, large-scale spreading ponds were maintained along the alignment in order to preconsolidate the soils before construction of the aqueduct.

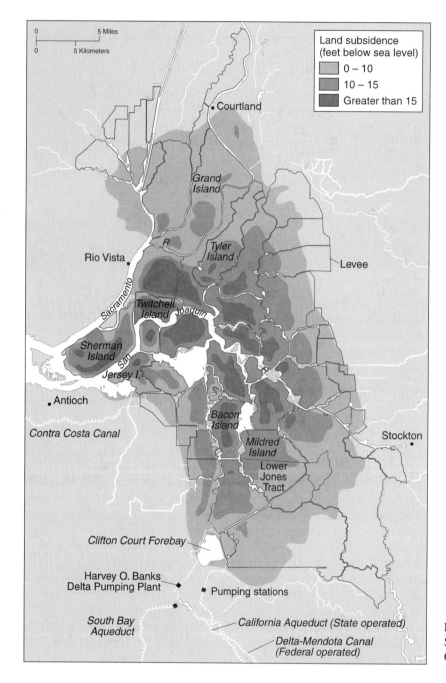

Figure 6.9.11. Land subsidence in the Sacramento–San Joaquin Delta, California (Galloway et al.[23]).

dissolved locally by groundwater. Eventually the ground surface sinks to form a cup-shaped depression. Over large areas of this type, a karstic sinkhole plain is formed with most of the drainage occurring in the subsurface.

New sinkholes often develop in regions where water tables have been lowered by pumping. For example, groundwater pumping from a dolomite aquifer for mine dewatering in the Far West Rand, South Africa, led to the formation of eight sinkholes larger than 50 m in diameter and deeper than 30 m within a period of 39 months.[18] Similarly, the massive Alabama sinkhole shown in Figure 6.9.12 appeared suddenly after a local decline of the water table. As

Figure 6.9.12. Sinkhole formed suddenly in December 1973 with a diameter of 130 m and a depth of 45 m in Shelby County, Alabama. This is one of more than 1,000 sinkholes that developed in Shelby County within a 15-year period. The concentration of sinkholes is attributed to a local lowering of the water table (courtesy U.S. Geological Survey).

a water table is lowered, removal of the buoyant support from the subsurface clay above the cavern together with increased velocities of infiltrating water are believed to be responsible for the cave-ins.

6.9.5 Crustal Uplift

The opposite of land subsidence, crustal uplift, can occur over large areas subject to heavy groundwater pumping. The tectonic uplift of land, involving an elastic expansion of the lithosphere, is caused by the removal of large masses of groundwater. The phenomenon has been noted in parts of Arizona, California, and Texas, where groundwater pumped from aquifers was removed by evapotranspiration by crops.[36] In the Santa Cruz River Basin, Arizona, a crustal uplift of 6 cm was observed between 1948 and 1967, when 43.5 billion tons of groundwater was pumped from an 8,070-km^2 area. This figure compares favorably with the 17-cm land surface depression, previously measured, resulting from the filling of Lake Mead on the Colorado River.

6.9.6 Monitoring of Land Subsidence

Several types of data are required in monitoring land subsidence. Wells are required to monitor water levels at multiple levels, and extensometer wells are required to measure sediment compaction. Borehole data, including inclinometer measurements, lithographic and geophysical data, and geotechnical data, are also collected. The water-level monitoring wells are constructed with multiple or nested piezometers at different depths, as illustrated in Figure 6.9.13, to monitor water levels in the multiple zones in the partially confined aquifer as a function of time. Submersible pressure transducers are installed to measure water levels in each of the nested piezometers.

Compaction is the decrease in the thickness of the subsurface sediments, whereas subsidence is the lowering of the land-surface elevation over a long period of time. The various

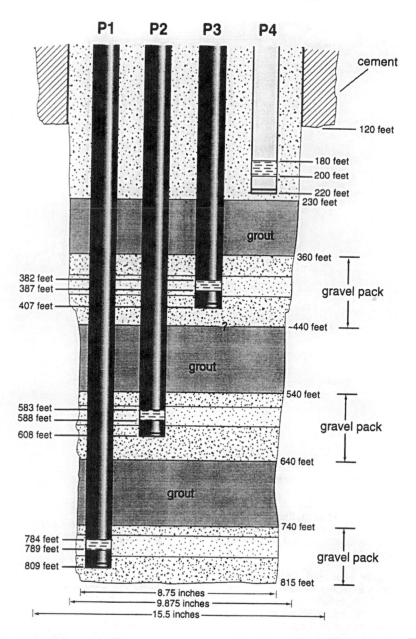

P1   P2   P3   P4

cement

120 feet

180 feet
200 feet
220 feet
230 feet

grout

382 feet
387 feet

407 feet

360 feet

gravel pack

?   ~440 feet

grout

583 feet
588 feet

608 feet

540 feet

gravel pack

640 feet

grout

784 feet
789 feet

809 feet

740 feet

gravel pack

815 feet

8.75 inches
9.875 inches
15.5 inches

NOT TO SCALE

Figure 6.9.13. Construction of nested piezometers P1 through P4 (Ikehara[37]).

sediment-sample properties measured in consolidation and permeability tests are listed in Table 6.9.1. Figure 6.9.14 illustrates the construction of an extensometer well and Figure 6.9.15 illustrates the instrumentation for the extensometer. Figure 6.9.16 shows the cumulative net sediment compaction superimposed on water levels for five water years (from Ikehara[37]). For this application in Yolo County, California, five piezometers and an extensometer were constructed in three boreholes.

Table 6.9.1 Key to Properties Measured in Consolidation and Permeability Tests on Sediment Samples

Time (m) . . . . from real time on first day of test (hours)

Sample thickness (c) . . . . . . (LVDT reading)(LVDT slope) + intercept from LVDT calibration

Void ratio, e (c) . . . . (height of sample–height of solids)/(height of solids)

Head difference (m) . . . . . . head drop across the sample

Flowrate, Q (m) . . . . flowrate used to conduct permeability test

Permeability (c) . . . . (Q/head difference)(sample thickness/cross-sectional area)

Vertical effective stress

(m) . . . . pressure on piston in contact with sample (lb/in^2)

(c) . . . . effective stress on sample (kPa)

Real vertical effective stress on the sample is about two times the measured effective stress because the area of the piston in contact with the sample is approximately half the area of the bottom of the loading chamber.

Compressibility (c) . . . . ($e_1 - e_2$)/(effective stress 1 – effective stress 2) [equivalent to the slope of the "void ratio versus effective stress" curve]

Specific storage (c) . . . . (compressibility)(specific gravity of water)/(1 + e) [see void ratio]

(c), calculated; kPa, kilopascal; 1b/in^2, pound per square inch; LVDT, linear variable differential transformer; (m), measured

Source: Ikehara[37]

6.10 EFFECTS OF GLOBAL CLIMATE CHANGE ON GROUNDWATER

The earth's temperature is affected by numerous influences, including (a) the incoming solar radiation that is absorbed by the atmosphere and the earth's surface, (b) the characteristics (emissivity) of the matter that absorbs the radiation, and (c) the part of the long-wave radiation emitted by the surface, absorbed by the atmosphere, and then re-emitted as long-wave radiation either in the upward or downward direction.

The so-called *greenhouse effect* is caused by the net change of the internal radiation balance of the atmosphere due to the continued increased emission of greenhouse gases, resulting in both the atmosphere and the earth's surface becoming warmer. The magnitude of the greenhouse effect is dependent on the composition of the atmosphere—with the most important factors being the concentrations of water vapor and carbon dioxide—and less importantly on certain trace gases, such as methane. There is mounting evidence that global warming is under way.

In general, the hydrologic effects are likely to influence water storage patterns throughout the hydrologic cycle and influence the exchange among aquifers, streams, rivers, and lakes. Chalecki and Glieck[8] provide a bibliography of the impacts of climate change on water resources in the United States.

The Pacific Institute has compiled a comprehensive online bibliography of peer-reviewed literature dealing with climate change and its effects on water resources and water systems. At last count, the bibliography included over 3,600 citations. This bibliography can be accessed at http://www.pacinst.org/resources/index.html. The International Panel on Climate Change (IPCC) has been active in compiling information concerning future climate change (IPCC[39–45]). These reports can be obtained from http://www.ipcc.ch.

The effects of climate change on groundwater sustainability include (Alley et al.[1]):

- Changes in groundwater recharge resulting from changes in average precipitation and temperature or in seasonal distribution of precipitation,

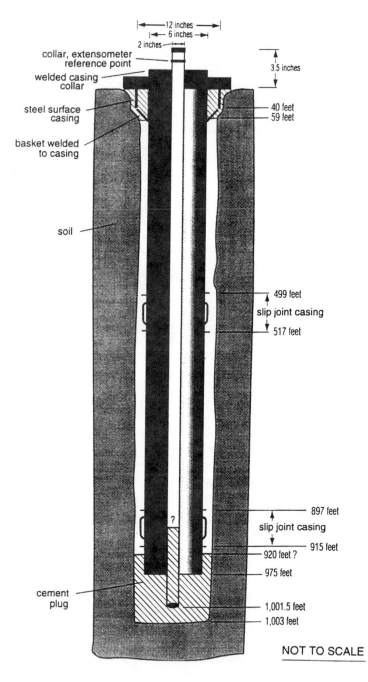

Figure 6.9.14. Construction of extensometer well (Ikehara[37]).

NOT TO SCALE

- More severe and longer lasting droughts, with the effects of drought illustrated in Figure 6.10.1,
- Changes in evapotranspiration resulting from changes in vegetation, and
- Possible increased demands of groundwater as a backup source of water supply.

Surficial aquifers are likely to be part of the groundwater system that is most sensitive to climate. These aquifers supply much of the flow to streams, lakes, wetlands, and springs. Because groundwater systems tend to respond more slowly to short-term variability in climate

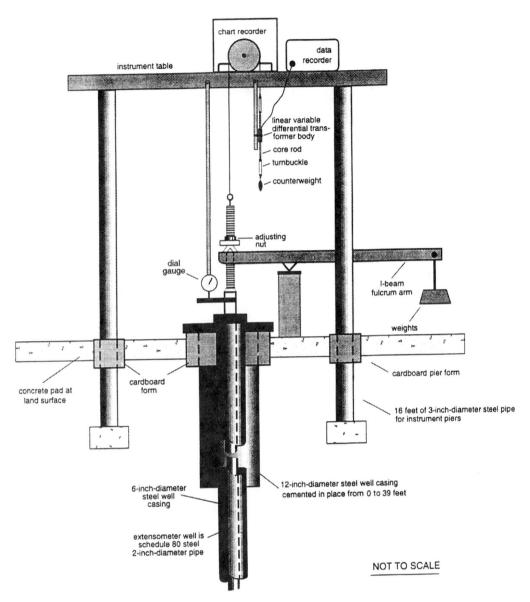

chart recorder

data recorder

instrument table

linear variable differential transformer body

core rod

turnbuckle

counterweight

adjusting nut

dial gauge

I-beam fulcrum arm

weights

concrete pad at land surface

cardboard form

cardboard pier form

16 feet of 3-inch-diameter steel pipe for instrument piers

12-inch-diameter steel well casing cemented in place from 0 to 39 feet

6-inch-diameter steel well casing

extensometer well is schedule 80 steel 2-inch-diameter pipe

NOT TO SCALE

Figure 6.9.15. Instrumentation of extensometer well (Ikehara[37]).

conditions than the response of surface water systems, the assessment of groundwater resources and related model simulations are based on average conditions, such as annual recharge and/or average annual discharge to streams. The use of average conditions may underestimate the importance of droughts (Alley et al.[1]).

The impacts of climate change on (a) specific groundwater basins, (b) the general groundwater recharge characteristics, and (c) groundwater quality have received little attention in the literature. Vaccaro[85] addressed the climate sensitivity of groundwater recharge, finding that a warmer climate (doubling CO_2) resulted in a relatively small sensitivity to recharge, which depended on land use. Sandstrom[72] studied a semiarid basin in Africa and concluded that a 15 percent reduction in rainfall could lead to a 45 percent reduction in groundwater

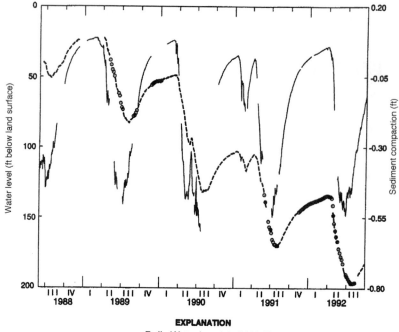

EXPLANATION

—— Daily Water Level At 0600 Hours

--- Daily Sediment Compaction At 0600 Hours

o   Instantaneous Sediment Compaction

Figure 6.9.16. Cumulative net sediment compaction superimposed on hydrograph of well 11N/1E-24Q6 (P3), water years 1988–92m in Yolo County, California (Ikehara[37]).

recharge. Sharma[76] and Green et al.[30] reported similar sensitivities of the effects of climate change on groundwater in Australia. Panagoulia and Dimou[62] studied the effect of climate change on groundwater–streamflow interactions in a mountainous basin in central Greece. They realized large impacts in the spring and summer months as a result of temperature-induced changes in snowfall and snowmelt patterns. Oberdorfer[61] looked at the impacts of climate change on groundwater discharge to the ocean using a simple water-balance model to study the effect of changes in recharge rates and sea level on groundwater resources and flows in a California coastal watershed. The impacts of sea-level rise on groundwater will include increased intrusion of salt water into coastal aquifers.

PROBLEMS

6.1.1 Develop a summary on the effects of groundwater development on groundwater flow to and from surface-water bodies. Refer to the Web site for the *U. S. Geological Survey Circular* 1186, *Sustainability of Ground-Water Resources*, by Alley et al.,[1] http://water.usgs.gov/pubs/circ/circ1186/html.

6.1.2 What has caused the reversal of declining groundwater levels in the Chicago area? Start with Avery,[3] *U.S. Geological Survey Fact Sheet* 222–95.

6.9.1 Write a report on the effect of land subsidence in the San Joaquin Valley, California. What were the major causes and

effects of the subsidence? Start your search for information with the *U. S. Geological Survey Circular* 1182, *Land Subsidence in the United States*, by Galloway et al.[23] and Carpenter[7], http://water.usgs.gov/pubs/circ/circ1182/html.

6.9.2 Write a report on the effect of land subsidence in south–central Arizona. What were the major causes and effects of the subsidence? Start your search for information with the *U.S. Geological Survey Circular* 1182, *Land Subsidence in the United States*, by Galloway et al.[23] and Galloway and Riley[24], http://water.usgs.gov/pubs/circ/circ1182/html.

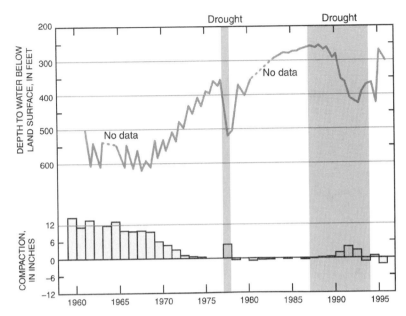

Figure 6.10.1. Effects of drought on groundwater levels and associated subsidence in the San Joaquin Valley, California (modified from Galloway and Riley,[24] as presented in Alley et al.[1]). The San Joaquin Valley is a major agricultural area that produces a large fraction of the fruits, nuts, and vegetables in the United States. Groundwater withdrawals during the 1930s to early 1960s caused water-level declines of tens to hundreds of feet in much of the valley. The water-level declines resulted in compaction of the alluvial deposits and extensive land subsidence. Subsidence in excess of one foot has affected more than 5,200 square miles in the San Joaquin Valley, representing perhaps the largest anthropogenic change in land-surface elevation in the world. Importation of surface water, beginning in the 1960s, led to a decrease in groundwater withdrawals, which in turn led to rising groundwater levels and at least a temporary end to further subsidence. During severe droughts in 1976–77 and 1987–93, deliveries of imported water were decreased. More groundwater was pumped to meet water demands, resulting in a decline in the water table and a renewal of compaction and land subsidence.

REFERENCES

1. Alley, W. M., T. E. Reilly, and O. L. Franke, *Sustainability of Ground-Water Resources, U.S. Geological Survey Circular* 1186, http://water.usgs.gov/pubs/circ/circ1186, U.S. Geological Survey, Denver CO, 1999.

2. Amer. Soc. Civil Engrs., *Ground water management,* Manual Engrng. Practice 40, New York, 216 pp., 1972.

3. Avery, C. F., *Reversal of Declining Ground-Water Levels in the Chicago Area, U.S. Geological Fact Sheet* 222-95, 1995.

4. Barnes, B. S., The structure of discharge-recession curves, *Trans. Amer. Geophysical Union,* v. 20, pp. 721–725, 1939.

5. Bouwer, H., Predicting reduction in water losses from open channels by phreatophyte control, *Water Resources Research,* v. 11, pp. 96–101, 1975.

6. Bredehoeft, J. D., Response of well-aquifer systems to earth tides, *Jour. Geophysical Research,* v. 72, pp. 3075–3087, 1967.

7. Carpenter, M. C., South-Central Arizona, in Land Subsidence in the United States, D. Galloway, D. Jones, and S. Ingebritsen, eds., *U.S. Geolog-ical Survey Circular* 1182, http://water.usgs.gov/pubs/circ/circ1182, Denver, CO, 1999.

8. Chalecki, L. H., and P. H. Gleick, A comprehensive bibliography of the impacts of climate change and variability on water resources of the United States, *Jour. American Water Resources Assn.,* v. 35, pp. 1657–1665, 1999.

9. Chow, V. T., D. R. Maidment, and L. W. Mays, *Applied Hydrology,* McGraw-Hill, New York, 1988.

10. Clark, W. E., Computing the barometric efficiency of a well, *Jour. Hydraulics Div.,* Amer. Soc. Civil Engrs., v 93, no. HY4, pp. 93–98, 1967.

11. Clawges, R. M., and C. V. Price, *Digital Data Sets Describing Principal Aquifers, Surficial Geology, and Ground-Water Regions of the Conterminous United States, U.S. Geological Survey Open-File Report* 99-77, http://water.usgs.gov/pubs/ofr/ofr99-77, 1999.

12. Cohen, P., et al., *An atlas of Long Island's water resources,* Bull. 62, New York Water Resources Comm., Albany, 117 pp., 1968.

13. Cooper, H. H., Jr., and M. I. Rorabaugh, *Ground-Water Movements and Bank Storage Due to Flood Stages in Surface Streams, U.S. Geological Survey Water-Supply Paper* 1536-J, pp. 343–366, 1963.

14. Cooper, H. H., Jr., et al., The response of well-aquifer systems to seismic waves, *Jour. Geophysical Research,* v. 70, pp. 3915–3926, 1965.

15. Da Costa, J. A., *Effect of Hegben Lake Earthquake on Water Levels in Wells in the United States, U.S. Geological Survey Professional Paper* 435, pp. 167–178, 1964.

16. Domenico, P. A., and M. D. Mifflin, Water from low-permeability sediments and land subsidence, *Water Resources Research,* v. 1, pp. 563–576, 1965.

17. Farvolden, R. N., Geologic controls on ground-water storage and base flow, *Jour. Hydrology,* v. 1, pp. 219–249, 1963.

18. Foose, R. M., Sinkhole formation by groundwater withdrawal: Far West Rand, South Africa, *Science,* v. 157, pp. 1045-1048, 1967.

19. Fox, D. J., Man-water relationships in metropolitan Mexico, *Geogr. Review,* v. 55, pp. 523–545, 1965.

20. Franke, O. L., *Double-Mass-Curve Analysis of the Effects of Sewering on Groundwater Levels on Long Island, New York, U.S. Geological Survey Prof. Paper* 600-B, pp. 205–209, 1968.

21. Freeze, R. A., Role of subsurface flow in generating surface runoff, *Water Resources Research,* v. 8, pp. 609–623, 1272–1283, 1972.

22. Gabrysch, R. K., and C. W. Bonnet, *Land–surface subsidence in the Houston–Galveston region, Texas,* Rept. 188, Texas Water Dev. Board, Austin, 19 pp., 1975.

23. Galloway, D., D. Jones, and S. Ingebritsen, eds., *Land Subsidence in the United States, U.S. Geological Survey Circular* 1182, http://water.usgs.gov/pubs/circ/circ1182, Denver, CO, 1999.

24. Galloway, D., and F. S. Riley, San Joaquin Valley, California, in *Land Subsidence in the United States,* D. Galloway, D. Jones, and S. Ingebritsen, eds., *U.S. Geological Survey Circular* 1182, http://water.usgs.gov/pubs/circ/circ1182, Denver, CO, 1999.

25. Gambolati, G., and R. A. Freeze, Mathematical simulation of the subsidence of Venice, *Water Resources Research,* v. 9, pp. 721–733, 1973; v. 10, pp. 563–577, 1974.

26. Gardner, W. R., and M. Fireman, Laboratory studies of evaporation from soil columns in the presence of a water table, *Soil Sci.,* v. 85, pp. 244–249, 1958.

27. Gatewood, J. S., et al., *Use of Water by Bottom-Land Vegetation in Lower Safford Valley, Arizona, U.S. Geological Survey Water-Supply Paper* 1103, 210 pp., 1950.

28. Gilliland, J. A., A rigid plate model of the barometric effect, *Jour. Hydrology,* v. 7, pp. 233–245, 1969.

29. Gleick, P. H., et al., *Water: The Potential Consequences of Climate Variability and Change for the Water Resources of the United States, A Report of the National Water Assessment Group, for the U.S. Global Change Research Program, Pacific Institute for Studies in Development, Environment, and Security,* Oakland, CA, September 2000.

30. Green, T. R., B. C. Bates, P. M. Fleming, S. P. Charles, and M. Taniguchi, Simulated Impacts of Climate Change on Groundwater Recharge in the Subtropics of Queensland, Australia, *Subsurface Hydrological Responses to Land Cover and Land Use Changes,* Kluwer Academic Publishers, Norwell, MA, U.S.A., pp. 187–204, 1997.

31. Gregg, D. O., An analysis of ground-water fluctuations caused by ocean tides in Glynn County, Georgia, *Ground Water,* v. 4, no. 3, pp. 24–32, 1966.

32. Hall, F. R., Base-flow recession—A review, *Water Resources Research,* v. 4, pp. 973–983, 1968.

33. Healy, J. H., et al., The Denver earthquakes, *Science,* v. 161, pp. 1301–1310, 1968.

34. Heath, R. C., *Basic Ground-Water Hydrology, U.S. Geological Survey Water Supply Paper* 2220, http://water.usgs.gov/pubs/wsp/wsp2220, Denver, CO, 1998.

35. Hellwig, D. H. R., Evaporation of water from sand, *Jour. Hydrology,* v. 18, pp. 93–118, 1973.

36. Holzer, T. J., Elastic expansion of the lithosphere caused by groundwater depletion, *Jour. Geophysical Research,* v. 84, pp. 4689–4698, 1979.

37. Ikehara, M. E., *Data from the Woodland Land Subsidence Monitoring Station, Yolo County, California, Water Years 1988–92, U.S. Geological Survey Open-File Report* 94-494, Sacramento, CA, 1995.

38. Ineson, J., Form of ground-water fluctuations due to nuclear explosions, *Nature,* v. 198, pp. 22–23, 1963.

39. Intergovernmental Panel on Climate Change (IPCC), Climate Change 1995: The Science of Climate Change: Contribution of Working Group I to the Second Assessment Report of the Intergovernmental Panel on Climate Change, Cambridge University Press, Cambridge, New York, 1996.

40. Intergovernmental Panel on Climate Change (IPCC), Climate Change 1995: Impacts, Adaptations, and Mitigation of Climate Change: Scientific-Technical Analysis: Contribution of Working Group II to the Second Assessment Report of the Intergovernmental Panel on Climate Change, Cambridge University Press, Cambridge, New York, 1996.

41. Intergovernmental Panel on Climate Change (IPCC), Hydrology and Freshwater Ecology, in Climate Change 1995: Impacts, Adaptations, and Mitigation of Climate Change: Contribution of Working Group II to the Second Assessment Report of the Intergovernmental Panel on Climate Change, Cambridge University Press, Cambridge, New York, 1996.

42. Intergovernmental Panel on Climate Change (IPCC), Regional Impacts of Climate Change: An Assessment of Vulnerability, Cambridge University Press, Cambridge, New York, 1998.

43. Intergovernmental Panel on Climate Change (IPCC), Climate Change 2001: The Scientific Basis, http://www.ipcc.ch, 2001.

44. Intergovernmental Panel on Climate Change (IPCC), Climate Change 2001: Impacts, Adaptation, and Vulnerability, http://www.ipcc.ch/pub.online.htm, 2001.

45. Intergovernmental Panel on Climate Change (IPCC), Climate Change 2001: http://www.ipcc.ch, 2001.

46. Intl. Assoc. Sci. Hydrology, *Land subsidence,* Publ. nos. 88 and 89 (2 vols.), 661 pp., 1969.

47. Jacob, C. E., Fluctuations in artesian pressure produced by passing railroad trains as shown in a well on Long Island, New York, *Trans. Amer. Geophysical Union,* v. 20, pp. 666–674, 1939.

48. Jacob, C. E., On the flow of water in an elastic artesian aquifer, *Trans. Amer. Geophysical Union,* v. 21, pp. 574–586, 1940.

49. Johnston, R. H., *Base Flow as an Indicator of Aquifer Characteristics in the Coastal Plain of Delaware, U.S. Geological Survey Professional Paper* 750-D, pp. 212–215, 1971.

50. Keppel, R. V., and K. G. Renard, Transmission losses in ephemeral stream beds, *Jour. Hydraulics Div.,* Amer. Soc. Civil Engrs., v. 88, no. HY3, pp. 59–68, 1962.

51. Klein, M., and P. Kaser, *A statistical analysis of ground-water levels in twenty selected observation wells in Ohio,* Tech. Rept. 5, Ohio Dept. Natural Resources, Div. of Water, Columbus, 124 pp., 1963.

52. Knisel, W. G., Jr., Baseflow recession analysis for comparison of drainage basins and geology, *Jour. Geophysical Research,* v. 68, pp. 3649–3653, 1963.

53. Kunkle, G. R., *Computation of Groundwater Discharge to Streams during Floods, or to Individual Reaches during Baseflow, by Use of Specific Conductance, U.S. Geological Survey Professional Paper* 525-D, pp. 207–210, 1965.

54. Lofgren, B. E., *Analysis of Stresses Causing Land Subsidence, U.S. Geological Survey Prof. Paper* 600-B, pp. 219–225, 1968.

55. Lofgren, B. E., Land subsidence due to the application of water, in D. J. Varnes, and G. Kiersch, eds., *Reviews in Engineering Geology,* v. 2, Geol. Soc. Amer., Boulder, CO, pp. 271-303, 1969.

56. Meinzer, O. E., *Ground Water in the United States, U.S. Geological Survey Water-Supply Paper* 836-D, pp. 157–232, 1939.

57. Meyboom, P., Estimating ground-water recharge from stream hydrographs, *Jour. Geophysical Research,* v. 66, pp. 1203–1214, 1961.

58. Minshall, N. E., Precipitation and base flow variability, *Intl. Assoc. Sci. Hydrology Publ.* 76, pp. 137–145, 1967.

59. Norris, S. E., and H. B. Eagon, Jr., Recharge characteristics of a watercourse aquifer system at Springfield, Ohio, *Ground Water,* v. 9, no. 1, pp. 30–41, 1971.

60. Norum, D. I., and J. N. Luthin, The effects of entrapped air and barometric fluctuations in the drainage of porous mediums, *Water Resources Research,* v. 4, pp. 417–424, 1968.

61. Oberdorfer, J. A., Numerical Modeling of Coastal Discharge: Predicting the Effects of Climate Change, in *Groundwater Discharge in the Coastal Zone,* R. W. Buddemeier, ed., Proceedings of an International Symposium, Moscow, pp. 85–91, July 1996.

62. Panagoulia, D., and G. Dimou, Sensitivities of groundwater-streamflow interactions to global climate change, *Hydrological Sciences Jour.,* v. 41, pp. 781–796, 1996.

63. Peck, A. J., The water table as affected by atmospheric pressure, *Jour. Geophysical Research,* v. 65, pp. 2383–2388, 1960.

64. Pinder, G. F., and J. F. Jones, Determination of the ground-water component of peak discharge from the chemistry of total runoff, *Water Resources Research,* v. 5, pp. 438–445, 1969.

65. Poland, J. F., Subsidence and its control, in *Underground Waste Management and Environmental Implications, Amer. Assoc. Petr. Geologists Memoir* 18, pp. 50–71, 1972.

66. Poland, J. F., and G. H. Davis, Land subsidence due to withdrawal of fluids, in D. J. Varnes and G. Kiersch, eds., *Reviews in Engineering Geology,* v. 2, Geol. Soc. Amer., Boulder, CO, pp. 187–269, 1969.

67. Poland, J. F., et al., *Studies in Land Subsidence, U.S. Geological Survey Professional Paper* 437-A to 437-H, 1964 to 1975.

68. Rasmussen, W. C., and G. E. Andreasen, *Hydrologic Budget of the Beaverdam Creek Basin, Maryland, U.S. Geological Survey Water-Supply Paper* 1472, 106 pp., 1959.

69. Ripple, C. D., et al., *Estimating Steady-State Evaporation Rates from Bare Soils under Conditions of High Water Table, U.S. Geological Survey Water-Supply Paper* 2019-A, 39 pp., 1972.

70. Robinson, E. S., and R. T. Bell, Tides in confined well-aquifer systems, *Jour. Geophysical Research,* v. 76, pp. 1857–1869, 1971.

71. Robinson, T. W., Earth-tides shown by fluctuations of water-levels in wells in New Mexico and Iowa, *Trans. Amer. Geophysical Union,* v. 20, pp. 656–666, 1939.

72. Sandstrom, K., Modeling the effects of rainfall variability on groundwater recharge in semi-arid Tanzania, *Nordic Hydrology,* v. 26, pp. 313–320, 1995.

73. Schicht, R. J., and W. C. Walton, *Hydrologic budgets for three small watersheds in Illinois,* Rept. Inv. 40, Illinois State Water Survey, Urbana, 40 pp., 1961.

74. Schneider, R., *Correlation of Ground-Water Levels and Air Temperatures in the Winter and Spring in Minnesota, U.S. Geological Survey Water-Supply Paper* 1539-D, 14 pp., 1961.

75. Singh, K. P., Some factors affecting baseflow, *Water Resources Research,* v. 4, pp. 985–999, 1968.

76. Sharma, M. L., Impact of Climate Change on Groundwater Recharge, Proceedings of the Conference on Climate and Water, Volume I, Helsinki, Finland, Valton Painatuskeskus, pp. 511–520, September 1989.

77. Shelton, M. J., and L. B. James, Engineer-geologist team investigates subsidence, *Jour. Pipeline Div.,* Amer. Soc. Civil Engrs., v. 85, no. PL2, 18 pp., 1959.

78. Smedema, L. B., and P. J. Zwerman, Fluctuations of the phreatic surface: 1. Role of entrapped air under a temperature gradient, *Soil Sci.,* v. 103, pp. 354–359, 1967.

79. Stone, M. C., R. H. Hotchkiss, C. M. Hubbard, T. A. Fontaine, L. O. Mearns, and J. G. Arnold, Impacts of climate change on Missouri River basin water yield, *Jour. of the American Water Resources Assn.,* v. 37(5): pp. 1119–1129, October 2001.

80. Taylor, C. J., and W. M. Alley, *Ground-Water-Level Monitoring and the Importance of Long-Term Water-Level Data, U.S. Geological Survey Circular 1217,* http://water.usgs.gov/pubs/circ/circ1217, Denver, CO, 2001.

81. Todd, D. K., Ground-water flow in relation to a flooding stream, *Proc. Amer. Soc. Civil Engrs.,* v. 81, sep. 628, 20 pp., 1955.

82. Troxell, H. C., The diurnal fluctuation in the ground-water and flow of the Santa Ana River and its meaning, *Trans. Amer. Geophysical Union,* v. 17, pp. 496–504, 1936.

83. Tschinkel, H. M., Short-term fluctuation in streamflow as related to evaporation and transpiration, *Jour. Geophysical Research,* v. 68, pp. 6459–6469, 1963.

84. Turk, L. J., Diurnal fluctuations of water tables induced by atmospheric pressure changes, *Jour. Hydrology,* v. 26, pp. 1–16, 1975.

85. Vaccaro, J. J., Sensitivity of groundwater recharge estimates to climate variability and change, Columbia Plateau, Washington, *Jour. Geophysical Research,* v. 97 (D3), pp. 2821–2833, 1992.

86. Vacher, H. L., Hydrology of small oceanic islands—Influence of atmospheric pressure on the water table, *Ground Water,* v. 16, pp. 417–423, 1978.

87. Veihmeyer, F. J., and F. A. Brooks, Measurement of cumulative evaporation of bare soil, *Trans. Amer. Geophysical Union,* v. 35, pp. 601–607, 1954.

88. Visocky, A. P., Estimating the ground-water contribution to storm runoff by the electrical conductance method, *Ground Water,* v. 8, no. 2, pp. 5–10, 1970.

89. Visocky, A. P., *Water-Level Trends and Pumpage in the Deep Bedrock Aquifers in the Chicago Region, 1991-1995, U.S. Geological Survey Circular* 182, Illinois State Water Survey, Champaign, IL, 1997.

90. Vorhis, R. C., Interpretation of hydrologic data resulting from earthquakes, *Geologische Rundschau*, v. 43, pp. 47–52, 1955.

91. Vorhis, R. C., Earthquake-induced water-level fluctuations from a well in Dawson County, Georgia, *Seismological Soc. Amer. Bull.*, v. 54, pp. 1023–1133, 1964.

92. Vorhis, R. C., *Hydrologic Effects of the Earthquake of March 27, 1964, Outside Alaska, U.S. Geological Survey Professional Paper* 544-C, 54 pp., 1967.

93. Werner, P. W., and D. Noren, Progressive waves in non-artesian aquifers, *Trans. Amer. Geophysical Union*, v. 32, pp. 238-244, 1951.

94. Werner, P. W., and K. J. Sundquist, On the ground-water recession curve for large watersheds, *Intl. Assoc. Sci. Hydrology Publ.* 33, pp. 202–212, 1951.

95. White, W. N., *A Method of Estimating Ground-Water Supplies Based on Discharge by Plants and Evaporation from Soil, U.S. Geological Survey Water-Supply Paper* 659, pp. 1–105, 1932.

96. Winter, T. C., J. W. Harvey, O. L. Franke, and W. M. Alley, *Groundwater and Surface Water: A Single Source, U.S. Geological Survey Circular* 1139, http://water.usgs.gov/pubs/circ/circ1139, Denver, CO, 1998.

Chapter 7

Quality of Groundwater

It is now recognized that the quality of groundwater is just as important as its quantity. All groundwater contains salts in solution that are derived from the location and past movement of the water. The quality required of a groundwater supply depends on its purpose; thus, needs for drinking water, industrial water, and irrigation water vary widely. To establish quality criteria, measures of chemical, physical, biological, and radiological constituents must be specified, as well as standard methods for reporting and comparing results of water analyses. Dissolved gases in groundwater can pose hazards if their presence goes unrecognized. The uniformity of groundwater temperature is advantageous for water supply and industrial purposes, and underlying saline groundwaters are important because they offer potential benefits.

7.1 NATURAL GROUNDWATER QUALITY

The groundwater in natural systems generally contains less than 1,000 mg/l dissolved solids, unless groundwater has (1) encountered a highly soluble mineral, such as gypsum, (2) been concentrated by evapotranspiration, or (3) been geothermally heated (Graham et al.[18]). Natural groundwater generally acquires dissolved constituents by dissolution of aquifer gases, minerals, and salts. Consequently, soil zone and aquifer gases and the most soluble minerals and salts in an aquifer generally determine the chemical composition of groundwater in an aquifer.

Table 7.1.1 gives examples of natural groundwater chemistry. Most groundwater is recharged through a soil zone containing partial pressures of carbon dioxide gas that are higher than the atmosphere. Recently recharged groundwater therefore generally contains high inorganic carbon concentrations. Minor and trace element compositions of natural groundwater depend on the availability of minor and trace elements in easily soluble phases or on sorption sites, and the redox state of the water in the aquifer (Graham et al.[18]). Anthropogenic sources can create significant levels of trace elements in the atmosphere. Recharge from precipitation introduces these trace elements into the groundwater.

Knowledge of the natural groundwater quality can provide important insights into the nature of the resource. Evaluation of the natural chemical and isotopic compositions of groundwater can provide inferences of the reactions that produce natural water chemistry and the recharge, movement, mixing, and discharge of groundwater. The chemistry of natural groundwater flow systems can be used to determine (1) geochemical reactions that produce observed water chemistry and changes in observed water chemistry, (2) groundwater flow paths, (3) groundwater mixing, and (4) groundwater ages and flow rates.

Arsenic, a naturally occurring element in rocks, soils, and water in contact with them, is a human health concern because it can contribute to skin, bladder, and other cancers (National Research Council[28]). Figure 7.1.1 illustrates the levels of arsenic concentrations in groundwa-

Table 7.1.1 Examples of Natural Groundwater Chemistry[*]

Site	Temperature	pH	Ca	Mg	Na	K	HCO₃	SO₄	Cl	SiO₂
				Carbonate rock aquifer						
Cold Creek Spring[a]	10	7.6	1.72	0.64	0.07	0.01	4.79	0.09	0.03	0.11
Corn Creek Spring[a]	21	7.7	1.14	1.34	0.27	0.05	4.67	0.18	0.20	0.28
Indian Springs[a]	26	7.4	1.15	0.92	0.18	0.03	3.97	0.16	0.10	0.20
				Carbonate rock aquifer with evaporite mineral						
Lewiston Big Spring[b]	10.6	7.58	1.87	1.15	0.10	0.02	3.31	1.46	0.05	—
Hanover Flowing Well[b]	20.4	7.63	2.10	1.19	0.12	0.03	3.42	1.77	0.04	—
Vanek Warm Spring[b]	19.6	7.40	3.25	1.65	0.16	0.03	3.53	3.44	0.07	—
Landusky Spring[b]	20.4	7.24	6.50	4.08	1.79	0.24	3.81	10.11	0.54	—
				Granitic rock aquifer						
Ephemeral Springs[c]	—	6.2	0.08	0.03	0.13	0.03	0.33	0.01	0.01	0.27
Perennial Springs[c]	—	6.8	0.26	0.08	0.26	0.04	0.90	0.03	0.03	0.41
Shallow Well 2[d]	8.0	7.89	0.85	0.19	0.54	0.04	2.34	0.09	0.10	0.21
				Rhyolitic rock aquifer						
Lower Indian Spring[e]	21	7.9	0.15	0.04	2.48	0.04	2.07	0.18	0.42	0.80
Crystal Springs[e]	24	7.7	0.55	0.15	2.18	0.09	2.33	0.23	0.59	0.75
S. Brown Well[f]	10.6	6.8	0.50	0.16	1.02	0.10	1.54	0.14	0.23	0.93
N. Brown Well[f]	14.7	7.5	0.87	0.23	2.09	0.17	2.46	0.54	0.71	1.03

[*] All analyses are in millimoles per liter (mmol/l) and temperature in degrees Celsius. Note that mmol/l times molecular weight equals mg/l.
Source: Graham et al.[18]

[a] Winograd and Pearson[49]
[b] Plummer et al.[32]
[c] Garrels and MacKenzie[17]
[d] Nordstrom et al.[29]
[e] White[46]
[f] Thomas et al.[43]

ter of the United States (also refer to Welch et al.[45]). Generally, arsenic concentrations in groundwater are the highest in the western part of the United States, with parts of the midwest and northeast having some locations with arsenic concentrations that exceed 10 μg/L, which is the provisional World Health Organization limit[52] for arsenic in drinking water.

7.2 SOURCES OF SALINITY

All groundwater contains salts in solution; reported salt contents range from less than 25 mg/l in a quartzite spring to more than 300,000 mg/l in brines[47]. The type and concentration of salts depend on the environment, movement, and source of the groundwater. Ordinarily, higher concentrations of dissolved constituents are found in groundwater than in surface water because of the greater exposure to soluble materials in geologic strata. Soluble salts in groundwater originate primarily from solution of rock materials. Bicarbonate, usually the primary anion in groundwater, is derived from carbon dioxide released by organic decomposition in the soil. Salinity varies with specific surface area of aquifer materials, solubility of minerals, and contact time; values tend to be highest where movement of groundwater is least. Hence, salinity generally increases with depth. A common geochemical sequence in groundwater includes bicarbonate waters near ground surface varying to chloride waters in the deepest portions of formations.

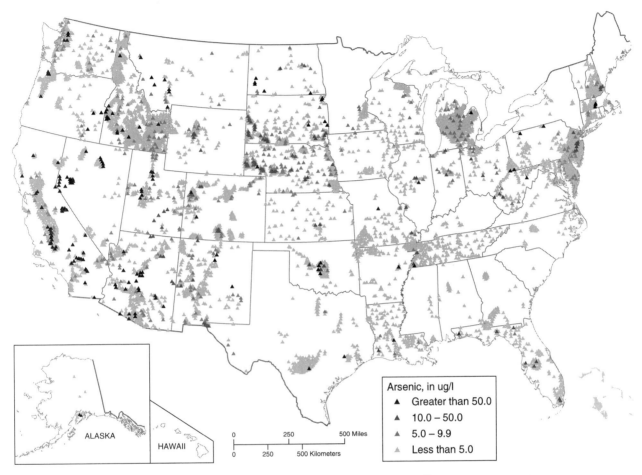

Figure 7.1.1. Arsenic concentrations in groundwater of the United States (Welch et al.[45]).

Precipitation reaching the earth contains only small amounts of dissolved mineral matter. Once on earth, the water contacts and reacts with the minerals of the soil and rocks. The quantity and type of mineral matter dissolved depend on the chemical composition and physical structure of the rocks as well as the hydrogen-ion concentration (pH) and the redox potential (Eh) of the water.[5] Carbon dioxide in solution, derived from the atmosphere and from organic processes in the soil, assists the solvent action of water as it moves underground.[21] The *geochemical cycle of surface water and groundwater* shown in Fig. 7.2.1 illustrates the principal chemical changes involved in water as it travels through the hydrologic cycle from precipitation to groundwater.

In areas recharging large volumes of water underground, such as alluvial streams, channels, or artificial recharge areas, the quality of the infiltrating surface water can have a marked effect on that of the groundwater. Locally, absorbed gases of magmatic origin contribute dissolved mineral products to groundwater; mineralized thermal springs are an excellent example. Connate water is usually highly mineralized because it is derived from water originally entrapped in sedimentary strata since the time of deposition.*

---

*It should be recognized that most connate water has been altered chemically by various chemical and physical processes and therefore does not necessarily represent the original water of deposition.

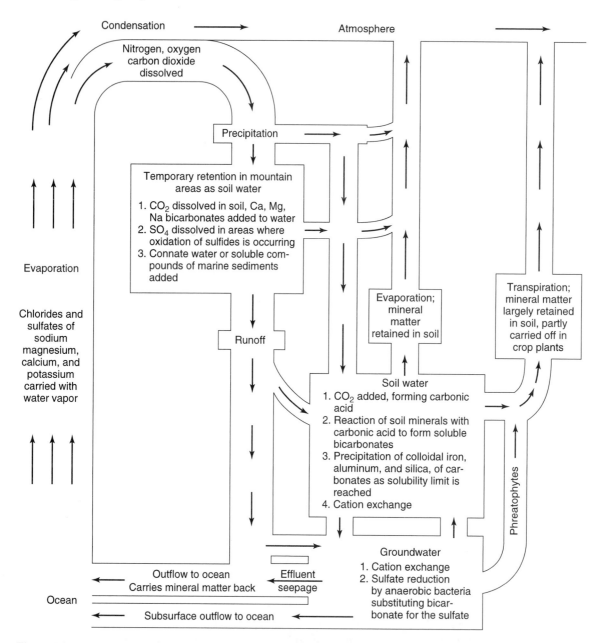

Figure 7.2.1. Geochemical cycle of surface water and groundwater (after *USGS Water-Supply Paper* 1469).

Salts are added to groundwater passing through soils by soluble products of soil weathering and of erosion by rainfall and flowing water. Excess irrigation water percolating to the water table may contribute substantial quantities of salt. Water passing through the root zone of cultivated areas usually contains salt concentrations several times that of the applied irrigation water. Increases result primarily from the evapotranspiration process, which tends to concentrate salts in drainage waters. In addition, soluble soil materials, fertilizers, and selective absorption of salts by plants will modify salt concentrations of percolating waters. Factors governing the increase include soil permeability, drainage facilities, amount of water applied, crops, and climate. Thus, high salinities may be found in soils and groundwater of arid climates where leaching by rainwater is not effective in diluting the salt solutions. Similarly, poorly drained areas, particularly basins having interior drainage, often contain high salt concentrations. Also, some regions contain remnants of sedimentary deposition under saline waters; the designation *badlands* implies in part the lack of productivity resulting from excess salt contents of the soil and water.

Groundwater passing through igneous rocks dissolves only very small quantities of mineral matter because of the relative insolubility of the rock composition. Percolating rainwater contains carbon dioxide from the atmosphere, which increases the solvent action of the water. The silicate minerals of igneous rocks result in silica being added to the groundwater.

Sedimentary rocks are more soluble than igneous rocks.* Because of their high solubility, combined with their great abundance in the earth's crust, they furnish a major portion of the soluble constituents of groundwater. Sodium and calcium are commonly added cations; bicarbonate and sulfate are corresponding anions. Chloride occurs to only a limited extent under normal conditions; important sources of chloride, however, are from sewage, connate water, and intruded seawater. Occasionally nitrate is an important natural constituent; high concentrations may indicate sources of past or present pollution. In limestone terrains, calcium and bicarbonate ions are added to the groundwater by solution.

A summary of the natural sources and concentrations of the principal chemical constituents found in groundwater is presented in Table 7.2.1 together with their effects on the usability of the water. A comprehensive discussion of the chemical constituents of groundwater can be found in Hem.[21]

Table 7.2.1 Principal Chemical Constituents in Groundwater—Their Sources, Concentrations, and Effect on Usability (modified from C. N. Durfer, and E. Baker, *USGS Water-Supply Paper* 1812, 1964)

Constituent	Major natural sources	Concentration in natural water	Effect on usability of water
Silica (SiO_2)	Feldspars, ferromagnesium and clay minerals, amorphous silica, chert, opal	Ranges generally from 1.0 to 30 mg/l, although as much as 100 mg/l is fairly common; as much as 4,000 mg/l is found in brines.	In the presence of calcium and magnesium, silica forms a scale in boilers and on steam turbines that retards heat; the scale is difficult to remove. Silica may be added to soft water to inhibit corrosion of iron pipes.
Iron (Fe)	Igneous rocks: amphiboles, ferromagnesian micas, ferrous sulfide (FeS), ferric sulfide or iron pyrite (FeS_2), magnetite (Fe_3O_4) Sandstone rocks: oxides, carbonates, and sulfides or iron clay minerals	Generally less than 0.50 mg/l in fully aerated water. Groundwater having a pH less than 8.0 may contain 10 mg/l; rarely as much as 50 mg/l may occur. Acid water from thermal springs, mine wastes, and industrial wastes may contain more than 6,000 mg/l.	More than 0.1 mg/l precipitates after exposure to air; causes turbidity, stains plumbing fixtures, laundry, and cooking utensils; imparts objectionable tastes and colors to foods and drinks. More than 0.2 mg/l is objectionable for most industrial uses.
Manganese (Mn)	Manganese in natural water probably comes most often from soils and sediments. Metamorphic and sedimentary rocks and mica biotite and amphibole hornblende minerals contain large amounts of manganese.	Generally 0.20 mg/l or less. Groundwater and acid mine water may contain more than 10 mg/l.	More than 0.2 mg/l precipitates upon oxidation; causes undesirable tastes, deposits on foods during cooking; stains plumbing fixtures and laundry; fosters growths in reservoirs, filters, and distribution systems. Most industrial users object to water containing more than 0.2 mg/l.

(continues)

*Important mineral sources in sedimentary rocks are feldspar, gypsum, and forms of calcium carbonate.

Table 7.2.1 (*continued*) Principal Chemical Constituents in Groundwater—Their Sources, Concentrations, and Effect on Usability (modified from C. N. Durfer, and E. Baker, *USGS Water-Supply Paper* 1812, 1964)

Constituent	Major natural sources	Concentration in natural water	Effect on usability of water
Calcium (Ca)	Amphiboles, feldspars, gypsum, pyroxenes, aragonite, calcite, dolomite, clay minerals	Generally less than 100 mg/l; brines may contain as much as 75,000 mg/l.	Calcium and magnesium combine with bicarbonate, carbonate, sulfate, and silica to form a heat-retarding, pipe-clogging scale in boilers and in other heat-exchange equipment. Calcium and magnesium combine with ions of fatty acid in soaps to form soapsuds; the more calcium and magnesium, the more soap required to form suds. A high concentration of magnesium has a laxative effect, especially on new users of the supply.
Magnesium (Mg)	Amphiboles, olivine, pyroxenes, dolomite, magnesite, clay minerals	Generally less than 50 mg/l; ocean water contains more than 1,000/mg/l, and brines may contain as much as 57,000 mg/l.	
Sodium (Na)	Feldspars (albite); clay minerals; evaporites, such as halite (NaCl) and mirabilite ($Na_2SO_4 \cdot 10H_2O$); industrial wastes	Generally less than 200 mg/l; about 10,000 mg/l in seawater; about 25,000 mg/l in brines.	More than 50 mg/l sodium and potassium in the presence of suspended matter causes foaming, which accelerates scale formation and corrosion in boilers. Sodium and potassium carbonate in recirculating cooling water can cause deterioration of wood in cooling towers. More than 65 mg/l of sodium can cause problems in ice manufacture.
Potassium (K)	Feldspars (orthoclase and microcline), feldspathoids, some micas, clay minerals	Generally less than about 10 mg/l; as much as 100 mg/l in hot springs; as much as 25,000 mg/l in brines.	
Carbonate (CO_3)	Limestone, dolomite	Commonly less than 10 mg/l in groundwater. Water high in sodium may contain as much as 50 mg/l of carbonate.	Upon heating, bicarbonate is changed into steam, carbon dioxide, and carbonate. The carbonate combines with alkaline earths—principally calcium and magnesium—to form a crust-like scale of calcium carbonate that retards flow of heat through pipe walls and restricts flow of fluids in pipes. Water containing large amounts of bicarbonate and alkalinity is undesirable in many industries.
Bicarbonate (HCO_3)		Commonly less than 500 mg/l; may exceed 1,000 mg/l in water highly charged with carbon dioxide.	
Sulfate (SO_4)	Oxidation of sulfide ores; gypsum; anhydrite	Commonly less than 300 mg/l except in wells influenced by acid mine drainage. As much as 200,000 mg/l in some brines.	Sulfate combines with calcium to form an adherent, heat-retarding scale. More than 250 mg/l is objectionable in water in some industries. Water containing about 500 mg/l of sulfate tastes bitter; water containing about 1,000 mg/l may be cathartic.
Chloride (Cl)	Chief source is sedimentary rock (evaporites); minor sources are igneous rocks	Commonly less than 10 mg/l in humid regions but up to 1,000 mg/l in more arid regions. About 19,300 mg/l in seawater; as much as 200,000 mg/l in brines.	Chloride in excess of 100 mg/l imparts a salty taste. Concentrations greatly in excess of 100 mg/l may cause physiological damage. Food processing industries usually require less than 250 mg/l. Some industries—textile processing, paper manufacturing, and synthetic rubber manufacturing—desire less than 100 mg/l.

(continues)

Table 7.2.1 (*continued*) Principal Chemical Constituents in Groundwater—Their Sources, Concentrations, and Effect on Usability (modified from C. N. Durfer, and E. Baker, *USGS Water-Supply Paper* 1812, 1964)

Constituent	Major natural sources	Concentration in natural water	Effect on usability of water
Fluoride (F)	Amphiboles (hornblende), apatite, fluorite, mica	Concentrations generally do not exceed 10 mg/l. Concentrations may be as much as 1,600 mg/l in brines.	Fluoride concentration between 0.6 and 1.7 mg/l in drinking water has a beneficial effect on the structure and resistance to decay of children's teeth. Fluoride in excess of 1.5 mg/l in some areas causes "mottled enamel" in children's teeth. Fluoride in excess of 6.0 mg/l causes pronounced mottling and disfiguration of teeth.
Nitrate (NO_3)	Atmosphere; legumes, plant debris, animal excrement	Commonly less than 10 mg/l.	Water containing large amounts of nitrate (more than 100 mg/l) is bitter tasting and may cause physiological distress. Water from shallow wells containing more than 45 mg/l has been reported to cause methemoglobinemia in infants. Small amounts of nitrate help reduce cracking of high-pressure boiler steel.
Dissolved solids	The mineral constituents dissolved in water constitute the dissolved solids.	Commonly contains less than 5,000 mg/l; some brines contain as much as 300,000 mg/l.	More than 500 mg/l is undesirable for drinking and many industrial uses. Less than 300 mg/l is desirable for dyeing of textiles and the manufacture of plastics, pulp paper, and rayon. Dissolved solids cause foaming in steam boilers; the maximum permissible content decreases with increases in operating pressure.

An important source of salinity in groundwater in coastal regions is airborne salts originating from the air–water interface over the sea. Detailed studies on a worldwide basis[13] and for Israel[25] suggest that salts are deposited on land both by precipitation and by dry fallout. Chloride deposition in coastal areas has been calculated to range from 4 to 20 kg Cl/ha. The deposition decreases inland, varying exponentially with distance from the sea. Thus, Israeli water measurements yielded the relation

$$N = 110 \, e^{-0.0133d} \tag{7.2.1}$$

where *N* is the annual amount of chloride precipitation in kg/ha and *d* is the distance from the sea in kilometers. In arid regions, where surface runoff is small and evapotranspiration is large, airborne salt deposition becomes intensified several fold in groundwater.

7.3 MEASURES OF WATER QUALITY

The chemical characteristics of groundwater are determined by the chemical and biological reaction in the zones through which the water moves, as illustrated in Figure 7.3.1. In specifying the quality characteristics of groundwater, chemical, physical, and biological analyses are normally required. The characteristics of water that affect water quality depend both on substances dissolved in water and on certain properties. Natural inorganic constituents commonly dissolved in water that are most likely to affect water use include bicarbonate,

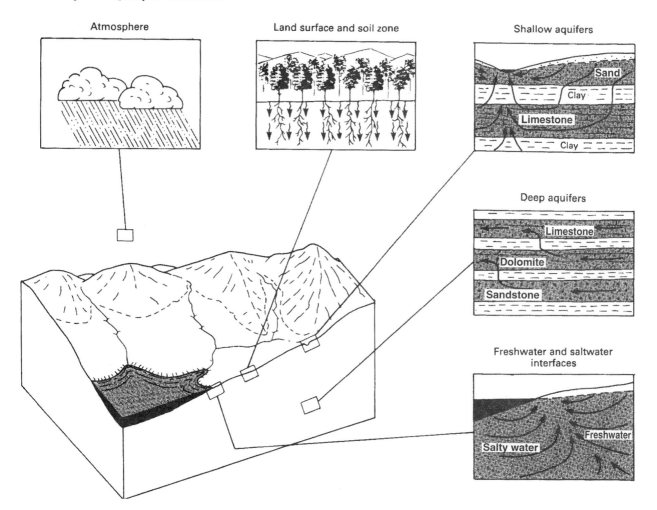

Figure 7.3.1. The chemical characteristics of groundwater are determined by the chemical and biological reactions in the zones through which the water moves (Heath[20]).

carbonate, calcium, magnesium, chloride, flouride, iron, manganese, sodium, and sulfate. See Heath[20] for more details. Table 7.3.1 provides a list of characteristics of water that affect water quality. A complete chemical analysis of a groundwater sample includes the determination of the concentrations of the inorganic constituents present; organic and radiological parameters are normally of concern only where human-induced pollution affects quality (see Chapter 8). Dissolved salts in groundwater of normal salinity occur as dissociated ions; in addition, other minor constituents are present and reported in elemental form. The analysis also includes measurement of pH and specific electrical conductance. Depending on the purpose of a water quality investigation, partial analyses of only particular constituents will sometimes suffice.

Properties of groundwater evaluated in a physical analysis include temperature, color, turbidity, odor, and taste. Biological analysis includes tests to detect the presence of coliform bacteria, which indicate the sanitary quality of water for human consumption. Because certain coliform organisms are normally found in intestines of humans and animals, the presence of these in groundwater is tantamount to its contact with sewage sources.

Table 7.3.1 Characteristics of Water that Affect Water Quality

Characteristic	Principal cause	Significance	Remarks
Hardness	Calcium and magnesium dissolved in the water	Calcium and magnesium combine with soap to form an insoluble precipitate (curd) and thus hamper the formation of a lather. Hardness also affects the suitability of water for use in the textile and paper industries and certain others, and in steam boilers and water heaters.	USGS classification of hardness (mg/L as $CaCO_3$: 0–60: Soft 61–120: Moderately hard 121–180: Hard More than 180: Very hard
pH for hydrogen-ion activity	Dissociation of water molecules and of acids and bases dissolved in water	The pH of water is a measure of its reactive characteristics. Low values of pH, particularly below pH 4, indicate a corrosive water that will tend to dissolve metals and other substances that it contacts. High values of H, particularly above pH 8.5, indicate an alkaline water that, on heating, will tend to form a scale. The pH significantly affects the treatment and use of water.	pH values: less than 7, water is acidic; value of 7, water is neutral; more than 7, water is basic.
Specific electrical conductance	Substances that form ions when dissolved in water	Most substances dissolved in water dissociate into ions that can conduct an electrical current. Consequently, specific electrical conductance is a valuable indicator of the amount of material dissolved in water. The larger the conductance, the more mineralized the water.	Conductance values indicate the electrical conductivity, in micromhos, of 1 cm^3 of water at a temperature of 25°C.
Total dissolved solids	Mineral substances dissolved in water	Total dissolved solids is a measure of the total amount of minerals dissolved in water and is, therefore, a very useful parameter in the evaluation of water quality. Water containing less than 500 mg/L is preferred for domestic use and for many industrial processes.	USGS classification of water based on dissolved solids (mg/L): Less than 1,000: Fresh 1,000–3,000: Slightly saline 3,000–10,000: Moderately saline 10,000–35,000: Very saline More than 35,000: Briny

Source: Heath[20]

7.4 CHEMICAL ANALYSIS

Once a sample of groundwater has been analyzed in a laboratory, methods for reporting water analyses must be considered. From an understanding of expressions and units for describing water quality, standards can be established so that analyses can be interpreted in terms of the ultimate purpose of the water supply. In a chemical analysis of groundwater, concentrations of different ions are expressed by weight or by chemical equivalence. Total dissolved solids can be measured in terms of electrical conductance. These and other measures of chemical quality are described in the following sections.

7.4.1 Concentrations by Weight

Concentrations of the common ions found in groundwater are reported by weight-per-volume units of milligrams per liter (mg/l) or micrograms per liter (mg/L). The total ionic concentration (or total dissolved solids) is also reported in this manner. The units milligrams per liter have replaced parts per million; however, they are numerically equivalent up to a concentration of dissolved solids of about 7,000 mg/l.

7.4.2 Chemical Equivalence

Positively charged cations and negative anions combine and dissociate in definite weight ratios. By expressing ion concentrations in equivalent weights, these ratios are readily determined because one equivalent weight of a cation will exactly combine with one equivalent weight of an anion. The combining weight of an ion is equal to its formula weight divided by its charge. When the concentration in milligrams per liter is divided by the combining weight, an equivalent concentration expressed in milliequivalents per liter (meq/l) results. Table 7.4.1 lists the reciprocals of combining weights of cations and anions; concentrations in milligrams per liter can be converted to milliequivalents per liter by multiplying by the appropriate conversion factor. For undissociated species with zero charge, of which silica is an example in groundwater quality, an equivalent weight cannot be computed.

Table 7.4.1 Conversion Factors for Chemical Equivalence (after Hem[21]) (concentrations in mg/1 times the conversion factor yields concentration in meq/1)

Chemical Constituent	Conversion Factor
Aluminum (Al^{+3})	0.11119
Ammonium (NH_4^+)	0.05544
Barium (Ba^{+2})	0.01456
Beryllium (Be^{+3})	0.33288
Bicarbonate (HCO_3^-)	0.01639
Bromide (Br^-)	0.01251
Cadmium (Cd^{+2})	0.01779
Calcium (Ca^{+2})	0.04990
Carbonate (CO_3^{-2})	0.03333
Chloride (Cl^-)	0.02821
Cobalt (Co^{+2})	0.03394
Copper (Cu^{+2})	0.03148
Fluoride (F^-)	0.05264
Hydrogen (H^+)	0.99209
Hydroxide (OH^-)	0.05880
Iodide (I^-)	0.00788
Iron (Fe^{+2})	0.03581
Iron (Fe^{+3})	0.05372
Lithium (Li^+)	0.14411
Magnesium (Mg^{+2})	0.08226
Manganese (Mn^{+2})	0.03640
Nitrate (NO_3^-)	0.01613
Nitrite (NO_2^-)	0.02174
Phosphate (PO_4^{-3})	0.03159
Phosphate (HPO_4^{-2})	0.02084
Phosphate ($H_2PO_4^-$)	0.01031
Potassium (K^+)	0.02557
Rubidium (Rb^+)	0.01170
Sodium (Na^+)	0.04350
Strontium (Sr^{+2})	0.02283
Sulfate (SO_4^{-2})	0.02082
Sulfide (S^{-2})	0.06238
Zinc (Zn^{+2})	0.03060

In application, therefore, it may be expected that of the total dissolved solids in a groundwater sample, the sum of the cations and the sum of the anions when expressed in milliequivalents per liter will be equal. If the chemical analysis of the various ionic constituents indicates a difference from this balance, it may be concluded either that there are other undetermined constituents present or that errors exist in the analysis.

7.4.3 Total Dissolved Solids by Electrical Conductance

A rapid determination of total dissolved solids can be made by measuring the electrical conductance of a groundwater sample. Conductance is preferred rather than its reciprocal, resistance, because it increases with salt content. Specific electrical conductance defines the conductance of a cubic centimeter of water at a standard temperature of 25°C; an increase of 1°C increases conductance by about 2 percent.

Specific conductance is measured in microsiemens/cm (μS/cm). The unit microsiemens/cm is equivalent to micromhos/cm. Because the definition of specific conductance already specifies the dimensions to which the measurement applies, the length in the units is often omitted in practice.

Because natural water contains a variety of ionic and undissociated species, conductance cannot be simply related to total dissolved solids. However, conductance is easily measured and gives results that are convenient as a general indication of dissolved solids. An approximate relation for most natural water[26,34] in the range of 100 to 5,000 μS/cm leads to the equivalencies 1 meq/l of cations = 100 μS/cm and 1 mg/l = 1.56 μS/cm.

7.4.4 Hardness

Hardness results from the presence of divalent metallic cations, of which calcium and magnesium are the most abundant in groundwater.* These ions react with soap to form precipitates and with certain anions present in the water to form a scale. Because of their adverse action with soap, hard waters are unsatisfactory for household cleansing purposes; hence, water-softening processes for removal of hardness are needed.

The hardness in water is derived from the solution of carbon dioxide, released by bacterial action in the soil, in percolating rainwater.[36] Low pH conditions develop and lead to the solution of insoluble carbonates in the soil and in limestone formations to convert them into soluble bicarbonates. Impurities in limestone, such as sulfates, chlorides, and silicates, become exposed to the solvent action of water as the carbonates are dissolved so that they also pass into solution. Thus, hard water tends to originate in areas where thick topsoils overlie limestone formations. A map of hardness of groundwater in the United States is shown in Figure 7.4.1.

Hardness H_T is customarily expressed as the equivalent of calcium carbonate. Thus,

$$H_T = Ca \times \frac{CaCO_3}{Ca} + Mg \times \frac{CaCO_3}{Ca} \qquad (7.4.1)$$

where H_T, Ca, and Mg are measured in milligrams per liter and the ratios in equivalent weights. Equation 7.4.1 reduces to

$$H_T = 2.5\,Ca + 4.1\,Mg \qquad (7.4.2)$$

The degree of hardness in water is commonly based on the classification listed in Table 7.4.2.

*The terms *hard* and *soft* as applied to water date from Hippocrates (460–354 B.C.), the father of medicine, in his treatise on public hygiene, *Air, Water and Places:* "Consider the waters which the inhabitants use, whether they be marshy and soft, or hard and running from elevated and rocky situations, and then if saltish and unfit for cooking . . . for water contributes much to health."[6]

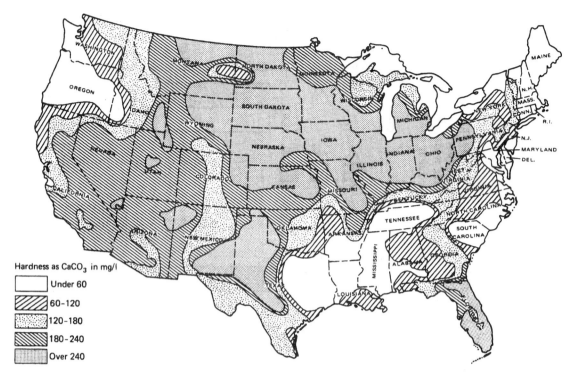

Figure 7.4.1. Hardness of groundwater in the United States. Areas delineated represent average conditions on a generalized basis (after Ackerman and Löf, *Technology in American Water Development,* Resources for the Future, copyright © 1959 by The Johns Hopkins University Press).

Table 7.4.2 Hardness Classification of Water (after Sawyer and McCarty[36])

Hardness, mg/l as $CaCO_3$	Water class
0–75	Soft
75–150	Moderately hard
150–300	Hard
Over 300	Very hard

7.5 GRAPHIC REPRESENTATIONS

Tables showing results of analyses of chemical quality of groundwater may be difficult to interpret, particularly when more than a few analyses are involved. To overcome this, graphic representations are useful for display purposes, for comparing analyses, and for emphasizing similarities and differences. Graphs can also aid in detecting the mixing of water of different compositions and in identifying chemical processes occurring as groundwater moves. A variety of graphic techniques have been developed for showing the major chemical constituents; some of the more useful graphs are described and illustrated in the following paragraphs.

Figure 7.5.1 illustrates vertical bar graphs, widely used in the United States for portraying chemical quality. Each analysis appears as a vertical bar having a height proportional to the total concentration of anions or cations, expressed in milliequivalents per liter. The left half of a bar represents cations, and the right half anions. These segments are divided horizontally to show the concentrations of major ions or groups of closely related ions and identified by

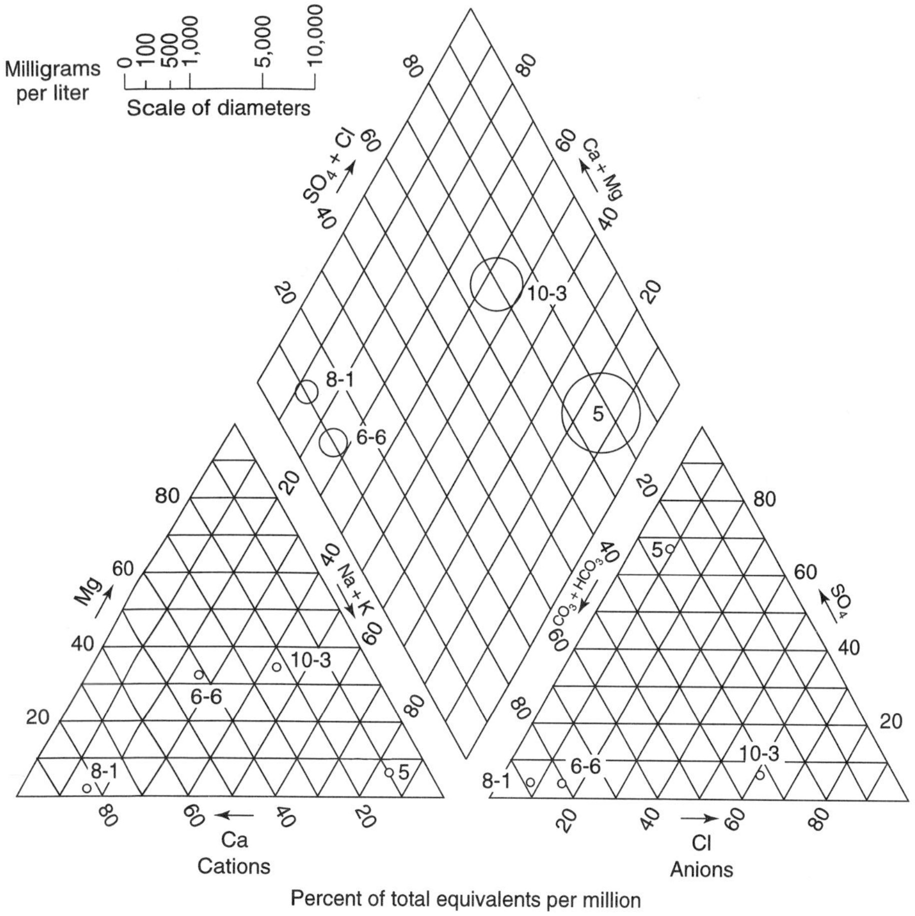

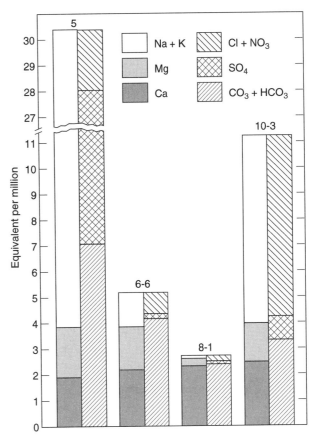

Figure 7.5.1. Vertical bar graphs for representing analyses of groundwater quality (after Hem[21]).

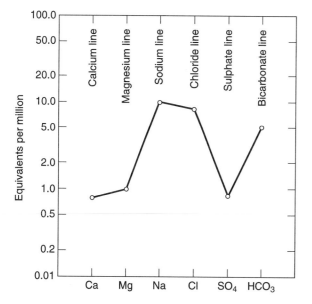

Figure 7.5.5. Trilinear diagram for representing analyses of groundwater quality (after Hem[21]).

Figure 7.5.6. Schoeller semi-logarithmic diagram for representing analyses of groundwater quality (after Schoeller[38]).

distinctive shading patterns.* The reference number of the analysis is shown at the top of the bar. This standard bar graph can be modified to include hardness and silica by additional segments.[21]

Another method for plotting chemical quality with radiating vectors is shown in Figure 7.5.2. The lengths of the six vectors represent ionic concentrations in milliequivalents per liter.

Pattern diagrams, first suggested by Stiff,[41] for representing chemical analyses by four parallel axes are illustrated by Figure 7.5.3. Concentrations of cations are plotted to the left of a vertical zero axis and anions to the right; all values are in milliequivalents per liter. The resulting points, when connected, form an irregular polygonal pattern; waters of a similar quality define a distinctive shape.

Figure 7.5.4 indicates circular diagrams of water quality with a special scale for the radii so that the area of a circle is proportional to the total ionic concentration of the analysis. Sectors within a circle show the fractions of the different ions expressed in milliequivalents per liter.

One of the most useful graphs for representing and comparing water quality analyses is the trilinear diagram by Piper,[30] shown in Figure 7.5.5. Here cations, expressed as percentages of total cations in milliequivalents per liter, plot as a single point on the left triangle; anions, similarly expressed as percentages of total anions, appear as a point in the right triangle. These two points are then projected into the central diamond-shaped area parallel to the upper edges

---

*It should be noted that the water quality diagrams in Figures 7.5.1–7.5.6 all depict the same four water analyses so as to facilitate comparisons. Also, equivalents per million is identical to meq/1.

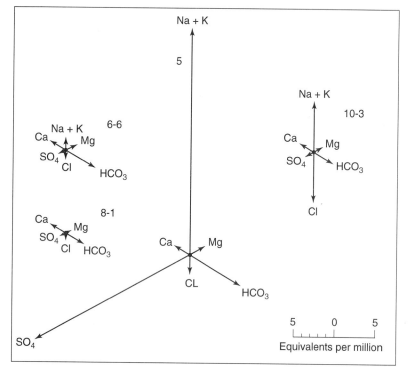

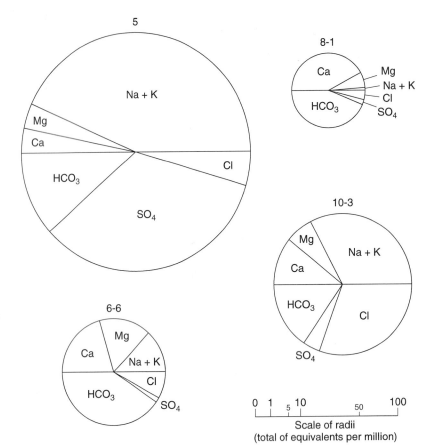

Figure 7.5.2. Vector diagrams for representing analyses of groundwater quality (after Hem[21]).

Figure 7.5.4. Circular diagrams for representing analyses of groundwater quality (after Hem[21]).

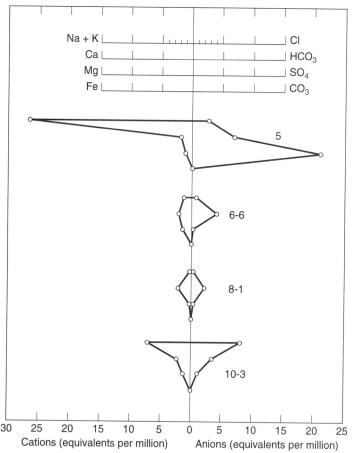

Figure 7.5.3. Pattern diagrams for representing analyses of groundwater quality (after Hem[21]).

of the central area. This single point is thus uniquely related to the total ionic distribution; a circle can be drawn at this point with its area proportional to the total dissolved solids. The linear diagram conveniently reveals similarities and differences among groundwater samples—those with similar qualities will tend to plot together as groups. Further, simple mixtures of two source waters can be identified; for example, an analysis of any mixture of two waters will plot on the straight line AB on the diagram, where A and B are the positions of the analyses of the two component waters.

In Europe, the semilogarithmic diagram developed by Schoeller[38] is employed widely for comparing groundwater analyses. Here the principal ionic concentrations, expressed in milliequivalents per liter, are plotted on six equally spaced logarithmic scales in the arrangement shown by Figure 7.5.6. The points thus plotted are joined by straight lines. This type of graph shows not only the absolute value of each ion but also the concentration differences among various groundwater analyses. Because of the logarithmic scale, if a straight line joining the points A and B of two ions in one water sample is parallel to another straight line joining the points A′ and B′ of the same two ions in another water sample, the ratio of the ions in both analyses is equal.[*]

―――――――――

[*]It should also be mentioned that the Schoeller diagram can be adapted to determine the degree of saturation of $CaCO_3$ and $CaSO_4$ in groundwater.[38]

7.6   PHYSICAL ANALYSIS

In a physical analysis of groundwater, temperature is reported in degrees Celsius and must be measured immediately after collecting the sample. Color in groundwater may be due to mineral or organic matter in solution and is reported in mg/l by comparison with standard solutions. *Turbidity* is a measure of the suspended and colloidal matter in water, such as clay, silt, organic matter, and microscopic organisms. Measurements are often based on the length of a light path through the water which just causes the image of a flame of a standard candle to disappear. The natural filtration produced by unconsolidated aquifers largely eliminates turbidity, but other types of aquifers can produce turbid groundwater. Tastes and odors may be derived from bacteria, dissolved gases, mineral matter, or phenols. These characteristics are subjective sensations that can be defined only in terms of the experience of a human being. Quantitative determinations of taste and odor have been developed based on the maximum degree of dilution that can be distinguished from taste free and odor free water.[2]

7.7   BIOLOGICAL ANALYSIS

As mentioned before, bacteriological analysis is important for detecting biological pollution of groundwater. Most pathogenic bacteria found in water are indigenous to the intestinal tract of animals and humans, but isolating them from natural water is difficult in the laboratory. Because bacteria of the coliform group are relatively easy to isolate and identify, standard tests to determine their presence or absence in a water sample are taken as a direct indication of the safety of the water for drinking purposes. Coliform test results are reported as the most probable number (MPN) of coliform group organisms in a given volume of water. By analysis of a number of separate portions of a water sample, the MPN is computed from probability tables for this purpose.

7.8   GROUNDWATER SAMPLES

In sampling groundwater for quality analysis, Pyrex glass or polyethylene bottles are generally satisfactory.[33] Volumes of one or two liters are usually sufficient for a normal routine chemical analysis. After rinsing the bottle with the water being sampled, the sample is then collected and securely sealed. The water should be stored in a cool place and transferred promptly to a laboratory for analysis. Samples should be taken from a well only after it has been pumped for some time; otherwise, nonrepresentative samples of stagnant or polluted water may be obtained. With each sample, a record should be made of well location, depth of sample, size of casing, date, water temperature, odor, color, turbidity, and operating conditions of the well immediately prior to the sampling. To analyze for organic and radiological constituents, special sampling and storage techniques are required.

The shorter the time that elapses between collection of a sample and its analysis, generally the more reliable the analytic results will be. For certain constituents and physical values, immediate analysis in the field is required to obtain dependable results; determinations of temperature, pH, alkalinity, and dissolved gases should always be carried out in the field because changes are inevitable by the time samples reach a laboratory. Storage of samples prior to analysis can also affect results. Cations such as Fe, Cu, Al, Mn, Cr, and Zn are subject to loss by adsorption or ion exchange on the walls of glass containers.[33]

Finally, it should be noted that samples taken from a well penetrating stratified aquifers can yield solute concentrations that differ significantly from those occurring in the individual layers. Under these conditions it is possible to obtain water meeting specified quality criteria, whereas in individual strata, concentrations could be entirely unacceptable.

7.9 WATER QUALITY CRITERIA

Whether a groundwater of a given quality is suitable for a particular purpose depends on the criteria or standards of acceptable quality for that use. Quality limits of water supplies for drinking water, industrial purposes, and irrigation apply to groundwater because of its extensive development for these purposes. Table 7.9.1 lists federal laws in the United States that regulate the quality of water.

7.9.1 Drinking Water Standards

Most drinking water supplies in the United States conform to standards established by the U.S. Environmental Protection Agency. The National Primary Drinking Water Standards Maximum Contaminant Levels (MCLs) and the Secondary Maximum Contaminant Levels (SMCLs) are provided at the following Internet address: http://www.epa.gov/safewater.

7.9.2 Industrial Water Criteria

It should be apparent that the quality requirements of waters used in different industrial processes vary widely. Thus, makeup water for high-pressure boilers must meet extremely exacting criteria, whereas water of as low a quality as seawater can be satisfactorily employed for cooling of condensers. Even within each industry, fixed criteria cannot be established; instead, only recommended limiting values or ranges can be stated.[3, 27] Salinity, hardness, and silica are three parameters that usually are important for industrial water.

Of almost equal importance for industrial purposes as quality of a water supply is the relative constancy of the various constituents. It is often possible to treat a poor quality water or adapt to it so that it is suitable for a given process, but if the quality fluctuates widely, continued attention and expense may be involved. Fluctuations of water temperature can be equally troublesome. From this standpoint, groundwater supplies are preferred to surface water supplies, which commonly display seasonal variations in chemical and physical quality. As a result, an adequate groundwater supply of suitable quality often becomes a primary consideration in selecting new industrial plant locations.

Table 7.9.1 Laws Regulating Water Quality

Federal laws in the United States that regulate the quality of water
National Environmental Policy Act of 1969 (P.L. 91–190)
Federal Water Pollution Control Act of 1972 (P.L. 92–500)
Clean Water Act Amendments of 1977 (P.L. 95–217)
Safe Drinking Water Act of 1974 (P.L. 93–523) and amendments
Resource Conservation and Recovery Act of 1976 (RCRA) (P.L. 94–580)
Comprehensive Environmental Response, Compensation and Liability Act of 1980 (CERCLA) (P.L. 96–510)
Superfund Amendments and Reauthorization Act of 1986 (SARA)
Surface Mining Control and Reclamation Act (SMCRA) (P.L. 95–87)
Uranium Mill Tailings Radiation and Control Act of 1978 (UMTRCA) (P.L. 95–604 as amended by P.L. 95–106 and P.L. 97–415)
Toxic Substances Control Act 1988 (TOSCA) (P.L. 94–469 as amended by P.L. 97–129)
Federal Insecticide, Fungicide and Rodenticide Act (FIFRA) (P.L. 92–516 as amended by P.L. 94–140, P.L. 95–396, P.L. 96–539, and P.L. 98–201)
Oil Pollution Act (OPA) of 1990

7.9.3 Irrigation Water Criteria

The suitability of groundwater for irrigation is contingent on the effects of the mineral constituents of the water on both the plant and the soil.[34, 48] Salts may harm plant growth physically by limiting the uptake of water through modification of osmotic processes, or chemically by metabolic reactions, such as those caused by toxic constituents. Effects of salts on soils, causing changes in soil structure, permeability, and aeration, indirectly affect plant growth. Specific limits of permissible salt concentrations for irrigation water cannot be stated because of the wide variations in salinity tolerance among different plants; however, field-plot studies of crops grown on soils that are artificially adjusted to various salinity levels provide valuable information relating to salt tolerance.

An important factor realting crop growth to water quality is drainage. If a soil is open and well drained, crops may be grown on it with the application of generous amounts of saline water; but, on the other hand, a poorly drained area combined with application of good-quality water may fail to produce as satisfactory a crop. Poor drainage permits salt concentrations in the root zone to build up to toxic proportions. Today, the necessity of adequate drainage is clearly recognized in order to maintain a favorable salt balance—where the total dissolved solids brought to the land annually by irrigation water is less than the total solids carried away annually by drainage water. It is believed that this factor accounted for the failure of many of the elaborate irrigation systems of historical times.

In place of rigid limits of salinity for irrigation water, quality is commonly expressed by classes of relative suitability. Most classification systems include limits on specific conductance (expressing total dissolved solids), sodium content, and boron concentration.

Sodium concentration is important in classifying an irrigation water because sodium reacts with soil to reduce its permeability (see following section). Soils containing a large proportion of sodium with carbonate as the predominant anion are termed *alkali soils*; those with chloride or sulfate as the predominant anion are *saline soils*. Ordinarily, either type of sodium-saturated soil will support little or no plant growth. Sodium content is usually expressed in terms of percent sodium (also known as sodium percentage and soluble-sodium percentage), defined by

$$\% \, Na = \frac{(Na + K)100}{Ca + Mg + Na + K} \qquad (7.9.1)$$

where all ionic concentrations are expressed in milliequivalents per liter. The Salinity Laboratory of the U.S. Department of Agriculture[34] recommends the sodium adsorption ratio (SAR) because of its direct relation to the adsorption of sodium by soil. It is defined by

$$SAR = \frac{Na}{\sqrt{(Ca + Mg)/2}} \qquad (7.9.2)$$

where the concentrations of the constituents are expressed in milliequivalents per liter.

Boron is necessary in very small quantities for normal growth of all plants, but in larger concentrations it becomes toxic. Quantities needed vary with the crop type; sensitive crops require minimum amounts whereas tolerant crops will reach maximum growth on several times these concentrations.

In investigations relating to quality of water for irrigation and the salinity of the soil solution, particularly regarding sampling programs, it is important to take cognizance of the salt distribution within the soil. As an illustration, the salt distribution under irrigated cotton plants is shown in Figure 7.9.1. It is apparent that the leaching effect of the irrigation water in the furrow, together with the movement toward the plants on the ridges, creates wide variations of salt concentration within short distances.

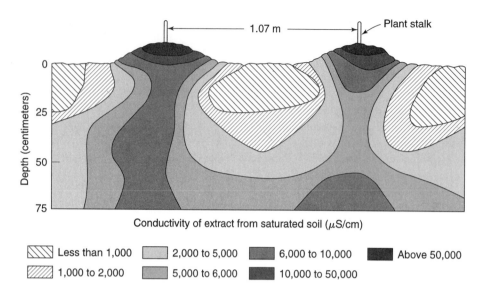

Figure 7.9.1. Salt distribution under furrow-irrigated cotton for soil initially salinized to 0.2 percent salt (3,100 μS/cm) and irrigated with water of medium salinity (after Wadleigh and Fireman;[44] reproduced by permission of the Soil Science Society of America).

7.10 CHANGES IN CHEMICAL COMPOSITION

As groundwater moves underground, it tends to develop a chemical equilibrium by chemical reactions with its environment. Effects of some equilibria in groundwater have important applications—for example, artificial recharge, movement of pollutants, and clogging of wells.

Chemical precipitation may remove ions in solution by forming insoluble compounds. Precipitation of calcium carbonate and release of dissolved carbon dioxide may result from a decrease in pressure and/or an increase in temperature. Ferrous iron in solution oxidizes on exposure to air and is deposited as ferric hydroxide.[22]

Ion exchange involves the replacement of ions adsorbed on the surface of fine-grained materials in aquifers by ions in solution. Because the exchange involves principally cations (sodium, calcium, and magnesium), the process is known as *base*, or *cation*, *exchange*. The direction of the exchange is toward an equilibrium of bases present in the water and on the finer materials of the aquifer. Base exchange is known to soften groundwater naturally and to produce, in coastal regions where seawater has entered an aquifer, groundwater having a quality other than a simple mixture of the two source waters (see Chapter 14).

Base exchange causes changes in the physical properties of soils. When high-sodium water is applied to a soil, the number of sodium ions combined with the soil increases, while an equivalent quantity of calcium or other ions is displaced. These reactions cause deflocculation and reduction of permeability. In the opposite case, where calcium is the dominant cation, the exchange occurs in the reverse direction, creating a flocculated and more permeable soil. The advantage of adding gypsum ($CaSO_4$) to a soil is that by base exchange the soil texture and drainability can be improved.

Chemical reduction of oxidized sulfur ions to sulfate ions or to the sulfide state occurs frequently in groundwater. The reaction is believed to take place in the presence of certain bacteria.[21] Typically, waters experiencing sulfate reduction have high bicarbonate and carbon dioxide content and contain hydrogen sulfide.

The equilibrium achieved by the various chemical reactions described above tends to produce a quality that remains stable with time because of the slow movement of groundwater and its long residence time within a given geologic formation. In general, quality variations are more noticeable in shallow aquifers where seasonal variations in recharge and discharge create corresponding fluctuations in salinity. Interestingly, freezing of shallow groundwater in

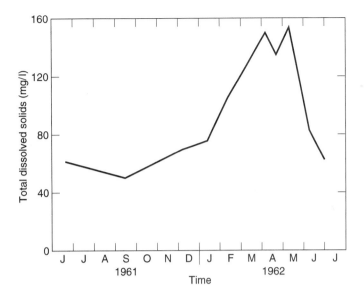

Figure 7.10.1. Variation in chemical quality of shallow groundwater at a site in northwestern Alaska due to seasonal effects of freezing (after Fuelner and Schupp[15]).

arctic regions can also cause seasonal changes in chemical quality, as shown in Figure 7.10.1. Here a marked increase in salinity during winter months results from freezing, which not only reduces the diluting effect of recharge from precipitation, but also selectively concentrates the mineral content in the unfrozen groundwater.

Finally, it should be noted that human acts can markedly change the natural quality of groundwater, usually for the worse. Figures 7.10.2–7.10.7 provide examples of these changes. The resulting degradation of groundwater has become an important international concern; Chapter 8 treats the subject of groundwater pollution.

7.11 DISSOLVED GASES

Although it is not generally recognized, most groundwater contains dissolved gases derived from natural sources. Those involved in the normal geochemical cycle of groundwater include the atmospheric gases: carbon dioxide (CO_2), oxygen (O_2), and nitrogen (N_2). Others derived from underground biochemical processes include the flammable gases methane (CH_4) and hydrogen sulfide (H_2S). With its distinctive rotten-egg odor, hydrogen sulfide is readily detected at concentrations of less than 1 mg/l and rarely accumulates to dangerous proportions.

Methane, which is colorless, tasteless, and odorless, occurs frequently in groundwater and can be a much more serious problem. The gas is a decomposition product of buried plant and animal matter in unconsolidated and geologically young deposits.[*] The minimum concentration of methane in water sufficient to produce an explosive methane–air mixture above the water from which it bubbles out of solution depends on the temperature, pressure, quantity of water pumped, and volume of air into which the gas evolves. Theoretically, water containing as little as 1 to 2 mg/l of methane can produce an explosion in a poorly ventilated air space. Fires and explosions in well pits, basements, and water tanks have occurred from methane emitted by groundwater; safety measures include analyses to detect the presence of the gas, aeration of water before use, and adequate ventilation where the water is being used. It is possible for people to suffocate in dug wells and pump pits where high methane concentrations form.

---

[*] For example, a buried oak log, only 1 ft in diameter and 5 ft long, can generate enough methane when it decomposes underground to form an explosive mixture in 400 to 1200 m³ of air.[4]

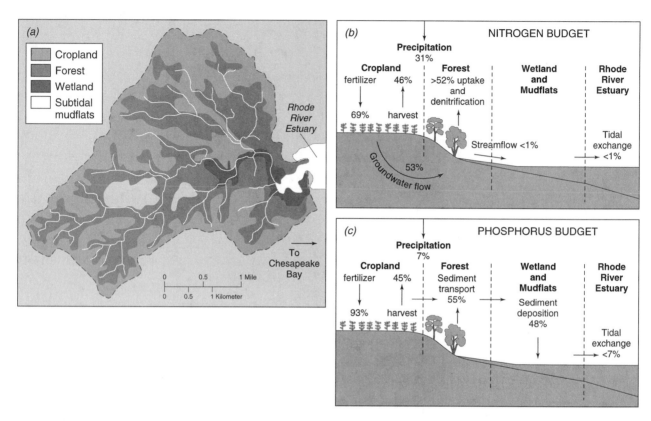

Figure 7.10.2. Forests and wetlands separate cropland from streams in the Rhode River watershed in Maryland (*a*). More than half of the nitrogen applied to cropland is transported by groundwater toward riparian forests and wetlands (*b*). More than half of the total phosphorus applied to cropland is transported by streams to wetlands and mudflats, where most is deposited in sediments (*c*) (modified from Correll,[11] as presented in Winter et al.[51]). (Reprinted by permission of the Estuarine Research Federation)

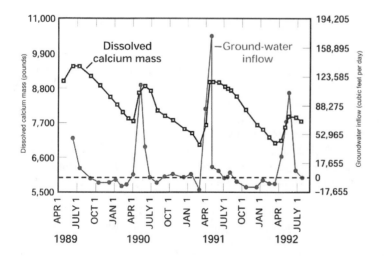

Figure 7.10.3. A large input of groundwater during spring supplies the annual input of calcium to Nevins Lake in the Upper Peninsula of Michigan (modified from Krabbenhoft,[24] as presented in Winter et al.[51]).

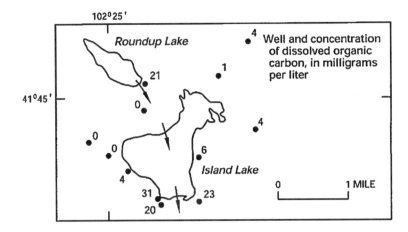

Figure 7.10.4. Seepage from lakes in the sand-hills of Nebraska causes plumes of dissolved organic carbon to be present in groundwater on the downgradient sides of the lakes (modified from J. W. LaBaugh,[23] as presented in Winter et al.[51]). (Reprinted with permission of Elsevier Science-NL, Amsterdam, The Netherlands)

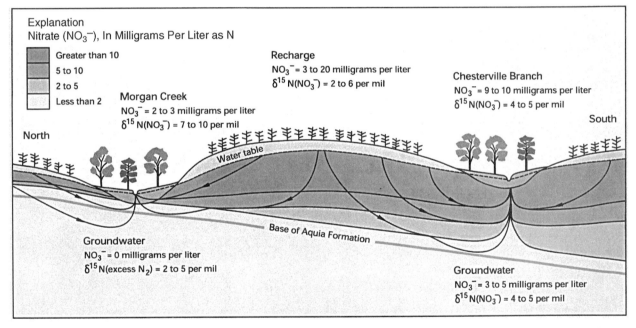

Figure 7.10.5. Denitrification had a greater effect on groundwater discharging to Morgan Creek than to Chesterville Branch in Maryland because a larger fraction of the local flow system discharging to Morgan Creek penetrated the reduced calcareous sediments near or below the bottom of the Aquia Formation than the flow system associated with the Chesterville Branch (modified from J. K. Bolke,[7] as presented in Winter et al.[51]).

7.12 TEMPERATURE

Variations in solar energy received at the earth's surface create periodicities, both diurnal and annual, in temperature below ground surface. The insulating qualities of the earth's crust rapidly damp the large temperature range found at ground surface so that only shallow groundwater displays any appreciable fluctuation in temperature.[19] In marked contrast to the large seasonal variation of surface water temperatures (except in tropical regions), groundwater temperatures tend to remain relatively constant—an important advantage for drinking water and industrial uses.

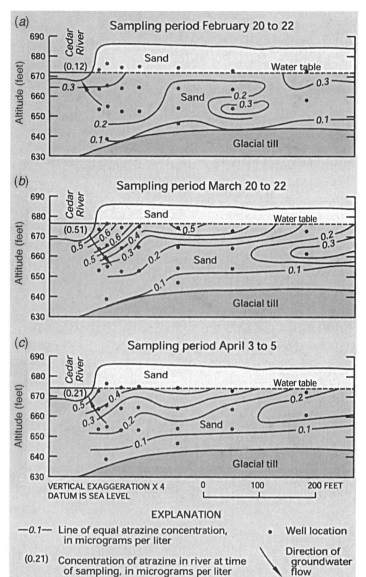

Figure 7.10.6. Concentrations of atrazine increased in the Cedar River in Iowa following applications of the chemical on agricultural areas upstream from a study site. During high streamflow (*b*), the contaminated river water moved into the alluvial aquifer as bank storage, contaminating groundwater. After the river level declined (*c*), part of the contaminated ground water returned to the river (modified from Squillace,[39] as presented in Winter et al.[51]).

An example of the variability of groundwater temperature in relation to that of a surface water body is shown by Figure 7.12.1 for the Schenectady, New York, well field adjoining the Mohawk River. The annual variation in groundwater temperature is only a small fraction of that observed in the river water. Furthermore, the tongue of larger temperature variation pointing toward the well field indicates the principal flow path of groundwater from the river to the Schenectady well field through a permeable zone of the shallow aquifer.[*]

---

[*]It should be noted generally that the velocity of groundwater can have an appreciable influence on the distribution of underground temperatures. Where sufficient field data are available, it may be possible to calculate flow velocities from temperature measurements and even to determine aquifer permeability from a combination of water level and temperature measurements.[37, 40]

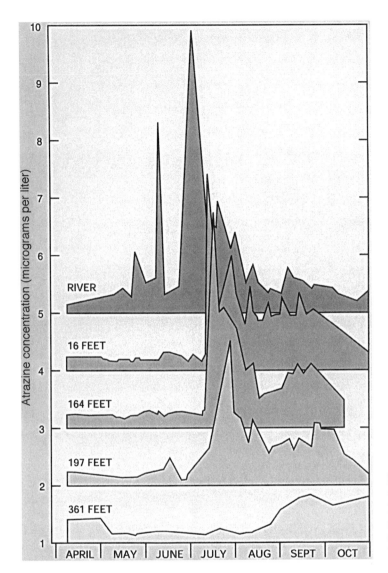

Figure 7.10.7. Pumping of municipal water-supply wells near Lincoln, Nebraska, has induced Platte River water contaminated with atrazine to flow into the aquifer. Distances shown are from river to monitoring well (modified from Duncan,[12] as presented in Winter et al[51]). (Used with permission)

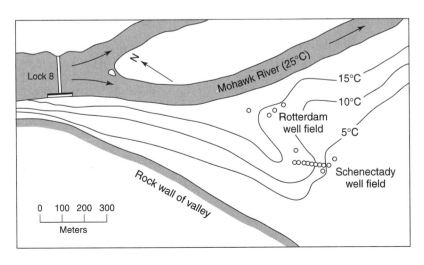

Figure 7.12.1. Map of annual variation in groundwater temperature near Mohawk River, New York. Numbered lines show locations of equal annual groundwater temperature change in °C; annual river water change was 25°C (after Winslow[50]).

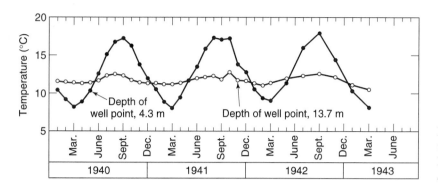

Figure 7.12.2. Monthly groundwater temperatures in two nearby wells on Long Island, New York (after *New York State Water Resources Comm. Bull.* 55).

Assuming that ground surface temperature is a sinusoidal function of time, the temperature T_z at a depth z can be given by

$$T_z = T_0 + Ae^{-z\sqrt{\pi/\alpha\tau}} \sin\left(\frac{2\pi t}{\tau} - z\sqrt{\pi/\alpha\tau}\right) \qquad (7.12.1)$$

where T_0 is the mean ground surface temperature, A is the amplitude of the surface temperature variation, τ is the oscillation period (one day or one year), α is the thermal diffusivity of the subsurface material (approximately 0.005 cm^2/sec), and t is time. The depth of nearly uniform temperature occurs at about 10 m in the tropics and increases to about 20 m in polar regions,[47] although influences such as rock type, elevation, precipitation, cloudiness, and wind can produce significant local deviations. Illustrative of annual fluctuations and the effect of depth are the temperatures in two nearby shallow wells shown in Figure 7.12.2.

Shallow groundwater temperatures can also be influenced by the type of overlying surface environment. Thus, a study on Long Island, New York,[31] showed that the mean annual temperature and the annual range in temperature of groundwater were larger beneath cleared areas than beneath wooded areas. The difference can be attributed to the absence of shade and the lack of an insulating layer of organic material on the ground in the cleared areas.

Below the zone of surface influence, groundwater temperatures increase approximately 2.9°C for each 100 m of depth in accordance with the geothermal gradient of the earth's crust.[*] Thus, groundwater pumped from deep wells is appreciably warmer than shallow groundwater. For example, if shallow groundwater has a temperature of 15°C, that obtained from an aquifer at 500 m might be expected to have a temperature near 30°C.

In the United States it has been found that the temperature of groundwater occurring at a depth of 10 to 20 m will generally be about 1° to 2°C higher than the local mean annual air temperature.[10] On this basis a groundwater temperature map, as shown in Figure 7.12.3, can be constructed from climatological data. As extremes, groundwater temperatures may vary from below freezing to above the boiling point in geothermal areas and in superheated water emerging from geysers.

7.13 SALINE GROUNDWATER

Many groundwaters contain dissolved salts in such concentrations as to make them unusable for ordinary water supply purposes. *Saline groundwater* is a general term referring to any groundwater containing more than 1,000 mg/l total dissolved solids. Various classification

---

[*]Most geothermal gradients measured in the United States fall within the range 1.8–3.6° C per 100 m. The heat flux from the interior of the earth is estimated to average 1.3 x 10^{-6} cal/cm^{2}sec.

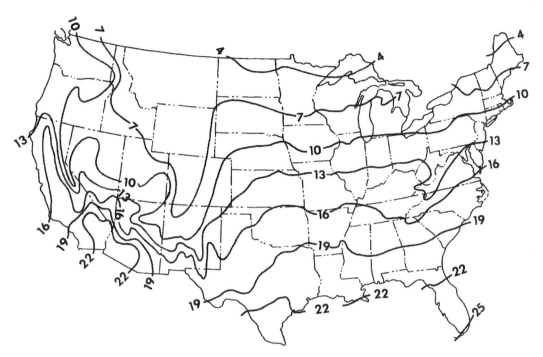

Figure 7.12.3. Approximate temperature of groundwater, in degrees Celsius, in the conterminous United States at depths of 10 to 25 m (Heath[20]).

schemes based on dissolved solids have been proposed; the simplicity of the one shown in Table 7.13.1 makes it particularly convenient.

Table 7.13.1 Classification of Saline Groundwater (after Carroll[9])

	Total dissolved solids (mg/l)
Fresh water	0–1,000
Brackish water	1,000–10,000
Saline water	10,000–100,000
Brine	>100,000

A map of the United States showing the minimum depth to aquifers containing groundwater exceeding 1,000 mg/l is given in Figure 7.13.1. Approximately two-thirds of the country is known to contain such waters; blank areas indicate areas where either well yields are less than 40 m^3/day or no data on saline groundwater are available. Four types of occurrence are recognized: connate water, intruded seawater (see Chapter 14), water salinized by contact with soluble salts in the formation where it is situated, and water in regions with shallow water tables where evapotranspiration concentrates the salts in solution.

Although saline groundwater has traditionally been regarded as an undesirable resource, modern technological advances may reverse this role. Rapid advances in desalination techniques suggest that saline groundwater may be a potentially important water supply source where shortages are imminent. Industrial use of saline groundwater for cooling purposes has yet to be exploited fully; high costs, involving deep wells, corrosion control, and disposal, have hindered such developments. Saline aquifers also serve useful purposes for temporary storage of fresh water and energy in the form of hot water and for disposal of wastewater (see Chapter 13).

Figure 7.13.1. Depth to saline groundwater in the United States (after Feth et al.[14]).

PROBLEMS

7.1.1 Develop a report on the (proposed) standard for arsenic. Discuss the technologies and costs for removal of arsenic. What importance does this have in groundwater studies? You may want to start with the following Web site: http://www.epa.gov/safewater/arsenic.html.

7.1.2 Develop a report on the connection between surface-water quality and groundwater quality in a karst aquifer. You might want to start at the Web site for the *U.S. Geological Circular* 1186.[1]

7.4.1 In *U.S. Geological Survey Water Resources Investigations Report 98-4133, Estimating Ground-Water Exchange with Lakes Using Water-Budget and Chemical Mass-Balance Approaches for Ten Lakes in Ridge Areas of Polk and Highland Counties, Florida,* by Sacks et al,[35] a chemical mass balance was used to estimate the groundwater exchange with lakes. This report is available at http://fl.water.usgs.gov. Develop a discussion of the method used to perform the chemical mass balance. How do the results compare with using the water-budget analysis?

7.9.1 Go to the U.S. EPA Web site for Drinking Water Standards and print out the standards. Discuss one organic and one inorganic chemical, defining the health effects of each, the source of

the contaminant, and the MCL. The Web site is http://www.epa.gov/safewater/standards.html.

7.9.2 Develop a short report on how the Federal Water Pollution Control Act of 1972 relates to the water quality of groundwater.

7.9.3 Develop a short report on how the Resource and Recovery Act of 1976 relates to the water quality of groundwater.

7.9.4 Develop a short report on how the Surface Mining Control and Reclamation Act relates to the water quality of groundwater.

7.9.5 Develop a short report on how the Toxic Substance Control Act relates to the water quality of groundwater.

7.9.6 Develop a short report on how the Superfund Amendments and Reauthorization Act relates to the water quality of groundwater.

7.9.10 Discuss the laws and regulations that relate to groundwater quantity and quality in your state or region.

7.10.1 Go to the U.S. Geological Web site http://water.usgs.gov/pubs. Find a water quality study on an area near you and summarize the water quality of the area.

7.10.2 Go to the U.S. Geological Survey Web site http://water. usgs.gov/pubs/circ/circ1207/index.html and develop a discussion of the water quality in southern Florida.

7.10.3 Discuss the concentrations of total phosphrous (TP) at the Southern Florida National Water Quality Assessment (NAWQA) Program sites. What is the major source(s) of the high TP? The program Web site is http://water.usgs.gov/nawqa.

7.10.4 Develop a report on the water quality factors affecting groundwater sustainability. You might want to start your search for information in *U.S. Geological Survey Circular* 1186, *Sustainability of Ground-Water Resources,* found at http://water.usgs.gov/pubs/circ/circ1186.[1]

REFERENCES

1. Alley, W. M., T. E. Reilly, and O. L. Franke, *Sustainability of Ground-Water Resources, U.S. Geological Survey Circular* 1186, http://water.usgs.gov/pubs/circ/circ1186, U.S. Geological Survey, Denver, CO, 1999.

2. Amer. Public Health Assoc., Amer. Water Works Assoc., and Water Pollution Control Fed., *Standard Methods for the Examination of Water and Wastewater,* 14th ed., Amer. Public Health Assoc., Washington, DC, 1200 pp., 1975.

3. Amer. Soc. Testing Matls., *Manual on Industrial Water and Industrial Waste Water,* 2nd ed., Philadelphia, 992 pp., 1966.

4. Anon., Gas in ground water, *Jour. Amer. Water Works Assoc.,* v. 61, pp. 413–414, 1969.

5. Back, W., and B. B. Hanshaw, Chemical geohydrology, in *Advances in Hydroscience*, V. T. Chow, ed., v. 2, Academic Press, New York, pp. 49–109, 1965.

6. Baker, M. N., *The Quest for Pure Water,* Amer. Water Works Assoc., New York, 527 pp., 1948.

7. Bolke, J. K., and J. M. Denver, Combined use of ground-water dating, chemical, and isotopic analyses to resolve the history and fate of nitrate contamination in two agricultural watersheds, Atlantic coastal plain, Maryland, *Water Resources Research,* v. 31, no. 9, pp. 2319–2337, 1995.

8. Carpentar, M. C., South-Central Arizona, in *Land Subsidence in the United States*, D. Galloway, D. Jones, and S. Ingebritsen, eds., *U.S. Geological Survey Circular* 1182, http://water.usgs.gov/pubs/circ/circ1182, Denver, CO, 1999.

9. Carroll, D., *Rainwater as a Chemical Agent of Geologic Processes—A Review, U.S. Geological Survey Water-Supply Paper* 1535-G, 18 pp., 1962.

10. Collins, W. D., *Temperature of Water Available for Industrial Use in the United States, U.S. Geological Survey Water-Supply Paper* 520-F, pp. 97–104, 1925.

11. Correll, D. L., T. E. Jordan, and D.E. Weller, Nutrient flux in a landscape—Effects of coastal land use and terrestial community mosaic on nutrient transport to coastal waters, *Estuaries,* v. 15, no. 4, pp. 431–442, 1992.

12. Duncan, D., D. T. Pederson, T. R. Shepherd, and J. D. Carr, Atrazine used as a tracer of induced recharge, *Ground Water Monitoring Review,* v. 11, no. 4, pp. 144–150, 1991.

13. Eriksson, E., Atmospheric transport of oceanic constituents in their circulation in nature, *Tellus,* v. 11. pp. 1–72, 1959.

14. Feth, J. H., et al., *Preliminary Map of the Conterminous United States Showing Depth to and Quality of Shallowest Ground Water Containing More than 1000 Parts per Million Dissolved Solids, U.S. Geological Survey Hydrologic Inv. Atlas* HA-199, 31 pp., 1965.

15. Feulner, A. J., and R. G. Schupp, *Seasonal Changes in the Chemical Quality of Shallow Ground Water in Northwestern Alaska, U.S. Geological Survey Professional Paper* 475-B, pp. 189–191, 1963.

16. Focazio, M. J., A. H. Welch, S. A. Watkins, D. R. Helsel, and M. A. Horn, *A Retrospective Analysis on the Occurrence of Arsenic in Ground Water Resources of the United States and Limitations in Drinking-Water Resources of the United States and Limitations in Drinking-Water-Supply Characterizations, U.S. Geological Survey Water-Resources Investigations Report* 99-4279, 1999.

17. Garrels, R. M. and F. T. MacKenzie, Origin of the chemical compositions of some springs and lakes, *Advances in Chemistry* Series, No. 67, American Chemical Society, Chapter 10, pp. 222–242, 1967.

18. Graham, M. J., J. A. Thomas, and F. B. Metting, Groundwater, in *Water Resources Handbook*, L.W. Mays, ed., McGraw-Hill, New York, 1996.

19. Heath, R. C., *Seasonal Temperature Fluctuations in Surficial Sand near Albany, New York, U.S. Geological Survey Professional Paper* 475-D, pp. 204–208, 1964.

20. Heath, R. C., *Basic Ground-Water Hydrology, U.S. Geological Survey Water-Supply Paper* 2220, http://water.usgs.gov/pubs/wsp/wsp2220, Denver, CO, 1998.

21. Hem, J. D., *Study and Interpretation of the Chemical Characteristics of Natural Water,* 2nd ed., *U.S. Geological Survey Water-Supply Paper* 1473, 363 pp., 1970.

22. Hem, J. D., et al., *Chemistry of Iron in Natural Water, U.S. Geological Survey Water-Supply Paper* 1459, 268 pp., 1962.

23. LaBaugh, J. W., Limnological characteristics of selected lakes in the Nebraska sandhills, U.S.A., and their relation to chemical characteristics of adjacent groundwater, *Journal of Hydrology,* v. 86, pp. 279–298, Elsevier Science-NL, Amsterdam, The Netherlands, 1986.

24. Krabbenhoft, D. P., and Webster, K. E., Transient hydrogeological controls on the chemistry of a seepage lake, *Water Resources Research,* v. 31, no. 9, pp. 2295–2305, 1995.

25. Loewengart, S., Airborne salts—The major source of the salinity of waters in Israel, *Bull. Research Council Israel,* v. 10G, pp. 183–206, 1961.

26. Logan, J., Estimation of electrical conductivity from chemical analysis of natural waters, *Jour. Geophysical Research,* v. 66, pp. 2479–2483, 1961.

27. McKee, J. E., and H. W. Wolf, eds., *Water quality criteria,* Publ. no. 3-A, California State Water Resources Control Board, Sacramento, 548 pp., 1963.

28. National Research Council, *Arsenic in Drinking Water,* National Academy Press, Washington, DC, 1999.

29. Nordstrom, D. K., J. W. Ball, R. J. Donahoe, and D. Whittemore, Groundwater chemistry and water–rock interactions stripa, *Geochim. Cosmochim. Acta,* v. 53, pp. 1727–1740, 1989.

30. Piper, A. M., A graphic procedure in the geochemical interpretation of water-analyses, *Trans. Amer. Geophysical Union,* v. 25, pp. 914–928, 1944.

31. Pluhowski, E. J., and I. H. Kantrowitz, *Influence of Land-Surface Conditions on Ground-Water Temperatures in Southwestern Suffolk County, Long Island, New York, U.S. Geological Survey Professional Paper* 475-B, pp. 186–188, 1963.

32. Plummer, L. N., J. F. Busby, R. W. Lee, and B. B. Hanshaw, Geochemical modeling of the Madison aquifer in parts of Montana, Wyoming, and South Dakota, *Water Resources Research,* v. 26, no. 9, pp. 1981–2014, 1990.

33. Rainwater, F. H., and L. L. Thatcher, *Methods for Collection and Analysis of Water Samples, U.S. Geological Survey Water-Supply Paper* 1454, 301 pp., 1960.

34. Richards, L. A., ed., *Diagnosis and Improvement of Saline and Alkali Soils,* Agric. Handbook 60, U.S. Dept. Agric., Washington, DC, 160 pp., 1954.

35. Sacks, L. A., A. Swancar, and T. M. Lee, *Estimating Ground-Water Exchange with Lakes Using Water-Budget and Chemical Mass-Balance Approaches for Ten Lakes in Ridge Areas of Polk and Highlands Counties, Florida, U.S. Geological Survey Water Resources Investigations Report* 98-4133, Tallahassee, FL, 1998.

36. Sawyer, C. N., and P. L. McCarty, *Chemistry for Sanitary Engineers,* 2nd ed., McGraw-Hill, New York, 518 pp., 1967.

37. Schneider, R., *An Application of Thermometry to the Study of Ground Water, U.S. Geological Survey Water-Supply Paper* 1544-B, 16 pp., 1962.

38. Schoeller, H., *Les eaux souterraines,* Masson & Cie, Paris, 642 pp., 1962.

39. Squillace, P. J., E. M. Thurman, and E. T. Furlong, Groundwater as a nonpoint source of atrazine and deethylatrazine in a river during base flow conditions, *Water Resources Research,* v. 29, no. 6, pp. 1719–1729, 1993.

40. Stallman, R. W., *Computation of Ground-Water Velocity from Temperature Data, U.S. Geological Survey Water-Supply Paper* 1544-H, pp. 36–46, 1963.

41. Stiff, H. A., Jr., The interpretation of chemical water analysis by means of patterns, *Jour. Petr. Technology,* v. 3, no. 10, pp. 15–17, 1951.

42. Taylor, C. J., and W. M. Alley, *Ground-Water-Level Monitoring and the Importance of Long-Term Water-Level Data, U.S. Geological Survey Circular* 1217, http://water.usgs.gov/pubs/circ/circ1217, Denver, CO, 2001.

43. Thomas, J. M., A. H. Welch, and A. M. Preissler, Geochemical evolution of ground water in Smith Creek Valley—A hydrologic ally closed basin in Central Nevada, U.S.A., *Appl. Geochem.,* v. 4, pp. 493–510, 1989.

44. Wadleigh, C. H., and M. Fireman, Salt distribution under furrow and basin irrigated cotton and its effect on water removal, *Soil Sci. Soc. Amer. Proc.,* v. 13, pp. 527–530, 1948.

45. Welch, A. H., S. A. Watkins, D. R. Helsel, and M. J. Focazio, *Arsenic in Ground-Water Resources of the United States, U.S. Geological Survey Fact Sheet* FS-063-00, May 2000.

46. White, A. F., *Geochemistry of Groundwater Associated with Tuffaceous Rocks, Oasis Valley, Nevada, U.S. Geological Survey Professional Paper* 712-E, Washington, DC, 1979.

47. White, D. E., et al., *Data of Geochemistry—Chemical Composition of Subsurface Waters,* 6th ed., *U.S. Geological Survey Professional Paper* 440-F, 67 pp., 1963.

48. Wilcox, L. V., *Classification and Use of Irrigation Waters,* U.S. Dept. Agric. Circular 969, Washington, DC, 19 pp., 1955.

49. Winograd, I. J., and F. J. Pearson, Jr., Major Carbon-14 anomaly in a region carbonate aquifer: Possible evidence for megascale channeling, south Central Great Basin, *Water Resources Research,* v. 12, no. 6, pp. 1125–1143, 1976.

50. Winslow, J. D., *Effect of Stream Infiltration on Ground-Water Temperatures near Schenectady, N.Y., U.S. Geological Survey Professional Paper* 450-C, pp. 125–128, 1962.

51. Winter, T. C., J. W. Harvey, O. L. Franke, and W. M. Alley, *Groundwater and Surface Water: A Single Source, U.S. Geological Survey Circular* 1139, http://water.usgs.gov/pubs/circ/circ1139, Denver, CO, 1998.

52. World Health Organization, *Arsenic in drinking water,* Fact sheet 210, http://www.who.int/inf-fs/en/fact210.html, 1999.

Chapter 8

Pollution of Groundwater

Groundwater pollution may be defined as the artificially induced degradation of natural groundwater quality. Pollution can impair the use of water and can create hazards to public health through toxicity or the spread of disease.[*] Most pollution originates from the disposal of wastewater following the use of water for any of a wide variety of purposes. Thus, a large number of sources and causes can modify groundwater quality, ranging from septic tanks to irrigated agriculture. In contrast with surface water pollution, subsurface pollution is difficult to detect, and is even more difficult to control, and may persist for decades. With the growing recognition of the importance of underground water resources, efforts are increasing to prevent, reduce, and eliminate groundwater pollution.

8.1 POLLUTION IN RELATION TO WATER USE

The possible pollutants in groundwater are virtually limitless. The sources and causes of groundwater pollution are closely associated with human use of water. A complex and interrelated series of modifications to natural water quality is created by the diversity of human activities impinging on the hydrologic cycle.

The principal sources and causes of groundwater pollution can be categorized— municipal, industrial, agricultural, and miscellaneous. Most pollution stems from disposal of wastes on or into the ground. Methods of disposal include placing wastes in percolation ponds, on the ground surface (spreading or irrigation), in seepage pits or trenches, in dry streambeds, in landfills, into disposal wells, and into injection wells.

All sources and causes of pollution can be classified as to their geometry. A point source originates from a singular location, a line source has a predominantly linear alignment, and a diffuse source occupies an extensive area that may or may not be clearly defined.

Groundwater contamination scenarios can be segregated into two categories: *point sources* and *nonpoint* (area, or distributed) *sources*. Point sources include storage tanks (both

---

[*]One need not go far back in history to find instances of major diseases transmitted by groundwater. A classic example was the outbreak of cholera in London, England, in 1854. Dr. John Snow, while investigating the epidemic, noted with considerable astuteness that more than 500 persons died from cholera in ten days within 250 yards of a public water supply well; this led to his conclusion:

> The result of the inquiry then was, that there had been no particular outbreak or increase of cholera, in this part of London, except among the persons who were in the habit of drinking the water of the above-mentioned pump-well. I had an interview with the Board of Guardians of St. James parish on the evening of Thursday, 7th September, and represented the above circumstances to them. In consequence of what I said, the handle of the pump was removed on the following day (On the Mode of Communication of Cholera, 1855).

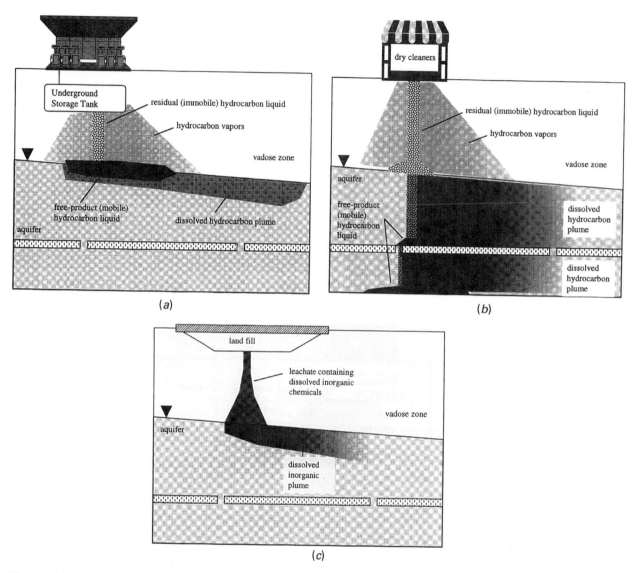

Figure 8.1.1. Example contaminant release scenarios: (*a*) LNAPL release, (*b*) DNAPL release, (*c*) dissolved contaminant release (Johnson[51]).

underground and above-ground), landfills, pipeline releases, chemical manufacturing locations, petroleum refining locations, wood treating facilities, and many others are considered to be point sources. Nonpoint (area or distributed) sources include agricultural activities.

Point-source groundwater contamination problems can be divided further into three main categories: (1) light nonaqueous phase liquids (LNAPLs), (2) dense nonaqueous phase liquids (DNAPLs), and (3) inorganics and other dissolved constituents. Example release scenarios for LNAPLs, DNAPLS, and dissolved inorganics are illustrated in Figure 8.1.1.

In their pure liquid form, LNAPLs are less dense than water and DNAPLs are denser than water. LNAPL sites are caused by the release of petroleum products or crude oil (e.g. service stations, refineries, and pipeline spills). DNAPL sites have been caused by dry cleaning, aviation, automobile, and electric circuit board operations, which historically have used chlorinated solvents, such as trichloroethylene (TCE) and perchloroethylene (PCE). Inorganics and other dissolved constituents include metals and salts. Sources that have added contaminants to groundwater in dissolved form include but are not limited to mining operations, electroplating operations, leaking wastewater treatment facilities, and landfills.

In the following sections, the principal sources and causes of pollution are briefly described with regard to their occurrence and their effects on groundwater quality. Considerable confusion exists in the literature over the distinction between *pollution* and *contamination*. Here *pollution* shall signify any degradation of natural water quality, while *contamination* shall be reserved for pollution that constitutes a hazard to human health.

8.2 MUNICIPAL SOURCES AND CAUSES

8.2.1 Sewer Leakage

Sanitary sewers are intended to be watertight; however, in reality, leakage of sewage into the ground is a common occurrence, especially from old sewers. Leakage may result from poor workmanship, defective sewer pipe, breakage by tree roots, ruptures from heavy loads or soil slippage, fractures from seismic activity, loss of foundation support, shearing due to differential settlement at manholes, and infiltration causing sewage flow into abandoned sewer laterals.[60] Because suspended solids in sewage tend to clog sewer cracks and because the surrounding soil tends to become clogged due to anaerobic conditions, leakage from minor sewer openings is often small.

Sewer leakage can introduce high concentrations of BOD (biochemical oxygen demand), COD (chemical oxygen demand), nitrate, organic chemicals, and possibly bacteria into groundwater. Where sewers serve industrial areas, heavy metals such as arsenic, cadmium, chromium, cobalt, copper, iron, lead, manganese, and mercury may enter the wastewater.

8.2.2 Liquid Wastes

Wastewater in an urban area may originate from domestic uses (see Figure 8.2.1), industries, or storm runoff. Most of this highly variable mix of waters receives some degree of treatment and is then discharged into surface waters. There is an increasing trend for treated wastewater to be recharged into the ground where it mingles with naturally occurring groundwater and subsequently becomes available for reuse.

Land application of municipal effluent is accomplished by one of three methods (see Figure 8.2.2): irrigation, infiltration–percolation, or overland flow. The selection of a method at a given site is primarily governed by the drainability of the soil, because this property determines the allowable liquid loading rate. In irrigation systems, wastewater is applied by spraying, ridge and furrow (see Chapter 13), and flooding; some water is lost by evapotranspiration. For the infiltration–percolation method, effluent is applied by spreading in basins or by spraying; almost all of the water so applied reaches the groundwater. In the overland flow technique, wastewater is sprayed over the upper reaches of sloped terraces and allowed to flow across a vegetated surface to runoff collection ditches; percolation to groundwater is minor in overland flow because surface runoff and evapotranspiration account for most of the applied water.

Municipal wastewaters can introduce bacteria, viruses, and inorganic and organic chemicals into groundwater. Where the recharged water is later extracted for potable use, concerns exist regarding health aspects of this reclaimed water, particularly involving viruses, trace elements and heavy metals, and stable organics. Furthermore, chlorination of wastewater effluent can produce additional potential pollutants.

Shallow wells are widely employed to place surface runoff and sometimes treated municipal wastewater underground in freshwater aquifers (see Chapter 13). Such *disposal*

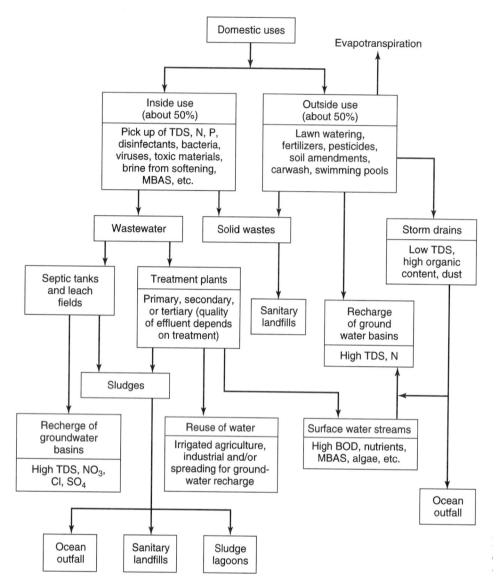

Figure 8.2.1. Domestic uses of water and their effects on water quality (after Hassan[45]).

wells have been criticized from a health standpoint because of the potential for pollutants to be released directly into an aquifer. The problem is most critical where disposal wells are near pumping wells (Figure 8.2.3) and where the beneficial effects of water passing through fine-grained materials may be absent, such as in basalt and limestone aquifers.

8.2.3 Solid Wastes

The land disposal of solid wastes creates a potential important source of groundwater pollution. A *landfill* may be defined as any land area serving as a depository of urban, or municipal, solid waste. Many landfills are simply refuse dumps; only a fraction can be regarded as sanitary landfills, indicating that they were designed and constructed according to engineering specifications. Leachate from a landfill can pollute groundwater if water moves through the fill material.[110] Possible sources of water include precipitation, surface water infiltration,

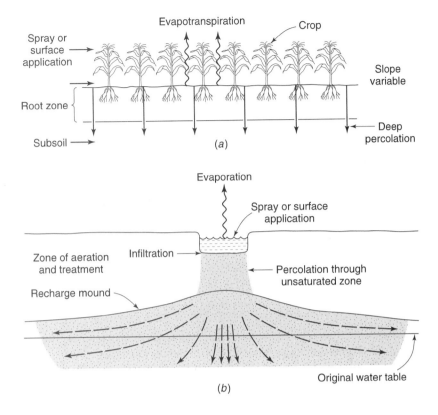

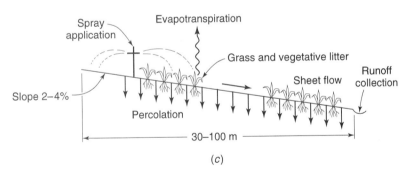

Figure 8.2.2. Methods of land application of municipal wastewaters. (*a*) Irrigation (*b*) Infiltration–percolation (*c*) Overland flow (after U.S. Environmental Protection Agency[96])

percolating water from adjacent land, and groundwater in contact with the fill. Ordinary mixed refuse usually has a moisture content less than that of field capacity; therefore, leachate from a landfill can be minimized if water from the above sources can be kept from the fill material. In a properly constructed sanitary landfill, any leachate generated can be controlled and prevented from polluting groundwater. In addition, it is assumed that the landfill is properly located, operated, and monitored.

The problem of pollution from landfills is greatest where high rainfall and shallow water tables occur. Important pollutants frequently found in leachate include BOD, COD, iron, manganese, chloride, nitrate, hardness, and trace elements. Hardness, alkalinity, and total dissolved solids are often increased, while generation of gases, such as methane, carbon dioxide, ammonia, and hydrogen sulfide, are further by-products of landfills.

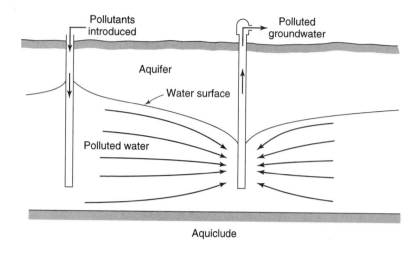

Figure 8.2.3. Diagram showing movement of pollutants from a disposal well to a nearby pumping well (after Deutsch[22]).

8.3 INDUSTRIAL SOURCES AND CAUSES

8.3.1 Liquid Wastes

The major uses of water in industrial plants are for cooling, sanitation, and manufacturing and processing. The quality of the wastewater varies with type of industry and type of use. A generalized flow diagram of industrial water use and its effects on water quality is shown in Figure 8.3.1. Cooling water that is softened before use to inhibit scale formation produces wastewaters with salts and heat as important pollutants. Groundwater pollution can occur where industrial wastewaters are discharged into pits, ponds, or lagoons, thereby enabling the wastes to migrate down to the water table.

Cooling water is sometimes recharged underground through shallow disposal wells because its quality, except for the addition of heat, may be unimpaired.[83] The disposal of hazardous and toxic industrial wastes is sometimes accomplished by means of deep injection wells that place the fluids into saline water formations far below developed freshwater aquifers.[53, 107]

8.3.2 Tank and Pipeline Leakage

Underground storage and transmission of a wide variety of fuels and chemicals are common practices for industrial and commercial installations. These tanks and pipelines are subject to structural failures so that subsequent leakage becomes a source of groundwater pollution. Petroleum and petroleum products are responsible for much of the pollution. Leakage is particularly frequent from gasoline station and home fuel oil tanks.[51, 63, 64, 76, 108] An immiscible liquid, such as oil, leaking underground moves downward through permeable soils until it reaches the water table. Thereafter it spreads to form a layer on top of the water table and migrates laterally with groundwater flow (see Figure 8.3.2a). Liquid radioactive wastes are sometimes stored in underground tanks; leakage from such installations, which has occurred, can cause serious pollution problems in local groundwater. Pumping wells can be used to create drawdown cones to contain the leakage, as shown in Figure 8.3.2b, and to remove a mixture of gasoline and water.

8.3.3 Mining Activities

Mines can produce a variety of groundwater pollution problems.[7, 26, 68] Pollution depends on the material being extracted and the milling process. Coal, phosphate, and uranium mines are

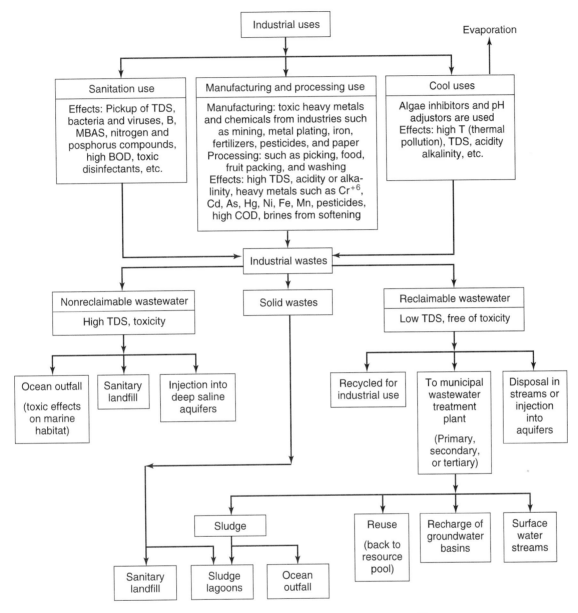

Figure 8.3.1. Industrial uses of water and their effects on water quality (after Hassan[45]).

major contributors; metallic ores for production of iron, copper, zinc, and lead are also important; stone, sand, and gravel quarries, although numerous, are chemically much less important. Both surface and underground mines invariably extend below the water table so that dewatering to expand mining is common. Water so pumped may be highly mineralized and is frequently referred to as *acid mine drainage*. Typical characteristics include low pH and high iron, aluminum, and sulfate.[67]

Coal deposits are often associated with pyrite (FeS_2). This is stable for conditions below the water table, but if the water table is lowered, oxidation occurs. Oxidation of pyrite followed by contact with water produces ferrous sulfate ($FeSO_4$) and sulfuric acid (H_2SO_4) in solution;

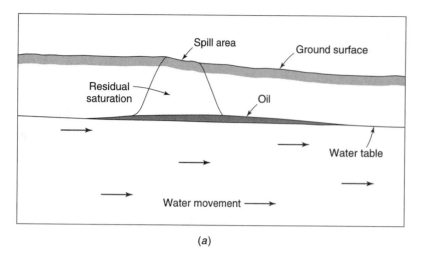

(a)

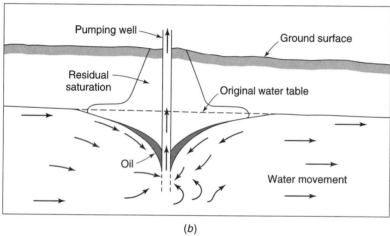

(b)

Figure 8.3.2. Oil pollution of groundwater and control by a pumping well. (*a*) Fluid oil floating on water table and elongating in direction of groundwater flow (*b*) Oil on water table contained in cone of depression created by pumping well (after Engrng. and Tech. Research Comm.[27])

groundwater intermingling with this water will have a reduced pH and an increase in iron and sulfate contents. Pollution of groundwater can also result from the leaching of old mine tailings and settling ponds; therefore, pollution problems can be associated with both active and abandoned mines.

8.3.4 Oil-Field Brines

The production of oil and gas is usually accompanied by substantial discharges of wastewater in the form of brine.[52] Constituents of brine include sodium, calcium, ammonia, boron, chloride, sulfate, trace metals, and high total dissolved solids. In the past, oil-field brine disposal was handled by discharge to streams or "evaporation ponds." In both instances, brine-polluted aquifers became commonplace in oil production areas as the infiltrating water reached the underlying groundwater. Today, such disposal methods are prohibited by most regulatory agencies; however, regulation is often ineffective, so that many brine-affected areas remain and will persist for years into the future. Figure 8.3.3 provides a graphic example of aquifer pollution from an oil-field brine disposal pit.

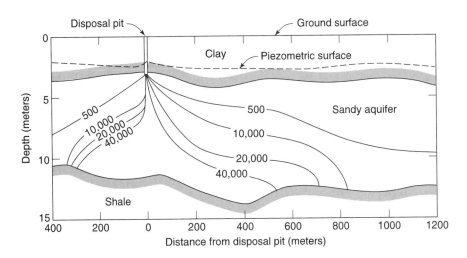

Figure 8.3.3. Distribution of saline water in confined aquifer resulting from an oil-field brine disposal pit in southwestern Arkansas. Numbered lines are isochlors in mg/l (after Fryberger[36]).

8.4 AGRICULTURAL SOURCES AND CAUSES

8.4.1 Irrigation Return Flows

Approximately one-half to two-thirds of the water applied for irrigation of crops is consumed by evapotranspiration; the remainder, termed *irrigation return flow*, drains to surface channels or joins the underlying groundwater. Irrigation increases the salinity of irrigation return flow from three to ten times that of the applied water. The degradation results from the addition of salts by dissolution during the irrigation process, from salts added as fertilizers or soil amendments, and from the concentration of salts by evapotranspiration (see Figure 8.4.1). Principal cations include calcium, magnesium, and sodium; major anions include bicarbonate, sulfate, chloride, and nitrate. Because irrigation is the primary use for water in arid and semiarid regions, irrigation return flow can be the major cause of groundwater pollution in such regions.

8.4.2 Animal Wastes

Where animals are confined within a limited area, as for beef or milk production, large amounts of wastes are deposited on the ground. Thus, for the 120 to 150 days that a beef animal remains in a feedlot, it will produce over a half-ton of manure on a dry-weight basis. With thousands of animals in a single feedlot, the natural assimilative capacity of the soil can become overtaxed. Storm runoff in contact with the manure carries highly concentrated pollutants to surface and subsurface waters. Animal wastes may transport salts, organic loads, and bacteria into the soil. Nitrate-nitrogen is the most important persistent pollutant that may reach the water table.[1, 39, 79]

8.4.3 Fertilizers and Soil Amendments

When fertilizers are applied to agricultural land, a portion usually leaches through the soil and to the water table. The primary fertilizers are compounds of nitrogen, phosphorus, and potassium. Phosphate and potassium fertilizers are readily adsorbed on soil particles and seldom constitute a pollution problem. But nitrogen in solution is only partially used by plants or adsorbed by the soils, and it is the primary fertilizer pollutant. Fertilizers are extensively used and will undoubtedly increase in the future.

Soil amendments are applied to irrigated lands to alter the physical or chemical properties of the soil. Lime, gypsum, and sulfur are widely used for this purpose; substantial amounts of these soil amendments may eventually leach to the groundwater, thereby increasing its salinity.

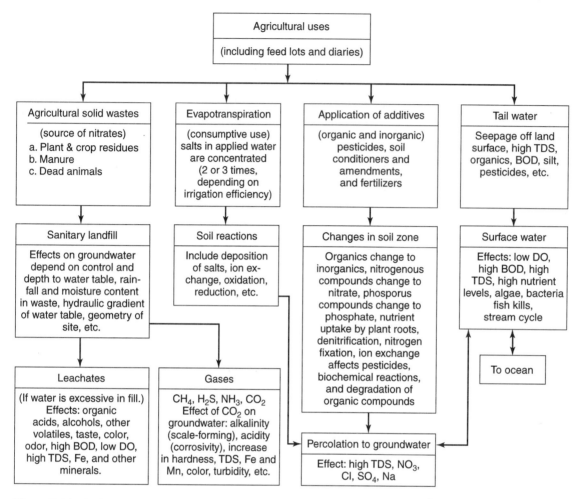

Figure 8.4.1. Agricultural uses of water and their effects on water quality (after Hassan[45]).

8.4.4 Pesticides, Insecticides, and Herbicides

Pesticides can be significant in agricultural areas as a diffuse source of groundwater pollution.[57, 94, 104] The terms *pesticide*, *insecticide*, and *herbicide* are broadly interpreted here to embrace any chemical applied to control, destroy, or mitigate pests. The presence of these materials in groundwater, even in minute concentrations, can have serious consequences in relation to the potability of the water. The impact of these chemicals on groundwater quality depends on the properties of these chemicals' residue, rainfall or irrigation rates, and soil characteristics.[91] Most pesticides are relatively insoluble in water, while others are readily adsorbed by soil particles or are subject to microbial degradation.

8.5 MISCELLANEOUS SOURCES AND CAUSES

8.5.1 Urbanization

Shallow groundwater quality can be affected by residential and commercial development as illustrated in Figure 8.5.1. The U.S. Geological Survey's National Water Quality Assessment (NAWQA) program (http://water.usgs.gov/nawqa/) seeks to determine how shallow groundwater quality is affected by development (Squillace and Price[85]). Residential developments have

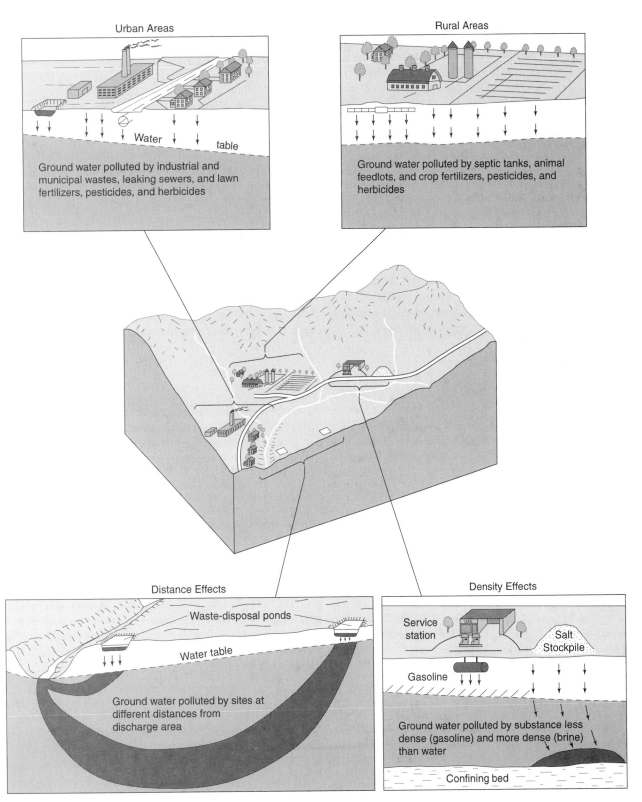

Figure 8.5.1. Groundwater pollution occurs in both urban and rural areas and is affected by differences in chemical composition, biological and chemical reactions, density, and distance from discharge areas (Heath[46]).

taken up large tracts of land, and as a consequence, have widespread influence on the quality of water that recharges aquifers and discharges into streams, lakes, and wetlands. Studies of various urbanizing areas have been performed in the past by Mullaney and Grady,[70] Stackelberg,[86] and Thomas.[89] The study by Thomas[89] of residential areas near Detroit, Michigan, concluded that the young and shallow waters have significantly higher median concentrations of nitrate, chloride, dissolved solids, sodium, calcium, and potassium than old and deep waters.

8.5.2 Spills and Surface Discharges

Liquids discharged onto the ground surface in an uncontrolled manner can migrate downward to degrade groundwater quality. At industrial sites, causal activities may include boilovers, losses during transfers of liquids, leaks from pipes and valves, and inadequate control of wastes and storm runoff. Washing aircraft with solvents and spills of fuel at airports can form a layer of hydrocarbons floating on the water table. Pollution can also occur from the intermittent dumping of fluids on the ground, especially near gasoline stations, small commercial establishments, and construction sites. It has been estimated that millions of gallons of automobile waste oil are discharged on the ground surface annually.[67] Finally, accidents involving above-ground pipelines, storage tanks, railroad cars, and trucks can release large quantities of a pollutant at a particular site. Hazardous and flammable liquids are often flushed by water from highways; this action may actually aid in transporting the pollutant to the water table.

8.5.3 Stockpiles

Solid materials are frequently stockpiled near industrial plants, construction sites, and large agricultural operations. These may be raw materials awaiting use, or they may be solid wastes placed for temporary or permanent storage. Precipitation falling on unsheltered stockpiles causes leaching into the soil to occur; this may transport heavy metals, salts, and other inorganic and organic constituents as pollutants to the groundwater.

8.5.4 Septic Tanks and Cesspools

Another potential source of groundwater pollution is septic tanks and cesspools.[13] Commercial establishments, hospitals, industrial plants, and resorts employ septic tanks in areas where community sewer systems are not available. A septic tank is a watertight basin intended to separate floating and settleable solids from the liquid fraction of domestic sewage and to discharge this liquid with its dissolved and suspended solids into the biologically active zone of the soil mantle through a subsurface percolation system such as a tile field, a seepage bed, or a sand-covered sand filter (see Figure 8.5.2). A cesspool is a large buried chamber with porous walls designed to receive and percolate raw sewage. Domestic sewage adds minerals to groundwater. Bacteria and viruses are normally removed by the soil system; phosphorus is generally retained by the soil, but significant quantities of nitrogen can be added to groundwater.[78]

8.5.5 Roadway Deicing

Pollution of groundwater also results from the application of deicing salts to streets and highways in winter.[31, 88] Urban areas with winter temperatures below freezing are affected the most. Salt reaches the groundwater in solution after spreading on roadways and also from stockpiles. Salt application rates range from 2 to 11 m-tons per single-lane kilometer per winter season. Sodium chloride accounts for most applications, with calcium chloride the remainder. A steady growth in roadway deicing has stemmed from the demand to maintain streets and

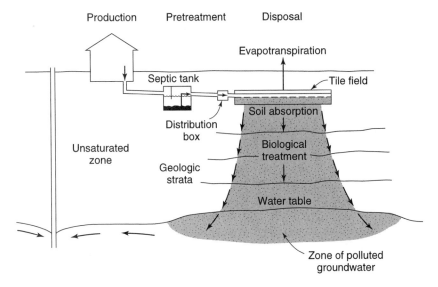

Figure 8.5.2. Disposal of household wastes through a conventional septic tank system (after Miller[67]).

highways in safe driving condition throughout the winter. The salts have produced widespread degradation of groundwater quality and have also hastened the corrosion of wells.

Illustrative of the effect is the dramatic increase in chloride measured in a well in Burlington, Massachusetts, shown in Figure 8.5.3. When chloride exceeded the 250 mg/l drinking water standard in 1970, local use of deicing chemicals on city streets was banned.

8.5.6 Saline Water Intrusion

Salt water may invade freshwater aquifers to create point or diffuse pollution sources. In coastal aquifers, seawater is the pollutant, while in inland aquifers underlying saline water may be responsible. Because of the worldwide importance of saline water intrusion, its occurrence, mechanisms, and control methods are discussed more fully in Chapter 14.

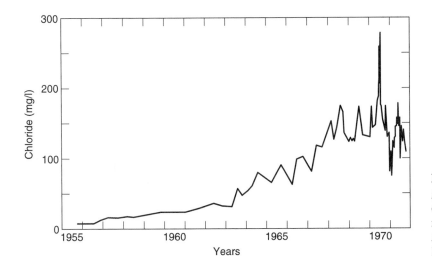

Figure 8.5.3. Time variation of chloride in groundwater at Burlington, Massachusetts, due to local roadway deicing (after Terry,[88] reprinted with permission from *Road Salt, Drinking Water, and Safety*, copyright 1974, Ballinger Publishing Company).

8.5.7 Interchange through Wells

Because wells form highly permeable vertical connections between aquifers, they can serve as avenues for groundwater pollution where inadequate attention is given to the proper construction, sealing, or abandonment of wells. Pollution occurs where there is incomplete hydraulic separation within a well and where a vertical difference in hydraulic head exists between two aquifers. Two flow conditions with a well connecting an unconfined aquifer with a confined aquifer are illustrated in Figure 8.5.4. In addition, if the top of a well is not constructed so as to divert surface water away from the well, it can admit a wide variety of surface pollutants. Although governmental regulations in most locations require plugging of abandoned wells, many such wells remain unplugged and serve as ongoing interchanges between aquifers.

8.5.8 Surface Water

Polluted surface water bodies that contribute to groundwater recharge become sources of groundwater pollution. The recharge may occur naturally from a losing stream, or it may be induced by a nearby pumping well, as indicated in Figure 8.5.5. Because many municipal water supply wells are located adjacent to rivers to ensure adequate flows, they can serve as important groundwater pollution mechanisms. Figure 8.5.6 illustrates the various pathways for the movement of contaminated groundwater into a well when the source of contaminants is at or near the water table.

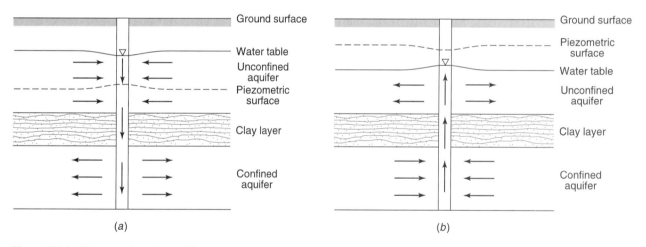

Figure 8.5.4. Diagrams showing aquifer leakage by vertical movement of water through a nonpumping well. (*a*) Water table above the piezometric surface (*b*) Piezometric surface above the water table (after Todd et al.[93])

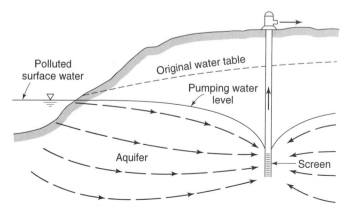

Figure 8.5.5. Diagram showing how polluted water can be induced to flow from a surface stream to a pumping well (after Deutsch[22]).

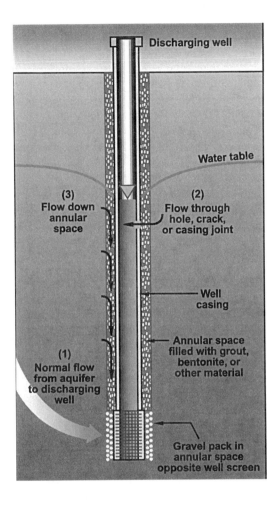

Figure 8.5.6. Pathways for movement of contaminated groundwater into a well, when the source of contaminants is at or near the water table.[32]

Analytical techniques for estimating areas contributing recharge to discharging wells, including numerical simulation, consider only pathway (1), which is the normal flow of groundwater from the aquifer to the discharging well. Pathways (2) (flow through a hole, crack, or casing joint) and (3) (flow down the annular space) can permit entry of adjacent ground water into the well from any depth above the well screen. The prevention of contamination along these pathways depends on the design and construction of the well and on maintenance of the well through time. Unfortunately, no analytical tools or simple predictors are available to identify if or when these pathways are or will become active (Franke et al.[32]).

8.6 ATTENUATION OF POLLUTION

Pollutants in groundwater tend to be removed or reduced in concentration with time and with distance traveled. Mechanisms involved include filtration, sorption, chemical processes, microbiological decomposition, and dilution. The rate of pollution attenuation depends on the type of pollutant and on the local hydrogeologic situation. Attenuation mechanisms tend to localize groundwater pollution near its source; they also are responsible for the interest in groundwater recharge as a water reclamation technique.

8.6.1 Filtration

Filtration removes suspended materials; hence, this action is most important at ground surface where polluted surface water is infiltrating into the ground. During groundwater flow, filtration can remove particulate forms of iron and manganese as well as precipitates formed by chemical reactions.

8.6.2 Sorption

Sorption serves as a major mechanism for attenuating groundwater pollution. Clays, metallic oxides and hydroxides, and organic matter function as sorptive materials. Most pollutants can be sorbed under favorable conditions, with the general exception of chloride, and to a lesser extent, nitrate and sulfate. The sorption process depends on the type of pollutant and the physical and chemical properties of both the solution and the subsurface materials; a substantial clay content in the strata above the water table is a key factor. The sorptive capacity of geologic materials is finite for most inorganic substances; however, for biodegradable substances such as bacteria and ammonia, the sorptive capacity may be renewed indefinitely.

8.6.3 Chemical Processes

Precipitation in groundwater can occur where appropriate ions are in solution in sufficient quantities. The most important precipitation reactions for the major constituents involve calcium, magnesium, bicarbonate, and sulfate. Trace elements having important precipitation potential include arsenic, barium, cadmium, copper, cyanide, fluoride, iron, lead, mercury, molybdenum, radium, and zinc. In arid regions, where moisture in the nearsurface zone may be minimal, chemical precipitation becomes a major attenuation mechanism.

In the zone above the water table, oxidation of organic matter acts as an important attenuation mechanism. Complex organic compounds are oxidized stepwise to more simple organic compounds until CO_2 and H_2O are formed along with numerous inorganic ions and compounds. Both oxidation and reduction reactions can occur underground in conjunction with other mechanisms, leading to precipitates, deposits of insoluble trace metals, and gases. Volatilization and loss as a gas apply particularly to reactions involving nitrate and sulfate. Radioactive decay, based on the half-life of a radioisotope, acts as an attenuation mechanism for radioactive pollutants.

8.6.4 Microbiological Decomposition

Most pathogenic microorganisms in the soil do not flourish in the soil and hence are subject to ultimate destruction, the timing of which depends on different species and environmental conditions. Bacteria and viruses as particulate matter suspended in water tend to move slower through a porous media than water. Field studies indicate that these pathogens are largely removed by passage through as little as one meter of soil, provided reasonable amounts of silt and clay are present.[25, 81]

8.6.5 Dilution

Pollutants in groundwater flowing through porous media tend to become diluted in concentration due to hydrodynamic dispersion occurring on both microscopic and macroscopic scales (see Chapter 3). These mixing mechanisms produce a longitudinal and lateral spreading of a pollutant within the groundwater so that the volume affected increases and the concentration decreases with distance traveled. Dilution is the most important attenuation mechanism for pollutants after they reach the water table.

In summary, attenuation above the water table is generally effective for most pollutants with the exception of the major inorganic constituents. Other exceptions include boron and tritium, as well as some trace elements and organic chemicals. Maximum attenuation requires an adequate distance to the water table and the presence of fine-grained geologic materials, such as silt and clay. Without these conditions, almost any pollutant can be introduced directly into the saturated zone. Pollutant attenuation below the water table occurs more slowly with dilution serving as the principal mechanism.

8.7 DISTRIBUTION OF POLLUTION UNDERGROUND

In a large region or basin, bodies of polluted groundwater can be visualized as innumerable scattered dots on a map, while in some areas diffuse sources, such as from irrigation return flows, would create areas with appreciable horizontal extent. Entry of pollutants into shallow aquifers occurs by percolation from ground surface, through wells, from surface waters, and by saline water intrusion. The extent of pollution in groundwater from a point source decreases as pollutants move away from the source until a harmless or very low concentration level is reached. Because each constituent of a pollution source may have a different attenuation rate, the distance to which pollutants travel will vary with each quality component.

A hypothetical example of a waste-disposal site is shown in Figure 8.7.1 with groundwater flowing toward a river. Zones *A, B, C, D,* and *E* represent essentially stable limits for different contaminants resulting from the steady release of wastes of unchanging composition. Pollutants, once entrained in the saturated groundwater flow, tend to form *plumes* (analogous to smoke from a smokestack as it drifts downwind in the atmosphere) of polluted water extending downstream from the pollution source until they attenuate to a minimum quality level. Only zone *E* reaches the river in Figure 8.7.1 and is subsequently diluted by surface water.

The shape and size of a plume depend on the local geology, the groundwater flow, the type and concentration of pollutant, the continuity of waste disposal, and any human modifications of the groundwater system, such as pumping wells.[61] Where groundwater is moving relatively rapidly, a plume from a point source tends to be long and thin, such as that shown in Figure 8.7.2, but where the flow rate is low, the pollutant tends to spread more laterally to form a wider plume. Irregularly shaped plumes can be created by local influences such as pumping wells and nonuniformities in permeability.

Plumes tend to become stable areas if there is a constant input of waste into the ground. This occurs for two reasons: enlargement as pollutants continue to be added at a point source is counterbalanced by attenuation mechanisms, or the pollutant reaches a location of groundwater discharge, such as a stream, and emerges from the underground. When a waste is first released into groundwater, the plume expands until a quasi-equilibrium stage is reached.

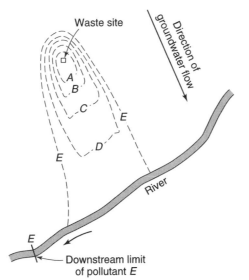

Figure 8.7.1. Plan view of an unconfined aquifer showing areal extent to which various pollutants of mixed wastes at a disposal site disperse and move to insignificant levels (after LeGrand[61]).

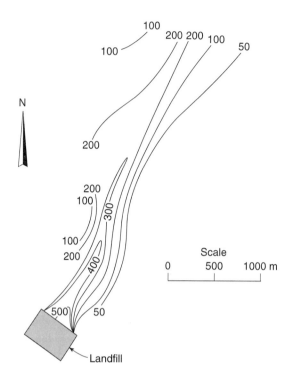

Figure 8.7.2. Plume of groundwater pollution from a landfill near Munich, West Germany. Numbered lines are isochlors in mg/l (after Cole[18]).

If sorption is important, a steady inflow of pollution will cause a slow expansion of the plume as the earth materials within it reach a sorption capability limit.

An approximately stable plume will expand or contract generally in response to changes in the rate of waste discharge. Figure 8.7.3 shows changes in plumes that can be anticipated from variations in waste inputs.

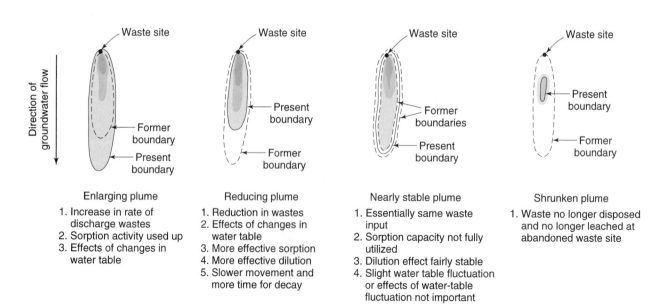

Figure 8.7.3. Changes in groundwater pollution plumes and causal factors (after LeGrand[61]).

An important aspect of groundwater pollution is the fact that it may persist underground for years, decades, or even centuries. This is in marked contrast to surface water pollution. Reclaiming polluted groundwater is usually much more difficult, time consuming, and expensive than reclaiming polluted surface water. Underground pollution control is achieved primarily by regulating the pollution source, and secondarily by physically entrapping and, when feasible, removing the polluted water from the underground.

8.7.1   Hanford Site in Richland, Washington

The Hanford site (see Figure 8.7.4) is considered to be one of the largest groundwater contamination problems in the United States by the Department of Energy (DOE) (see Graham et al.[41]). Tritium and nitrates from the reprocessing of fuels have contaminated tens of square miles of the aquifer underlying the site (Freshley and Thorne[35]). The spread of the tritium plume with time is shown in Figure 8.7.5. Other chemicals and radionuclides have also been disposed at the Hanford site and are found in the groundwater (Dirkes et al.[23]). Figure 8.7.6 shows the distribution of carbon tetrachloride in the groundwater under one area of the site.

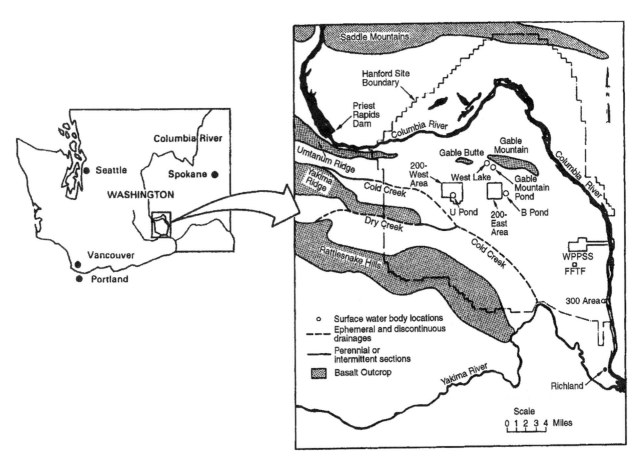

Figure 8.7.4. Location of the Hanford site in Richland, Washington (Freshley and Graham[34]).

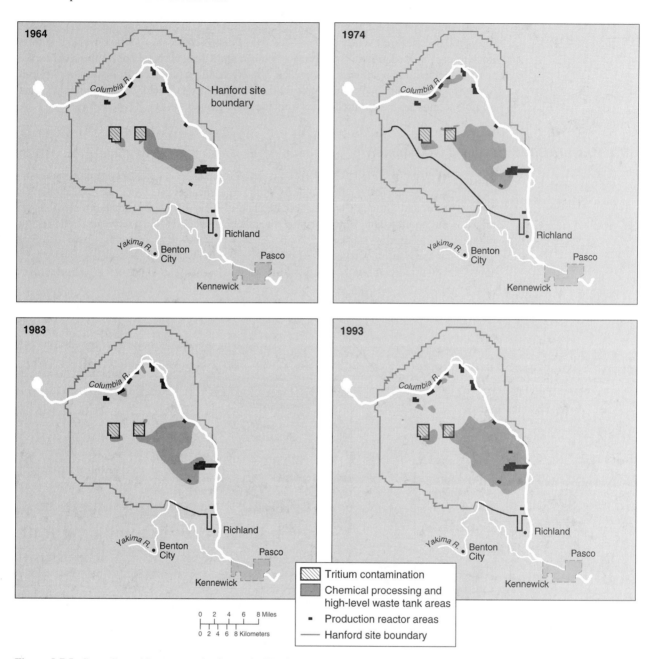

Figure 8.7.5. Spreading tritium contamination at the Hanford Site in Washington. The shaded areas on these maps show how tritium contamination in concentrations above safe drinking-water standards has spread over time (DOE[95]).

8.8 MASS TRANSPORT OF POLLUTANTS

8.8.1 Transport Processes

The law of conservation (advective–dispersive equation) for solute transport in saturated media is derived following Ogata,[74] Bear,[8] and Freeze and Cherry.[33] The conservation of mass for solutes in porous media considers the flux of solute into and out of a fixed elemental volume (see Figure 8.8.1) as follows:

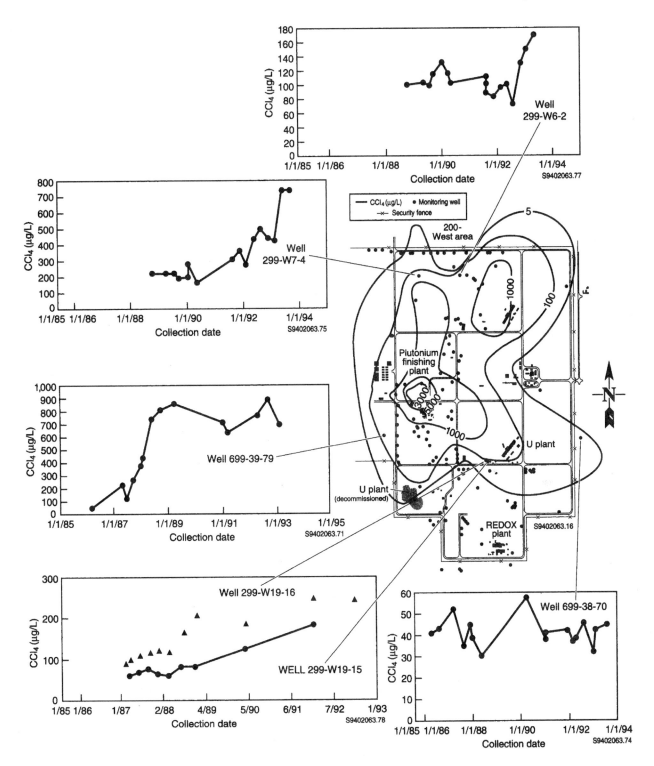

Figure 8.7.6. Distribution of carbon tetrachloride in groundwater under the 200-West area of the Hanford site (Dirkes et al.[23]).

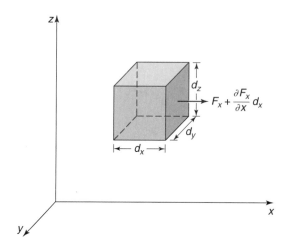

Figure 8.8.1. Elemental control volume for derivation of the conservation of mass showing the flux (Freeze and Cherry[33]).

$$
\begin{bmatrix}
\text{net rate of} \\
\text{change of mass} \\
\text{of solute within} \\
\text{the element}
\end{bmatrix}
=
\begin{bmatrix}
\text{flux of} \\
\text{solute out} \\
\text{of the} \\
\text{element}
\end{bmatrix}
-
\begin{bmatrix}
\text{flux of} \\
\text{solute into} \\
\text{the} \\
\text{element}
\end{bmatrix}
\pm
\begin{bmatrix}
\text{loss or gain} \\
\text{of solute mass} \\
\text{due to} \\
\text{reactions}
\end{bmatrix}
\tag{8.8.1}
$$

Solutes can be considered in two broad classes: conservative and reactive. Conservative solutes are nonreactive with the soil and/or native groundwater and do not undergo biological or radiaoactive decay. Chloride ion is an example of a nonreactive or conservative solute.

Advection and hydrodynamic dispersion are the physical processes that control flux into and out of the elemental volume of porous media. *Advection* is the transport of solute by the flowing groundwater (bulk groundwater flow), expressed by Darcy's law. *Hydrodynamic dispersion* results from mechanical mixing (dispersivity) and molecular diffusion.

Mechanical dispersion occurs when contaminated groundwater mixes with noncontaminated groundwater, resulting in a dilution of the contaminate, which is referred to as *dispersion*. Figure 8.8.2 illustrates longitudinal and transverse spreading due to mechanical dispersion. Mixing that occurs along the streamline is *longitudinal dispersion*. Dispersion normal to the longitudinal dispersion is *lateral (transverse) dispersion*. Mechanical dispersion in the transverse direction is a much weaker process than in the longitudinal direction; however, the coefficients of longitudinal and transverse dispersion are nearly equal at low velocities when molecular diffusion is the dominant dispersive mechanism.

Figure 8.8.3 compares streamlines and groundwater age distribution with and without dispersion in a regional layered aquifer system. The effect of not considering dispersion on groundwater age is clearly indicated by the darker areas as compared to the same areas that are much lighter for advection–dispersion.

Diffusion is the flux of solute from a zone of higher concentration to a zone of lower concentration as a result of Brownian motion of ionic and molecular species. *Fick's law* describes the diffusion of a solute for a steady-state condition:

$$
F = -D \frac{dC}{dx}
\tag{8.8.2}
$$

where F is the mass flux of solute per unit area per unit time ($M/L^2/T$); D is the diffusion coefficient (L^2/T); C is the solute concentration (M/L^3); and dC/dx is the concentration gradient ($M/L^3/L$). The negative sign indicates the movement from greater to lesser concentration. Diffusion coefficients for the major ions in groundwater (Na^+, K^+, Mg^{2+}, Ca^{2+}, Cl^-, HCO^-, SO_4^{2-})

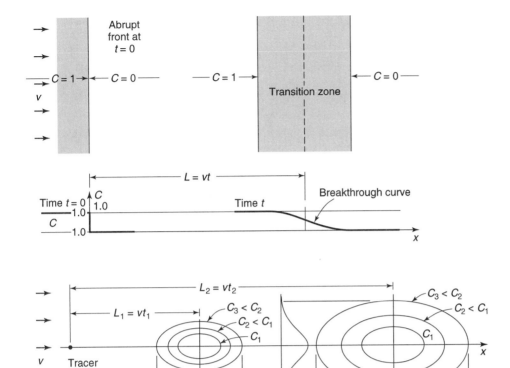

Figure 8.8.2. Longitudinal and transverse spreading due to mechanical dispersion (from Bear and Verruijt[9] with permission).

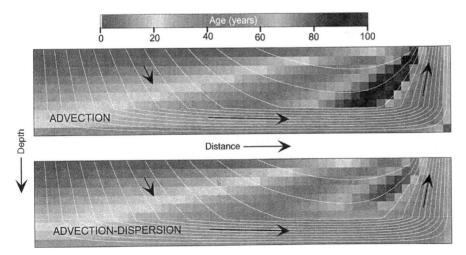

Figure 8.8.3. Streamlines and groundwater age distributions with and without dispersion in a regional layered aquifer system (Goode[40]).

are in the range of 1×10^{-9} to 2×10^{-9} m^2/s at 25°C (Robinson and Stokes[80]). The coefficients are temperature dependent and at 5°C the coefficients are about 50% smaller.

Fick's second law describes the change of concentration over time inside the element (control volume) subject to diffusion flux,

$$\frac{\partial C}{\partial t} = D * \frac{\partial^2 C}{\partial x^2} \tag{8.8.3}$$

where $\partial C/\partial t$ is the change in concentration with time. The above expressions of Fick's first and second laws are one-dimensional, but can be expressed more generally for three dimensions.

Diffusion in porous media does not proceed at the rate it can in water because ions follow longer paths caused by the presence of particles in the solid matrix and because of adsorption on the solids. To account for these phenomena, an apparent (or effective) diffusion coefficient for nonabsorbed species in porous media flow is defined as

$$D* = \omega D \tag{8.8.4}$$

where ω is an empirical coefficient (<1) that takes into account the effect of the solid phase of the porous media on the diffusion. Freeze and Cherry[33] suggest using the above effective diffusion coefficient with ω ranging from 0.5 to 0.01, to account for the tortuosity of the flow path.

Because the processes of molecular diffusion and mechanical dispersivity cannot be separated in groundwater flow, the coefficient of hydrodynamic dispersion is used to account for both. The longitudinal coefficient of hydrodynamic dispersion, D_L, is expressed as

$$D_L = \alpha_L \bar{v} + D* \tag{8.8.5}$$

where α_L is the dynamic longitudinal dispersivity, a characteristic property of the porous medium; $\bar{v}$ is the average linear groundwater velocity; and $D*$ is the molecular diffusion (referred to as the effective diffusion coefficient or the apparent diffusion coefficient).

A rough approximation of α_L based on averaging published date (Gelhar et al.[38]) is

$$\alpha_L = 0.1L \tag{8.8.6}$$

where L is the length of the flow path (m). For lengths less than 3,500 m, Neuman[72] gave

$$\alpha_L = 0.0175L^{1.46} \tag{8.8.7}$$

For transverse (lateral) dispersivity, α_T is typically 1/10 to 1/100 of the longitudinal dispersivity α_L. Xu and Eckstein[109] used a statistical study to develop the following relationship

$$\alpha_L = 0.83(\log L)^{2.414} \tag{8.8.8}$$

where L is in ft or m and α_L is in ft or m.

8.8.2 Advection–Dispersion Equation for Solute Transport in Saturated Porous Media

Consider the elemental volume (Figure 8.8.1) of homogeneous porous media of porosity n. F_x represents the total mass of solute per unit cross-sectional area transported in the x-direction per unit time. The mass of solute is transported in the x-direction by advection and by dispersion expressed respectively as

$$\text{Mass transported by advection} = \bar{v}_x n C \, dA \tag{8.8.9}$$

$$\text{Mass transported by dispersion} = n D_x 2C/2x \, dA \tag{8.8.10}$$

where dA is the elemental cross-sectional area of the cubic element and D_x is the dispersion coefficient in the x-direction defined by

$$D_x = \alpha_x \bar{v}_x + D* \tag{8.8.11}$$

where α_x is the dynamic dispersivity and $\alpha_x v_x$ is the *mechanical dispersion*.

F_x is now represented as

$$F_x = \bar{v}_x nC - nD_x \frac{\partial C}{\partial x} \qquad (8.8.12)$$

with the negative sign for the dispersive term indicating that the contaminant (solute) moves toward the zone of lower concentration. In a similar manner, F_y and F_z are respectively

$$F_y = \bar{v}_y nC - nD_y \frac{\partial C}{\partial y} \qquad (8.8.13)$$

$$F_z = \bar{v}_z nC - nD_z \frac{\partial C}{\partial z} \qquad (8.8.14)$$

The total solute entering (flux entering) the cubic element is

$$F_{\text{entering}} = F_x \, dz \, dy + F_y \, dz \, dx + F_z \, dx \, dy \qquad (8.8.15)$$

and the total solute leaving (flux leaving) the cubic element is

$$F_{\text{leaving}} = \left(F_x + \frac{\partial F_x}{\partial x} dx \right) dy \, dz + \left(F_y + \frac{\partial F_y}{\partial y} dy \right) dz \, dx + \left(F_z + \frac{\partial F_z}{\partial z} dz \right) dx \, dy \quad (8.8.16)$$

Partial terms indicate the spatial change of the solute mass in the respective direction.

For a nonreactive dissolved substance, the flux into the element and the flux out of the element is equal to the net rate of change of mass of solute within the element, defined as

$$\Delta F = -n \frac{\partial C}{\partial t} dx \, dy \, dz \qquad (8.8.17)$$

Combining the above three expressions (Equations 8.8.15–8.8.17) and simplifying gives

$$\frac{\partial F_x}{\partial x} + \frac{\partial F_y}{\partial y} + \frac{\partial F_z}{\partial z} = -n \frac{\partial C}{\partial t} \qquad (8.8.18)$$

Substituting Equations 8.8.12–8.8.14 into Equation 8.8.18 gives

$$\left[\frac{\partial}{\partial x}\left(D_x \frac{\partial C}{\partial x} \right) + \frac{\partial}{\partial y}\left(D_y \frac{\partial C}{\partial y} \right) + \frac{\partial}{\partial z}\left(D_z \frac{\partial C}{\partial z} \right) \right] - \left[\frac{\partial}{\partial x}(\bar{v}_x C) + \frac{\partial}{\partial y}(\bar{v}_y C) + \frac{\partial}{\partial z}(\bar{v}_z C) \right] = \frac{\partial C}{\partial t}$$
$$(8.8.19)$$

For a homogeneous medium with v steady and uniform in space and time, in which the dispersion coefficients D_x, D_y, and D_z do not vary through space, then the above Equation 8.8.19 simplifies to

$$\left[D_x \frac{\partial^2 C}{\partial x^2} + D_y \frac{\partial^2 C}{\partial y^2} + D_z \frac{\partial^2 C}{\partial z^2} \right] - \left[\bar{v}_x \frac{\partial C}{\partial x} + \bar{v}_y \frac{\partial C}{\partial y} + \bar{v}_z \frac{\partial C}{\partial z} \right] = \frac{\partial C}{\partial t} \qquad (8.8.20)$$

For one dimension, the conservation of mass (advection–dispersion equation) is

$$D_x \frac{\partial^2 C}{\partial x^2} - \bar{v}_x \frac{\partial C}{\partial x} = \frac{\partial C}{\partial t} \qquad (8.8.21)$$

which can also be expressed along a flowline by using L for x where L is the coordinate direction along the flowline. D_L is the longitudinal coefficient of hydrodynamic dispersion and $\bar{v}_L$ is the average linear velocity along the flowline.

8.8.3 Analytical Procedure

The analytical solution to Equation 8.8.21 is (see Ogata[74] for details)

Initial condition	$C(x, 0) = 0$	$x \geq 0$
Boundary condition	$C(0, t) = C_0$	$t \geq 0$
Boundary condition	$C(\infty, t) = 0$	$t \geq 0$

$$C(x,t) = \frac{C_0}{2}\left[\text{erfc}\left(\frac{x - vt}{2\sqrt{D_L t}}\right) + \exp\left(\frac{vx}{D_L}\right)\text{erfc}\left(\frac{x + vt}{2\sqrt{D_L t}}\right)\right] \qquad (8.8.22)$$

where x is the distance from the injection point.

The argument of the exponential $(\bar{v}_L x/D_L)$ is the Peclet number, $P_e = \bar{v}_L x/D_L$, which is a measure of the ratio of the rate of transport by advection to the rate of transport by diffusion. Large Peclet numbers $(P_e > 100)$ indicate that advection dominates. When advection dominates, the second term (with the exponent) on the right-hand side becomes negligible. Other books that discuss contaminant transport include: Bear,[8] Bear and Verruijt,[9] Bedient et al.,[10] Charbeneau,[14] Delleur,[19] de Marsily,[21] Domenico and Schwartz,[24] Fetter,[30] Freeze and Cherry,[33] Javandel et al.,[49] and Palmer.[77]

EXAMPLE 8.8.1

The objective of this example is to illustrate the use of Equation 8.8.22 to compute the concentration of a pollutant as a function of time and distance from a point or line source in an aquifer with known properties. The aquifer properties are: hydraulic conductivity = 2.5×10^{-5} m/s; hydraulic gradient = 0.001; effective porosity = 0.25; and an effective diffusion coefficient = 0.75×10^{-9} m²/s. A chloride solution with a concentration of 600 mg/l penetrates (enters) the aquifer along a line source. Determine the chloride concentration at a distance of 25 m from the source after one year, two years, and four years.

SOLUTION

Step 1: Compute the pore velocity (average linear velocity) using Darcy's law:

$$\bar{v}_L = \frac{Ki}{n_e} = \frac{(2.5 \times 10^{-5})0.001}{0.25} = 1 \times 10^{-7} \text{ m/s}$$

Step 2: Compute the longitudinal dispersivity using the approximation of Neuman,[72] Equation 8.8.7:

$$\alpha_L = 0.0175L^{1.46}$$

$$= 0.0175(25)^{1.46}$$

$$= 1.92 \text{ m}$$

Using the approximation by Xu and Eckstein,[109] Equation 8.8.8, we have

$$\alpha_L = 0.83(\log L)^{2.414}$$

$$= 0.83(\log 25)^{2.414}$$

$$= 1.86 \text{ m}$$

Step 3: Compute the coefficient of longitudinal mechanical dispersion–diffusion (coefficient of longitudinal hydrodynamic dispersion) using Equation 8.8.11 where $\alpha_L = 1.86$ m:

$$D_L = \alpha_L v_L + D*$$

$$= 1.86 \times (1 \times 10^{-7}) + 0.75 \times 10^{-9}$$

$$= 1.9 \times 10^{-7} \text{ m}^2/\text{s}$$

Step 4: Use Equation 8.8.22 to compute the concentration for times of $t = 1$ year $= 60$ s/min $\times$ 1,440 min/day $\times$ 365 days/yr $= 3.15 \times 10^7$ s; $t = 2$ years $= 6.31 \times 10^7$ s; and $t = 4$ years $= 12.6 \times 10^7$ s; $\bar{v}_L = 1 \times 10^{-7}$ m/s; $C_0 = 600$ mg/l; $x = L = 25$ m; $D_L = 1.9 \times 10^{-7}$ m^2/s.

$$\text{For } t = 1 \text{ year: } C(25\text{m}, 1 \text{ yr}) = 0.0 \text{ mg/l}$$

$$\text{For } t = 2 \text{ years: } C(25\text{m}, 2 \text{ yr}) = 0.037 \text{ mg/l}$$

$$\text{For } t = 4 \text{ years: } C(25\text{m}, 4 \text{ yr}) = 21.6 \text{ mg/l}$$

■

8.8.4 Transport of Reactive Pollutants

This section considers the transport of solutes that behave as nonreactive, but also have the added influence of chemical reactions. *Sorption* is the exchange of molecules and ions between the solid phase and the liquid phase, including both adsorption and desorption. *Adsorption* is the attachment of molecules and ions from the solute to the rock material, causing a decrease of concentration of the solute. This is also referred to as causing a *retardation* of the contaminant transport. *Desorption* is the release of molecules and ions from the solid phase to the solute. A *sorption isotherm* is the relationship between the solute concentration in the adsorbed phase and in the water phase.

The one-dimensional transport, Equation 8.8.16, for advection–dispersion can be extended to include the effects of retardation of solute transportation through sorption, chemical reaction, biological transformations, or radioactive decay by including source–sink term. The form of the one-dimensional equation that includes retardation, in a homogeneous saturated media due to adsorption, adds a source–sink term:

$$D_L \frac{\partial^2 C}{\partial l^2} - \bar{v}_L \frac{\partial C}{\partial l} + \frac{\rho_b}{n} \frac{\partial S}{\partial t} = \frac{\partial C}{\partial t} \tag{8.8.23}$$

(dispersion term)   (advection term)   (reaction term)

where ρ_b is the bulk mass density of the porous medium, n is the porosity, and S is the mass of chemical constituent adsorbed on a unit mass of the solid part of the porous medium. $\partial S/\partial t$ is the rate at which the constituent is adsorbed [M/MT] and $(\rho_b/n)(\partial S/\partial t)$ is the change in concentration in the groundwater caused by adsorption or desorption [M/(L^{3}T)].

Adsorption relationships that can be plotted as straight-line graphical representations on log–log paper are referred to as *Freundlich isotherm*, expressed as

$$\log S = b \log C + \log K_d$$
$$S = K_d C^b \tag{8.8.24}$$

where S is the mass of solute species adsorbed or precipitated on the solids per unit bulk dry mass of the porous medium, C is the solute concentration, and K_d and b are coefficients. These coefficients depend on the solute species, nature of the porous medium, and other conditions of the system. b is the slope of the log–log relationship, so if it has a value of 1, then the isotherm is a *linear isotherm,* so that $K_d = dS/dC$ and is referred to as the *distribution coefficient.* K_d is a representation of the partitioning between liquid and solids only if the reactions that cause the partitioning are fast and reversible, if the isotherm is linear. Many contaminants that are of interest in groundwater studies meet these requirements.

The advance of the contaminant front is retarded as a result of the transfer by adsorption (or other chemical processes) of the contaminant mass from the pore water to the solid part of the porous medium. For a fast and reversible adsorption with linear isotherm, the partitioning of the contaminant can be described by the distribution coefficient and the *retardation* of the front relative to the bulk mass of water as it passes through the soil as described by the retardation factor, R_a, expressed as the *retardation equation,*

$$R_a = 1 + \frac{(1-n)\rho_b}{n} K_d \tag{8.8.25}$$

where n is the porosity and ρ_b is the bulk mass density of the soil, gm/cm^3; and K_d is the distribution coefficient for the solute with the soil. Also $1/\theta = (n-1)/n$ where θ is the volumetric content of the soil, which is dimensionless. The retardation factor ranges from 1 to 10,000. A reactive solute will travel at a slower rate than the groundwater flow because of the adsorption. The velocity of the solute front v_c (where the concentration is 0.5 of the original concentration, $C/C_0 = 0.5$) is

$$v_c = \bar{v}/R_a \tag{8.8.26}$$

so that the reciprocal of the retardation factor is referred to as the *relative velocity*.

The amount (or degree) of contaminant adsorbed by the solids is a function of the concentration in solution, $S = f(C)$, so that

$$-\frac{\partial S}{\partial t} = \frac{\partial S}{\partial C} \cdot \frac{\partial C}{\partial t} \tag{8.8.27}$$

and

$$-\frac{\rho_b}{n} \cdot \frac{\partial S}{\partial t} = \frac{\rho_b}{n} \cdot \frac{\partial S}{\partial C} \cdot \frac{\partial C}{\partial t} \tag{8.8.28}$$

Using $K_d = dS/dC$ and the above relation, the governing equation (8.8.16) can be expressed in terms of the retardation factor to express the one-dimensional advection–dispersion equation as

$$\frac{\partial C}{\partial t} = -\frac{v}{R_a} \frac{\partial C}{\partial x} + \frac{D_L}{R_a} \frac{\partial^2 C}{\partial x^2} \tag{8.8.29}$$

which defines the change in storage of contaminant in the control volume. The first term on the right-hand side represents the retarded advective inflow–outflow and the second term represents the retarded diffusion and dispersion. Van Genuchten[106] presented solutions of the advection/dispersion/adsorption equation.

8.9 MONITORING GROUNDWATER QUALITY

To protect a groundwater resource against pollution, a water quality monitoring program—defined as a scientifically designed surveillance system of continuing measurements, observation, and evaluations—is necessary. Monitoring methods may include not only sampling and analyses of groundwater quality, but also determination of groundwater levels and flow directions, measurements of moisture in the unsaturated zone, geophysical surveys, evaluations of wastes and other materials contributing to subsurface pollution, testing of pipelines and tanks, and aerial surveillance.[28, 69, 73]

Facilities that are regulated under the Resource Conservation and Recovery Act (RCRA) commonly require four types of monitoring programs (U.S. EPA[97, 100, 101]). (1) *Detection monitoring* programs detect an impact to groundwater quality. (2) *Assessment monitoring* programs assess the nature and extent of contaminants and collect data for designing remediation. (3) *Corrective action monitoring* programs assess the impact of a particular remediation design. (4) *Performance monitoring* programs evaluate the effectiveness of an element of a groundwater remediation system. As pointed out by Houlihan and Lucia,[48] these four monitoring programs may also be useful in non-RCRA programs such as at Superfund sites, at sites being voluntarily remediated, and at legally required remediation sites. Figure 8.9.1 shows the sequence of activities for groundwater monitoring.

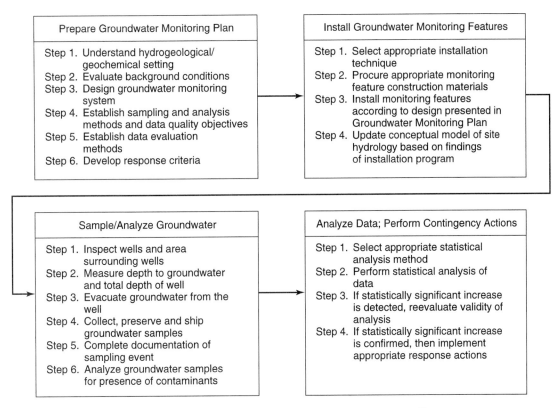

Figure 8.9.1. Sequence of activities for groundwater monitoring (Houlihan and Lucia[48]).

In developing a groundwater-monitoring program, federal and state regulations must be observed. Houlihan and Lucia[48] provided the following questions as a guide to identifying applicable groundwater monitoring regulations for a particular site or operation:

- What is the activity that could have an adverse impact on groundwater quality? Are there related activities that could also have an adverse impact on groundwater quality?
- What specific requirements do the regulations contain for planning groundwater monitoring, sampling groundwater, analyzing groundwater samples, and evaluating data from groundwater sampling?
- Do the regulations contain specific requirements for action in the event that impacts to groundwater quality are identified during the monitoring program?

Once the applicable regulatory requirements have been identified, the overall strategy of the groundwater monitoring program should be defined prior to preparing the groundwater monitoring plan. This will help in guiding the development of the monitoring plan. Figures 8.9.2 and 8.9.3 illustrate the development of a groundwater monitoring strategy. Potential sources of contamination and the aquifers of concern should be characterized before developing a monitoring strategy. The selection of target monitoring zones cannot be made until the source and the aquifer of concern have been evaluated through a hydrogeologic evaluation.

A groundwater-monitoring plan is required on a site-by-site basis under the RCRA and the Comprehensive Emergency Response and Compensation Liability Act (CERCLA), and for other regulated facilities under various state and local programs. These plans are developed to describe each aspect of monitoring and to control the monitoring activities so that the overall goals of the strategy are fulfilled. Table 8.9.1 is an example outline for a groundwater monitoring plan.

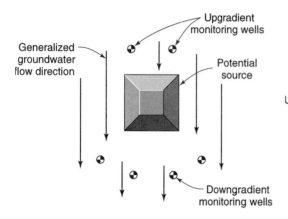

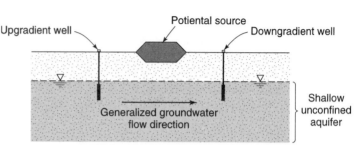

Figure 8.9.2. Plan view of typical unconfined aquifer groundwater monitoring system (Houlihan and Lucia[48]).

Figure 8.9.3. Section of typical unconfined aquifer groundwater monitoring system (Houlihan and Lucia[48]).

Table 8.9.1 Sample Outline—Groundwater Monitoring Plan

1. Introduction
2. Site Conditions
    2.1 Site Topography and Surface-Water Hydrology
    2.2 Site Geology
    2.3 Site Hydrogeology
3. Groundwater Monitoring System
    3.1 Introduction
    3.2 Existing Groundwater Monitoring Wells
    3.3 Construction of New Groundwater Monitoring Wells
    3.4 Development of New Groundwater Monitoring Wells
4. Groundwater Sampling Procedures
    4.1 Introduction
    4.2 Health and Safety Procedures
    4.3 Recordkeeping
    4.4 Equipment Use, Cleaning, Decontamination
    4.5 Preparation for Sampling
    4.6 Groundwater Sample Collection Procedures
    4.7 Groundwater Sample Handling and Preservation
    4.8 Shipment of Groundwater Samples
5. Laboratory Groundwater Analytical Program
6. Quality Assurance and Quality Control Procedures
    6.1 Data Quality Objectives
    6.2 Field Sampling QA/QC Requirements
    6.3 Laboratory QA/QC Requirements
7. Evaluation of Groundwater Data
    7.1 Introduction
    7.2 Groundwater Quality Data Validation
    7.3 Evaluation of Static Groundwater Elevations
    7.4 Statistical Analysis Procedures for Groundwater Quality Data
    7.5 Response Actions
    7.6 Reporting Requirements for Groundwater Quality Data
8. References

Source: Houlihan and Lucia[48]

The National Water-Quality Assessment (NAWQA) Program has developed groundwater data-collection protocols and procedures for the selection, installation, and documentation of related data (Lapham et al.[59]) and also for the collection and documentation of water quality samples and related data (Koterba et al.[58]). Three types of studies are defined in the NAWQA program: (1) *study-unit (or subunit) surveys,* designed to obtain occurrence and distribution data on a variety of analytes; (2) *land-use studies;* and (3) *flowpath studies,* which assess spatial differences and possibly temporal variability in each of a selected number of analytes among wells located in different parts of a local groundwater flow system. Table 8.9.2 provides a summary of required, recommended, and optional water-quality constituents measured in the three studies.

Table 8.9.2 Summary of Current (1995) Required, Recommended, and Optional Water-Quality Constituents to Be Measured in the Three National Water-Quality Assessment Program Groundwater Components of the Occurrence and Distribution Assessment

Water-quality constituent or constituent class	Study-unit survey	Land-use studies	Flowpath studies[1]	Method[2]
Field measurements				
—Temperature	Req	Req	Req	Field
—Specific electrical conductance	Req	Req	Req	Field
—pH	Req	Req	Req	Field
—Dissolved oxygen	Req	Req	Req	Field
—Acid neutralizing capacity (ANC) (unfiltered sample)[3]	Rec	Rec	Rec	Field incremental
—Alkalinity (filtered sample)[3]	Req	Req	Req	Field incremental
—Turbidity[4]	Rec	Rec	Rec	Field
Major inorganics	Req	Req	Req	NWQL SC2750
Nutrients	Req	Req	Req	NWQL SC2752
Filtered organic carbon	Req	Req	Opt	NWQL SC2085
Pesticides	Req	Req	Opt	NWQL SC2001/2010 NWQL SC2050/2051
Volatile organic compounds (VOCs)	Req	Req or Opt[5]	Req or Opt[6]	NWQL SC 2090
Radon	Req	Req or Rec[7]	Req or Rec[6]	NWQL LC 1369
Trace elements[4]	Opt	Opt	Opt	NWQL SC 2703
Radium	Opt	Opt	Opt	NWQL-Opt
Uranium	Opt	Opt	Opt	NWQL-Opt
Tritium, tritium–helium, chlorofluorocarbons (CFCs)[8]	Rec	Rec	Rec	NWQL LC1565 (tritium)
Environmental isotopes[9]	Rec	Rec	Rec	NWQL-Opt

[Required water-quality constituents to be measured for the Occurrence and Distribution Assessment are determined partly by the water-quality topics of national interest selected for National Assessment. Topics selected for National Assessment (1994) are nutrients, pesticides, and volatile organic compounds. The topics selected can change over time. Quality-control samples also are required—types of quality control samples depend on study component. Req, Required; Rec, Recommended; Opt, Optional; NWQL, National Water-Quality Laboratory; SC, Schedule; LC, Laboratory Code]

[1]Selection of constituents for measurement in flowpath studies is determined by flowpath study objectives. During at least the first round of sampling, however, the broad range of constituents measured in study-unit surveys and land-use studies would be measured.

[2]Schedules and laboratory codes listed are required for study units that began their intensive phase in 1991 or 1994, and apply until changed by National Program directive. Schedules for radium and uranium can be selected by the study unit, but require NAWQA Quality-Assurance Specialist approval. A detailed discussion is found in Koterba et al.[58]

[3]ANC (formerly referred to as unfiltered alkalinity) is measured on an unfiltered sample. Alkalinity is measured on a filtered sample. Study unit could have collected ANC, alkalinity, or both to date.

[4]Turbidity measurements are required whenever trace-element samples are collected to evaluate potential colloidal contributions to measured concentrations of iron, manganese, and other elements.

[5]VOCs are required at all urban land-use study wells, but optional in agricultural land-use studies. If VOCs are chosen as part of an agricultural land-use study, then they should be measured in at least 20 of the land-use study wells.

(continues)

Table 8.9.2 (*continued*) Summary of Current (1995) Required, Recommended, and Optional Water-Quality Constituents to Be Measured in the Three National Water-Quality Assessment Program Groundwater Components of the Occurrence and Distribution Assessment

[6]VOCs are required at all urban flowpath wells for at least the first round of sampling. If VOCs are measured in an agricultural land-use study, then they should be measured at all flowpath-study wells within that land-use study for at least the first round of sampling.

[7]Radon is required at any land-use or flowpath study well if that well also is part of a study-unit survey; otherwise, radon collection is recommended for land-use or flowpath study wells located in likely source areas.

[8]Collection of tritium, tritium–helium, chlorofluorocarbons (CFCs), and/or other samples for dating groundwater is recommended, depending on hydrogeologic setting. For tritium methods, see NWQL catalog; for CFCs, see Office of Water Quality Technical Memorandum No. 95.02 (unpublished document located in the USGS Office of Water Quality, MS 412, Reston, VA 22092).

[9]For a general discussion of the use of environmental isotopes in groundwater studies, see Alley.[3]

The study-unit survey can involve data collection from as many as 100 to 120 wells associated with multiple land uses, with wells distributed among several subunit surveys, each of about 30 wells (Koterba et al.[58] and Lapham et al.[59]). The flowpath studies used for water-quality data collection involve about 20 wells with most of the wells in a single aquifer that underlies a single land use. Figure 8.9.4 shows the general design of a monitoring well in

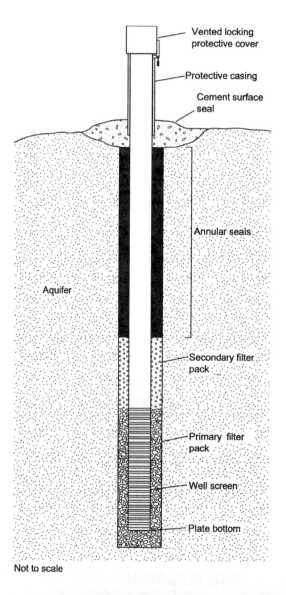

Figure 8.9.4. General design of a monitoring well in unconsolidated deposits for National Water-Quality Assessment Program Land-Use and Flowpath Studies (Koterba et al.[58]).

Vented locking protective cover

Protective casing

Cement surface seal

Annular seals

Aquifer

Secondary filter pack

Primary filter pack

Well screen

Plate bottom

Not to scale

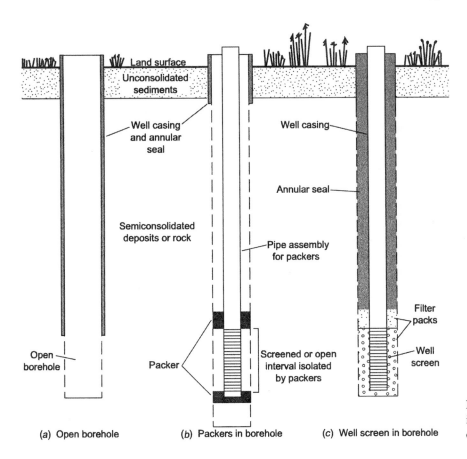

Figure 8.9.5. Examples of monitoring-well designs in semiconsolidated deposits and in rock (Lapham et al.[59]).

unconsolidated deposits for the land-use and flowpath studies. Figure 8.9.5 shows examples of monitoring-well designs in semiconsolidated deposits and in rock. Figure 8.9.6 shows examples of three well-cluster designs. Figure 8.9.7 shows the schematic setup for well purge and sample collection. The American Society for Testing Materials (ASTM[5]) and the U.S. Environmental Protection Agency (EPA)[97] also provide standard approaches for design of groundwater monitoring wells.

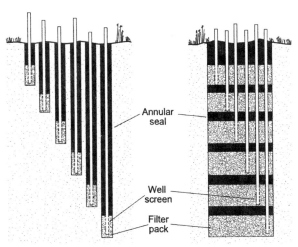

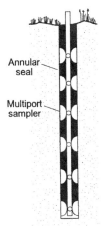

(a) Monitoring wells with short screens, each installed in its own borehole

(b) Multiple monitoring wells with short screens installed in a single borehole

(c) A series of multiport samplers installed in a single borehole

Figure 8.9.6. Examples of three well-cluster designs (Koterba et al.[58]).

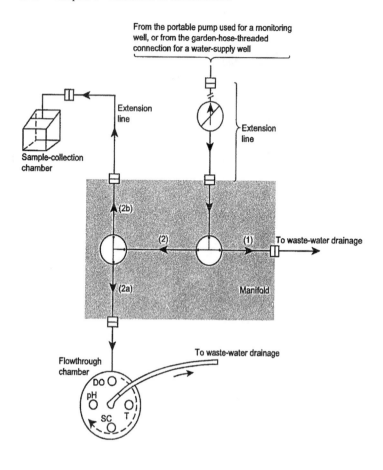

From the portable pump used for a monitoring
well, or from the garden-hose-threaded
connection for a water-supply well

Extension
line

Sample-collection
chamber

Extension
line

(2b)

(2)

(1)   To waste-water drainage

(2a)   Manifold

To waste-water drainage

Flowthrough
chamber

DO

pH

SC   T

	Rigid-wall Teflon tubing		Antibacksiphon
	Quick connection		Three-way Teflon flow valve

Flow direction, at different times

(1)	During initial purge stage	
(2)	During intermediate and final stages	
(2a)	To obtain most field measurements	
(2b)	To obtain turbidity samples, and at end of purge to route flow for collection	

Field sensors (flowthrough chamber)

DO	Dissolved-oxygen sensor
T	Temperature sensor
pH	pH sensor
SC	Specific electrical conductance sensor

Figure 8.9.7. Schematic of equipment setup for well purge and sample collection (Koterba et al.[58]).

8.10 REMEDIATION OF CONTAMINATED GROUNDWATER

8.10.1 Remediation Goals

Needs for remediation of contaminated groundwater can be established using the results of a groundwater assessment or monitoring program. *Remediation* is a broad term that refers to the reduction of risk caused by exposure to contaminated groundwater. Alternatively, remediation seeks to protect human health and the environment and to restore groundwater to beneficial uses where practicable, ideally to restore groundwater to drinking water standards. In the United States, cleanup goals under CERCLA and RCRA generally are set at drinking water standards. Maximum Contaminant Levels (MCLs) have been established under the Safe Drinking Water Act. Cleanup requirements may vary for groundwater that is not used for drinking water.

To determine an appropriate strategy to manage contaminated groundwater, site conditions must be evaluated and remediation goals must be defined. One of the most important steps in a remediation program is to define goals, such as (NRC[71]):

- *Complete restoration*, which involves removal of all contaminants from the contaminated aquifer;
- *Nondegradation*, which involves removal of contaminants that exceed either the detection limits of available analytical equipment or background concentrations;
- *Remediation to health-based standards*, which involves removal of contaminants that are present at a concentration that could cause adverse health effects (some examples of health-based standards are maximum contaminant levels (MCLs), alternate concentration limits (ACLs), and local or federal drinking water standards);
- *Remediation to the limits of technology-based standards*, which involves use of the best available technology to remove as much of the contaminants as possible;
- *Partial-use restrictions* (or *institutional controls*), such as legal restrictions on the use of groundwater in areas where groundwater has been contaminated, or physical barriers (e.g., fences) to prevent access to contaminated media;
- *Containment*, which involves the use of engineered systems for preventing migration of the contaminants to locations where receptors could be exposed to the contaminants.

Table 8.10.1 lists advantages and disadvantages of each of the above remediation goals. Once a remediation goal is selected, a remedy must be implemented to achieve the goal. Three major remedies are (1) containment of groundwater, (2) groundwater extraction, and (3) treatment of contaminated groundwater.

Table 8.10.1 Advantages and Disadvantages of Cleanup Goals

Goal	Advantages	Disadvantages
Complete restoration	Eliminates all risk	Likely impossible
Nondegradation	Reduction of contaminants to lowest level measurable	Extremely difficult, expensive, and time-consuming for many contaminants and hydrogeologic settings
Health-based standards	Designed to prevent measurable impacts to human health or environment	Difficult to defineand may not accurately address all possible health impacts of exposure to contaminated groundwater
Technology-based standards	Allows treatment to the best capabilities of current technology	May not reduce risk to a level that is protective of human health and the environment
Partial-use restrictions	Prevents contact between contaminants and receptors in a cost-effective manner	Leaves contaminants that could cause risk if partial use restrictions are ineffective
Containment	Relatively predictable and reliable; typically less costly than other remediation approaches	Leaves contamination that could migrate if containment system fails

(Left margin annotations: Increasing flexibility; Increasing cost; Increasing protection)

Adapted from National Research Council[71] with permission

8.10.2 System Design

Design of an efficient, cost-effective groundwater remediation system requires (Houlihan and Lucia[47]):

- A comprehensive understanding of the nature and extent of contamination;
- Remediation goals;
- A careful evaluation of remedial technologies and their abilities to meet the objectives.

The steps in a groundwater remediation system design are as follows (Houlihan and Lucia[47]):

- Step 1—Define the problem.
- Step 2—Define the goal of the groundwater remediation (see Table 8.10.2).
- Step 3—Screen candidate remedies (see Table 8.10.3) for a summary of commonly used groundwater remediation technologies.
  - Prepare preliminary conceptual design for each candidate remedy.
  - Evaluate effectiveness of conceptual designs.
  - Evaluate cost of remedies.
- Step 4—Prepare detailed design.
- Step 5—Implement the design.
- Step 6—Confirm the effectiveness of the design.

8.10.3 Hydraulic Containment of Groundwater

Hydraulic containment can be used at sites where the contaminant source cannot be removed, such as at landfills or in bedrock with DNAPLs. The most widely used types of hydraulic containment are physical barriers and hydraulic barriers. *Physical barriers* are vertical features

Table 8.10.2 Candidate Groundwater Remediation Goals and Corresponding Remediation Approaches

Remediation goal	Examples of candidate remediation approaches
Groundwater restoration	Natural (intrinsic) remediation
	Bioremediation
	Combinations of conventional remediation approaches
Nondegradation	Source removal
	Conventional pump-and-treat
	Bioremediation SVE/AS
Return aquifer to health-based standards	Source removal
	Conventional pump-and-treat
	Bioremediation SVE/AS
Apply technology-based standards	Phytoremediation
	Electrokinetics
	Solvent extraction
	Thermal desorption
Implement restricted-use policies	Deed restrictions
	Local ordinances
Containment	Physical barriers
	Hydraulic barriers
	Capping

Adapted from National Research Council[71] with permission

Table 8.10.3 Summary of Commonly Used Groundwater Remediation Technologies

Alternative technology	Residual groundwater concentration	Residual sorbed concentration in source area	Cleanup time[a]	Number of peer-reviewed publications[b]
Source Remediation				
Conventional pump-and-treat	Low to medium	Medium to high	Long	Some
Vacuum extraction and bioventing	NA	Low to medium	Short	Some
Air sparging (vertical or horizontal wells)	Low to medium	Low to medium	Short to medium	Limited
In-situ bioremediation—hydrocarbons	Low to medium	Low to high	Short to medium	Some
In-situ bioremediation—chlorinated solvents	Low to medium	Low to high	Medium to long	Some
Cosolvent and surfactant flushing	Low to medium	Low to medium	Short to medium	Some
Steam stripping	Low to medium	Low to medium	Short	Some
In-situ thermal desorption	Low to medium	Low to medium	Short	Some
In-situ chemical oxidation	Medium (?)	Medium to high (?)	Medium (?)	Limited
In-situ bioremediation—metals	Low to medium	Low to high	Medium to high	Limited
Intrinsic bioremediation	Low to medium	Low to high	Long	Limited
Plume Remediation				
Conventional pump-and-treat	Low	Medium to high	Long	Many
Air sparging (vertical or horizontal wells)	Low to medium	Low to medium	Medium to long	Limited
In-situ bioremediation—hydrocarbons	Low to medium	Low to high	Medium to long	Some
In-situ bioremediation—chlorinated solvents	Low to medium	Low to high	Medium to long	Some
In-situ reactive barriers	Low	NA	Long	Limited
Intrinsic bioremediation	Low to medium	Low to medium	Long	Limited

Note: A "low" residual concentration and "medium" cleanup time reflect relatively good performance, while a "high" residual concentration and "long" cleanup time reflect much less effective performance. "NA" denotes that the technology is not applicable to this situation. Cost of the technology and feasibility for different situations are not addressed in this table but should be carefully evaluated before selecting a technology.

[a] Because few cases of achieving cleanup goals have been reported, these qualitative assessments reflect the judgment of National Research Council.

[b] "Limited" indicates that very little information about this technology is available in peer-reviewed publications, while "some" indicates a greater availability of information.

Adapted from National Research Council[71] with permission, as presented in Houlihan and Lucia[47]

(consisting of a variety of different materials, as listed in Table 8.10.4) in the ground that are barriers to the flow of groundwater. They can be *nonselective physical barriers,* which obstruct the flow of all groundwater, or *selective physical barriers,* which obstruct only the migration of target contaminants. Cutoff walls, which limit migration of groundwater by forming a physical barrier to the flow, are the oldest and most commonly used. Soil–bentonite slurry walls are the most commonly used in geotechnical and environmental remediation projects (Houlihan and Lucia[47]).

Hydraulic barriers create a depression in the piezometric surface of groundwater, acting as a barrier beyond which groundwater within the zone of influence of the barrier should not flow. They can be formed using trenches, wells, or other methods that remove groundwater. Other options for hydraulic containment include horizontal drains, geomembranes, and well points.

Table 8.10.4 Types of Physical Barriers

Barrier type	Width (ft)	Maximum depth (apx) (ft)	Unit cost ($/SF)	Production rate per 10 hrs (SF)
Soil bentonite	2–3	80	2–8	2,500–15,000
Cement bentonite	2–3	80	5–18	1,000–8,000
Biopolymer drain	2–3	70	7–25	1,500–5,000
Deep mixing	2.5	90	6–15	1,000–8,000
DM structural	2.5	90	15–30	1,000–3,000
Jet grouting	1.5–3	200	30–80	300–2,500
Grout curtain	One row	200	40–100	200–1,000
Sheet piling	One sheet	150	15–40	2,000–15,000
Geochemical barrier	Varies	Hundreds	8–400	100–1,000

Source: Rumer and Mitchell[82]

8.10.4 Groundwater Extraction Systems

Groundwater extraction systems, the most common being extraction well systems and extraction trench systems, are a combination of subsurface components and above-ground components. The subsurface components are the wells or trenches, which provide access to the contaminated water, and the above-ground components are used to regulate and monitor the extraction process. A typical groundwater extraction well system is shown is Figure 8.10.1 and a typical groundwater extraction trench system is shown in Figure 8.10.2.

8.10.5 Treatment of Contaminated Groundwater

A large number of groundwater treatment technologies have been developed as described in U.S. EPA.[103] These technologies can be classified as *in-situ* or *ex-situ*, as summarized in Table 8.10.5. *In-situ treatment techniques* either render a contaminant nontoxic through treatment or enhance extraction of the contaminant from the aquifer. *Ex-situ treatment techniques* treat groundwater that has been extracted from the aquifer. These are summarized in Table 8.10.6.

In-situ treatment technologies are generally designed to perform one or more of the following functions (Johnson[51]):

- Remove contaminant source zone
- Restore aquifer
- Prevent or minimize continued contaminant migration

A single technology rarely accomplishes all three of these goals.

Houlihan and Lucia[47] discuss in-situ treatment techniques as subdivided into biological remediation (bioremediation) techniques, volatization processes, and chemical and physical processes. *Bioremediation* is the biological degradation of contaminants using naturally occurring microbes in the soil. *Volatization processes* transfer the chemical from the liquid state to the gaseous state (vapor phase), because vapor phase contaminants are typically easier to remediate: Aquifer materials are significantly more permeable to vapors than to liquids, making vapors easier to remove from the aquifer. Johnson[51] subdivides in-situ techniques into the following three categories (see Figure 8.10.3 for most common technologies used):

- *Source zone treatment technologies* :
  - Free-product recovery
  - Excavation and disposal or above-ground treatment
  - In-situ soil venting

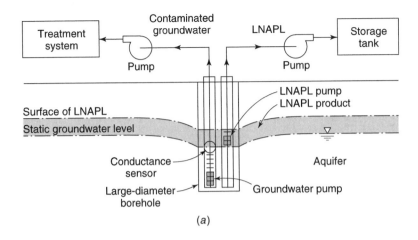

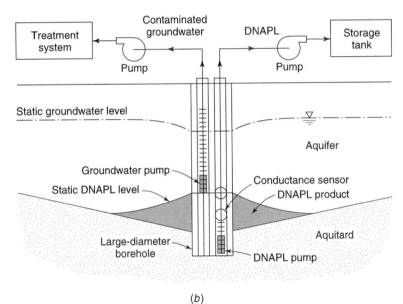

Figure 8.10.1. (*a*) LNAPL dual-extraction system (*b*) DNAPL dual-extraction system (from Houlihan and Lucia[47]).

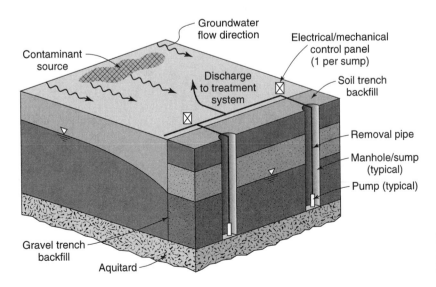

Figure 8.10.2. Typical groundwater extraction trench system (Houlihan and Lucia[47]).

Table 8.10.5 Summary of Candidate Treatment Technologies

In situ	Process description
Bioremediation	Biological degradation of contaminants using naturally occurring microbes in soil
Soil vapor extraction	Volatization of contaminants that are present in the vadose zone
Air sparging	Volatization of contaminants in the saturated zone
Permeable reaction barriers	Physical or chemical treatment in a trench
Vacuum vapor extraction	Volatization, within a well, of contaminants from saturated zone
Density driven convection	Enhanced bioremediation using single-well driven convection system in aquifer.

Ex situ	Process description
Bioreactor	Biological degradation of contaminants (activated sludge, fixed-film biological reactor, biophysical treatment)
Slurry-phase biological treatment	Variation on bioreactor in which contaminants are treated in a slurry form
Air stripping	Volatization of contaminants
Carbon adsorption	Adsorption of contaminants to activated carbon
Ion exchange	Exchange-type attachment of contaminants to ion-exchange resin
Alkaline precipitation	Alteration of water quality (usually pH adjustment) such that concentration exceeds the compound's solubility limit, causing precipitation
Membrane	Separation of solids from water using membranes (reverse osmosis, ultrafiltration)
Wetlands treatment	Uptake of contaminants by wetland features
Electrokinetic decontamination	Desorption of contaminants by "acidic front" of groundwater caused by hydrolysis of the groundwater

Source: Houlihan and Lucia[47]

- Bioventing
- In-situ air sparging
- Enhanced in-situ soil venting with soil heating and/or soil fracturing
- In-situ vitrification
- Phytoremediation
- Groundwater pump and treatment systems
- *Aquifer restoration techniques*:
  - Groundwater pump-and-treat systems
  - Natural attenuation
  - In-situ air sparging
  - Enhanced bioremediation
- *Contaminant migration prevention* (see Figure 8.10.4):
  - Natural attenuation
  - Groundwater pump-and-treat systems
  - In-situ reaction walls
  - In-situ air sparging curtains
  - Infiltration barriers
  - In-situ contaminant options (grout walls, sheet piling walls, etc.)

Source zone treatment technologies treat the cause of the groundwater contamination so that they target the removal and destruction of the residual contaminants in the soil that are the

Table 8.10.6 Summary of Selected Ex-Situ Groundwater Treatment Technologies

Groundwater treatment technology	Representative examples	Residual streams	Status of technology
Organic contaminants			
Air stripping	Packed towers, surface or diffused aeration removal of volatile compounds; soil venting	Air stream with VOCs	Commercial
Liquid-phase	GAC removal of broad spectrum of VOCs	GAC for regeneration or disposal	Commercial
Steam stripping	Packed tower with steam stripping, removal of low volatile organics	Recovered solvent	Some commercial
Membranes	Ultrafiltration for removal of selected organics	Concentrated brine side stream	Commercial
Oxidation	Ozone/UV, or ozone/H_2O_2, destruction of chlorinated organics	None	Some commercial in development stages
Activated sludge	Oxygen or air biological oxidations for removal/destruction of degradable organics	Sludge	Commercial
Fixed-film biological reactors	Fixed-film fluidized bed, for oxidation of less degradable organics	Sludge	Commercial
Biophysical	Powdered carbon, with activated sludge, treatment of high strength waste waters	Powdered carbon and bacterial	Commercial, PACT process
Inorganic contaminants			
Alkaline precipitation	Heavy metals removal	Hazardous sludge	Commercial
Coagulation	Ferric sulfate or alum for heavy metals removal	Hazardous sludge	Commercial
Ion exchange	Heavy metals; nitrate	Regeneration stream	Commercial
Adsorption	Selenium removal on activated alumina	Regeneration stream	Commercial
Filtration	Removal of clays, other particulates	Backwash wastes	Commercial
Reduction	SO_2 reduction of CR(VI)	Sludge	Commercial
Membranes	Reverse osmosis, ultrafiltration for removal of metals, other ions	Concentrated liquid waste	Commercial, new membranes are under development
Oxidation	Fe(II) and Mn(II)	Sludge	Commercial

Source: USEPA[102]

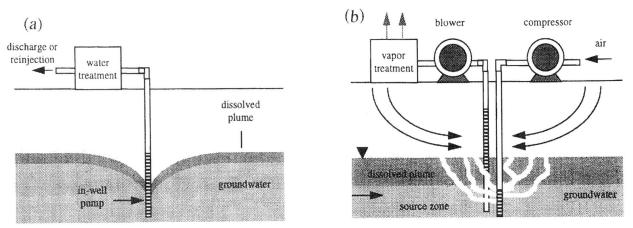

Figure 8.10.3. Example aquifer restoration process schematics. (*a*) Groundwater pump-and-treat (*b*) In-situ air sparging (Johnson[51])

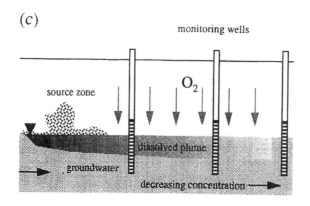

Figure 8.10.3 (*continued*). Example aquifer restoration process schematics. (*c*) Natural attenuation (Johnson[51])

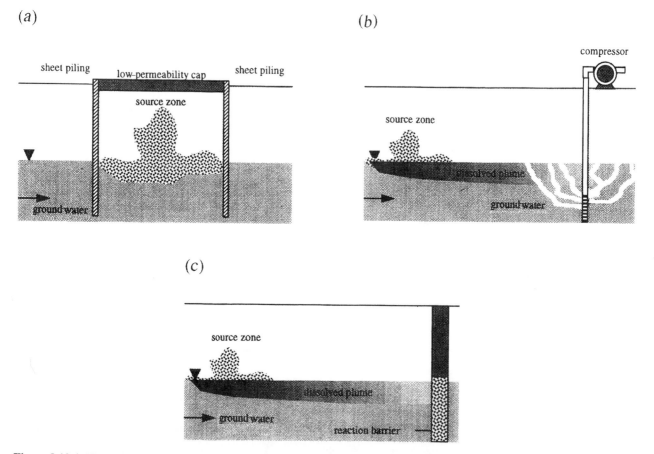

Figure 8.10.4. Example contaminant migration technology processes schematics. (*a*) Physical barriers (*b*) In-situ air sparging curtain (*c*) Reaction barrier (Johnson[51])

source. Aquifer restoration technologies are employed before, during, or after source zone treatment to target the treatment of dissolved contaminant plumes. Contaminant migration prevention technologies are employed at sites where the source zone location is unknown and/or there are no practicable source zone and aquifer restoration options in order to minimize future impacts of contaminants on groundwater.

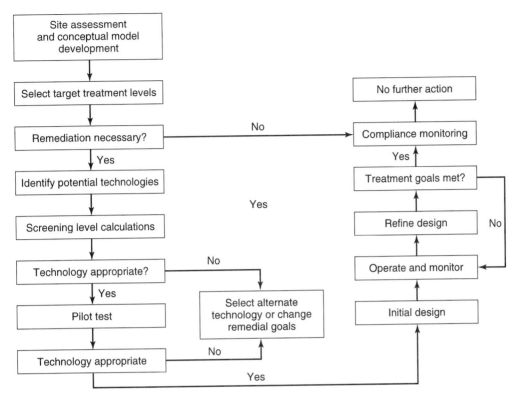

Figure 8.10.5. Generalized technology selection and design flowchart (Johnson[51]).

8.11 CONVENTIONAL PUMP-AND-TREATMENT SYSTEMS

8.11.1 Pump-and-Treat Remediation Strategies

Groundwater pump-and-treatment systems are most often used for hydraulic containment purposes. Conventional pump-and-treat systems (see Figure 8.11.1) extract the contaminated groundwater and treat it on the surface before discharging or reinjection. As described in Section 8.10, to determine an appropriate strategy, first an evaluation of site conditions followed by a definition of the remediation goals must be performed. Conflicting objectives for restoration may generally include: (1) reduce contaminant concentrations to cleanup standards, (2) maximize mass removal, (3) minimize cleanup time, and (4) minimize costs.

The strategies for managing groundwater contamination using pump-and-treat technology include (Cohen et al.[17]): (1) hydraulic/physical containment, (2) groundwater quality restoration, and (3) mixed objective strategies. The selected management strategy depends on site-specific hydrogeology and contaminant conditions, and remediation goals. Figure 8.11.2 illustrates several groundwater contamination management strategies using pump-and-treat technology.

The use of pump-and-treatment for the hydraulic containment of dissolved containment by pumping groundwater from wells or drains is illustrated in Figure 8.11.3. Groundwater cleanup is typically more difficult to achieve than hydraulic containment; however, for sites where the contaminative source has been removed or contained, pump-and-treat technology may be used for cleanup of a dissolved plume. The use of pump-and-treat technology for aquifer restoration generally combines hydraulic containment with more aggressive manipulation of the groundwater.

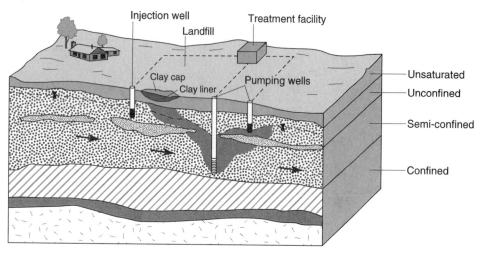

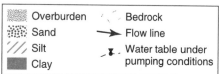

Figure 8.11.1. Example of a pump-and-treat system (after Mercer et al.,[65] as presented in Cohen et al.[17]).

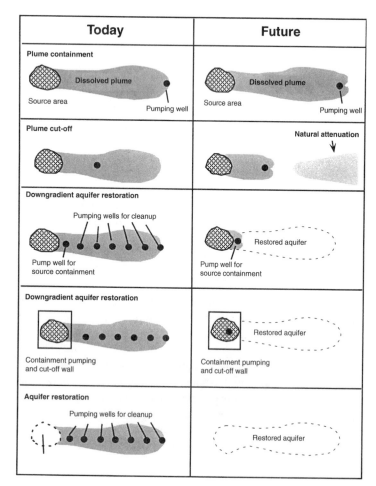

Figure 8.11.2. Several groundwater contamination management strategies using pump-and-treat technology (after NRC,[71] Cherry et al.,[15] as presented in Cohen et al.[17]).

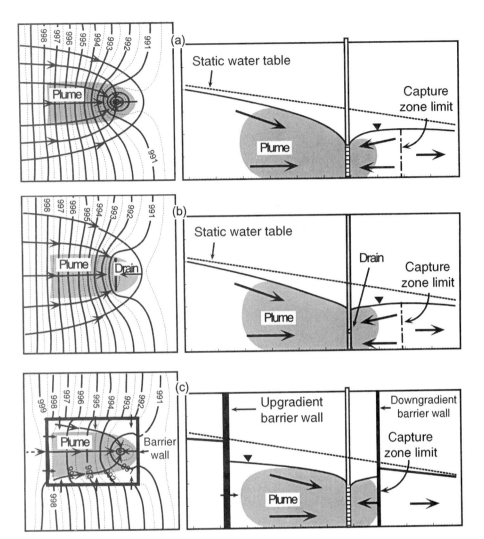

Figure 8.11.3. Examples of hydraulic containment in plan view and cross section using an extraction well (a), a drain (b), and a well within a barrier wall (c) (Cohen et al.[17]).

8.11.2 Characterizing Sites for Pump and Treatment Design

Site characterization should have as the main goal to obtain sufficient data to select and design a remedy (NRC[71]). Investigations are made of (Cohen et al.[17]) the following:

- Nature, extent, and distribution of contaminants in source areas and down-gradient plumes;
- Potential receptors and risks posed by contaminated groundwater; and
- Hydrogeologic and contaminant properties that affect containment, restoration, and system design in different site areas.

Figure 8.11.4 illustrates the types of data used to develop a site conceptual model for remedy assessment. Important goals for contaminant characterization data include (Cohen et al.[17]):

1. Delineating contaminant source areas and release characteristics;
2. Defining the nature and extent (horizontal and vertical) of contamination;

Contaminant Sources

- Location and characteristics of continuing near-surface releases and sources

- Location of subsurface sources (NAPL pools, residual NAPL, metal precipitates, etc.)

Nature and Extent of Contamination

- Spatial distribution of subsurface contaminants

- Types and concentrations of contaminants

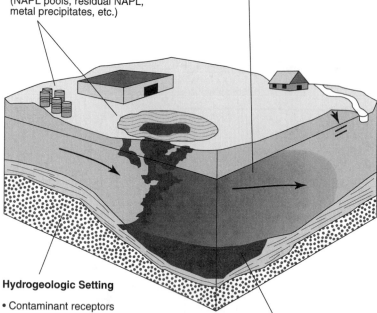

Hydrogeologic Setting

- Contaminant receptors

- Description of regional and site geology

- Stratigraphy (thickness and lateral extent of units, preferential pathways, etc.)

- Depth to ground water

- Hydraulic gradients

- Hydraulic conductivity, storage coefficient, porosity distribution

- Temporal variation in water levels

- Groundwater recharge and discharge

- Groundwater/surface water interactions

Restoration Potential

- Estimates of contaminant mass

- Temporal trend in contaminant concentration

- Sorption data

- Contaminant transformation processes and rates

- Contaminant migration rates

- NAPL properties

- Other characteristics that affect transport and fate

Figure 8.11.4. Types of data used to develop a site conceptual model for remedy assessment (modified from U.S. EPA[99] as presented in Cohen, et al.[17]).

3.   Characterizing contaminant transport pathways, processes, and rates;

4.   Estimating risks associated with contaminant transport; and

5.   Assessing aquifer restoration potential.

A site conceptual model is used to formulate remedial strategies, restoration and/or containment.

Many systems will operate for decades in order to clean up and contain contaminated groundwater, because of the slow contaminant transport and interphase transfer. This results in the use of a phased and integrated approach to site characterization as illustrated in Figure 8.11.5. The performance of the pump-and-treat systems is assessed by measuring hydraulic

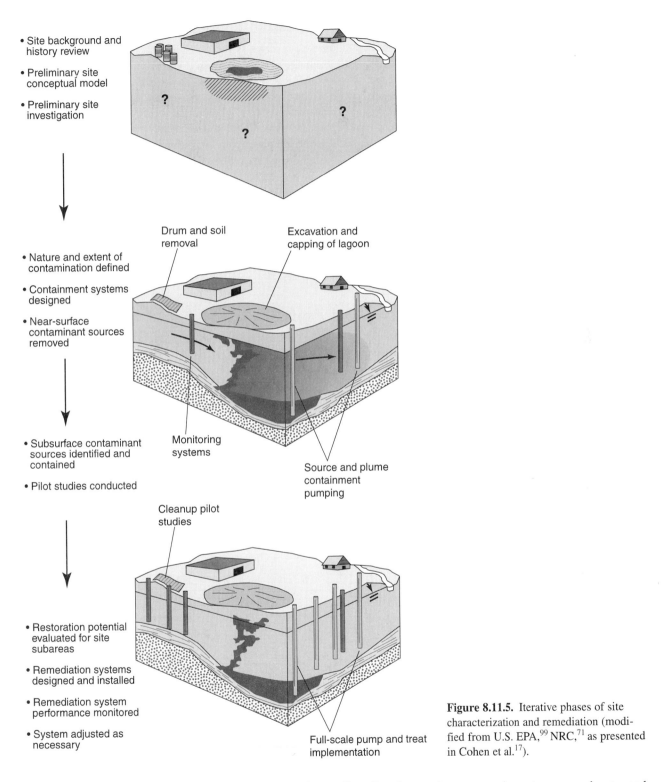

- Site background and history review
- Preliminary site conceptual model
- Preliminary site investigation

Drum and soil removal

Excavation and capping of lagoon

- Nature and extent of contamination defined
- Containment systems designed
- Near-surface contaminant sources removed

Monitoring systems

Source and plume containment pumping

- Subsurface contaminant sources identified and contained
- Pilot studies conducted

Cleanup pilot studies

- Restoration potential evaluated for site subareas
- Remediation systems designed and installed
- Remediation system performance monitored
- System adjusted as necessary

Full-scale pump and treat implementation

Figure 8.11.5. Iterative phases of site characterization and remediation (modified from U.S. EPA,[99] NRC,[71] as presented in Cohen et al.[17]).

heads and gradients, groundwater flow directions and rates, pumping rates, pumped water and treatment system effluent quality, and contaminated distributions in groundwater and porous media (see Cohen et al.[16, 17]).

8.11.3 Capture Zone Analysis

The *capture zone* of an extraction well or drain is the portion of the subsurface containing groundwater that ultimately will discharge to the well or drain (see Figures 8.11.3 and 8.11.6). The capture zone of a well is not coincident with the drawdown zone of influence, as illustrated in Figure 8.11.6. The shape of the capture zone depends on the natural hydraulic gradient as well as the pumping rate and transmissivity, whereas the drawdown zone of influence depends largely on the pumping rate and transmissivity alone. There have been many models developed to determine capture zones, ranging from simple graphical and semianalytical methods (Javandel and Tsang,[50] Blanford et al.,[11] Strack et al.,[87] and Haitjema et al.[43]) to numerical models for groundwater flow and contaminant transport (see Chapter 9) to optimization models that link numerical models (see Chapter 10). Refer to van der Heijde and Elnawawy[105] for a summary of models.

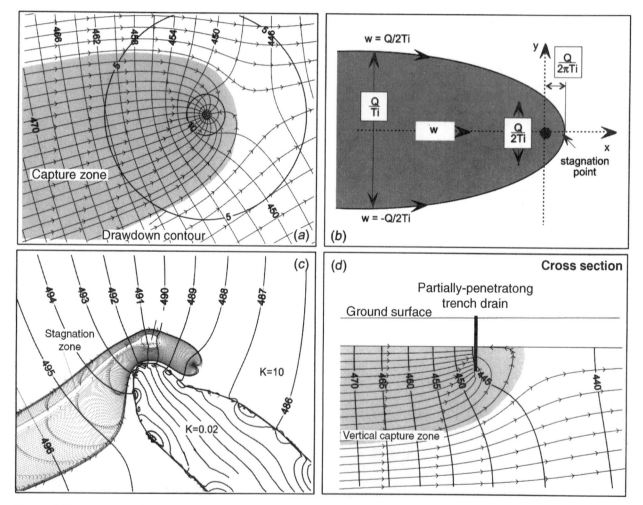

Figure 8.11.6. (*a*) Illustration of drawdown contours (i.e., zone of influence) and the capture zone of a single pumping well in a uniform medium. Equations for the dividing streamlines ($w = Q/2Ti$) that separate the capture zone of a single well from the rest of an isotropic, confined aquifer with a uniform regional hydraulic gradient are given in (*b*), where T = transmissivity, Q = pumping rate, and i = initial uniform hydraulic gradient. Simplified capture zone analysis methods may provide misleading results when applied to more complex problems, such as those dealing with heterogeneous media, as depicted in (*c*), where K = relative hydraulic conductivity, and three-dimensional flow (*d*) (Cohen et al.[17]).

EXAMPLE 8.11.1

Determine the maximum width of the capture zone for a well that is pumping 2,200 m³/day from a confined aquifer with a hydraulic conductivity of 1,200 m/day, an initial hydraulic gradient of 0.0008, and a saturated thickness of 35 m. What is the distance from the well to the stagnation point?

SOLUTION

Maximum width of the capture zone (w) is computed using the following equation from Figure 8.11.6:

$$w = Q/(2Ti) = Q/(2Kbi) = 2200 \text{ m}^3/\text{day}/[2(1200 \text{ m/day})(35 \text{ m})(0.0008)] = 32.7 \text{ m}$$

Distance from the well to the stagnation point (refer to Figure 8.11.6) is computed using

$$X = Q/(2\pi Kbi) = w/\pi = 32.7 \text{ m}/3.14 = 10.4 \text{ m} \qquad \blacksquare$$

8.11.4 Extraction/Injection Scheme Design

The design of pump-and-treat systems involves optimizing well locations, depths, and injection/extraction rates to maintain an effective hydraulic sweep through the contamination zone, minimize stagnation zones, flush pore volumes through the system, and contain contaminated groundwater. Figure 8.11.7 illustrates a conceptual modeling analysis of an idealized site using three pumping strategies for plume management. The medium is uniform and has a single nondegrading contaminant with continuing release and linear equilibrium sorption.

The three strategies are *down-gradient pumping, source control with down-gradient pumping,* and *source control with mid-plume and down-gradient pumping.* The down-gradient pumping results in expansion of the plume, making it more difficult to achieve cleanup. The other two strategies are more effective, with the source control clearly being important in preventing continued off-site migration. The third strategy with source control, mid-plume, and down-gradient pumping reduces the flow path and travel time of the contaminants to extraction wells and diminishes the impact of processes that cause tailing (Faust et al.[29]). Harvey et al.[44] have modeled the effects of pulsed pumping, namely alternating pumping and resting periods, as shown in Figure 8.11.8. Pulsed pumping has been studied as a means to address tailing and flush stagnation zones. Various pumping schemes have been reported in the literature (Gailey and Gorelick,[37] Haggerty and Gorelick,[42] Satkin and Bedient[84]). Marquis and Dineen[62] studied the relative merits of various methods, including pump-and-treat, in-situ bioremediation, and others. Bumb et al.[12] discuss the design of groundwater extraction/reinjection systems using the MODFLOW model (see Chapter 9).

PROBLEMS

8.1.1 Go to the U.S. EPA Web site for the Toxic Release Inventory (TRI) program and find out about toxic chemical releases in your zip code. The Web site is http://www.epa.gov/tri.

8.1.2 Go to the U.S. EPA Web site for the Drinking Water Contaminant Candidate List (CCL) and print out the list. Write a short discussion on one of the microbiological contaminants and one of the chemical contaminants. The Web site is http://www.epa.gov/OGWDW/ccl/cclfs.html.

8.1.3 Go to the U.S. EPA Web site for Superfund and find a fact sheet on a Superfund site in your state. Write a summary of the site, identifying the location, contaminants of concern, source of contamination, techniques being used for remediation, and the status of the remediation process. The Web site is http://www.epa.gov/superfund.

8.1.4 Go to the U.S. EPA Web site for the Integrated Risk Information System (IRIS) and develop a report on what can be found at the Web site. What is news on the site? The Web site is http://www.epa.gov/iris.

8.1.5 Go to the U.S. EPA Web site for the National Drinking Water Contaminant Occurrence Database (NCOD) and perform a search to see if there have been any violations by your local water source and what contaminants were found in the water source. The Web site is http://www.epa.gov/safewater/data/ncod.html.

8.8.1 Resolve Example 8.8.1 with a hydraulic gradient of 0.003 m/m. All other data remain the same.

8.8.2 Resolve Example 8.8.1 with a hydraulic conductivity of 5.0 × 10⁻⁵ m/s. All other data remain the same.

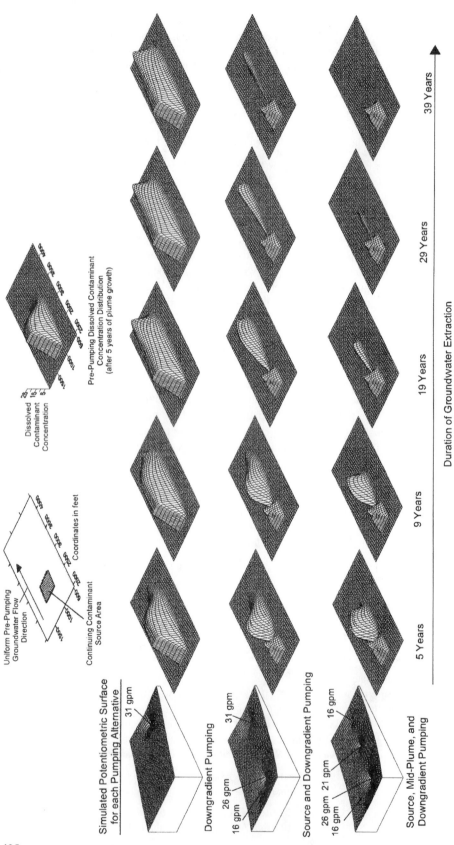

Figure 8.11.7. Results of FTWORK (Faust et al.[29]) simulation analysis of three pump-and-treat alternatives for an idealized site (with uniform media, linear equilibrium sorption, and a single nondegrading contaminant) showing dissolved contaminant concentrations with time of pumping (as presented in Cohen et al.[17]).

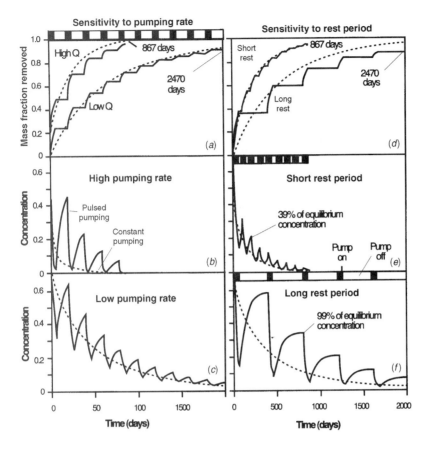

Figure 8.11.8. Effects of varying pulsed pumping parameters (after Harvey et al.[44]). The fraction of total mass removed with time is shown in (a) and (d); pumping well concentrations are shown in (b), (c), (e), and (f). Dashed lines represent equivalent constant pumping rates. Black bars atop of figures represent pumping periods and white bars represent rest periods (as presented in Cohen et al.[17]).

8.8.3 Develop the relationship (table and graph) of the dimensionless concentration (C/C_0) versus distance (0 m to 600 m) and consider times ranging from 1 to 60 years. $v = 0.04$ m/d and dispersion coefficient $D = 0.4$ m²/d.

8.8.4 An effluent with a concentration of a solute of 1,250 mg/L is leaking from a landfill into an aquifer with a hydraulic gradient of 9.8 m/day, a gradient of 0.004, and an effective porosity of 0.15. What are the solute concentrations in monitoring wells 25 m and 100 m from the landfill 300 days after the leak began? Assume that the effective diffusion coefficient D* is 0.0 and use the Xu and Eckstein[109] method to compute the longitudinal dispersivity.

8.9.1 Find the ASTM[5] standards for groundwater monitoring wells and write a brief description of the designs.

8.9.2 Write a description of the National Water-Quality Assessment (NAWQA) Program.

8.9.3 What are the selection, design, and documentation criteria for wells used in the NAWQA Program?

8.10.1 Write a description of the RCRA and CERCLA programs.

8.10.2 Write a description of the cleanup goals of the RCRA and CERCLA programs.

8.11.1 Determine the maximum width of the capture zone for a well that is pumping 2,000 m³/day from a confined aquifer with hydraulic conductivity of 1,500 m/day, an initial hydraulic gradient of 0.0009, and a saturated thickness of 35 m. What is the distance from the well to the stagnation point?

8.11.2 Determine the maximum width of the capture zone for a well that is pumping 30,000 ft³/day from a confined aquifer with a hydraulic conductivity of 95 ft/day, an initial hydraulic gradient of 0.004, and a saturated thickness of 50 ft. What is the distance from the well to the stagnation point?

REFERENCES

1. Adriano, D. C., et al., Nitrate and salt in soils and ground waters from land disposal of dairy manure, *Soil Sci. Soc. Amer. Proc.*, v. 35, pp. 759–762, 1971.

2. Ahlfeld, D. P., and C. S. Sawyer, Well location in capture zone design using simulation and optimization techniques, *Ground Water,* v. 28, no. 4, pp. 507–512, 1990.

3. Alley, W. M., ed., *Regional Ground-Water Quality,* Van Nostrand Reinhold, New York, 1993.

4. Alley, W. M., T. E. Reilly, and O. L. Franke, *Sustainability of Ground-Water Resources, U.S. Geological Survey Circular* 1186, U.S. Geological Survey, Denver, CO, 1999.

5. American Society for Testing and Materials (ASTM), *Standard practice for design and installation of ground water monitoring wells in aquifers,* Standard D5092-90, Philadelphia, 1990.

6. American Society for Testing and Materials (ASTM), *Standard guide for sampling groundwater monitoring wells,* Standard K4448-45a; ASTM Standards on Environmental Sampling, Philadelphia, 1995.

7. Barnes, I., and F. E. Clarke, *Geochemistry of Ground Water in Mine Drainage Problems, U.S. Geological Survey Professional Paper* 473-A, 6 pp., 1964.

8. Bear, J., *Dynamics of Fluids in Porous Media,* American Elsevier, New York, 1972.

9. Bear, J., and A. Verruijt, *Modeling Groundwater Flow and Pollution,* Reidel, Dordrecht, The Netherlands, 1987.

10. Bedient, P. B., H. S., Rifai, and C. J. Newell, *Groundwater Contamination: Transport and Remediation,* Prentice Hall, Upper Saddle River, NJ, 1994.

11. Blandford, T. N., P. S. Huyakorn, and Y. Wu, *WHPA—A Modular Semi-Analytical Model for the Delineation of Wellhead Protection Areas,* U.S. EPA Office of Drinking Water and Ground Water, Washington, DC, 1993.

12. Bumb, A. C., J. T. Mitchell, and S. K. Gifford, Design of a groundwater extraction/reinjection system at a superfund site using MODFLOW, *Ground Water,* v. 35, no. 3, pp. 400–408, 1997.

13. Canter, L. W., and R. C. Knox, *Septic Tank System Effects on Ground Water Quality,* Lewis Publishers, Chelsea, MI, 1986.

14. Charbeneau, R. J., *Groundwater Hydraulics and Pollutant Transport,* Prentice Hall, Upper Saddle River, NJ, 2000.

15. Cherry, J. A., S. Feenstra, and D. M. Mackay, Developing rational goals for in situ remedial technologies, Subsurface Restoration, *Conference Proceedings, Third International Conference on Ground Water Quality Research,* National Center for Ground Water Research, Dallas, TX, June, 1992.

16. Cohen, R. M., A. H. Vincent, J. W. Mercer, C. R. Faust, and C. P. Spalding, *Methods for Monitoring Pump-and-Treat Performance,* EPA/600/R-94/123, U.S. EPA, ORD, R. S. Kerr Environmental Research Laboratory, Ada, OK, 1994.

17. Cohen, R. M., J. W. Mercer, R. M. Greenwald, and M. S. Beljin, *Design Guidelines for Conventional Pump-and-Treatment Systems,* EPA/540/S-97/504, U.S. Environmental Protection Agency, 1997.

18. Cole, J. A., ed., *Groundwater Pollution in Europe,* Water Information Center, Port Washington, NY, 347 pp., 1975.

19. Delleur, J. W., ed., *The Handbook of Groundwater Engineering,* CRC Press, Boca Raton, FL, 1999.

20. Delleur, J. W., Elementary groundwater flow and transport process, in *The Handbook of Groundwater Engineering,* J. W. Delleur, ed., CRC Press, Boca Raton, FL, 1999.

21. de Marsily, G., *Quantitative Hydrogeology—Groundwater Hydrology for Engineers,* Academic Press, Orlando, FL, 1986.

22. Deutsch, M., *Groundwater Contamination and Legal Controls in Michigan, U.S. Geological Survey Water-Supply Paper* 1691, 79 pp., 1963.

23. Dirkes, R. L., R. W. Hanf, and R. K. Woodruff, *Hanford site environmetal report for calendar year 1993,* PNL-9823, Batelle Pacific Northwest Laboratory, Richland, WA, 1994.

24. Domenico, P. A., and F. W. Schwartz, *Physical and Chemical Hydrogeology,* John Wiley & Sons, NY, 1990.

25. Drewry, W. A., and R. Eliassen, Virus movement in ground water, *Jour. Water Poll. Control Fed.,* v. 40, pp. 257–272, 1968.

26. Emrich, G. H., and G. L. Merritt, Effects of mine drainage on ground water, *Ground Water,* v. 7, no. 3, pp. 27–32, 1969.

27. Engineering and Technical Research Committee, *The migration of petroleum products in soil and ground water,* Publ. 4149, Amer. Petroleum Inst., 35 pp., 1972.

28. Everett, L. G., et al., *Monitoring Groundwater Quality: Methods and Costs,* Rept. EPA-600/4-76-023, U.S. Environmental Protection Agency, Las Vegas, 140 pp. 1976.

29. Faust, C. R., P. N. Sims, C. P. Spalding, P. F. Anderson, B. H. Lester, M. G. Shupe, and A. Harrover, FTWORK: *Groundwater flow and solute transport in three dimensions,* Version 2.8, Geo Trans, Inc., Sterling, VA, 1993.

30. Fetter, C. W., *Contaminant Hydrogeology,* 2nd ed., Prentice Hall, Upper Saddle River, NJ, 1999.

31. Field, R., et al., *Water Pollution and Associated Effects from Street Salting,* Rept. EPA-R2-73-257, U.S. Environmental Protection Agency, Cincinnati, 48 pp., 1973.

32. Franke, O. L., T. E. Reilly, D. W. Pollock, and J. W. LaBaugh, *Estimating Areas Contributing Recharge to Wells: Lessons from Previous Studies, U.S. Geological Survey Circular* 1174, 1998.

33. Freeze, R. A., and J. A. Cherry, *Groundwater,* Prentice Hall, Englewood Cliffs, NJ, 1979.

34. Freshley, M. D., and M. J. Graham, *Estimation of ground-water travel time at the Hanford site: Description, past work, and future need,* PNL-6328, Pacific Northwest Laboratory, Richland, WA, 1988.

35. Freshley, M. D., and P. D. Thorne, *Ground-water contribution to DOE from past Hanford operations,* PNWD-1974 HEDR, Battelle Pacific Northwest Laboratories, Richland, WA, 1992.

36. Fryberger, J. S., Investigation and rehabilitation of a brine-contaminated aquifer, *Ground Water,* v. 13, no. 2, pp. 155–160, 1975.

37. Gailey, R. M., and S. M. Gorelick, Design of optimal, reliable, plume capture schemes: Application to the Gloucester landfill ground-water contamination problem, *Ground Water,* v. 31, no. 1, pp. 107–114, 1993.

38. Gelhar, L. W., C. Welty, and K. R. Rehfeldt, A critical review of data on field-scale dispersion in aquifers, *Water Resources Research,* v. 28, no. 7, pp. 1955–1974, 1992.

39. Gilham, R. W., and L. R. Webber, Nitrogen contamination of groundwater by barnyard leachates, *Jour. Water Poll. Control Fed.,* v. 41. no. 10, pp. 1752–762, 1969.

40. Goode, D. J., *Simulating Contaminant Alternatives, Double-Porosity Exchange, and Water Age in Aquifers Using MOC3D, U.S. Geological Survey Fact Sheet* 086-99, 1999.

41. Graham, M. J., J. A. Thomas, and F. B. Metting, Groundwater, in *Water Resources Handbook,* L. W. Mays, ed., McGraw-Hill, New York, 1996.

42. Haggerty, R., and S. M. Gorelick, Design of multiple contaminant remediation: Sensitivity to rate-limited mass transport, *Water Resources Research,* v. 30, pp. 435–446, 1994.

43. Haitjema, H. M., J. Wittman, V. Kelson, and N. Bauch, *WhAEM: Program Documentation for the Wellhead Analytical Element Model,* EPA/600/R-94/210, U.S. EPA, ORD, R. S. Kerr Environmental Research Laboratory, Ada, OK, 1994.

44. Harvey, C. F., R. Haggerty, and S. M. Gorelick, Aquifer remediation: A method for estimating mass transfer rate coefficients and an evaluation of pulsed pumping, *Water Resources Research,* v. 30, no. 7, pp. 1979–1991, 1994.

45. Hassan, A. A., Water quality cycle—Reflection of activities of nature and man, *Ground Water,* v. 12, no. 1, pp. 16–21, 1974.

46. Heath, R. C., *Basic Ground-Water Hydrology, U.S. Geological Survey Water-Supply Paper* 2220, 1998.

47. Houlihan, M. F., and P. C. Lucia, Remediation of groundwater, in *The Handbook of Groundwater Engineering,* J. W. Delleur, ed., CRC Press, Boca Raton, FL, 1999.

48. Houlihan, M. F., and P. C. Lucia, Groundwater monitoring, in *The Handbook of Groundwater Engineering,* J. W. Delleur, ed., CRC Press, Boca Raton, FL, 1999.

49. Javandel, I., C. Doughty, and C. F. Tsang, *Groundwater Transport: Handbook of Mathematical Models,* Water Resources Monograph 10, American Geophysical Union, Washington, DC, 1984.

50. Javandel, I., and C. F. Tsang, Capture-zone type curves: A tool for aquifer cleanup, *Ground Water,* v. 24, pp. 616–625, 1986.

51. Johnson, P. C., Hydraulic design for groundwater contamination, in *Hydraulic Design Handbook,* L. W. Mays, ed., McGraw-Hill, New York, 1999.

52. Karubian, J. F., *Polluted Groundwater: Estimating the Effects of Man's Activities,* Rept. EPA 600/4-74-002, U.S. Environmental Protection Agency, Washington, DC, 1974.

53. Kaufman, M. I., Subsurface wastewater injection, Florida, *Jour. Irrig. Drain. Div.,* Amer. Soc. Civil Engrs., v. 99, no. IR1, pp. 53–70, 1973.

54. Keely, J. F., *Performance Evaluation of Pump-and-Treat Remediations,* EPA/540/4-89/005, U.S. EPA, ORD, R. S. Kerr Environmental Research Laboratory, Ada, OK, 1989.

55. Keely, J. F., and K. Boateng, Monitoring well installation, purging, and sampling techniques—Part 1: Conceptualizations, *Ground Water,* v. 25, no. 3, pp. 300–313, 1987.

56. Keely, J. F. and K. Boateng, Monitoring well installation, purging, and sampling techniques—Part 2: Case Histories, *Ground Water,* v 25, no. 4, pp. 300–313, 1987.

57. Kolpin, D. W., J. E. Barbash, and R. J. Gilliom, Pesticides in groundwater of the United States, 1992–1996, *Ground Water,* v. 38, no. 6, pp. 858–863, 2000.

58. Koterba, M. T., F. D. Wilde, and W. W. Lapham, *Ground-Water Data-Collection Protocols and Procedures for the National Water-Quality Assessment Program: Collection and Documentation of Water-Quality Samples and Related Data, U.S. Geological Survey Open-File Report* 95-399, Reston, VA, 1995.

59. Lapham, W. W., F. D. Wilde, and M. T. Koterba, *Ground-Water Data-Collection Protocols and Procedures for the National Water-Quality Assessment Program: Selection, Installation, and Documentation of Wells, and Collection of Related Data, U.S.* Geological Survey Open-File Report 95-398, Reston, VA, 1995.

60. LeGrand, H. E., Environmental framework of ground-water contamination, *Ground Water,* v. 3, no. 2, pp. 11–15, 1965.

61. LeGrand, H. E., Patterns of contaminated zones of water in the ground, *Water Resources Research,* v. 1, pp. 83–95, 1965.

62. Marquis, Jr., S. A., and D. Dineen, Comparison between pump and treat, biorestoration, and biorestoration/pump and treat combined: Lessons from computer modeling, *Ground Water Monitoring and Remediation,* v. 14, no. 2, pp. 105–119, 1994.

63. Matis, J. R., Petroleum contamination of ground water in Maryland, *Ground Water,* v. 9, no. 6, pp. 57–61, 1971.

64. McKee, J. E. et al., Gasoline in groundwater, *Jour. Water Poll. Control Fed.,* v. 44, no. 2, pp. 293–302, 1972.

65. Mercer, J. W., D. C. Skipp, and Griffin, *Basics of Pump-and-Treatment Groundwater Remediation,* EPA/600/8-90/003, U.S. Environmental Protection Agency, R. S. Kerr Environmental Research Laboratory, Ada, OK, 1990.

66. Meyer, C. F., ed., *Polluted Groundwater: Some Causes, Effects, Controls, and Monitoring,* Rept. EPA-600/4–73–0016, U.S. Environmental Protection Agency, Washington, DC, 282 pp., 1973.

67. Miller, D. W., ed., *Waste Disposal Effects on Ground Water,* Premier Press, Berkeley, CA, 512 pp., 1980.

68. Mink, L. L. et al., Effect of early day mining operations on present day water quality, *Ground Water,* v. 10, no. 1, pp. 17–26, 1972.

69. Morrison, R. D., *Ground Water Monitoring Technology: Procedures, Equipment, and Applications,* Timco Manufacturing, Praire du Sac, WI, 1983.

70. Mullaney, J. R., and S. J. Grady, *Hydrogeology and Water Quality in a Surficial Aquifer Underlying an Urban Area, Manchester, Connecticut, U.S. Geological Survey Water Resources Investigation Report* 97-4195, 1997.

71. National Research Council (NRC), Water Science and Technology Board, *Alternatives for Ground Water Cleanup,* Washington, DC, 1994.

72. Neuman, S. P., Universal scaling of hydraulic conductivities and dispersivities in geologic media, *Water Resources Research,* v. 26, no. 8, pp. 1749–1758, 1990.

73. Nielson, D. M., ed., *Practical Handbook of Ground-Water Monitoring,* Lewis Publishers, Chelsea, MI, 1991.

74. Ogata, A., *Theory of Dispersion in a Granular Medium, U.S. Geological Survey Professional Paper* 411-I, U.S. Government Printing Office, Washington, DC, 1970.

75. Ogata, A., and R. B. Banks, *A Solution of the Differential Equation of Longitudinal Dispersion in Porous Media, U.S. Geological Survey Professional Paper* 411-A, U.S. Government Printing Office, Washington, DC, 1961.

76. Osgood, J. L., Hydrocarbon dispersion in ground water: Significance and characteristics, *Ground Water,* v. 12, no. 6, pp. 427–438, 1974.

77. Palmer, C. M., *Contaminant Hydrogeology,* Lewis Publishers, Boca Raton, FL, 1996.

78. Quan, E. L., et al., Subsurface sewage disposal and contamination of ground water in East Portland, Oregon, *Ground Water,* v. 12, no. 6, pp. 356–368, 1974.

79. Robbins, J. W. D., and G. J. Kriz, Relation of agriculture to groundwater pollution: a review, *Trans. Amer. Soc. Agric. Engrs.,* v. 12, pp. 397–403, 1969.

80. Robinson, R. A., and R. H. Stokes, *Electrolyte Solutions,* 2nd ed., Butterworth, London, 1965.

81. Romero, J. C., The movement of bacteria and virus through porous media, *Ground Water,* v. 8, no. 2, pp. 37–48, 1970.

82. Rumer, R. R., and J. K. Mitchell, *Assessment of barrier containment technologies — A comprehensive treatment for environmental remediation applications,* National Technical Information Services, Springfield, VA, 1995.

83. Sasman, R. T., Thermal pollution of ground water by artificial recharge, *Water and Sewage Works,* v. 119, no. 12, pp. 52–55, 1972.

84. Satkin, R. L., and P. B. Bedient, Effectiveness of various aquifer restoration schemes under variable hydrogeologic conditions, *Ground Water,* v. 26, no. 4, pp. 488–498, 1988.

85. Squillace, P. J., and C. V. Price, *Urban Land-Use Study Plan for the National Water-Quality Assessment Program, U.S. Geological Survey Open-File Report* 96-217, 1996.

86. Stackelberg, P. E., *Relation Between Land Use and Quality of Shallow, Intermediate, and Deep Ground Water in Nassau and Suffolk Counties, Long Island, New York, U.S. Geological Survey Water Resources Investigations Report* 94-4080, 1995.

87. Strack, O. D. L., E. I. Anderson, M. Bakker, W. C. Olsen, J. C. Panda, R. W. Pennings, and D. R. Steward, *CZAEM User's Guide: Modeling Capture Zones of Ground-Water Using Analytic Elements,* EPA/600/R-94/174, Office of Research and Development, Cincinnati, OH, 1994.

88. Terry, R. C., Jr., *Road Salt, Drinking Water, and Safety,* Ballinger, Cambridge, MA, 161 pp., 1974.

89. Thomas, M. A., The effect of residential development on groundwater quality near Detroit, Michigan, *Jour. American Water Resources Assn.,* v. 36, no. 5, pp. 1023–1038, October 2000.

90. Tinlin, R. M., ed., *Monitoring Groundwater Quality: Illustrative Examples,* Rept. EPA-600/4-76-036, U.S. Environmental Protection Agency, Las Vegas, 81 pp., 1976.

91. Todd, D. K., and D. E. O. McNulty, *Polluted Groundwater,* Water Information Center, Port Washington, NY, 179 pp., 1976.

92. Todd, D. K. et al., A groundwater quality monitoring methodology, *Jour. Amer. Water Works Assoc.,* v. 68, pp. 586–593, 1976.

93. Todd, D. K. et al., *Monitoring Groundwater Quality: Monitoring Methodology,* Rept. EPA-600/4-76-026, U.S. Environmental Protection Agency, Las Vegas, 154 pp., 1976.

94. U.S. Department of Agriculture Agricultural Research Service, Pesticide properties database (http://www.arsusda.gov/acsl/ppbd.html), 1995.

95. U.S. Department of Energy (DOE), *Closing the Circle on the Splitting of the Atom, The Environmental Legacy of Nuclear Weapons Productions in the United States and What the Department of Energy is Doing About It,* U.S. Department of Energy, Office of Environmental Management, Washington, DC, 1995.

96. U.S. Environmental Protection Agency, *Land Application of Wastewater,* Rept. EPA 903/9-75-017, Philadelphia, 94 pp., 1975.

97. U.S. Environmental Protection Agency (EPA), *RCRA Ground-Water Monitoring,* technical enforcement guidance document, Office of Waste Programs Enforcement and Office of Solid Waste and Emergency Response, OSWER-9950.1, Washington, DC, 1986.

98. U.S. Environmental Protection Agency (EPA), *Handbook of Suggested Practices for the Design and Installation of Ground-Water Monitoring Wells,* PB90-159-807, Washington, DC, 1989.

99. U.S. Environmental Protection Agency (EPA), *Guidance for Evaluating the Technical Impracticability of Ground-Water Restoration,* Directive 9234.2-25, Interim Final, U.S. EPA, Office of Solid Waste and Emergency Response, Washington, DC, 1993.

100. U.S. Environmental Protection Agency (EPA), *Subsurface Characterization and Monitoring Techniques: A Desk Reference Guide, Volume I: Solids and Ground Water,* Appendices A and B, EPA/625/R-93-003a, Washington, DC, 1993.

101. U.S. Environmental Protection Agency (EPA), *Subsurface Characterization and Monitoring Techniques: A Desk Reference Guide, Volume II: The Vadose Zone, Field Screening and Analytical Methods,* Appendices C and D, EPA/625/R-93-003b, Washington, DC, 1993.

102. U.S. Environmental Protection Agency (EPA), *Superfund Innovative Technology Evaluation Program, Technology Profiles,* 7th ed., EPA/540/R-94/526, Washington, DC, 1994.

103. U.S. Environmental Protection Agency (EPA), *Ground-Water and Leachate Treatment Systems,* EPA/625/R-94/005, Washington, DC, 1995.

104. U.S. Geological Survey, *The Quality of Our Nation's Waters—Nutrients and Pesticides, U.S. Geological Survey Circular* 1225, 1999.

105. Van der Heijde, P. K. M., and O. A. Elnawawy, *Compilation of ground-water models,* EPA/600/R-93/118, U.S. EPA, ORD, R. S. Kerr Environmental Research Laboratory, Ada, OK, 1993.

106. van Genuchten, M. T., Analytical solutions for chemical transport with simultaneous adsorption, zero-order production and first order decay, *Journal of Hydrology,* v. 49, pp. 213–233, 1981.

107. Warner, D. L., and J. H. Lehr, *An Introduction to the Technology of Subsurface Wastewater Injection,* Rept. EPA-600/2-77-240, U.S. Environmental Protection Agency, Ada, OK, 344 pp., 1977.

108. Williams, D. E., and D. G. Wilder, Gasoline pollution of a groundwater reservoir—A case history, *Ground Water,* v. 9, no. 6, pp. 50–56, 1971.

109. Xu, M., and Y. Eckstein, Use of weighted least squares method in evaluation of the relationship between dispersion and field scale, *Ground Water,* v. 33, no. 6, pp. 905–908, 1995.

110. Zanoni, A. E., Ground water pollution and sanitary landfills—A critical review, *Ground Water,* v. 10, no. 1, pp. 3–16, 1972.

Chapter 9

Groundwater Flow Modeling Techniques

Groundwater is essentially a hidden resource; therefore, studies of groundwater under both natural and artificial boundary conditions have employed modeling techniques. Fortunately, during the past several decades, computer simulation models for analyzing flow and solute transport in groundwater systems have played an increasing role in the evaluation approaches to groundwater development and management.

9.1 WHY DEVELOP GROUNDWATER MODELS?

Alley et al.[1] define groundwater models:

> Groundwater models are an attempt to represent the essential features of the actual groundwater system by means of a mathematical counterpart. The underlying philosophy is that an understanding of the basic laws of physics, chemistry, and biology that describe groundwater flow and transport and an accurate description of the specific system under study will enable a quantitative representation of the cause and effect relationships for that system. Quantitative understanding of cause and effect relationships enables forecasts to be made for any defined set of conditions.

Konikow and Bredehoeft[31] point out that such forecasts, which are usually outside the range of observed conditions, typically are limited by uncertainties due to sparse and inaccurate data, poor definition of stresses acting on the system, and errors in system conceptualization. Even though forecasts of future events using models may be imprecise, they do represent the best available information at a given time that can be used in decision making. Simulation models can be used as learning tools to identify additional data needed in order to better define and understand groundwater systems. These models have the capability to test and quantify the consequences of various errors and uncertainties in the information needed to determine the cause and effect relationships and related model-based forecasts. In the process of forecasting, this model capability may be the most important aspect to define the uncertainties of the forecasts, allowing water managers to evaluate the significance of their decisions. Also this process may allow the evaluation of unexpected consequences of their decisions.

Groundwater modeling has become a very important process in managing groundwater resources. It is interesting to note that each of the classic texts by Walton,[59] Freeze and Cherry,[11] and Todd,[57] dated respectively 1970, 1979, and 1980, had very little material dealing with groundwater modeling using the computer. The past two decades have resulted in tremendous changes in our use of the computer for groundwater management.

9.2 TYPES OF GROUNDWATER MODELS

Todd[57] classified groundwater models as porous media models (sand tank models and transparent models), analog models (viscous fluid models, membrane models, Moire pattern models, thermal models, and blotting paper models), electric analog models (conductive solid models, resistance–capacitance networks, and resistance networks), and digital computer models (finite difference models, finite element models, and hybrid computer models). Prickett[49, 50] provided a detailed list of references on the various models that existed during that time, listing applications of various types of groundwater models, including aquifer features, purpose of the models, and schematic sketches of the models.

A great deal has changed in groundwater modeling since the 1970s, when the numerical (or mathematical) modeling approach was a relatively new field. It was not extensively pursued until the mid-1960s, when digital computers with adequate capacity became generally available (Appel and Bredehoeft).[5] During the 1970s, many articles in the literature, government publications, and books, such as Peaceman,[43] Pinder and Gray,[45] Prickett and Lonnquist,[51] Remson et al.,[54] Rushton and Redshaw,[55] Texas Water Development Board,[56] and Trescott et al.,[58] appeared. Presently, the use of mathematical models is the state-of-the-art in practice for groundwater modeling.

Groundwater models (for the purposes of this book) are physically based mathematical models derived from Darcy's law and the law of conservation of mass.[64] Various established solution techniques based upon either finite difference or finite element approximations, or a combination of both, are available for solving the governing equations of the model. The accuracy of the solutions (model predictions) is dependent upon the reliability of the estimated model parameters and the accuracy of the prescribed boundary conditions.

Groundwater models can be classified in several ways: steady state or transient; confined, unconfined, or a combination of confined and unconfined; one-dimensional, two-dimensional, quasi–three-dimensional, or three-dimensional. They can be solved using *finite difference methods* or *finite element methods* or a combination of these. The major emphasis in this chapter is on the finite difference method.

The finite difference method requires a rectangular element shaped discretization of the aquifer and the finite element method consists of a triangular discretization as illustrated in Figure 9.2.1. The size and shape of the finite elements are arbitrary, typically being triangular or quadrilateral. In fact, the elements can be disordered and nonuniform and should be the smallest where flow is concentrated, such as near a well. Also, it is easy to define the boundaries of

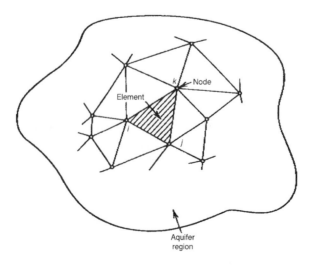

Figure 9.2.1. Example of a triangular finite element within an aquifer (after Prickett[49]).

irregularly shaped aquifers and to ensure that node points coincide with monitoring wells or various types of geographic features. The mathematical basis for finite element methods is more complex than for the finite difference method. Galerkin's method is the most commonly used solution method for the finite element method. Detailed descriptions, while not discussed in this book, of the finite element method can be found elsewhere in the literature.[23, 54, 62]

9.3 STEPS IN THE DEVELOPMENT OF A GROUNDWATER MODEL

Anderson and Woessner[4] presented the following modeling protocol for groundwater modeling:

1. Determine the purpose of the model.
2. Develop a conceptual model.
3. Select the governing equations and computer code.
4. Model design—put the conceptual model into a form suitable for modeling.
5. Calibration—establish that the model can reproduce field-measured heads and flows.
6. Calibration sensitivity analysis—determine the effect of uncertainty on the calibrated model.
7. Model verification—use the calibrated parameter values and stresses to reproduce a second set of field data in order to establish greater confidence in the model.
8. Prediction to quantify the response of the system to future events.
9. Predictive sensitivity analysis to quantify the effect of parameter uncertainty.
10. Presentation of modeling design and results.
11. Post-audit analysis in the future with new field data.
12. Model redesign using insight from the post-audit analysis.

See Hill[25] for more insight on effective model calibration.

9.4 SIMULATION OF TWO-DIMENSIONAL GROUNDWATER SYSTEMS

The first finite difference groundwater models that found fairly widespread usage were developed for two-dimensional (horizontal) flow. These included models such the ones developed by Prickett and Lonnquist,[51] Trescott et al.,[58] and the Texas Water Development Board.[56] Even though the use of these models has been superceded by the U.S. Geological Survey MODFLOW model (which can be used for either two- or three-dimensional modeling), it is worthwhile to describe the two-dimensional framework as a precursor to a description of the more complicated three-dimensional models.

9.4.1 Governing Equations

Darcy's law relates the Darcy flux v with dimension (L/T) to the rate of headloss per unit length of porous medium $\partial h / \partial l$, as discussed in Chapter 3. The negative sign indicates that the total head is decreasing in the direction of flow because of friction. This law applies to a cross section of porous medium, which is large compared to the cross section of individual pores and grains of the medium. At this scale, Darcy's law describes a steady uniform flow of constant velocity, in which the net force on any fluid element is zero. For unconfined saturated flow, the two forces are gravity and friction. Darcy's law (Equation 3.1.5) can also be expressed in terms of the transmissivity for confined conditions as

$$v = -\frac{T}{b}\frac{\partial h}{\partial l} \tag{9.4.1}$$

or for unconfined conditions as

$$v = -\frac{T}{h}\frac{\partial h}{\partial l} \tag{9.4.2}$$

where b is the thickness of the confined aquifer and h is the saturated thickness of the unconfined aquifer.

Considering two-dimensional (horizontal) flow, a general flow equation can be derived by considering flow through a rectangular element (control volume), as shown in Figure 9.4.1. The flow components ($q = Av$) for the four sides of the element are expressed using Darcy's law where $A = \Delta x \cdot h$ for unconfined conditions and $A = \Delta x \cdot b$ for confined conditions so that

$$q_1 = -T_{x_{i-1,j}}\Delta y_j\left(\frac{\partial h}{\partial x}\right)_1 \tag{9.4.3a}$$

$$q_2 = -T_{x_{i,j}}\Delta y_j\left(\frac{\partial h}{\partial x}\right)_2 \tag{9.4.3b}$$

$$q_3 = -T_{y_{i,j+1}}\Delta x_i\left(\frac{\partial h}{\partial y}\right)_3 \tag{9.4.3c}$$

$$q_4 = -T_{y_{i,j}}\Delta x_i\left(\frac{\partial h}{\partial y}\right)_4 \tag{9.4.3d}$$

where $T_{x_{i,j}}$ is the transmissivity in the x flow direction from element (i, j) to element ($i + 1, j$). The terms $(\partial h/\partial x)_1$, $(\partial h/\partial x)_2$, . . . , define the hydraulic gradients at the element sides 1, 2, ...

The rate at which water is stored or released in the element over time is

$$q_5 = S_{i,j}\Delta x_i\Delta y_i\frac{\partial h}{\partial t} \tag{9.4.4}$$

in which $S_{i,j}$ is the storage coefficient for element (i, j). In addition, the flow rate q_6 for constant net withdrawal or recharge from the element over time interval Δt is considered as

$$q_6 = q_{i,j,t} \tag{9.4.5}$$

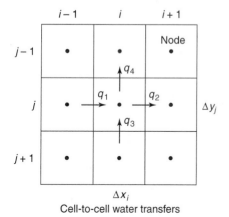

Δy_j

Δx_i

Cell-to-cell water transfers

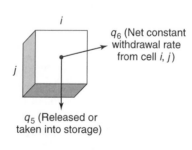

q_6 (Net constant withdrawal rate from cell i, j)

q_5 (Released or taken into storage)

Figure 9.4.1. Finite difference grid showing the control volume for an element of a two-dimensional (horizontal) flow. The node within each grid (cell) is the point where the head is calculated.

in which $q_{i,j,t}$ has a positive value for pumping, whereas it has a negative value for recharge.

By continuity the flow into and out of a grid or cell is

$$q_1 - q_2 + q_3 - q_4 = q_5 + q_6 \tag{9.4.6}$$

Substituting in Equations 9.4.3 and 9.4.5 gives

$$-T_{x_{i-1,j}} \Delta y_j \left(\frac{\partial h}{\partial x}\right)_1 + T_{x_{i,j}} \Delta y_j \left(\frac{\partial h}{\partial x}\right)_2 - T_{y_{i,j+1}} \Delta x_i \left(\frac{\partial h}{\partial y}\right)_3 + T_{y_{i,j}} \Delta x_i \left(\frac{\partial h}{\partial y}\right)_4$$
$$= S_{i,j} \Delta x_i \Delta y_i \frac{\partial h}{\partial t} + q_{i,j,t} \tag{9.4.7}$$

Dividing Equation 9.4.7 by $\Delta x \, \Delta y$ and simplifying for constant transmissivities in the x and y directions yields

$$-T_x \left[\frac{\left(\frac{\partial h}{\partial x}\right)_1 - \left(\frac{\partial h}{\partial x}\right)_2}{\Delta x_i}\right] - T_y \left[\frac{\left(\frac{\partial h}{\partial y}\right)_3 - \left(\frac{\partial h}{\partial y}\right)_4}{\Delta y_j}\right] = S_{i,j} \frac{\partial h}{\partial t} + \frac{q_{i,j,t}}{\Delta x_i \Delta y_j} \tag{9.4.8}$$

For Δx and Δy infinitesimally small, the terms in brackets [ ] become second derivatives of h. Then Equation 9.4.8 reduces to

$$T_x \frac{\partial^2 h}{\partial x^2} + T_y \frac{\partial^2 h}{\partial y^2} = S \frac{\partial h}{\partial t} + W \tag{9.4.9}$$

which is the general partial differential equation for unsteady flow in the horizontal direction in which $W = q_{i,j,t}/\Delta x_i \Delta y_j$ is a sink term with dimensions (L/T).

In the more general case for unsteady, two-dimensional heterogeneous anisotropic flow, Equation 9.4.9 is expressed as

$$\frac{\partial}{\partial x}\left(T_x \frac{\partial h}{\partial x}\right) + \frac{\partial}{\partial y}\left(T_y \frac{\partial h}{\partial y}\right) = S \frac{\partial h}{\partial t} + W \tag{9.4.10a}$$

or more simply

$$\frac{\partial}{\partial x_i}\left(T_{i,j} \frac{\partial h}{\partial x_j}\right) = S \frac{\partial h}{\partial t} + W \quad i,j = 1,2 \tag{9.4.10b}$$

9.4.2 Finite Difference Equations

The partial derivative expressions for Darcy's law, (Equations 9.4.3a–d), can be expressed in finite difference form for time t in Equation 9.4.7 using

$$\left(\frac{\partial h}{\partial x}\right)_1 = \left(\frac{h_{i-1,j,t} - h_{i,j,t}}{\Delta x_i}\right) \tag{9.4.11a}$$

$$\left(\frac{\partial h}{\partial x}\right)_2 = \left(\frac{h_{i,j,t} - h_{i+1,j,t}}{\Delta x_i}\right) \tag{9.4.11b}$$

$$\left(\frac{\partial h}{\partial y}\right)_3 = \left(\frac{h_{i,j+1,t} - h_{i,j,t}}{\Delta y_j}\right) \tag{9.4.11c}$$

$$\left(\frac{\partial h}{\partial y}\right)_4 = \left(\frac{h_{i,j,t} - h_{i,j-1,t}}{\Delta y_j}\right) \tag{9.4.11d}$$

and the time derivative in Equation 9.4.10 is

$$\frac{\partial h}{\partial t} = \left(\frac{h_{i,j,t} - h_{i,j,t-1}}{\Delta t}\right) \tag{9.4.12}$$

Substituting Equations 9.4.11 and 9.4.12 into Equation 9.4.7 yields

$$-T_{x_{i-1,j}} \Delta y_j \left(\frac{h_{i-1,j,t} - h_{i,j,t}}{\Delta x_i}\right) + T_{x_{i,j}} \Delta y_j \left(\frac{h_{i,j,t} - h_{i+1,j,t}}{\Delta x_i}\right)$$

$$-T_{y_{i,j+1}} \Delta x_i \left(\frac{h_{i,j+1,t} - h_{i,j,t}}{\Delta y_j}\right) + T_{y_{i,j}} \Delta x_i \left(\frac{h_{i,j,t} - h_{i,j-1,t}}{\Delta y_j}\right) \tag{9.4.13}$$

$$-S_{i,j} \Delta x_i \Delta y_j \left(\frac{h_{i,j,t} - h_{i,j,t-1}}{\Delta t}\right) - q_{i,j,t} = 0$$

which can be further simplified to

$$A_{i,j}h_{i,j,t} + B_{i,j}h_{i-1,j,t} + C_{i,j}h_{i+1,j,t} + D_{i,j}h_{i,j+1,t} +$$

$$E_{i,j}h_{i,j-1,t} + F_{i,j,t} = 0 \tag{9.4.14}$$

where

$$A_{i,j} = \left[T_{x_{i-1,j}}\frac{\Delta y_j}{\Delta x_i} + T_{x_{i,j}}\frac{\Delta y_j}{\Delta x_i} + T_{y_{i,j+1}}\frac{\Delta x_i}{\Delta y_j} + T_{y_{i,j}}\frac{\Delta x_i}{\Delta y_j} - S_{i,j}\frac{\Delta x_i \Delta y_j}{\Delta t}\right] \tag{9.4.15a}$$

$$B_{i,j} = -T_{x_{i-1,j}}\frac{\Delta y_j}{\Delta x_i} \tag{9.4.15b}$$

$$C_{i,j} = -T_{x_{i,j}}\frac{\Delta y_j}{\Delta x_i} \tag{9.4.15c}$$

$$D_{i,j} = -T_{y_{i,j+1}}\frac{\Delta x_i}{\Delta y_j} \tag{9.4.15d}$$

$$E_{i,j} = -T_{y_{i,j}}\frac{\Delta x_i}{\Delta y_j} \tag{9.4.15e}$$

$$F_{i,j,t} = -S_{i,j}\frac{\Delta x_i \Delta y_j}{\left(\frac{h_{i,j,t} - h_{i,j,t-1}}{\Delta t}\right)} - q_{i,j,t} \tag{9.4.15f}$$

The coefficients $A_{i,j}$, $B_{i,j}$, $C_{i,j}$, $D_{i,j}$, $E_{i,j}$, and $F_{i,j,t}$ are linear functions of the thickness of cell (i, j) and the thickness of one of the adjacent cells. For artesian conditions, this thickness is a known constant, so if cell (i, j) and its neighbors are artesian, Equation 9.4.14 is linear for all t. For unconfined (water table) conditions, the thickness of cell (i, j) is $h_{i,j,t} - BOT_{i,j}$, where

$BOT_{i,j}$ is the average elevation of the bottom of the aquifer for cell (i, j). Then for unconfined conditions, Equation 9.4.14 involves products of heads and is nonlinear in terms of the heads.

9.4.3 Solution

An *iterative alternating direction implicit (IADI) procedure* can be used to solve the set of equations. The IADI procedure involves reducing a large set of equations to several smaller sets of equations. A smaller set of equations is generated by writing Equation 9.4.14 for each cell or element in a column but assuming that the heads for the nodes on the adjacent columns are known. The unknowns in this set of equations are the heads for the nodes along the column. The heads for the nodes along adjoining columns are not considered unknowns. This set of equations is solved by Gauss elimination and the process is repeated until each column is treated. The next step is to develop a set of equations along each row, assuming the heads for the nodes along adjoining rows are known. The set of equations for each row is solved and the process is repeated for each row in the finite difference grid.

9.4.4 Case Study

Once the sets of equations for the columns and the sets of equations for the row have been solved, one "iteration" has been completed. The iteration process is repeated until the procedure converges. Once convergence is accomplished, the terms $h_{i,j}$ represent the heads at the end of the time step. These heads are used as the beginning heads for the following time step. For a more detailed discussion of the IADI procedure, see references 44, 51, or 62. Two-dimensional finite difference models for groundwater flow are by Prickett and Lonnquist[51] and by Trescott et al.[58]

Two studies utilizing a two-dimensional finite difference groundwater model have been made of the Edwards (Balcones Fault Zone) Aquifer. First, this aquifer has been modeled using the GWSIM groundwater simulation model developed by the Texas Water Development Board.[56] GWSIM is a finite difference simulation model that uses the IADI method similar to the model by Prickett and Lonnquist.[51] The finite difference grid for the Edwards Aquifer is shown in Figure 9.4.2, which has 856 active cells to describe the aquifer. Second, Wanakule[61] has modeled a smaller portion of the Edwards aquifer called the Barton Springs—Edwards Aquifer in Austin, Texas. This application used a finite difference grid system (Figure 9.4.3) containing 330 cells whose dimensions varied from 0.379×0.283 mi^2 to 0.95×1.51 mi^2.

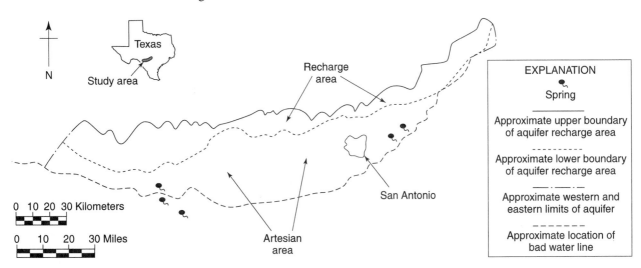

Figure 9.4.2a. Edwards (Balcones Fault Zone) Aquifer, San Antonio, TX, region.[56]

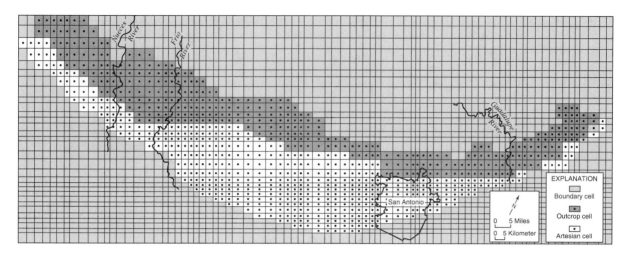

Figure 9.4.2b. Cell map used for the digital computer model of the Edwards (Balcones Fault Zone) Aquifer (after Klempt et al.[30]).

The above two case studies are for the purposes of illustrating the two-dimensional finite difference method for horizontal flows. The present day modeling of the Edwards Aquifer is being done using the MODFLOW model that is introduced in the following sections of this chapter.

9.5 THREE-DIMENSIONAL GROUNDWATER FLOW MODEL

9.5.1 Derivation of Finite Difference Equations

The partial differential equation for transient three-dimensional groundwater flow in a heterogeneous and anisotropic medium (provided the principal axes of hydraulic conductivity are aligned with the coordinate directions), for a confined or unconfined aquifer, is expressed as

$$\frac{\partial}{\partial x}\left(K_{xx}\frac{\partial h}{\partial x}\right) + \frac{\partial}{\partial y}\left(K_{yy}\frac{\partial h}{\partial y}\right) + \frac{\partial}{\partial z}\left(K_{zz}\frac{\partial h}{\partial z}\right) - W = S_s\frac{\partial h}{\partial t} \qquad (9.5.1)$$

where K_{xx}, K_{yy}, and K_{zz} are the hydraulic conductivities along the x, y, and z coordinate axes parallel to the major axes of hydraulic conductivities; h is the potentiometric head; W is a volumetric flux per unit volume representing sources (W is negative) and/or sinks (W is positive) of water; S_s is the specific storage of the porous medium; and t is time. K_{xx}, K_{yy}, K_{zz}, and S_s are functions of space (x, y, z) and W is a function of space and time (x, y, z, t).

Equation 9.5.1, together with specification of flow and/or head conditions at the boundaries of an aquifer system and specification of initial head conditions, constitutes a mathematical representation of a groundwater flow system. The solution of Equation 9.5.1 requires the use of a numerical method such as the finite difference method. The application of finite differences replaces the continuous system described by Equation 9.5.1 with a finite set of discrete points in space and time, such that the partial derivatives are replaced by terms calculated from the head differences calculated at these points. A system of simultaneous linear algebraic difference equations results from this process, which is solved for the heads at specific points and times that constitute the approximation to the time-varying head distribution.

Figure 9.5.1 illustrates a three-dimensional hypothetical aquifer system discretized into a mesh of blocks called *cells*, described by rows, columns, and layers. The hypothetical system is discretized into five rows, nine columns, and five layers. An *i, j, k* indexing system is used

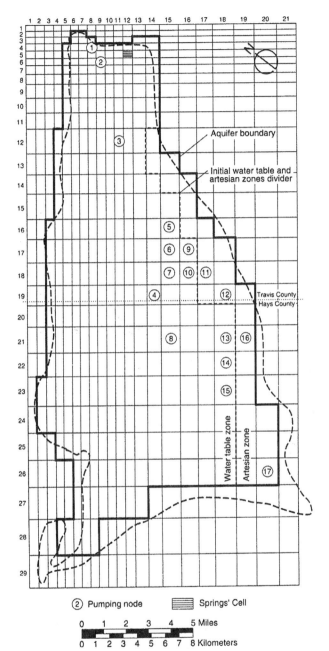

(2) Pumping node   ▦ Springs' Cell

0 1 2 3 4 5 Miles
0 1 2 3 4 5 6 7 8 Kilometers

Figure 9.4.3. Pumping locations used in Barton Springs–Edwards Aquifer model (after Wanakule[61]).

to reference the cells. Layers generally refer to horizontal geohydrologic units or intervals, so that the k index denotes changes along the vertical (z-direction in Cartesian coordinates). The layers are numbered from the top down so that an increment in the k index corresponds to a decrease in elevation. Rows are considered parallel to the x axis and columns are parallel to the x axis.

A *node* is a point within each cell (grid) where the head is calculated. Figure 9.5.2 illustrates block-centered and point-centered grid systems. Whichever grid system is used, the

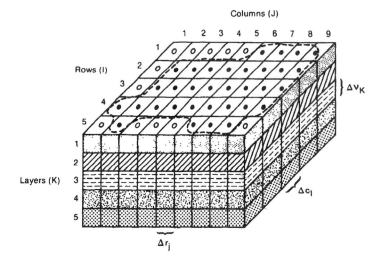

Figure 9.5.1. A discretized hypothetical aquifer system (McDonald and Harbaugh[39]).

spacing of nodes should be selected so that the hydraulic properties are generally uniform over the extent of the cell. The finite difference equations developed herein are applicable for either grid system. Figure 9.5.3 illustrates cell (i, j, k) and the six adjacent cells that are used to derive the finite difference equation for the cell.

Figure 9.5.4 illustrates the flow into cell (i, j, k) from cell $(i, j - 1, k)$. The effective hydraulic conductivity for the entire region between the nodes is denoted as $KR_{i,j-1/2,k}$. The subscript $j - 1/2$ does not relate to a specific point between the nodes but is used to relate to the region from cell $(i, j - 1, k)$ to cell (i, j, k). The effective hydraulic conductivity is computed as the harmonic mean, so that for flow into cell (i, j, k) from cell $(i, j - 1, k)$, the *effective hydraulic conductivity* is

$$KY_{i-1/2,j,k} = \frac{\left(\Delta y_{i-1} + \Delta y_i\right)/2}{\left(\dfrac{\Delta y_{i-1}/2}{KY_{i-1,j,k}} + \dfrac{\Delta y_i/2}{KY_{i,j,k}}\right)} = \frac{KY_{i-1,j,k}KY_{i,j,k}\left(\Delta y_{i-1} + \Delta y_i\right)}{KY_{i,j,k}\Delta y_{i-1} + KY_{i-1,j,k}\Delta y_i} \qquad (9.5.2)$$

If Δy is constant, then Equation 9.5.2 is expressed as

$$KY_{i-1/2,j,k} = \frac{2KY_{i,j,k}KY_{i-1,j,k}}{KY_{i,j,k} + KY_{i-1,j,k}} \qquad (9.5.3)$$

Flows are considered positive if they are entering cell (i, j, k) and the negative sign incorporated into Darcy's law is dropped from all terms. The flow into cell (i, j, k) from cell $(i, j - 1, k)$ is expressed by Darcy's law as (McDonald and Harbaugh[39])

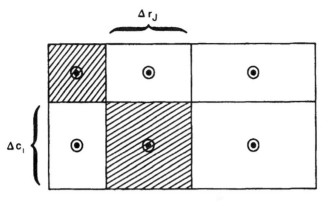

Block-Centered Grid System

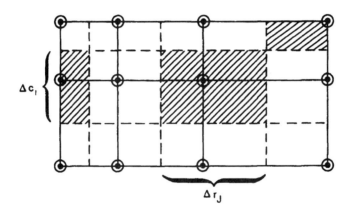

Point-centered grid system

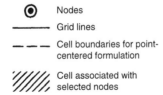

⊙ Nodes

─── Grid lines

─ ─ ─ Cell boundaries for point-
centered formulation

//// Cell associated with
selected nodes

Figure 9.5.2. Grids showing
the difference between block-
centered and point-centered
grids (McDonald and
Harbaugh[39]).

$$q_{i,j-1/2,k} = \mathrm{KY}_{i,j-1/2,k} \Delta x_i \Delta z_k \left(\frac{h_{i,j-1,k} - h_{i,j,k}}{\Delta y_{j-1/2}} \right) \qquad (9.5.4)$$

where $q_{i,j-1/2,k}$ is the volumetric fluid discharge through the face between cell $(i, j - 1, k)$ and cell (i, j, k); $h_{i,j,k}$ and $h_{i,j-1,k}$ are the heads at the respective nodes; $\Delta x_i \Delta z_k$ is the area of the cell face normal to the row direction; and $\Delta y_{j-1/2}$ is the distance between nodes (i, j, k) and $(i, j - 1, k)$. Equation 9.5.4 defines the flow for a one-dimensional steady-state flow through a block of aquifer extending from node $(i, j -1, k)$ to node (i, j, k), and having cross-sectional area $\Delta x_i \Delta z_k$.

Darcy's law is applied in a similar manner to describe the flow through the remaining five sides of cell (i, j, k) to obtain the following (McDonald and Harbaugh[39]):

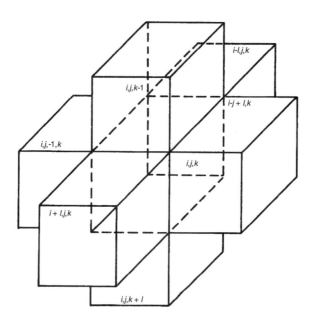

Figure 9.5.3. Cell (i, j, k) and indices for the six adjacent cells (McDonald and Harbaugh[39]).

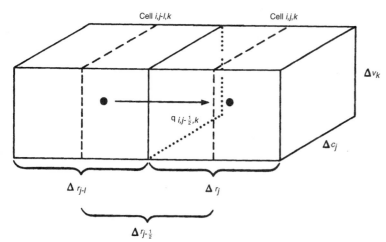

Figure 9.5.4. Flow into cell (i, j, k) from cell $(i, j - 1, k)$ (McDonald and Harbaugh[39]).

Flow cell (i, j, k) to cell $(i, j + 1, k)$

$$q_{i,j+1/2,k} = KY_{i,j+1/2,k} \Delta x_i \Delta z_k \left(\frac{h_{i,j+1,k} - h_{i,j,k}}{\Delta y_{j+1/2}} \right) \tag{9.5.5}$$

Flow cell (i, j, k) to cell $(i + 1, j, k)$

$$q_{i+1/2,j,k} = KX_{i+1/2,j,k} \Delta y_j \Delta z_k \left(\frac{h_{i+1,j,k} - h_{i,j,k}}{\Delta x_{i+1/2}} \right) \tag{9.5.6}$$

Flow cell $(i - 1, j, k)$ to cell (i, j, k)

$$q_{i-1/2,j,k} = KX_{i-1/2,j,k} \Delta y_j \Delta z_k \left(\frac{h_{i-1,j,k} - h_{i,j,k}}{\Delta x_{i-1/2}} \right) \tag{9.5.7}$$

Flow cell (i, j, k) to cell $(i, j, k + 1)$

$$q_{i,j,k+1/2} = \text{KZ}_{i,j,k+1/2}\Delta y_j \Delta x_i \left(\frac{h_{i,j,k+1} - h_{i,j,k}}{\Delta z_{k+1/2}} \right) \qquad (9.5.8)$$

Flow cell $(i, j, k - 1)$ to cell (i, j, k)

$$q_{i,j,k-1/2} = \text{KZ}_{i,j,k-1/2}\Delta y_j \Delta x_i \left(\frac{h_{i,j,k-1} - h_{i,j,k}}{\Delta z_{k-1/2}} \right) \qquad (9.5.9)$$

Equations 9.5.4 through 9.5.9 describe the steady-state, one-dimensional flow through each side of cell (i, j, k) in terms of the heads, grid dimensions, and hydraulic conductivity. The notation can be simplified by defining a constant referred to as the *hydraulic conductance* or the *conductance* by combining the grid dimensions and hydraulic conductivity. *Conductance* is the product of the effective hydraulic conductivity and cross-sectional area of flow divided by the length of the flow path (distance between nodes). The hydraulic conductance $\text{CY}_{i,j-1/2,k}$ in row i and layer k between nodes $(i, j - 1, k)$ and (i, j, k) is

$$\text{CY}_{i,j-1/2,k} = \text{KY}_{i,j-1/2,k}\,\Delta x_i \Delta z_k / \Delta y_{j-1/2} \qquad (9.5.10)$$

where $\text{CY}_{i,j-1/2,k}$ is the conductance in row i and layer k between nodes $(i, j-1, k)$ and (i, j, k) [L^2/T]. Equations 9.5.4 through 9.5.9 are expressed with the conductance as

$$q_{i,j-1/2,k} = \text{CY}_{i,j-1/2,k}\,(h_{i,j-1,k} - h_{i,j,k}) \qquad (9.5.11)$$

$$q_{i,j+1/2,k} = \text{CY}_{i,j+1/2,k}\,(h_{i,j+1,k} - h_{i,j,k}) \qquad (9.5.12)$$

$$q_{i-1/2,j,k} = \text{CX}_{i-1/2,j,k}\,(h_{i-1,j,k} - h_{i,j,k}) \qquad (9.5.13)$$

$$q_{i+1/2,j,k} = \text{CX}_{i+1/2,j,k}\,(h_{i+1,j,k} - h_{i,j,k}) \qquad (9.5.14)$$

$$q_{i,j,k-1/2} = \text{CZ}_{i,j,k-1/2}\,(h_{i,j,k-1} - h_{i,j,k}) \qquad (9.5.15)$$

$$q_{i,j,k+1/2} = \text{CZ}_{i,j,k+1/2}\,(h_{i,j,k+1} - h_{i,j,k}) \qquad (9.5.16)$$

The flows from features or processes external to the aquifer, such as streams, drains, areal recharge, evapotranspiration, or wells, are described through additional terms. These flows may be dependent on the head in the receiving cell but independent of all other heads in the aquifer, or they may be entirely independent of head in the receiving cell.

The rate at which water is stored or released in the cell, $\Delta V/\Delta t$ over time Δt, is

$$\frac{\Delta V}{\Delta t} = \text{SS}_{i,j,k}\Delta x_i \Delta y_j \Delta z_k \frac{\Delta h_{i,j,k}}{\Delta t} \qquad (9.5.17)$$

where $\text{SS}_{i,j,k}$ is the specific storage of cell (i, j, k); $\Delta h_{i,j,k}/\Delta t$ is the finite difference approximation for the derivative of head with respect to time; and $\Delta x_i \Delta y_j \Delta z_k$ is the volume of cell (i, j, k). The time derivative of head is expressed in terms of specific heads and time by defining the head difference as $\Delta h = h_{i,j,k}^m - h_{i,j,k}^{m-1}$ and the time interval as $\Delta t = t_m - t_{m-1}$:

$$\frac{\Delta h}{\Delta t} = \frac{h_{i,j,k}^m - h_{i,j,k}^{m-1}}{t_m - t_{m-1}} \qquad (9.5.18)$$

The rate at which water is stored in or released from cell (i, j, k) is expressed as

$$\frac{\Delta V}{\Delta t} = \text{SS}_{i,j,k}\left(\Delta x_i \Delta y_j \Delta z_k\right)\frac{h_{i,j,k}^m - h_{i,j,k}^{m-1}}{t_m - t_{m-1}} \qquad (9.5.19)$$

External flows (sources and stresses) from outside the aquifer into each cell, such as recharge, streams, and flow out of each cell—such as evapotranspiration and well pumpage for each individual cell—must also be taken into account. The total external flow for cell (i, j, k) is designated $W_{i,j,k}$, which is the combination of source and/or stress terms for an individual cell expressed as

$$W_{i,j,k} = \sum_{n=1}^{N} p_{i,j,k,n} h_{i,j,k} + \sum_{n=1}^{N} q_{i,j,k,n} = P_{i,j,k} h_{i,j,k} + Q_{i,j,k} \qquad (9.5.20)$$

where N is the number of external sources or stresses affecting cell (i, j, k), and $p_{i,j,k,n}$ and $q_{i,j,k,n}$ are constants that describe the individual external sources or stresses. These constants for the MODFLOW model are described in more detail in subsection 9.6.3.

The continuity equation for cell (i, j, k) is expressed as

$$
\begin{aligned}
q_{i,j-1/2,k} + q_{i,j+1/2,k} + q_{i-1/2,j,k} &+ q_{i+1/2,j,k} + q_{i,j,k-1/2} \\
+ q_{i,j,k+1/2} + W_{i,j,k} &= SS_{i,j,k} \frac{\Delta h_{i,j,k}}{\Delta t} \left(\Delta x_i \Delta y_j \Delta z_k \right)
\end{aligned}
\qquad (9.5.21)
$$

Equations 9.5.11–9.5.16, 9.5.19, and 9.5.20 are substituted into 9.5.21, resulting in the following finite difference approximation for cell (i, j, k):

$$
\begin{aligned}
CY_{i,j-1/2,k}(h_{i,j-1,k} - h_{i,j,k}) &+ CY_{i,j+1/2,k}(h_{i,j+1,k} - h_{i,j,k}) + CX_{i-1/2,j,k}(h_{i-1,j,k} - h_{i,j,k}) \\
+ CX_{i+1/2,j,k}(h_{i+1,j,k} - h_{i,j,k}) &+ CZ_{i,j,k-1/2}(h_{i,j,k-1} - h_{i,j,k}) + CZ_{i,j,k+1/2}(h_{i,j,k+1} - h_{i,j,k}) \\
+ P_{i,j,k} h_{i,j,k} + Q_{i,j,k} &= SS_{i,j,k}(\Delta x_i \Delta y_j \Delta z_k)(\Delta h_{i,j,k}/\Delta t)
\end{aligned}
\qquad (9.5.22)
$$

The flow terms can be expressed in terms of h^m at time t_m as

$$
\begin{aligned}
CY_{i,j-1/2,k}\left(h_{i,j-1,k}^m - h_{i,j,k}^m\right) &+ CY_{i,j+1/2,k}\left(h_{i,j+1,k}^m - h_{i,j,k}^m\right) + CX_{i-1/2,j,k}\left(h_{i-1,j,k}^m - h_{i,j,k}^m\right) \\
+ CX_{i+1/2,j,k}\left(h_{i+1,j,k}^m - h_{i,j,k}^m\right) &+ CZ_{i,j,k-1/2}\left(h_{i,j,k-1}^m - h_{i,j,k}^m\right) + CZ_{i,j,k+1/2}\left(h_{i,j,k+1}^m - h_{i,j,k}^m\right) \\
+ P_{i,j,k} h_{i,j,k}^m + Q_{i,j,k} &= SS_{i,j,k}\left(\Delta x_i \Delta y_j \Delta z_k\right) \frac{\left(h_{i,j,k}^m - h_{i,j,k}^{m-1}\right)}{t_m - t_{m-1}}
\end{aligned}
\qquad (9.5.23)
$$

This equation can be rearranged so that all terms of h^m containing heads at the current time step are grouped on the left-hand side of the equation and all remaining known terms independent of heads at the current time step are on the right-hand side. The resulting equation is

$$
\begin{aligned}
CZ_{i,j,k-1/2}\left(h_{i,j,k-1}^m\right) + CX_{i-1/2,j,k}\left(h_{i-1,j,k}^m\right) &+ CY_{i,j-1/2,k}\left(h_{i,j-1,k}^m\right) \\
+ \left(-CZ_{i,j,k-1/2} - CX_{i-1/2,j,k} - CY_{i,j-1/2,k} \right. &- CY_{i,j+1/2,k} \\
\left. -CZ_{i+1/2,j,k} - CZ_{i,j,k+1/2} + HCOF_{i,j,k} \right) &h_{i,j-1,k}^m \\
+ CY_{i,j+1/2,k}\left(h_{i,j+1,k}^m\right) + CX_{i+1/2,j,k}\left(h_{i+1,j,k}^m\right) &+ CZ_{i,j,k+1/2}\left(h_{i,j,k+1}^m\right) = RHS_{i,j,k}
\end{aligned}
\qquad (9.5.24)
$$

where

$$HCOF_{i,j,k} = P_{i,j,k} - \frac{SC1_{i,j,k}}{(t_m - t_{m-1})} \qquad \left[L^2/T \right]$$

$$RHS_{i,j,k} = -Q_{i,j,k} - \frac{SC1_{i,j,k} h_{i,j,k}^{m-1}}{(t_m - t_{m-1})} \qquad \left[L^3/T \right]$$

and

$$SC1_{i,j,k} = SS_{i,j,k}\left(\Delta x_i \Delta y_j \Delta z_k\right) \quad \left[L^2\right]$$

Expressing Equation 9.6.24 for each variable-head cell in the aquifer results in a system of equations of the following matrix form:

$$[A]\{h\}=\{q\} \tag{9.5.25}$$

where [A] is a matrix of the coefficients of heads from the left side of Equation 9.5.24 for all active nodes in the mesh; {h} is the vector of unknown heads at the end of time step m for all nodes in the mesh; and {q} is the vector of constant terms, the right side of Equation 9.5.24, for all nodes of the mesh.

9.5.2 Simulation of Boundaries

In order to simulate boundary conditions, cells are grouped as constant-head cells and inactive (or no-flow) cells. *Constant-head cells* have a specified value through all time steps of the simulation. *Inactive cells* do not allow flow into or out of the cell. The remaining cells of a finite difference grid system (mesh) are variable head cells. The finite difference Equation 9.5.24 is formulated for each variable-head cell in the mesh. The system of finite difference equations is solved simultaneously for each time step in the simulation.

The constant-head and no-flow cells are used to represent conditions of various hydrologic conditions, as illustrated in Fig. 9.5.5. Even though the aquifer is irregular in shape the mesh defining the aquifer is always rectangular in outline. No-flow cells are used to delete the portion of the array of cell beyond the aquifer boundary. Figure 9.5.5 shows constant-head cells along one section of the boundary to simulate direct contact with a major surface water body. Boundary conditions, such as areas of constant flow or areas where inflow varies with head, can be simulated as an external source or through a combination of external source and a no-flow cell.

9.5.3 Vertical Discretization

The range of vertical discretization of an aquifer cross section (see Fig. 9.5.6) can be visualized simply as an extension of areal discretization in an effort to represent individual aquifers or

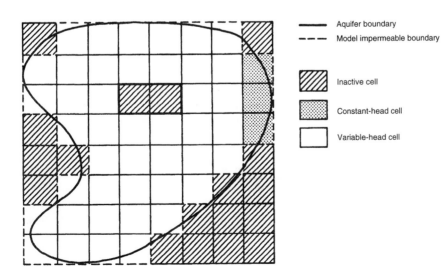

Figure 9.5.5. Discretized aquifer showing boundaries and cell designations (McDonald and Harbaugh[39]).

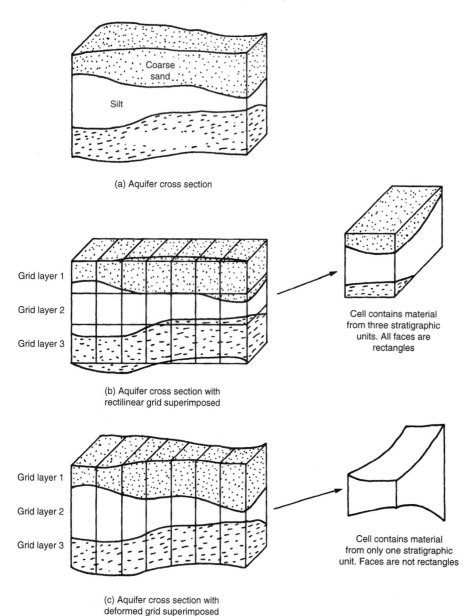

(a) Aquifer cross section

Cell contains material
from three stratigraphic
units. All faces are
rectangles

(b) Aquifer cross section with
rectilinear grid superimposed

Cell contains material
from only one stratigraphic
unit. Faces are not rectangles

(c) Aquifer cross section with
deformed grid superimposed

Figure 9.5.6. Schemes of vertical discretization.[39]

permeable zones by individual layers of the model. The simpler approach (Figure 9.5.6*b*) leads to a rigid superposition of an orthogonal three-dimensional mesh on the geohydrologic system. In the second approach (Figure 9.5.6*c*), the model layer thickness is considered variable in order to simulate the varying thickness of the geohydrologic units leading to a deformed mesh.

The advantage of the second approach is that the uniform hydraulic properties of a cell are better approximated when the model layers conform to the geohydrologic units. Also, greater accuracy is expected if model layers correspond to intervals within which vertical headloss is negligible, which is more likely using the approach shown in Figure 9.5.6*c*. However, the deformed mesh of Figure 9.5.6*c* fails to conform to the assumptions of the model equations. As an example, cells may no longer have rectangular sides and the major axes of hydraulic conductivity may not be aligned with the model axis, both introducing errors.

In practice, vertical discretization schemes turn out to be a combination of the viewpoints illustrated in Figure 9.5.6 (McDonald and Harbaugh[39]). Figure 9.5.7 illustrates a system consisting of two sand layers separated by a clay layer. Each of the units is of equal thickness so that each of the units could be represented by a single layer without deformation of the mesh; however, the flow pattern is much more complicated as illustrated by the flow lines. In this case, several layers must be used to represent each layer. Figure 9.5.8 illustrates another case of a sand and clay system in which pumpage from the sands is sustained partially by vertical flow of water released from storage in the clay. Several model layers would be required to determine the pattern of storage release in the clay as shown in the figure. In the case that the storage is released only in the sands, the flow would be nearly horizontal with the flow in the clay nearly vertical. A single layer could be used for each sand unit and the clay layer could be represented by a vertical conductance between sand layers, which is sometimes referred to as a quasi–three-dimensional approach. Each of these vertical-discretization approaches leads to a set of model equations of the form of Equation 9.5.24.

9.5.4 Hydraulic Conductance Equations

Conductance has previously been defined as the product of hydraulic conductivity and cross-sectional area of flow divided by the length of the flow path. Start with Darcy's law defining

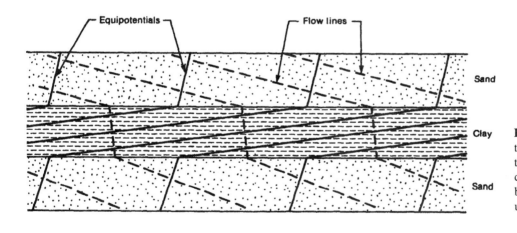

Figure 9.5.7. Possible pattern of flow in a cross section consisting of two high conductivity units separated by a low conductivity unit.[39]

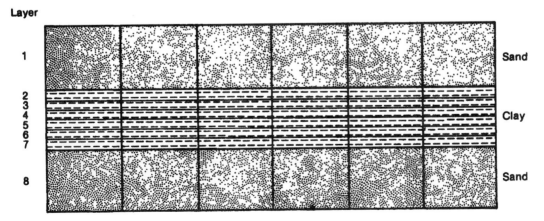

Figure 9.5.8. A cross section in which a low conductivity unit is represented by six model layers.[39]

one-dimensional flow in a prism of porous medium (Figure 3.1.1) as $Q = -KA(h_2 - h_1)/L$; where Q is the volumetric flow, K is the hydraulic conductivity, A is the cross-sectional area perpendicular to flow, $h_1 - h_2$ is the head difference across the prism parallel to flow, and L is the length of the prism parallel to the flow path. Then the hydraulic conductance is defined as $C = KA/L$ and Darcy's law can be written as $Q = C(h_1 - h_2)$. Another form of the hydraulic conductance for horizontal flow in a prism is $C = TB/L$ where T is the transmissivity in the direction of flow L^2/T and B is the width of the prism.

Hydraulic conductance is defined for a particular prism of material and for a particular direction of flow. Anisotropic media are characterized by three principal directions of hydraulic conductivity so that the hydraulic conductances of a prism differ in the three directions.

Now consider a prism of porous medium consisting of two or more subprisms in a series, as shown in Figure 9.5.9. Knowing the conductance in each subprism, we can compute a hydraulic conductance representing the entire prism, referred to as the *equivalent hydraulic conductance*. The equivalent hydraulic conductance for the entire prism is $C = Q/(h_A - h_B)$. The total head loss (referring to Figure 9.5.9) is

$$\sum_{i=1}^{n} \Delta h_i = h_A - h_B$$

(9.5.26)

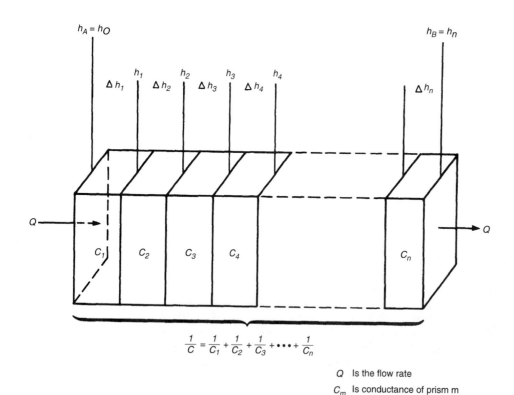

$$\frac{1}{C} = \frac{1}{C_1} + \frac{1}{C_2} + \frac{1}{C_3} + \cdots + \frac{1}{C_n}$$

Q Is the flow rate
C_m Is conductance of prism m
h_m Is head at the right side of prism m
Δh_m Is the head change across prism m

Figure 9.5.9. Calculation of conductance through several prisms in a series.[39]

And substituting for head change across each section using Darcy's law then leads to

$$\sum_{i=1}^{n} \frac{q_i}{C_i} = h_A - h_B \tag{9.5.27}$$

Because flow is one dimensional and it is assumed that there is no accumulation or depletion in storage; then all q_i are equal to the total flow Q. Then

$$Q\sum_{i=1}^{n} \frac{1}{C_i} = h_A - h_B \quad \text{and} \quad \frac{h_A - h_B}{Q} = \sum_{i=1}^{n} \frac{1}{C_i} \tag{9.5.28}$$

From $C = Q/(h_A - h_B)$ and Equation 9.5.28, the equivalent hydraulic conductance is

$$\frac{1}{C} = \sum_{i=1}^{n} \frac{1}{C_i} \tag{9.5.29}$$

which states that for a set of hydraulic conductances arranged in series, the inverse of the equivalent conductance equals the sum of the inverses of the individual conductances. For two sections, the equivalent hydraulic conductance is $C = C_1 C_2/(C_1 + C_2)$ which explains the form of Equation 9.5.3.

9.6 MODFLOW-2000 DESCRIPTION

9.6.1 Model Introduction

MODFLOW-2000 is a computer program that simulates one-, two-, or three-dimensional groundwater flow using a finite difference solution of the model formulation presented in Section 9.5 of this chapter, MODFLOW, the original computer program, was first published by McDonald and Harbaugh.[38] This program was developed using a modular structure for ease in adapting the code for particular applications. The second version of MODFLOW, documented in McDonald and Harbaugh,[39] was referred to as MODFLOW-88. The third version was documented by Harbaugh and McDonald[21, 22] and was referred to as MODFLOW-96. The most recent version, MODFLOW-2000, documented in Harbaugh et al.,[20] includes many capabilities added since the original computer program. This section briefly describes these capabilities.

MODFLOW-2000 can simulate steady and nonsteady flow in an irregularly shaped flow system in which aquifer layers can be confined, unconfined, or a combination of confined and unconfined. Flow from external stresses, such as flow to wells, areal recharge, evapotranspiration, flow to drains, and flow through river beds, can be simulated. The hydraulic conductivities or transmissivities for any layer may differ spatially and be anisotropic (restricted to having the principal directions aligned with the grid axes), and the storage coefficient may be heterogeneous. Specified head and specified flux boundaries can be simulated, as can a head-dependent flux across the model's outer boundary, which allows water to be supplied by a boundary block in the modeled area at a rate proportional to the current head difference between a source of water outside the modeled area and the boundary block. MODFLOW-2000 has been expanded to simulate solute transport and parameter estimation.

The MODFLOW-2000 computer program is divided into a main program and a series of independent subroutines called modules. The modules are grouped into packages that deal with a single aspect of the simulation. The packages are listed and briefly described in Table 9.6.1. Individual packages may or may not be required, depending on the problem being

Table 9.6.1 Processes, Packages, and Additional Capabilities of MODFLOW-2000 (Version 1.1, 1/17/2001)

Processes	Packages	Additional Capabilities
GWF1—Ground-Water Flow Process	BAS6—Basic Package	HYDMOD—Hydrograph Option[17]
SEB1—Sensitivity Process	BCF6—Block-Centered Flow Package[15, 21, 22, 40]	
OBS1—Observation Process	LPF1—Layer-Property Flow Package	
PES1—Parameter-Estimation Process	RIV6—River Package	
	DRN6—Drain Package	
	WEL6—Well Package	
	GHB6—General Head Boundary Package	
	RCH6—Recharge Package	
	EVT6—Evapotransipiration Package	
	CHD6—Time-Variant Specified-Head Package[36]	
	HFB6—Horizontal Flow Barrier Package[28]	
	SIP5—Strongly Implicit Procedure Package	
	SOR5—Slice Successive Over-Relaxation Package	
	PCG2—Version 2 of Preconditioned Conjugate Gradient Package[24]	
	DE45—Direct Solver[19]	
	STR6—Streamflow-Routing Package[53]	
	ADV2—Advective-Transport Observation Package[2]	
	RES1—Reservoir Package[10]	
	FHB1—Flow and Head Boundary Package[35]	
	IBS6—Interbed Storage (Subsidence) Package[36]	
	HUF1—Hydrogeologic-Unit Flow Package[3]	
	LAK3—Lake Package[41]	
	ETS1—Evapotranspiration with a Segmented Function Package[6]	
	DRT1—Drains with Return Flow Package[6]	

solved. The basic documentation for MODFLOW is contained in Harbaugh and McDonald,[21, 22] Harbaugh et al.,[20] Hill et al.,[26] and McDonald and Harbaugh.[39]

9.6.2 Space and Time Discretization

The physical size of the finite difference grid is provided through input to the program. In all versions of MODFLOW, the finite difference grid is assumed to be rectangular horizontally and can be distorted vertically as illustrated in Figure 9.6.1. The horizontal grid dimensions are specified by the cell widths DELR and DELC (Figure 9.6.1a). Columns are numbered starting from the left side of the grid and rows are numbered starting from the upper edge (plan view) of the grid. All cells in a column have the same width and all cells in a row have the same width. Layers are numbered starting from the top layer down (see Figure 9.6.1b). The elevation of the top of Layer 1 and the bottom elevation of each layer for each cell are used to determine the thickness of each cell. A confining bed through which only vertical flow exists can be simulated below each layer except the bottom layer. This simulation of confining beds is referred to as the quasi–three-dimensional (quasi-3D) approach.[39]

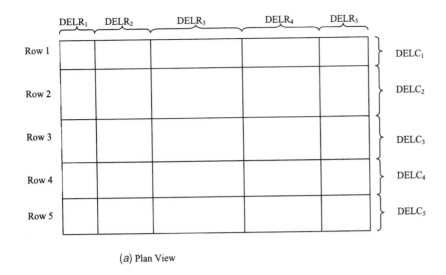

Row 1
Row 2
Row 3
Row 4
Row 5

DELR₁ DELR₂ DELR₃ DELR₄ DELR₅

DELC₁
DELC₂
DELC₃
DELC₄
DELC₅

(a) Plan View

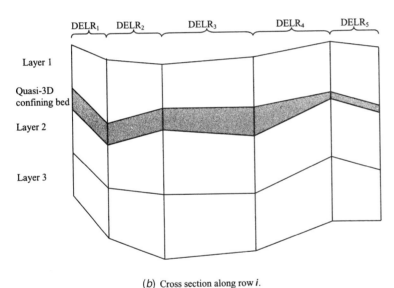

DELR₁ DELR₂ DELR₃ DELR₄ DELR₅

Layer 1
Quasi-3D confining bed
Layer 2
Layer 3

(b) Cross section along row *i*.

Figure 9.6.1. Finite difference grid with (*a*) plan view and (*b*) cross-sectional view (Harbaugh et al.[20]).

Time discretization is dealt with through time steps, which are grouped into stress periods. For each stress period, the user specifies the total length (PERLEN), the number of time steps (NSTP), and the multiplier for the length of the time step $n-1$ times TSMULT. This procedure develops a geometric series with the length of the first time step given as

$$\Delta t_1 = \text{PERLEN}\left(\frac{\text{TSMULT}-1}{\text{TSMULT}^{\text{NSTP}}-1}\right) \qquad (9.6.1)$$

9.6.3 External Sources and Stresses

9.6.3.1 Rivers and Streams

The effects of flow between surface water features and groundwater systems are modeled by adding terms for the seepage to or from the surface water feature to the finite difference model

Equation 9.5.24. Flow between a stream and the groundwater system, $W_{i,j,k}$ = QRIV, is expressed as

$$QRIV = CRIV(HRIV - h_{i,j,k}), \quad h_{i,j,k} > RBOT$$

$$QRIV = CRIV(HRIV - RBOT), \quad h_{i,j,k} \leq RBOT$$

(9.6.2)

where CRIV is the hydraulic conductance defined as *KLB/M*, in which K is the hydraulic conductance of the streambed layer, L is the length of the stream reach contained within the cell, B is the width of the stream reach, and M is the thickness of the streambed material; HRIV is the head in the stream; $h_{i,j,k}$ is the head at the node in the cell underlying the stream reach; and RBOT is the elevation of the bottom of the streambed layer.

9.6.3.2 Areal Recharge

Areal recharge, which most commonly occurs as a result of precipitation percolating into the groundwater system, $W_{i,j,k}$ = $QR_{i,j}$, is expressed as

$$QR_{i,j} = I_{i,j} \times DELR_j \times DELC_i$$

(9.6.3)

where $QR_{i,j}$ is the recharge flow rate applied to horizontal cell (i, j), expressed as water volume per unit time, and $I_{i,j}$ is the recharge flux (length per unit time) applicable to the area $DELR_j DELC_i$.

9.6.3.3 Drains

Buried and open drains are shown in Figure 9.6.2. The flow into a drain is illustrated in the cross-section through cell (i, j, k) in Figure 9.6.3. Flow in the drain is assumed to be only partially full, so the head within the drain is approximately equal to the median drain elevation, $d_{i,j,k}$. The head, $h_{i,j,k}$, is the average head for the cell. There are three processes of flow into the drain: convergent flow through the drain, flow through immediately around the drain with a different hydraulic conductivity, and flow through the wall of the drain. Each of these processes creates head losses that are assumed to be proportional to the discharge, $QD_{i,j,k}$, with the total head loss $h_{i,j,k} - d_{i,j,k}$. The drain function, $W_{i,j,k}$ = $QD_{i,j,k}$, is expressed as

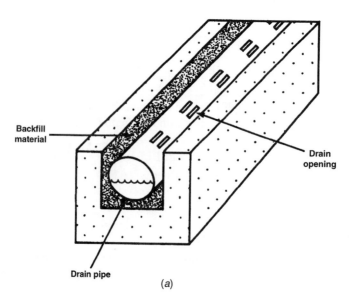

Backfill material

Drain opening

Drain pipe

(a)

Figure 9.6.2. Factors affecting head loss immediately around a drain: (*a*) buried drain pipe in backfilled ditch and (*b*) open drain (McDonald and Harbaugh[39]).

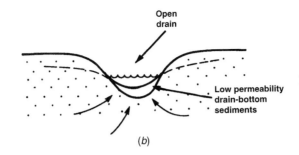

Figure 9.6.2 (continued). Factors affecting head loss immediately around a drain: (*a*) buried drain pipe in backfilled ditch and (*b*) open drain (McDonald and Harbaugh[39]).

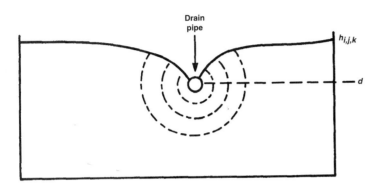

Figure 9.6.3. Cross section through cell (*i, j, k*) illustrating head loss in convergent flow into drain. (McDonald and Harbaugh[39]).

$$QD_{i,j,k} = CD_{i,j,k} (h_{i,j,k} - d_{i,j,k}) \qquad \text{for } h_{i,j,k} > d_{i,j,k} \tag{9.6.4}$$

$$QD_{i,j,k} = 0 \qquad \text{for } h_{i,j,k} \leq d_{i,j,k} \tag{9.6.5}$$

where $CD_{i,j,k}$ is a lumped or equivalent conductance describing all of the head loss between the drain and the cell (see McDonald and Harbaugh[39]). See Banta[6] for more information on modeling drains with return flows.

9.6.3.4 Evapotranspiration

To simulate the effects of plant transpiration and direct evaporation of water from the saturated groundwater, the volumetric rate of flow from a given cell is required. The volumetric rate of evaporation, $W_{i,j,k} = Q_{ETi,j,k}$, is the product of the loss rate per unit area of the cell and the horizontal surface area, $DELR_j$ $DELC_i$, given as

$$Q_{ETi,j} = R_{ETi,j} \times DELR_j \times DELC_i \tag{9.6.6}$$

where $R_{ETi,j}$ is the rate of loss per unit surface area of water table due to evapotranspiration, given as

$$
\begin{aligned}
R_{ETi,j} &= R_{ETMi,j} & h_{i,j,k} &> h_{si,j} \\
R_{ETi,j} &= 0 & h_{i,j,k} &< h_{si,j} - d_{i,j} \\
R_{ETi,j} &= R_{ETMi,j} \left[\frac{h_{i,j,k} - \left(h_{si,j} - d_{i,j}\right)}{d_{i,j}} \right] & \left(h_{si,j} - d_{i,j}\right) &\leq h_{i,j,k} \leq h_{si,j}
\end{aligned}
\tag{9.6.7}
$$

in which $h_{si,j}$ is the water table elevation at which the maximum value of evapotranspiration loss occurs, $R_{ETMi,j}$ is the maximum possible value of $R_{ETi,j}$, and $d_{i,j}$ is the cutoff or extinction depth so that when the distance between $h_{si,j}$ and $h_{i,j,k}$ exceeds $d_{i,j}$, then evapotranspiration ceases. See Banta[6] for modeling evapotranspiration with a segmented function.

9.6.3.5 General Head Boundary

The flow $Q_{bi,j,k}$ into or out of a cell from an external source with a known head is proportional to the difference between the head in cell, $h_{i,j,k}$, and the head assigned to the external source, $h_{bi,j,k}$, as

$$Q_{bi,j,k} = C_{bi,j,k} (h_{bi,j,k} - h_{i,j,k}) \tag{9.6.8}$$

where $C_{bi,j,k}$ is the conductance between the external source and cell (i, j, k).

9.6.3.6 Wells

For injection wells, the source strength is

$$W_{i,j,k} = \text{QW}_{i,j,k} (t_m) \tag{9.6.9}$$

where $\text{QW}_{i,j,k}(t_m)$ is the well injection discharge during the stress period at time t_m.

9.6.4 Hydraulic Conductance—Layer-Property Flow Package (LPF)

9.6.4.1 Horizontal Hydraulic Conductance

MODFLOW solves the finite difference Equations 9.5.24 using equivalent hydraulic conductances between nodes of adjacent cells, as explained in Section 9.5.4. CR (along rows) and CC (along columns) are the horizontal equivalent hydraulic conductance terms between adjacent horizontal nodes. The LPF Package in MODFLOW-2000 reads data defining the horizontal hydraulic conductivity for individual cells and uses this information to compute the equivalent hydraulic conductance between nodes, using one of three methods.

The first method assumes a constant transmissivity (hydraulic conductivity times thickness) within a cell, but allows a discrete change in transmissivity at the boundary between the cells. Because it is at the center of the cells, the hydraulic conductance between nodes is the equivalent hydraulic conductance of two half cells in series, as shown in Figure 9.6.4. Using $C = C_1 C_2/(C_1 + C_2)$, we obtain the conductance CR given in Figure 9.6.4 and the conductance CC as

$$\text{CR}_{i,j+\frac{1}{2},k} = \frac{\left[\dfrac{\text{TR}_{i,j,k}\text{DELC}_i}{\left(\frac{1}{2}\right)\text{DELR}_j}\right]\left[\dfrac{\text{TR}_{i,j+1,k}\text{DELC}_i}{\left(\frac{1}{2}\right)\text{DELR}_{j+1}}\right]}{\dfrac{\text{TR}_{i,j,k}\text{DELC}_i}{\left(\frac{1}{2}\right)\text{DELR}_j} + \dfrac{\text{TR}_{i,j+1,k}\text{DELC}_i}{\left(\frac{1}{2}\right)\text{DELR}_{j+1}}} \tag{9.6.10}$$

where $\text{TR}_{i,j,k}$ is the transmissivity in the row direction at cell (i, j, k). This approach for calculating interblock conductance is the *harmonic mean method*. $\text{CC}_{i+1/2,k}$ is computed in the same manner. The equations for CR and CC are used unless the transmissivities of both cells are zero, resulting in an equivalent hydraulic conductance of zero.

Goode and Appel[15] describe three alternative approaches for calculating horizontal hydraulic conductances. One method assumes that the transmissivity varies linearly between nodes, referred to as the *logarithmic-mean interblock transmissivity method*. The second method assumes a flat, homogeneous aquifer with a water table defining the *arithmetic-mean interblock transmissivity method*. The third method assumes a flat aquifer with a water table in which hydraulic conductivity varies linearly between nodes defining the *arithmetic-mean thickness and logarithmic-mean hydraulic conductivity method*. The first and third methods are also included in the LPF package in MODFLOW-2000. Harbaugh[18] describes a generalized

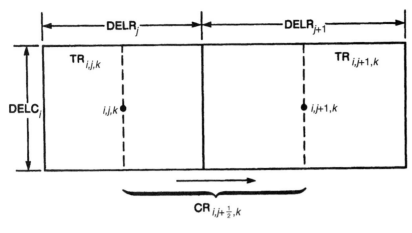

$$\frac{1}{CR_{i,j+\frac{1}{2},k}} = \frac{1}{\left(\frac{TR_{i,j,k}DELC_i}{\left(\frac{DELR_j}{2}\right)}\right)} + \frac{1}{\left(\frac{TR_{i,j+1,k}DELC_i}{\left(\frac{DELR_{j+1}}{2}\right)}\right)}$$

$$CR_{i,j+\frac{1}{2},k} = 2\,DELC \times \frac{TR_{i,j,k}\,TR_{i,j+1,k}}{TR_{i,j,k}\,DELR_{j+1} + TR_{i,j+1,k}\,DELR_j}$$

$TR_{i,j,k}$ Is transmissivity in the row direction in cell (i,j,k)

$CR_{i,j+\frac{1}{2},k}$ Is conductance in the row direction between nodes (i,j,k) and $(i,j+1,k)$

Figure 9.6.4. Calculation of conductance between nodes using transmissivity and dimensions of cells (McDonald and Harbaugh[39]).

finite-difference interblock transmissivity for unconfined aquifers and for aquifers having smoothly varying transmissivity.

9.6.4.2 Vertical Hydraulic Conductance

The vertical hydraulic conductance is calculated assuming that nodes are in the center of cells and discrete changes in vertical hydraulic conductivity occur at the boundaries of the layers, as illustrated in Figure 9.6.5. Using Equation 9.5.29 with the hydraulic conductance for each half cell results in

$$\frac{1}{CV_{i,j,k+\frac{1}{2}}} = \frac{1}{\dfrac{DELR_j DELC_i VK_{i,j,k}}{\left(\frac{1}{2}\right)THICK_{i,j,k}}} + \frac{1}{\dfrac{DELR_j DELC_i VK_{i,j,k+1}}{\left(\frac{1}{2}\right)THICK_{i,j,k+1}}} \qquad (9.6.11)$$

where $VK_{i,j,k}$ is the vertical hydraulic conductivity of cell (i,j,k) and $THICK_{i,j,k}$ is the saturated thickness of cell (i, j, k). Simplifying Equation 9.6.11 results in

$$\frac{1}{CV_{i,j,k+\frac{1}{2}}} = \frac{DELR_j DELC_i}{\dfrac{\left(\frac{1}{2}\right)THICK_{i,j,k}}{VK_{i,j,k}} + \dfrac{\left(\frac{1}{2}\right)THICK_{i,j,k+1}}{VK_{i,j,k+1}}} \qquad (9.6.12)$$

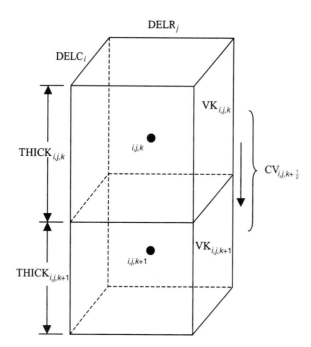

Figure 9.6.5. Calculation of vertical conductance between two nodes (Harbaugh et. al.[20]).

The calculation of vertical conductance between two nodes with a semiconfining unit in between (Figure 9.6.6) assumes that the semiconfining layer makes no measurable contribution to the horizontal conductance or the storage capacity of either model layer. Under such assumptions, the confining bed restricts vertical flow between the cells. A quasi–three-dimensional approach (McDonald and Harbaugh[39]) is used. In this approach, three intervals (lower half of upper aquifer, semiconfining unit, and the upper half of the lower aquifer) are represented in the summation of conductance between the nodes. Using Equation 9.5.29, we obtain for the vertical hydraulic conductance

$$\frac{1}{CV_{i,j,k+\frac{1}{2}}} = \frac{1}{\dfrac{DELR_j DELC_i VK_{i,j,k}}{\left(\frac{1}{2}\right)THICK_{i,j,k}}} + \frac{1}{\dfrac{DELR_j DELC_i VKCB_{i,j,k}}{THICK_{CB}}} + \frac{1}{\dfrac{DELR_j DELC_i VK_{i,j,k+1}}{\left(\frac{1}{2}\right)THICK_{i,j,k+1}}}$$

$$(9.6.13)$$

where $VKCB_{i,j,k}$ is the hydraulic conductivity of the semiconfining unit between cells (i, j, k) and $(i, j, k + 1)$ and $THICK_{CB}$ is the thickness of the semiconfining unit. Equation 9.6.13 can be simplified to

$$CV = \frac{DELR_j\, DELC_i}{\dfrac{\left(\frac{1}{2}\right)THICK_{i,j,k}}{VK_{i,j,k}} + \dfrac{THICK_{CB}}{VKCB_{i,j,k}} + \dfrac{\left(\frac{1}{2}\right)THICK_{i,j,k+1}}{VK_{i,j,k+1}}} \qquad (9.6.14)$$

9.6.4.3 Vertical Flow Calculation under Dewatered Conditions

A modification to the vertical flow calculation is required when a cell in a convertible layer is unconfined and the cell immediately above is fully or partially saturated (see Figure 9.6.7). This modification (vertical flow correction) is applied to every iteration of the solution process.

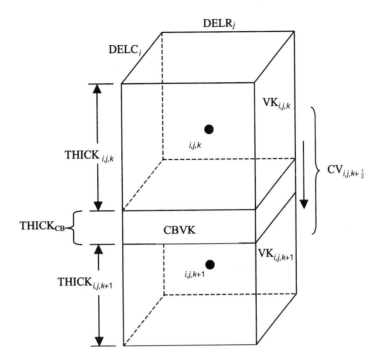

Figure 9.6.6.
Calculation of vertical conductance between two nodes with a semi-confining unit between.[20]

The actual flow between the nodes in the upper and lower cells is the flow through the confining bed, $q_{i,j,k+1/2}$ computed using

$$q_{i,j,k+\frac{1}{2}} = CV_{i,j,k+\frac{1}{2}} \left(TOP_{i,j,k+1} - h_{i,j,k} \right)$$ (9.6.15)

so that the flow is downward from cell (i, j, k) to cell $(i, j, k + 1)$ but is not dependent on $h_{i,j,k+1}$. Equation 9.6.15 could be substituted into the finite difference Equation 9.4.24; however, the

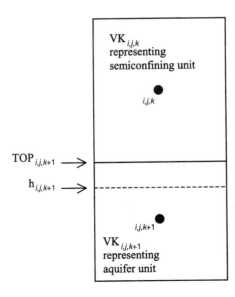

Figure 9.6.7. Situation in which a correction is required to limit the downward flow into cell $(i, j, k + 1)$, as a result of partial desaturation of the cell.[20]

matrix A in Equation 9.4.25 would be asymmetric, generating problems in the solution process. Alternatively, in MOFLOW-2000, the flow is computed as $q_{i,j,k+1/2} = CV_{i,j,k+1/2}(h_{i,j,k+1}, -h_{i,j,k})$ so that this term remains on the left side of Equation 9.4.24. Then a correction term q_c is added to the right side to compensate for allowing the computed flow to remain on the left side of Equation 9.4.24,

$$q_c^n = CV_{i,j,k+\frac{1}{2}}\left(h_{i,j,k+1}^{n-1} - TOP_{i,j,k+1}\right)$$

$$(9.6.16)$$

where q_c^n is the value of q_c added to the right side of Equation 9.5.24 in the nth iteration and $h_{i,j,k+1}^{n-1}$ is the head from the preceding iteration.

9.6.5 Solver Packages

The four solver packages in the MODFLOW-2000 are (1) the strongly implicit procedure (SIP),[39] (2) the slice successive over-relaxation (SOR),[39] (3) the preconditioned conjugate gradient (PCG),[24] and (4) the direct solution based on alternating diagonal ordering (DE4).[19] A detailed discussion of these various solvers is beyond the scope of this book. Interested readers should refer to the noted references.

9.6.6 Telescopic Mesh Refinement

Telescopic mesh refinement in groundwater flow modeling is an approach to use a larger (regional) encompassing model to define the boundary conditions and model parameters for a smaller (local) embedded model.[34] (Also refer to Buxton and Reilly[8] and Ward et al.[63] for more detail and applications of the approach.) This method is used when a detailed model is needed for a relatively small area as compared to the entire aquifer system, such the analysis of well fields[42] and contaminant plumes.[63] Essentially, the approach can be used to construct a detailed model from a larger (regional) groundwater flow model and can be used repeatedly to construct successively smaller (local) embedded models, hence the name "telescopic" mesh refinement.

Leake and Claar[34] discuss two basic approaches to construct data for local groundwater models. The first approach is to use a telescopic mesh refinement program such as MODTMR (*MOD*FLOW *T*elescopic *M*esh *R*efinement), which constructs the data set for the perimeter boundary conditions of a local model using cell-by-cell terms and head values in a regional model. This model can construct embedded-model data sets for most of the MODFLOW packages. A second approach is to use a manual-gridding process, a geographic information system (GIS), or other data processing programs to construct local-model data sets from grid-independent data. The first approach can provide maximum consistency between the local and regional model and would require less effort than the second approach, which has the main advantage in that local-model data sets are most consistent with locations and distributions of aquifer properties and sources and sinks contained in the grid-independent data sets. Two other programs have been developed by the U.S. Geological survey. TMRDIFF (*Tele*scopic *M*esh *R*efinement *DIFF*erence) constructs files with comparisons of computed heads or drawdowns in the local model with computed heads or drawdowns in the regional model. RIVGRD constructs MODFLOW input data files for the River Package, the Drain Package, the General-Head Boundary Package, and the Stream Package.[53]

Figure 9.6.8 illustrates the telescopic mesh refinement modeling approach. Figures 9.6.9 and 9.6.10 illustrate the application of this approach to a hazardous waste site in the Miami River Valley near Dayton, Ohio. Ward et al.[63] used a nested series of three models designed on the regional, local, and site scales. Note that the grid for the site model is oriented differently than the local scale and regional models.

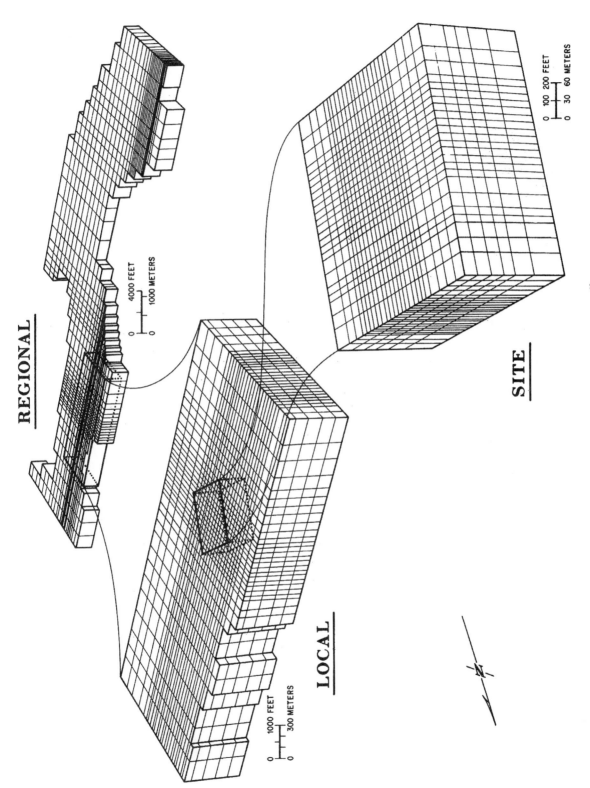

Figure 9.6.8. Conceptual diagram of the telescopic mesh refinement modeling approach (Ward et al.[63]).

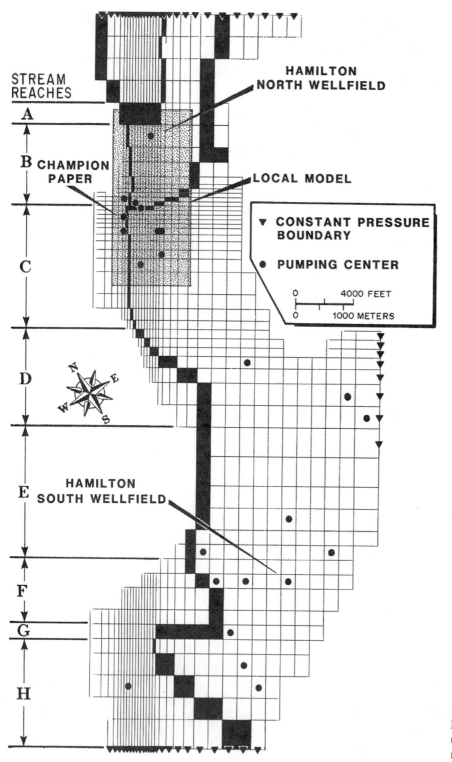

Figure 9.6.9. Finite difference grid used for the regional scale flow model of the Great Miami River Valley-Fill Aquifer.[63]

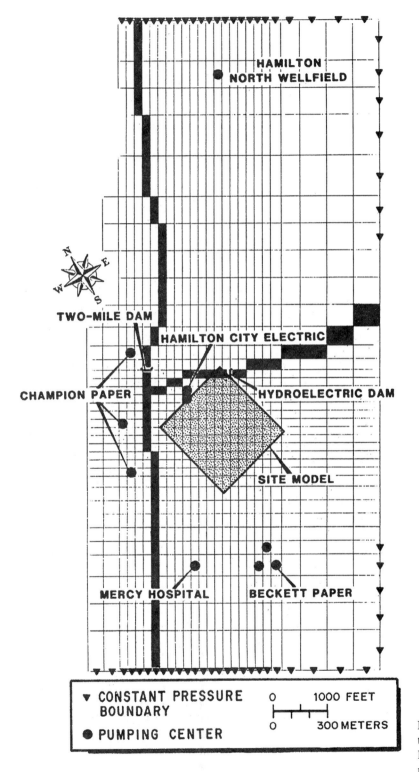

Figure 9.6.10. Finite difference grid used for the local-scale flow model of the Great Miami River Valley-Fill Aquifer. Relationship to regional-scale model is shown in Figure 9.6.9.[63]

MODFLOW is used widely in the United States and throughout the world by consultants and government agencies. The U.S. Geological Survey, as well as many other federal and state government agencies, use this program extensively and have applied it to many modeling applications. A few of the many applications by the U.S. Geological Survey include Belitz and Phillips,[7] Grubbs,[16] and Prince et al.[52]

The hydrologic budget for Lake Five-O in northwestern Florida (location shown in Figure 1.6.4) by Grubbs[16] was introduced in Chapter 1. The data-collection sites for the study area are shown in Figure 1.6.5. A three-dimensional model was developed by Grubbs[16] to represent the Lake Five-O groundwater system under steady-state and transient conditions.

9.7.1 Finite Difference Grid and Boundary Conditions

A finite difference grid was developed to approximate the geometry of the major lithologic contacts and aquifer boundaries. The areal discretization (81 rows and 57 columns) and boundary conditions for the models are illustrated in Fig. 9.7.1. All the rows in the grid have a width of 20 m. Columns 4 through 1 progressively increase in size westward from 30 to 100 m (expansion factor of 1.5). Columns 7 through 57 have a constant width of 20 m. The vertical discretization and boundary conditions are illustrated in Fig. 9.7.2. In the vertical, the sediments from the land surface to the uppermost part of the Upper Floridan aquifer were modeled using seven horizontal layers of varying thickness. Layer 1 was modeled as a water table layer; layers 2–6 were modeled as confined layers; and layer 7 was modeled as a specified boundary representing the Upper Floridan Aquifer.

Figure 9.7.2 presents the calibrated values of the hydraulic conductivity for the parameter zones used to represent variations in hydraulic properties. A single value or limited number of values of the hydraulic properties were assigned to parameter zones representing the lithologic units in order to approximate the surficial aquifer and intermediate confining unit. The following parameter zones were used by Grubbs:[16]

Basal clay zone—to represent the dense, low-permeability clay at the base of the intermediate confining unit;

Plateau zone—to represent increased anisotropy in the upper 5 to 10 m of the intermediate confining unit in the plateau area;

Transition zone—to represent the geometry of the contact between the surficial aquifer and intermediate confining unit where this contact dips toward the breaches in the intermediate confining unit;

Lakebed sediment zone—to represent lower permeability sediments (relative to the surficial aquifer) in surficial aquifer cells contiguous with and beneath the lake in model layers 3 and 4;

Lake Five-O—represented by a zone of highly conductive material in layers 1–4 and for transient simulations. The storage properties of the lake zone are identical to those of water.

The within-layer variations in horizontal hydraulic conductivity were accounted for by using equivalent hydraulic conductivity values computed using a weighted mean algorithm (Freeze and Cherry[11]). The transmissivity, T, of cells in the transition zone were computed using

$$T = LK_{h,eq} = L\left\{ \frac{\left[(K_{h,SA})\dfrac{L}{2}\right]}{L} + \frac{\left[(K_{h,ICU})\dfrac{L}{2}\right]}{L} \right\}$$

$$= L\left(\frac{K_{h,SA}}{2} + \frac{K_{h,ICU}}{2} \right) \tag{9.7.1}$$

where L is the layer thickness, $K_{h,eq}$ is the equivalent horizontal hydraulic conductivity, and $K_{h,SA}$ and $K_{h,ICI}$ are respectively the horizontal hydraulic conductivities for the surficial aquifer and intermediate confining unit. The vertical flow between model layers was simulated using the leakance parameter (vertical hydraulic conductivity divided by flow path distance). In the cases when the vertical hydraulic conductivity varied within a model layer or across contiguous model layers, the leakance values were calculated with equivalent vertical hydraulic conductivities using the harmonic mean algorithm of Equation 9.5.2.

9.7.2 Model Calibration and Sensitivity Analysis

Four steady-state models were calibrated to the hydrologic conditions on four separate dates in 1988 to 1990 in order to satisfy one or more of the following objectives (see Grubbs[16]): (1) Refine the premodeling or prior estimates of hydraulic properties and their spatial distribution, (2) Provide initial conditions for transient simulations, and (3) Provide flow fields necessary for evaluation of flow paths and residence times of groundwater inflow into the lake. Calibration of the steady-state models was performed by comparing simulated and observed heads, and by comparing simulated lake seepage with minimum leakage estimates from hydrologic budget calculations (see Section 1.6 for the hydrologic budget and see Grubbs[16] for more detail on the calibration).

Calibration-sensitivity analysis of the steady-state model was performed by varying model input variables (within probable ranges) and comparing model outputs (simulate heads, lake inflow, and lake leakage) to known conditions. The root-mean-square error (RMSE) of the simulated heads to changes in horizontal hydraulic conductivity, leakance, recharge, and anisotropy

(continues)

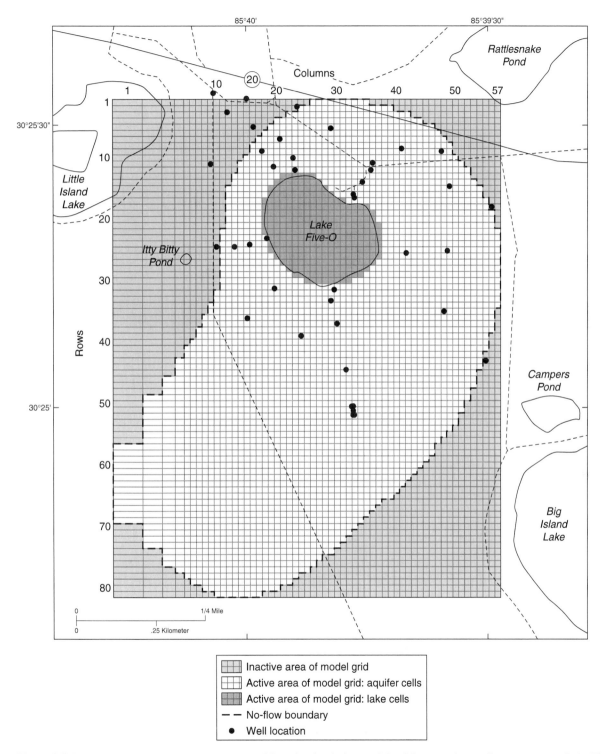

Figure 9.7.1. Areal discretization and boundary conditions for simulation models of the groundwater flow system near Lake Five-O (Grubbs[16]).

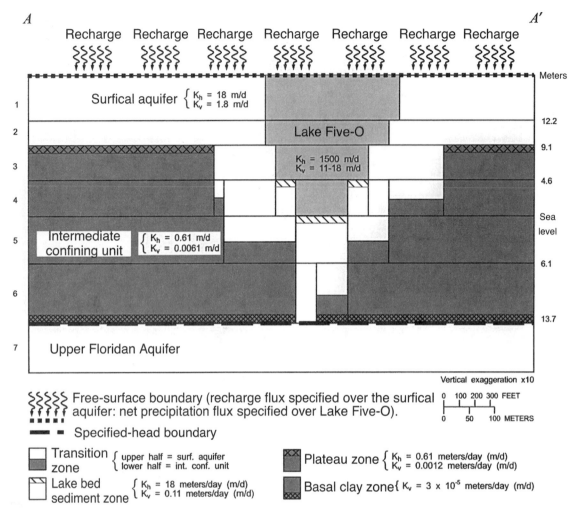

Figure 9.7.2. Vertical discretization, boundary conditions, and calibrated values of hydraulic conductivity for simulation models of the groundwater flow system near Lake Five-O(Grubbs[16]). Cross-section A-A' location is shown in Figure 1.6.5.

9.7 CASE STUDY   *Using MODFLOW: Lake Five-O, Florida (continued)*

---

are shown in Figures 9.7.3 and 9.7.4. The steep slope of the head-sensitivity curves (change from calibrated values versus RMSE) illustrate the sensitivity of the model to these variables. Head response was most sensitive to changes in recharge, horizontal hydraulic conductivity, and anisotropy of the surficial aquifer, and in horizontal hydraulic conductivity of the intermediate confining unit.

Transient model calibration determined storage properties and temporal changes in recharge rates. Hydraulic conductivities determined from the calibrated steady-state model calibrations were used in the transient model. The simulation period was divided into 43 stress periods. Assigned to each stress period were

recharge rates, net precipitation rates, and beginning and ending heads to each stress period. First the specific yield in layer 1 and the specific storage in layers 2–6 were adjusted to simulate the slope of well hydrographs. Recharge was negligible during the calibration period selected because of below-normal rainfall, which made it possible to calibrate specific yield and specific storage independent of recharge. Sensitivity of the transient model to specific yield and specific storage is illustrated in Figure 9.7.5. The model is more sensitive to changes in specific yield than to changes in specific storage.

(continues)

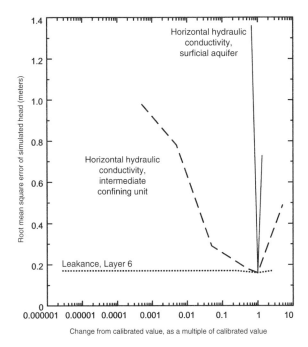

Figure 9.7.3. Sensitivity of steady-state models of groundwater flow system near Lake Five-O to changes in horizontal hydraulic conductivity and leakance of layer 6 (Grubbs[16]).

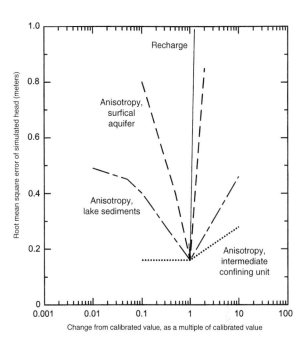

Figure 9.7.4. Sensitivity of steady-state models of groundwater flow system near Lake Five-O to changes in recharge and anisotropy (Grubbs[16]).

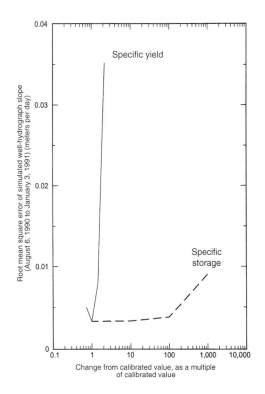

Figure 9.7.5. Sensitivity of transient model of the groundwater flow system near Lake Five-O to changes in aquifer storage properties (Grubbs[16]).

9.7 CASE STUDY   *Using MODFLOW: Lake Five-O, Florida (continued)*

The calibration of recharge consisted of adjusting the recharge function to reproduce well hydrographs and monthly estimates of the net groundwater flow. The recharge function used by Grubbs[16] is

$$R_{\Delta t} = S_y \left[\int_{\Delta t} \dot{h}(t)\,dt + \overline{h}_{recess}\Delta t \right] \qquad (9.7.2)$$

where $R_{\Delta t}$ is the recharge over time interval Δt, S_y is the specific yield, and $\overline{h}_{recess}$ is the absolute value of the average change in water table during a given hydrograph recession under suitably dry conditions. Simulated and measured heads are shown in

Figure 9.7.6. Simulated and computed net groundwater flows to Lake Five-O are shown in Figure 9.7.7.

9.7.3 Model Results

Grubbs[16] concluded that the model simulations indicated that the groundwater system is the dominant source of water for Lake Five-O and that the groundwater system is the dominant sink for water leaving Lake Five-O. In fact, the simulated groundwater inflow and leakage were approximately four and five times larger than precipitation inputs and evaporative losses, respectively, during the calendar years 1989–1990. Spatial distributions of simulated groundwater inflow and leakage showed little variation between wet and dry periods.

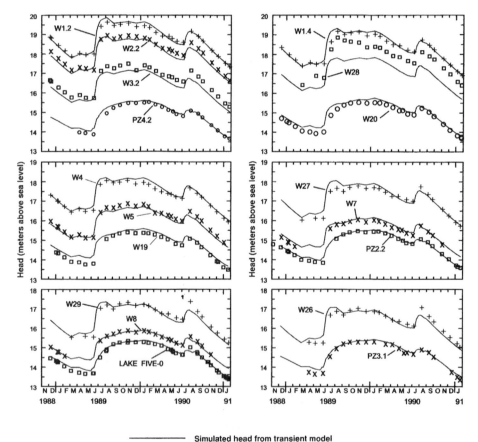

Simulated head from transient model
+ x □ ○   Observed head in indicated wells
W8   well number

Figure 9.7.6. Simulated and observed heads for Lake Five-O and adjacent groundwater system, December 12, 1988, through January 22, 1991 (Grubbs[16]).

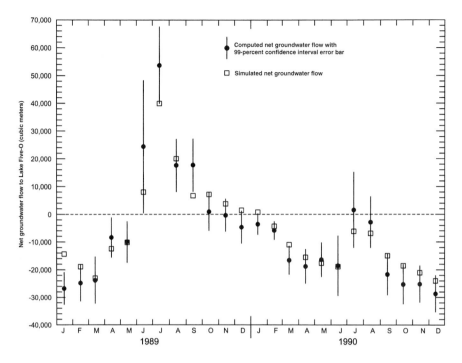

Figure 9.7.7. Simulated and computed monthly net groundwater flow to Lake Five-O, 1989–90 (Grubbs[16]).

9.8 PARTICLE TRACKING—MODPATH

9.8.1 What Is Particle Tracking?

Particle tracking is used to trace out flow paths, or path lines, by tracking the movement of infinitely small imaginary particles placed in the flow field.[4] Particle tracking computer programs, such as the MODPATH,[46–48] are postprocessors to MODFLOW. MODPATH accepts the head distribution from MODFLOW and uses it to compute the velocity distribution,[12] which is then used to trace out path lines. Particle tracking can be used to help visualize the flow field and to track contaminant paths. Anderson and Woessner[4] also point out that particle tracking analysis should be used routinely with groundwater flow modeling to detect conceptual errors that cannot be detected solely by examining the head distribution. Particle tracking analysis can be used to evaluate the effects of different boundary conditions and can show the location of recharge and discharge more clearly than results from groundwater flow models.

9.8.2 Particle Tracking Analysis—An Application

Grubbs[16] used the particle tracking code by Pollock[47] to evaluate the groundwater path lines (flow paths) and residence times for hypothetical parcels of groundwater (particles) that discharge to Lake Five-O in northwestern Florida. Figures 9.8.1 and 9.8.2 illustrate the results of the particle tracking. Figure 9.8.1 indicates that most of the surrounding groundwater basin contributes to flow into the lake. According to Grubbs,[16] the *residence times* (travel time of the inflow particles from recharge at the water table to discharge at the lake) generally ranged from 0.6 to 9 years, with mean residence times ranging from 3 to 6 years. Also, there was very little variation in these residence times between high and low water conditions.

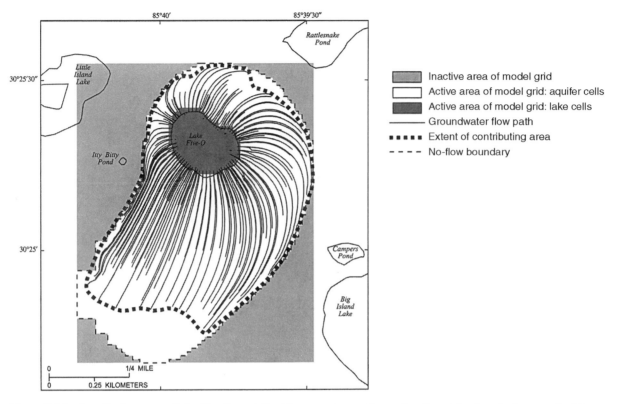

Figure 9.8.1. Contributing area to Lake Five-O, as defined by groundwater flow paths of particles that discharge to the lake (Grubbs[16]).

The flow paths in Figure 9.8.2 illustrate the vertical character of the groundwater flow within and between the surficial aquifer and intermediate confining unit. Flow lines are deflected downward at two stagnation points, a very small distance outside the lakebed. These are locations where the head in the groundwater flow is the same as the altitude of the lake surface. At these two points, groundwater flow into the lake and leakage from the lake to the groundwater occur below these points. According to Grubbs,[16] the location of these two points did not change much from low to high water conditions. Flow paths in Figure 9.8.2 indicate that most all of the groundwater flow near Lake Five-O occurs within the surficial aquifer. In addition the groundwater that discharges into the lake does not move through the intermediate confining unit.

9.9 EXAMPLE APPLICATIONS AND INPUT OF MODFLOW

EXAMPLE 9.9.1

TWO-DIMENSIONAL STEADY-STATE PROBLEM

Consider the aquifer (uniformly discretized into cells of 500 ft × 500 ft) shown in Figure 9.9.1, with specified head boundaries along row 1 and along column 7. Think of this as two rivers intersecting perpendicularly in the northeastern corner of the modeled groundwater system. A single aquifer is being modeled so that only one layer is required. The aquifer is treated as a confined aquifer because of its relative thickness and because it does not experience large changes in saturated thickness. The transmissivity of the aquifer is 500 ft²/d and it recharges at a rate of 0.001 ft/d. A well is located at row 5, column 3, which discharges at a rate of 8,000 ft³/d. Heads for all the active cells are set at 10 ft. A time step of 365 days is to be used.

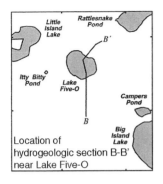

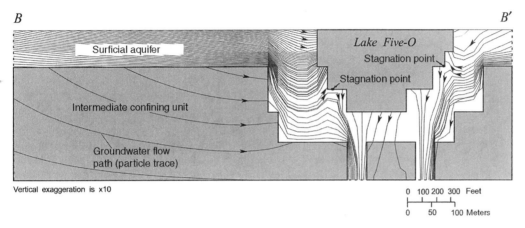

Figure 9.8.2. Particle traces projected onto hydrogeologic section B–B' near Lake Five-O (Grubbs[16]).

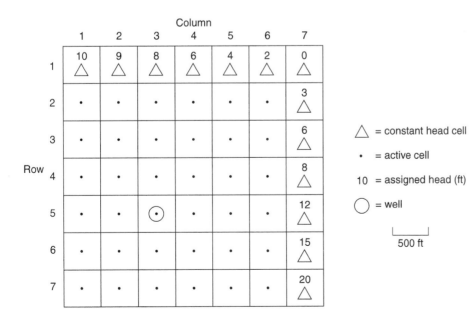

Figure 9.9.1. Example 9.9.1 aquifer discretization with 500 ft × 500 ft cells.

The MODFLOW model input data is presented in Table 9.9.1 (Appendix B) with brief explanations of the input. The strongly implicit procedure (SIP) solution technique is used and a maximum of 50 iterations (MXITER) is specified. The number of iteration parameters (NPARM) is 5, the acceleration number (ACCL) is 1, the head change criterion is 0.01, IPCALC = 1, WSEED = 0, and IPRSIP = 1. See the MODFLOW input manual for detailed descriptions of the input.

SOLUTION

Output data are given in Table 9.9.2 (Appendix B).   ■

EXAMPLE 9.9.2

STEADY-STATE APPLICATION COMPARING ANALYTICAL AND NUMERICAL SOLUTIONS

This example uses MODFLOW model for a steady-state problem in a confined aquifer. The aquifer is the same as described in Example 4.2.2:

$$K = 16.43 \text{ m/day}$$

$$b = 10 \text{ m}$$

$$r_w = 0.5 \text{ m}$$

$$Q = 425 \text{ m}^3/\text{day}$$

The radius of influence of pumping is at 300 m from the pumping well. Construct a numerical model under steady-state conditions using MODFLOW, calculate the drawdown curve from $r = 0$ m to $r = 300$ m and compare it with the analytical (Thiem) solution.

SOLUTION

The input data are presented in Table 9.9.3 (Appendix B). To represent the radius of influence given in the original problem, a 600 m × 600 m one-layer mesh is constructed with 19 rows and 19 columns in MODFLOW model. The well is placed at the center point (300 m, 300 m). An increasing grid spacing expansion gives better results than a regular grid spacing model. The grid spacing scheme is given in Table 9.9.4 (Appendix B) and is shown in Figure 9.9.2.

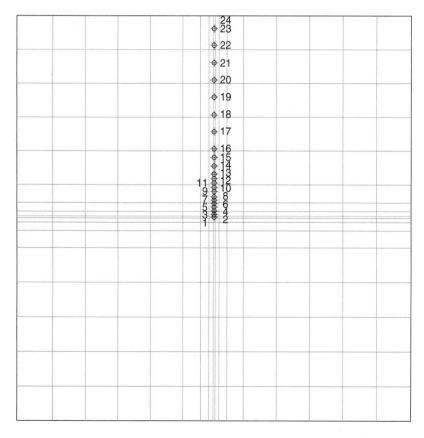

Figure 9.9.2. Grid-1 scheme used for Example 9.9.2.

The layer type is 0: *confined*. The *Transmissivity* flag is set to *User-specified*. The boundaries of the model are defined as *fixed-head boundaries*. A homogeneous transmissivity of 164.3 m/day is specified over the model domain. To calculate the drawdown curve, 27 boreholes are defined along a radial direction from $r = 0.5$ m to $r = 300$ m. The pumping well is placed at the center of the domain, row 10, column 10, with a pumping rate of 425 m³/day. Simulation flow type is set to steady-state. See Table 9.9.5 (Appendix B) for output.

The resulting drawdown curves from the analytical (Thiem) and numerical (MODFLOW) solutions are given in Table 9.9.6 (Appendix B) and plotted in Figure 9.9.3.

The MODFLOW results compare well to the analytical solution. The numerical results of the given problem are within 0.05 m of the analytic results except near the pumping well (0.5 m $< r < 8$ m). An exact comparison is not attained because of the use of a discrete rather than continuous spatial domain. It should also be remembered that the radius of influence in the original problem, $r = 300$ m, is approximated by setting the boundaries of the square model domain as fixed-head boundaries. ■

EXAMPLE 9.9.3

TRANSIENT APPLICATION COMPARING ANALYTICAL AND NUMERICAL SOLUTIONS

Using the data given in Example 4.4.4, construct a one-layer model in MODFLOW and calculate the drawdown vs. time curve 50 m away from the pumping well for $t = 0$ to $t = 1$ day. Compare the numerical results to the analytical solution of Theis.

SOLUTION

The aquifer given in Example 4.4.4 is confined with the following supplemental data:

$$b = 12 \text{ m}$$

$$K = 25 \text{ m/day}$$

$$S = 0.0001$$

$$Q = 300 \text{ m}^3/\text{day}$$

The exact analytical solution of the problem (using Theis) is also given below for comparison. Two different grid spacing schemes are used in the MODFLOW model listed in Table 9.9.7 (Appendix B).

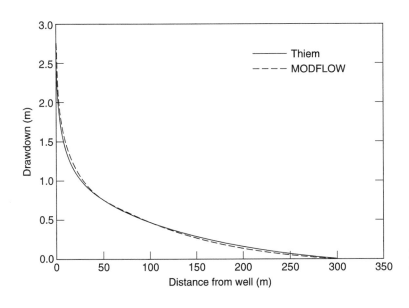

Figure 9.9.3. Example 9.9.2: Comparison of distance versus drawdown for the Thiem equation and MODFLOW.

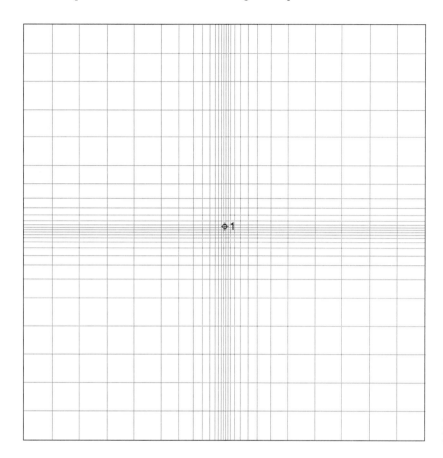

Figure 9.9.4. Grid-2 model scheme for Example 9.9.3.

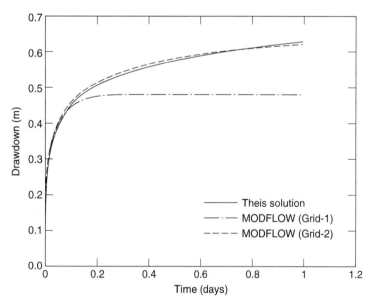

Figure 9.9.5. Example 9.9.3: Comparison of time-drawdown for the Theis equation and MODFLOW.

Grid-1 covers a square area of 2,000 m × 2,000 m and consists of 19 rows and 19 columns. An increasing grid spacing is used as shown in Table 9.9.7. Grid-2 (Figure 9.9.4) is an expanded version of Grid-1 with 29 rows and 29 columns covering an area of 5,000 m × 5,000 m. The simulation time is set to one day with

20 time steps and a time step expansion factor of 1.3. The SIP package is used as the model solver with five iteration parameters, 0.0001 as the closure criterion, and the maximum number of iterations is set to 50.

The resulting drawdown curves 50 m away from the pumping well using the Theis solution and MODFLOW model are illustrated in Figure 9.9.5. The artificial fixed head boundaries of the Grid-1 model start to affect the drawdown curve at 50 m approximately 0.1 day after the pumping starts. To meet the requirement of an infinite aerial extent aquifer, inherent in the Theis solution, the fixed head boundaries are moved farther away from the pumping well in the Grid-2 model. With this modification, the effect of the boundaries are not observed at 50 m for the given time period ($0 < t < 1$). The MODFLOW results using Grid-2 compare well to the analytic solution. The numerical results are generally within 1 cm of the analytic solution. ∎

9.10 SOLUTE TRANSPORT MODELING—MOC3D

9.10.1 Solute Transport Equation

The solute transport equation[32] is

$$\frac{\partial(\varepsilon C)}{\partial t} + \frac{\partial(\rho_b \overline{C})}{\partial t} + \frac{\partial}{\partial x_i}(\varepsilon C V_i) - \frac{\partial}{\partial x_i}\left(\varepsilon D_{ij}\frac{\partial C}{\partial x_j}\right)$$
$$- \sum C'W + \lambda(\varepsilon C + \rho_b \overline{C}) = 0 \tag{9.10.1}$$

(summation over repeated indices is understood), where C is volumetric concentration (mass of solute per unit volume of fluid, ML^{-3}), ρ_b is the bulk density of the aquifer material (mass of solids per unit volume of aquifer, ML^{-3}), $\overline{C}$ is the mass concentration of solute sorbed on or contained within the solid aquifer material (mass of solute per unit mass of aquifer material, MM^{-1}), ε is the effective porosity (dimensionless), V is a vector of interstitial fluid velocity components (LT^{-1}), D is a second-rank tensor of dispersion coefficients (L^2T^{-1}), W is a volumetric fluid sink ($W < 0$) or fluid source ($W > 0$) rate per unit volume of aquifer (T^{-1}), C' is the volumetric concentration in the sink/source fluid (ML^{-3}), λ is the decay rate (T^{-1}), t is time (T), and x_i are the Cartesian coordinates (L).

The terms controlling sorption are combined into a single parameter—the retardation factor (R_f), which is assumed to be constant in time because on a linear isotherm, $C/\overline{C}$ is constant. The *retardation factor* is defined as

$$R_f = 1 + \frac{\rho_b \overline{C}}{\varepsilon C}. \tag{9.10.2}$$

A number of the assumptions that follow are made in the development of the governing equations.[32]

1. Darcy's law is valid, and hydraulic head gradients are the only significant driving mechanism for fluid flow.
2. The hydraulic conductivity of the aquifer system is constant with time. Also, if the system is anisotropic, it is assumed that the principal axes of the hydraulic-conductivity tensor are aligned with the coordinate system of the grid, so that the cross-derivative terms of the hydraulic-conductivity tensor are eliminated.
3. Gradients of fluid density, viscosity, and temperature do not affect the velocity distribution.
4. Chemical or biological reactions do not affect the fluid or aquifer properties.
5. The dispersivity coefficients are constant over a flow time step, and the aquifer is isotropic with respect to longitudinal dispersivity.

The solution of the transport equation requires knowledge of the velocity (or specific discharge) field. After the head distribution has been calculated for a given time step or steady-state flow condition, the specific discharge across every face of each finite difference cell within the transport subgrid is calculated using a finite difference approximation.[32]

The mass-tracking algorithm requires that the seepage velocity at any point within a cell be defined to compute advective transport. Seepage velocity is calculated at points within a finite difference cell based on linearly interpolated estimates of specific discharge at those points divided by the effective porosity of the cell.[32]

9.10.2 MOC3D Model

MOC3D is a general-purpose computer model developed by the U.S. Geological Survey for simulation of three-dimensional solute transport in groundwater (see references 13, 14, 15, 23, 27, 29, 32, and 41 for more detail). This model computes changes in concentration of a single dissolved chemical constituent over time that are caused by advective transport, hydrodynamic dispersion (including both mechanical dispersion and diffusion), mixing (or dilution) from fluid sources, and mathematically simple chemical reactions (including linear sorption, which is represented by a retardation factor, and decay). MOC3D can be used to simulate groundwater age transport and the effects of double porosity and zero-order growth/loss. The model uses the method of characteristics to solve the transport equation on the basis of the hydraulic gradients that are computed with MODFLOW for a given time step. Particle tracking is used to represent advective transport and explicit finite difference methods are used to calculate the effects of other processes.

The notation and conventions used in MOC3D to describe the grid and to number nodes are illustrated in Figures 9.10.1 and 9.10.2. The indexing notation used is consistent with that used in MODFLOW.

Application of MOC3D is the third step in a logical sequence of (1) flow, (2) advective transport, and (3) solute transport simulation as illustrated in Figure 9.10.3. After the flow

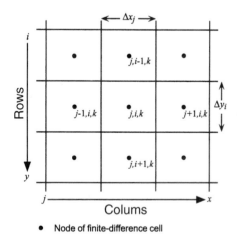

- Node of finite-difference cell
$\Delta x_j = \Delta r_j$ = cell dimension in row direction
$\Delta y_i = \Delta c_i$ = cell dimension in column direction

Figure 9.10.1. Notation used to label rows, columns, and nodes within one layer (k) of a three-dimensional, block-centered, finite difference grid for MOC3D.[32]

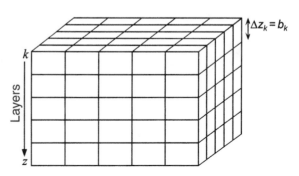

Figure 9.10.2. Representative three-dimensional grid for MOC3D illustrating notation for layers.[32]

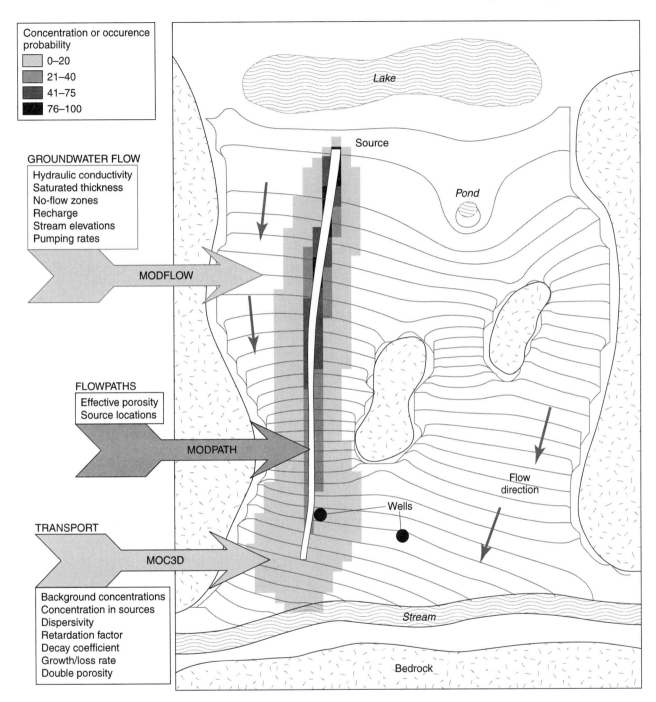

Figure 9.10.3. Typical input and illustration of output for simulation of groundwater flow and solute transport using MODFLOW, MODPATH, and MOC3D (Goode[14]).

model (MODFLOW) is constructed, advective transport can be modeled by particle tracking using MODPATH.[48] Because travel times are needed, the effective porosity must be specified. The solute transport as a result of advection and dilution (mixing) can be modeled using MOC3D using specified initial and source concentrations. Optional input on dispersion and reactions allows effects of these processes to be modeled. The simulation of solute transport with dispersion using MOC3D can enhance particle-tracking analyses by approximating uncertainty in source–area delineation and potential contamination advection.[13, 14]

Figure 9.10.3 shows nine closely spaced pathlines of groundwater flow from a continuous source of contamination after 20 years. These were calculated using MODFLOW and MODPATH. MOC3D was used to simulate the dispersed plume from the same source location. This dispersed plume can be interpreted as the probability that the contaminant will be detected at each location within 20 years. As pointed out by Goode,[13] the dispersion coefficient approximately accounts for the uncertainty in predicted flowpaths caused by imperfect knowledge of aquifer properties, recharge rates, stream levels, and other factors. Detailed field characterization can reduce uncertainties and make the contaminant occurrence probability map more accurate. These approaches can also be used to delineate probabilistic source areas (capture zones) for pumping wells using MOC3D.

An example of a MOC3D application is illustrated in Figure 9.10.4, which uses time- and space-varying decay coefficients to approximate biodegradation of benzene at a Laurel Bay, South Carolina research site. Benzene attenuation depends on decay rates that are generally higher in aerobic (oxygen rich) zones and lower in anaerobic (zero oxygen). Anaerobic biodegradation rates are dependent on the specific terminal-electron-accepting process (TEAP) controlling the redox state. Figure 9.10.4a shows the TEAP zones. Also in the groundwater at this site is methyl *tert*-butyl ether (MTBE). This is also a gasoline-derived contaminant, but is essentially nonreactive during transport. Source concentrations are the same for

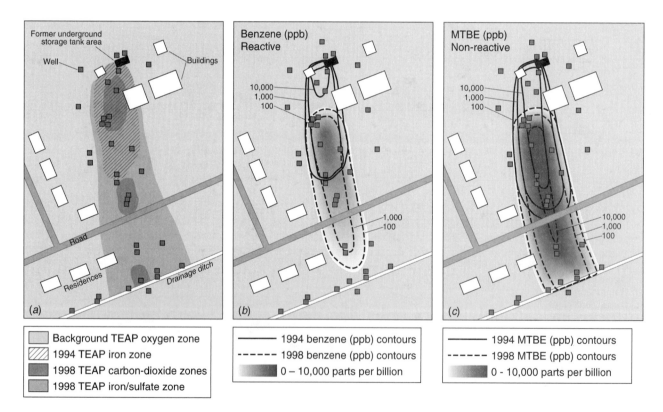

Figure 9.10.4. MOC3D simulation of benzene and MTBE at the Laurel Bay, SC, research site, based on field and lab data of Landmeyer[33] and others (1998): (*a*) Location map with anaerobic terminal-electron-accepting-process (TEAP) zones in 1994 (hatched) and 1998 (shaded); (*b*) Benzene in 1994 and 1998; (*c*) MTBE in 1994 and 1998. In (*b*) and (*c*), the 1998 shaded maps range from 0 to 10,000 parts per billion (ppb); contours show 100, 1,000, and 10,000 ppb concentration levels (as presented in Goode[14]).

both contaminants; therefore, differences in simulated concentrations result from differences in decay coefficients. Outside the TEAP zones fast aerobic degradation is taking place, except at the early time when microbes are assumed to be in low concentrations. This is illustrated by the differences in the 1994 and 1998 zones (see Figure 9.10.4a). Figure 9.10.4b show the results of MOC3D simulations of the benzene migration for 1994 to 1998 that has temporally and spatially varying decay coefficients. Figure 9.10.4c shows the results of MOC3D simulations of the MBTE for 1994 to 1998.

9.11 GROUNDWATER MODELING SOFTWARE SUPPORT

9.11.1 U.S. Geological Survey

The U.S. Geological Survey provides access to public-domain software, such as MODFLOW, MODPATH, MOC3D, and several other simulations models such as SUTRA for two- and three-dimensional variably saturated flow, solute or energy transport and several other computer models. ZONEBUDGET is a program for computing water budgets for MODFLOW. Access this software at http://water.usgs.gov/software/ground_water.html.

9.11.2 U.S. EPA Center for Exposure Assessment Modeling (CEAM)

The Center for Subsurface Modeling Support (CSMoS) provides public domain groundwater and vadose zone modeling software and services to public agencies and private companies. CSMoS is located at the National Risk Management Laboratory (NRML), U.S. EPA Center for Ground-Water Research, Ada, Oklahoma. Its primary aims are to provide direct technical support to EPA and state decision makers in subsurface modeling applications and to support the groundwater models and databases resulting from the research at NRML. A major focus of CSMoS is the coordination of the use of models for risk assessment, site characterization, remedial activities, wellhead protection, and Geographic Information Systems (GIS) application. The Web address for CSMoS is http://www.epa.gov/ada/csmos.html.

9.11.3 International Groundwater Modeling Center (IGWMC)

The International Ground Water Modeling Center (IGWMC) provides information, education, and research for groundwater modeling. It is located at the Colorado School of Mines, Golden, Colorado. IGWMC advises on groundwater modeling problems, distributes software, organizes short courses and workshops, and provides technical assistance on problems related to groundwater modeling. The Web address for IGWMC is http://www.mines.edu/igwmc/about.

9.11.4 Processors for MODFLOW

Because of the wide usage of the MODFLOW model, several pre and post-processing models have been developed. Some of the processors are PMWIN, Visual MODFLOW, MODFLOW-GUI PIE, and VIEWLOG. These processors can be located easily on the Internet by performing a search for their names.

PMWIN (see Chiang and Kinzelbach[9]) is a simulation package for pre- and post-processing that includes MODFLOW, MT3D, MT3DMS, PMPATH, UCODE, and PEST-LITE.

Visual MODFLOW, developed by Waterloo Hydrogeologic, Inc. of Ontario, Canada, is a pre-and post-processor for the MODFLOW and MODPATH models. This allows the user to solve the flow model and to perform particle tracking and velocity vector analysis. The program is completely graphical, allowing the user to input all necessary input parameters, run the program, and visualize the analysis results. The Visual MODFLOW package includes the

MODFLOW and MODPATH models and can import previously created MODFLOW data files. VIEWLOG is also a pre-and post-processor for the MODFLOW and MODPATH models.

MODFLOW-GUI PIE is a U.S. Geological Survey public domain GIS pre-and post-processor graphical-user interface for preparing MODFLOW, MOC3D, MODPATH, and ZONEBDGT input and viewing model output for use within Argus ONE. MODFLOW GUI PIE is used to graphically input GIS data, run MODFLOW, and visualize the results from within Argus ONE, which is an easy-to-use graphical workplace.

PROBLEMS

9.2.1 Explain the concepts of physical scale models, analog models, and mathematical models.

9.2.2 Discuss the concepts of porous media models, analog models, and electric analog models.

9.2.3 Discuss the theory upon which viscous fluid models are based.

9.2.4 Explain how electric analog models are used to model groundwater flow.

9.2.5 Explain how sand tank models are used to model groundwater flow.

9.2.6 Explain the concept of membrane models for groundwater flow.

9.2.7 Explain how Hele-Shaw models have been used in groundwater flow modeling.

9.2.8 Discuss in general the difference between the finite difference and finite element methods of groundwater modeling.

9.3.1 Discuss in detail the processes of calibration, calibration sensitivity analysis, and model verification and how they work together to develop a model.

9.3.2 Discuss in detail the processes of prediction and predictive sensitivity analysis in model development.

9.3.3 Develop a brief discussion of the various ASTM standards for groundwater flow modeling.

9.4.1 Show the development of Equation 9.4.10a.

9.5.1 Derive Equation 9.5.2 for the effective hydraulic conductivity.

9.5.2 Derive Equation 9.5.22, which is the finite difference approximation of the continuity for cell (i, j, k).

9.6.1 Derive Equation 9.6.10 for the calculation of conductance between nodes.

9.6.2 Derive Equation 9.6.12 for the calculation of vertical conductance between nodes.

9.9.1 Run the MODFLOW model for the problem in Example 9.9.1 using a time step of one day and a time step multiplier of 1.0. Compare the results with those obtained in Example 9.9.1 and justify your conclusions.

9.9.2 Run the MODFLOW model for the problem in Example 9.9.1 using a time step of one day and a time step multiplier of 1.0. Use an initial head condition of 1,000 ft in the active part of the grid. Compare the results with those obtained in Example 9.9.1 and justify your conclusions.

9.9.3 Run the MODFLOW model for the system in Example 9.9.1 in a transient mode. Five time steps are to be used with a time step multiplier of 1.5 and a stress period length of 365 days specified in the BASIC package. Print the mass balance (budget) and head distributions for all five time steps using the OUTPUT CONTROL PACKAGE. Use a specific storage of 0.01.

9.9.4 Modify the data set from Problem 9.9.3 to use the OUTPUT CONTROL PACKAGE to print out the model-wide mass balance and to save the cell-by-cell budgets for the BCF, WELL, and RECHARGE packages at time step 1. Use the hydraulic heads generated for time step 1 to manually compute the model-wide rate components into storage, out of storage, well discharge, out of constant heads, into constant heads, and recharge. Use Darcy's law to compute constant head flux. Also use the definition of the storage coefficient to determine the rate change in storage. Compare your results to those computed by the model.

9.9.5 Use the input file developed for Problem 9.9.4 and run it to obtain the cell-by-cell budgets. Compare the model-generated values to the Problem 9.9.4 hand calculations.

9.9.6 Use the model developed in Problem 9.9.1 and eliminate the well in row 5, column 3. Perform a steady-state simulation for a recharge of 0.001 ft/d and a transmissivity of 500 ft^2/d.

9.9.7 Use the model developed in Problem 9.9.1 and eliminate the well in row 5, column 3. Perform a steady-state simulation for a recharge of 0.001 ft/d and a transmissivity of 50 ft^2/d. Compare the hydraulic heads to those generated in Problem 9.9.6.

9.9.8 Use the model developed in Problem 9.9.1 and eliminate the well in row 5, column 3. Perform a steady-state simulation for a recharge of 0.0001 ft/d and a transmissivity of 50 ft^2/d. Compare the hydraulic heads to those generated in Problem 9.9.6.

9.9.9 Rerun the steady-state model (one stress period and one time step of one day) developed in Problem 9.9.6 with the option in the BCF package invoked to print out the individual specified head fluxes.

9.9.10 Run the steady-state model (one stress period and one time step of one day) developed in Problem 9.9.9 with a well in row 5, column 3 pumping at a rate of $-8,000$ ft^3/d. Print out the individual specified head fluxes and compare these to those generated in Problem 9.9.9.

9.9.11 Develop a drawdown model using the parameters and stresses of Problem 9.9.10. Use an initial head of 0.0, a recharge rate of 0.0, and specified heads of 0.0 along row one and column 7. Run the steady-state model with one stress period and one time step of one day, printing out the individual specified head fluxes. On a node-by-node basis, add the heads generated in Problem 9.9.9 and the solution generated in this problem. Compare these heads to those generated in Problem 9.9.10.

9.9.12 Resolve Problem 9.9.11 with the well rate doubled. Compare the heads of Problem 9.9.11 with those generated in this problem.

9.9.13 Resolve Example 9.9.2, doubling the transmissivity. All other data remain the same. Compare the results with those reported in Example 9.9.2.

9.9.14 Resolve Example 9.9.2 using a 30 by 30 grid. Compare the results with those presented in Example 9.9.2.

REFERENCES

1. Alley, W. M., T. E. Reilly, and O. L. Franke, *Sustainability of Ground-Water Resources, U.S. Geological Survey Circular* 1186, 1999.

2. Anderman, E. R., and M. C. Hill, *Advective-Transport Observation (ADV) Package, a Computer Program for Adding Advective-Transport Observations of Steady-State Flow Fields to the Three-Dimensional Ground-Water Flow Parameter-Estimation Nodel MODFLOWP, U.S. Geological Survey Open-File Report* 97-14, 67 pp, 1997.

3. Anderman, E. R., and M. C. Hill, *MODFLOW-2000, The U.S. Geological Survey Modular Groundwater Model—Documentation of the Hydrogeologic-Unit Flow (HUF) Package, U.S. Geological Survey Open-File Report* 00-342, 89 pp, 2000.

4. Anderson, M. P., and W. W. Woessner, *Applied Groundwater Modeling: Simulation of Flow and Advective Transport*, Academic Press, San Diego, CA, 1992.

5. Appel, C. A., and J. D. Bredehoeft, *Status of Groundwater Modeling in the U.S. Geological Survey, U.S. Geological Survey Circular* 737, 1976.

6. Banta, E. R., *MODFLOW-2000, The U.S. Geological Survey Modular Groundwater Model—Documentation of Packages for Simulating Evapotranspiration with a Segmented Function (ETS1) and Drains with Return Flow (DRT1), U.S. Geological Survey Open-File Report* 00-466, 127 pp, 2000.

7. Belitz, K., and S. P. Phillips, *Numerical Simulation of Ground-Water Flow in the Central Part of the Western San Joaquin Valley, California, U.S. Geological Survey Water-Supply Paper* 2396, 69 pp, 1993.

8. Buxton, H., and T. E. Reilly, *A Technique for Analysis of Ground-Water Systems of Regional and Subregional Scales Applied to Long Island, New York, U.S. Geological Survey Water-Supply Paper* 2310, pp. 129–142, 1986.

9. Chiang, W. H., and W. Kinzelbach, *3D-Groundwater Modeling with PMWIN, A Simulation System for Modeling Groundwater Flow and Pollution*, Springer, Berlin, 2000.

10. Fenske, J. P., S. A. Leake, and D. E. Prudic, *Documentation of a Computer Program (RES1) to Simulate Leakage from Reservoirs Using Modular Finite Difference Groundwater Flow Model (MODFLOW), U.S. Geological Survey Open-File Report* 96-364, 51 pp, 1996.

11. Freeze, R. A., and J. A. Cherry, *Groundwater*, Prentice Hall, Englewood Cliffs, NJ, 1979.

12. Goode, D. J., Particle velocity interpolation in block-centered finite difference groundwater flow models, *Water Resources Research*, v. 26, no. 5, pp. 925–940, 1990.

13. Goode, D. J., Age, Double Porosity, and Simple Reaction Modifications for the MOC3D Ground-Water Transport Model, *U.S. Geological Survey Water-Resources Investigations Report* 99-4041, 1999.

14. Goode, D. J., *Simulating Contaminant Attenuation, Double-Porosity Exchange, and Water Age in Aquifers Using MOC3D, U.S. Geological Survey Fact Sheet* 086-99, April 1999.

15. Goode, D. J., and C. A. Appel, *Finite Difference Interblock Transmissivity for Unconfined Aquifers and for Aquifers Having Smoothly Varying Transmissivity, U.S. Geological Survey Water-Resources Investigations Report* 92-4124, 79 pp, 1992.

16. Grubbs, J. W., *Evaluation of Ground-Water and Hydrologic Budget for Lake Five-O, a Seepage Lake in Northwestern Florida, U.S. Geological Survey, Water-Resources Investigations Report* 94-4145, Tallahassee, FL, 1995.

17. Hanson, R. T., and S. A. Leake, *Documentation for HYMOD, A Program for Extracting and Processing Time-Series Data from the U.S. Geological Survey's Modular Three-Dimensional Finite Difference Ground-Water Flow Model, U.S. Geological Survey Open-File Report* 98-564, 57 pp, 1999.

18. Harbaugh, A. W., *A Generalized Finite-Difference Interblock Transmissivity for Unconfined Aquifers and for Aquifers Having Smoothly Varying Transmissivity, U.S. Geological Survey Open-File Report* 91-494, 60 pp, 1992.

19. Harbaugh, A. W., *Direct Solution Package Based on Alternating Diagonal Ordering for the U.S. Geological Modular Finite-Difference Groundwater Flow Model, U.S. Geological Survey Open-File Report* 95-288, 46 pp, 1995.

20. Harbaugh, A. W., E. R. Banta, M. C. Hill, and M. G. McDonald, *MODFLOW-2000: The U.S. Geological Survey Modular Ground-Water Model—User Guide to Modularization Concepts and the Ground-Water Flow Process, U.S. Geological Survey Open-File Report* 00-92, 121 pp, Reston, VA, 2000.

21. Harbaugh, A. W., and M. G. McDonald, *User's Documentation for MODFLOW-96, An Update to the United States Geological Survey Modular Finite Difference Groundwater Flow Model, U.S. Geological Survey Open-File Report* 96-485, 56 pp, 1996.

22. Harbaugh, A. W., and M. G. McDonald, *Programmer's Documentation for MODFLOW-96, An Update to the U.S. Geological Modular Finite Difference Groundwater Flow Model, U.S. Geological Open File Report* 96-486, 220 pp, 1996.

23. Heberton, C. I., T. F. Russell, L. F. Konikow, and G. Z. Hornberger, *A Three-Dimensional Finite Volume Eulerian-Lagrangian Localized Adjoint Method (ELLAM) for Solute Transport Modeling, U.S. Geological Survey Water Resources Investigations Report* 00-4087, Reston, VA, 2000.

24. Hill, M. C., *Preconditioned Conjugate-Gradient 2 (PCG2), a Computer Program for Solving Ground-Water Flow Equations, U.S. Geological Survey Water-Resources Investigations Report* 90-4048, 43 pp, 1990.

25. Hill, M. C., *Methods and Guidelines for Effective Model Calibration*, U.S. Geological Survey Water-Resources Investigations Report 98-4005, 90 pp, 1998.

26. Hill, M. C., E. R. Banta, A. W. Harbaugh, and E. R. Anderman, *MODFLOW-2000, The U.S. Geological Survey Modular Ground-Water Model—User Guide to the Observation, Sensitivity, and Parameter-Estimation Processes and Three Post-Processing Programs*, U.S. Geological Survey Open-File Report 00-184, 210 pp, 2000.

27. Hornberger, G. Z., L. F. Konikow, and P. T. Harte, *Simulating Horizontal-Flow Barriers Using the MODFLOW Ground-Water Transport Process*, U.S. Geological Survey Open-File Report 02-52, 2002.

28. Hsieh, P. A., and J. R. Freckleton, *Documentation of a Computer Program to Simulate Horizontal-Flow Barriers Using the U.S. Geological Survey Modular Three-Dimensional Finite Difference Ground-Water Flow Model*, U.S. Geological Survey Open-File Report 92-477, 32 pp, 1993.

29. Kipp, K. L., L. F. Konikow, and G. Z. Hornberger, *An Implicit Dispersive Transport Algorithm for the U.S. Geological Survey MOC3D Solute-Transport Model*, U.S. Geological Survey Water-Resources Investigations Report 98-4234, 1998.

30. Klempt, W. B., T. R. Knowles, G. R. Elder, and T. W. Sieh, Groundwater Resources and Model Applications for the Edwards (Balcones Fault Zone) Aquifer in the San Antonio Region, Texas, Report 239, Texas Department of Water Resources, Austin, TX, October 1979.

31. Konikow, L. F., and J. D. Bredehoeft, Groundwater models cannot be validated, *Advances in Water Resources*, v. 15, pp. 75–83, 1992.

32. Konikow, L. F., D. J. Goode, and G. Z. Hornberger, *A Three-Dimensional Method-of-Characteristics Solute-Transport Model (MOC3D)*, U.S. Geological Survey Water Resources Investigations Report 96-4267, 87 pp, 1996.

33. Landmeyer, J. E., F. H. Chapelle, P. M. Bradley, J. F. Pankow, C. D. Church, and P. G. Tratnyek, Fate of MTBE relative to benzene in a gasoline-contamination aquifer (1993–1998), *Ground Water Monitoring and Remediation*, v. 18, No. 4, pp. 93–102, 1998.

34. Leake, S. A., and D. V. Claar, *Procedures and Computer Programs for Telescopic Mesh Refinement Using MODFLOW*, U.S. Geological Survey Open-File Report 99-238, 53 pp, 1999.

35. Leake, S. A., and M. R. Lilly, *Documentation of a Computer Program (FHB1) for Assignment of Transient Specified-Flow and Specified-Head Boundaries in Applications of the Modular Finite Difference Groundwater Flow Model (MODFLOW)*, U.S. Geological Survey Open-File Report 97-571, 50 pp, 1997.

36. Leake, S. A., and D. E. Prudic, *Documentation of a Computer Program to Simulate Aquifer-System Compaction Using the Modular Finite Difference Ground-Water Flow Model*, U.S. Geological Survey Techniques of Water-Resources Investigations, Book 6, Chap. A2, 68 pp, 1991.

37. Liggett, J. A., and P. L-F., Liu, *The Boundary Integral Equation Method for Porous Media Flow*, Allen and Unwin, 255 pp, 1983.

38. McDonald, J. M., and A. W. Harbaugh, *A Modular Three-Dimensional Finite Difference Groundwater Flow Model*, U.S. Geological Survey Open-File Report 83-875, 528 pp, 1984.

39. McDonald, J. M., and A. W. Harbaugh, A modular three-dimensional finite difference groundwater flow model, *Techniques of Water Resources Investigations of the United States Geological Survey*, Book 6, pp. 586, 1988.

40. McDonald, M. G., A. W. Harbaugh, B. R. Orr, and D. J. Ackerman, *A Method of Converting No-Flow Cells to Variable-Head Cells for the U.S. Geological Survey Finite Difference Groundwater Flow Model*, U.S. Geological Survey Open-File Report 91-536, 99 pp, 1992.

41. Merritt, M. L., and L. F. Konikow, *Documentation of a Computer Program to Simulate Lake-Aquifer Interaction Using MODFLOW Groundwater Flow Model and the MOC3D Solute-Transport Model*, U.S. Geological Survey Water-Resources Investigations Report 00-4167, 146 pp, 2000.

42. Miller, R. T., and C. I. Voss, Finite difference grid for a doublet well in an anisotropic aquifer, *Ground Water*, v. 24, no. 4, pp. 490–496, 1987.

43. Peaceman, D. W., *Fundamentals of Numerical Reservoir Simulation*, Elsevier Scientific Publishing Company, Oxford, UK, 1977.

44. Peaceman, D. W., and H. H. Rachford, Jr., The numerical solution of parabolic and elliptic differential equations, *Jour. Soc. of Industrial and Applied Mathematics*, v. 3, pp. 28–41, 1955.

45. Pinder, G. F., and W. G. Gray, *Finite Element Simulation in Surface and Subsurface Hydrology*, Academic Press, San Diego, CA, 295 pp, 1977.

46. Pollock, D. W., Semianalytical computation of path lines for finite difference models, *Ground Water*, v. 26, no. 6, pp. 743–750, 1988.

47. Pollock, D. W., *Documentation of Computer Programs to Compute and Display Pathlines Using Results from the U.S. Geological Survey Modular Three-Dimensional Finite-Difference Ground-Water Flow Model*, U.S. Geological Survey Open-File Report 89-381, 188 pp, 1989.

48. Pollock, D. W., *User's guide for MODPATH/MODPATH-PLOT, Version 3: A Particle Tracking Post-Processing Package for MODFLOW, the U.S. Geological Survey Finite-Difference Ground-Water Flow Model*, U.S. Geological Survey Open-File Report 94-464, 234 pp, 1994.

49. Prickett, T. A., Modeling techniques for groundwater evaluation, in V. T. Chow, ed., *Advances in Hydroscience*, Academic Press, v. 10, pp. 1–143, 1975.

50. Prickett, T. A., Ground-water computer models—State of the art, *Ground Water*, v. 17, pp. 167–173, 1979.

51. Prickett, T. A., and Lonnquist, C. G., Selected digital computer techniques for groundwater resources evaluation, *Bulletin No. 55*, Illinois State Water Survey, Urbana, IL, 1971.

52. Prince, K. R., O. L. Franke, and T. E. Reilly, *Quantitative Assessment of the Shallow Ground-Water Flow System Associated with Conneeetquot Brook, Long Island, New York*, U.S. Geological Survey Water-Supply Paper 2309, 28 pp, 1988.

53. Prudic, D. E., *Documentation of a Computer Program to Simulate Stream-Aquifer Relations Using a Modular, Finite-Difference, Groundwater Flow Model*, U.S. Geological Survey Open-File Report 88-729, 113 pp, 1989.

54. Remson, I., G. W. Hornberger, and F. J. Molz, *Numerical Methods in Subsurface Hydrology*, Wiley-Interscience, New York, 389 pp, 1971.

55. Rushton, K. R., and S. C. Redshaw, *Seepage and Groundwater Flow—Numerical Analysis by Analog and Digital Methods*, John Wiley & Sons, New York, 1979.

56. Texas Water Development Board, GWSIM groundwater simulation program, program document and user's manual, UM S7405, Austin, TX, 1974.

57. Todd, D. K., *Groundwater Hydrology*, 2nd ed., John Wiley & Sons, New York, 1980.

58. Trescott, P. C., G. F. Pinder, and S. P Larson, Finite-difference model for aquifer simulation in two dimensions with results of numerical experiments, in *U S. Geological Survey Techniques of Water Resources Investigations*, Book 7, C1, U.S. Geological Survey, Reston, VA, 1976.

59. Walton, W.C., *Groundwater Resource Evaluation*, McGraw-Hill, New York, 1970.

60. Walton, W. C., *Numerical Groundwater Modeling: Flow and Contaminant Migration*, Lewis Publishers, Boca Raton, FL, 272 pp, 1989.

61. Wanakule, N., *Optimal Groundwater Management Models for the Barton Springs-Edwards Aquifer*, Edwards Aquifer Research and Data Center, San Marcos, TX, 1989.

62. Wang, H. F., and M. P. Anderson, *Introduction to Groundwater Modeling: Finite Difference and Finite Element Models*, W. H. Freeman, San Francisco, CA, 1982.

63. Ward, D. S., D. R. Buss, J. W. Mercer, and S. S. Hughes, Evaluation of a groundwater corrective action at the Chem-Dyne hazardous waste site using telescope mesh refinement modeling approach, *Water Resources Research*, v. 23, no. 4, pp. 603–617, 1987.

64. Yeh, W. W. G., Groundwater Systems, Ch. 16, in *Water Resources Handbook*, L. W. Mays, ed., McGraw-Hill, New York, 1996.

Chapter 10

Management of Groundwater

Maximum development of groundwater resources for beneficial use involves planning in terms of an entire groundwater basin. Because a basin is a large natural underground reservoir, it follows that utilization of groundwater by one landowner affects the water supply of all other landowners. Management objectives must be selected in order to develop and operate the basin. These involve not only geologic and hydrologic considerations but also economic, legal, political, and financial aspects. Typically, optimum economic development of water resources in an area requires an integrated approach that coordinates the use of both surface water and groundwater resources. After evaluation of total water resources and preparation of alternative management plans, action decisions can then be made by appropriate public bodies or agencies.

10.1 CONCEPTS OF BASIN MANAGEMENT

The management of a groundwater basin implies a program of development and utilization of subsurface water for some stated purpose, usually of a social or economic nature. In general, the desired goal is to obtain the maximum quantity of water to meet predetermined quality requirements at least cost.* Because a groundwater basin can be visualized as a large natural underground reservoir, it follows that extraction of water by wells at one location influences the quantity of water available at other locations within the basin.

Groundwater is extracted from the ground just as are other minerals, such as oil, gas, or gold. Water typically carries a special constraint: It is regarded as a renewable natural resource. Thus, when a water well is drilled, people presume that production of water will continue indefinitely with time. In effect, this can occur only if there exists a balance between water recharged to the basin from surface sources and water pumped from within the basin by wells.

Development of water supplies from groundwater begins typically with a few pumping wells scattered over a basin. With time, more wells are drilled and the rate of extraction increases. As wells become more numerous, development of the basin reaches and exceeds its natural recharge capability. Continued development thereafter without a management plan could eventually deplete the groundwater resource.

By regulating inflow to and outflow from the basin, an underground reservoir can be made to function beneficially and indefinitely just as a surface water reservoir. The increasing demand for water in the United States and throughout the world has produced the realization that the vast underground reservoirs formed by aquifers constitute invaluable water storage facilities; proper management of them, therefore, has become a matter of considerable interest. Some of the pros and cons of subsurface and surface reservoirs are summarized in Table 10.1.1.

---

*As Bear and Levin[16] succinctly stated: "The basic idea is to regard the aquifer as a system which has to be operated in an optimal manner."

Table 10.1.1 Advantages and Disadvantages of Subsurface and Surface Reservoirs (after U.S. Bureau of Reclamation[89])

Subsurface reservoirs	Surface reservoirs
Advantages	Disadvantages
1. Many large-capacity sites available	1. Few new sites available
2. Slight to no evaporation loss	2. High evaporation loss even in humid climate
3. Require little land area	3. Require large land area
4. Slight to no danger of catastrophic structural failure	4. Ever-present danger of catastrophic failure
5. Uniform water temperature	5. Fluctuating water temperature
6. High biological purity	6. Easily contaminated
7. Safe from immediate radioactive fallout	7. Easily contaminated by radioactive material
8. Serve as conveyance systems—canals or pipelines across lands of others unnecessary	8. Water must be conveyed
Disadvantages	Advantages
1. Water must be pumped	1. Water may be available by gravity flow
2. Storage and conveyance use only	2. Multiple use
3. Water may be mineralized	3. Water generally of relatively low mineral content
4. Minor flood control value	4. Maximum flood control value
5. Limited flow at any point	5. Large flows
6. Power head usually not available	6. Power head available
7. Difficult and costly to investigate, evaluate, and manage	7. Relatively easy to evaluate, investigate, and manage
8. Recharge opportunity usually dependent on surplus surface flows	8. Recharge dependent on annual precipitation
9. Recharge water may require expensive treatment	9. No treatment required of recharge water
10. Continuous expensive maintenance of recharge areas or wells	10. Little maintenance required of facilities

Forecasts of future water demand suggest that mismanagement—or lack of management—of major groundwater basins cannot be permitted if adequate ongoing water supplies are to be provided. The management objective consists of providing an economic and continuous water supply to meet a usually growing demand from an underground water resource of which only a small portion is perennially renewable.

10.1.1 Managing Groundwater Resources

The demand for groundwater in recent decades has led to development of a variety of strategies for managing subsurface water and making efficient use of the available underground storage space. These may involve shifting of local water sources, changing pumping patterns, limiting pumpage, artificial recharge, conjunctive use of groundwater and surface water, and reuse of wastewater (Galloway[34]). The key concern is to maintain a sustainable long-term yield from aquifers (Alley et al.[9]). The consequences of overdraft in terms of adverse impacts on water quantity, quality, land subsidence, and water rights have been mentioned already. What is often overlooked in large-scale groundwater pumping is the fact that surface water depletions can occur far removed from the responsible wells. This situation, usually unbeknownst to pumpers and unrecognized by water managers, has become commonplace in the United States and poses an ongoing threat to our nation's rivers (Glennon[36]).

It is important to recognize on a regional basis that groundwater and surface water constitute a single unified source. What is surface water today can be groundwater tomorrow and vice versa. It follows that a depletion of one can affect the other. Failure to realize this basic concept regionally has served as the cause of several interstate water rights litigations. Illustrative is the basin of the Republican River, draining from Colorado to Nebraska and into Kansas. The area, centered in the corn and wheat belt of the Midwest, derives almost all of its water for irrigation from the extensive and underlying High Plains Aquifer. Since World War

II, the availability of deep-well turbine pumps, and later the innovation of center-pivot irrigation systems, stimulated groundwater production. In the drier western portion of the basin, water levels declined substantially, thereby reducing base flow in all tributaries. At the same time, conservation practices such as terracing of fields and tail-water ponds reduced summer thunderstorm runoff. The cumulative effect has been to deplete a majority of the river flow reaching Kansas. Thus, although all water was used beneficially, a maldistribution has resulted in the absence of a management approach that considered regional needs.

10.2 GROUNDWATER BASIN INVESTIGATIONS AND DATA COLLECTION

Ideally, before groundwater is developed in a basin, an investigation of the underground water resources should be made. In practice this rarely occurs; instead, a study is usually initiated either after extensive development with a view toward further development or after overdevelopment when a problem threatening the water supply appears imminent. Investigations are seldom concerned with simply locating groundwater supplies. More commonly the concerns involve evaluating the quantity and quality of groundwater resources already known to exist or determining the impact of human plans or activities on the quantity and quality of groundwater. Figure 10.2.1 illustrates the sequence of activities preceding the start of a groundwater management investigation.

Groundwater management studies are usually undertaken by local government agencies. Four levels of study are generally recognized, although not all are required.[10] In brief, these include:

1. *Preliminary Examination*—Based largely on judgment by experienced personnel, this study identifies the management possibilities of meeting a defined need for a specified area.

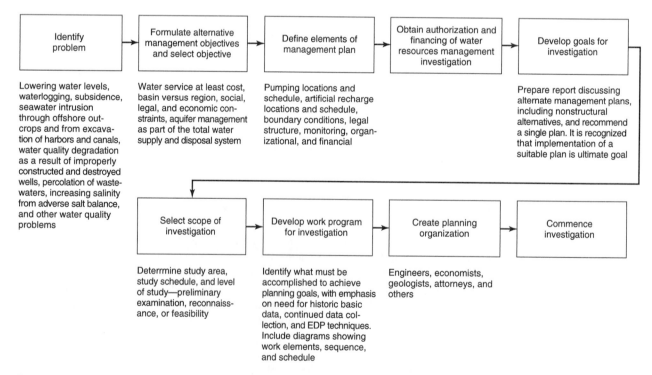

Figure 10.2.1. Sequence of activities preceding start of a groundwater management investigation (after Amer. Soc. Civil Engrs.[10]).

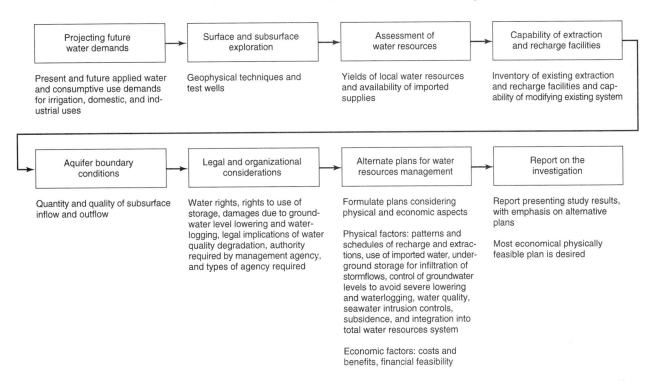

Figure 10.2.2. Sequence of activities during a feasibility investigation for groundwater management (after Amer. Soc. Civil Engrs.[10]).

2. *Reconnaissance*—This study considers possible alternatives in the formulation of a water management plan to meet a defined need for an area, including estimates of benefits and costs. The investigation draws on available data and generally necessitates a minimum of new data collection.

3. *Feasibility*—This study requires detailed engineering, hydrogeologic, and economic analyses together with cost and benefit estimates to ensure that the selected project is an optimum development. The sequence of activities normally involved in a feasibility investigation is outlined in Figure 10.2.2. Typically, the investigation concludes with a report recommending approval and funding for the project.

4. *Define Project*—This investigation involves planning studies necessary for defining specific features of the selected project. The completed report forms the basis for starting final design and preparation of plans and specifications.

The types of data and the tasks involved in the physical portion of a reconnaissance or feasibility study for groundwater management are outlined in the sections below. Table 1.8.1 presents the principal types of data and data compilations.

10.2.1 Topographic Data

Contour maps, aerial photographs, and benchmarks related to a leveling network are basic requirements. They are directly applicable for locating and identifying wells, measuring groundwater levels, conducting crop and land use surveys, and plotting areal data.

10.2.2   Geologic Data

Surface and subsurface geologic mapping provides the framework for the occurrence and movement of groundwater and hence is essential for feasibility studies. Subsurface information is gained from a drilling program, including classification and analysis of well logs, and geophysical surveys (see Chapter 12). As part of the drilling program, pumping tests of wells are conducted to evaluate storage coefficients and transmissivities of aquifers, while samples of groundwater are collected and analyzed for quality. From interpretation of subsurface geologic data, principal aquifers and their extent are mapped together with regions of confined and unconfined groundwater. Location of faults, dikes, and other structures that may significantly affect groundwater is also a part of the geologic program.

10.2.3   Hydrologic Data

The principal purpose of hydrologic data collection is to evaluate the equation of hydrologic equilibrium. The following outline summarizes types of basic data required and methods of their analysis:

Surface Inflow and Outflow; Imported and Exported Water. These quantities are measurable by standard hydrographic and hydraulic procedures. Where complete data on surface flows to and from the basin are not available, supplemental stream gauging stations should be installed.

Precipitation. Records of precipitation in the area should be assembled. Gauges should be well distributed over the basin to provide a good estimate of the annual precipitation from the isohyetal or Thiessen methods. If gauges are not so located, supplemental stations should be established.

Consumptive Use. All water, surface and subsurface, released into the atmosphere by processes of evaporation and transpiration is consumptive use, or evapotranspiration. To compute this discharge from a given basin, it is first necessary to make a land use, or cultural, survey to yield the amount of each type of water-consuming area. Aerial photographs are helpful for this task. Unit values of consumptive use must then be determined. For crops and native vegetation, methods based on available heat (such as the Thornthwaite or Blaney–Criddle method) are generally satisfactory. For water surfaces, local evaporation records should be employed. Urban and industrial areas require careful estimates from samples of representative areas using metered deliveries and sewage outflows. Multiplying the unit value of consumptive use by the corresponding acreage gives the water consumption for each area; the sum of these products yields the total consumptive use over the basin.

Changes in Surface Storage. These can be computed directly from changes in water levels of surface reservoirs and lakes.

Changes in Soil Moisture. The moisture content of the soil can be measured by devices embedded in the soil or by a neutron probe (see Chapter 2). In practice, however, the variability of soil moisture both in time and place makes it difficult to obtain an accurate basin-wide measurement. The problem can be minimized by selecting periods of storage change in which the amount of water in unsaturated storage at the beginning and end of the period is nearly equal. In irrigated areas period limits should correspond to the beginning or end of the irrigation season.

Changes in Groundwater Storage. Changes in groundwater storage can be determined from geologic data on aquifers and measurement of groundwater levels. Antecedent information on groundwater levels, pumping records, pumping tests, and artificial recharge should be collected. Specific yields of unconfined aquifers are determined by laboratory tests of samples and/or by classifications of well logs; storage coefficients are best determined from pumping tests of wells.

Select a grid of measuring wells distributed over the basin. Supplement with test holes where required. Water levels in these wells should be measured under conditions as nearly static as possible, preferably after the season of heavy draft and again after the season of recharge. A few control wells should be equipped with automatic water-level recorders or have their water levels measured monthly to facilitate detailed study of groundwater fluctuations. A basin map showing lines of equal change in groundwater level is then prepared. The product of change in water level times storage coefficient times area gives the change of groundwater storage for each aquifer within the basin.

Subsurface Inflow and Outflow. These items of the equation are the most difficult to evaluate because they cannot be directly measured. Often one of them, or the difference, is calculated by being the only unknown in the equation. From geologic investigation it may be found that either subsurface inflow or outflow is lacking, or both. Many times after study, subsurface inflow may be estimated to equal that of subsurface outflow so that the items cancel.

Difficulties arise in the case of underground flows from one basin to another. The direction of flow can be established from water table or piezometric gradients. Knowing groundwater slopes and transmissivities, subsurface flows can be computed from Darcy's law. Where surface streams and subsurface drainage systems control groundwater levels, better estimates of subsurface flow are usually possible because more data are available.

10.3 YIELD

10.3.1 Alternative Basin Yields

The maximum quantity of water that is actually available from a groundwater basin on a perennial basis is limited by the possible deleterious side effects that can be caused by pumping and by the operation of the basin. As a result, several concepts of basin yield are generally recognized.[10] These are briefly defined in the following subsections together with comments as to their consequences.

Mining Yield

If groundwater is withdrawn at a rate exceeding the recharge, a *mining yield* exists.[29] As a consequence, this yield must be limited in time until the aquifer storage is depleted. Many groundwater basins today are being mined; if mining continues, the local economy served by this pumping may change, evolving into other forms that use less water or involve importations of water into the basin. The Salt River Valley of southern Arizona and the High Plains of western Texas are classic examples of such situations.

Various valid arguments, economic and other, have been advanced to justify mining of groundwater. One is that water in storage is of no value unless it is used.[75] In arid areas, such as the Sahara Desert, where groundwater represents the only available water resource, almost any development of groundwater constitutes a mining yield. But the needs are there and the benefits are great, and so such exploitation will continue. With proper management plus water conservation, such groundwater resources can be made to last from several decades to a few centuries.

Perennial Yield

The *perennial yield* of a groundwater basin defines the rate at which water can be withdrawn perennially under specified operating conditions without producing an undesired result.* An

---

*In the past the term *safe yield*, implying a fixed quantity of extractable water basically limited to the average annual basin recharge, has been widely used. The term has now fallen into disfavor because a never-changing quantity of available water depending solely on natural water sources and a specified configuration of wells is essentially meaningless from a hydrologic standpoint.

undesired result is an adverse situation, such as (1) progressive reduction of the water resource, (2) development of uneconomic pumping conditions, (3) degradation of groundwater quality, (4) interference with prior water rights, or (5) land subsidence caused by lowered groundwater levels.[56, 105] (Evaluation of perennial yield is discussed in a subsequent section.) Any draft in excess of perennial yield is referred to as *overdraft*. Existence of overdraft implies that continuation of present water management practices will result in significant negative impacts on environmental, social, and/or economic conditions.

A schematic diagram of a groundwater basin developed to less than perennial yield is shown in Figure 10.3.1a. Here a portion of the natural recharge is lost by subsurface outflow from the basin. But Figure 10.3.1b suggests a minimum perennial yield situation in which extractions balance recharge so that no groundwater is lost.

Deferred Perennial Yield

The concept of a deferred perennial yield consists of two different pumping rates. The initial rate is larger and exceeds the perennial yield, thereby reducing the groundwater level. This planned overdraft furnishes water from storage at low cost and without creating any undesirable effects. In fact, reducing storage eliminates wasteful subsurface outflow of groundwater and losses to the atmosphere by evapotranspiration from high water table areas. After the groundwater level has been lowered to a predetermined depth, a second rate, comparable to that of

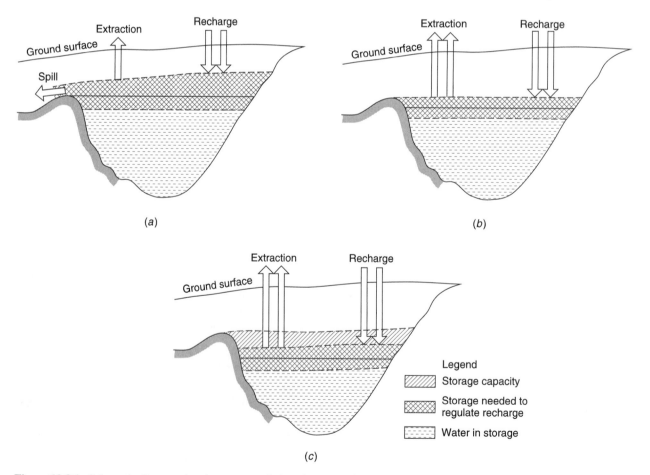

Figure 10.3.1. Schematic diagram showing storage relations in a groundwater basin for three stages of development. (*a*) Less than perennial yield (*b*) Minimum perennial yield (*c*) Increased perennial yield (after Peters[66])

perennial yield, is established so that a balance of water entering and leaving the basin is maintained thereafter. With a larger available storage volume, more water can be recharged and a larger perennial yield can be obtained. Figure 10.3.1c indicates this situation schematically.

Maximum Perennial Yield

The *maximum perennial yield*, as the name suggests, means the maximum quantity of groundwater perennially available if all possible methods and sources are developed for recharging the basin. In effect, this quantity depends on the amount of water economically, legally, and politically available to the organization or agency managing the basin. Clearly, the more water that can be recharged both naturally and artificially to a basin, the greater the yield.

To achieve the maximum perennial yield, the aquifer should be managed as a unit. Thus, efficient and economic production of water requires that all pumping, importations, and distributions of water be done for the benefit of the largest manageable system. Where surface water is available in addition to groundwater, these two sources are operated conjunctively. Such a conjunctive use scheme provides a larger and more economic yield of water than can be obtained from the two sources operated independently. The limit to such an operation is governed by the ability to import and distribute water and also by the storage available for surface water and groundwater.

10.3.2 Evaluation of Perennial Yield

Consideration of the above definitions of perennial yield reveals that there can be more than one "undesired result" from pumping a groundwater basin, that perennial yield may be limited to an amount less than the net amount of water supplied to the basin, and that perennial yield can vary with different patterns of recharge, development, and use of water in a basin.

If groundwater is regarded as a renewable natural resource, then only a certain quantity of water may be withdrawn annually from a groundwater basin. The maximum quantity of water that can be extracted from an underground reservoir, and still maintain that supply unimpaired, depends on the perennial yield. Overdraft areas constitute the largest potential groundwater problem in the United States.[88] Until overdrafts are reduced to perennial yields in these basins, permanent damage or depletion of groundwater supplies must be anticipated.

Factors Governing Perennial Yield

Determination of the perennial yield of a groundwater basin requires analysis of the undesired results that may accrue if the extraction rate is exceeded. The recharge[*] criterion (progressive reduction of the water resource) is the most important because exceeding this factor is normally responsible for introducing other undesired results. Water supplied to a basin may be limited either by the storage volume of the underground basin or by the rate of water movement through the basin from the recharge area to the withdrawal area. The quantity concept is usually applicable to unconfined aquifers where supply and disposal areas are close together, whereas the rate concept applies more to confined aquifers where supply and disposal areas are widely separated.

Economic considerations can govern perennial yield in basins where the cost of pumping groundwater becomes excessive. Excessive costs may be associated with lowered groundwater levels, necessitating deepening wells, lowering pump bowls, and installing larger pumps. Where pumpage is largely for irrigation, power costs, crop prices, or government farm subsidies may establish an economic limit for pumping groundwater; alternatively, other uses that can support higher pumping costs may evolve.

---

[*]Here, *recharge* refers to water reaching the saturated zone of an aquifer, where it is available for extraction.

Water quality can govern perennial yield if draft on a basin produces groundwater of inferior quality. Possibilities include: (1) pumping in a coastal aquifer could induce seawater intrusion into the basin (see Chapter 14); (2) lowered groundwater levels could lead to pumping of underlying connate brines; (3) polluted water from nearby areas might be drawn into a pumped aquifer. A quality limitation on perennial yield depends on the minimum acceptable standard of water quality, which in turn depends on the use made of the pumped water. Therefore, by lowering the quality requirement, the perennial yield can be increased.

Legal considerations affect perennial yield if pumpage interferes with prior water rights.[61] Finally, if pumpage is responsible for land subsidence, a limitation on perennial yield can result.

Calculation of Perennial Yield

In general, the basin recharge criterion will govern perennial yield because, as mentioned earlier, one or more of the other undesired results will often be induced by pumpage exceeding this rate. Quantitative determination of perennial yield where recharge is the limiting factor can be made under specified conditions if adequate knowledge of the hydrology of the basin is available. Methods are based on the equation of hydrologic equilibrium or approximations thereto.[57] Basically, this implies that perennial yield is defined in terms of a rate at which groundwater can be withdrawn from a basin over a representative time period without producing a significant change in groundwater storage.

Variability of Perennial Yield

It is important to recognize that perennial yield of a groundwater basin tends to vary with time. Any quantitative determination is based on specified conditions, either existing or assumed, and any changes in these conditions will modify the perennial yield. This fact applies to the degree and pattern of groundwater development within a basin as well as to the other factors that govern safe yield.

Investigations of the availability of groundwater within a basin are typically not initiated until basin development has produced an overdraft. Yet this is almost necessary in order to obtain a reasonable estimate for perennial yield. In a virgin basin, where a balance exists between natural inflow and outflow and there is no pumping, the absence of hydrogeologic data may not justify the cost of a management investigation. Similarly, estimating future perennial yield of a basin under greater development than at present requires careful evaluation of all items in the equation of hydrologic equilibrium.

Perennial yield may vary with the level of groundwater within a basin. Thus, if levels are lowered, subsurface inflow will be increased and subsurface outflow will be decreased, recharge from losing streams will be increased and discharge from gaining streams will be decreased, and uneconomic evapotranspiration losses will be reduced. Conversely, a rise in water levels will have the opposite effects. Therefore, where recharge is sufficient, the greater the utilization of underground water, the larger the perennial yield. The maximum perennial yield will be controlled by economic or legal constraints.

An unconfined basin fed by an adequate recharge source can increase its perennial yield, not only by increasing pumpage but also by rearrangement of the pumping pattern. If the concentration of wells is shifted to near the recharge source, greater inflow can be induced. The rearrangement has the additional advantage that a greater supply may be obtained without necessarily increasing pumping lifts. For example, in the cross section shown in Figure 10.3.2a, it is assumed that the stream is the principal recharge source. By moving the well field nearer to the stream as in Figure 10.3.2b, the water table slope is increased and a greater yield for equal pumping depths results.

For a confined aquifer with its recharge area located some distance from the pumping area, the rate of flow through the aquifer will govern the perennial yield. In large confined

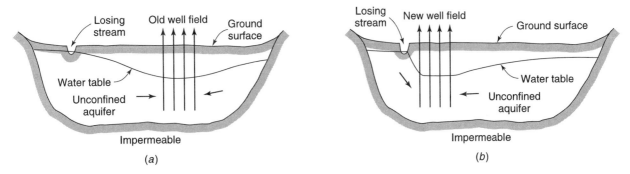

Figure 10.3.2. Example of increased groundwater yield for same pumping depths obtained by shifting wells nearer to a recharge source.

aquifers, pumpage of water from storage can be carried on for many years without establishing an equilibrium with basin recharge. Although the slope of the piezometric surface will increase, the permeability of the aquifer is seldom sufficient to maintain a compensating flow into the basin.[88]

Besides operational changes, perennial yield can also vary due to gradual and subtle modifications occurring within a basin. Changes in vegetation and even in crops, particularly where root depth is affected, may influence surface infiltration and subsequent percolation to the water table. Urbanization of an area, accompanied by greater surface runoff and installation of sewer systems, can be expected to reduce recharge. Changes in the purpose of pumping groundwater, such as from irrigation to municipal or industrial use, may—from an economic viewpoint—permit greater pumping lifts; consequently, perennial yield can be increased. Other economic factors include changes in value of irrigated crops, increased efficiency of new wells and pumps, treatment to meet revised water quality standards, and power costs.

10.4 CONJUNCTIVE USE AND WATERSHED MANAGEMENT

In basins approaching full development of water resources, optimal beneficial use can be obtained by *conjunctive use*, which involves the coordinated and planned operation of both surface water and groundwater resources to meet water requirements in a manner whereby water is conserved.[*] The basic difference between the usual surface water development with its associated groundwater development and a conjunctive operation of surface water and groundwater resources is that the separate firm yields of the former can be replaced by the larger and more economic joint yields of the latter.

The concept of conjunctive use of surface water and groundwater is predicated on surface reservoirs impounding streamflow, which is then transferred at an optimum rate to groundwater storage. Surface storage in reservoirs behind dams supplies most annual water requirements, while the groundwater storage can be retained primarily for cyclic storage to cover years of subnormal precipitation. Thus, groundwater levels would fluctuate, being lowered during a cycle of dry years and being raised during an ensuing wet period. Figure 10.4.1 depicts how groundwater levels might vary under such a system of conjunctive use.

During periods of above-normal precipitation, surface water is utilized to the maximum extent possible and also artificially recharged into the ground to augment groundwater storage and raise groundwater levels (see Chapter 13). Conversely, during drought periods, limited

---

[*]Coordinated use of surface water and groundwater does not preclude importing water, as required, to meet growing needs. In fact, to store and distribute additional water economically may require more intensive use of groundwater storage space.

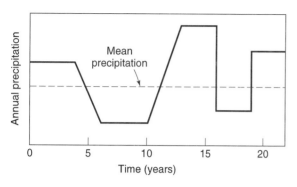 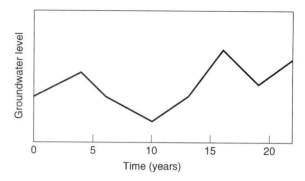

Figure 10.4.1. Illustrative example of variation in groundwater levels in relation to annual precipitation under conjunctive use management.

surface water resources are supplemented by pumping groundwater, thereby lowering groundwater levels. The feasibility of the conjunctive-use approach depends on operating a groundwater basin over a range of water levels; that is, there must be space to store recharged water, and, in addition, there must be water in storage for pumping when needed.

Management by conjunctive use requires physical facilities for water distribution, for artificial recharge, and for pumping. The procedure does require careful planning to optimize use of available surface water and groundwater resources. Such operations can be complex and highly technical; they require competent personnel, detailed knowledge of the hydrogeology of the basin, records of pumping and recharge rates, and continually updated information on groundwater levels and quality. A schematic diagram of a systematic approach for a conjunctive use analysis is illustrated in Figure 10.4.2.

A conjunctive use management study requires data on surface water resources, groundwater resources, and geologic conditions; data on water distribution systems, water use, and wastewater disposal are also necessary.[45, 67] Figure 10.4.3 shows a simplified flowchart of the various phases and steps involved for a basin management study in California. This suggests the diversity of data and effort required in order to determine an optimal basin management plan. It should be noted from Figures 10.4.2 and 10.4.3 that mathematical models are usually incorporated in such studies (see Chapter 9). A basin model simulates the responses of a basin to variations in variables such as natural and artificial recharge and pumping so that the best operating procedures for basin management can be practiced. In effect, this will optimize the water supply obtained from the basin.[28, 40, 72]

Because every water development project is unique, it is impossible to generally present economic considerations for conjunctive operations and have them apply specifically to any given situation. Nevertheless, the advantages and disadvantages, mostly economic, are summarized in Table 10.4.1. The tabulation compares a conjunctive use operation relative to development of surface-water resources only, assuming irrigation to be the principal water use in a semiarid region.

Total usable water supply can be increased by coordinated operation of surface and underground water resources. With an optimum coordinated operation, the unit cost of water supply storage and distribution can be minimized. The basic principles of groundwater basin operation that will produce an optimum water resources management scheme include the following, as reported by Fowler:[32]

1.   The surface and underground storage capacities must be integrated to obtain the most economical utilization of the local storage resources and the optimum amount of water conservation.

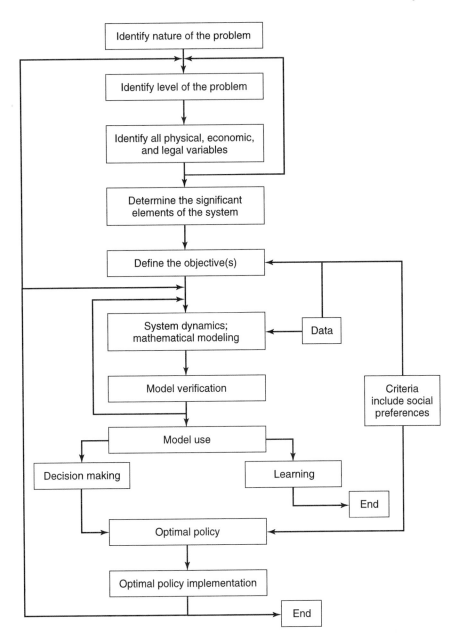

Figure 10.4.2. Schematic diagram of a systematic approach for studying conjunctive use problems (after Maknoon and Burges,[55] reprinted from *Journal American Water Works Association,* Vol. 70, by permission of the Association; copyright © 1978 by American Water Works Association, 6666 West Quincy Avenue, Denver, CO 80235).

2. The surface distribution system must be integrated with the groundwater basin transmission characteristics to provide the minimum cost distribution system.

3. An operating agency must be available with adequate power to manage surface-water resources, groundwater recharge sites, surface-water distribution facilities, and groundwater extractions.

The procedure for developing a sound conjunctive-use operation within a basin requires estimation of the various elements of water supply and distribution. The optimum use of surface-water and groundwater resources is determined for assumed conditions, usually those during the most critical drought period of record. There are many article in the literature addressing conjunctive management (particularly modeling), including: Basagaoglu and

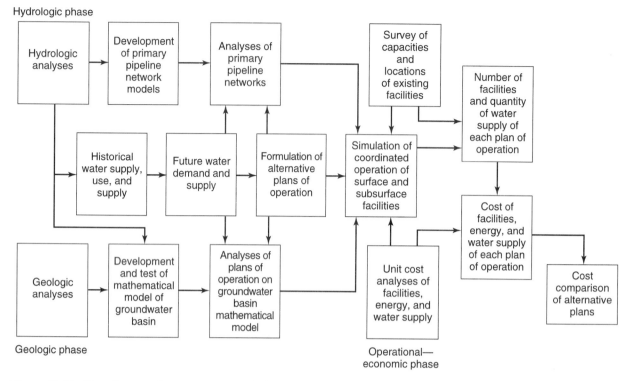

Figure 10.4.3. Flow diagram of a management study for the San Gabriel Valley, California, groundwater basin (after Amer. Soc. Civil Engrs.[10]).

Mariño;[15] Bredehoeft and Young;[19] Galloway et al.;[34] Haimes and Dreizen;[39] LaBolle, Ahmed, and Fogg;[47] Reichard;[68] Willis and Yeh;[102] Winter et al.;[103] and Young and Bredehoeft.[106]

Table 10.4.1 Conjunctive Use of Surface Water and Groundwater Resources
(after Clendenen[24])

Advantages	Disadvantages
1. Greater water conservation	1. Less hydroelectric power
2. Smaller surface storage	2. Greater power consumption
3. Smaller surface distribution system	3. Decreased pumping efficiency
4. Smaller drainage system	4. Greater water salination
5. Reduced canal lining	5. More complex project operation
6. Greater flood control	6. More difficult cost allocation
7. Ready integration with existing development	7. Artificial recharge is required
8. Stage development facilitated	8. Danger of land subsidence
9. Smaller evapotranspiration losses	
10. Greater control over outflow	
11. Improvement of power load and pumping plant use factors	
12. Less danger from dam failure	
13. Reduction in weed seed distribution	
14. Better timing of water distribution	

10.5 GROUNDWATER MANAGEMENT: WATER LAWS AND POLICIES

10.5.1 Water Law and Policy

Water-management decisions and policies should have water laws as the fundamental basis. In the United States, water law has two basic functions: (1) creation of supplemental private property rights in scarce resources and (2) imposition of public interest limitation on private use. For our purposes water law is divided into *surface water law* and *groundwater law*. Surface water law is further categorized into riparian law and appropriation law. *Riparian law* is based on the riparian doctrine, which states that the right to use water is considered real property, although the water itself is not property of the landowner (Wehmhoefer[97]). *Appropriation law* states that the allocation of water rests on the proposition that the beneficial use of water is the basis, measure and limit of the appropriative right: the first in time is prior in right. See Tarlock[83] for further details on concepts of water law.

In the western United States, surface water policy generally follows this doctrine of "first in time, first in right." In order to appropriate water, the user need only demonstrate availability of water in the source of supply, show an intent to put the water to beneficial use, and give priority to more senior permit holders during times of shortage (Schmandt et al.[76]). Beneficial use of water under the law includes domestic consumption, livestock watering, irrigation, mining, power generation, municipal use, and other circumstances. The states of Arizona and New Mexico follow the appropriation law of surface water, and in California and Texas the appropriation doctrine and the riparian doctrine coexist.

Groundwater allocation is handled quite differently and is typically divided into common law or statutory law. Common law doctrines include the overlaying rights doctrines of absolute ownership, reasonable use, and correlative rights. These doctrines give equal rights to all landowners overlaying an aquifer. Arizona, California, and Texas have adopted these principles for groundwater allocation.

The above surface and groundwater laws serve as the basis for individual state water policies. The burden of developing water policies lies upon the state. This is often achieved by the state proposing a water project and securing federal funds for the construction. It is also up to the states to agree on apportionment in interstate waters; if states cannot agree, then the courts will intervene and settle the dispute by decree. The federal government gets involved in such disputes only where interstate waters, federal lands, or Indian reservations are concerned.

10.5.2 Arizona's Groundwater Management Code

Throughout Arizona's history, water policy has been directed at supporting growth of its population and major revenue-producing activities. Starting with mining, ranching, and farming, with the gradual shift to municipal and industrial uses, the water policy of the state has been directed at obtaining imported supplies. This has been an effort to augment what has appeared to be an insufficient and indigenous resource. Waterstone[96] points out that the "state's water policies have led to the protracted exercise to capture and secure the Central Arizona Project (CAP), the ongoing infatuation with weather and watershed manipulation, the current experimentation with groundwater recharge and effluent use, and the recent spate of purchases of remote water farms."

In Arizona, the state's water policy and management focused more on surface water than groundwater prior to 1980, when the Groundwater Management Code was developed; thereafter, the emphasis has been on groundwater. In regards to surface water, Arizona law defines surface water as "the waters of all sources, flowing in streams, canyons, ravines or other natural channels, or in definite underground channels, whether perennial or intermittent, flood, waste, or surplus water, and of lakes, ponds and springs on the surface." These surface waters are subject to the *doctrine of prior appropriation* (ADWR[12]). In Arizona, surface water rights are obtained by filing an application with the Department of Water Resources for a permit to appropriate surface water. Once the permit is issued and the water is actually put to beneficial use, proof of that use is made to the Department and a certificate of water right is issued to the applicant. Once a certificate is issued, the use of the water is subject to all prior appropriations.

Since water law in the state of Arizona has changed substantially over the years, Arizona is now conducting a general adjudication of water rights in certain parts of the state. *Adjudications* are court determinations of the status of all state law rights to surface water and all claims based upon federal law within the river systems. These adjudications will provide a comprehensive way to identify and rank the rights to the use of water in some areas. The adjudications will also quantify the water rights of the federal government and the Indian reservations within Arizona.

In Arizona, groundwater problems arise from the overdrafting of water from the aquifers. Groundwater overdrafts cause many problems, such as increased well pumping costs and water quality issues. In areas of severe groundwater depletion, the earth's surface may also subside, causing cracks or fissures that can damage roads or building foundations. In order to manage groundwater pumping in Arizona, the Arizona Groundwater Management Code was developed in 1980 as state legislation. The Arizona Groundwater Management Code was named one of the nation's ten most innovative programs in state and local government by the Ford Foundation in 1986. This achievement came from the cooperation of Arizonans working together and compromising when necessary in order to protect the future of the state's water supply.

The Groundwater Management Code has three primary goals (ADWR[12]):

1. control the severe overdraft currently occurring in many parts of the state;

2. provide a means to allocate the state's limited groundwater resources to most effectively meet the changing needs of the state; and

3. augment Arizona's groundwater through water supply development.

In order to achieve these goals, the code set up a comprehensive management environment and established the Arizona Department of Water Resources.

There are three levels of water management outlined by the code. Each level is based on different groundwater conditions. The lowest level applies statewide, and includes general groundwater provisions. The next level applies to Irrigation Non-Expansion Areas (INAs), and the highest level applies to Active Management Areas (AMAs), where groundwater depletion is the highest. The boundaries that divide the INAs and AMAs are determined by groundwater basins and not by political jurisdiction. The main purpose of groundwater management is to determine who may pump groundwater and how much may be pumped. This includes identifying existing water rights and providing new ways for nonirrigation water users to initiate new withdrawals. In an AMA or INA, new irrigation users are not allowed. Even with the original publicity and enthusiasm, many people now feel that the efforts under the Groundwater Management Code have been very costly with very little savings in water, making the success questionable. For more information, refer to the Arizona Department of Water Resources Web site at http://www.water.az.gov/adwr.

10.5.3 Texas Groundwater Law

As a comparison to Arizona's water law, it is worthwhile to look at Texas. A large amount of Texas' water supply is from groundwater resources, which has resulted in a severe depletion of some of the major aquifers in the state. It is surprising that a state such as Texas that depends on groundwater so much has such an ill-defined groundwater plan. The Texas groundwater law is based on the "absolute ownership rule," which states, "percolating waters are the private property of the landowner" (Schmandt et al.[76]). This means that a landowner has the right to pump water from a groundwater deposit beneath his land at any rate, as long as the withdrawal does not maliciously harm his neighbor. The groundwater law excludes underground streams, where the classification of an underground stream is still a matter of debate. This lack of regulation has caused many problems in the state of Texas, such as overdrafting, subsidence, and water quality issues. The Texas Water Development Board has conducted efforts to develop the best management techniques for regional groundwater planning.

10.6 CASE STUDY   *Groundwater Management: Examples*

10.6.1 *Edwards Aquifer Management*

The Edwards Aquifer of Texas extends in a narrow band from Hays to Kinney Counties in the south–central part of Texas, as illustrated in Figures 2.6.2 and 9.4.2a. The aquifer is approximately 175 miles in length and ranges in width from 5 to 30 miles. The aquifer has a shallow outcrop area where recharge occurs, and a deeper (800–1,200 feet) confined or artesian area through which most of the flow moves in an east–northeasterly direction (TWBD[87]). Figure 9.4.2a illustrates the outcrop area (recharge area) in the aquifer.

This aquifer is highly transmissive and accepts an average of about 642,000 acre-feet of recharge annually, and discharges an equivalent amount of water through various springs and wells. Large fissures and cracks decrease the amount of filtration through the aquifer, which provides a continuous threat of contamination of the aquifer through the recharge process. The water quality in the aquifer generally contains less than 500 mg/L of total dissolved solids, with salinity increasing with depth. This aquifer is the source of 246,000 acre-feet for municipal supplies annually, 79,000 acre-feet of irrigation water, 29,000 acre-feet of industrial water, and 13,000 acre-feet of water for other purposes.

The Edwards Aquifer is managed by three different entities: the Edwards Aquifer Authority, the Barton Springs/Edwards Aquifer Conservation District, and the Texas Natural Resource Conservation Commission. The principal uses of water from the Edwards Aquifer are for municipal and irrigation purposes. San Antonio, Texas, with a population of over 1.5 million people, obtains its entire municipal water supply from the aquifer and is one of the largest cities in the world that relies solely on groundwater for its supply. The aquifer also supplies industrial users with water in both the San Antonio and Austin, Texas regions. The Barton Springs/Edwards Aquifer Conservation District is the aquifer management entity for the Austin region. A comprehensive plan, developed by the district, outlines programs and activities covering areas of concern, such as water level and water quality monitoring, well construction and spacing standards, production regulations, and public education in conservation and protection of groundwater.

The management of the Edwards Aquifer in the San Antonio region has been under controversy for many years. After a severe drought from 1950 to 1957, the water levels lowered enough that Comal Springs located in Comal County went dry. This event caused the Texas legislature to create the Edwards Underground Water District. The district, including five counties that are affected by the aquifer, was to conserve, protect, and monitor recharging, as well as prevent waste and pollution of the aquifer. However, management efforts failed as bitter conflicts sparked between the various users of the aquifer. In 1993, for fear of federal intervention, the Texas legislature enacted a management plan for the aquifer. The plan abolished the district and created the Edwards Aquifer Authority. It called for the Authority to implement a comprehensive plan for the aquifer that regulated pumpage while taking into consideration the needs of all entities involved that rely on the aquifer. It was also important for the plan to ensure a successful balance to maintain the delicate relationship between spring flows and the environment. In essence, the Authority was created to regulate the withdrawal of groundwater from the aquifer and ensure that the spring flows do not drop below the jeopardy level. In order to achieve this, the Edwards Aquifer Authority changed the rules from "rule of capture" to a system of permits to be awarded and managed by the Authority. The objective was to maintain a sufficient level of spring flows at Comal and San Marcos Springs for the purpose of protecting endangered species.

Some important management issues for the Edwards Aquifer include the following:

1. Establishing a level of groundwater withdrawals to ensure adequate water levels and at least minimum spring flows;

2. Developing a plan for the acceptable and equitable distribution of pumping among the numerous users involved;

3. Securing additional supplies to meet projected demands in excess of the allowable pumping limits.

10.6.2 *High Plains Aquifer: Conjunctive Water Use on the High Plains*

The High Plains Aquifer (see Section 2.9) is a 174,000-square-mile-area of flat to gently rolling terrain including parts of Colorado, Kansas, Nebraska, New Mexico, Oklahoma, South Dakota, Texas, and Wyoming (see Figures 2.9.2, 2.9.3, and 2.9.5). The High Plains have moderate precipitation with a low natural recharge rate to the groundwater system. The region is underlain by the High Plains Aquifer, a water table aquifer of unconsolidated alluvial deposits that form the water table aquifer. The High Plains Aquifer consists largely of the Ogallala Aquifer, which covers more than 35,000 square miles of the High Plains.

Irrigation water pumped from the aquifer has made the High Plains one of the United States' most important agricultural areas.

(continues)

10.6 CASE STUDY   *Groundwater Management: Examples(continued)*

In 1949, the groundwater pumpage for irrigation was approximately 480 million cubic feet per day. By 1980 the pumpage had reached approximately 2,150 million cubic feet per day (U.S. Geological Survey[91]). Because of major water level declines, by 1990 the pumpage declined to about 1,870 million cubic feet per day (McGuire and Sharpe[62]). According to Alley et al.,[9] pumping from this aquifer has resulted in the largest decrease in storage of any major aquifer in the United States. Much of the predevelopment saturated thickness has been dewatered (refer to Figure 2.9.5). The water table continues to decline in much of the High Plains Aquifer; however, monitoring of water levels in wells indicates an overall reduced rate of decline (McGuire and Sharpe[62]). The decline is the result of improved irrigation and cultivation practices, decreases in irrigated acreage, and above-normal precipitation.

The Ogallala Aquifer is primarily composed of sand, gravel, clay, and silt deposits. The groundwater moves slowly through the aquifer in a southeastward direction. The water quality is generally fresh; however, both dissolved solids and chloride concentrations in the aquifer increase from north to south. These dissolved solids' concentrations range from about 300 to more than 3,000 mg/L. In some areas the aquifer contains fluoride and nitrate concentrations which exceed the U.S. Environmental Protection Agency and the Texas Department of Health primary standards for fluoride and nitrate.

More water is pumped from the Ogallala Aquifer in Texas than from any other aquifer in the region. The total pumpage from the aquifer in 1994 was approximately 5.9 million acre feet, and 96% of the water pumped was used for irrigation. In addition to irrigation, many communities use the Ogallala Aquifer as their only source of drinking water. Recharge rates cannot keep up with the large amounts of withdrawal. The combination of these activities has significantly reduced the water level in the aquifer. The water level is so low that the cost of pumping has become increasingly high. The previous lack of water management in the region has added to this problem of groundwater depletion.

In order to better manage the aquifer, the High Plains Ogallala Area Water Management Plan (HPOAWMP) was developed. From 1990 to 2005, the HPOAWMP will conduct a study of the aquifer and will reflect the analyses and provide management recommendations for the 47-county area in multiple states. The study consists of individual water assessments by each municipality that examine the present and potential water needs. It will also include water conservation measures, reuse, new project development, and improved system operations.

(continues)

Figure 10.6.1. Playa Lakes on the Texas High Plains

Playa lakes like these can be found throughout the High Plains between Lubbock and Amarillo. Studies suggest that playa lakebeds can be modified to increase recharge to the Ogallala Aquifer. The Texas High Plains are covered with more than 17,000 shallow, naturally occurring depressions called *playa lakes*. The lakes capture an average of 1.8 to 5.7 million AF of water annually. The characteristics of individual playas vary tremendously. Some cover only a small area while others can be as large as 200 acres. Most of the playas are 2 to 10 feet deep when full of water. There are an average of one to two playas per square mile in the Southern High Plains.

In the spring and fall, intense thunderstorms frequently fill the playas. However, little infiltration occurs, 50% to 80% of the water is lost to evaporation and winds, and minimal groundwater recharge occurs. Almost all the lakes have clay bottoms (photo courtesy of Lloyd Urban, Texas Tech University).

According to the Ogallala plan, for the sake of consistency, examination of the full range of alternatives, and taking account of individual area differences, it is a joint goal of the regional management plan to:

1. have some commonality in the range of various management tools examined for each user class and geographical area;

2. identify some common regional management proposals (such as conservation plans) and best-use techniques that would be appropriate for all user classes and areas to implement; and

3. consider each area's uniqueness in selecting an appropriate array of these tools for developing a customized plan for each management area or municipality that is consistent with the planning methodologies and goals used for the overall region.

In order to execute the plan, a management team is in charge of the above tasks and identifies any special studies that are necessary. These studies may include identifying and evaluating drought management aspects, developing appropriate groundwater models, addressing water quality issues, performing an economic analysis of water management alternatives, and developing and evaluating new management strategies. The interconnections between surface water and groundwater and the relationship to watersheds are much less direct and obvious in the High Plains Aquifer as compared to the Edwards Aquifer in the San Antonio region. Texas law is clear concerning the use and ownership of percolating groundwater under the absolute ownership doctrine as established in a 1904 lawsuit (*Houston and T.C. Ry. Co.* v. *East*). Landowners can pump and use groundwater beneath their property with little regulation. According to Templer and Urban,[84] conjunctive water use is widely practiced concerning water from (1) groundwater in the dwindling Ogallala Aquifer, (2) runoff water accumulating in the thousands of small playa lakes that dot the High Plains (see Figure 10.6.1), and (3) the meager flow of streams in the area. Water from the playa lakes is used mainly to supplement irrigation water pumped from the Ogallala.

Further information on the High Plains aquifer is found in Alley and Schefter,[8] Grubb,[38] Longley and Jordan,[51] and Smerdon and Jordan.[80] For more general discussions of water development in the southwestern United States, refer to Baker et al.,[13] Bowden,[18] Espeland,[31] Fradkin,[33] Ingram,[44] Killgore and Eaton,[46] Reisner,[69] Shupe and Folk-Williams,[77] and Smith.[81]

10.7 GROUNDWATER MANAGEMENT USING MODELS

10.7.1 What Are Groundwater Management Models?

Aquifer simulation models (discussed in Chapter 9) have been used to examine the effects of various groundwater management strategies. Use has been primarily of the "case study" or "what if" type. The analyst specifies certain quantities and the model predicts the technical and perhaps economic consequences of this choice. The analyst evaluates these consequences and uses judgment and intuition to specify the next case. *Groundwater management models* may be used for optimization or as combined simulation–optimization. Reviews of groundwater management models and applications can be found in Ahfeld and Heidari,[4] Ahfeld and Mulligan,[5] Gorelick,[37] Wagner,[92] Willis and Yeh,[102] and Yeh.[104]

Yeh[104] reviewed *inverse solution procedures* for parameter identification, groundwater-management models, and optimal experimental design methodologies. Optimization methods have been developed extensively over the past three decades for inverse solution procedures, but used only sparingly in applications. Groundwater management applications have focused on explicitly combining simulation and optimization, resulting in so-called *simulation management models*. Groundwater management models can be developed and used very effectively for groundwater management purposes.

An example of a groundwater management model is one that is used to identify the least-cost management strategy to meet specified hydraulic and water quality restrictions in an aquifer. Typically, total pumping is used as a surrogate for total cost. The restrictions dictate the allowable heads, gradients, velocities, and concentrations through space and time. The management strategy defines the well locations and rates to minimize pumping.

10.7.2 Optimization Methods

Optimization procedures use mathematical expressions to describe a system and its response to the system inputs for various parameters. *Constraints* are used to define the limits of the design variables, and performance is evaluated through the use of *objective functions.* An *optimization problem* may be formulated in a general framework as

$$\text{Optimize } f(\mathbf{x}) \qquad (10.7.1)$$

subject to the constraints

$$\mathbf{G(x)} = \mathbf{0} \qquad (10.7.2)$$

and bound constraints on the decision variables

$$\mathbf{x}_l < \mathbf{x} < \mathbf{x}_u \qquad (10.7.3)$$

where $\mathbf{x}_l$ and $\mathbf{x}_u$ are the vectors of lower and upper bounds, respectively, on the decision variables $\mathbf{x}$. A set of values of the decision variables that simultaneously satisfies the constraints is a *feasible solution.* A feasible region is the region of feasible solutions defined by the constraints. An optimal solution is a set of values of the decision variables that satisfies the constraints and provides an optimal value of the objective function.

Depending upon the nature of the objective function and the constraints, an optimization problem can be classified as linear or nonlinear. *Linear programming (LP)* problems consist of a linear objective function and all the constraints are linear. *Nonlinear programming (NLP)* problems have a nonlinear objective function and/or nonlinear constraints. Linear programming (LP) models have been extensively applied to optimal resource allocation problems.

The general form of an LP model can be expressed as

$$\text{Max (or Min) } X_0 = \Sigma c_j X_j \qquad (10.7.4)$$

which is the linear objective function subject to the linear constraints

$$\Sigma a_{ij} X_j < b_i \qquad (10.7.5)$$

$$X_j < 0 \qquad (10.7.6)$$

The c's are the objective function coefficients, the a's are the technological coefficients, and the b's are the right-hand side (RHS) coefficients. The well-known algorithm for solving LP problems is the *simplex method.* References for the simplex method include Hillier and Lieberman,[42] Mays and Tung,[60] and Taha.[82]

10.7.3 Types of Groundwater Management Models

A classification of groundwater management models is presented in Figure 10.7.1. Two basic categories are (a) *hydraulic management models* that are aimed at managing pumping and recharge and (b) *policy evaluation models* that also can consider the economics of water allocations. Both categories of models have been developed based upon four major approaches: (1) embedding approach, (2) optimal control approach, (3) response matrix approach, and (4) heuristic optimization-simulation approach. In addition, the policy evaluation models have been developed using hierarchical approaches.

The *embedding approach* incorporates the equations of the simulation model (represented as a set of difference equations) directly into the optimization problem to be solved. This method has limited applications and is mostly used in groundwater hydraulic management. The optimization problem quickly becomes too large to solve by available algorithms when a large-scale aquifer (especially an unconfined aquifer) is considered. Unconfined aquifers result in nonlinear programming problems. Previous work based on this approach includes Aguado

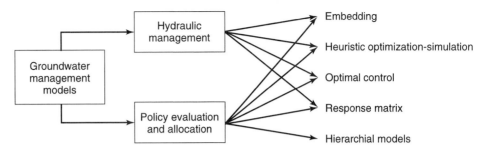

Figure 10.7.1. Classification of optimization models for groundwater management.

et al.,[2] Aguado et al.,[3] Willis and Newman,[101] Aguado and Remson,[1] Remson and Gorelick,[71] and Willis and Liu.[100]

The *optimal control approach*, based upon concepts from optimal control (Mays[59]), couples optimization techniques with a groundwater simulator to implicitly, repeatedly solve the governing equations of groundwater flow. This methodology could be thought of as a variation of the embedding approach. The state variables that represent heads and the control variables that represent pumpages are implicitly related through the simulator. The simulator equations are used to express the state variables in terms of the control variables, yielding a much smaller reduced optimization problem that must be solved many times. Wanakule, Mays, and Lasdon[95] presented a general groundwater management model, based upon a nonlinear programming code GRG2 (Lasdon and Warren[48]) interfaced with a groundwater simulation model GWSIM (Texas Water Development Board[85]), that can be used to solve both hydraulic management problems and groundwater policy evaluation (allocation) problems. Other applications of optimal control include Andricevic;[11] Chang, Shoemaker, and Liu;[23] Culver and Shoemaker;[25, 26] Georgakakos and Vlatsa;[35] Lee and Kitanidis;[50] and Whiffen and Shoemaker.[98]

The *response matrix approach* generates a unit response matrix by solving the simulation model several times, each with unit pumpage at a single pumping node. Superposition is used to determine total drawdowns. This yields a smaller optimization problem, but the method has two major limitations. It is exact only for a confined aquifer but has good accuracy for an unconfined aquifer with relatively small drawdowns compared to the aquifer thickness. A drawdown correction method may be used to improve accuracy for an unconfined aquifer with larger drawdowns, but acceptable accuracy cannot be guaranteed (Heidari[41]). In addition, the response matrix must be recomputed when exogenous factors such as aquifer boundary conditions or potential well locations change. An alternative is to treat these factors as decision variables, but then more variables and constraints are included in the optimization problem. Work stemming from this approach includes that by Ahfeld and Sawyer[6], Ahfeld et al.[7], Barlow et al.[14], Maddock,[52, 53] Maddock and Haimes,[54] Morel-Seytoux and Daly,[64] Morel-Seytoux et al.,[65] Heidari,[41] Illangasekare and Morel-Seytoux,[43] Reichard,[68] and Willis.[99]

More recently, less conventional *heuristic optimization methods* have been developed, including neural networks (Rogers and Dowla[74]), genetic algorithms (McKinney and Lin[63], Ritzel et al.[73]), and simulated annealing (Dougherty and Marryott,[30] Marryott, Dougherty, and Stollar,[58] and Skaggs et al.[78,79]). These methods use a simulation model interfaced with a heuristic optimization procedure such as genetic algorithms or simulated annealing approaches. Figure 10.7.2 illustrates the interfacing of such a procedure. These procedures have the advantage of being able to handle nonlinearities, but can be computationally time consuming because of the many times that the simulation model must be solved during the heuristic search procedure.

Groundwater policy evaluation and allocation models are used for water allocation plans involving economic management objectives subject to institutional policies as constraints in

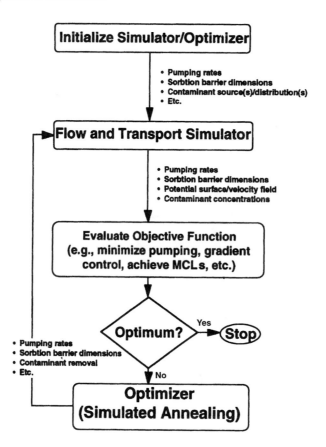

Figure 10.7.2.
Flowchart of ground-
water management
methodology.

addition to the hydraulic management constraints. These types of models have been applied to transient aquifer problems that consider the agricultural economy in response to institutional policies and to conjunctive use of surface water–groundwater problems. Figure 10.7.1 illustrates five types of approaches for policy evaluation and allocation problems. The response matrix approach is used for problems considering hydraulic-economic response (Gorelick[37]). Linked simulation–optimization models use the results of an external groundwater simulation model as input to a series of subarea economic optimization models (Gorelick[37]). Examples of the linked simulation–optimization models include Young and Bredehoeft,[106] Daubert and Young,[27] and Bredehoeft and Young.[19] Hierarchical models use subarea decomposition and a response matrix approach (Haimes and Dreizen,[39] Bisschop et al.[17]).

10.8 GROUNDWATER MANAGEMENT MODELING: HYDRAULIC MANAGEMENT MODELS

10.8.1 Steady-State One-Dimensional Problems for Confined Aquifers

Consider a confined aquifer with flow in one dimension and fixed-head boundaries as shown in Figure 10.8.1 with pumping wells that are fully penetrating. The governing equation for steady-state flow is

$$\frac{\partial^2 h}{\partial x^2} = \frac{W}{T_x}$$

(10.8.1)

Using a central differencing scheme, we can write Equation 10.8.1 in finite difference form as

$$\frac{h_{i+1} - 2h_i + h_{i-1}}{(\Delta x)^2} = \frac{W_i}{T_x} \tag{10.8.2}$$

Aguado et al.[2] formulated the following type of linear programming model for determining the optimal steady-state pumpage from a one-dimensional confined aquifer with fixed head boundaries. The objective function for optimization problem can be stated as

$$\text{Maximize } Z = \sum_{i \in I} h_i \tag{10.8.3}$$

subject to Equation 10.8.2 for each well and the following additional constraints

$$\sum_{i \in I} W_i \geq W_{\min} \tag{10.8.4}$$

$$h_i \geq 0 \quad i \in I \tag{10.8.5a}$$

$$W_i \geq 0 \quad i \in I \tag{10.8.5b}$$

where I is the set of wells and $W_{\min}$ is the minimum total production rate for the wells. The unknowns in this problem are h and W. Once the model is solved the pumpage can be determined from $W = q_i/\Delta x_i^2$. The head maintenance objective of Equation 10.8.3 is practical for managing some aquifers; however, other types of objective functions could be used, such as minimizing pumping costs. The above model formulation considers negligible well diameters and negligible well losses.

EXAMPLE 10.8.1

Develop an LP model for determining the optimal (maximize heads) steady-state pumpage of the one-dimensional confined aquifer shown in Figure 10.8.1. The wells are equally spaced at a distance of Δx apart with constant head boundaries h_0 and h_5.

SOLUTION

The objective function is simply

$$\text{Maximize } Z = h_1 + h_2 + h_3 + h_4$$

subject to the following finite difference equations

$$-2h_1 + h_2 - \frac{(\Delta x)^2}{T} W_1 = -h_0$$

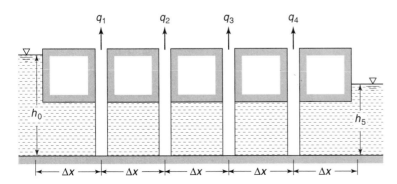

Figure 10.8.1. Confined one-dimensional aquifer.

$$h_1 - 2h_2 + h_3 - \frac{(\Delta x)^2}{T} W_2 = 0$$

$$h_2 - 2h_3 + h_4 - \frac{(\Delta x)^2}{T} W_3 = 0$$

$$h_3 - 2h_4 - \frac{(\Delta x)^2}{T} W_4 = -h_5$$

and the constraint on the production rate:

$$W_1 + W_2 + W_3 + W_4 \geq W_{min}$$

$$h_i \geq 0 \quad i = 1, \ldots, 4$$

$$W_i \geq 0 \quad i = 1, \ldots, 4$$

The unknowns in this LP model are $h_1, \ldots, h_4$ and $W_1, \ldots, W_4$. Additional constraints can be used to force the heads to be decreasing in the direction of flow, which are

$$h_4 \geq h_5; \qquad h_3 - h_4 \geq 0 \qquad h_2 - h_3 \geq 0 \qquad h_1 - h_2 \geq 0 \qquad \text{and } h_1 \leq h_0 \qquad \blacksquare$$

10.8.2 Steady-State One-Dimensional Problems for Unconfined Aquifers

A one-dimensional unconfined aquifer is shown in Figure 10.8.2 with constant head boundaries and fully penetrating wells that are equally spaced. The governing equation for steady-state flow is

$$\frac{\partial}{\partial x}\left(T_x \frac{\partial h}{\partial x} \right) = W \tag{10.8.6}$$

where $T_x = Kh$ so that

$$\frac{d^2 h^2}{dx^2} = \frac{2W}{K} \tag{10.8.7}$$

In order to simplify the notation, the substitution $w = h^2$ can be made to linearize the problem so that the finite difference expression can be written as

$$\frac{d^2 w}{dx^2} = \frac{w_{i+1} - 2w_i + w_{i-1}}{(\Delta x)^2} = \frac{2W_i}{K} \tag{10.8.8}$$

assuming that the hydraulic conductivity K is constant throughout the aquifer. The governing Equation 10.8.8 for each well is now linear. Aguado et al.[2] formulated the following linear programming model for determining the optimal steady-state pumpage from a one-dimensional unconfined aquifer:

$$\text{Maximize } Z = \sum_{i \in I} w_i \tag{10.8.9}$$

subject to Equation 10.8.8 for each well and

$$\sum_{i \in I} W_i \geq W_{min} \tag{10.8.10}$$

$$w_i \geq 0 \quad i \in I \tag{10.8.11a}$$

$$W_i \geq 0 \quad i \in I \tag{10.8.11b}$$

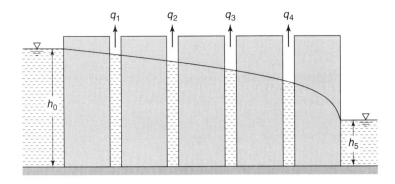

Figure 10.8.2. Unconfined one-dimensional aquifer.

The unknowns in the LP model are w_i and W_i. The heads h_i can be determined from $h_i = \sqrt{w_i}$ once the LP model has been solved for the unknowns.

EXAMPLE 10.8.2

Develop an LP model for determining the optimal steady-state pumpage of the one-dimensional unconfined aquifer in Figure 10.8.2 to maximize heads. The wells are equally spaced at a distance of Δx apart with constant head boundaries h_0 and h_5.

SOLUTION

The objective function is simply

$$\text{Maximize } Z = w_1 + w_2 + w_3 + w_4$$

subject to the following finite difference equations:

$$-2w_1 + w_2 - \frac{2(\Delta x)^2}{K} W_1 = -w_0$$

$$w_1 - 2w_2 + w_3 - \frac{2(\Delta x)^2}{K} W_2 = 0$$

$$w_2 - 2w_3 + w_4 - \frac{2(\Delta x)^2}{K} W_3 = 0$$

$$w_3 - 2w_4 - \frac{2(\Delta x)^2}{K} W_4 = -w_5$$

and the constraint on the production rate:

$$W_1 + W_2 + W_3 + W_4 \geq W_{\min}$$

$$w_i \geq 0 \quad i = 1, \ldots, 4$$

$$W_i \geq 0 \quad i = 1, \ldots, 4$$

The unknowns in this LP model are $w_1, \ldots, w_4$ and $W_1, \ldots, W_4$. ∎

10.8.3 Steady-State Two-Dimensional Model for Confined Aquifers

The governing steady-state two-dimensional equation for a homogeneous confined aquifer is

$$\frac{\partial^2 h}{\partial x^2} + \frac{\partial^2 h}{\partial y^2} = \frac{W}{T} \tag{10.8.12}$$

for which $T_x = T_y = T$. Using central differences, we can express Equation 10.8.12 in finite difference form as

$$\frac{h_{i+1,j} - 2h_{i,j} + h_{i-1,j}}{(\Delta x)^2} + \frac{h_{i,j+1} - 2h_{i,j} + h_{i,j-1}}{(\Delta y)^2} = \frac{W_{i,j}}{T} \qquad (10.8.13)$$

which can be reduced for $\Delta x = \Delta y$ to

$$h_{i+1,j} - 4h_{i,j} + h_{i-1,j} + h_{i,j+1} + h_{i,j-1} = \frac{(\Delta x)^2 W_{i,j}}{T} \qquad (10.8.14)$$

An LP model for the optimal steady state pumpage from a two-dimensional confined aquifer can be formulated as

$$\text{Maximize } Z = \sum_{i,j \in I} h_{i,j} \qquad (10.8.15)$$

subject to Equation 10.8.14 for each cell and

$$\sum_{i,j \in I} W_{i,j} \geq W_{\min} \qquad (10.8.16)$$

$$h_{i,j} \geq 0 \qquad (10.8.17a)$$

$$W_{i,j} \geq 0 \qquad (10.8.17b)$$

where I represents the set of pumping wells. The unknowns in the LP model are $h_{i,j}$ for all the cells and $W_{i,j}$ for the pumping cells.

EXAMPLE 10.8.3

Develop an LP model for determining the optimal steady-state pumping from the two-dimensional confined aquifer shown in Figure 10.8.3. This aquifer has constant (fixed) heads along the aquifer boundaries. It has three pumping cells, (2, 2), (3, 2) and (3, 3), as shown in Figure 10.8.3.

SOLUTION

The objective function is simply

$$\text{Maximize } Z = h_{2,2} + h_{3,2} + h_{3,3}$$

subject to the finite difference Equation 10.8.14 written for each cell in the aquifer. The finite difference equations for cell (1, 1) is

$$h_{2,1} - 4h_{1,1} + h_{0,1} + h_{1,2} + h_{1,0} = 0$$

$W_{1,1} = 0$ because there is no pumping and heads $h_{0,1}$ and $h_{1,0}$ are known constant heads, so this constraint can be written as

$$h_{2,1} - 4h_{1,1} + h_{1,2} = -h_{0,1} - h_{1,0}$$

with the known values on the RHS. The finite difference equations for pumping cell (2, 2) is

$$h_{3,2} - 4h_{2,2} + h_{1,2} + h_{2,3} + h_{2,1} - \frac{(\Delta x)^2}{T} W_{2,2} = 0$$

The finite difference equation can be written for each of the remaining cells in the aquifer. The pumpage constraint Equation 10.8.16 is simply

$$W_{2,2} + W_{3,2} + W_{3,3} \geq W_{\min}$$

∎

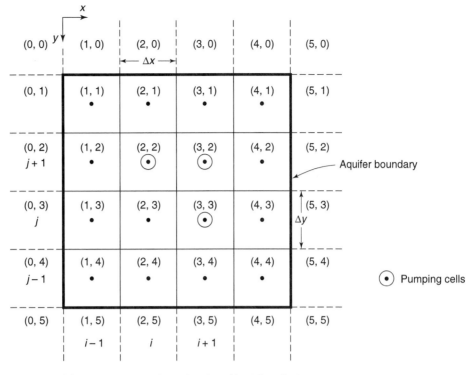

Figure 10.8.3. Confined two-dimensional aquifer (plan view).

10.8.4 Transient One-Dimensional Problem for Confined Aquifers

The governing equation for transient, one-dimensional confined problems has the form

$$T\frac{\partial^2 h}{\partial x^2} = S\frac{\partial h}{\partial t} + W \tag{10.8.18}$$

Using the Crank–Nicholson scheme (Remson et al.[70]), we can express the finite difference approximation of the second-order derivative in Equation 10.8.18 as

$$\frac{\partial^2 h}{\partial x^2} = \frac{1}{2}\left[\frac{h_{i+1,t} - 2h_{i,t} + h_{i-1,t}}{(\Delta x)^2} + \frac{h_{i+1,t-1} - 2h_{i,t-1} + h_{i-1,t-1}}{(\Delta x)^2}\right] \tag{10.8.19}$$

The finite difference equation for Equation 10.8.18 is determined using Equation 10.8.19 and the finite difference approximation for $\partial h/\partial t = (h_{i,t} - h_{i,t-1})/\Delta t$,

$$T\left[\frac{h_{i+1,t} - 2h_{i,t} + h_{i-1,t}}{2(\Delta x)^2} + \frac{h_{i+1,t-1} - 2h_{i,t-1} + h_{i-1,t-1}}{2(\Delta x)^2}\right] - S\left[\frac{h_{i,t} - h_{i,t-1}}{\Delta t}\right] - \frac{W_{i,t} + W_{i,t-1}}{2} = 0 \tag{10.8.20}$$

which can be simplified to

$$h_{i+1,t} - 2h_{i,t} + h_{i-1,t} + h_{i+1,t-1} - 2h_{i,t-1} + h_{i-1,t-1} - \frac{2(\Delta x)^2}{T(\Delta t)}S[h_{i,t} - h_{i,t-1}] - \frac{(\Delta x)^2}{T}\left(W_{i,t} + W_{i,t-1}\right) = 0 \tag{10.8.21}$$

Aguado et al.[2] proposed an LP model for transient one-dimensional flow with an objective of maximizing the sum of heads in the last time step, τ, that is,

$$\text{Maximize } Z = \sum_i h_{i,\tau} \qquad (10.8.22)$$

The constraints are Equation 10.8.21 written for each cell along with

$$\sum_i W_{i,t} \geq W_{\min,t} \qquad t = 1,\ldots,\tau \qquad (10.8.23)$$

EXAMPLE 10.8.4

Develop an LP model for the transient, one-dimensional confined aquifer in Figure 10.8.1 considering two time periods. The head boundaries at h_0 and h_5 are known for time $t = 0$, $t = 1$, and $t = 2$.

SOLUTION

The objective function is simply

$$\text{Maximize } Z = h_{1,2} + h_{2,2} + h_{3,2} + h_{4,2}$$

and the constraint for well $i = 1$ and time period $t = 1$ is

$$h_{2,1} - 2h_{1,1} + h_{0,1} + h_{2,0} - 2h_{1,0} + h_{0,0} - \frac{2(\Delta x)^2}{T(\Delta t)} S\left[h_{1,1} - h_{1,0}\right] - \frac{(\Delta x)^2}{T}\left(W_{1,1} + W_{1,0}\right) = 0$$

in which $h_{0,1}$ is the constant boundary and $W_{1,0}$, $h_{2,0}$, $h_{1,0}$, and $h_{0,0}$ represent the known initial conditions. This constraint can be rearranged with only the unknowns on the left-hand side:

$$h_{2,1} - 2h_{1,1} - \frac{2(\Delta x)^2}{T(\Delta t)} Sh_{1,1} - \frac{(\Delta x)^2}{T} W_{1,1} = -h_{0,1} - h_{2,0} + 2h_{1,0} - h_{0,0} - \frac{2(\Delta x)^2}{T(\Delta t)} Sh_{1,0} + \frac{(\Delta x)^2}{T} W_{1,0}$$

The constraint for well $i = 2$ and time period $t = 2$ is

$$h_{3,2} - 2h_{2,2} + h_{1,2} + h_{3,1} - 2h_{2,1} + h_{1,1} - \frac{2(\Delta x)^2}{T(\Delta t)} S\left[h_{2,2} - h_{2,1}\right] - \frac{(\Delta x)^2}{T}\left(W_{2,2} + W_{2,1}\right) = 0$$

The finite difference equations are written for each well at each time period.

The pumpage constraints are

$$W_{1,1} + W_{2,1} + W_{3,1} + W_{4,1} \geq W_{\min,1}$$

and

$$W_{1,2} + W_{2,2} + W_{3,2} + W_{4,2} \geq W_{\min,2} \qquad \blacksquare$$

10.8.5 Steady-State Two-Dimensional Problem for Unconfined Aquifers

This type of problem may be typical for dewatering of a construction or mining site (see Figure 10.8.4). Aguado et al.[2] presented an LP model for solving this problem. The governing equation for a homogeneous and isotropic aquifer is

$$\frac{\partial^2 h^2}{\partial x^2} + \frac{\partial^2 h^2}{\partial y^2} = \frac{2W}{K} \qquad (10.8.24)$$

or

$$\frac{\partial^2 w}{\partial x^2} + \frac{\partial^2 w}{\partial y^2} = \frac{2W}{K} \qquad (10.8.25)$$

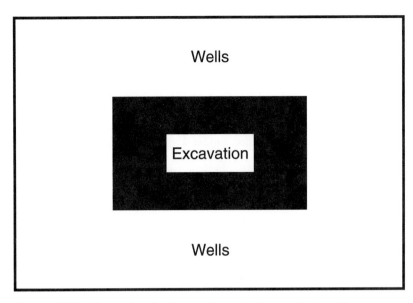

Figure 10.8.4.   Excavation site for two-dimensional unconfined aquifer.

where $w = h^2$. With the central differencing scheme of Equation 10.8.2, the finite difference representation of Equation 10.8.25 is

$$\frac{w_{i+1,j} - 2w_{i,j} + w_{i-1,j}}{(\Delta x)^2} + \frac{w_{i,j+1} - 2w_{i,j} + w_{i,j-1}}{(\Delta y)^2} = \frac{2W_{i,j}}{K} \tag{10.8.26}$$

For the purposes of this modeling effort, both the heads and the pumpages are to be determined so that the unknowns are w and W, then Equation 10.8.26 is

$$\left[\frac{1}{(\Delta x)^2}\right]w_{i+1,j} + \left[\frac{1}{(\Delta x)^2}\right]w_{i-1,j} - \left[\frac{2}{(\Delta x)^2} + \frac{2}{(\Delta y)^2}\right]w_{i,j} + \left[\frac{1}{(\Delta y)^2}\right]w_{i,j+1} + \left[\frac{1}{(\Delta y)^2}\right]w_{i,j-1} - \frac{2}{K}W_{i,j} = 0$$

$$\tag{10.8.27}$$

An LP model for the dewatering problem could be stated to minimize the amount of steady-state pumpage required to maintain head levels below specified levels. The LP model statement is

$$\text{Minimize } Z = \sum_{i,j \in I} W_{i,j} \tag{10.8.28}$$

subject to Equation 10.8.27 written for each cell, and the maximum allowable head requirement in the excavation site

$$w_{i,j} \leq w_r \tag{10.8.29}$$

for $c, j \in$ excavation site with $w_r = h_r^2$ and nonnegativity constraints

$$w_{i,j} \geq 0 \tag{10.8.30a}$$

$$W_{i,j} \geq 0 \qquad i,j \in I \tag{10.8.30b}$$

The number of wells and their locations are affected by the well spacing in the two-dimensional grid.

10.9 POLICY EVALUATION AND ALLOCATION MODELS: RESPONSE MATRIX APPROACH

The *response matrix approach* uses a groundwater simulation model external to the optimization model to develop unit responses. A unit response describes the influence of a pulse stimulus (such as a unit pumpage or injection over a time period) at a selected location (well) or cell upon the hydraulic heads for the other locations (wells) or cells throughout an aquifer. The response matrix consists of all the unit responses. Maddock[52] derived a function relating the drawdown in a confined aquifer to pumpages through the use of a unit response function, β. This function, also referred to as an *algebraic technological function*, defines the drawdown $s_{k,n}$ in the kth cell at the end of the nth time period:

$$s_{k,n} = \sum_{p=1}^{n} \sum_{j=1}^{J} \beta_{k,j,n,p} \, q_{j,p} \qquad (10.9.1)$$

where the unit response function $\beta_{k,j,n,p}$ is the change in drawdown (unit drawdown) in the kth cell at the end of the nth time period due to a unit pumpage from the jth cell (j may equal k) during the pth time period; $q_{j,p}$ is the quantity pumped from j during the pth time period; and J is the number of cells.

Equation 10.9.1 is based on the assumption that (1) flow is horizontal only, (2) wells are fully penetrating, (3) the transmissivity and storage coefficient may be nonhomogeneous, and (4) the pumpage is constant over a time period but can vary for different time periods. The unit response function, β, can be calculated analytically or numerically. To numerically determine β, a separate groundwater simulation is required for each pumped well or cell. For each simulation, unit pumpage over a time period is assumed and the simulation determines the response or drawdown at the other well locations or cells, for this unit pulse. Maddock[53] extended the idea of the algebraic technological function to unconfined aquifers.

A groundwater management model by Heidari[41] was formulated based upon the response function approach as follows:

$$\text{Maximize } Z = \sum_{j=1}^{J} \sum_{n=1}^{N} q_{j,n} \qquad (10.9.2)$$

subject to:

a. satisfying the governing equation of flow through the response matrix

$$s_{k,n} = \sum_{p=1}^{n} \sum_{j=1}^{J} \beta_{k,j,n,p} \, q_{j,p} \quad k = 1, \ldots, J$$
$$n = 1, \ldots, N \qquad (10.9.3)$$

b. pumpage cannot exceed $\bar{q}_{j,n}$, which is the smaller of the appropriated right or its capacity

$$0 \le q_{j,n} \le \bar{q}_{j,n} \qquad j = 1, \ldots, J$$
$$n = 1, \ldots, N \qquad (10.9.4)$$

c. drawdown at each well or cell cannot exceed an upper limit, $\bar{s}_{k,n}$

$$0 \le s_{j,n} \le \bar{s}_{k,n} \qquad k = 1, \ldots, J$$
$$n = 1, \ldots, N \qquad (10.9.5)$$

d.   demand, Q_n, for each time period n should be satisfied

$$\sum_{j=1}^{J} q_{j,n} \geq Q_n \quad n = 1, \ldots, N \tag{10.9.6}$$

The upper limit on drawdown $\bar{s}_{k,n}$ can be defined as a fraction (γ) of the saturated thickness b_k of the aquifer at the kth well (or cell), so that

$$\bar{s}_{k,n} = \gamma \cdot b_k \tag{10.9.7}$$

Actually, constraint Equations 10.9.3 and 10.9.5 can be combined. Heidari[41] applied the above LP model to the Pawnee Valley in Kansas to determine optimal pumpage for policies with and without net appropriations as constraints.

EXAMPLE 10.9.1

Develop a policy evaluation model using LP for a simple confined aquifer defined by four cells with a pumping well in each cell. A total of four time periods are to be considered. The unit response function $\beta_{k,j,n,p}$ is known and the objective is to maximize pumpage.

SOLUTION

$$\text{Maximize } Z = \sum_{j=1}^{4} \sum_{n=1}^{4} q_{j,n}$$

$$= q_{1,1} + q_{1,2} + q_{1,3} + q_{1,4} + \ldots + q_{4,1} + q_{4,2} + q_{4,3} + q_{4,4}$$

subject to:

a.   satisfying governing equations of flow through the response matrix

Well 1 ($k = 1$)

($n = 1$)

$$s_{1,1} = \sum_{p=1}^{n=1} \sum_{j=1}^{4} \beta_{k,j,n,p} \, q_{j,p}$$

$$= \beta_{1,1,1,1} q_{1,1} + \beta_{1,2,1,1} q_{2,1} + \beta_{1,3,1,1} q_{3,1} + \beta_{1,4,1,1} q_{4,1}$$

($n = 2$)

$$s_{1,2} = \sum_{p=1}^{n=2} \sum_{j=1}^{4} \beta_{k,j,n,p} \, q_{j,p}$$

$$= \beta_{1,1,2,1} q_{1,1} + \beta_{1,2,2,1} q_{2,1} + \beta_{1,3,2,1} q_{3,1} + \beta_{1,4,2,1} q_{4,1}$$
$$+ \beta_{1,1,2,2} q_{1,2} + \beta_{1,2,2,2} q_{2,2} + \beta_{1,3,2,2} q_{3,2} + \beta_{1,4,2,2} q_{4,2}$$

($n = 3$)

$$s_{1,3} = \sum_{p=1}^{n=3} \sum_{j=1}^{4} \beta_{k,j,n,p} \, q_{j,p}$$

$$= \beta_{1,1,3,1} q_{1,1} + \beta_{1,2,3,1} q_{2,1} + \beta_{1,3,3,1} q_{3,1} + \beta_{1,4,3,1} q_{4,1}$$
$$+ \beta_{1,1,3,2} q_{1,2} + \beta_{1,2,3,2} q_{2,2} + \beta_{1,3,3,2} q_{3,2} + \beta_{1,4,3,2} q_{4,2}$$
$$+ \beta_{1,1,3,3} q_{1,3} + \beta_{1,2,3,3} q_{2,3} + \beta_{1,3,3,3} q_{3,3} + \beta_{1,4,3,3} q_{4,3}$$

($n = 4$)

$$s_{1,4} = \sum_{p=1}^{n=4} \sum_{j=1}^{4} \beta_{k,j,n,p}\, q_{j,p}$$

$$= \beta_{1,1,4,1} q_{1,1} + \beta_{1,2,4,1} q_{2,1} + \beta_{1,3,4,1} q_{3,1} + \beta_{1,4,4,1} q_{4,1}$$
$$+ \beta_{1,1,4,2} q_{1,2} + \beta_{1,2,4,2} q_{2,2} + \beta_{1,3,4,2} q_{3,2} + \beta_{1,4,4,2} q_{4,2}$$
$$+ \beta_{1,1,4,3} q_{1,3} + \beta_{1,2,4,3} q_{2,3} + \beta_{1,3,4,3} q_{3,3} + \beta_{1,4,4,3} q_{4,3}$$
$$+ \beta_{1,1,4,4} q_{1,4} + \beta_{1,2,4,4} q_{2,4} + \beta_{1,3,4,4} q_{3,4} + \beta_{1,4,4,4} q_{4,4}$$

Well 2 ($k = 2$)

($n = 1$)

$$s_{2,1} = \sum_{p=1}^{n=1} \sum_{j=1}^{4} \beta_{k,j,n,p}\, q_{j,p}$$

$$= \beta_{2,1,1,1} q_{1,1} + \beta_{2,2,1,1} q_{2,1} + \beta_{2,3,1,1} q_{3,1} + \beta_{2,4,1,1} q_{4,1}$$

($n = 2$)

$$s_{2,2} = \sum_{p=1}^{n=2} \sum_{j=1}^{4} \beta_{k,j,n,p}\, q_{j,p}$$

$$= \beta_{2,1,2,1} q_{1,1} + \beta_{2,2,2,1} q_{2,1} + \beta_{2,3,2,1} q_{3,1} + \beta_{2,4,2,1} q_{4,1}$$
$$+ \beta_{2,1,2,2} q_{1,2} + \beta_{2,2,2,2} q_{2,2} + \beta_{2,3,2,2} q_{3,2} + \beta_{2,4,2,2} q_{4,2}$$

The remaining governing constraints are written in a similar manner.

b. pumpage cannot be greater than the appropriated right

$$0 \le q_{1,1} \le \bar{q}_{1,1}$$
$$0 \le q_{1,2} \le \bar{q}_{1,2}$$
$$\vdots$$
$$0 \le q_{2,1} \le \bar{q}_{2,1}$$
$$\vdots$$
$$0 \le q_{3,1} \le \bar{q}_{3,1}$$
$$\vdots$$
$$0 \le q_{4,1} \le \bar{q}_{4,1}$$

c. drawdown at each well cannot exceed an upper limit

$$0 \le s_{1,1} \le \bar{s}_{1,1}$$
$$0 \le s_{1,2} \le \bar{s}_{1,2}$$
$$\vdots$$
$$0 \le s_{2,1} \le \bar{s}_{2,1}$$
$$\vdots$$
$$0 \le s_{3,1} \le \bar{s}_{3,1}$$
$$\vdots$$
$$0 \le s_{4,1} \le \bar{s}_{4,1}$$

d. demand requirement

$$q_{1,1} + q_{2,1} + q_{3,1} + q_{4,1} \geq Q_1$$

$$q_{1,2} + q_{2,2} + q_{3,2} + q_{4,2} \geq Q_2$$

$$q_{1,3} + q_{2,3} + q_{3,3} + q_{4,3} \geq Q_3$$

$$q_{1,4} + q_{2,4} + q_{3,4} + q_{4,4} \geq Q_4$$ ■

10.10 OPTIMAL CONTROL GROUNDWATER MANAGEMENT MODELING

The general groundwater management problem (GGMP) can be expressed mathematically as follows:

$$\text{Optimize } Z = f(\boldsymbol{h}, \boldsymbol{q}) \tag{10.10.1}$$

subject to

a.   the general groundwater flow constraints Equation 9.4.10a or b or 9.5.1

$$g(\boldsymbol{h}, \boldsymbol{q}) = 0 \tag{10.10.2}$$

b.   the simple bounds

$$\underline{q} \leq q \leq \bar{q} \tag{10.10.3}$$

$$\underline{h} \leq h \leq \bar{h} \tag{10.10.4}$$

c.   other constraint sets, such as demand constraints

$$w(\boldsymbol{h}, \boldsymbol{q}) \leq 0 \tag{10.10.5}$$

where ($^-$) represents an upper bound and (_) represents a lower bound.

Both head h and pumpage or recharge q are vectors of decision variables that have maximum dimensions equal to the product of the number of active nodes within the aquifer boundary and time steps. Fixed pumpages or recharges are considered to be constants. By convention, available pumpages have a positive value and the elements of q have a negative value where there is available recharge. Usually the number of variable pumpages and/or recharges (hereafter the terms pumpages that refer to q will imply both pumpages and/or recharges) is small and results in a much smaller dimension of q than h.

The objective function Equation 10.10.1, which may be either maximization (e.g., sum of heads) or minimization (e.g., minimize pumpage), can be a linear or nonlinear function. Also, it may be nonseparable or contain only terms of pumpages or heads, but has to be differentiable. Constraint Equation 10.10.2 represents a system of equations governing groundwater flow, which are finite difference or simulator equations when q is unknown. The upper ($\bar{q}$) and lower ($\underline{q}$) bounds on pumpages physically may or may not exist. Unlike pumpage, the lower bound on heads ($\underline{h}$) can be viewed as the bottom elevation of the aquifer while the upper head bound ($\bar{h}$) can be regarded as ground surface elevations for the unconfined cells. In addition to constraint Equations 10.10.2–10.10.4, constraint Equation 10.10.5 may be included to impose restrictions such as water demands, operating rules, budgetary limitations, and so forth.

The above general optimization model, Equations 10.10.1–10.10.5, has been solved through the development of several simulation optimization groundwater management models based upon optimal control approaches (Mays[59]). These include models developed by Culver and Shoemaker[25] and Wanakule, Mays, and Lasdon.[95]

10.11 CASE STUDIES   *Groundwater Management Modeling Applications*

10.11.1 Optimal-Control Model for Barton Springs-Edwards Aquifer, Texas

Wanakule[93] developed a groundwater management (simulation–optimization) model for determining the optimal yield under various management policies of the Barton Springs-Edwards Aquifer, Austin, Texas (see Figure 9.4.3). The model (Wanakule et al.[94, 95]) used the Texas Water Development Board (TWDB) finite difference GWSIM groundwater simulation model (TWDB[86]) interfaced with the GRG2 nonlinear programming optimizer (Lasdon et al.[49]). Figure 9.4.3 shows the finite difference grid system containing 330 cells whose dimensions varied from 0.379 × 0.283 mi^2 to 0.95 × 1.51 mi^2. Figure 10.11.1 shows the 1981 water level contours.

The objective of the management model was to maximize the pumpage of all existing wells in 1981. This mathematical model can be stated as follows:

Maximize total pumpage of all existing wells in 1981 subject to these constraints:

1) The groundwater hydraulics of the Edwards Aquifer for the 1981 hydrologic conditions including recharge.

2) Recreation constraints for maintaining the spring flows at Barton Springs.

Dewatered areas in the eastern part of the aquifer had to be considered due to the high aquifer bottom elevations. Two different cases were presented based upon the size of the dewatered zones, as shown in Figure 10.11.2. The case 002, which allows for a larger dewatered zone (5.095 mi^2), resulted in an optimal yield of 8,999 acre-feet. Case 003 with the smaller dewatered zone (4.233 mi^2) resulted in a smaller yield of 4,928 acre-feet as compared to the 1981 pumpage of 2,666 acre-feet. This example

(continues)

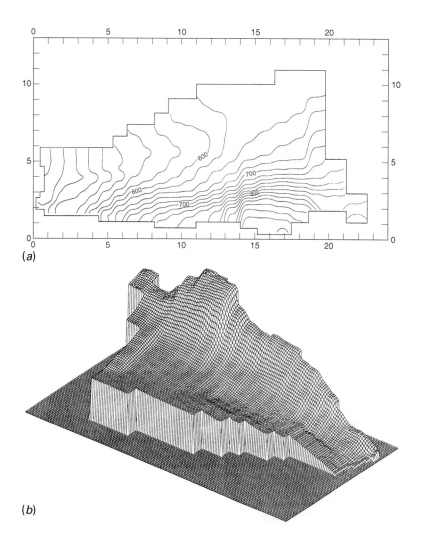

(a)

(b)

Figure 10.11.1. (*a*) 1981 Water level contours in the Barton Springs-Edward Aquifer at 20-ft. intervals.
(*b*) Perspective block diagram of 1981 water levels viewed from the east side of the aquifer (Wanakule[93]).

illustrates the use of a simulation–optimization model to define optimal pumpages to maximize yield and maintain spring flows.

10.11.2 Heuristic Optimization–Simulation Model for Groundwater Remediation Design: N-Springs Site, Hanford, Washington

Skaggs et al.[78, 79] developed a heuristic optimization–simulation model for groundwater remediation design, based upon a simulated (enhanced) annealing algorithm with a groundwater simulation model (see Figure 10.7.2). The groundwater management model was applied to the N-Springs site located at Hanford, Washington, adjacent to the Columbia River (see Figure 10.11.3).

N-Springs is the location of the 1301-N Liquid Waste Disposal Facility (LWDF), which was used for disposal of radiological-contaminated water associated with pass-through cooling, spent fuel storage, and other activities related to operation of the N Reactor (U.S. DOE[90]).

The LWDF is a long, zigzagging trench running parallel to and about 800 ft from the Columbia River. From 1963 until 1985, the LWDF received discharges of water containing both radiological and nonradiological contaminants at a rate of approximately 2,100 gal/min. Almost immediately after receiving water, mobile contaminants from the LWDF were observed in springs along the

(*continues*)

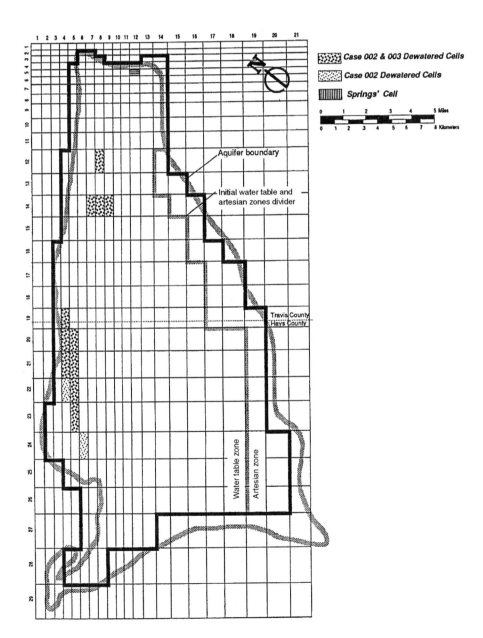

Figure 10.11.2. Dewatered area for Cases 002 and 003 (Wanakule[93]).

| 10.11 CASE STUDIES | *Groundwater Management Modeling Applications (continued)* |

banks of the Columbia River. Beginning in 1980, ^{90}Sr was detected in the groundwater near the river. It is estimated that 2,997 Ci of ^{90}Sr was discharged to the facilities at N-Springs, of which ~46 Ci was released to the Columbia River and 1,085 Ci has decayed away. The majority of the radiological inventory remaining at the N-Springs area is located in the sediments underlying the LWDF, including 75.5 Ci within the unconfined aquifer, of which it is estimated that 75 Ci is adsorbed by the sediments and 0.5 Ci is dissolved in the groundwater.

Contours depicting the areal concentration distribution of ^{90}Sr in the groundwater at N-Springs are shown in Figure 10.11.4. The plume covers approximately 55 acres and has maximum concen-

trations on the order of 4,000 pCi/L. The highest concentrations are shown to be immediately beneath and down gradient (i.e., toward the river) from the LWDF.

The modeled area is that used by the U.S. DOE[90] analysis as shown in Figure 10.11.3, encompassing the 8,000 ft by 12,000 ft study area. The prevailing hydraulic gradient is almost due north, from the lower right-hand corner of the model area to the upper left-hand corner. Over the study area the gradient varies from 0.0005 to 0.003, producing average groundwater velocities between 0.1 to 2 ft/day in the vicinity of the LWDF. The U.S. DOE[90] study was based on a two-dimensional finite-difference

(continues)

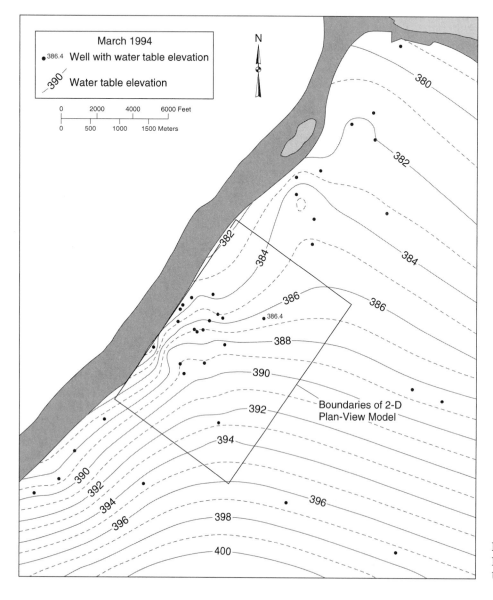

Figure 10.11.3. Plan view of N-Springs study area (from U.S. DOE[90]).

code with the 8,000 ft by 12,000 ft study area discretized using a variable grid spacing. The flow model assumed steady-state conditions and transport was determined using particle tracking based on computed groundwater pathlines and accounting for retardation and radioactive decay. The site was discretized using a 35 by 68 element grid. In the vicinity of the ^{90}Sr plume (from 3,000 ft to 9,000 ft in the y direction and 0 ft to 3,000 ft in the x direction) a subgrid, regularly spaced with element dimensions of 100 ft by 100 ft, was used. Outside the plume area, element dimensions increased from 250 ft to 1,500 ft. An overlay of the subgrid over the model area is shown in Figure 10.11.5.

The management problem under consideration for the N-Springs area is selection of a least-cost containment system that would prevent any release of ^{90}Sr into the Columbia River for a ten-year period. Factors evaluated include construction costs for a barrier and/or well(s), treatment costs for any extracted contaminated groundwater, plus maintenance and operating costs for both extraction and injection wells. In addition, the net groundwater withdrawal should be zero (i.e., total extraction rate equals total

injection rate). Twenty-eight candidate well locations with twelve possible flow rates for each well were considered: five injection rates, zero, and six extraction rates. Similar to the U.S. DOE study, four possible barrier configurations were considered: no barrier and candidate lengths of 2,000 ft, 3,000 ft, and 3,800 ft.

The optimal solution obtained consisted of a 3,000-ft barrier with a single withdrawal well and a single injection well. The computed contaminant release to the river is virtually zero for the ten-year period. The optimal solution had a total cost of $17.6 million as compared to the U.S. DOE[90] solution costing $23.1 million.

10.11.3   Groundwater Management Model Using Response Matrix Approach: Santa Clara-Calleguas Basin, California

Reichard[68] presented an application of a response matrix model for groundwater–surface water management to identify efficient strategies for meeting water demand (with uncertain surface water

(continues)

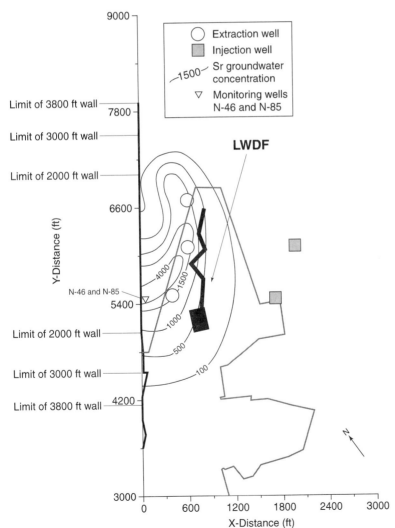

Figure 10.11.4. Areal distribution of ^{90}Sr in the groundwater at the N-Springs area.[79]

10.11 CASE STUDIES   *Groundwater Management Modeling Applications (continued)*

---

supplies) and controlling regional water quality (seawater intrusion) in the Santa Clara-Calleguas Basin in southern California, shown in Figure 10.11.6. The study focused on the issues relating to groundwater use and operation of the facilities under the control of the United Water Conservation District (UWCD). An efficient strategy was assumed to be one that met demands and water quality constraints while minimizing the need for supplemental water, imposed water use reductions, or changes from current regional pumping patterns. The management agency (Fox Canyon Groundwater Management Agency (GMA) created by the state of California to manage endangered groundwater resources) must supply water to demand sectors, using a combination of water delivered by pipelines and distributed pumpage from wells in an upper and lower aquifer. Water delivered by pipelines includes river diversions, water pumped from central production wells, and supplemental

water. Supplemental water was any additional source of surface water, including capturing additional surface water in the basin and importation. Surface water that was not delivered directly to water users could be artificially recharged. The surface water availability from the river diversion is uncertain.

10.11.3.1 Description of Study Area

In order to apply the groundwater management model, the Santa Clara-Calleguas Basin was subdivided into 13 water demand sectors, with the Oxnard Plain and Las Posas subbasins further subdivided, as shown in Figure 10.11.6. Surface water is diverted from the Santa Clara River, with the goal of capturing as much of the river water as possible, subject to the water rights. The major regional water quality issue addressed in the modeling effort is

(continues)

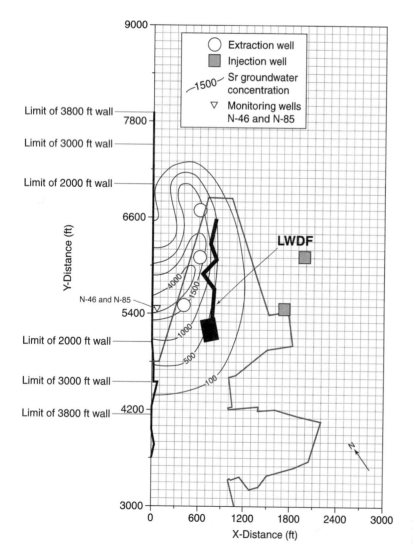

Figure 10.11.5. Overlay of 100 ft by 100 ft subgrid on the study area.[79]

seawater intrusion, illustrated in Figure 10.11.7 with the high chloride plumes with concentrations greater than 500 mg/L. The water management facilities for the Santa Clara-Calleguas Basin are shown in Figure 10.11.8.

The groundwater basin is approximately 300 mi^2 (800 km^2) in area, with at least five different aquifers having been identified in the Santa Clara-Calleguas Basin, grouped into an upper aquifer system (Oxnard and Mugu aquifers) and a lower system aquifer system (Hueneme, Fox Canyon, and Grimes Canyon). The upper system is more transmissive and more easily recharged of the two aquifer systems. For purposes of modeling, two layers were used in the Oxnard Plain (sectors 1–4 in Figure 10.11.6) and one layer elsewhere. Only the lower aquifer was assumed to be

present in the Pleasant Valley, Santa Rosa, Las Posas subbasins (sectors 5, 6, and 7–9, respectively, in Figure 10.11.6).

The issues of the study are related to groundwater use and operation of facilities under the control of the United Water Conservation District (UWCD). During the 1984–1989 period, annual use of groundwater in the basin plus water delivered by pipelines by UWCD was approximately 360 ft^3/s (10.3 m^3/s), of which 80 % was for agriculture. The GMA has the responsibility for developing and implementing a plan to control seawater intrusion. They also have the authority to limit pumping and collect fees within sectors 1–9. GMA strategy has been to increase artificial recharge, decrease pumpage from the upper aquifer, and increase pumpage from the lower aquifer system.

(*continues*)

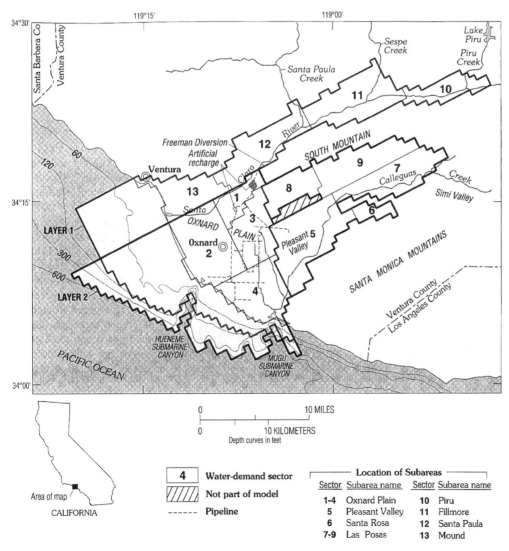

Figure 10.11.6. Santa Clara-Calleguas Basin (Reichard[68]).

10.11.3.2 Groundwater Model

The groundwater study area model was developed by the California Department of Water Resources[21, 22] and MODFLOW was used to perform the groundwater flow computations to develop the response functions (see Reichard[68]) for the optimization model. Two layers were used to represent the upper and lower aquifers, with 1,044 active cells in the upper aquifer and 1,106

active cells in the lower aquifer. Model cells are 2,640 ft × 2,640 ft (805 m × 805 m). The modeled groundwater system assumes heterogeneous, isotropic confined flow in both layers.

Offshore boundaries represented outcrops of the two aquifers and were modeled as specified-head boundaries. Drain boundaries were used to represent zones of rising water along the Santa Clara River (see Figure 10.11.7). All other boundaries were no-flow

(continues)

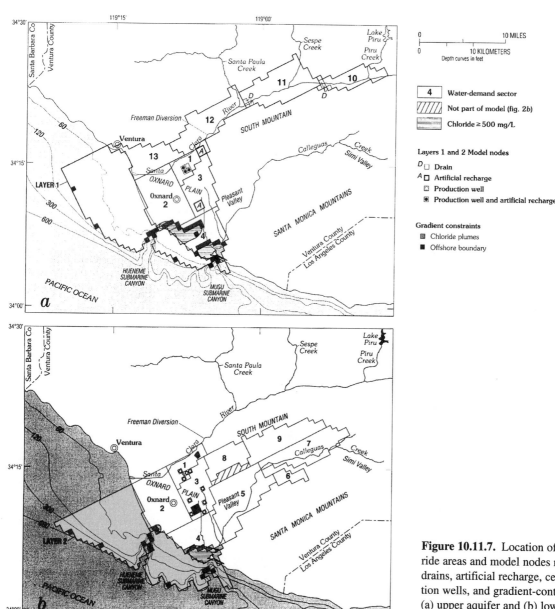

Figure 10.11.7. Location of high-chloride areas and model nodes representing drains, artificial recharge, central production wells, and gradient-control locations: (a) upper aquifer and (b) lower aquifer (Reichard[68]).

boundaries, representing either the contact between consolidated and unconsolidated materials, the presence of the faults, or the pinching of the upper aquifer.[68] Natural recharge, from stream-flow and precipitation, was determined from modified estimates from water-budgets. Many other details of the model development can be found in Reichard.[68]

10.11.3.3 Model Formulation

The statement of the mathematical optimization model is as follows:

Objectives

Minimize the relative amount of supplemental water obtained for a specified water use (i.e., minimize the quantity of supplemental water as a fraction of surface water diverted from the Santa Clara River).

or

Minimize the imposed water use reduction for a specified amount of supplemental water.

or

Minimize the expected sum of squared deviations from current average pumping rates for a specified reduction in water use.

Subject to the constraints:

1. Availability of supplemental water
2. Water level constraints
3. Hydraulic gradient constraints
4. Demand requirements

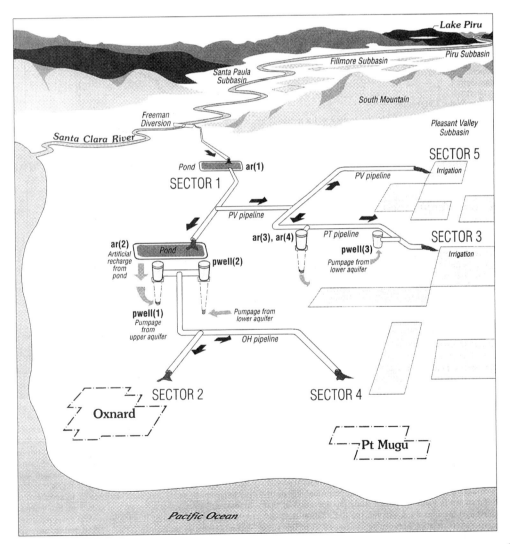

Figure 10.11.8. Schematic of water management facilities in the Santa Clara-Calleguas Basin (Reichard[68]).

5. Continuity on water routing including operation of pipelines, artificial recharge facilities, and the central production wells for every year

6. Artificial recharge and pipeline capacity constraints

7. Groundwater hydraulics constraints

The model planning period is T years. Constraints on the water levels and hydraulic gradients are imposed at the end of T years. Groundwater modeling simulations were used to develop the response functions used in the groundwater hydraulic constraints.

The nonlinear optimization was solved using MINOS through the GAMS modeling system (Brooke et al.[20]).

Application of the groundwater management model indicated that a large quantity of supplemental water or significant pumpage reductions, particularly in the lower aquifer, may be required to control seawater intrusion with current facilities. The supplemental water would be delivered directly to users through pipelines and artificially recharged. The current artificial recharge facilities had been valuable and new artificial recharge facilities would be beneficial.

PROBLEMS

10.2.1 A pumping test is a frequently used field technique to determine the properties of aquifers. It is done by pumping a well at a constant rate and observing the drawdown of the piezometric surface or water table in an observation well at some distance from the pump well. Then an appropriate analytical solution for groundwater flow is used to determine the transmissivity and storage coefficient. Therefore, aquifer properties so determined will be valid in the vicinity of the pumping well when flow satisfies the conditions used in developing the analytical solution. Utilize the Cooper–Jacob equation to determine the optimal transmissivity (T) and storage coefficient (S) that minimizes the sum of square of errors between the observed and calculated drawdowns (listed in the table below) at two observation wells over time.

10.2.2 Resolve Problem 10.2.1 to minimize the sum of absolute error between the observed and calculated drawdowns at two observation wells.

10.5.1 Develop a report describing the groundwater laws in your state or region of your country. How have these laws promoted or hindered the development of groundwater?

10.8.1 Set up and solve the linear programming model to determine the optimal pumpage for the one-dimensional confined

aquifer in Figure 10.8.1 where $\Delta x = 80$ m, $T = 1000$ m^2/day, $W_{min} = 500$ m/day, $h_0 = 40$ m, and $h_5 = 35$ m. Solve this model for h_1, ..., h_4 and W_1, ..., W_4.

10.8.2 Set up the LP model to determine the optimal pumpage for the one-dimensional confined aquifer in Figure 10.8.1 where $\Delta x = 10$ ft, $T = 10,000$ ft^2/day, $h_0 = 100$ ft, $h_4 = 110$ ft, and $W_{min} = 100$ ft/day.

10.8.3 Solve the LP model set up in Problem 10.8.2.

10.8.4 Set up and solve the LP model to determine the optimal pumpage for the one-dimensional unconfined aquifer in Figure 10.8.2 where $\Delta x = 80$ m, $K = 150$ m/day, $W_{min} = 500$ m/day, $h_0 = 40$ m, and $h_5 = 35$ m.

10.8.5 Set up the complete LP model for the two-dimensional confined aquifer described in Example 10.8.3.

10.8.6 Set up and solve the LP model for the two-dimensional confined aquifer shown in Figure 10.8.3 to determine the optimal pumpage from cells (2, 2), (3, 2) and (3, 3); $\Delta x = 1,000$ ft, $T = 10,000$ ft^2/day, $W_{min} = 2$ ft/day. The boundary condition is a constant head of 20 ft.

10.8.7 Set up the complete LP model for the transient, one-dimensional confined aquifer problem in Example 10.8.4.

10.8.8 Solve the model for the transient, one-dimensional confined aquifer developed in Problem 10.8.7 for $\Delta x = 100$ ft, $\Delta t = 1$ day, $T = 10,000$ ft^2/day, $S = 0.001$, $h_0 = 120$ ft, $h_5 = 105$ ft, and $W_{min} = 200$ ft/day. At $t = 0$, $h_1 = 118.5$ ft, $h_2 = 116$ ft, $h_3 = 113$ ft, $h_4 = 103.5$ ft, and $W_{1,0} = W_{2,0} = W_{3,0} = W_{4,0} = 0$.

10.9.1 Develop a groundwater allocation model for an irrigation district to determine the optimal cropping pattern and groundwater allocation that maximizes the net return from agricultural production. The irrigation district has the responsibility for determining the optimal cropping pattern (acreage devoted to each crop) with the district's authority. Water is pumped from an extensive confined aquifer system for which a steady-state model is considered for the two-dimensional response of the aquifer. The profit per acre for a crop is (the unit revenue of a crop times the yield per acre) minus (the unit cost of groundwater times the amount of groundwater applied per acre). The yield would be a

Drawdown Data for Problem 10.2.1

Time (days)	Drawdown (m)	
	Observation wells	
	Well 1 ($r = 100$ m)	Well 2 ($r = 200$ m)
0.001	0.087	0.015
0.005	0.200	0.100
0.01	0.252	0.147
0.05	0.370	0.270
0.1	0.435	0.320
0.5	0.555	0.450
1	0.610	0.500
5	0.745	0.630
10	0.805	0.690

nonlinear function defined by a production function. Define the objective function, constraints, parameters, and decision variables. What would be the solution methodology?

10.9.2 Develop a conjunctive groundwater and surface water planning model that can optimally distribute, over time, the water resources for various water demands. The decisions to be made are the groundwater and surface water allocation in each planning period. The objective function is to maximize the net discounted benefits from operating the system over the planning horizon, including costs for capital, operation, and maintenance. Constraints would include (a) conservation or balance equations for the surface water system, (b) capacity limitations of the surface water system, (c) groundwater system flow equations to define aquifer response, and (d) bound constraints on groundwater levels. Others may be required such as for artificial recharge, pumping schedules. Use GW_{ij}^t and SW_{kj}^t to represent, respectively, groundwater allocated from groundwater source i to demand j in time period t and surface water allocated from source k to demand j in time period t. Define the objective function, constraints, parameters, and decision variables. What would be the solution methodology?

10.9.3 In the planning stage of groundwater management, applications of analytical solutions of groundwater flow can be made. Those analytical equations can be used to calculate the unit response matrix in groundwater management models. Consider an undeveloped, homogenous, confined groundwater aquifer (see Figure 10.P.1) in which there are three potential pump wells and five control points at which the drawdowns are to be observed. For steady-state management, the Thiem equation

$$s_{kl} = \frac{\ln(r_{01} / r_{kl})}{2\pi T} q_l$$

can be used in which s_{kl} = drawdown at control point k resulting from a pumpage of q_l at the well location l; r_{01} = radius of influence of potential production well; r_{kl} = distance between control

point k and well location l; and T = aquifer transmissivity. Assume that the radius of influence of all pump wells is 700 feet (213 meters) and the aquifer transmissivity is 5,000 gallon/day/ft (0.0007187 m²/sec). Based on the information about the maximum allowable drawdown and pump capacities given in Figure 10.P.1, formulate an LP model to determine the optimal pumpage for each well.

10.9.4 Refer to the hypothetical groundwater system as shown in Figure 10.P.1. Consider now that the management is to be made for three periods of 50 days each. This is a management problem for transient groundwater systems which can be approximately described by the Theis equation or Cooper–Jacob equation.

(a) Using the Cooper–Jacob, equation define the unit response function.

(b) Assume that the storage coefficient of the aquifer is 0.002 and transmissivity is 5,000 ft²/day (465 m²/day). Further, the maximum allowable drawdowns (in ft) at each control point are given as

Period	Control point				
	1	2	3	4	5
1	5	5	8	5	5
2	8	8	10	8	8
3	10	10	15	10	10

Develop the LP model and solve it for maximizing total pumpage for all three 50-day periods.

10.9.5 Assume that the transmissivity of the hypothetical aquifer (Figure 10.P.1) is a lognormal random variable with mean 5,000 gallon/day/ft and coefficient of variation of 0.3. Convert the LP model developed in Problem 10.9.1 into a chance-constrained model such that the specified drawdown at the five control points will not be exceeded with 95 percent reliability. Also, solve the model.

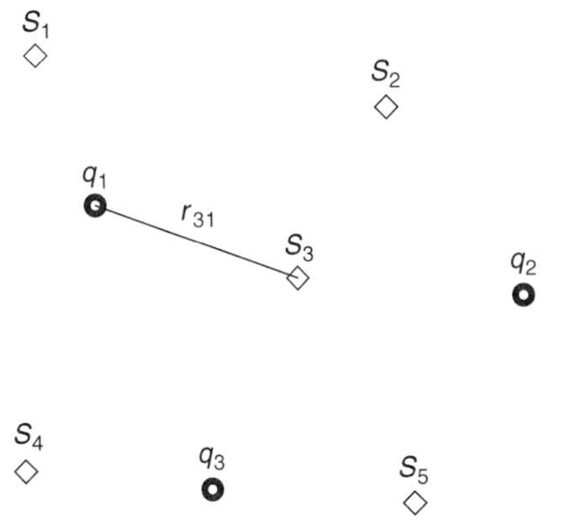

Distance (ft.) between potential pump wells and control points

Pumping well	Control points					Pump capacity (gpd)
	1	2	3	4	5	
1	158	381	158	255	430	200,000
2	515	255	292	474	158	200,000
3	447	447	200	200	200	200,000
Maximum allowable drawdown (ft)	7	7	15	7	7	

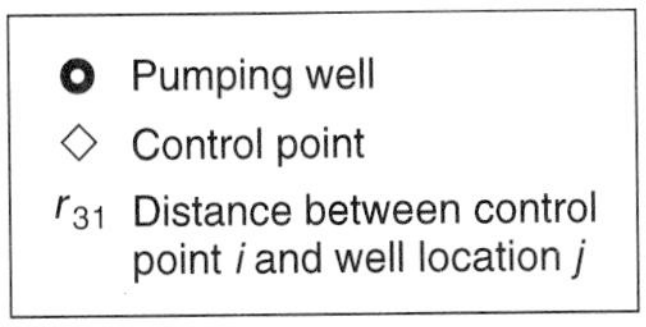

- ● Pumping well
- ◇ Control point
- r_{31} Distance between control point i and well location j

Figure 10.P.1. Location of pumping wells and control points for a hypothetical groundwater basin (Problems 10.9.3–10.9.5).

REFERENCES

1. Aguado, E., and I. Remson, Ground-water management with fixed charges, *Jour. Water Resources Planning Mgmt. Div., Amer. Soc. Civil Engrs.,* v. 106, pp. 375–382, 1980.

2. Aguado, E., I. Remson, M. F. Pikul, and W. A. Thomas, Optimum pumping for aquifer dewatering, *Jour. Hydraulics Div., Amer. Soc. Civil Engrs.,* v. 100, pp. 860–877, 1974.

3. Aguado, E., N. Sitar, and I. Remson, Sensitivity analysis in aquifer studies, *Water Resources Research,* v. 13, pp. 733–737, 1977.

4. Ahfeld, D. P., and M. Heidari, Applications of optimal hydraulic control to groundwater systems, *Jour. Water Resources Planning and Mgt.,* v. 120, no. 3, May/June 1994.

5. Ahfeld, D. P., and A. Mulligan, *Optimal Management of Flow in Groundwater Systems,* Academic Press, 2000.

6. Ahlfeld, D. P., and C. S. Sawyer, Well location in capture zone design using simulation and optimization techniques, *Ground Water,* v. 28, no. 4, pp. 507–512, 1990.

7. Ahlfeld, D. P., R. H. Page, and G. F. Pinder, Optimal ground-water remediation methods applied to a Superfund site: From formulation to implementation, *Ground Water,* v. 33, no.1, pp. 58–70, Jan.–Feb. 1995.

8. Alley, W. M., and J. E. Schefter, External effects of irrigator's pumping decisions, High Plains Aquifer, *Water Resources Research,* v. 23, no. 7, pp. 1123–1130, 1987.

9. Alley, W. M., T. E., Reilly, O. L. Franke, *Sustainability of Groundwater Resources, U.S. Geological Survey Circular* 1186, Denver, CO, 1999.

10. Amer. Soc. Civil Engrs., *Ground Water Management,* Manual Engrng. Practice 40, 216 pp., 1972.

11. Andricevic, R., Coupled withdrawal and sampling designs for groundwater supply models, *Water Resources Research,* v. 29, no. 1, pp. 5–16, 1993.

12. Arizona Department of Water Resources (ADWR), Overview of Arizona's groundwater management code, http://www.adwr.state.az.us/Azwaterinfo, 1998.

13. Baker, T. L., R. Rae, J. E. Minor, and S. V. Connor, *Water for the Southwest: Historical Survey and Guide to Historic Sites,* ASCE Historical Publication No. 3, Amer. Soc. Civil Engrs., New York, 1973.

14. Barlow, P. M., B. J. Wagner, and K. Belitz, Pumping strategies for management of a shallow water table: The value of the simulation–optimization approach, *Ground Water,* v. 34, no. 2, March/April, 1996.

15. Basagaoglu, H., and M. A. Mariño, Joint management of surface and ground water supplies, *Ground Water,* v. 37, no. 2, pp. 214–222, 1999.

16. Bear, J., and O. Levin, The optimal yield of an aquifer, *Intl. Assoc. Sci. Hydrology Publ.* 72, pp. 401–412, 1967.

17. Bisschop, J. W., J. H. Candler, and G. T. O'Mara, The Indian Basin model: A special application of two-level linear programming, *Math Program,* v. 20, pp. 30–38, 1982.

18. Bowden, C., *Killing the Hidden Waters,* University of Texas Press, Austin, TX, 1977.

19. Bredehoeft, J. D., and R. A. Young, Conjunctive use of groundwater and surface water for irrigated agriculture: Risk aversion, *Water Resources Research,* AGU, v. 19, pp. 1111–1121, 1983.

20. Brooke, A., D. Kendrick, and A. Meerhaus, *GAMS: A User's Guide,* Scientific Press, Redwood City, CA, 1988.

21. California Department of Water Resources, *Mathematical modeling of water quality for water resources management, 1. Development of the ground water quality model,* District report, Southern District, Los Angeles, CA, 1974.

22. California Department of Water Resources, *Compilation of technical information for the Ventura County cooperative investigation,* Vol. II, Sacramento, 1975.

23. Chang, L. C., C. A. Shoemaker, and P. L. F. Liu, Optimal time-varying pumping rates for groundwater remediation: Application of a constrained optimal control algorithm, *Water Resources Research,* v. 28, no. 12, pp. 3157–3173, 1992.

24. Clendenen, F. B., Economic utilization of ground water and surface water storage reservoirs, Paper presented before meeting of Amer. Soc. Civil Engrs., San Diego, Feb. 1955.

25. Culver, T. B., and C. A. Shoemaker, Dynamic optimal control for groundwater remediation with flexible management periods, *Water Resources Research,* v. 28, no. 3, pp. 629–641, March 1992.

26. Culver, T. B., and C. A. Shoemaker, Optimal control for groundwater remediation by differential dynamic programming with quasi-Newton approximations, *Water Resources Research,* v. 29, no. 4, pp. 823–831, 1993.

27. Daubert, J. T., and R. A. Young, Groundwater development in western river basins: Large economic gains with unseen costs, *Ground Water,* v. 20 pp. 80–85, 1982.

28. Domenico, P. A., Economic aspects of conjunctive use of water, Smith Valley, Nevada, USA, *Intl. Assoc. Sci. Hydrology Publ 72,* pp. 474–482, 1967.

29. Domenico, P. A. et al., Optimal ground-water mining, *Water Resources Research,* v. 4, pp. 247–255, 1968.

30. Dougherty, D. E., and R. A. Marryott, Optimal Groundwater Management, 1. Simulated Annealing, *Water Resources Research,* v. 27, pp. 2493–2508, 1991.

31. Espeland, W. N., *The Struggle for Water: Politics, Rationality, and Identity in the American Southwest,* University of Chicago Press, Chicago, IL, 1998.

32. Fowler, L. C., Ground-water management for the nation's future—ground-water basin operation, *Jour. Hydraulics Div., Amer. Soc. Civil Engrs.,* v. 90, no. HY4, pp. 51–57, 1964.

33. Fradkin, P.L., *A River No More: The Colorado River and the West,* University of Arizona Press, Tucson, AZ, 1981.

34. Galloway, D. L., W. M. Alley, P. M. Barlow, T. E. Reilly, P. Tucci, *Evolving Issues and Practices in Managing Ground-Water Resources: Case Studies on the Role of Science, U.S. Geological Survey Circular* 1247, 2003.

35. Georgakakos, A. P., and D. A. Vlatsa, Stochastic control of ground water systems, *Water Resources Research,* v. 27, no. 8, pp. 2077–2090, 1991.

36. Glennon, R., *Water Follies: Groundwater Pumping and the Fate of America's Rivers,* Island Press, Washington, D.C., 2002.

37. Gorelick, S. M., A review of distributed parameter groundwater management modeling methods, *Water Resources Research,* v. 19, pp. 305–319, 1983.

38. Grubb, H. W., The Edwards Aquifer: Conflicts surrounding use of a regional water resource, *Water Resources Update,* Universities Council on Water Resources, Issue No. 106, Carbondale, IL, Winter 1997.

39. Haimes, Y. Y., and Y. C. Dreizen, Management of groundwater and surface water via decomposition, *Water Resources Research,* AGU, v. 13, pp. 69–77, 1977.

40. Hartman, L. M., Economics and ground-water development, *Ground Water,* v. 3, no. 2, pp. 4–8, 1965.

41. Heidari, M., Application of linear systems theory and linear programming to groundwater management in Kansas, *Water Resources Bull,* v. 18, pp. 1003–1012, 1982.

42. Hillier, F. S., and G. J. Lieberman, *Introduction to Operations Research,* 7th ed., McGraw-Hill, New York, 2001.

43. Illangasekare, T., and H. J. Morel-Seytoux, Stream-aquifer influence coefficients as tools for simulation and management, *Water Resources Research,* v. 18, pp. 168–176, 1982.

44. Ingram, H. et al., Water scarcity and the politics of plenty in the Four-Corner states, *Western Political Quarterly,* September 1979.

45. Johnson, A. I., et al., Symposium on planning and design of ground-water data programs, *Water Resources Research,* v. 8, pp. 177–241, 1972.

46. Killgore, M. K., and D. J. Eaton, *NAFTA Handbook for Water Resource Managers and Engineers,* Amer. Soc. Civil Engrs., New York, 1995.

47. LaBolle, E. M., A. A. Ahmed, and G. E. Fogg, Review of the integrated groundwater and surface-water model (IGSM), *Ground Water,* v. 41, no. 2, pp. 238–246, 2003.

48. Lasdon, L. S., and A. D. Warren, *GRG2 User's Guide,* Department of General Business, University of Texas, Austin, 1989.

49. Lasdon, L. S., A. D. Warren, A. Jain, and M. Ratner, Design and testing of a generalized reduced gradient code for nonlinear programming, *Assoc. Comput. Mach. Trans. Math. Software,* v. 4, pp. 34–50, 1978.

50. Lee, S. L., and P. K. Kitanidis, Optimization estimation and scheduling in aquifer remediation with incomplete information, *Water Resources Research,* v. 27, no. 9, pp. 2203–2217, 1991.

51. Longley, G., and W. R. Jordan, Management of the Edwards Aquifer region: How the Endangered Species Act influenced action, *Water Resources Update,* Universities Council on Water Resources, Issue No. 106, Carbondale, IL, Winter 1996.

52. Maddock, T., III, Algebraic technological function for a simulation model, *Water Resources Research,* v. 8, pp. 129–134, 1972.

53. Maddock, T., III, Nonlinear technological functions for aquifers whose transmissivities vary with drawdown, *Water Resources Research,* v. 10, pp. 877–881, 1974.

54. Maddock, T., III, and Y. Y. Haimes, A tax system for groundwater management, *Water Resources Research,* v. 11, pp. 7–14, 1975.

55. Maknoon, R., and S. J. Burges, Conjunctive use of ground and surface water, *Jour. Amer. Water Works Assoc.,* v. 70, pp. 419–424, 1978.

56. Mann, J. F., Jr., Factors affecting the safe yield of ground-water basins, *Trans. Amer. Soc. Civil Engrs.,* v. 128, Pt. III, pp. 180–190, 1963.

57. Mann, J. F., Jr., Ground-water management in the Raymond Basin, California, *Engrng. Geol. Case Histories 7,* Geol. Soc. Amer., pp. 61–74, 1969.

58. Marryott, R. A., D. E. Dougherty, and R. L. Stollar, Optimal groundwater management, 2. Application of simulated annealing to a field-scale contamination site, *Water Resources Research,* v. 29, pp. 847–860, 1993.

59. Mays, L. W., *Optimal Control of Hydrosystems,* Marcel Dekker, New York, 1997.

60. Mays, L. W., and Y. K. Tung, *Hydrosystems Engineering and Management,* McGraw-Hill, New York, 1992.

61. McCleskey, G. W., Problems and benefits in ground-water management, *Ground Water,* v. 10, no. 2, pp. 2–5, 1972.

62. McGuire, V. L., and J. B. Sharp, *Water-Level Changes in the High Plains Aquifer—Predevelopment to 1995, U.S. Geological Survey Water Resources Investigations Report* 97-4081, 1997.

63. McKinney, D. C., and M. D. Lin, Genetic algorithm solution of groundwater management models, *Water Resources Research,* v. 30, no. 6, pp. 1897–1906, 1994.

64. Morel-Seytoux, H. J., and C. J. Daly, A discrete kernel generator for stream-aquifer studies, *Water Resources Research,* v. 11, pp. 253–260, 1975.

65. Morel-Seytoux, H. J., G. Peters, R. Young, and T. Illangasekare, Groundwater modeling for management, International Symposium on Water Resource Systems, Water Resour. Dev. and Training Cent., Univ. of Roorkee, Roorkee, India, 1980.

66. Peters, H. J., Groundwater management, *Water Resources Bull,* v. 8, pp. 188–197, 1972.

67. Pfannkuch, H. O., and B. A. Labno, Design and optimization of ground-water monitoring networks for pollution studies, *Ground Water,* v. 14, pp. 455–462, 1976.

68. Reichard, E. G., Groundwater-surface water management with stochastic surface water supplies: A simulation optimization approach, *Water Resources Research,* v. 31, no. 11, pp. 2845–2865, November 1995.

69. Reisner, M., *Cadillac Desert,* Viking Penguin, New York, 1986.

70. Remson, I., G. M. Hornberger, and F. J. Molz, *Numerical Methods in Subsurface Hydrology,* Wiley-Interscience, New York, 1971.

71. Remson, I., and S. M. Gorelick, Management models incorporating groundwater variables, in *Operation Research in Agriculture and Water Resources,* D. Yaron and C. S. Tapiero, eds., North-Holland, Amsterdam, 1980.

72. Renshaw, E. F., The management of ground water reservoirs, *Jour. Farm Economics,* v. 45, pp. 285–295, 1963.

73. Ritzel, B. J., J. W. Eheart, and S. Ranjithan, Using genetic algorithms to solve a containment problem, *Water Resources Research,* v. 30, pp. 1589–1603, 1994.

74. Rogers, L. L., and F. U. Dowla, Optimization of groundwater remediation using artificial neural networks with parallel solute transport modeling, *Water Resources Research,* v. 30, pp. 457–481, 1994.

75. Sasman, R. T., and R. J. Schicht, To mine or not to mine groundwater, *Jour. Amer. Water Works Assoc.,* v. 70, pp. 156–161, 1978.

76. Schmandt, J., E. T. Smerdon, and J. Clarkson, *State Water Policies, A Study of Six States,* Praeger, New York, 1988.

77. Shupe, S. J., and J. Folk-Williams, *The Upper Rio Grande: A Guide to Decision-Making,* Western Network, Santa Fe, NM, 1988.

78. Skaggs, R. L., L. W. Mays, and L. W. Vail, Simulated annealing with memory and directional search for ground water remediation design, *Jour. Amer. Water Resources Assn.,* v. 37, no. 4, pp. 853–866, August 2001.

79. Skaggs, R. L., L. W. Mays, and L. W. Vail, Application of enhanced annealing to ground water remediation design, *Jour. Amer. Water Resources Assn.,* v. 37, no. 4, pp. 867–875, August 2001.

80. Smerdon, E. T., and W. R. Jordan, eds., *Issues in groundwater management,* Center for Research in Water Resources, University of Texas at Austin, Austin,TX, 1985.

81. Smith, Z. A., ed., *Water and the Future of the Southwest,* University of New Mexico Press, Albuquerque, NM, 1989.

82. Taha, H. D., *Operations Research: An Introduction,* 7th ed., Macmillan Publishing, New York, 1987.

83. Tarlock, D., Water law, in *Water Resources Handbook,* L. W. Mays, ed., McGraw-Hill, New York, 1996.

84. Templer, O., and L.V. Urban, Conjunctive use of water on the Texas High Plains, *Water Resources Update,* Universities Council on Water

Resources, Issue No. 116, Integrated Water Management, pp. 102–108, Carbondale, IL, Winter 1997.

85. Texas Water Development Board, GWSIM—Groundwater simulation program, *Program Document and User's Manual,* UM 57405, Austin, TX, 1974.

86. Texas Water Development Board (TWDB), *Water for Texas,* Austin, TX, August 1997.

87. Texas Water Development Board, GWSIM II—Groundwater simulation program, *Program Document and User's Manual,* UM16, Austin, TX, 1978.

88. Thomas, H. E., *The conservation of ground water,* McGraw-Hill, New York, 327 pp., 1951.

89. U.S. Bureau of Reclamation, *Ground water manual,* U.S. Dept. Interior, 480 pp., 1977.

90. U.S. Department of Energy, Modeling evaluation of N-Springs barrier and pump-and-treat system, DOE/RL-94-132, Rev. 0, Richland, WA, 1995.

91. U.S. Geological Survey, *National Water Summary 1983—Hydrologic Events and Issues, U.S. Geological Survey Water-Supply Paper* 2250, 1984.

92. Wagner, B. J., Recent advances in simulation-optimization groundwater management modeling, U.S. National Report to IUGG, 1991–1994, *Rev. Geophys.,* v. 33, Suppl., 1995.

93. Wanakule, N., Optimal groundwater management models for the Barton Springs-Edwards Aquifer, Report EARDC R1-89, Edwards Aquifer Research and Data Center (EARDC), Southwest Texas State University, San Marcos, TX, March 1989.

94. Wanakule, N., L. W. Mays, and L. S. Lasdon, Development and testing of a model for determining optimal plumping and recharge of large-scale aquifers, Technical Report CRWR-217, Center for Research in Water Resources, The University of Texas, Austin, TX, October 1985.

95. Wanakule, N., L. W. Mays, and L. S. Lasdon, Optimal management of large-scale aquifers: Methodology and application, *Water Resources Research,* v. 22, no. 4, pp. 447–465, April 1986.

96. Waterstone, M., Of dogs and tails: Water policy and social policy in Arizona, *Water Resources Bull,* American Water Resources Association, v. 28, no. 3, pp. 479–486, 1992.

97. Wehmhoefer, R. A., Chapter 2 in *Water and the Future of the Southwest,* Z. A. Smith, ed., University of New Mexico Press, Albuquerque, NM, 1989.

98. Whiffen, G. J., and C. A. Shoemaker, Nonlinear weighted feedback control of groundwater remediation under uncertainty, *Water Resources Research,* v. 29, no. 9, pp. 3277–3289, 1993.

99. Willis, R., A unified approach to regional groundwater management, in *Groundwater Hydraulics, Water Resour. Monogr.* 9, J. S. Rosenshein and G. D. Bennett, eds., AGU, Washington, DC, 1984.

100. Willis, R., and P. Liu, Optimization model for ground-water planning, *Jour. Water Resour. Plan. Manag. Div.,* Amer. Soc. Civil Engrs., v. 110, pp. 333–347, 1984.

101. Willis, R., and B. A. Newman, Management model for groundwater development, *Jour. Water Resour. Plan. Manag. Div..* Amer. Soc. Civil Engrs., v. 13, pp. 159–171, 1977.

102. Willis, R., and W. W.-G. Yeh, *Groundwater Systems Planning and Management,* Prentice-Hall, Englewood Cliffs, NJ, 1987.

103. Winter, T. C., J. W. Harvey, O. L. Franke, and W. M. Alley, *Ground Water and Surface Water: A Single Source, U.S. Geological Survey Circular* 1139, Denver, CO, 1998.

104. Yeh, W.-G., Groundwater systems, in *Water Resources Handbook,* L. W. Mays, ed., McGraw-Hill, New York, 1996.

105. Young, R. A., Safe yield of aquifers: An economic reformulation, *Jour. Irrig. Drain. Div.,* Amer. Soc. Civil Engrs., v. 96, no. IR4, pp. 377–385, 1970.

106. Young, R. A., and Bredehoeft, J. D., Digital computer simulation for solving management problems of conjunctive groundwater and surface water systems, *Water Resources Research,* v. 8, pp. 533–556, 1972.

Chapter **11**

Surface Investigations of Groundwater

Although groundwater cannot be seen on the earth's surface, a variety of techniques can provide information concerning its occurrence and—under certain conditions—even its quality from surface or above-surface locations. Surface investigations of groundwater are seldom more than partially successful in that results usually leave the hydrogeologic picture incomplete; however, such methods are normally less costly than subsurface investigations. Geologic methods, involving interpretation of geologic data and field reconnaissance, represent an important first step in any groundwater investigation. Remote sensing from aircraft or satellite has become an increasingly valuable tool for understanding subsurface water conditions. Finally, geophysical techniques, especially electrical resistivity and seismic refraction methods, provide only indirect indications of groundwater so that underground hydrologic data must be inferred from surface data. Techniques have been developed for sensing buried wastes and waste migration.[3] Correct interpretation requires supplemental data from subsurface investigations (described in Chapter 12) to substantiate surface findings.

11.1 GEOLOGIC METHODS

Geologic studies enable large areas to be rapidly and economically appraised on a preliminary basis as to their potential for groundwater development. A geologic investigation begins with the collection, analysis, and hydrogeologic interpretation of existing topographic maps, aerial photographs, geologic maps and logs, and other pertinent records. This should be supplemented, when possible, by geologic field reconnaissance and by evaluation of available hydrologic data on streamflow and springs; well yields; groundwater recharge, discharge, and levels; and water quality. Such an approach should be regarded as a first step in any investigation of subsurface water because no expensive equipment is required; furthermore, information on geologic composition and structure defines the need for field exploration by other methods.

Knowledge of the depositional and erosional events in an area may indicate the extent and regularity of water-bearing formations. The type of rock formation will suggest the magnitude of water yield to be expected; one formation may be adequate for a domestic supply but entirely unsatisfactory for an industrial or municipal supply. Stratigraphy and geologic history of an area may reveal aquifers beneath unsuitable upper strata, the continuity and interconnection of aquifers, or important aquifer boundaries. The nature and thickness of overlying beds, as well as the dip of water-bearing formations, will enable estimates of drilling depths to be made. Similarly, confined aquifers may be noted and the possibility of flowing wells or low pumping lifts foretold. Landforms can often reveal near-surface unconsolidated formations serving as aquifers, such as glacial outwash, eskers, terraces, and sand dunes. Faults, which may form impermeable barriers to subsurface flow, frequently can be mapped from surface traces.

The fundamental relationships between geology and groundwater are presented elsewhere: geologic formations as aquifers in Chapter 2, the quality of groundwater as affected by geologic sources in Chapter 7, and geologic logs in Chapter 12.

11.2 REMOTE SENSING

Photographs of the earth taken from aircraft or satellite at various electromagnetic wavelength ranges can provide useful information regarding groundwater conditions.[5] The technology of remote sensing has developed rapidly in recent years, while its application to water resources is still being discovered. [20, 21, 27] Furthermore, the ready availability of photographs from commercial firms and government agencies has stimulated their use.

Stereoscopic examination of black-and-white aerial photographs has gained steadily in importance. Observable patterns, colors, and relief make it possible to distinguish differences in geology, soils, soil moisture, vegetation, and land use.[3] Thus, photogeology can differentiate between rock and soil types and indicate their permeability and areal distribution—and hence areas of groundwater recharge and discharge.[37] Maps classifying an area into good, fair, and poor groundwater yields can be prepared.[20] Table 11.2.1 summarizes the role aerial photographs can play as interpretive aids in groundwater studies.

Aerial photographs also reveal fracture patterns in rocks, which can be related to porosity, permeability, and ultimately well yield.[25, 39] Springs and marshy areas indicate relatively shallow depths to groundwater. Hydrobotanical studies of vegetation in photographs can be productive.[9] Phreatophytes, which transpire water from shallow water tables, define depths to groundwater; Figure 11.2.1 illustrates vegetation on an alluvial fan. Halophytes, plants with a high tolerance for soluble salts, and white efflorescences of salt at ground surface indicate the presence of shallow brackish or saline groundwater. Xerophytes, desert plants subsisting on minimal water, suggest a considerable depth to the water table.

Other nonvisible portions of the electromagnetic spectrum hold promise for a whole array of imaging techniques that can contribute to hydrogeologic surveys. *Infrared imagery*, which records differences in apparent surface temperatures, enables information on soil moisture, groundwater circulation, and faults functioning as aquicludes to be obtained.[36] Near-infrared imaging has outlined seepage patterns from canals.[1] One of the most interesting results of infrared aerial imaging has been mapping coastal submarine springs—both hot and cold—in regions of basalt or limestone.[14] Figure 11.2.2 shows swirls of groundwater, colder than seawater, emerging around the island of Hawaii.* *Radar imagery* can provide information on the presence of moisture on or at shallow depths below ground surface. Finally, low-frequency electromagnetic aerial surveys have outlined buried channels and zones of seawater intrusion.[32]

---

*The detection of submarine springs should prove most beneficial in the eastern Mediterranean area where such springs are common in karstic limestone. One such spring 2 km offshore of Chekka, Lebanon, discharges an estimated 17 m^3/sec at a depth of 43 m and produces potable water at the sea surface.

Table 11.2.1 Surficial Features Identified on Aerial Photographs that Aid in Evaluating Groundwater Conditions (after Heath and Trainer, *Introduction to Ground-Water Hydrology,* John Wiley, New York, 1968, and Mollard[27])

Topography
   Appraisal of regional relief setting
   Appraisal of local relief setting
Phreatophytes and aquatic plants
Geologic land forms likely to contain relatively permeable strata
   Modern alluvial terraces and floodplains
   Stratified valley-fill deposits in abandoned meltwater and spillway channels
   Glacial outwash and glacial deltas
   Kames and kame–moraine complexes
   Eskerine–kame complexes
   Alluvial fans
   Beach ridges
   Partly drift-filled valleys marked by a chain of elongate closed depressions
   Largely masked bedrock valleys cutting across modern valleys, indicated by local nonslumping of
      weak shale strata in valley sides
   Local drift-filled valleys in extensive bedrock-exposed terrains
   Sand dunes assumed to overlie sandy glaciofluvial sediments
Lakes and streams
   Drainage density of stream network
   Localized gain or loss of streamflow
   Nearby small perennial and intermittent lakes (e.g., lakes in outwash, elongate saline lakes in inactive
      drainage systems)
   Perennial rivers and larger creeks in valleys having inactive floodplains
   Small intermittent drainages (including misfit creeks in abandoned glacial spillways and meltwater
      channels)
   No defined drainage channel in former glacial spillways and meltwater channels
Moist depressions and seepages
   Moist depressions, marshy environments, and seepages (significance depends on interpretation of
      associated phenomena)
   String of alkali flats or lakes (playas, salinas) along inactive drainage systems
   Salt precipitates (e.g., salt crusts), localized anomalous-looking "burnout" patches in the soil, and
      vegetation associated with salt migration and accumulation
Springs (types tentatively inferred from aerial photographs)
   Depression springs (where land surface locally cuts the water table or the upper surface of the zone
      of saturation)
   Contact springs (permeable water-bearing strata overlying relatively impermeable strata—usually
      along the sides of valleys that cut across the interface between different strata)
   Artesian springs occurring on undulating upland till plains (permeable water-bearing bed between
      relatively impermeable confining beds, with enough head to discharge water at the ground surface)
   Artesian springs occurring on or near the base of hillsides, valley slopes, and local scarps
   Springs where the type could not be reasonably inferred from aerial photographs
Artificial water features
   Wells
   Developed springs
   Reservoirs
   Canals

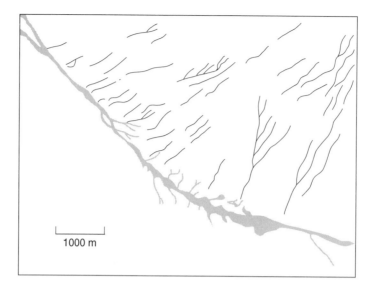

Figure 11.2.1. Tracing of an aerial photograph showing a strip of phreatophytes along the toe of an alluvial fan in a desert area. There should be a good supply of shallow groundwater along the upslope portion of the strip (after Mann[26]).

Figure 11.2.2. Infrared aerial photograph showing submarine springs emerging along the coast of Hilo Bay, Hawaii. Dark ocean areas contain cooler fresh water (courtesy U.S. Geological Survey).

11.3 GEOPHYSICAL EXPLORATION

Geophysical exploration is the scientific measurement of physical properties of the earth's crust for investigation of mineral deposits or geologic structure.[11, 18] With the discovery of oil by geophysical methods in 1926, economic pressures for locating petroleum and mineral deposits stimulated the development and improvement of many geophysical methods and equipment. Application to groundwater investigations was slow because the commercial value of oil overshadows that of water. In recent years, however, refinement of geophysical techniques—as well as an increasing recognition of the advantages of the methods for groundwater study—has changed the situation. Today, many organizations concerned with groundwater employ geophysical methods. The methods are frequently inexact or difficult to interpret, and they are most useful when supplemented by subsurface investigations.

Geophysical methods detect differences, or anomalies, of physical properties within the earth's crust. Density, magnetism, elasticity, and electrical resistivity are properties most commonly measured; these are described in the following sections. Experience and research have enabled pronounced differences in these properties to be interpreted in terms of geologic structure, rock type and porosity, water content, and water quality.

11.4 ELECTRICAL RESISTIVITY METHOD

The electrical resistivity of a rock formation limits the amount of current passing through the formation when an electrical potential is applied. It may be defined as the resistence in ohms between opposite faces of a unit cube of the material. If a material of resistance R has a cross-sectional area A and a length L, then its resistivity can be expressed as

$$\rho = \frac{RA}{L} \qquad (11.4.1)$$

Units of resistivity are ohm-m^2/m, or simply ohm-m.

Resistivities of rock formations vary over a wide range, depending on material, density, porosity, pore size and shape, water content and quality, and temperature.[23] There are no fixed limits for resistivities of various rocks; igneous and metamorphic rocks yield values in the range 10^2 to 10^8 ohm-m; sedimentary and unconsolidated rocks, 10^0 to 10^4 ohm-m. Figure 11.4.1 provides a representative guide to electrical resistivity ranges of various sediments and rocks. In relatively porous formations, the resistivity is controlled more by water content and quality within the formation than by the rock resistivity. For aquifers composed of unconsolidated materials, the resistivity decreases with the degree of saturation and the salinity of the groundwater. Clay minerals conduct electrical current through their matrix; therefore, clayey formations tend to display lower resistivities than do permeable alluvial aquifers.[*]

Actual resistivities are determined from apparent resistivities, which are computed from measurements of current and potential differences between pairs of electrodes placed in the ground surface. The procedure involves measuring a potential difference between two electrodes (*P* in Figure 11.4.2) resulting from an applied current through two other electrodes (*C* in Figure 11.4.2) outside but in line with the potential electrodes. If the resistivity is uniform in the subsurface zone beneath the electrodes, an orthogonal network of circular arcs will be formed by the current and equipotential lines, as shown in Figure 11.4.2. The measured

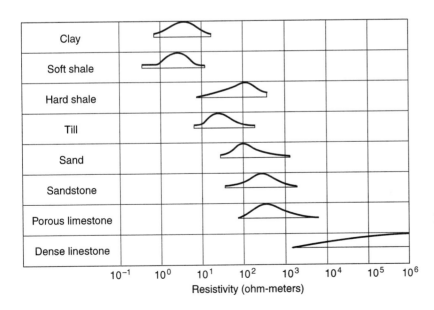

Figure 11.4.1. Representative ranges of electrical resistivity for various sediments and rocks. Values assume presence of fresh groundwater; saline water will shift values at least an order of magnitude to the left (after Amer. Soc. Civil Engrs.[2]).

---

[*]Clay and till when wet typically have low resistivities of 5–30 ohm-m, whereas wet sand and gravel have resistivities five to ten times higher; therefore, relatively high resistivity zones are of interest as shallow aquifers.

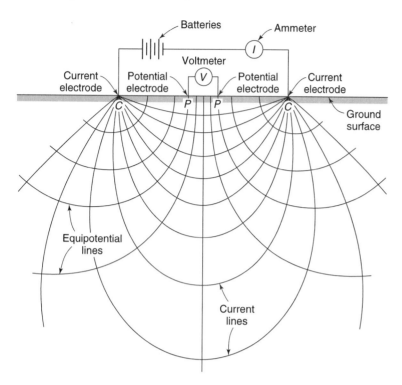

Figure 11.4.2. Electrical circuit for resistivity determination and electrical field for a homogeneous subsurface stratum.

potential difference is a weighted value over a subsurface region controlled by the shape of the network. Thus, the measured current and potential differences yield an apparent resistivity over an unspecified depth. If the spacing between electrodes is increased, a deeper penetration of the electric field occurs and a different apparent resistivity is obtained. In general, actual subsurface resistivities vary with depth; therefore, apparent resistivities will change as electrode spacings are increased, but not in a like manner. Because changes of resistivity at great depths have only a slight effect on the apparent resistivity compared to those at shallow depths, the method is seldom effective for determining actual resistivities below a few hundred meters.

Electrodes consist of metal stakes driven into the ground.* In practice, various standard electrode spacing arrangements have been adopted; most common are the Wenner and Schlumberger arrangements.

The Wenner[43] arrangement, shown in Figure 11.4.3a, has the potential electrodes located at the one-third points between the current electrodes. The apparent resistivity is given by the ratio of voltage to current times a spacing factor. For the Wenner arrangement, the apparent resistivity

$$\rho_a = 2\pi a \frac{V}{I} \tag{11.4.2}$$

where a is the distance between adjacent electrodes, V is the voltage difference between the potential electrodes, and I is the applied current.

The Schlumberger arrangement, shown in Figure 11.4.3b, has the potential electrodes close together. The apparent resistivity is given by

$$\rho_a = \pi \frac{(L/2)^2 - (b/2)^2}{b} \frac{V}{I} \tag{11.4.3}$$

---

*For the potential electrodes, porous cups filled with a saturated solution of copper sulfate are sometimes employed to inhibit electric fields from forming around them.

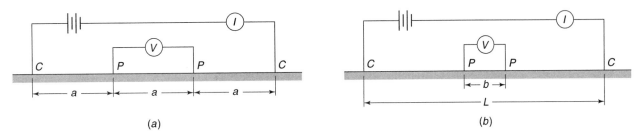

Figure 11.4.3. Common electrode arrangements for resistivity determination. (*a*) Wenner[43] (*b*) Schlumberger[10]

where L and b are the current and potential electrode spacings, respectively. Theoretically, $L \gg b$, but for practical application good results can be obtained if $L \geq 5b$.[10]

When apparent resistivity is plotted against electrode spacing (a for Wenner, and $L/2$ for Schlumberger) for various spacings at one location, a smooth curve can be drawn through the points. The interpretation of such a resistivity-spacing curve in terms of subsurface conditions is a complex and frequently difficult problem. The solution can be obtained in two parts: (1) interpretation in terms of various layers of actual (as distinguished from apparent) resistivities and their depths, and (2) interpretation of the actual resistivities in terms of subsurface geologic and groundwater conditions. Part 1 can be accomplished with theoretically computed resistivity-spacing curves of two-, three-, and four-layer cases for various ratios of resistivities.[5] Curves and explanations of curve-matching techniques have been published for the Wenner configuration[31, 42] and the Schlumberger configuration.[10, 35] Part 2 depends on supplemental data. Comparing actual resistivity variations with depth to data from a nearby logged test hole enables a correlation to be established with subsurface geologic and groundwater conditions. This information can then be applied for interpretation of resistivity measurements in surrounding areas.

Figure 11.4.4 illustrates the interpretation of a two-layer situation from measurements with the Schlumberger electrode spacings. The field curve, plotted on logarithmic transparent paper to the same scale as published master curves, is superposed on the two-layer master set. By keeping the coordinate axes parallel, the sheet is moved until a best fit of the field and theoretical curves is obtained. The abscissa of the cross, which is the origin of the theoretical curve, equals the thickness of the first layer, while the ordinate of the cross defines the actual resistivity ρ_1 of the first layer. The asymptote of the end of the curve with the largest spacing defines the actual resistivity ρ_2 of the second layer. Physically, such a curve might represent a clay layer overlying a sandy aquifer at a depth of 14 m.

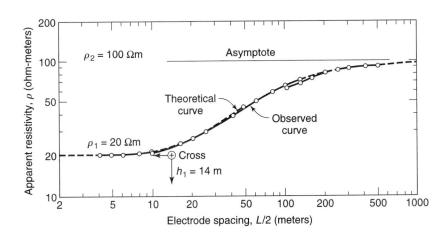

Figure 11.4.4. Interpretation of a two-layer electrical resistivity measurement from Schlumberger electrode spacings (after Zohdy et al.[44]).

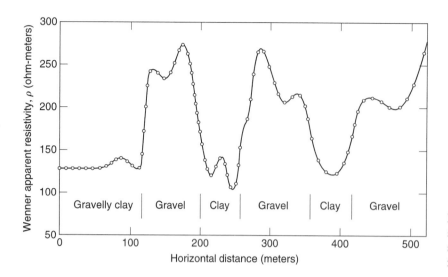

Figure 11.4.5. Horizontal profile by surface resistivity measurements over a shallow gravel deposit in California and its interpretation (after Zohdy et al.[44]).

Resistivity surveys can cover vertical variations (soundings) at selected locations by varying electrode spacings.* More generally, they are conducted to obtain horizontal profiles of apparent resistivity or apparent-resistivity maps of an area by adopting a constant electrode spacing. Figure 11.4.5 shows a horizontal resistivity profile across a shallow gravel deposit together with its geologic interpretation. Areal resistivity changes can be interpreted in terms of aquifer limits and changes in groundwater quality, whereas sounding surveys may indicate aquifers, water tables, salinities, impermeable formations, and bedrock depths.

Any factors that disturb the electric field in the vicinity of electrodes may invalidate the resistivity measurements. These may include lateral geologic inhomogeneities; in addition, buried pipelines, cables, and wire fences are common hazards. Advances to these methods have been reported by Radstake et al.[36] and Goyal et al.[14]

Of all surface geophysical methods, electric resistivity has been applied most widely for groundwater investigations. Its portable equipment and ease of operation facilitate rapid measurement. The method frequently aids in planning efficient and economic test drilling programs.[33] It is especially well adapted for locating subsurface saltwater boundaries because the decrease in resistance when salt water is encountered becomes apparent on a resistivity-spacing curve. Where subsurface conditions are relatively homogeneous, the technique can be employed to detect the water table as the top of a relatively conductive layer. In California, locations of highly permeable zones for groundwater recharge were aided by resistivity measurements.[33] The method has also been employed for delineating geothermal areas and estimating aquifer permeability.[17, 23]

An important new application of resistivity surveys involves defining areas and magnitudes of polluted groundwater. Results correlate best with groundwater samples where a highly conductive pollutant, such as soluble salt, is moving in a relatively shallow zone with uniform geologic conditions.[7] Studies of pollution from landfills, wastewater disposal, industrial wastes, and acid mine drainage have demonstrated the feasibility of the technique.

---

*It is often assumed that a given electrode spacing represents the depth of resistivity measurement. Although this rule of thumb is untrue, the greater the current electrode separation, the greater the amount of current that penetrates a given depth.[46]

11.5 SEISMIC REFRACTION METHOD

The *seismic refraction method* involves the creation of a small shock at the earth's surface, either by the impact of a heavy instrument or by a small explosive charge, and measuring the time required for the resulting sound, or shock, wave to travel known distances. Seismic waves follow the same laws of propagation as light rays and may be reflected or refracted at any interface where a velocity change occurs. Seismic reflection methods provide information on geologic structure thousands of meters below the surface, whereas seismic refraction methods—of interest in groundwater studies—go only about 100 meters deep.[34] The travel time of a seismic wave depends on the media through which it is passing; velocities are greatest in solid igneous rocks and least in unconsolidated materials.

Characteristic seismic velocities for a variety of geologic materials are shown in Figure 11.5.1; these can be employed to identify the nature of alluvium or bedrock. In coarse alluvial materials, seismic velocity increases markedly from unsaturated to saturated zones; consequently, the depth to water table can be mapped, often to an accuracy of 10 percent, where

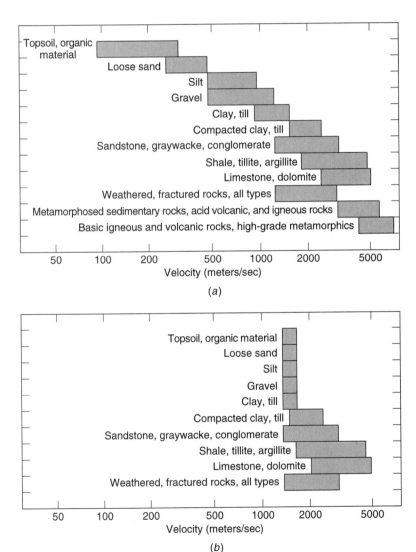

Figure 11.5.1. Seismic velocity of geologic materials. (*a*) Unsaturated materials (*b*) Saturated materials (after Amer. Soc. Civil Engrs.[2])

geologic conditions are relatively uniform. Changes in seismic velocities are governed by changes in elastic properties of the formations. The greater the contrast of these properties, the more clearly the formations and their boundaries can be identified. In sedimentary rocks, the texture and geologic history are more important than the mineral composition. Porosity tends to decrease wave velocity, but water content increases it.

For consolidated formations with a uniform distribution of small pores, such as a sandstone, velocity and porosity can be related by[19, 41]

$$\frac{1}{v} = \frac{\alpha}{v_L} + \frac{1-\alpha}{v_S} \tag{11.5.1}$$

where v is the measured velocity, v_L is the velocity in the liquid saturating the rock,* v_S is the velocity of the solid rock matrix, and α is porosity of the rock.

A spherical wave expands outward from a shock point, as shown in Figure 11.5.2a. It travels at a speed governed by the material through which it is passing. Assume, for example, a homogeneous unconsolidated material with a water table; when the wave reaches the water table it will travel along the interface. As it travels, a series of waves is propagated back into the unsaturated layer. Positions of the wave front drawn at intervals of a few milliseconds in Figure 11.5.2a illustrate this refraction. At any location on the surface, the first wave will arrive either directly from the shot point or from a refracted path. By measuring the time interval of the first arrival at varying distances from the shot point, a time-distance graph can be plotted. For the horizontal two-layer case, the depth H to the water table can be computed from the velocities v_1 and v_2 and the distance s to the intersection on the graph, as shown in Figure 11.5.2b. The equation is

$$H = \frac{s}{2}\sqrt{\frac{v_2 - v_1}{v_2 + v_1}} \tag{11.5.2}$$

s is the distance from the shockpoint to the point at which the direct wave and the refracted wave arrive simultaneously.

Alternatively, the depth to the lower layer, H, can be determined using

$$H = \frac{T_i}{2}\frac{v_1 v_2}{\sqrt{v_2^2 - v_1^2}} \tag{11.5.3}$$

where T_i is the intercept time, determined by projecting the second-line segment (for $v_2 = 2,000$ m) backward to the time axis.

Multilayered problems can be solved in a similar manner, often aided by nomographs. Different surface elevations, sloping formations, faults, and changes in the interfacial configuration require special analysis.[34, 44] Computational procedures are described in textbooks of geophysics.[11, 16, 18]

For a three-layered seismic refraction case with $v_1 < v_2 < v_3$, the thickness of the first layer, H_1, is computed using Equations 11.5.2 or 11.5.3, and the thickness of the second layer, H_2, is computed using

$$H_2 = \frac{1}{2}\left(T_{i_2} - 2H_1\frac{\sqrt{v_3^2 - v_1^2}}{v_3 v_1}\right)\left(\frac{v_2 v_3}{\sqrt{v_3^2 - v_2^2}}\right) \tag{11.5.4}$$

The field procedure for seismic refraction investigations has been simplified with the help of compact and efficient instruments. A small charge of dynamite is placed in a hand-augered

---

*Seismic velocity in water under typical groundwater conditions approximates 1,460 m/s.

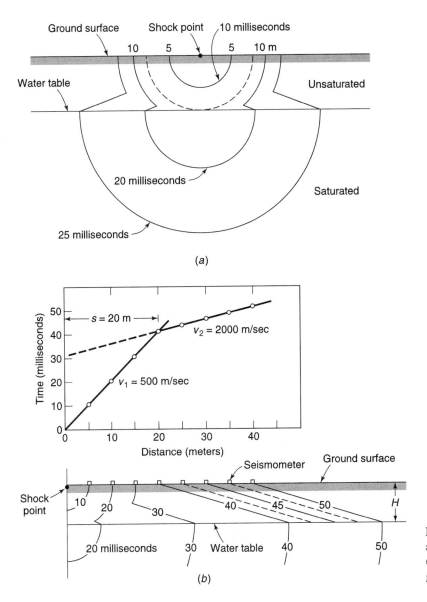

Figure 11.5.2. Seismic refraction method applied to determine depth to water table. (*a*) Wave front advance (*b*) Time–distance graph

hole about one meter deep, and the hole is backfilled. Seismometers, also known as *geophones*, *detectors*, or *pickups*, are spaced in a line from the shock point 3 to 15 m apart. They receive the shock wave and convert the vibration into electric impulses. An electric circuit connects the seismometers to an amplifier and a recording oscillograph, which automatically records the instant of firing and the various first arrivals of the shock wave. Depth determinations to 60 to 100 m are typical with this equipment, although satisfactory work to depths of 300 m has been accomplished. For investigations of depths less than about 20 m, a sledgehammer blow on the ground surface can produce a recordable shock wave.

Interpretation of seismic refraction data assumes homogeneous layers bounded by interfacial planes. Where no distinct boundary exists, but rather a gradual transition zone, a curve replaces the break in slope on the time–distance graph. Fortunately, water tables approximate planes, so that many of the problems imposed by irregular configurations of geologic structure are avoided.[12] Efficient application of the method requires skill in proper interpretation in terms of rock materials, depths, and irregularities. Other knowledge of subsurface conditions aids in proper analysis of field records. The actual presence of groundwater is difficult to determine without supplemental information because velocities overlap in saturated and unsaturated zones. Seismic velocities must increase with depth in order to obtain satisfactory results; as a result, a dense layer overlying an unconsolidated aquifer can mask the presence of the aquifer.

The seismic refraction method when applied in applicable areas, can rapidly and economically eliminate areas unfavorable for test drilling. It is not readily adapted to small areas. Minimum distances of a few hundred meters are needed for seismic profiles in different directions. Local noise or vibrations from sources such as highways, airports, and construction sites interfere with seismic work.

The seismic method, because it requires special equipment and trained technicians for operation and interpretation, has been applied to only a relatively limited extent for groundwater investigations. It is commonly employed to map cross sections of alluvial valleys so that variations in thickness of unconfined aquifers can be determined. The successful work of Linehan and Keith[26] in locating groundwater supplies in New England provides notable examples of the applicability of the seismic method. Lateral seismic velocity variations in unsaturated sediments can represent lithology differences that correlate with water well yields.

11.6 GRAVITY AND MAGNETIC METHODS

The *gravity method* measures differences in density on the earth's surface that may indicate geologic structure. Because the method is expensive and because differences in water content in subsurface strata seldom involve measurable differences in specific gravity at the surface, the method has little application to groundwater prospecting. Under special geologic conditions, such as a large buried valley, the gross configuration of an aquifer can be detected from gravity variations.

The *magnetic method* enables magnetic fields of the earth to be mapped. Indirect information pertinent to groundwater studies, such as dikes that form aquifer boundaries or limits of a basaltic flow, has been obtained using the method.[44]

PROBLEMS

11.2.1 Perform an Internet search to determine the latest advances made in remote sensing.

11.4.1 Refer to the references by Radstake et al.[36] and Goyal et al.[14] to determine what advances have been made in the area of electrical resistivity methods.

11.4.2 Perform an Internet search to determine what advances have been made in electrical resistivity methods.

11.5.1 Perform an Internet search to determine what advances have been made in seismic refraction methods.

11.5.2 For a three-layer seismic problem with the first arrival times measured in both the forward and reverse directions, explain how to determine the seismic velocities for each layer to find the thickness of the layers and the critical angle between layer 1 and layer 2.

11.5.3 Given the first arrival time results for a two-layer seismic refraction problem, explain how to determine v_1, v_2, the angle of refraction, and H.

11.5.4 Refraction seismology has been very successful in the delineation of bedrock beneath a landfill. Study "A Refraction Statics Method for Mapping Bedrock" and write a summary of the methodology.[40]

11.6.1 Gravity surveys have been used to delineate faults in buried bedrock valleys. Study "The Successful Use of Microgravity Profiling to Delineate Faults in Buried Bedrock Valleys" and write a summary of the methodology.[1]

11.6.2 Gravity measurements have been used to determine the specific yield of an alluvial aquifer. Study "Measurements of Aquifer-Storage Change and Specific Yield Using Gravity Surveys" and write a summary of the methodology.[35]

REFERENCES

1. Allen, D. M., and F. A. Michel, The successful use of microgravity profiling to delineate faults in buried bedrock valleys, *Ground Water,* v. 34, no. 6, pp. 1132–1140, 1996.

2. American Soc. Civil Engrs., *Ground Water Management,* Manual Engrng. Practice 40, New York, 216 pp., 1972.

3. Avery, T. E., *Interpretation of Aerial Photographs,* Burgess Publishing, Minneapolis, 392 pp., 1977.

4. Benson, R., R. A. Glaccum, and M. R. Noel, *Geophysical Techniques for Sensing Buried Wastes and Waste Migration,* National Water Well Association, Dublin, OH, 1988.

5. Bhattacharya, P. K., and H. P. Patra, *Direct Current Geoelectric Sounding—Principles and Interpretation,* Elsevier, Amsterdam, 135 pp., 1968.

6. Bowden, L. W., and E. L. Pruit, eds., *Manual of Remote Sensing, Vol. II, Interpretation and Applications,* Amer. Soc. Photogrammetry, Falls Church, VA, pp. 869–2144, 1975.

7. Bugg, S. F., and J. W. Lloyd, A study of fresh water lens configuration on the Cayman Islands using resistivity methods, *Quarterly Jour. Engrng. Geol.,* v. 9, pp. 291–302, 1976.

8. Chadwick, D. G., and L. Jensen, *The detection of magnetic fields caused by groundwater and the correlation of such fields with water dowsing,* Utah Water Research Lab., Logan, 57 pp., Jan. 1971.

9. Chikishev, A. G., ed., *Plant Indicators of Soils, Rocks, and Subsurface Waters,* Consultants Bureau, New York, 210 pp., 1965.

10. Compagnie Générale de Géophysique, *Master Curves for Electrical Sounding,* 2nd rev. ed., European Assoc. Exploration Geophysicists, The Hague, Netherlands, 49 pp., 1963.

11. Dobrin, M. B., *Introduction to Geophysical Prospecting,* 3rd ed., McGraw-Hill, New York, 630 pp., 1976.

12. Duguid, J. O., Refraction determination of water table depth and alluvium thickness, *Geophysics,* v. 33, pp. 481–488, 1968.

13. Fischer, W. A., et al., *Fresh-Water Springs of Hawaii from Infrared Images,* U.S. Geological Survey Hydrologic Atlas 218, 1966.

14. Goyal, V. C., Sri Niwas, and P. K. Gupta, Theoretical evaluation of modified Weiner array for shallow resistivity exploration, *Ground Water,* v. 29, no. 4, pp. 582–586, 1991.

15. Grant, F. S., and G. F. West, *Interpretation Theory in Applied Geophysics,* McGraw-Hill, New York, 583 pp., 1965.

16. Griffiths, D. H., and R. F. King, *Applied Geophysics for Engineers and Geologists,* Pergamon, Oxford, 223 pp., 1965.

17. Hatherton, T., et al., Geophysical methods in geothermal prospecting in New Zealand, *Bull. Volcanol.,* v. 29, pp. 485–498, 1966.

18. Hearst, J. R., P. H. Nelson, and F. L. Paillet, *Well Logging for Physical Properties,* 2nd ed., John Wiley & Sons, New York, 2000.

19. Howe, R. H., et al., Application of air photo interpretation in the location of ground water, *Jour. Amer. Water Works Assoc.,* vol. 48, pp. 1380–1390, 1956.

20. Idso, S. B., et al., Detection of soil moisture by remote surveillance, *Amer. Sci.,* vol. 63, pp. 549–557, 1975.

21. Inland Waters Branch, *Instrumentation and observation techniques,* Proc. Hydrology Symposium no. 7, Dept. of Energy, Mines and Resources, Ottawa, 343 pp., 1969.

22. Keller, G. V., and F. C. Frischknecht, *Electrical Methods in Geophysical Prospecting,* Pergamon, Oxford, 517 pp., 1966.

23. Kelly, W. E., Geoelectric sounding for estimating aquifer hydraulic conductivity, *Ground Water,* v. 15, pp. 420–425, 1977.

24. Lattman, L. H., and Parizek, R. R., Relationships between fracture traces and the occurrence of groundwater in carbonate rocks, *Jour. Hydrology,* v. 2, pp. 73–91, 1964.

25. Linehan, D., and S. Keith, Seismic reconnaissance for ground-water development, *Jour. New England Water Works Assoc.,* v. 63, pp. 76–95, 1949.

26. Mann, J. F., Jr., Estimating quantity and quality of ground water in dry regions using airphotos, *Intl. Assoc. Sci. Hydrology Publ.* 44, pp. 125–134, 1958.

27. Mollard, J. D., The role of photo-interpretation in finding groundwater sources in Western Canada, Proc. 2nd Seminar on Air Photo Interpretation in the Development of Canada, The Queen's Printer, Ottawa, pp. 57–75, 1968.

28. Mooney, H. M., and W. W. Wetzel, *The Potentials About a Point Electrode and Apparent Resistivity Curves for a Two-, Three-, and Four-Layered Earth,* Univ. Minnesota Press, Minneapolis, 146 pp. + set of curves, 1956.

29. Morley, L. W., ed., *Mining and groundwater geophysics/1967,* Econ. Geol. Rept. 26, Geological Survey of Canada, Ottawa, 722 pp., 1970.

30. Morris, D. B., The application of resistivity methods to ground-water exploration of alluvial basins in semi-arid areas, *Jour. Instn. Water Engrs.,* v. 18, pp. 59–65, 1964.

31. Musgrave, A. W., ed., *Seismic Refraction Prospecting,* Soc. Explor. Geophysists, Tulsa, 604 pp., 1967.

32. Orellana, E., and H. M. Mooney, *Master Tables and Curves for Vertical Electrical Sounding over Layered Structures,* Interciencia, Madrid, 150 pp., 66 tables, 1966.

33. Page, L. M., Use of the electrical resistivity method for investigating geologic and hydrologic conditions in Santa Clara County, California, *Ground Water,* v. 6, no. 5, pp. 31–40, 1968.

34. Pluhowski, E. J., *Hydrologic Interpretations Based on Infrared Imagery of Long Island, New York,* U.S. Geological Survey Water-Supply Paper 2009-B, 20 pp., 1972.

35. Pool, D. R., and J. H. Eychaner, Measurements of aquifer-storage change and specific yield using gravity surveys, *Ground Water,* v. 33, no. 3, pp. 425–432, 1995.

36. Radstake, R., et al., Applications of forward modeling resistivity profiles, *Ground Water,* v. 20, no. 1, pp. 13–17, 1991.

37. Ray, R. G., *Aerial Photographs in Geologic Interpretation and Mapping,* U.S. Geological Survey Professional Paper 373, 227 pp., 1960.

38. Setzer, J., Hydrologic significance of tectonic fractures detectable on airphotos, *Ground Water,* v. 4, no. 4, pp. 23–27, 1966.

39. Stollar, R., and P. Roux, Earth resistivity surveys—A method for defining ground-water contamination, *Ground Water,* v. 13, no. 2, pp. 145–150, 1975.

40. Taucher, P., and B. N. Fuller, A refraction statistics method for mapping bedrock, *Ground Water,* v. 32, no. 6, pp. 895–904, 1994.

41. Wyllie, M. R. J., *The Fundamentals of Well Log Interpretation,* 3rd ed., Academic Press, New York, 238 pp., 1963.

42. Van Nostrand, R. G., and K. L. Cook, *Interpretation of Resistivity Data, U.S. Geological Survey Professional Paper* 499, 310 pp., 1966.

43. Wenner, F., A method of measuring earth-resistivity, *Bull. Bureau Standards,* v. 12, Washington, DC, pp. 469–478, 1916.

44. Zohdy, A. A., G. P. Eaton, and D. R. Mabey, *Application of Surface Geophysics to Ground-Water Investigations, U.S. Geological Survey Techniques of Water-Resources Investigations,* Chap. D1, Bk. 2, 116 pp., 1974.

Chapter 12

Subsurface Investigations of Groundwater

Detailed and comprehensive study of groundwater and conditions under which it occurs can be made only by subsurface investigations. Whether the information needed concerns an aquifer (its location, thickness, composition, permeability, and yield) or groundwater (its location, movement, and quality), quantitative data can be obtained from subsurface examinations. It should be emphasized that all work classed as subsurface investigations is conducted entirely by personnel on the surface who operate equipment extending underground. Test drilling furnishes information on substrata in a vertical line from the surface. Geophysical logging techniques provide information on physical properties of geologic formations, water quality, and well construction. Books dealing with the topic of subsurface investigations include those by Bassiouni[1], Heath et al.,[14] Kearey and Brooks,[18] Keys,[19, 20] Vogelsang,[44] and Ward.[46]

12.1 TEST DRILLING

Test drilling of small-diameter holes to ascertain geologic and groundwater conditions is useful in verifying other means of investigation and to obtain assurance of underground conditions prior to well drilling. Many times, if a test hole proves fruitful, it is redrilled or reamed to a larger diameter to form a pumping well. Test holes also serve as observation wells for measuring water levels or for conducting pumping tests.

Almost any well-drilling method can be employed for test drilling; however, in unconsolidated formations, cable tool and hydraulic rotary methods are most common (see Chapter 5). The former is slower but provides more accurate samples from the bailer; the latter is faster, but it is sometimes difficult to determine the exact character of the formations. This fact is particularly true where fine-grained materials are encountered because these mix with the drilling fluid. For great depths and fairly uniform sands, the hydraulic rotary method is quicker and cheaper. Accurate samples can be obtained by pulling the drill stem and using a sampler at the bottom of the hole, or if intact cores are desired in exploratory drilling, a hollow-stem cutter head can be affixed to cut cylindrical cores. For test holes in soft ground and shallow depths, drilling with an auger is quick and economic. Jetting has proved to be an economic method of drilling shallow, small-diameter holes for investigational purposes. The rapidity of the jetting operation combined with its lightweight portable equipment gives it important advantages, but the lack of good samples is a disadvantage. The choice of method for test drilling depends on what information is necessary, the type of material encountered, drilling depth, and location.

12.1.1 Geologic Log

A geologic log is constructed from sampling and examination of well cuttings collected at frequent intervals during the drilling of a well or test hole. Such logs furnish a description of the geologic character and thickness of each stratum encountered as a function of depth, thereby enabling aquifers to be delineated. Figure 12.1.1 shows a geologic log for a well in unconsolidated alluvium.

Considering all types of logs, the geologic log is probably the most important, but preparation of a good geologic log can be difficult. One problem is that well cuttings are small and mixed with mud. Particularly in rotary drilling, the drilling mud masks the presence of material in the silt and clay ranges. Another problem stems from the fact that most geologic logs are prepared by well drillers, who are busy with many activities during drilling operations and sometimes lack formal geologic training; therefore, logs often tend to be prepared in a perfunctory manner. But experienced drillers, who recognize the value of geologic logs, make diligent efforts to prepare them carefully and completely.

It is good practice to store samples of well cuttings systematically. These not only permit detailed geologic logs to be prepared but also enable grain size analyses and correlations with other nearby wells to be made after drilling is finished. Figure 12.1.2 shows well cuttings collected by a bucket sampler during drilling; they have been arranged in order on the ground, bagged, and labeled for storage.

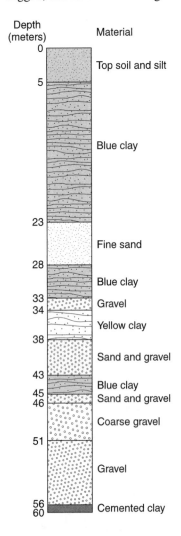

Figure 12.1.1. Geologic log of a well prepared from well cuttings.

Figure 12.1.2. Well cuttings, taken at regular intervals from a rotary-drilled well in Joufrah, Libya, being tagged and labeled for subsequent storage in compartmented boxes (photo by David K. Todd).

12.1.2 Drilling-Time Log

A drilling-time log is a useful supplement to test drilling. It consists of an accurate record of the time, in minutes and seconds, required to drill each unit depth of the hole.[23] The technique is most practical with hydraulic rotary drilling although it is applicable to other methods as well. Because the texture of a stratum being penetrated largely governs the drilling rate, a drilling-time log may be readily interpreted in terms of formation types and depths. A portion of a drilling time log obtained by the hydraulic rotary method is shown in Figure 12.1.3 together with the log of the test hole based on cuttings.

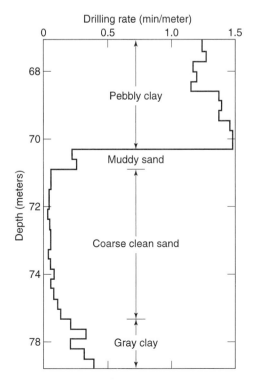

Figure 12.1.3 Drilling-time log and strata penetrated (after Kirby[23]; courtesy Johnson Div., UOP Inc.).

12.2 WATER LEVEL MEASUREMENT

One of the most common measurements in groundwater investigations is the determination of the depth to groundwater. In both existing and new wells, such data are needed to define groundwater flow directions, changes in water levels over time, and effects of pumping tests.

A simple and accurate method for obtaining water depth is lowering a steel tape into a well. By adding chalk to the end of the tape, the length of submersion becomes apparent, thus giving the distance from the top of the well to the water surface.

For repeated measurements and for depths exceeding 50 m, an *electric water-level sounder* is preferable. A sounder consists of a battery, a voltmeter, a calibrated two-wire cable on a reel, and an electrode. When the electrode contacts water, the circuit is completed and the voltmeter shows a deflection. The depth is read directly from graduations along the cable.

Another widely employed technique is the *air-line method.*[11] A small-diameter tube is placed in the annular space between the pump column and the casing. Commonly, the tube is fastened to the pump column so that the two are installed simultaneously. The tube extends below the water surface and is connected to a tire pump or small air compressor with a pressure gauge. Air is pumped into the tube until a maximum pressure is observed; this pressure, converted to depth of water, indicates the distance from the lower end of the tube to the water surface. Although the air-line method is less accurate than the above methods, it is especially applicable in pumped wells where water splash and turbulence may invalidate other techniques.

A unique and convenient method for measuring water levels in deep wells is the *rock technique* developed by Stewart.[41] He determined empirically the time required for a common 1.55-cm glass marble or a standard BB (air rifle shot) to fall to the water surface plus the time for the sound of the splash to return to ground surface. Measuring the elapsed time by stopwatch, the depth to water can be read directly from Table 12.2.1. For water depths exceeding 57 m, the sphere reaches a constant terminal velocity; therefore, for depths greater than those listed in Table 12.2.1, the equation

$$d = 27.3\,t - 47.6 \tag{12.2.1}$$

Table 12.2.1 Depth to Water Surface in a Well as a Function of the Time Interval of a Falling Sphere (after Stewart[41])

Time (s)	Distance (m)	Time (s)	Distance (m)	Time (s)	Distance (m)
0.0	0.0	1.4	9.6	2.8	33.3
0.1	0.0	1.5	11.0	2.9	35.4
0.2	0.2	1.6	12.4	3.0	37.6
0.3	0.4	1.7	13.9	3.1	39.8
0.4	0.8	1.8	15.3	3.2	42.1
0.5	1.2	1.9	16.8	3.3	44.4
0.6	1.8	2.0	18.4	3.4	46.9
0.7	2.4	2.1	20.0	3.5	49.2
0.8	3.1	2.2	21.7	3.6	51.6
0.9	4.0	2.3	23.5	3.7	54.1
1.0	4.9	2.4	25.4	3.8	56.7
1.1	5.9	2.5	27.3	3.9	59.1
1.2	7.1	2.6	29.3	4.0	61.6
1.3	8.3	2.7	31.3	4.1	64.3

can be employed where *d* is the depth to water in meters and *t* is the time interval in seconds. The method is accurate to within 1.5 m.[*]

Automatic water level recorders or *pressure transducers* are installed in observation wells where short-term fluctuations are of interest, such as near intermittently operating wells or for pumping tests. A typical recorder consists of a float and counterweight, a gear linkage that rotates a chart drum, and a recording pen driven across the chart by a clock mechanism.

Where multiple aquifers exist with differing water levels, individual observation wells screened in only one aquifer are often drilled. Alternatively, individual small piezometer tubes extending down to the levels of the various aquifers are placed inside a large single perforated casing. The casing is backfilled with sand and sealed by grout between adjoining aquifers.

12.3 GEOPHYSICAL LOGGING

Geophysical logging involves lowering sensing devices in a borehole and recording a physical parameter that may be interpreted in terms of formation characteristics; groundwater quantity, quality, and movement; or physical structure of the borehole.[21, 31] A wide variety of logging techniques are available; Table 12.3.1 lists the types of information that can be obtained from various logging techniques described in this chapter.

Table 12.3.1 Summary of Logging Applications to Groundwater Hydrology (after Keys and MacCrary[21])

Required information	Possible logging techniques
Lithology and stratigraphic correlation of aquifers and associated rocks	Resistivity, sonic, or caliper logs made in open holes; radiation logs made in open or cased holes
Total porosity or bulk density	Calibrated sonic logs in open holes; calibrated neutron or gamma-gamma logs in open or cased holes
Effective porosity or true resistivity	Calibrated long-normal resistivity logs
Clay or shale content	Natural gamma logs
Permeability	Under some conditions, long-normal resistivity logs
Secondary permeability—fractures, solution openings	Caliper, sonic, or television logs
Specific yield of unconfined aquifers	Calibrated neutron logs
Grain size	Possible relation to formation factor derived from resistivity logs
Location of water level or saturated zones	Resistivity, temperature, or fluid conductivity logs; neutron or gamma-gamma logs in open or cased holes
Moisture content	Calibrated neutron logs
Infiltration	Time-interval neutron logs
Dispersion, dilution, and movement of waste	Fluid conductivity or temperature logs; natural gamma logs for some radioactive wastes
Source and movement of water in a well	Fluid velocity or temperature logs
Chemical and physical characteristics of water, including salinity, temperature, density and viscosity	Calibrated fluid conductivity or temperature logs; resistivity logs
Construction of existing wells, diameter and position of casing, perforations, screens	Gamma-gamma, caliper, casing, or television logs
Guide to screen setting	All logs providing data on the lithology, water-bearing characteristics, and correlation and thickness of aquifers
Cementing	Caliper, temperature, or gamma-gamma logs; acoustic logs for cement bond
Casing corrosion	Under some conditions, caliper, casing, or television logs
Casing leaks and/or plugged screen	Fluid velocity logs

[*]Table 12.2.1 applies only to the two specified spheres; ordinary pebbles give erratic results because of their irregular shapes. Stewart[41] also pointed out that BBs make a sharp "ping" sound, while marbles cause a short "blurred" sound. Furthermore, he noted, BBs are deflected by spiderwebs, but marbles are not.

Geophysical logs furnish continuous records of subsurface conditions that can be correlated from one well to another.[16] They serve as valuable supplements to geologic logs. Data from geophysical logs can be digitized, stored on magnetic tape, or transmitted by radio or telephone for interpretation. Graphic displays of log data permit rapid visual interpretations and comparisons in the field so decisions regarding completion and testing of wells can be made immediately.

The application of geophysical logging to groundwater hydrology lags far behind its comparable use in petroleum exploration. It is doubtful if more than a few percent of the new water wells drilled each year are logged by geophysical equipment. The primary reason for this is cost. Most water wells are shallow, small-diameter holes for domestic water supply; logging costs would be relatively large and usually unnecessary. But for deeper and more expensive wells, such as for municipal, irrigation, or injection purposes, logging can be economically justified in terms of improved well construction and performance.

Another deterrent to geophysical logging is the lack of experience among drillers, engineers, and geologists in the interpretation of logs. As logging techniques become more sophisticated, the data they produce become more complex. The interpretation of many logs is more of an art than a science; log responses are governed by numerous environmental factors, making quantitative analysis difficult. In general, best results are obtained with experience and with supplemental hydrogeologic information.

In the following sections, the geophysical logging techniques that are most important in groundwater hydrology are described. Emphasis is on concepts and applications. Figure 12.3.1 is a schematic diagram showing several of the logs and their typical relative responses in various unconsolidated and consolidated geologic formations.

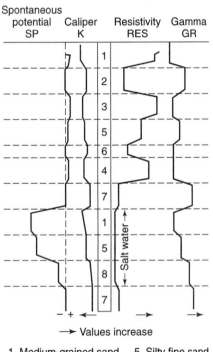

→ Values increase

1 Medium-grained sand   5 Silty fine sand
2 Boulder clay (till)   6 Brown coal
3 Coarse sand   7 Clay
4 Fine sand   8 Clayey silt

(a)

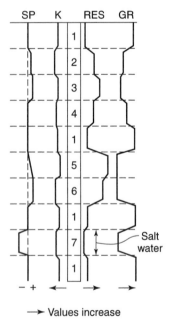

→ Values increase

1 Claystone   5 Sandstone
2 Marly limestone   6 Sandstone
3 Limestone   7 Sandstone
4 Marlstone   6 but with
                  salt water

(b)

Figure 12.3.1. Schematic diagram of various geophysical logs showing their relative responses in (*a*) unconsolidated rocks and (*b*) consolidated rocks (after Brown et al.,[6] reproduced by permission of UNESCO).

12.4 RESISTIVITY LOGGING

Within an uncased well, current and potential electrodes can be lowered to measure electric resistivities of the surrounding media and to obtain a trace of their variation with depth. The result is a resistivity (or electric) log. Such a log is affected by fluid within a well, by well diameter, by the character of surrounding strata, and by groundwater.

Of several possible methods for measuring underground resistivities, the multielectrode method is most commonly employed, because it minimizes effects of the drilling fluid and well diameter and also makes possible a direct comparison of several recorded resistivity curves.[25] Four electrodes, two for emitting current and two for potential measurement, constitute the system. Recorded curves are termed *normal* or *lateral,* depending on the electrode arrangement, as shown in Figure 12.4.1. In the normal arrangement, the effective spacing is considered to be the distance *AM* (Figure 12.4.1) and the recorded curve is designated *AM.* Sometimes a long normal curve (*AM'*) is recorded based on the same electrode arrangement as the normal but with a larger *AM* distance (Figure 12.4.1). The spacing for lateral (*AO*) curves is taken as the distance *AO*, measured between *A* and a point midway between the electrodes *M* and *N* (Figure 12.4.1*c*). Boundaries of formations having different resistivities are located most readily with a short electrode spacing, whereas information on fluids in thick permeable formations can be best obtained with long spacings.

An electric log of a well usually consists of vertical traverses that record the short and long normals, the lateral, and the spontaneous potential curves (see following section). An illustration of an electric log is given in Figure 12.4.2. Accurate interpretation of resistivity logs is difficult, requires careful analysis, and is best done by specialists.[18, 26, 47]

Resistivity curves indicate the lithology of rock strata penetrated by the well and enable fresh and salt waters to be distinguished in the surrounding material.[22, 24] In old wells, exact

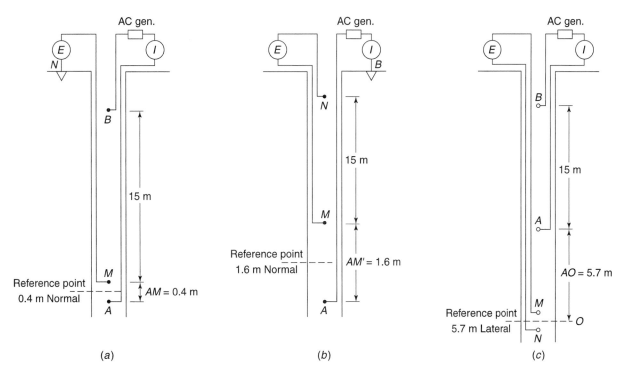

Figure 12.4.1 Typical electrode arrangements and standardized distances for resistivity logs. (*a*) Short normal (*b*) Long normal (*c*) Lateral (after Keys and MacCrary[21])

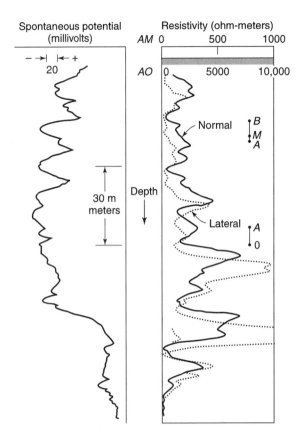

Figure 12.4.2. Spontaneous potential and resistivity logs of a well (courtesy Schlumberger Well Surveying Corp.).

locations of casings can be determined. Resistivity logs may be used to determine specific resistivities of strata, or they may indicate qualitatively changes of importance. As mentioned in the previous chapter, resistivity of an unconsolidated aquifer is controlled primarily by porosity, packing, water resistivity, degree of saturation, and temperature. Although specific resistivity values cannot be stated for different aquifers, on a relative basis, shale, clay, and salt-water sand give low values; freshwater sand moderate to high values; and cemented sandstone and nonporous limestone high values (see Figure 11.4.1). Casings and metallic objects will indicate very low resistivities. Correlation of rock samples, taken from wells during drilling, with resistivity curves furnishes a sound basis for interpretation of curves measured in nearby wells without available samples.

Resistivity of groundwater depends on ionic concentration and ionic mobility of the salt solution. This mobility is related to the molecular weight and electrical charge, so that differences exist for various compounds. For example, the ion mobility of a sodium chloride solution is several times that of a comparable calcium carbonate solution. Relationships between resistivity and total dissolved solids for several salt solutions and natural groundwaters are shown in Figure 12.4.3.

As the temperature of groundwater increases, it has a greater ionic mobility, associated with a decrease in viscosity. Hence, an inverse relation exists between resistivity and temperature. The relation, expressed as a correction factor, is shown in Figure 12.4.4. Resistivity at the measurement temperature when multiplied by the correction factor for that temperature yields the resistivity at the standard temperature of 25°C.

One of the most common uses of an electric log is to determine the proper place to set well screens. A log provides a basis for selecting proper lengths of screens and for setting them

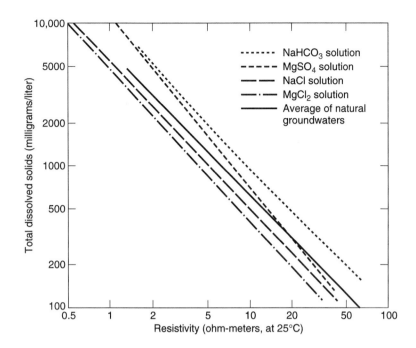

Figure 12.4.3. Resistivity–concentration curves for various salt solutions and natural groundwaters (after *Agric. Handbook* 60, U.S. Dept. of Agriculture).

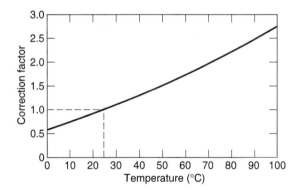

Figure 12.4.4. Correction factor to convert resistivities at other temperatures to resistivities at 25°C (after Jones and Buford[17]).

opposite the best formations. Because of this application, many well drillers have their own "loggers" for this purpose.

Investigations of Louisiana aquifers by Jones and Buford[17] and later by Turcan[42] have extended the applicability of resistivity logs to the estimation of groundwater quality. First, a *field-formation factor F* for an aquifer is determined from previous data by

$$F = \frac{\rho_0}{\rho_w} \tag{12.4.1}$$

where ρ_0 is the resistivity of the saturated aquifer and ρ_w is the resistivity of the groundwater in the aquifer.[*] Second, specific conductance is related to chloride content or total dissolved

---

[*]It should be noted that the resistivity of groundwater is the reciprocal of its specific conductance (see Chapter 7). The relation has the form

$$\rho_w = 10^4/E_c$$

where ρ_w is in ohm-m and E_c is specific conductance in μS/cm.

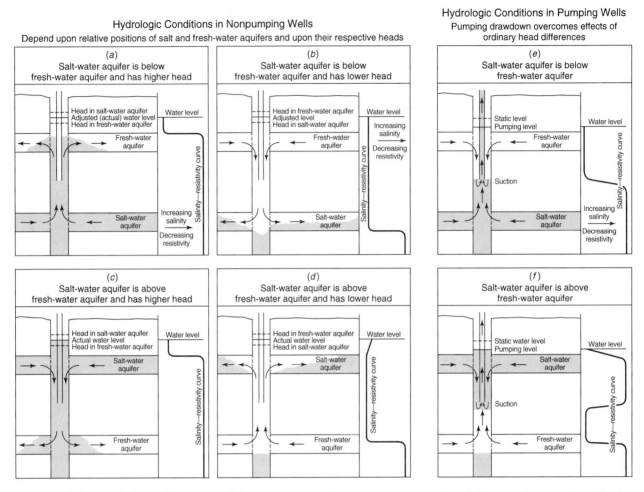

Figure 12.4.5. Hydrologic conditions and resistivity curves for wells penetrating two aquifers of different salinities (after Poland and Morrison[36]).

solids for the aquifer, as in Figure 12.4.3. Finally, with these relationships known, ρ_0 is read from the long-normal resistivity curve in an aquifer; this enables ρ_w and then the salinity of the groundwater to be calculated. The method yields best results in uniform clastic aquifers such as sand and sandstone, consisting mainly of intergranular pores saturated with water.

Another application of long-normal curves has been suggested by Croft[8] to estimate permeability. A value of F is determined as above; then from a previously established relationship, permeability is calculated directly from F. Again, the method should be limited to clastic rock formations.

Resistivity logs can also aid in identifying wells that intersect both fresh and saline zones. Circulation within such a well under non-pumping conditions depends on the relative hydrostatic heads, water densities, aquifer locations and thicknesses, and the physical structure and condition of the well. Various hydrologic conditions for pumping and nonpumping wells are shown diagrammatically in Figure 12.4.5 together with corresponding resistivity curves. Resistivity logs are also employed for locating aquifers, determining bed sequences, correlating aquifers, and estimating changes in groundwater quality.

12.5 SPONTANEOUS POTENTIAL LOGGING

The *spontaneous potential method* measures natural electrical potentials found within the earth.* Measurements, usually in millivolts, are obtained from a recording potentiometer connected to two like electrodes. One electrode is lowered in an uncased well and the other is connected to the ground surface, as illustrated by electrodes *M* and *N* in Figure 12.4.1. The potentials are primarily produced by electrochemical cells formed by the electrical conductivity differences of drilling mud and groundwater where boundaries of permeable zones intersect a borehole. In some instances, electrokinetic effects of fluids moving through permeable formations are also responsible for spontaneous potentials. Therefore, potential logs indicate permeable zones but not in absolute terms; they can also aid in determining casing lengths[10] and in estimating total dissolved solids in groundwater.[45] Where no sharp contrasts occur in permeable zones, as often happens in shallow alluvial formations, potential logs lack relief and contribute little. In urban and industrial areas, spurious earth currents may occur, such as from electric railroads, which interfere with potential logging.

Potential values range from zero to several hundred millivolts. By convention, potential logs are read in terms of positive and negative deflections from an arbitrary baseline, usually associated with an impermeable formation of considerable thickness. The sign of the potential depends on the ratio of the salinity (or resistivity) of the drilling mud to the formation water.[9]

Spontaneous potentials resulting from electrochemical potentials can be expressed by

$$SP = -(64.3 + 0.239T)\log\frac{\rho_f}{\rho_w} \tag{12.5.1}$$

where ρ_f is the drilling fluid resistivity in ohm-m, ρ_w is the groundwater resistivity in ohm-m, and T is the borehole temperature in °C. Therefore, for measured SP, ρ_f, and T values, the resistivity and hence salinity of groundwater can be determined.[21, 45] It should be noted, however, that the formula applies only where the groundwater is very saline, NaCl is the predominant salt, and the drilling mud contains no unusual additives.

In practice, potential and resistivity logs are usually recorded together as shown in Figure 12.4.2. The two logs often indicate the same subsurface conditions and thereby supplement each other; however, occasionally the two types of logs will furnish information not available directly from either alone.

12.6 RADIATION LOGGING

Radiation logging, also known as *nuclear* or *radioactive logging*, involves the measurement of fundamental particles emitted from unstable radioactive isotopes. Logs having application to groundwater are natural gamma, gamma-gamma, and neutron. These are promising but not widely used hydrogeologic tools. An important advantage of these logs over most others is that they may be recorded in either cased or open holes that are filled with any fluid.

12.6.1 Natural-Gamma Logging

Because all rocks emit natural-gamma radiation, a record of this constitutes a natural-gamma log. The radiation originates from unstable isotopes of potassium, uranium, and thorium. In general, the natural-gamma activity of clayey formations is significantly higher than that of

*Potentials are also referred to as *self-potentials* or simply *SP*.

quartz sands and carbonate rocks. The most important application to groundwater hydrology is identification of lithology, particularly clayey or shale-bearing sediments, which possess the highest gamma intensity.[28, 34] Because most of the gamma rays detected originate within 15–30 cm of the borehole wall, logs run before and after well development can reveal zones where clay and fine-grained material were removed. Borehole dimensions and fluid, casing, and gravel pack all exert minor influences on gamma probe measurements.

Figure 12.6.1 shows the natural-gamma log of a test hole in unconsolidated sediments together with its geologic interpretation.

12.6.2 Gamma-Gamma Logging

Gamma radiation originating from a source probe and recorded after it is backscattered and attenuated within the borehole and surrounding formation constitutes a gamma-gamma log. The source probe generally contains cobalt-60 or cesium-137, which is shielded from a sodium iodide detector built into the probe. Primary applications of gamma-gamma logs are for identification of lithology and measurement of bulk density and porosity of rocks. The porosity α can be determined by

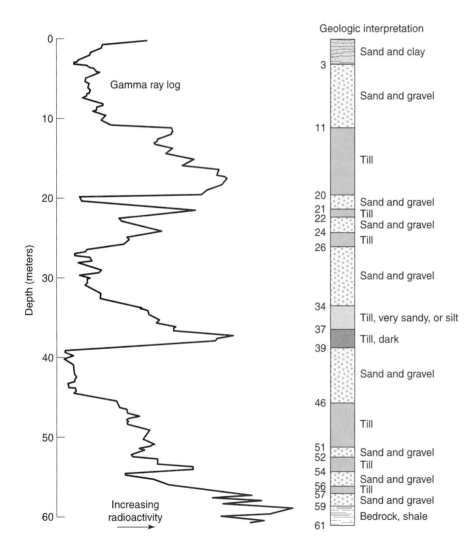

Figure 12.6.1. Natural-gamma log of a test hole in Moraine City, Ohio, together with its geologic interpretation (after *USGS Water-Supply Paper* 1808).

$$\alpha = \frac{\rho_G - \rho_B}{\rho_G - \rho_F} \qquad (12.6.1)$$

where ρ_G is grain density (obtained from cuttings or cores), ρ_B is bulk density (measured from a calibrated log), and ρ_F is the fluid density. Also, within the same geologic formation it should be possible to derive specific yield from the difference in bulk density measured above and below a water table. Finally, gamma-gamma logs can assist in locating casing, collars, grout, and zones of hole enlargement. As with natural-gamma logs, borehole and fluid conditions affect readings.

12.6.3 Neutron Logging

Neutron logging is accomplished by a neutron source and detector arranged in a single probe, which produces a record related to the hydrogen content of the borehole environment. In most formations the hydrogen content is directly proportional to the interstitial water; therefore, neutron logs can measure moisture content above the water table and porosity below the water table.[33] Neutrons have a relative mass of 1 and no electric charge; therefore, the loss of energy when passing through matter is by elastic collisions. Neutrons are slowed most effectively by collisions with hydrogen because the nucleus of a hydrogen atom has approximately the same mass as a neutron.[*] Several designs of neutron probes are currently available, utilizing sources of beryllium combined with radium-226, plutonium-239, or americium-241. Probes for measurement of soil moisture (moisture meters) are compact and designed to fit snugly in a small-diameter access tube for accurate quantitative results. For porosity determination in large-diameter holes, larger probes are employed. By measuring moisture contents above and below the water table, the specific yield of unconfined aquifers can be determined.[27] The lateral penetration of neutron logs is in the range 0.2–0.6 m. Neutron log results are influenced by hole size; therefore, in large uncased holes information on hole diameter is required for proper interpretation.

Figure 12.6.2 shows a neutron log of a shallow well in unconsolidated alluvium together with the geologic log. This log is calibrated in moisture content as a percentage of bulk volume. Note that the capillary fringe is apparent above the water table as well as high porosity clay layers in the saturated zone.

12.7 TEMPERATURE LOGGING

A vertical traverse measurement of groundwater temperature in a well can be readily obtained with a recording resistance thermometer. Such data can be of value in analyzing subsurface conditions.[3, 38] Ordinarily, temperatures will increase with depth in accordance with the geothermal gradient, amounting to roughly 3°C for each 100 m in depth. Departures from this normal gradient may provide information on circulation or geologic conditions in the well.[4, 40] Abnormally cold temperatures may indicate the presence of gas or, in deep wells, may suggest recharge from ground surface. Likewise, abnormally warm water may occur from water of deep-seated origin. Temperatures may indicate waters from different aquifers intersected by a well. In a few instances, temperature logs have aided the location of the approximate top of new concrete behind a casing,[2] because the heat generated during setting produces a marked temperature increase of the water within the casing.

---

[*]Dynamically, the slowing of a neutron by a hydrogen atom is analogous to a golf ball losing energy upon collision with another golf ball but rebounding elastically with little energy loss from a large mass such as a concrete wall.

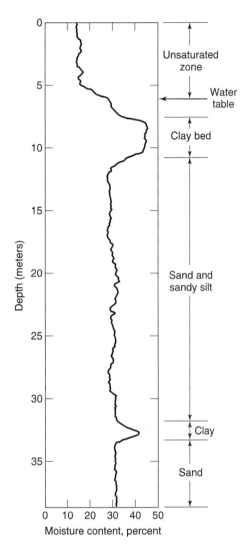

Figure 12.6.2. Neutron log of a shallow well in unconsolidated alluvium near Garden City, Kansas, together with its geologic interpretation (courtesy U.S. Geological Survey).

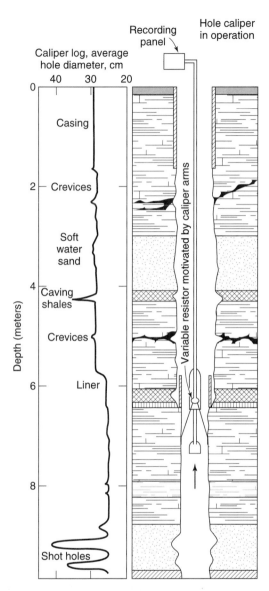

Figure 12.8.1. Hole caliper and corresponding caliper log (after Bays and Folk[2]).

12.8 CALIPER LOGGING

A caliper log provides a record of the average diameter of a borehole. Caliper tools are designed either with arms hinged at the upper end and pressed against the hole wall by springs or with bow springs fastened at both ends. These logs aid in the identification of lithology and stratigraphic correlation, in the location of fractures and other rock openings,[30] and in correcting other logs for hole-diameter effects. During well construction, caliper logs indicate the size of casing that can be fitted into the hole and enable the annular volume for gravel packing to be calculated. Other applications include measuring casing diameters in old wells and locating swelling and caving zones. A hole caliper and the resulting log are shown in Figure 12.8.1.

12.9 FLUID-CONDUCTIVITY LOGGING

A continuous record of the conductivity of fluid in a borehole is a *fluid-conductivity log*. The probe measures the AC-voltage drop across two closely spaced electrodes and is governed by the resistivity of the fluid between the electrodes. Fluid resistivity is generally measured in ohm-m; its reciprocal, conductivity, is measured in μS/cm (see Equation 12.4.2). Use of the term *fluid-conductivity log* avoids confusion with a resistivity log, which measures rock and fluid conditions outside a borehole. Temperature logs should be made in conjunction with fluid-conductivity logs so that values can be corrected to a standard temperature.

Fluid-conductivity logs enable saline water zones to be located, furnish information on fluid flow within a well, and provide a means to extrapolate water-sample data from a well. Figure 12.9.1 illustrates how a fluid-conductivity log can define the location and transition zone of saline water underlying fresh water within a well.

12.10 FLUID-VELOCITY LOGGING

Measurement of fluid movement within a borehole constitutes a fluid-velocity log. Such data reveal strata contributing water to a well, flow from one stratum to another within a well, hydraulic differences between aquifers intersected by a well, and casing leaks. Several flowmeter designs have been developed for boreholes; it is important that they be compact and sensitive to small water movements and directions.[32] Figure 12.10.1 illustrates a fluid-velocity log for a pumping well.

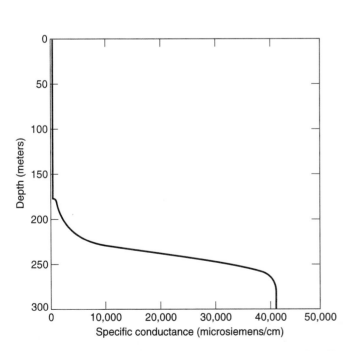

Figure 12.9.1. Fluid conductivity log in basalt at Honolulu, Hawaii, showing fresh water overlying saline water and separated by a transition zone (courtesy Honolulu Board of Water Supply).

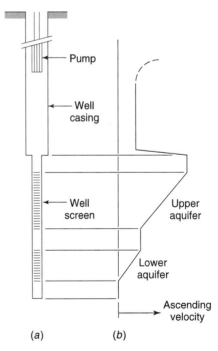

Figure 12.10.1. Example of a fluid-velocity log for a well tapping two confined aquifers. (*a*) Well (*b*) Fluid-velocity log

12.11 MISCELLANEOUS LOGGING TECHNIQUES

12.11.1 Television Logging

A convenient tool with increasing use is a television camera lowered in a well. Specially designed wide-angle cameras, typically less than 7 cm in diameter, are equipped with lights and provide continuous visual inspection of a borehole; with videotape a record of the interior can be preserved. Among the variety of applications are locating changes in geologic strata, pinpointing large pore spaces, inspecting the condition of the well casing and screen, checking for debris in wells, locating zones of sand entrance, and searching for lost drilling tools.[7, 12] Photographs taken within a well at close intervals, termed a photolog,[24] can be employed for the same purposes. As an example, photographs of cased and uncased portions of a well in Honolulu appear in Figure 12.11.1.

12.11.2 Acoustic Logging

Acoustic, or *sonic, logging* measures the velocity of sound through the rock surrounding an uncased, fluid-filled hole. Sound velocity in rock is governed by the velocity of the rock matrix and the fluid filling the pore space (see Chapter 11); therefore, the greater the porosity, the closer the measured sound velocity approaches that of the fluid. Chief applications of the acoustic log include determining the depth and thickness of porous zones,[35] estimating porosity, identifying fracture zones, and determining the bonding of cement between the casing and the formation.

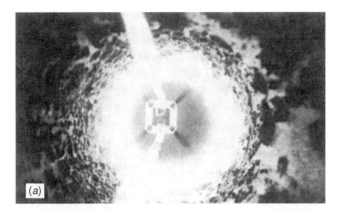

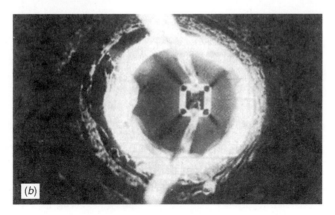

Figure 12.11.1. Photographs inside a water supply well in Honolulu, Hawaii. (*a*) Tuberculated condition within a wrought iron casing 83 years old (*b*) Fractured and cavernous lava formation below the casing (courtesy Honolulu Board of Water Supply)

12.11.3 Casing Logging

A casing-collar locator is a useful device for recording locations of casing collars, perforations, and screens.[21] The instrument consists of a magnet wrapped with a coil of wire; voltage fluctuations caused by changes in the mass of metal cutting the lines of flux from the magnet are recorded to form the log.

12.12 OTHER SUBSURFACE METHODS

Besides the array of logging techniques described in this chapter, it should be recognized that other subsurface methods can yield important information about hydrogeologic conditions. These have been described in earlier chapters and by way of summary are listed here:

1. Tracer tests for groundwater flow (Chapter 3)
2. Groundwater level measurements for flow directions and aquifer conditions (Chapter 3)
3. Pumping tests of wells for aquifer characteristics (Chapter 4)
4. Groundwater level fluctuation measurements for aquifer characteristics (Chapter 6)
5. Groundwater samples for water quality determination (Chapter 7)

12.13 CASE STUDY *Oasis Valley, Nevada*

This case study is reported by Robledo et al.[37] In 1997, the U.S. Geological Survey installed a network of wells in the Oasis Valley near Beatty, Nye County, Nevada, to develop a long-term groundwater-monitoring network in the Oasis Valley. The Oasis Valley is of concern to the U.S. Department of Energy because it is down gradient from the Pahute Mesa, one of the sites of intensive underground nuclear testing at the Nevada Test Site. The objective of the monitoring network is to define spatial and temporal changes in the chemical and isotopic character of the groundwater. Seven sites up gradient from the spring-discharge area in Oasis Valley were selected for installation of single or multiple wells. Twelve monitoring wells (in 11 boreholes) were installed, ranging in depth from 65 to 642 ft. Figure 12.13.1 show the typical construction of single- and multiple-well boreholes in the Oasis Valley.

The boreholes were geophysically logged to define the lithology, stratigraphy, and geohydrologic character of the rock units during drilling. The typical suite of logs for each borehole included a caliper log, natural-gamma log, conductivity log, resistivity log, resistance log, and spontaneous-potential log. Some of the boreholes also have short- and long-normal resistivity logs, lateral logs, acoustic velocity logs, fluid temperature logs, fluid resistivity logs, fluid specific conductance logs, acoustic televiewer logs, and heat-pulse flowmeter logs.

A borehole video log was run in well ER-OV-06a. Table 12.13.1 is a summary of the inflow and outflow zones and relative transmissivities in the borehole. Figure 12.13.2 shows flow-producing fractures or bedding planes in six water-bearing zones

that were identified in this borehole. Figure 12.13.3 shows the geologic and geophysical logs for the borehole. The six water-bearing zones in Figure 12.13.2 were identified with the aid of acoustic televiewer, temperature, caliper, and heat-pulse flowmeter logs (Hess[19]). The large scatter of the flowmeter log data in the interval where the flow was the greatest made it difficult to interpret the flowmeter log. Most of the data scatter of the flowmeter log was caused by the rugged borehole, as shown by the caliper log. This did not allow a good seal between aquifer zones. The temperature log is nearly isothermal in the interval between 210 and 340 ft, supporting the interpretation that the greatest ambient flow (both inflow and outflow) is in this interval. The flow is fast enough that it does not cool much as it travels up the borehole.

The differences in inflow rates of the water production zones are proportional to transmissivity and may therefore be used to determine relative transmissivities of the zones (Paillet[29]). Amount of flow entering from a bed or zone is proportional to the product of the transmissivity of the zone and the change in hydraulic head across the borehole wall. The relative transmissivity for a given zone can be determined using the following equation that describes the change in flow, ΔQ, for a given fracture zone,

$$\Delta Q = \Delta h \times T$$

where Δh is the change in gradient from the fracture zone to the borehole between ambient and injection conditions and T is the

(continues)

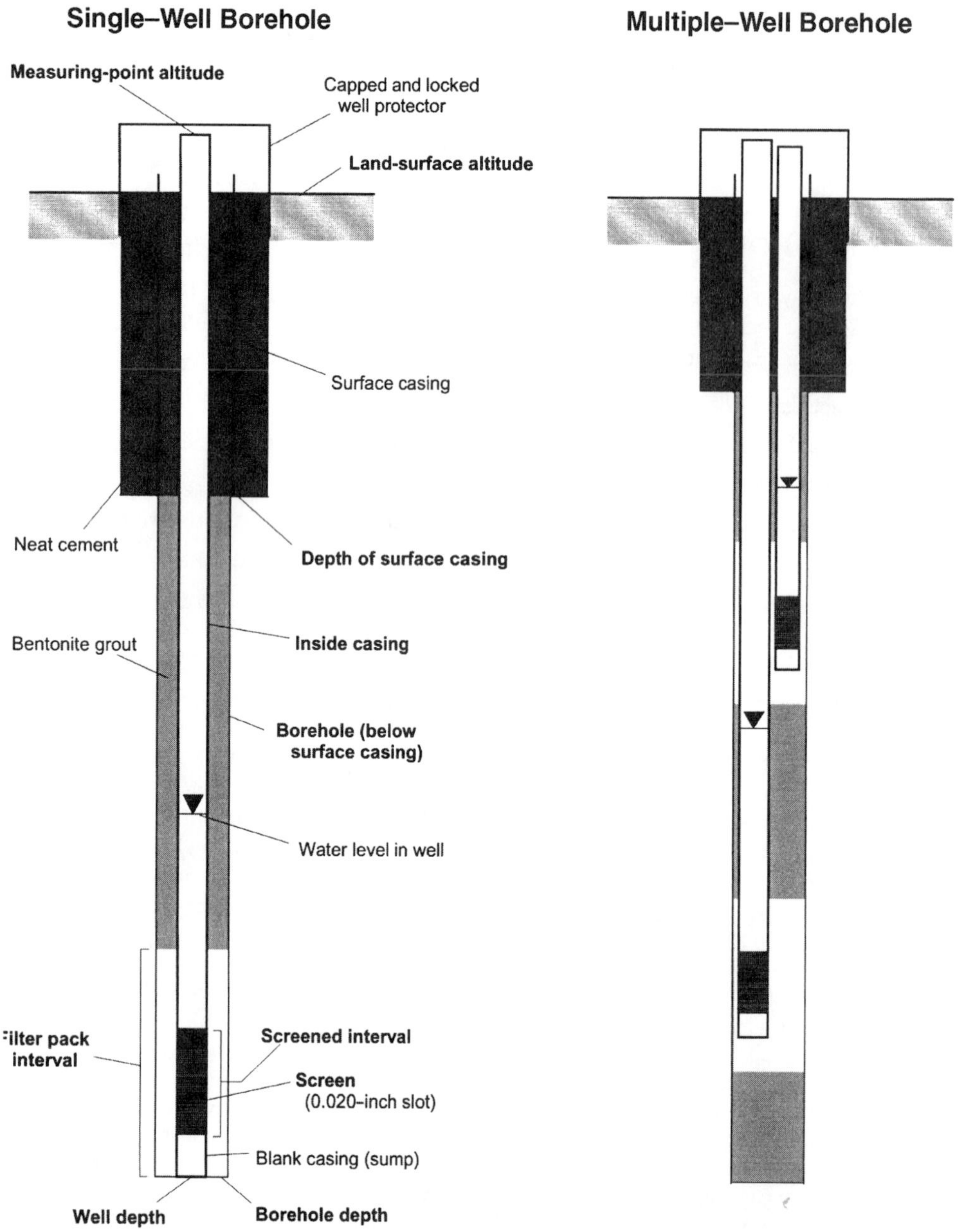

Single–Well Borehole

Multiple–Well Borehole

Figure 12.13.1. Typical construction of single- and multiple-well boreholes in Oasis Valley, Nevada (Robledo et al.[37]).

Table 12.13.1 Summary of Inflow and Outflow Zones and Relative Transmissivities in Borehole ER-OV-06a, Oasis Valley, Nevada

Zone	Depth of inflow or outflow zone (feet below land surface)	Probable inflow or outflow point (feet below land surface)	Inflow or outflow (gallons per minute)			Percent difference between ambient and injection inflow or outflow[2]	Relative transmissivity
			Ambient[1]	Injection[1]	Difference		
A	43.0–48.5	45	–0.10	–0.10	0.00	<1	Low
B	62.5–99.0	75	–.30	–.32	.02	5.1	Low
C	195.0–213.0	205	–.20	–.37	.17	43.6	High
D	325.5–346.0	340	.50	.34	.16	41.0	High
E	397.0–417.0	400	.06	.03	.03	7.7	Low
F	483.0–536.0	515	.04	.03	.01	2.6	Low
	Total		0.00	–.39	.39	100	

[1]By convention, inflow (water flowing into well) is shown as positive number and outflow is shown as negative number.

[2]Percent difference for zone is calculated using the following equation: $100 \times$ [(ambient inflow) – (injection inflow)]/(sum of differences between ambient and injection inflow for all zones). Precision implied by percent differences for zone exceeds intended accuracy.

Source: Robledo et al.[37]

12.13 CASE STUDY   *Oasis Valley, Nevada (continued)*

transmissivity of the fracture zone. Assuming that the change in head difference or gradient between the borehole and the aquifer at each zone is the same, then the change in flow is directly proportional to the difference in transmissivity of the fracture zones.

This difference between the inflow under two head conditions can be used to determine the relative transmissivity. Another good example case study is presented by Brendle[5] for the San Luis Valley, Colorado.

PROBLEMS

12.3.1 Refer to ASTM Reference D5753-95—Standard Guide for Planning and Conducting Borehole Geophysical Logging and write a summary of the guidelines.

12.13.1 Develop a detailed summary of the Oasis Valley, Nevada, case study. Refer to the report by Robledo et al.[37] for additional details. Your instructor may identify additional points to emphasize.

12.13.2 Discuss the methodologies and results of the geophysical logging procedures used in the *U. S. Geological Water Resources Investigation Report* 02-4058 by Brendle.[5] Your instructor may identify additional points to emphasize.

12.13.3 Perform a search of the ASTM standards and develop a list of the various standards that apply to subsurface investigations of groundwater. Briefly define how the standard applies.

12.13.4 Perform an Internet search to determine the most recent advances that have been made in logging techniques.

12.13.5 Develop a categorized list of the required information on the properties of rocks, fluid, wells, and groundwater systems. Then identify the available logging techniques that might be utilized for each. A starting point might be to refer to Keys and MacCary.[21]

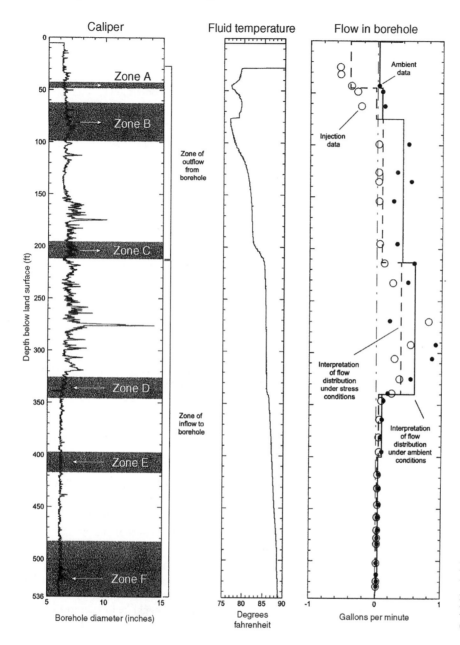

Figure 12.13.2. Caliper and fluid-temperature logs and heat-pulse flowmeter interpretation from well ER-OV-06a, Oasis Valley, Nevada (Robledo et al.[37]).

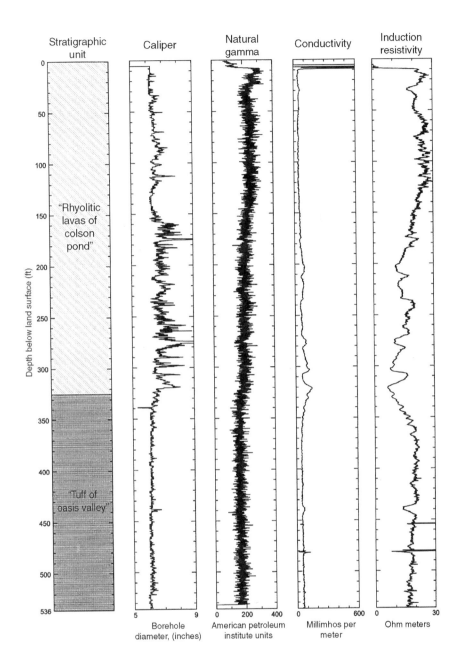

Figure 12.13.3. Geologic and geophysical logs from well ER-OV-06a, Oasis Valley, Nevada (Robledo et al.[37]).

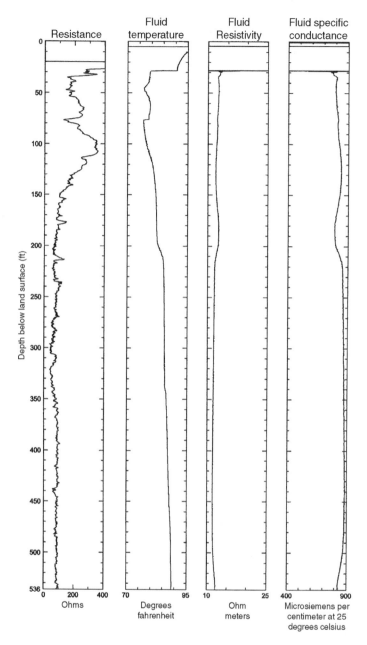

Figure 12.13.3 (continued). (Robledo et al.[37])

REFERENCES

1. Bassiouni, Z., *Theory, Measurement, and Interpretation of Well Logs,* Society of Petroleum Engineers, Houston, 1994.

2. Bays, C. A., and S. H. Folk, Developments in the application of geophysics to ground-water problems, Illinois Geological Survey Circ. 108, Urbana, 25 pp., 1944.

3. Birman, J. H., Geothermal exploration for ground water, *Geol. Soc. Amer. Bull.,* v. 80, pp. 617–630, 1969.

4. Bredehoeft, J. D., and I. S. Papadopulos, Rates of vertical ground water movement estimated from earth's thermal profile, *Water Resources Research,* v. 1, pp. 325–328, 1965.

5. Brendle, D. L., *Geophysical Logging to Determine Construction, Contributing Zones, and Appropriate Use of Water Levels Measured in Confined-Aquifer Network Wells, San Luis Valley, Colorado, 1998–2000, U.S. Geological Survey Water Resources Investigations Report* 02-4058, http://water.usgs.gov/pubs/wri/wrir02-4058, 2002.

6. Brown, R. H., et al., *Ground-water Studies,* Studies and Repts. in Hydrology 7, UNESCO, Paris, vars. pp., 1972.

7. Callahan, J. T., et al., Television—A new tool for the ground-water geologist, *Ground Water,* v. 1, no. 4, pp. 4–6, 1963.

8. Croft, M. G., *A Method for Calculating Permeability from Electric Logs, U.S. Geological Survey Professional Paper* 750-B, pp. B265–B269, 1971.

9. Doll, H. G., The S. P. log: Theoretical analysis and principles of interpretation, *Trans. Amer. Inst. Min. and Met. Engrs.,* v. 179, pp. 146–185, 1949.

10. Frimpter, M. H., Casing detector and self-potential logger, *Ground Water,* v. 7, no. 6, pp. 24–27, 1969.

11. Garber, M. S., and F. C. Koopman, Methods of measuring water levels in deep wells, *U.S. Geological Survey Techniques of Water-Resources Investigations,* Bk. 8, Chap. A1, 23 pp., 1968.

12. Gorder, Z. A., Television inspection of a gravel pack well, *Jour. Amer. Water Works Assoc.,* v. 55, pp. 31–34, 1963.

13. Guyod, H., and J. A. Pranglin, *Analysis Charts for the Determination of True Resistivity from Electric Logs,* Houston, TX, 202 pp., 1959.

14. Heath, J. R., P. H. Nelson, and F. L. Paillet, *Well Logging for Physical Properties: A Handbook for Geologists, and Engineers,* 2nd ed., John Wiley & Sons, New York, 2000.

15. Hess, A. E., Identifying hydraulically conductive features with a slow-velocity borehole flowmeter, *Canadian Geotechnical Journal,* v. 23, no. 1, pp. 69–78, 1986.

16. Jakosky, J. J., *Exploration Geophysics,* Trija, Los Angeles, 1195 pp., 1950.

17. Jones, P. H., and T. B. Buford, Electric logging applied to ground water exploration, *Geophysics,* v. 16, pp. 115–139, 1951.

18. Kearey, P., and M. Brooks, *An Introduction to Geophysical Exploration,* 2nd ed., Blackwell Scientific Publications, Oxford, 1991.

19. Keys, W. S., *Borehole Geophysics Applied to Ground-Water Investigations,* National Water Well Association, Dublin, OH, 1989.

20. Keys, W. S., Borehole geophysics applied to Ground-water investigations, Book 2, Chapter E2, *U.S. Geological Survey Techniques of Water-Resources Investigations Reports,* http://water.usgs.gov/pubs/twri/twri2-e2, 1990.

21. Keys, W. S., and L. M. MacCrary, Application of borehole geophysics water-resources investigations, Techniques of Water-Resources Investigations Reports, *U.S. Geological Survey Techniques of Water-Resources Invs.,* Bk. 2, Chap. E1, http://water.usgs.gov/pubs/twri/twri2-e1, 126 pp., 1971.

22. Keys, W. S., and L. M. MacCrary, *Location and Characteristics of the Interface between Brine and Fresh Water from Geophysical Logs of Boreholes in the Upper Brazos River Basin, Texas, U.S. Geological Survey Professional Paper* 809-B, 23 pp., 1973.

23. Kirby, M. E., Improve your work with drilling-time logs, *Johnson National Drillers Jour.,* v. 26, no. 6, pp. 6–7, 14, 1954.

24. Lao, C., et al., Application of electric well logging and other well logging methods in Hawaii, Tech. Rept. 21, Water Resources Research Center, Univ. of Hawaii, Honolulu, 108 pp., 1969.

25. LeRoy, L. W., *Subsurface Geologic Methods,* 2nd ed., Colorado School of Mines, Golden, CO, 1156 pp., 1951.

26. Lynch, E. J., *Formation Evaluation,* Harper and Row, New York, 422 pp., 1962.

27. Meyer, W. R., *Use of a Neutron Probe to Determine the Storage Coefficient of an Unconfined Aquifer, U.S. Geological Survey Professional Paper* 450E, pp. E174-E176, 1963.

28. Norris, S. E., The use of gamma logs in determining the character of unconsolidated sediments and well construction features, *Ground Water,* v. 10, no. 6, pp. 14–21, 1972.

29. Paillet, F. L., Flow modeling and permeability interpretation using borehole flow logs in heterogeneous fractured formations, *Water Resources Research,* v. 34, no. 5, pp. 997–1010, 1998.

30. Parizek, R. P., and S. H. Siddiqui, Determining the sustained yield of wells in carbonate and fractured aquifers, *Ground Water,* v. 8, no. 5, pp. 12–20, 1970.

31. Patten, E. P., Jr., and G. D. Bennett, *Application of Electrical and Radioactive Well Logging to Ground-Water Hydrology, U.S. Geological Survey Water-Supply Paper* 1544-D, 60 pp., 1963.

32. Patten, E. P., Jr., and G. D. Bennett, *Methods of Flow Measurement in Well Bores, U.S. Geological Survey Water-Supply Paper* 1544-C, 28 pp., 1962.

33. Peterson, F. L., *Neutron Well Logging in Hawaii,* Tech. Rept. 75, Water Resources Research Center, Univ. of Hawaii, Honolulu, 42 pp., 1974.

34. Pickell, J. J., and J. G. Heacock, Density logging, *Geophysics,* v. 25, pp. 891–904, 1960.

35. Pickett, G. R., The use of acoustic logs in the evaluation of sandstone reservoirs, *Geophysics,* v. 25, pp. 250–274, 1960.

36. Poland, J. F., and R. B. Morrison, An electrical resistivity-apparatus for testing well-waters, *Trans. Amer. Geophysical Union,* v. 21, pp. 35–46, 1940.

37. Robledo, A. R., P. L. Ryder, J. M. Fenelon, and F. L. Paillet, *Geohydrology of Monitoring Wells Drilled in Oasis Valley near Beatty, Nye County, Nevada, 1997, U.S. Geological Water-Resources Investigations Report* 98-4184, http://water.usgs.gov/pubs/wri/wri984184, 1998.

38. Schneider, R., *An Application of Thermometry to the Study of Ground Water, U.S. Geological Survey Water-Supply Paper* 1544-B, 16 pp., 1962.

39. Shuter, E., and W. E. Teasdale, Application of drilling, coring, and sampling techniques to test holes and wells, Bk. 2, Chapt. F1, *U.S. Geological Survey Techniques of Water-Resources Investigations Reports,* http://water.usgs.gov/pubs/twri/twri2-f1, 1989.

40. Sorey, M. L., Measurement of vertical groundwater velocity from temperature profiles in wells, *Water Resources Research,* v. 7, pp. 963–970, 1971.

41. Stewart, D. M., The rock and bong techniques of measuring water levels in wells, *Ground Water,* v. 8, no. 6, pp. 14–18, 1970.

42. Turcan, A. N., Jr., Calculation of water quality from electrical logs-theory and practice, Water Resources Pamphlet 19, Louisiana Geological Survey, Baton Rouge, 23 pp., 1966.

43. U. S. Geological Survey, A Guide to Safe Field Operations, *U.S. Geological Survey Open-File Report* 95-777, http://water.usgs.gov/pubs/of/of95-777, 1995.

44. Vogelsang, D., *Environmental Geophysics: A Practical Guide,* Springer Verlag, Berlin, 1995.

45. Vonhof, J. A., Water quality determination from spontaneous-potential electric log curves, *Jour. Hydrology,* v. 4, pp. 341–347, 1966.

46. Ward, S., ed., Geotechnical and environmental geophysics, vols. I, II, and III, *Investigations in Geophysics*, SEG Series, SEG, Tulsa, 1990.

47. Wyllie, M. R. J., *The Fundamentals of Well Log Interpretation,* 3rd ed., Academic Press, New York, 238 pp., 1963.

Chapter 13

Artificial Recharge of Groundwater

In order to increase the natural supply of groundwater, people artificially recharge groundwater basins. Artificial recharge may be defined as augmenting the natural movement of surface water into underground formations by some method of construction, by spreading of water, or by artificially changing natural conditions. A variety of methods have been developed, including water spreading, recharging through pits and wells, and pumping to induce recharge from surface water bodies.[5, 18, 32, 34, 66] The choice of a particular method is governed by local topographic, geologic, and soil conditions; the quantity of water to be recharged; and the ultimate water use. In special circumstances, land value, water quality, or even climate may be an important factor.

13.1 CONCEPT OF ARTIFICIAL RECHARGE

Artificial recharge projects are designed to serve one or more of the following purposes:

1. Maintain or augment the natural groundwater as an economic resource.
2. Coordinate operation of surface and groundwater reservoirs.
3. Combat adverse conditions such as progressive lowering of groundwater levels, unfavorable salt balance, or saline water intrusion.
4. Provide subsurface storage for local or imported surface waters.
5. Reduce or stop significant land subsidence.
6. Provide a localized subsurface distribution system for established wells.
7. Provide treatment and storage for reclaimed wastewater for subsequent reuse.
8. Conserve or extract energy in the form of hot or cold water.

Thus, in most situations, artificial recharge projects not only serve as water-conservation mechanisms but also assist in overcoming problems associated with overdrafts.[16] The role of artificial recharge in groundwater management is described in Chapter 9; control of seawater intrusion is described in Chapter 14.

To place water underground for future use requires that adequate amounts of water be obtained for this purpose. In some localities, storm runoff is collected in ditches, basins, or reservoirs for subsequent recharge. Elsewhere, recharge water is imported into a region by pipeline or aqueduct from a distant surface water source. A third possibility involves utilization of treated wastewater.

Recharging began in Europe early in the nineteenth century and in the United States near the end of the century; since then recharge installations have steadily increased throughout the world. Recharge basins form integral parts of many Swedish municipal water supply systems. Artificial recharge is widely practiced in Germany to meet industrial and municipal water demands. In the Netherlands, water supply systems for Amsterdam, Leiden, and The Hague include basins for recharging surface water into coastal sand dunes.[9]

Artificial recharge is accomplished by applying water to the soil for infiltration and downward movement through the unsaturated or "vadose" zone to the groundwater. Bouwer[14] defined artificial recharge systems according to permeable materials in which they can be placed, as illustrated in Figure 13.1.1. The surface system (*a*) requires soils that are sufficiently permeable, vadose zones that have no clay or other restricting layers, and aquifers that are unconfined. Artificial recharge can be accomplished with *excavated basins* (*b*) that are sufficiently deep to reach permeable material where permeable soils are not available at relatively small depths (e.g., 1 m). *Seepage trenches* (*c*) can be used if the permeable material is too deep for removal of overlying material, but is within trenchable depth (e.g., less than about 7 m). Trenches are also suitable in soils that are highly stratified with alternating layers of fine and coarse materials.[14] *Large-diameter wells, pits,* or *shafts* (*d*) in the vadose zone can, used when permeable material is too deep for trenches. These shafts can be drilled with bucket augers to a depth of about 50 m with a diameter of about 1 m.[14] *Recharge wells,* penetrating the aquifer (*e*), can be used in situations where permeable surface soils are not available, vadose zones are not sufficiently permeable to transmit water, or aquifers are confined.

13.2 RECHARGE METHODS

13.2.1 Methods

A variety of methods have been developed to recharge groundwater artificially. The most widely practiced methods can be described as types of *water spreading*—releasing water over the ground surface in order to increase the quantity of water infiltrating into the ground and then percolating to the water table. Although field studies of spreading have shown that many factors govern the rate at which water will enter the soil, from a quantitative standpoint, area of recharge and length of time that water is in contact with soil are most important. Spreading

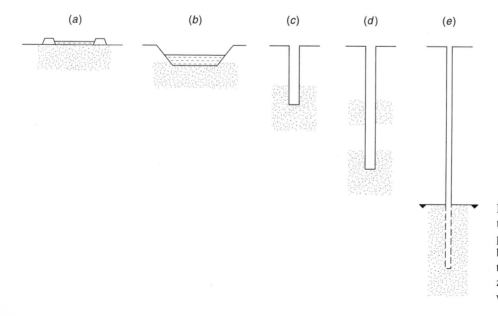

Figure 13.1.1. Recharge systems for increasingly deep permeable materials: surface basin (*a*), excavated basin (*b*), trench (*c*), shaft or vadose zone well (*d*), and aquifer well (*e*) (Bouwer[14]).

efficiency is measured in terms of the recharge rate, expressed as the velocity of downward water movement over the wetted area.

Spreading methods may be classified as basin, stream channel, ditch and furrow, flooding, and irrigation. These, together with techniques employing pits and recharge wells, are described in the following sections.

13.2.1.1 Basin Method

Water may be recharged by releasing it into basins formed by construction of dikes or levees or by excavation. Generally, basin sizes and shapes are adapted to land surface slope. Silt-free water aids in preventing sealing of basins during submergence. Most basins require periodic maintenance to improve infiltration rates by scarifying, disking, or scraping the bottom surfaces when dry. Where local storm runoff is being recharged, a single basin will normally suffice, but where streamflow is being diverted for recharge, a series of basins, often parallel to the natural stream channel, becomes advantageous.[8] Water from the stream is led by a ditch into the uppermost basin. As the first fills, it spills into the second, and the process is repeated through the entire chain of basins. From the lowest basin, any excess water is returned to the stream channel. Figure 13.2.1 illustrates a typical plan of a multiple-basin recharge project. This method permits water contact over 75 to 90 percent of the gross area.

Multiple basins provide for continuity of operation when certain basins are removed from service for drying and maintenance. Furthermore, where streamflow from storm runoff is being spread, upper basins can be reserved for settling silt. Figure 13.2.2 shows an aerial photograph of an extensive series of recharge basins in Los Angeles, California. Figure 13.2.3 on page 551 shows the recharge operations of Orange County Water District.

Basins, because of their general feasibility, efficient use of space, and ease of maintenance, are the most favored method of recharge. Long-time recharge rates vary widely.

Infiltration basins (SAT systems, see Section 13.4) are by far the most widely used method for groundwater recharge and municipal waste removal (U.S. Environmental Protection Agency[66]). Because of the economical attractiveness and low-cost maintenance, many SAT systems have been installed or are planned to be installed in the United States and other countries throughout the world. Los Angeles County implemented the use of SAT systems for groundwater recharge in the Montbello Forebay in the early 1960s. After over 20 years of operation and extensive study, no measurable impact on groundwater quality or human health could be established (Nellor et al.[47]). Bouwer[12, 13] studied SAT systems in the Phoenix metropolitan area with effluents from two different wastewater treatment plants. SAT systems have also been studied and used in Israel (Idelovitch et al.[33]) and in Australia (Ho et al.[31]). Wilson et al.[69] has studied water quality changes and the fate of disinfection byproducts at the Tucson

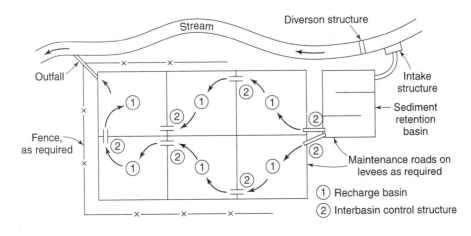

Figure 13.2.1. Typical plan of a multiple-basin recharge project diverting water from a stream (after Amer. Soc. Civil Engrs., *Ground Water Management*, Man. and Repts. on Engrng. Practice 40, 1972).

Figure 13.2.2. Aerial view of spreading basins adjoining the San Gabriel River, Los Angeles, California, and temporary finger dikes within the river channel (courtesy Los Angeles County Flood Control District).

Sweetwater Recharge Site. Arizona State University et al.[4] have performed extensive studies on the feasibility of a large-scale SAT project in Phoenix, Arizona.

13.2.1.2 Stream-Channel Method

Water spreading in a natural stream channel involves operations that will increase the time and area over which water is recharged from a naturally losing channel. This involves both upstream management of streamflow and channel modifications to enhance infiltration. Upstream reservoirs enable erratic runoff to be regulated and ideally limit streamflows to rates that do not exceed the absorptive capacity of downstream channels.

Improvements of stream channels may include widening, leveling, scarifying, or ditching to increase infiltration. In addition, low check dams and dikes can be constructed across a stream where a wide bottom occurs; these act as weirs and distribute the water into shallow ponds occupying the entire streambed (see Figure 13.2.4). These structures are normally

Figure 13.2.4. Channel spreading with rock-and-wire check dams in Cucamonga Creek near Upland, California (courtesy D. C. Muckel).

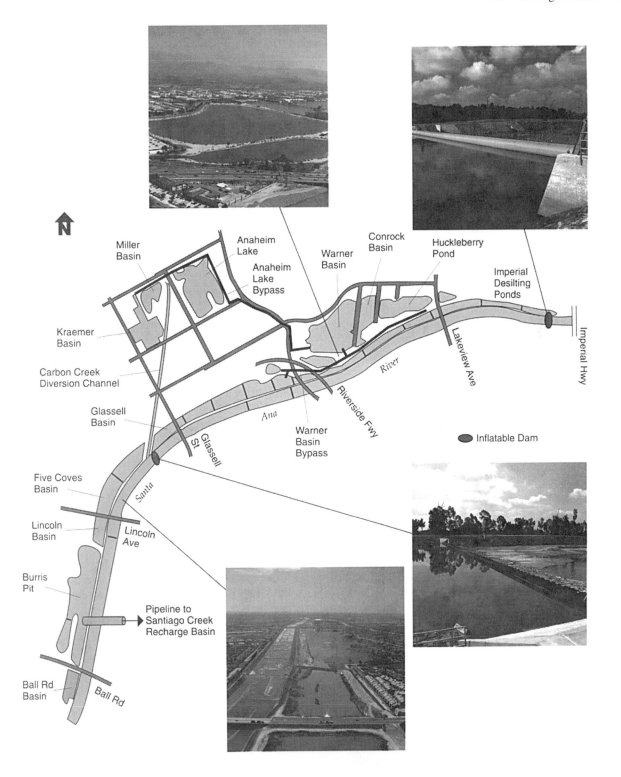

Figure 13.2.3. The efficiency of Orange County Water District's recharge operations has been greatly increased over the past seven years by a major capital improvement program. As a result, up to 400,000 acre-feet of water can percolate into the groundwater basin in a year's time (Orange County Water District[49]).

temporary, consisting of river-bottom material, and sometimes protected by vegetation, wire, or rocks. Such works quickly collapse when high streamflows occur; if permanent structures are placed in a channel, it is important that they do not create a flood hazard.

L-shaped finger levees can be seen in the stream channel in Figure 13.2.2, each of which impounds water. These are simply constructed annually by bulldozer at the end of the high-streamflow season.

Channel spreading can also be conducted without a specific spreading works. In streams with storage reservoirs primarily for flood control, releases of clear water may be entirely recharged into downstream reaches. A majority of the spreading works in and near Los Angeles County are part of an integrated water conservation and flood protection plan.

13.2.1.3 Ditch-and-Furrow Method

In the ditch-and-furrow method, water is distributed to a series of ditches, or furrows, that are shallow, flat-bottomed, and closely spaced to obtain maximum water-contact area. One of three basic layouts is generally employed: (1) contour, where the ditch follows the ground contour and by means of sharp switchbacks meanders back and forth across the land; (2) tree-shaped, where the main canal successively branches into smaller canals and ditches; and (3) lateral, where a series of small ditches extend laterally from the main canal.

Ditch widths range from 0.3 to 1.8 m. On very steep slopes, checks are sometimes placed in ditches to minimize erosion and to increase the wetted area.

Gradients of major feeder ditches should be sufficient to carry suspended material through the system. Deposition of fine-grained material clogs soil surface openings. Although a variety of ditch plans have been devised, a particular plan should be tailored to the configuration of the local area. A collecting ditch is needed at the lower end of the site to convey excess water back into the main stream channel.

The method is adaptable to irregular terrain but seldom provides water contact to more than about 10 percent of the gross area. Figure 13.2.5 shows typical spreading ditches on an alluvial plain.

13.2.1.4 Flooding Method

In relatively flat topography, water may be diverted to spread evenly over a large area. In practice, canals and earth-distributing gullies are usually needed to release the water at intervals over the upper end of the flooding area. It is desirable to form a thin sheet of water over the land, which moves at a minimum velocity to avoid disturbing the soil. Tests indicate that highest infiltration rates occur on areas with undisturbed vegetation and soil covering. Compared with other spreading methods, flood spreading costs least for land preparation. In order to control the water

Figure 13.2.5. Spreading ditches in Tujunga Wash, Los Angeles, California (courtesy City of Los Angeles Department of Water and Power).

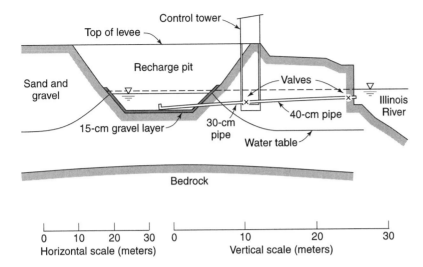

Figure 13.2.6. Cross section of a recharge pit at Peoria, Illinois (after Suter and Harmeson[61]).

at all times, embankments or ditches should surround the entire flooding area. To obtain maximum efficiency, a person should be on the grounds during flooding operations—frequently, movement of a few shovelsful of dirt will effectively increase the wetted area.

13.2.1.5 Irrigation Method

In irrigated areas, water is sometimes deliberately spread by irrigating cropland with excess water during dormant, winter, or nonirrigating seasons. The method requires no additional cost for land preparation because the distribution system is already installed. Even keeping irrigation canals full will contribute to recharge by seepage from the canals. Where a large portion of the water supply is pumped, the method has the advantage of raising the water table and consequently reducing power costs. Consideration needs to be given to the effects of the leaching action of the percolating water both in carrying salts from the root zone to groundwater and in removing soil nutrients, thereby possibly reducing crop yields.

13.2.1.6 Pit Method

A pit (see Figure 13.2.6) excavated into a permeable formation serves as an ideal facility for groundwater recharge. Because the cost of excavation and removal of material is high, use of abandoned excavations, such as gravel pits, is most economic. In areas where shallow subsurface strata, such as hardpans and clay layers, restrict the downward passage of water, pits can effectively reach materials with higher infiltration rates.[35, 41]

Besides the small capital cost of pits constructed for a purpose other than recharge, the steep sides provide a high silt tolerance. Silt usually settles to the bottom of the pit, leaving the walls relatively unclogged for continued infiltration of water.[8] Nonsilty water should be recharged whenever possible so as to minimize silt accumulation and periodic removal costs.[23] Attention to the geometry of a recharge pit is important in order to obtain the maximum infiltration rate.[20, 58]

Studies by the Illinois State Water Survey[29] of the filtration efficiency of coarse media resulted in the equation

$$SS_0 = 13.1H^{-0.25}d^{0.5}Q_0SS_i^{1.33} \tag{13.2.2}$$

where SS_0 is the suspended solids concentration (mg/l) transmitted through the filter layer, H is the filter layer thickness (cm), d is the mean diameter (mm) of particles forming the filter layer, Q_0 is the rate of recharge (m/day), and SS_i is suspended solids concentration of the recharged water.

13.2.1.7 Recharge Well Method

A *recharge well* may be defined as a well that admits water from the surface to freshwater aquifers. Recharge wells are also known as *disposal wells* and *drain wells*. They should be distinguished from *injection wells*, described in Chapter 8, which recharge brines and toxic industrial wastes to deep, saline-water aquifers. A recharge well's flow is the reverse of a pumping well, but its construction may or may not be the same. Well recharging is practical where deep, confined aquifers must be recharged, or where economy of space, such as in urban areas, is an important consideration.

If water is admitted into a well, a cone of recharge will be formed that is similar in shape but is the reverse of a cone of depression surrounding a pumping well. The equation for the curve can be derived in a similar manner to that for a pumping well (see Chapter 4). For a confined aquifer with water being recharged into a completely penetrating well at a rate Q_r, the approximate steady-state expression

$$Q_r = \frac{2\pi K b\left(h_w - h_0\right)}{\ln\left(r_0/r_w\right)} \tag{13.2.3}$$

is applicable. (Symbols are identified in Figure 13.2.7a.) For a recharge well penetrating an unconfined aquifer (see Figure 13.2.7b),

$$Q_r = \frac{\pi K\left(h_w^2 - h_0^2\right)}{\ln\left(r_0/r_w\right)} \tag{13.2.4}$$

By comparing the discharge equations for pumping and recharge wells, it might be anticipated that the recharge capacity would equal the pumping capacity of a well if the recharge cone has dimensions equivalent to the cone of depression. Field measurements, however, rarely support this reasoning; recharge rates seldom equal pumping rates. The difficulty lies in the fact that pumping and recharging differ by more than a simple change of flow direction.

As water is pumped from a well, fine material present in the aquifer is carried through the coarser particles surrounding the well and into the well. On the other hand, any silt carried by water into a recharge well is filtered out and tends to clog the aquifer surrounding the well.[51, 54] Similarly, recharge water may carry large amounts of dissolved air, tending to reduce the permeability of the aquifer by air binding. Recharge water may also contain bacteria, which can form growths on the well screen and the surrounding formation, thereby reducing the effective flow area. Chemical constituents of the recharge water may differ sufficiently from the normal

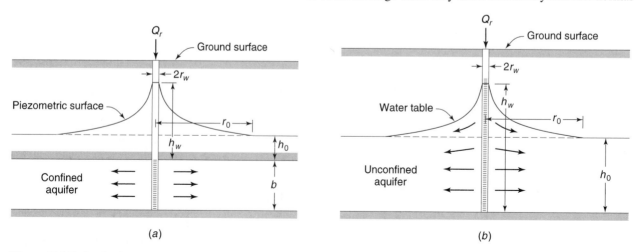

Figure 13.2.7. Radial flow from recharge wells penetrating (*a*) confined and (*b*) unconfined aquifers.

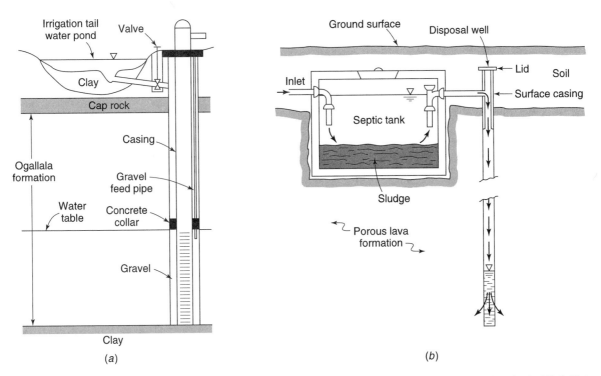

Figure 13.2.8. Examples of recharge well designs. (*a*) Combined irrigation and recharge well in alluvium in the High Plains of Texas (*b*) Recharge well for disposal of septic tank effluent into a lava formation in Central Oregon (after Hauser and Lotspeich[30] and Sceva.[56])

groundwater to cause undesired chemical reactions—for example, deflocculation caused by reaction of high-sodium water with fine soil particles. These factors all act to reduce recharge rates; as a result, well recharging has been limited to a few areas where experience, particularly with water treatment and redevelopment of wells, has shown the practicality of the method.[67]

It should be noted that supply wells can alternate as recharge wells, as shown in Figure 13.2.8*a* Recharge wells serve as convenient means for disposal of septic tank effluent, excess irrigation water, and surface runoff into the permeable volcanic terrains of the northwestern United States.[55, 64] A typical domestic wastewater disposal system including a recharge well is shown in Figure 13.2.8*b*.

An extensive series of recharge wells has been successfully operated since 1953 along the coasts of Los Angeles and Orange Counties, California. These wells create and maintain a pressure ridge of fresh water to control seawater intrusion (see Chapter 14). Experience gained on the project demonstrated that gravel-packed wells operate most efficiently; a typical dual-aquifer well is shown in Figure 13.2.9. Favorable recharge rates have been maintained by chlorination and deaeration of the water supply and by a comprehensive well-maintenance program involving periodic pumping of the wells. It was also found that a concrete seal should be provided on the outside of the casing where it passes through the impermeable zone above the confined alluvial aquifer to prevent upward movement of water. Figure 13.2.10 illustrates the Scottsdale Water Campus located in Scottsdale, Arizona, which has been in operation since 1999.

Finally, it should be mentioned that field and laboratory studies have demonstrated the feasibility of temporary storage of fresh water in saline water aquifers through wells first recharged and later pumped.[21, 37] The efficiency of the procedure increases with each recharge-storage-withdrawal cycle. The technique has application in flat coastal areas underlain by saline water aquifers where no surface reservoir sites are available to provide freshwater supplies on a year-round basis.

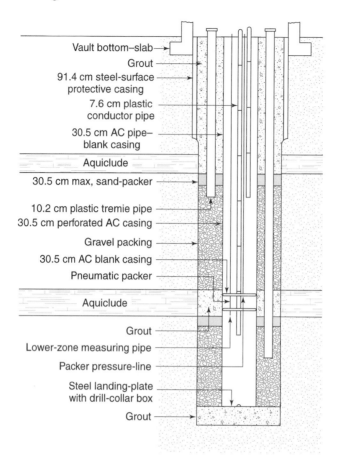

Vault bottom–slab
Grout
91.4 cm steel-surface protective casing
7.6 cm plastic conductor pipe
30.5 cm AC pipe– blank casing
Aquiclude
30.5 cm max, sand-packer
10.2 cm plastic tremie pipe
30.5 cm perforated AC casing
Gravel packing
30.5 cm AC blank casing
Pneumatic packer
Aquiclude
Grout
Lower-zone measuring pipe
Packer pressure-line
Steel landing-plate with drill-collar box
Grout

Figure 13.2.9. Design of a dual-aquifer recharge well for control of sea water intrusion, Los Angeles County, California (after Department of Economic and Social Affairs[19]).

13.2.1.8 Incidental Recharge

Incidental, or unplanned, recharge occurs where water enters the ground as a result of a human activity whose primary objective is unrelated to artificial recharge of groundwater.[19] Included in this category is water from irrigation, cesspools, septic tanks, water mains, sewers, landfills, waste-disposal facilities, canals, and reservoirs. The quantity of incidental recharge normally far exceeds that deliberately accomplished by artificial recharge projects. Because several of these sources introduce polluted waters into the underground, degradation of the quality of ground-water can occur (see Chapter 8).

13.2.2 Recharge Rates

The economy of water spreading hinges on maintenance of a high infiltration rate. Typical rate curves, however, show a pronounced tendency to decrease with time. Determining the cause of this decrease and how to counteract it has led to extensive research programs. A variety of soil and water treatments and operational methods have been undertaken to study the problem, from the 1960s[7, 11, 55] to more recent studies (Arizona State University et al.[4]).

Figure 13.2.11 shows a typical curve of recharge rate versus time. The initial decrease is attributed to dispersion and swelling of soil particles after wetting; the subsequent increase accompanies elimination of entrapped air by solution in passing water; the final gradual decrease results from microbial growths clogging the soil pores. Laboratory tests with sterile soil and water give nearly constant maximum recharge rates, thereby substantiating the effect of microbial growths.

(*a*) Outside view

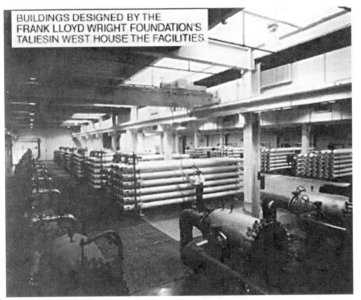

(*b*) Filtration system

Figure 13.2.10. Scottsdale Water Campus (courtesy of Black & Veach[10] and the City of Scottsdale, AZ).

This facility includes a 43-mgd conventional water treatment plant, a 10-mgd wastewater treatment plant, a 10-mgd dual membrane water reclamation plant, and groundwater recharge facilities.

The dual membrane system, consisting of microfiltration (MF) and reverse osmosis (RO), is being used to treat both surface water from the Colorado River and tertiary effluent prior to groundwater recharge. The water then percolates through the soil and eventually mixes with local groundwater. The city receives groundwater withdrawal credits based on the volume recharged from the Arizona Department of Water Resources within its service area.

The MF system treats both surface water and secondary treated sewage effluent.

After MF, the filtrate is further treated using RO for desalting before groundwater recharge. This is to prevent salt buildup in the groundwater.

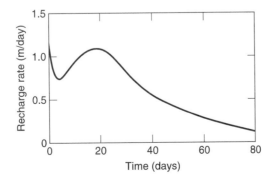

Figure 13.2.11. Typical time variation of recharge rate for water spreading on undisturbed soil (after Muckel[45]).

Recharge rates generally decrease as the mean particle size of soil on a spreading area decreases. Efforts to maintain soil pores free for water passage have led to additions of organic matter and chemicals to the soil as well as to growing vegetation on the spreading area. Chemical soil conditioners, which tend to aggregate the soil, show promise in soils of certain textures. Alternating wet and dry periods on a basin generally furnishes a greater total recharge than does continuous spreading, in spite of the fact that water is in contact with the soil for as little as one-half of the total time. Drying kills microbial growths, and this, combined with scarification of the soil surface, reopens the soil pores.[47]

Other factors contribute to recharge rates. Studies in small ponds have confirmed that infiltration rates are directly proportional to head of water. Where less pervious strata lie below the surface stratum, the recharge rate depends on the rate of subsurface lateral flow. Hence, spreading only in narrow, widely spaced strips recharges nearly as much water as spreading over an entire area.[20, 58] Water containing silt or clay is known to clog soil pores, leading to rapid reductions in recharge rates. Wave action in large, shallow ponds can stir bottom sediments and seal pores that would otherwise remain open. Water quality can be an important factor; thus, recharging water of high sodium content tends to deflocculate colloidal soil particles and thereby hinder water passage. Because a high water table limits the downward flow of recharged water, this surface should be at least 3 to 6 m below spreading surfaces.

Clogging effects can be modeled considering the hydraulic effects. The hydraulic characteristics on hydraulic conductivity K, water content ω, and pressure head h have been described by Kool and van Genuchten:[36]

$$\omega = \omega_r + \frac{\omega_s - \omega_r}{\left(1 + |\alpha h|^\beta\right)^\gamma} \tag{13.2.5}$$

and

$$K = K_c S_e^{1/2} (1 - (1 - S_e^{1/\gamma})^\gamma)^2 \tag{13.2.6}$$

where ω_s and ω_r are the saturated and residual water content; α, β, and γ are curve-fitting parameters; S_e is the degree of saturation, $S_e = (\omega - \omega_r)/(\omega_s - \omega_r)$, and K_c is the saturated hydraulic conductivity affected by the clogging effect, defined as

$$K_c = K_s f_{clog} \tag{13.2.7}$$

where K_s is the saturated hydraulic conductivity without clogging. The factor f_{clog} expresses the effect of surface clogging ($z = 0$) and biological clogging ($z \geq 0$).

General expressions for the clogging effect can be described using the affected hydraulic conductivity K_c, based on Ahuja[2] and Moore:[43]

$$K_c = K_f + (K_s - K_f) \exp(-\alpha_c J) \tag{13.2.8}$$

where K_f is the lowest hydraulic conductivity value, α_c is the coefficient related to characteristics of soil and clogging substance, and J is the cumulative influx of clogging substances as a function of time:

$$J = \int_0^t q_s \left(\partial_i c_1 + \delta_2 c_2\right) d\tau \tag{13.2.9}$$

where ∂_i is the percent of each clogging substance being intercepted by the soil surface, c_i denotes concentration of clogging substances i in ponding water, and $i = 1$ for suspended solids and $i = 2$ for algae.

The effect of biological clogging is described according to Taylor and Jaffe[65] by

$$f_{clog}(z,t) = \begin{cases} \exp\left(aB + bB^2\right) & B \leq B_0 \\ f_0 & B > B_0 \end{cases} \tag{13.2.10}$$

where a, b, f_0 and B_0 are empirical values and B is the biomass.

13.3 WASTEWATER RECHARGE FOR REUSE

In recent years, increasing attention has been focused on reuse of municipal wastewater. Almost all uses are nonpotable, such as for irrigation or industrial purposes, because of questionable health effects,[59] and one of the mostly widely favored is recharge to supplement groundwater resources.

The major concerns associated with recharge to provide high-quality potable or nonpotable water from wastewater effluent include organic carbon, nitrogen species, and pathogens. As an example, priority pollutants in biologically treated wastewater from the 91st Avenue Wastewater Treatment Plant in Phoenix, Arizona, have satisfied primary drinking water standards since strict industrial pretreatment requirements were enforced (Arizona State University et al.[4]). The levels of organic carbon, nitrogen species, and pathogens found in biologically treated wastewater tend to be at levels greater than those found in surface waters and provide serious concerns for reuse.

The organics in biologically treated wastewaters are analogous to the organics present in surface waters and are of concern just as with organics occurring naturally in both surface waters and groundwaters. This concern is the reaction of the organics with the commonly used disinfectant chlorine to produce disinfection byproducts. Both the removal of organic carbon during soil aquifer treatment and the effect on the formation of disinfection byproducts have been studied.

Nitrogen species, in particular nitrate-nitrogen, are one of the most common reasons that groundwaters do not meet drinking water standards. Nitrogen can be removed effectively before soil aquifer treatment to ensure compliance with nitrogen standards. Significant transformations of nitrogen occur during recharge to provide further removal of nitrogen species. A wide range of nitrogen characteristics in wastewater effluents has been studied to determine impacts on nitrogen transformations.[4] Table 13.3.1 presents the effluents studied and their wide range of characteristics.

The removal of all three major classes of pathogens (parasites, bacteria, and viruses) needs to be considered when wastewater is reclaimed. Pathogenic bacteria and viruses can be present in large numbers in biologically treated wastewater; however, disinfection prior to recharge is effective in eliminating pathogens. During previous studies on recharge, no enteric viruses or enteric bacteria have been found in aquifers that were recharged with wastewater. The study by Arizona State University et al.[4] particularly focused on examining the fate of *Cryptosporidium* during soil aquifer treatment and Hepatitis A, since these two pathogens are of concern and very little data exists on their fate during recharge.

Table 13.3.1 Typical Range of Water Quality Parameters (mg/L)

Effluent	Type	BOD	SS	NH_3-N	NO_x-N	Total N	TOC
Phoenix 91st Ave. WWTP effluent	Dechlorinated, denitrified	3–7	3–7	1–3	1–6	4–10	8–10
	Chlorinated, denitrified	3–7	3–7	1–3	1–6	4–10	8–10
	High rate activated sludge (chlorinated)	10–15	10–25	10–30	0–8	10–40	13–15
Tucson Roger Road WWTP effluent	Secondary trickling filter*	20–25	20–30	15–25	0–2	15–30	15–20
	Tertiary-pressure filter	10–15	5–10	15–25	0–2	15–30	10–15
	Primary	50–80	60–100	20–35	0–1	25–40	40–50

*O_3 secondary effluent was also studied

Source: Arizona State University et al.[4]

Recharge of wastewater (usually after secondary treatment) improves its quality by removal of physical, biological, and some chemical constituents; provides storage until subsequent reuse; reduces seasonal temperature variations; and dilutes the recharged water with native groundwater. Land application practices involve irrigation, spreading, overland flow, and recharge wells. Selection of a given system is governed by soil and subsurface conditions, climate, availability of land, and intended reuse of the wastewater.

Irrigation Method

Effluent can be applied by sprinklers or surface irrigation techniques to irrigate cropland. Application rates are low, ranging from 0.05 to 0.2 m/week; also, only the portion not consumed by plants percolates downward to the water table. In humid regions, low evapotranspiration rates plus dilution with rainwater contributes good-quality water to the groundwater; in arid regions, brackish water that degrades the groundwater can result.

Spreading Method

Recharging of effluent onto bare ground or native vegetation for infiltration and percolation functions as a tertiary treatment plant, producing reclaimed water for reuse. Application rates are high, ranging from 0.5 to 10 m/week, depending on local conditions. High-rate systems require deep permeable soils (sandy loams to loamy sands) and a water table that does not rise to ground surface. Flooding is conducted intermittently—for example, 2 to 14 days wet alternating with 5 to 20 days dry. Movement of effluent through the soil removes bacteria and viruses, almost all biochemical oxygen demand and suspended solids, up to 50 percent of nitrogen, and 60 to 95 percent of phosphorus. Because municipal use adds up to 300 mg/l of dissolved solids to water and this cannot be removed by recharging, wastewater can deteriorate the quality of groundwater unless adequate subsurface dilution is available.

Overland Flow Method

Where soils have low infiltration rates, such as clays and clay loams, wastewater is applied by irrigation or spray techniques to the upper end of sloping vegetated plots and allowed to flow in a shallow sheet to runoff collection ditches. Only a minor fraction of the applied water infiltrates; hence, this method contributes little to groundwater recharge.

Recharge Well Method

High-quality, tertiary-treated effluent can be placed underground through recharge wells. The high cost of recharging effluent into wells can be justified economically only where some special purpose, such as control of land subsidence, control of seawater intrusion, or delayed development of a costly alternative water supply source, can be served.[6, 57]

The removal of pollutants from secondary effluent by recharging depends on the time and distance of travel underground and on the type and properties of the soils and subsurface formations. In general, for percolation through fine-textured alluvial deposits, bacteria and viruses are removed, nitrogen is reduced, soluble salts are not removed, and trace elements and heavy metals may be reduced.

13.4 SOIL AQUIFER TREATMENT (SAT) SYSTEMS

13.4.1 What Are SAT Systems?

Artificial recharge of sewage effluent has an important role to play in water reuse. Treated effluent can be placed in recharge basins, allowing for infiltration into the ground for the recharge of aquifers. As the effluent moves through the soil and the aquifer, it undergoes sig-

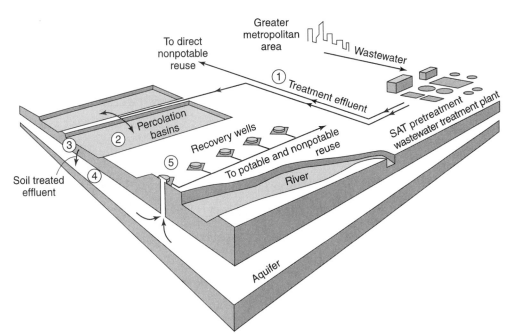

Figure 13.4.1a. An SAT system and its components (Arizona State University et al.[4])

Figure 13.4.1b. Avra Valley recharge facility near Tucson, Arizona (courtesy of Central Arizona Project, photo by M. Early).

nificant quality improvements through physical, chemical, and biological processes. Collectively, these processes and the water quality improvement obtained are called *soil aquifer treatment (SAT)*. SAT is premised on the infiltration of treated wastewater into the soil and percolation through the vadose zone. Improvements in water quality can occur due to many different mechanisms, including infiltration, biological degradation, physical adsorption, ion exchange, and precipitation.

An SAT system consists of five major components (see Figures 13.4.1a and b): (1) the pipeline that carries the treated effluent from the wastewater treatment plant, (2) percolation

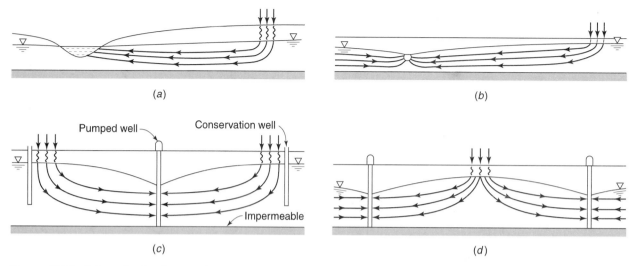

Figure 13.4.2. Schematic of recharge and recovery SAT systems: (*a*) with natural drainage of renovated water into stream, lake, or low area, (*b*) collection of renovated water by subsurface drain (*c*) infiltration areas in two parallel rows and line of wells midway between, and (*d*) infiltration areas in center surrounded by a circle of wells (Bouwer[14]).

(infiltration) basins where the treated effluent infiltrates into the ground, (3) the soil immediately below the infiltration basins (vadose zone), (4) the aquifer where water is stored for a long duration, and (5) the recovery well where water is pumped from the aquifer for potable or nonpotable reuse. Recharge systems for SAT can be designed as infiltration-recovery systems in which basically all effluent water is recovered, as shown by systems A, B, and C in Figure 13.4.2, or after blending with the native groundwater as shown by system D.

Figure 13.4.3 illustrates the SAT system dynamics including inputs (design inputs), the state of the system, and the output of the system. The inputs are the soil type, water quality, operation schedule, and environment. The state of the system, which controls the residence time of water in the vadose zone and the level of microbial activity, includes the soil moisture profile, level of oxygen in the system, algal growth, and the soil hydraulic conductivity. Of particular importance is the level of oxygen, which is related to the microorganism distribution and oxygen demanding substrates. Algal growth is related to clogging layer formation on the soil surface and effective soil hydraulic conductivity. Soil moisture is directly affected by soil type and the environment, and indirectly by the water quality, which affects algal growth and the soil hydraulic conductivity. Oxygen in the vadose zone is affected by the soil moisture profile, the water quality of treated effluent, the operation schedule, and the algal growth. Algal growth is affected by the water quality of the treated effluent, the operation schedule and the environment. Soil hydraulic conductivity is affected by the algal growth, the soil type, and the environment.

The *critical outputs of the system* are the total infiltration, the *total organic carbon removal efficiency*, the *nitrogen removal efficiency*, and the *pathogen removal efficiency*. The total organic carbon removal efficiency is primarily affected by the amount of oxygen in the system through the organic oxidation process and the presence of acclimated microorganisms. Nitrogen removal efficiency is dependent upon oxygen levels and the availability of biodegradable organic carbon through the nitrification–denitrification process. Pathogen removal efficiency is dependent upon the soil moisture profile and the soil hydraulic conductivity through the adsorption–inactivation process.

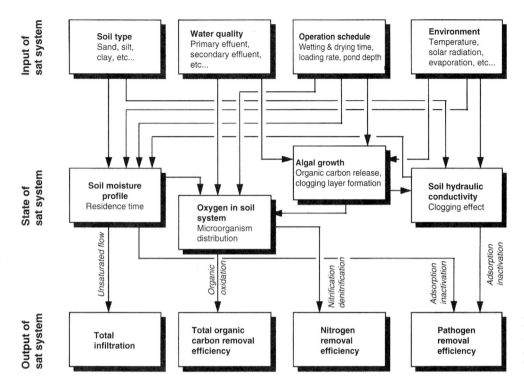

Figure 13.4.3. Soil aquifer treatment system dynamics (Arizona State University et al.[4]).

The major *purification processes* in the soil aquifer system are infiltration, chemical precipitation/dissolution, organic biodegradation, nitrification, denitrification, disinfection, ion exchange, and adsorption/desorption. SAT is a key component of overall water reuse strategies, providing for (1) mechanical filtration of suspended particles, (2) biologically mediated transformation of organics and nitrogen, and (3) physical–chemical retention of inorganic and organic dissolved constituents (e.g., phosphorus, potassium, trace elements) from biologically treated wastewater.

The removal of nitrogen, organic compounds, and biochemical oxygen demand (BOD) can be a continuous biological process that is sustainable. Removal of trace metals, phosphorus, and refractory organics by abiotic mechanisms can result in accumulation in the soil and eventual breakthrough. Bouwer[13] has estimated that SAT-related accumulation of trace metals and phosphorus could become a problem in future decades.

13.4.2 Operation of SAT Systems

Cyclic flooding and drying of the SAT infiltration basins is necessary both to improve infiltration rates and to control aerobic/anoxic conditions in the soil. Basins are flooded until infiltration rates decrease due to development of a surface-clogging layer. Most SAT-related treatment occurs in the upper strata of soil, near the clogging layer (see Figure 13.4.4). During normal conditions (native conditions), unsaturated porous soils (see Figure 13.4.5) contain voids consisting of water and air. Accumulation of biofilm, algae, and suspended solids reduce the void space as illustrated in Figure 13.4.6. The clogging layer development and reduced infiltration rates can dominate SAT system performance, adversely affecting the operation. Cyclic flooding/drying of the basins is necessary to renew hydraulic capacity. Cycle times also influence the transport of oxygen with depth in the vadose zone, causing cycle times to be critical for the control and efficiency of biological processes.

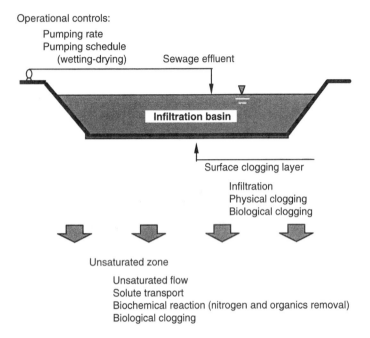

Operational controls:
  Pumping rate
  Pumping schedule
    (wetting-drying)   Sewage effluent

Infiltration basin

Surface clogging layer

  Infiltration
  Physical clogging
  Biological clogging

Unsaturated zone

  Unsaturated flow
  Solute transport
  Biochemical reaction (nitrogen and organics removal)
  Biological clogging

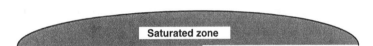

Saturated zone

Figure 13.4.4. Processes related to operation of SAT system. The saturated zone represents the point at which groundwater enters the water table and a mound develops (Arizona State University et al.[4]).

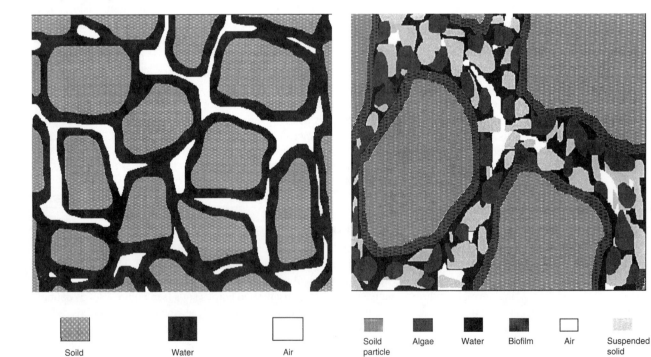

| Soild | Water | Air |

Figure 13.4.5. A cross section of soil in the vadose zone consisting of a matrix of soil, water, and air. Prior to SAT, the air spaces are void and allow for infiltration of water.[4]

| Soild particle | Algae | Water | Biofilm | Air | Suspended solid |

Figure 13.4.6. After SAT, the air spaces become filled with biofilm, algae, and SS. In addition, a mat of algae can cover the surface, greatly reducing infiltration rate.[4]

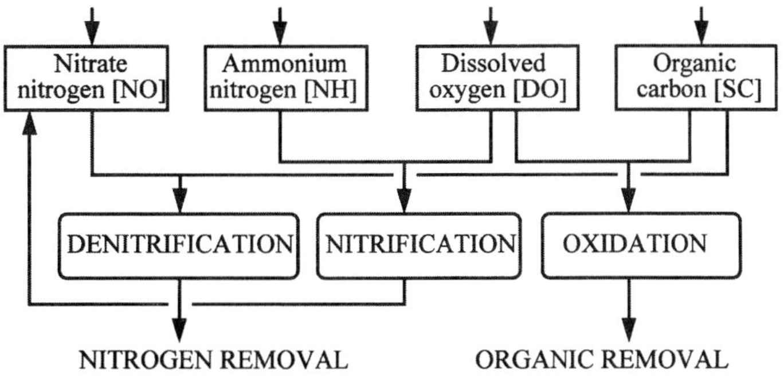

Figure 13.4.7. Flow diagram of the nitrogen cycle in an SAT system.

Removal of organic carbon and nitrogen in SAT systems are related (see Figure 13.4.7) and the *removal mechanisms* include *adsorption–desorption* and *biodegradation*. During the flooding period, dissolved oxygen in the vadose zone often becomes limited and the primary removal mechanism for organic carbon is adsorption, while ammonia is removed by ion exchange. Larger gravel and sands may hold larger air pockets, which increases aerobic biodegradation of carbon and nitrification of ammonia during flooding. Soils that contain more clays and silts have large adsorptive and ion-exchange capacities to remove organic carbon and ammonia. During drying, the soil in the vadose zone becomes aerobic; however, desiccation of the upper soil layer can decrease biological activity and hinder removal since most removal occurs in the upper soil layer. As the flooding cycle begins again, the nitrate that was produced from nitrification becomes soluble. Denitrification will occur if sufficient organic carbon and low dissolved oxygen are simultaneously present. Longer flooding cycles and/or the use of primary effluent tend to increase available organic carbon and increase total nitrogen removal.

13.4.3 Modeling SAT Systems

Abdulrazzak et al.[1] performed a combined field experiment and analytical approach to conclude that the Green–Ampt infiltration model provides adequate representation of the actual infiltration process for artificial recharge systems. Zomorodi[70] developed an infiltration equation that accounted for the change in infiltration rate due to clogging from turbid water and the concentration of suspended sediments. Using several simplified assumptions, the analysis concluded that a system of intermittent multibasin recharge is more efficient than either continuous or intermittent recharge on a single basin when clogging occurs. This work did not account for redistribution of water within the soil, and required many parameters to be defined.

Mushtaq et al.[46] developed optimization models for determining the operation cycles of recharge basins with the objective of maximizing the infiltration over many cycle times. They defined each *cycle time* as consisting of an *application period* (X), a *draining period* (Y), and a *drying period* (Z). During the application period, water is discharged into the basin. During the draining period, water is drained from the basin to begin the drying period. The Green–Ampt infiltration method was used to define the infiltration process. A kinematic wave model (see Section 3.11) was used to define the redistribution of the soil-moisture profile as a function of time as the soil moisture moves downward through the soil. The kinematic-wave-redistribution model developed by Charbeneau[17] was used to define the wetting front profile and water propagation (refer to Figure 3.11.1). Beginning with a zone of saturated soil, the soil-moisture wave propagates downward through the soil. The optimization model has the following assumptions: (1) infiltration model is one dimensional; (2) redistribution model is one dimensional; (3) interference between adjacent basins is neglected; (4) soil-moisture-redistribution model applies at the end of the drain time; (5) hydraulic conductivity remains constant; (6) clogging effects are ignored; and (7) water quality effects are ignored.

Tang et al.[63, 64] and Arizona State University et al.[4] developed a simulation model (NITRINFIL) to quantitatively describe the chemical, physical, and biological interactions that take place between wastewater constituents and the soil system. NITRINFIL (a modified version of the HYDRUS model by Kool and van Genuchten[36]) is a one-dimensional variably saturated flow and interactive multicomponent solute transport model for cyclic operation of soil aquifer systems. NITRINFIL consists of three major submodels, describing unsaturated flow (hydrological), solute transport, and solute biotransformation. The functions incorporated into the unsaturated flow hydraulics simulator are (1) ponding on the surface; (2) clogging effects (surface clogging and biological clogging); (3) infiltration; (4) solute transport process (dissolved oxygen, ammonium nitrogen, nitrite/nitrate nitrogen, biodegradable organic carbon, nonbiodegradable organic carbon, bacteria); (5) biochemical reaction process including nitrogen removal (nitrification and denitrification) and organics removal (oxidation); (6) temperature effects on unsaturated flow and biochemical reaction; and (7) system performance evaluation (removal efficiency of nitrogen and organic carbon and infiltration volume and infiltration volume and infiltration rate). Figures 13.4.8 and 13.4.9 are operation curves generated through a series of computer simulations of the hydrological submodel of NITRINFIL.

The NITRINFIL model was interfaced with an optimization procedure to develop an optimization model referred to as SATOM (SAT Operation Model) for use in determining optimal cycle times for SAT operation (Arizona State University et al.[4]).

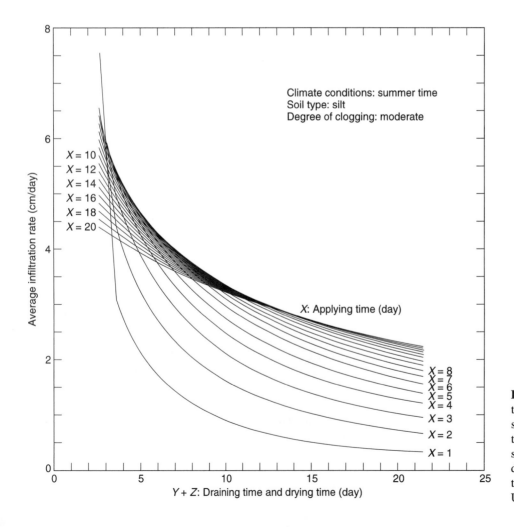

Figure 13.4.8. SAT operation curve showing relationships between average infiltration rate and operation schedule (applying time and drying time) during summer time (Arizona State University et al.[4]).

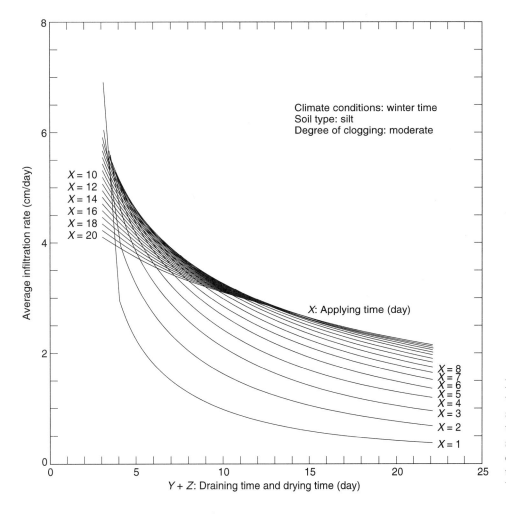

Figure 13.4.9. SAT operation curve showing relationships between average infiltration rate and operation schedule (applying time and drying time) during winter time (Arizona State University, et al.[4]).

13.5 RECHARGE MOUNDS

Groundwater mounds resulting from artificial recharge and stormwater infiltration basins have been studied by several investigators (Bouwer et al.,[14] Glover,[22] Griffen and Warrington,[24] Guo,[25, 26, 27] Hantush,[28] Marino,[39, 40] Molden et al.,[42] Morel-Seytoux et al.,[44] Ortiz et al.,[50] Rao and Sarma,[52] Rastogi and Pandy,[53] Sumner et al.,[60] Swamee and Ojha,[62] and Warner et al.[68]).

13.5.1 Perched Groundwater Mounds

A *perched mound* occurs when a mound is created above a restricting layer (see Figure 13.5.1). In the case of a zero pressure head for the water at the bottom of the restricting layer, the height of the perched mound above the restricting layer, L_p, can be calculated using (Bouwer et al.[14])

$$L_p = L_r \frac{\left[\dfrac{i}{K_r} - 1 \right]}{\left[1 - \dfrac{i}{K_s} \right]}$$

(13.5.1)

where L_r is the thickness of the restricting layer; i is the infiltration rate, K_r is the hydraulic conductivity of the restricting layer, and K_s is the hydraulic conductivity of soil above the restricting the layer.

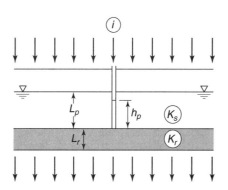

Figure 13.5.1. Geometry and symbols for perched mound above restricting layer in vadose zones (Bouwer[14]).

Often, the infiltration rate is much smaller than K_s because surface soils are finer textured than deeper soils. Also, there may be a clogging layer on the surface soil that reduces infiltration. The infiltration rate is often much larger than K_r. For these conditions, Equation 13.5.1 can be simplified to

$$L_p = i\frac{L_r}{K_r} \tag{13.5.2}$$

L_p should be small enough so that the top of the perched mound is deep enough to avoid reductions in infiltration rates. The above equations will overestimate the height of perched mounds above noncontinuous perching layers (lenses or strips), because they are not as high as above continuous perching layers with the same L_r and K_r because there is no lateral flow in the perched mound.

EXAMPLE 13.5.1	The average annual recharge over an area of 9.57 square miles is 0.52 ft/year. A restricting layer of 10 ft thickness, having $K_r = 0.026$ ft/day, is detected in the subsurface. If the hydraulic conductivity of the soil above the restricting layer is 5.6 ft/day, estimate the height of the perched mound.
SOLUTION	The given information in the problem statement is $i = 0.52$ ft/year $= 1.425 \times 10^{-3}$ ft/day; $K_s = 5.6$ ft/day; $K_r = 0.026$ ft/day; and $L_r = 10$ ft. Substituting this information into Equation 13.5.1 yields a negative height because the hydraulic conductivity of the so-called restricting layer is higher than the infiltration rate. In fact, the infiltration rate must be at least 0.026 ft/day or 9.49 ft/year in order for a perched mound above the restricting layer to form. ∎
EXAMPLE 13.5.2	If the same area in Example 13.5.1 is artificially recharged at a rate of 0.09 ft/day, estimate the height of the perched mound above the restricting layer.
SOLUTION	Again, substituting the given information into Equation 13.5.1 yields

$$L_p = L_r\frac{\left[\dfrac{i}{K_r}-1\right]}{\left[1-\dfrac{i}{K_s}\right]} = (10\text{ ft})\frac{\dfrac{0.09\text{ ft/day}}{0.026\text{ ft/day}}-1}{1-\dfrac{0.09\text{ ft/day}}{5.6\text{ ft/day}}} = (10\text{ ft})\frac{(3.46-1)}{(1-0.016)} = 25\text{ ft}$$

∎

13.5.2 Steady-State Equations for Groundwater Mounds

For *long strip basins* (length of at least five times the width) groundwater flow away from the strip can be approximated as linear horizontal flow (*Dupuit–Forchheimer flow*). Below an

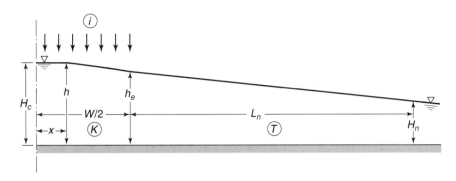

Figure 13.5.2. Geometry and symbols for groundwater mound below long infiltration area (strip) (Bouwer[14]).

infiltration area, lateral flow can be assumed to increase linearly with distance from the center (see Figure 13.5.2). Lateral flow can be assumed to be constant between the edge of the recharge system at a distance $W/2$ from the center and the constant control water table at a distance L_n from the edge (Figure 13.5.2). With these assumptions, Bouwer et al.[15] developed an equation for the ultimate rise of a groundwater mound below the center of the recharge strip in the case of equilibrium between recharge and pumping from the aquifer:

$$H_c - H_n = \frac{iW}{2T}\left(\frac{W}{4} + L_n\right) \qquad (13.5.3)$$

where H_c is the height of the groundwater mound in the center of the recharge area, H_n is the height of the groundwater table at the control area, i is the average infiltration rate in the recharge area (total recharge divided by total area), W is the width of the recharge area, L_n is the distance between the edge of the recharge area and the control area, and T is the transmissivity of the aquifer.

EXAMPLE 13.5.3

A recharge basin 300 m × 2,500 m is proposed for a new area. The recharge rate is 63,000,000 m³ per year. The depth to natural groundwater in the area is 46 m and the aquifer has a saturated thickness of 20 m. The transmissivity of the aquifer is 2,200 m²/day. If the maximum allowable rise of the recharge mound is 38 m, how far should the control area be established from the centerline of the recharge basin?

SOLUTION

Since $L/W = 2500$ m/300 m $= 8.3 > 5$, we can use Equation 13.5.3 developed for long strip basins. Given: $(H_c)_{max} = 20$ m $+ 38$ m $= 58$ m; $H_n = 20$ m; $W = 300$ m; $T = 2,200$ m²/day; and

$$i = \frac{63,000,000 \text{ m}^3/\text{year}}{(300 \text{ m} \times 2,500 \text{ m})} = 84 \text{ m/year} = 0.23 \text{ m/day},$$

$$H_c - H_n = \frac{iW}{2T}\left(\frac{W}{4} + L_n\right)$$

$$58 \text{ m} - 20 \text{ m} = \frac{(0.23 \text{ m/day})(300 \text{ m})}{(2)(2200 \text{ m}^2/\text{day})}\left(\frac{300 \text{ m}}{4} + L_n\right)$$

$$L_n = 2350 \text{ m}$$

Thus, the control area should be established 2,350 m + 150 m = 2,500 m = 2.5 km or closer from the centerline of the recharge basin. ∎

For a round or square recharge area, the groundwater flow is in a radial direction from the recharge area. Bouwer et al.[15] used radial flow theory to develop an equation for the equilibrium height of the mound below the center of the recharge system, above the constant groundwater table at a distance R_n from the center of the recharge system.

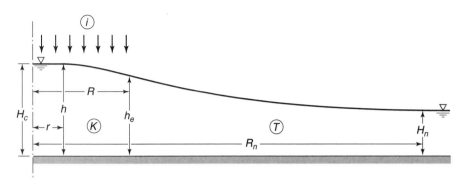

Figure 13.5.3. Geometry and symbols for groundwater mounds below round infiltration area (Bouwer[14]).

$$H_c - H_n = \frac{iR^2}{4T}\left(1 + 2\ln\frac{R_n}{R}\right) \tag{13.5.4}$$

where R is the radius or equivalent radius of the recharge area and R_n is the distance from the center of the recharge area to the water table control area (see Figure 13.5.3).

Equations 13.5.3 and 13.5.4 can be used to determine where groundwater should be recovered and to what depth groundwater levels should be pumped to keep the mound from rising too high. These equations can also be used to determine the dimensions of the recharge basins and to determine allowable recharge (infiltration) rates.

EXAMPLE 13.5.4

A circular recharge area with a radius of 500 m will be established over an unconfined aquifer where depth to groundwater is 73 m. If the estimated recharge rate is 0.14 m/day and the groundwater mound should be kept at least 10 m below the bottom of the recharge basin, determine the distance from the center of the recharge area to the control area where the original groundwater table is maintained. The transmissivity of the aquifer is 440 m²/day.

SOLUTION

The following information is given: $(H_c)_{max} - H_n = 73$ m $- 10$ m $= 63$ m; $i = 0.14$ m/day; $R = 500$ m; $T = 440$ m²/day. Equation 13.5.4 is used to compute R_n:

$$H_c - H_n = \frac{iR^2}{4T}\left(1 + 2\ln\frac{R_n}{R}\right)$$

$$63\text{ m} = \frac{(0.14\text{ m/day})(500\text{ m})^2}{(4)(440\text{ m}^2/\text{day})}\left(1 + 2\ln\frac{R_n}{500}\right)$$

$$R_n = 1478\text{ m}$$

Thus, the control area should be established approximately 1.5 km from the center of the recharge area.

13.5.3 Hantush Equation

Hantush[28] developed the following equation to determine the height of the water table (mound) as a function of location (x, y) and time t (see Figure 13.5.4):

$$h_{x,y,t} - H = \frac{v_a t}{4f}\{F[(W/2 + x)n,(L/2 + y)n] + F[(W/2 + x)n,(L/2 - y)n]$$
$$+ F[(W/2 - x)n,(L/2 + y)n] + F[(W/2 - x)n,(L/2 - y)n]\} \tag{13.5.5}$$

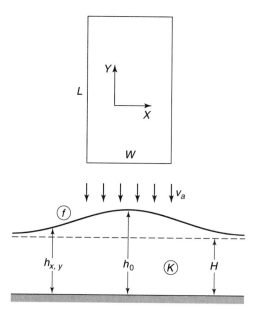

Figure 13.5.4. Geometry and symbols for recharge system and groundwater mound (Bouwer[14]).

where $h_{x,y,t}$ is the height of the water table above the impermeable layer at x, y, and time t; H is the original height of the water table above the impermeable layer; v_a is the arrival rate at the water table of water from the infiltration basin; t is the time since the start of the recharge; f is the fillable porosity $(1 > f > 0)$; L is the length of the recharge basin (in y direction); W is the width of the recharge basin (in x direction); $n = (4tT/f)^{-1/2}$; and $F(\alpha, \beta) = \int_0^1 erf\left(\alpha\tau^{-1/2}\right) \cdot erf\left(\beta\tau^{-1/2}\right)d\tau$. The values of $F(\alpha, \beta)$ are tabulated in Table 13.5.1, where $\alpha = (W/2 + x)n$ or $(W/2 - x)n$ and $\beta = (L/2 + y)n$ or $(L/2 - y)n$.

EXAMPLE 13.5.5

The average leakage from the artificially constructed Tempe Town Lake in Tempe, Arizona, is 1.5 ft/day. Depth to bedrock in the area is about 150 ft and the regional aquifer is 100 ft thick. The average transmissivity in the area is 10,000 ft²/day and the fillable porosity of the aquifer is 0.2. The geometry of the lake can be approximated by a rectangle that is 1,000 ft wide and 2 miles long. Assuming no natural gradient of groundwater and no mound control by groundwater pumping, how far away from the center of the lake must one move before the groundwater system doesn't feel the effect of the lake? How long does it take before the area becomes waterlogged?

SOLUTION

The answer to this problem is obtained by utilizing the Mathcad software and the exact solution proposed by Hantush. The most important assumption/limitation in the present solution is that there is no natural gradient of the regional groundwater. The application of the Hantush equation yielded the contour plots of groundwater levels (shown in Figure 13.5.5) at $t = 4, 6, 8, 12$ days. The maximum rise of the groundwater mound occurs directly below the center point of the lake and reaches 150 ft above the bedrock at about 12 days (i.e., reaches the ground surface or the bottom of the lake).

Figure 13.5.6 shows the propagation of the groundwater mound along the y-axis passing through the center point of the lake (i.e., $x = 0$) as well as the propagation along the x-axis (i.e., $y = 0$). It is difficult to determine the distance from the center of the lake where the mound rise is negligible since the solution is still transient when the area starts to become waterlogged. On the other hand, a 0.5-ft rise of the mound is observed at about 2,680 ft from the center of the lake along the x-direction at $t = 12$ days. Similarly, a 0.5-ft increase is observed 7,150 ft from the center point along the y-axis.

Groundwater mounding beneath the Tempe Town Lake has been controlled to minimize the water loss and the impact on regional groundwater levels by means of recovery wells located around the

Table 13.5.1 Values of the Function F (α, β) in Equation 13.5.5 for Different Values of α and β

α \ β	0.02	0.04	0.06	0.08	0.10	0.14	0.18	0.22	0.26	0.30	0.34	0.38	0.42	0.46	0.50	0.54	0.58	0.62
0.02	0.0041	0.0073	0.0101	0.0125	0.0146	0.0184	0.0216	0.0243	0.0267	0.0288	0.0306	0.0322	0.0337	0.0349	0.0361	0.0371	0.0380	0.0387
0.04	0.0073	0.0135	0.0188	0.0236	0.0278	0.0353	0.0416	0.0470	0.0518	0.0559	0.0596	0.0628	0.0657	0.0683	0.0705	0.0725	0.0743	0.0759
0.06	0.0101	0.0188	0.0266	0.0335	0.0398	0.0509	0.0602	0.0684	0.0754	0.0817	0.0871	0.0920	0.0963	0.1001	0.1035	0.1065	0.1091	0.1115
0.08	0.0125	0.0236	0.0335	0.0425	0.0508	0.0652	0.0776	0.0884	0.0978	0.1060	0.1133	0.1197	0.1254	0.1305	0.1350	0.1389	0.1425	0.1456
0.10	0.0146	0.0278	0.0398	0.0508	0.0608	0.0786	0.0939	0.1072	0.1188	0.1290	0.1381	0.1461	0.1532	0.1595	0.1650	0.1700	0.1744	0.1783
0.14	0.0184	0.0353	0.0509	0.0652	0.0786	0.1025	0.1232	0.1414	0.1573	0.1714	0.1839	0.1941	0.2048	0.2135	0.2212	0.2281	0.2343	0.2397
0.18	0.0216	0.0416	0.0602	0.0776	0.0939	0.1232	0.1490	0.1716	0.1916	0.2094	0.2251	0.2391	0.2515	0.2626	0.2724	0.2812	0.2890	0.2959
0.22	0.0243	0.0470	0.0684	0.0884	0.1072	0.1414	0.1716	0.1984	0.2222	0.2433	0.2621	0.2789	0.2938	0.3071	0.3189	0.3295	0.3389	0.3472
0.26	0.0267	0.0518	0.0754	0.0978	0.1188	0.1573	0.1916	0.2222	0.2494	0.2737	0.2954	0.3147	0.3320	0.3474	0.3612	0.3735	0.3844	0.3941
0.30	0.0288	0.0559	0.0817	0.1060	0.1290	0.1714	0.2094	0.2433	0.2737	0.3009	0.3252	0.3470	0.3665	0.3839	0.3995	0.4134	0.4257	0.4368
0.34	0.0306	0.0596	0.0871	0.1133	0.1391	0.1839	0.2251	0.2621	0.2954	0.3252	0.3520	0.3761	0.3976	0.4169	0.4341	0.4495	0.4633	0.4756
0.38	0.0322	0.0628	0.0920	0.1197	0.1461	0.1949	0.2391	0.2789	0.3147	0.3470	0.3761	0.4022	0.4256	0.4466	0.4654	0.4823	0.4973	0.5108
0.42	0.0337	0.0657	0.0963	0.1254	0.1532	0.2048	0.2515	0.2938	0.3320	0.3665	0.3976	0.4256	0.4508	0.4734	0.4937	0.5119	0.5281	0.5427
0.46	0.0349	0.0683	0.1001	0.1305	0.1595	0.2135	0.2626	0.3071	0.3474	0.3839	0.4169	0.4466	0.4734	0.4975	0.5161	0.5385	0.5559	0.5715
0.50	0.0361	0.0705	0.1035	0.1350	0.1650	0.2212	0.2724	0.3189	0.3612	0.3995	0.4341	0.4654	0.4937	0.5191	0.5420	0.5626	0.5810	0.5975
0.54	0.0371	0.0725	0.1065	0.1389	0.1700	0.2281	0.2812	0.3295	0.3735	0.4134	0.4495	0.4823	0.5119	0.5385	0.5626	0.5842	0.6036	0.6209
0.58	0.0380	0.0743	0.1091	0.1425	0.1744	0.2343	0.2890	0.3389	0.3844	0.4257	0.4633	0.4973	0.5281	0.5559	0.5810	0.6036	0.6238	0.6420
0.62	0.0387	0.0759	0.1115	0.1456	0.1783	0.2397	0.2959	0.3472	0.3941	0.4368	0.4756	0.5108	0.5427	0.5715	0.5975	0.6209	0.6420	0.6609
0.66	0.0394	0.0773	0.1136	0.1484	0.1718	0.2445	0.3020	0.3547	0.4027	0.4466	0.4865	0.5227	0.5556	0.5854	0.6122	0.6364	0.6582	0.6778
0.70	0.0401	0.0785	0.1154	0.1509	0.1849	0.2488	0.3075	0.3612	0.4104	0.4553	0.4962	0.5334	0.5672	0.5977	0.6254	0.6503	0.6728	0.6929
0.74	0.0406	0.0796	0.1117	0.1531	0.1876	0.2526	0.3123	0.3671	0.4172	0.4630	0.5048	0.5429	0.5774	0.6087	0.6371	0.6627	0.6857	0.7064
0.78	0.0411	0.0806	0.1185	0.1550	0.1900	0.2559	0.3166	0.3722	0.4232	0.4699	0.5125	0.5513	0.5865	0.6185	0.6475	0.6736	0.6972	0.7184
0.82	0.0415	0.0814	0.1198	0.1567	0.1921	0.2589	0.3203	0.3768	0.4286	0.4760	0.5192	0.5587	0.5946	0.6272	0.6567	0.6834	0.7074	0.7291
0.86	0.0419	0.0822	0.1209	0.1582	0.1940	0.2615	0.3237	0.3808	0.4333	0.4813	0.5252	0.5653	0.6017	0.6348	0.6648	0.6920	0.7165	0.7386
0.90	0.0422	0.0828	0.1219	0.1595	0.1957	0.2638	0.3266	0.3844	0.4374	0.4860	0.5305	0.5711	0.6080	0.6416	0.6721	0.6996	0.7245	0.7469
0.94	0.0425	0.0834	0.1228	0.1607	0.1971	0.2658	0.3292	0.3875	0.4411	0.4902	0.5351	0.5762	0.6136	0.6476	0.6784	0.7063	0.7316	0.7543
0.98	0.0428	0.0839	0.1236	0.1617	0.1984	0.2676	0.3314	0.3902	0.4442	0.4938	0.5392	0.5807	0.6184	0.6528	0.6840	0.7123	0.7378	0.7608
1.00	0.0429	0.0842	0.1239	0.1622	0.1990	0.2684	0.3324	0.3914	0.4457	0.4955	0.5410	0.5827	0.6206	0.6552	0.6865	0.7150	0.7406	0.7638
1.20	0.0437	0.0858	0.1263	0.1654	0.2030	0.2740	0.3396	0.4001	0.4558	0.5070	0.5540	0.5969	0.6362	0.6719	0.7044	0.7339	0.7605	0.7846
1.40	0.0441	0.0866	0.1275	0.1669	0.2049	0.2767	0.3431	0.4043	0.4608	0.5127	0.5603	0.6039	0.6438	0.6801	0.7132	0.7432	0.7704	0.7949
1.80	0.0444	0.0871	0.1283	0.1680	0.2062	0.2785	0.3454	0.4071	0.4641	0.5165	0.5645	0.6086	0.6489	0.6856	0.7190	0.7494	0.7769	0.8018
2.00	0.0444	0.0871	0.1284	0.1681	0.2064	0.2787	0.3457	0.4075	0.4645	0.5169	0.5651	0.6092	0.6495	0.6863	0.7198	0.7502	0.7778	0.8027
2.20	0.0444	0.0872	0.1284	0.1682	0.2065	0.2788	0.3458	0.4076	0.4646	0.5171	0.5653	0.6094	0.6497	0.6865	0.7200	0.7505	0.7781	0.8030
2.50	0.0444	0.0872	0.1284	0.1682	0.2065	0.2788	0.3458	0.4077	0.4647	0.5172	0.5653	0.6095	0.6498	0.6867	0.7202	0.7506	0.7782	0.8032
3.00	0.0444	0.0872	0.1284	0.1682	0.2065	0.2789	0.3458	0.4077	0.4647	0.5172	0.5654	0.6095	0.6499	0.6867	0.7202	0.7506	0.7782	0.8032

Table 13.5.1 (continued) Values of the Function F (α, β) in Equation 13.5.5 for Different Values of α and β

α \ β	0.62	0.66	0.70	0.74	0.78	0.82	0.86	0.90	0.94	0.98	1.00	1.20	1.40	1.80	2.00	2.20	2.50	3.00
0.02	0.0387	0.0394	0.0401	0.0406	0.0411	0.0415	0.0419	0.0422	0.0425	0.0428	0.0429	0.0437	0.0441	0.0444	0.0444	0.0444	0.0444	0.0444
0.04	0.0759	0.0773	0.0785	0.0796	0.0806	0.0814	0.0822	0.0828	0.0834	0.0839	0.0842	0.0858	0.0866	0.0871	0.0871	0.0872	0.0882	0.0882
0.06	0.1115	0.1136	0.1154	0.1171	0.1185	0.1198	0.1209	0.1219	0.1228	0.1236	0.1239	0.1263	0.1275	0.1283	0.1284	0.1284	0.1284	0.1284
0.08	0.1456	0.1484	0.1509	0.1531	0.1550	0.1567	0.1582	0.1595	0.1606	0.1617	0.1622	0.1654	0.1669	0.1680	0.1681	0.1682	0.1682	0.1682
0.10	0.1783	0.1818	0.1849	0.1876	0.1900	0.1921	0.1940	0.1957	0.1971	0.1984	0.1990	0.2030	0.2049	0.2062	0.2064	0.2065	0.2065	0.2065
0.14	0.2397	0.2445	0.2488	0.2526	0.2559	0.2589	0.2615	0.2638	0.2658	0.2676	0.2684	0.2740	0.2777	0.2785	0.2787	0.2788	0.2788	0.2788
0.18	0.2959	0.3020	0.3075	0.3123	0.3166	0.3203	0.3237	0.3266	0.3292	0.3314	0.3324	0.3396	0.3431	0.3454	0.3457	0.3454	0.3454	0.3454
0.22	0.3472	0.3547	0.3612	0.3671	0.3722	0.3768	0.3808	0.3844	0.3875	0.3902	0.3914	0.4001	0.4043	0.4071	0.4075	0.4076	0.4077	0.4077
0.26	0.3941	0.4027	0.4104	0.4172	0.4232	0.4286	0.4333	0.4374	0.4411	0.4442	0.4457	0.4558	0.4608	0.4641	0.4645	0.4646	0.4647	0.4647
0.30	0.4368	0.4466	0.4553	0.4630	0.4699	0.4760	0.4813	0.4860	0.4902	0.4938	0.4955	0.5070	0.5127	0.5165	0.5159	0.5171	0.5172	0.5172
0.34	0.4756	0.4865	0.4962	0.5048	0.5125	0.5192	0.5252	0.5305	0.5351	0.5392	0.5410	0.5540	0.5603	0.5645	0.5651	0.5653	0.5653	0.5654
0.38	0.5108	0.5227	0.5334	0.5429	0.5513	0.5587	0.5653	0.5711	0.5762	0.5807	0.5827	0.5969	0.6039	0.6086	0.6092	0.6094	0.6095	0.6095
0.42	0.5427	0.5556	0.5672	0.5774	0.5865	0.5946	0.6017	0.6080	0.6136	0.6184	0.6206	0.6362	0.6438	0.6489	0.6495	0.6497	0.6498	0.6499
0.46	0.5715	0.5854	0.5977	0.6087	0.6185	0.6272	0.6348	0.6416	0.6476	0.6528	0.6552	0.6719	0.6801	0.6856	0.6863	0.6865	0.6867	0.6867
0.50	0.5975	0.6122	0.6254	0.6371	0.6475	0.6567	0.6648	0.6721	0.6784	0.6840	0.6865	0.7044	0.7132	0.7190	0.7198	0.7200	0.7202	0.7202
0.54	0.6209	0.6364	0.6503	0.6627	0.6736	0.6834	0.6920	0.6996	0.7063	0.7123	0.7150	0.7379	0.7432	0.7494	0.7502	0.7505	0.7506	0.7506
0.58	0.6420	0.6578	0.6728	0.6857	0.6972	0.7074	0.7165	0.7245	0.7316	0.7378	0.7406	0.7605	0.7704	0.7799	0.7778	0.7781	0.7782	0.7782
0.62	0.6609	0.6778	0.6929	0.7064	0.7184	0.7291	0.7386	0.7469	0.7543	0.7608	0.7638	0.7846	0.7949	0.8018	0.8027	0.8030	0.8032	0.8032
0.66	0.6778	0.6953	0.7110	0.7250	0.7375	0.7486	0.7584	0.7671	0.7748	0.7816	0.7846	0.8064	0.8171	0.8243	0.8252	0.8255	0.8257	0.8257
0.70	0.6929	0.7110	0.7272	0.7417	0.7546	0.7660	0.7762	0.7852	0.7932	0.8002	0.8034	0.8259	0.8370	0.8445	0.8454	0.8458	0.8460	0.8460
0.74	0.7064	0.7250	0.7414	0.7566	0.7698	0.7816	0.7921	0.8014	0.8096	0.8168	0.8201	0.8434	0.8549	0.8627	0.8636	0.8640	0.8642	0.8642
0.78	0.7184	0.7375	0.7546	0.7698	0.7834	0.7956	0.8063	0.8159	0.8243	0.8317	0.8351	0.8591	0.8710	0.8789	0.8799	0.8803	0.8805	0.8805
0.82	0.7291	0.7486	0.7660	0.7816	0.7956	0.8080	0.8190	0.8288	0.8374	0.8450	0.8485	0.8731	0.8853	0.8935	0.8945	0.8949	0.8951	0.8951
0.86	0.7386	0.7584	0.7762	0.7921	0.8063	0.8190	0.8302	0.8402	0.8491	0.8569	0.8604	0.8855	0.8980	0.9065	0.9075	0.9079	0.9081	0.9081
0.90	0.7469	0.7671	0.7852	0.8014	0.8159	0.8288	0.8402	0.8504	0.8594	0.8674	0.8710	0.8966	0.9094	0.9180	0.9191	0.9195	0.9197	0.9197
0.94	0.7543	0.7748	0.7932	0.8096	0.8243	0.8374	0.8491	0.8594	0.8686	0.8767	0.8803	0.9064	0.9195	0.9282	0.9294	0.9298	0.9300	0.9300
0.98	0.7608	0.7816	0.8002	0.8168	0.8317	0.8450	0.8569	0.8674	0.8767	0.8849	0.8886	0.9151	0.9284	0.9373	0.9384	0.9389	0.9391	0.9391
1.00	0.7638	0.7846	0.8034	0.8201	0.8351	0.8485	0.8604	0.8710	0.8803	0.8886	0.8924	0.9191	0.9324	0.9414	0.9426	0.9430	0.9432	0.9433
1.20	0.7846	0.8064	0.8259	0.8434	0.8591	0.8731	0.8855	0.8966	0.9064	0.9151	0.9191	0.9472	0.9614	0.9709	0.9722	0.9726	0.9728	0.9729
1.40	0.7949	0.8171	0.8370	0.8549	0.8710	0.8853	0.8980	0.9094	0.9195	0.9284	0.9324	0.9614	0.9759	0.9858	0.9871	0.9875	0.9878	0.9878
1.80	0.8018	0.8243	0.8445	0.8627	0.8789	0.8935	0.9065	0.9180	0.9282	0.9373	0.9414	0.9709	0.9858	0.9959	0.9972	0.9972	0.9979	0.9980
2.00	0.8027	0.8252	0.8454	0.8636	0.8799	0.8945	0.9075	0.9191	0.9294	0.9384	0.9426	0.9722	0.9871	0.9972	0.9985	0.9990	0.9992	0.9994
2.20	0.8030	0.8255	0.8458	0.8640	0.8803	0.8949	0.9079	0.9195	0.9298	0.9389	0.9430	0.9726	0.9875	0.9977	0.9990	0.9995	0.9997	0.9998
2.50	0.8032	0.8257	0.8460	0.8642	0.8805	0.8951	0.9081	0.9197	0.9300	0.9391	0.9432	0.9728	0.9878	0.9979	0.9992	0.9997	1.0000	1.0000
3.00	0.8032	0.8257	0.8460	0.8642	0.8805	0.8951	0.9081	0.9197	0.9300	0.9391	0.9433	0.9729	0.9878	0.9980	0.9993	0.9998	1.0000	1.0000

Source: From Hantush[28]

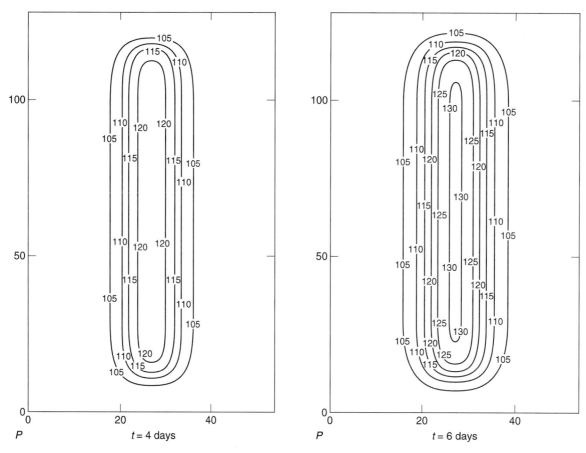

Figure 13.5.5. Growth of groundwater mound beneath Tempe Town Lake using Hantush Equation (Scale: 1 unit = 100 ft).

periphery of the lake since its construction. It should be noted that the solution presented here assumes no artificial mound control. ∎

EXAMPLE 13.5.6

A square recharge basin 800 m × 800 m is constructed over an area where an unconfined aquifer is found in the subsurface. The thickness of the aquifer is 50 m and the average transmissivity is 1,140 m²/day. The fillable porosity of the vadose zone is 0.25. If the aquifer is recharged at a rate of 25,000,000 m³ per year and the maximum permissible rise of the groundwater mound is 25 m above the original water table, how long does it take for the mound to rise to its maximum level? How much water can be stored underground until that time?

SOLUTION

We are interested only in the maximum rise of the mound. The given information is $W = 800$ m; $L = 800$ m; $x = 0$ m; $y = 0$ m; $f = 0.25$; $H = 50$ m; $T = 1,140$ m²/day and

$$v_a = \frac{25 \times 10^6 \text{ m}^3/\text{year}}{800 \text{ m} \times 800 \text{ m}} = 39.0625 \text{ m/year} = 0.107 \text{ m/day}$$

$$h_{0,0,t} = 50 \text{ m} + 25 \text{ m} = 75 \text{ m}$$

and

$$n = \left[\frac{4tT}{f}\right]^{-1/2} = \left[\frac{4(t)\left(1,140 \text{ m}^2/\text{day}\right)}{0.25}\right]^{-1/2} = 0.0074(t)^{-1/2}$$

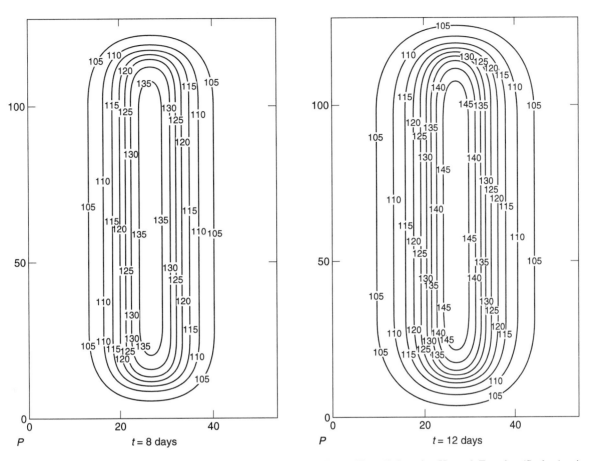

Figure 13.5.5 (continued). Growth of groundwater mound beneath Tempe Town Lake using Hantush Equation (Scale: 1 unit = 100 ft).

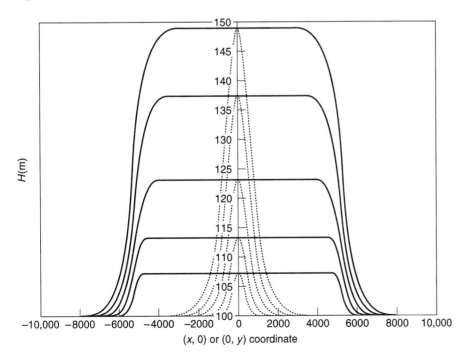

Figure 13.5.6. Propagation of groundwater mound along x- and y-axes
Solid line: Propagation along y-axis for $t = 1$ day, 2 days, 4 days, 8 days, and 12 days
Dashed line: Propagation along x-axis for $t = 1$ day, 2 days, 4 days, 8 days, and 12 days

Table 13.5.2 Results for Example 13.5.6

t (years)	n	$\alpha = \beta$	$F(\alpha, \beta)$	$h_{0,0,t}$ (m)
0.50	5.481E–04	0.22	0.1984	65.50
0.74	4.505E–04	0.18	0.1490	67.22
1.23	3.495E–04	0.14	0.1025	69.70
2.40	2.502E–04	0.10	0.0608	72.80
3.75	2.001E–04	0.08	0.0425	74.90
6.70	1.497E–04	0.06	0.0266	77.84
15.00	1.001E–04	0.04	0.0135	81.63

Substituting the above information into Equation 13.5.5 yields

$$h_{0,0,t} - H = \frac{v_a t}{4f}\left\{ F\left[(W/2+x)n,(L/2+y)n\right] + F\left[(W/2+x)n,(L/2-y)n\right] \right.$$

$$\left. + F\left[(W/2-x)n,(L/2+y)n\right] + F\left[(W/2-x)n,(L/2-y)n\right]\right\}$$

$$h_{0,0,t} - 50 \text{ m} = \frac{(0.107 \text{ m/day})(t)}{4(0.25)}\left\{4F(400n, 400n)\right\}$$

The values of $h_{0,0,t}$ can be found for a range of t values using Table 13.5.1 as shown in the following Table 13.5.2. Note that t values in the table are chosen such that no interpolation was necessary while using Table 13.5.1. The results show that the height of the mound would reach 75 m above the impermeable bed in about 3.75 years. A more accurate answer could be obtained using the integral expression for $F(\alpha, \beta)$ instead of Table 13.5.1, while the answer above is sufficient for practical purposes. Finally, the volume of water that can be stored underground in 3.75 years is given by

$$V = (25 \times 10^6 \text{ m}^3/\text{year})(3.75 \text{ year}) = 93.75 \text{ million m}^3 \qquad \blacksquare$$

13.6 STORMWATER INFILTRATION BASIN MOUND DEVELOPMENT

Groundwater mounds resulting from stormwater infiltration basins have been studied by several investigators (Griffen and Warrington,[24] Guo,[25, 26, 27] Marino,[39, 40] Molden et al.,[42] Morel-Seytoux et al.,[44] Ortiz et al.,[50] Sumner et al.,[60] and Swamee and Ojha[62]).

13.6.1 Potential Flow Model for a Trench

Guo[25, 26, 27] utilized potential flow theory to depict the seepage flow through soils, considering both vertical and lateral movements. Guo used a stream function for infiltrating flow under a trench, defined to satisfy two-dimensional continuity and irrotationality (Guo[25]) as $\Psi = (f/D)xy$. Here, f is the infiltration rate, D is the saturated depth, x is the horizontal distance from the central axis of the basin, and y is the vertical distance below the basin. Variables and parameters are illustrated in Figure 13.6.1. At a point $C(B, D)$ the stream function is $\Psi_c = q = fB$, where q is the infiltration volume rate and B is the half-width of the trench. In order to maintain continuity along the streamline, Ψ_c, between the basin and any cross section at a vertical distance y, the width of the wetting front, w, can be determined using $\Psi = (f/D)xy$, so that $w = BD/y$.

The velocity vectors of seepage flow are determined by taking the derivative of $\Psi = (f/D)xy$ with respect to x and y to obtain $u = (f/D)x$ and $v = -(f/D)y$. For steady-state conditions, the flow rate is constant at sections AC, PR, and RF, shown in Figure 13.6.1. When the soil suction is ignored, the hydraulic gradient at PR can be approximated as unity ($i = -1$, downward), so that from Darcy's law, the flow of recharge at PR is $v = q/W = K_i W/W = -K_y$, where K_y is the hydraulic conductivity in the vertical direction. Also at $y = H$, $v = -(f/D)H$ and also $v = -K_y$, so that $(f/D)H = K_y$. Solving for $D = (f/K_y)H$, then we find $D = \lambda_y H$, where $\lambda_y = f/K_y$.

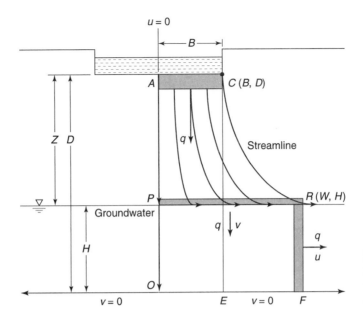

Figure 13.6.1. Potential flow pattern for infiltrating water under a trench (Guo[27]).

Streamlines of recharge flow near the mound are concentrated in the upper or active thickness of the aquifer, with much less flow and almost stagnant water in the deeper portion of the aquifer (Bouwer[14]). The active flow depth, H, below the initial groundwater table can be estimated by the *Dupuit–Forchheimer equation*,

$$q = \Psi_c = \frac{K_x}{2} \frac{\left(D^2 - H^2\right)}{W - B}$$ (13.6.1)

in which K_x is the hydraulic conductivity in the lateral direction.

Guo[25] reasoned that stormwater infiltration basins are small in size and are operated using short loading periods, so that the height of the water mound decays toward the initial groundwater table. Under this condition, Equation 13.6.1 can be approximated to the edge of the area of recharge, that is, point $R(W, H)$ in Figure 13.6.1. Equation 13.6.1 divided by the effective flow depth, H, yields a sectional velocity which must be equal to $u = (f/D)x$ at $x = W$; resulting in an effective thickness of the aquifer expressed as

$$\frac{H}{B} = \sqrt{\frac{2\lambda_x}{\lambda_y + 1}}$$ (13.6.2)

where

$$\lambda_x = \frac{f}{K_x}$$

Using $D = \lambda_y H$, we can compute the saturated depth:

$$\frac{D}{B} = \lambda_y \sqrt{\frac{2\lambda_x}{\lambda_y + 1}}$$ (13.6.3)

EXAMPLE 13.6.1

Consider an infiltration basin 20 ft by 180 ft with an average soil infiltration rate of 1.8 in/hr (3.6 ft/day) and a hydraulic conductivity of 1.8 ft/day. The subsurface medium is isotropic: $K_x = K_y = K$. Determine the saturated depth, D, and the effective flow depth, H, for a half-width of the trench of $B = 10$ ft. Also determine Z, the required distance to the groundwater table.

SOLUTION   The ratio of the average infiltratration rate to the hydraulic conductivity is $f_a/K = 2.0$. The subsurface medium is isotropic so that $\lambda_x = \lambda_y = f_a/K_x = f_a/K_y = 2.0$. From Equation 13.6.3, we have

$$\frac{D}{B} = \lambda_y \sqrt{\frac{2\lambda_x}{\lambda_y + 1}} = 2\sqrt{\frac{2(2.0)}{2.0 + 1}} = 2.309 \text{ and } D = B(D\ B) = (10\,\text{ft})(2.309) = 23.09\,\text{ft}.$$

The effective flow depth, H, is computed using Equation 13.6.2, so that

$$\frac{H}{B} = \sqrt{\frac{2\lambda_x}{\lambda_y + 1}} = \sqrt{\frac{2(2.0)}{2.0 + 1}} = 1.155 \text{ and } H = B(H\ B) = (10\ \text{ft})(1.155) = 11.55\ \text{ft}.$$

The required distance to the groundwater table is $Z = D - H = 23.09 - 11.55 = 11.54$ ft.   ∎

13.6.2 Potential Flow Model for Circular Basin

The flow pattern below a circular basin (see Figure 13.6.2), described by the stream function, $\Psi = \pi(f/D)r^2y$, satisfies the continuity and irrotationality equations for three-dimensional axisymmetric flow (Guo[25]). Streamlines below the basin are distributed as concentric circles with $\Psi = 0$ along the y-axis and $\Psi =$ infiltration volume rate at the circumference of the basin bottom (i.e., $C(r, y) = (R_0, D)$ in Figure 13.6.2). R_0 is the radius of the circular basin.

The infiltration volume rate released from the circular basin is $q = f\pi R_0^2$. To maintain continuity of flow, then $q = f\pi R_0^2 = \Psi = \pi(f/D)r^2y$, so solving for r gives $r = \sqrt{D/y}\,R_0$. Then at section PR in Figure 13.6.2, the radius of the area of recharge must pass point $R(R, H)$, expressed as $R = \sqrt{D/H}\,R_0$. Using the same approach as for the rectangular basin, the following relationships can be developed for the subsurface geometry,

$$\frac{H}{R_0} = \sqrt{\frac{\lambda_r \ln \lambda_y}{2(\lambda_y^2 - 1)}} \tag{13.6.4}$$

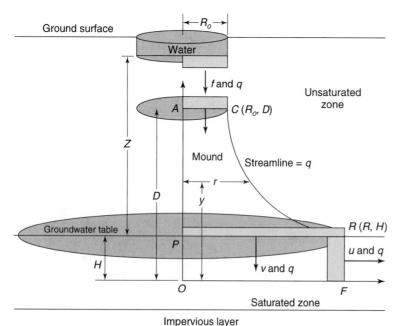

Figure 13.6.2. Flow pattern under circular basin (Guo[27]).

where

$$\lambda_r = \frac{f}{K_r}$$

and K_r is the hydraulic radius in the radial direction. Substituting

$$R = \sqrt{\frac{D}{H}} R_0$$

into Equation 13.6.4 gives the following expression for computing the saturation depth:

$$\frac{D}{R_0} = \lambda_y \sqrt{\frac{\lambda_r \ln \lambda_y}{2(\lambda_y^2 - 1)}} \qquad (13.6.5)$$

EXAMPLE 13.6.2

A stormwater infiltration basin has a radius of 72 ft. The basin is lined with loamy sand and the resulting average infiltration rate is 1.60 in/hr. Determine the required saturation depth, Y_0, between the bottom of the basin and the groundwater table if the hydraulic conductivity of the vadose zone is 1.6 ft/day. Assume an isotropic subsurface.

SOLUTION

The following information is given: $f = 1.60$ in/hr; $K_r = K_y = 1.6$ ft/day $= 0.8$ in/hr; and $R_0 = 72$ ft. The required saturation depth, Y_0, is equal to $(D - H)$. Thus, combining Equations 13.6.4 and 13.6.5 yields

$$\frac{Y_0}{R_0} = (\lambda_y - 1)\sqrt{\frac{\lambda_r \ln \lambda_y}{2(\lambda_y^2 - 1)}}$$

Since the subsurface is isotropic,

$$\lambda_r = \lambda_y = \frac{f}{K} = \frac{1.6 \text{ in/hr}}{0.8 \text{ in/hr}} = 2.0$$

Then

$$\frac{Y_0}{R_0} = (2.0 - 1)\sqrt{\frac{2(\ln 2)}{2(2^2 - 1)}} = 0.481$$

$$Y_0 = 0.481(72 \text{ ft}) = 34.6 \text{ ft}$$

Thus, the minimum required distance between the bottom of the basin and the groundwater table is about 35 ft, so that there will be a sufficient hydraulic gradient in the subsurface.   ■

EXAMPLE 13.6.3

The saturation depth, the distance between the bottom of the basin and the groundwater table, beneath a stormwater infiltration basin is 5.5 m. If the design infiltration rate from the basin is 5.13 cm/hr, determine and plot the required basin radius as a function of the hydraulic conductivity. Assume the hydraulic conductivity beneath the basin is isotropic.

SOLUTION

The required saturation depth Y_0 is given by $(D - H)$. As in the previous example, combining Equations 13.6.4 and 13.6.5 yields

$$\frac{Y_0}{R_0} = (\lambda_y - 1)\sqrt{\frac{\lambda_r \ln \lambda_y}{2(\lambda_y^2 - 1)}}$$

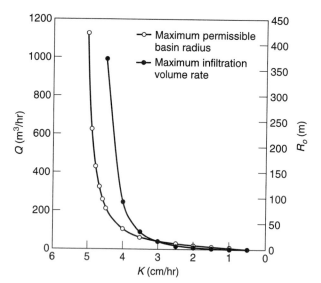

Figure 13.6.3. Results for Example 13.6.3

However, since the saturation depth is specified and fixed in this problem, the basin radius is the design parameter as a function of the unknown hydraulic conductivity. Thus, we must solve this relationship for R_0:

$$R_0 = \frac{Y_0}{\left(\lambda_y - 1\right)\sqrt{\dfrac{\lambda_r \ln \lambda_y}{2\left(\lambda_y^2 - 1\right)}}}$$

Assuming an isotropic formation

$$\lambda_r = \lambda_y = \frac{f}{K} = \frac{5.13 \text{ cm/hr}}{K}$$

and substituting into above the expression, we have

$$R_0 = \frac{5.5 \text{ m}}{\left(\dfrac{5.13}{K} - 1\right)\sqrt{\dfrac{\dfrac{5.13}{K} \ln \dfrac{5.13}{K}}{2\left(\left(\dfrac{5.13}{K}\right)^2 - 1\right)}}}$$

where K is in cm/hr and the basin radius, R_0, is in meters. A plot of R_0 versus K is shown in Figure 13.6.3.

If $K \geq 5.13$ cm/hr, this would be the case of a direct downward recharge to the groundwater. In fact, $(R_0)_{max} \rightarrow +\infty$ as $K \rightarrow 5.13$ cm/hr, as shown in Figure 13.6.3. On the other hand, when $K \geq 5.13$ cm/hr, there is a maximum permissible basin radius depending on the K value while the saturation depth is a fixed value. This constraint is to assure there will be sufficient hydraulic gradients beneath the basin to prevent the infiltration system from being backed up (Guo[26, 27]). Note that as the permissible basin radius decreases with decreasing K values, the corresponding infiltration volume rates decrease even more significantly. This shows that a system with a high f/K ratio, while the saturation depth is fixed, may prove to be very inefficient as a stormwater infiltration system. ∎

13.6.3 Mound Growth

The growth (before steady state is reached) of a mound for flow under an infiltration trench is depicted in Figure 13.6.4. To define the storage effect in the soil as a function of time, the continuity equation between the inflow at section AC and the outflow at section PR is

$$\frac{dV}{dt} = q - q_0$$

where V is the soil water storage volume, t is time, q is the infiltrating volumetric rate, and q_0 is the outflow rate to recharge. The incremental volume, dV, at depth y is

$$dV = S_y w \, dy = S_y \frac{D}{y} B dy$$

in which S_y is the soil yield. The outflow rate, q_0, at Section PR is defined using Darcy's law as

$$q_0 = S_y K_u W = S_y K_u \frac{D}{H} B$$

in which $K_u = \sqrt{K_r K_y}$ is the representative or equivalent hydraulic conductivity for the growth of the mound and

$$w = \frac{BD}{y}$$

The growth function of an infiltration trench can be developed by substituting the expression $dV = S_y(D/y)B \, dy$ into the continuity equation,

$$\frac{dV}{dt} = q - q_0$$

and integrating from the thickness of the aquifer, H, to the depth y to obtain

$$y = He^{-kt/D} \qquad (13.6.6)$$

in which k is a pseudo–hydraulic conductivity for the growth of a water mound, expressed as

$$k = \left(\frac{f}{S_y} - K_u \frac{D}{H} \right) \qquad (13.6.7)$$

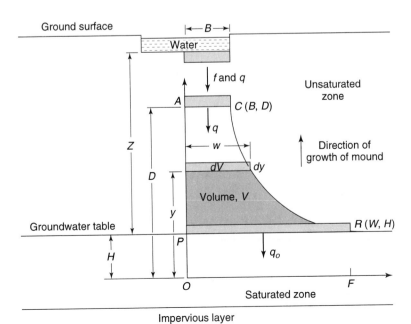

Figure 13.6.4. Growth of water mound underneath a trench (Guo[27]).

Equation 13.6.6 is an increasing function with depth y, which cannot exceed D in Equation 13.6.3 for an infiltration trench or Equation 13.6.5 for a circular infiltration basin. Equation 13.6.6 can be used to estimate the growth of a water mound for an elapsed time t. For circular infiltration basins, the same derivation procedure can be used to show that Equation 13.6.6 also applies. Generally, the exponent $(-kt/D)$ is small enough so that Equation 13.6.6 reduces to (Guo[26, 27])

$$y = H\left(1 - \frac{kt}{D}\right) \tag{13.6.8}$$

EXAMPLE 13.6.4

A circular stormwater infiltration basin has a radius of 29 m. The saturation depth (distance from the basin bottom to the impermeable bedrock) is 18.5 m and the saturated thickness of the unconfined aquifer beneath the basin is 13.5 m. The saturated hydraulic conductivity of the aquifer is 2.4 m/day. The infiltration rate from the basin is 1.25 cm/hr and the soil yield, S_y, is estimated as 0.25. Assuming a representative unsaturated hydraulic conductivity of 0.23 m/day, compute the pseudo-hydraulic conductivity.

SOLUTION

To determine the pseudo–hydraulic conductivity, we solve Equation 13.6.7 with $f = 1.25$ cm/hr $= 0.3$ m/day to obtain

$$k = \left(\frac{f}{S_y} - K_u \frac{D}{H}\right) = \left(\frac{0.3 \text{ m/day}}{0.25} - (0.23 \text{ m/day})\frac{(18.5 \text{ m})}{(13.5 \text{ m})}\right) = 0.8848 \text{ m/day}$$

To estimate the growth of the mound, we would solve Equation 13.6.6. ∎

13.6.4 Mound Recession

A mound begins to recede during the time of no infiltration from a trench or basin by gradually recharging the aquifer (see Figure 13.6.5). The volume release from the soil medium is equal to the volume of recharge, at Section PR to the groundwater table. The recession of a mound starts at the bottom. If the receding water front maintains the shape as the saturated, the volume released from the soil is estimated as

$$V_r = S_y \int_{y=y}^{y=D} w \, dy = S_y \int_{y=y}^{y=D} \frac{BD}{y} \, dy = S_y DB \ln\left(\frac{D}{y}\right) \tag{13.6.9}$$

Considering vertical flow through section PR, the volume of recharge to the groundwater table is

$$V_r = S_y K_u Wt = S_y K_u \frac{D}{H} Bt \tag{13.6.10}$$

Equating the above two expressions for the volume gives the following recession function for a mound:

$$y = De^{-(K_u t/H)} \tag{13.6.11}$$

For a small exponent, the recession function reduces to

$$y = D\left(1 - \frac{K_u t}{H}\right) \tag{13.6.12}$$

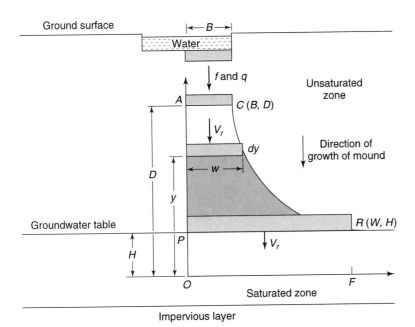

Figure 13.6.5. Recession of water mound (Guo[27]).

The unsaturated hydraulic conductivity, K_u, is the most sensitive parameter in using the above models to predict the growth and recession of a water mound. K_u varies with respect to the soil moisture content as discussed in Chapter 3. Guo[26] compared the results of using the recession function described above with results from using the unsaturated flow simulation model VS2D developed by Lappala et al.[38]

13.7 INNOVATIVE APPROACHES

Several innovative approaches employing artificial recharge are described by Alley et al[3] as follows:

13.7.1 Wildwood, New Jersey

Wildwood, New Jersey, is a resort town on a barrier island along the Atlantic Coast. As a resort community, it has a large influx of tourists in the summertime. The population can increase from about 5,000 during the winter to 30,000 at the height of the summer tourist season. The Wildwood Water Utility withdraws water from wells located about 5 miles inland from the barrier island. To supply water from these wells to meet the island's needs during the summer would require a large pumping facility and transmission lines that would be little used the rest of the year. To avoid these excessive costs, the utility injects groundwater into a shallow aquifer on the island during periods of low demand and withdraws the water in the summer by using dual injection and recovery wells. This system, operated since 1967, represents perhaps the oldest operational aquifer storage recovery (ASR) project in the United States. In ASR, water is injected underground, commonly into nonpotable or saline aquifers, where it forms a lens of good quality water for later recovery from the same well(s). Operation of a typical ASR installation is designed to smooth out annual variability in water demand by recharging aquifers during periods of low demand and recovering the water during periods of high demand. Advantages of ASR over other artificial recharge schemes are that it uses very little land (especially compared to surface spreading) and reduces the cost and maintenance of separate injection and recovery wells.

13.7.2 Orange County, California

Orange County, near Los Angeles, California, receives an average of only 13 to 15 inches of rainfall annually, yet sustains a population of approximately 2.5 million people. A groundwater basin that underlies the northwestern half of the county supplies about 75 percent of the total water demand. As the area developed from a thriving agricultural center into a highly urbanized area, increased demands for groundwater resulted in a gradual lowering of the water table below sea level and encroachment of saltwater from the Pacific Ocean. To prevent further saltwater intrusion and to replenish the groundwater supply, the Orange County Water District operates a hydraulic barrier system composed of a series of 23 multipoint wells that inject fresh water into the aquifer, blocking further passage of seawater. The source of injection water is a blended combination of deep well water and recycled secondary effluent. The recycled product meets drinking-water standards through advanced treatment processes of reverse osmosis and activated carbon adsorption.

13.7.3 Long Island, New York

Groundwater is the sole source of fresh water for the more than 3 million people who live on Long Island, outside the metropolitan New York City boundary. To help replenish the aquifer, as well as reduce urban flooding and control saltwater intrusion, more than 3,000 recharge basins dispose of storm runoff at an average rate of about 150 million gallons per day. Initially, many of these basins were abandoned gravel pits, but since 1936 urban developers are required to provide recharge basins with new developments. Practically all basins are unlined excavations in the upper glacial deposits and have areas from less than 0.1 to more than 30 acres.

13.7.4 Orlando, Florida

Large volumes of reclaimed water, which has undergone advanced secondary treatment, are reused through land-based applications in a 40-square-mile area near Orlando, Florida. These applications include citrus crop irrigation and artificial recharge to the surficial aquifer through rapid infiltration basins.

13.7.5 Dayton, Ohio

Dayton, Ohio, is heavily dependent upon groundwater to meet municipal and industrial water-supply needs. Nearly one-fourth of all groundwater used in Ohio is withdrawn from wells completed in a sole-source sand and gravel aquifer that underlies the Dayton metropolitan area. Much of the water is pumped from a 30- to 75-foot thick shallow aquifer that underlies the Mad River Valley. To ensure that groundwater levels are maintained high enough to allow for large drawdowns by high-capacity wells, an artificial recharge system has been in place since the 1930s. The source of recharge is streamflow diverted from the Mad River into a series of interconnected infiltration ditches and lagoons that occupy about 20 acres on Rohrers Island. To meet increasing demand, a new municipal well field was developed in the 1960s in a section of the aquifer north of Dayton. Here, water is pumped from the Great Miami River into a series of ponds and lagoons, some of which also serve as water hazards on a city-owned golf course. Recharge lagoons at both well fields are periodically drained, and accumulated muck and silt are excavated to maintain a high rate of infiltration into the underlying aquifer.

13.8 INDUCED RECHARGE

Direct methods of artificial recharge described above involve the conveyance of surface water to some point where it enters the ground. Distinguished from these is the method of induced

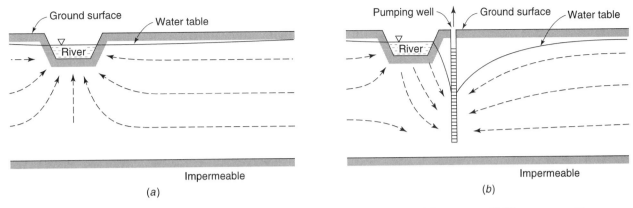

Figure 13.8.1. Induced recharge resulting from a well pumping near a river. (*a*) Natural flow pattern (*b*) Flow pattern with pumping well

recharge, accomplished by withdrawing groundwater at a location adjacent to a river or lake so that lowering of the groundwater level will induce water to enter the ground from the surface source. The schematic cross sections of a river valley in Figure 13.8.1 show flow patterns with and without induced infiltration from a stream. On the basis of this definition, wells located directly adjacent to and fed largely by surface water provide artificial recharge. The hydraulics of wells located near streams is described in Chapter 4; the construction and operation of wells, collectors, and galleries is described in Chapter 5.

Induced infiltration where supplied by a perennial stream ensures a continuing water supply even though overdraft conditions may exist in nearby areas supplied only by natural recharge. The method has proved effective in unconsolidated formations of permeable sand and gravel hydraulically connected between stream and aquifer. The amount of water induced into the aquifer depends on the rate of pumping, permeability, type of well, distance from surface stream, and natural groundwater movement. It is important that the velocity of the surface stream be sufficient to prevent silt deposition from sealing the streambed.

Studies of water quality have shown that induced recharge can furnish water free of organic matter and pathogenic bacteria. Because surface water commonly is less mineralized than groundwater, water obtained by induced infiltration, being a mixture of two water sources, possesses a higher quality than natural groundwater.

PROBLEMS

13.5.1 What should be the maximum recharge rate if the water table should be at least 1 m below the soil surface (bottom of basin) in order to maintain unimpeded infiltration rates? A restricting layer 3 m thick is detected in the vadose zone. The top of the restricting layer is 8.6 m below the bottom of the recharge basin. The hydraulic conductivities of the restricting layer and the soil above the restricting layer are 0.022 m/day and 7.8 m/day, respectively.

13.5.2 The artificially constructed Tempe Town Lake in Tempe, AZ, is approximately two miles long and 800–1,200 ft wide, covering 222 acres of surface area. The depth to bedrock under the lake is about 150 ft and the saturated thickness of the aquifer is 100 ft. The average transmissivity in the area is about 10,000

ft^2/day. If the leakage from the lake is about 1.5 ft/day and the maximum permissible rise of groundwater mound is 140 ft above the bedrock, how far should the control area be established from the centerline of the lake? Assume steady-state conditions and no natural groundwater gradient.

13.5.3 A square recharge basin 1 mile × 1 mile is proposed over a new area. The average transmissivity of the regional aquifer is about 8,500 ft^2/day and the unconfined aquifer is 150 ft thick. If the maximum allowable rise of the groundwater mound is 70 ft above the original water table, plot the permissible recharge (infiltration) rate versus R_n, the distance from the center of recharge area to the water table control area. Assume that the original groundwater table will be maintained along the perimeter of the control area.

13.5.4 How far should a water table control area be established from the recharge area in Example 13.5.6 in order to prevent the groundwater mound from rising more than 25 m above the original water table?

13.5.5 The rise of a groundwater mound directly below the center of gravity of a rectangular recharge area 0.5 km × 1.5 km is 41 m in three years. If the recharge rate is approximately 37 million m^3 per year and the fillable porosity of the vadose zone is 0.22, estimate the aquifer transmissivity beneath the recharge area.

13.5.6 A rectangular recharge area, 0.3 mi × 1.6 mi, recharges an unconfined aquifer at a rate of 2.25×10^9 ft^3 per year. The initial water table of the unconfined aquifer is 97 ft above the impermeable bedrock and the depth to the horizontal water table is 220 ft. The fillable porosity of the vadose zone is 0.25. The average transmissivity in the area is 14,500 ft^2/day. If the maximum permissible rise of the groundwater mound is 290 ft above the bedrock to prevent waterlogging, when and where should a water table control area be established? How much water can be stored underground during that time?

13.6.1 A new stormwater infiltration basin will be designed at a site where the average soil infiltration rate at the land surface is 1.27 cm/hr. The distance between the bottom of the basin and the groundwater table will be approximately 13 m. The hydraulic conductivity of the subsurface material is isotropic and estimated to be in the range of 0.80–1.00 cm/hr. Design a square infiltration basin that can sustain the design infiltration rate for the given range of hydraulic conductivity values.

13.6.2 Rework Example 13.6.2 if the hydraulic conductivity in the vadose zone were anisotropic. Assume $K_y = 0.5$ ft/day and $K_r = 1.6$ ft/day.

13.6.2 A circular stormwater infiltration system has a diameter of 70 ft. The hydraulic conductivity of the subsurface soil is

2.4 ft/day and the depth from the basin bottom to the groundwater table is 26 ft. Determine the maximum permissible infiltration rate through the basin bottom. Assume an istotropic subsurface medium.

13.6.3 A new stormwater infiltration basin will be constructed over an area where the depth to the groundwater table is about 6.8 m. Determine the required radius of a circular infiltration basin if the design infiltration rate is 4.82 cm/h and the hydraulic conductivity of the formation beneath the basin is 0.58 m/day. Assume the hydraulic conductivity of the formation is isotropic.

13.6.4 A new stormwater infiltration basin will be constructed at a site where the design volumetric infiltration rate based on the design rainfall event selected is 0.00625 m^3/s. The hydraulic conductivity of the medium beneath the basin is about 0.64 m/day. If the distance from the basin bottom to the groundwater level is 10.2 m, determine the required basin radius and the corresponding design infiltration rate that will prevent the infiltration system from being backed up.

13.6.5 Show that growth function for a mound below a circular infiltration can be described by Equation 13.6.6. To begin the derivation, the incremental water storage volume in the soil medium is

$$dV = S_r \pi r^2 dy = S_y \pi \frac{D}{y} R_0^2 dy$$

and the rate of recharge at Section PR in Figure 13.6.2 is

$$q_0 = S_y K_u \pi R^2 = S_y K_u \frac{D}{h} R_0^2$$

13.6.6 Show that the recession function for a mound from a circular infiltration basin is identical to the recession function for a mound from an infiltration trench.

REFERENCES

1. Abdulrazzak, M. J., A. Sorman, and A. Alhames, Techniques of artificial recharge from an ephemeral wadi channel under extreme arid conditions, Proceedings, International Symposium on Artificial Recharge of Groundwater, A. Z. Johnson and D. J. Finlayson, eds., Am. Soc. of Civil Engrs., New York, pp. 602–611, 1988.

2. Ahuja, L. R., J. W. Naney, and R. D. Williams, Estimating soil water characteristics from simpler properties or limited data, Soil Sci. Soc. Am. Proc., v. 45, pp. 1100–1105, 1985.

3. Alley, W. M., T. E. Reilly, and O. L. Franke, Sustainability of Groundwater Resources, U.S. Geological Survey Circular 1186, 1999.

4. Arizona State University, University of Arizona, and University of Colorado, Soil Treatability Pilot Studies to Design and Model Soil Aquifer Treatment Systems, AWWA Research Foundation and American Water Works Association, Denver, CO, 1998.

5. Asano, T., ed., Artificial Recharge of Groundwater, Butterworth Publishers, Stoneham, MA, 1985.

6. Baier, D. C., and G. W. Wesner, Reclaimed waste for groundwater recharge, Water Resources Bull., v. 7, pp. 991–1001, 1971.

7. Behnke, J. J., Clogging in surface spreading operations for artificial ground-water recharge, Water Resources Research, v. 5, pp. 870–876, 1969.

8. Bianchi, W. C., and D. C. Muckel, Ground-Water Recharge Hydrology, ARS 41–161, Agric. Research Service, U.S. Dept. Agric., 62 pp., 1970.

9. Biemond, C., Dune water flow and replenishment in the catchment area of the Amsterdam water supply, Jour. Instn. Water Engrgs., v. 11, pp. 196–213, 1957.

10. Black and Veach, www.bv.com/services/projects/water/scottsdale.asp.

11. Bouwer, H., Renovating municipal wastewater by high rate infiltration of groundwater recharge, Jour. Amer. Water Works Assoc., v. 66, pp. 159–162, 1974.

12. Bouwer, H., R. C. Rice, J. C. Lance, and R. G. Gilbert, Rapid infiltration research: The Flushing Meadows project, Arizona, Jour. Water Poll. Control Fed., v. 42, no. 10, pp. 2457–2470, 1980.

13. Bouwer, E. J., P. L. McCarty, H. R. Bouwer, and R. C. Rice, Organic contaminant behavior during rapid infiltration of secondary wastewater at the Phoenix 23rd Avenue project, Water Research, v. 18, pp. 463–472, 1984.

14. Bouwer, H., Artificial recharge of groundwater: Systems, design, and management, Ch. 24, *Hydraulic Design Handbook,* L. W. Mays, ed., McGraw-Hill, New York, 1999.

15. Bouwer, H., J. T. Back, and J. M. Oliver, Predicting infiltration and ground-water mounds for artificial recharge, *Jour. of Hydrol. Engrg,* v. 4, no. 4 pp. 350–357, Oct. 1999.

16. Brown, R. F., and D. C. Signor, Artifical recharge—State of the art, *Ground Water,* v. 12, pp. 152–160, 1974.

17. Charbeneau, R. J., Kinematic models for soil moisture and solute transport, *Water Resources Research,* v. 17, no. 3, pp. 600–706, 1984.

18. Crook, J., Water reclamation and reuse, Ch. 21, *Water Resources Handbook,* L. W. Mays, ed., McGraw-Hill, New York, 1996.

19. Dept. Econ. and Social Affairs, *Ground-water storage and artificial recharge,* Natural Resources/Water Ser. 2, United Nations, New York, 270 pp., 1975.

20. Dvoracek, M. J., and V. H. Scott, Ground-water flow characteristics influenced by recharge pit geometry, *Trans. Amer. Soc. Agric. Engrs.,* v. 6, pp. 262–265, 267, 1963.

21. Esmail, O. J., and O. K. Kimbler, Investigation of the technical feasibility of storing fresh water in saline aquifers, *Water Resources Research,* v. 3, pp. 683–695, 1967.

22. Glover, R. E., *Groundwater movement.* U.S. Bureau of Reclamation, Engineering Monograph, 31, 1964.

23. Goss, D. W., et al., Fate of suspended sediment during basin recharge, *Water Resources Research,* v. 9, pp. 668–675, 1973.

24. Griffin, D. M., Jr., and R. O. Warrington, Examination of 2-D ground-water recharge solution, *Jour. of Irrig. Drain. Engrg.,* v. 114, no. 4, pp. 691–704, Nov. 1988.

25. Guo, J. C. Y., Surface-subsurface model for trench infiltration basins, *Jour. of Water Res. Plan. Mgt.* v. 124, no. 5, pp. 280–284, Sept./Oct. 1998.

26. Guo, J. C. Y., Design of circular infiltration basin under water mound effects, *Jour. of Water Res. Plan. Mgt.,* Jan 2001.

27. Guo, J. C. Y., Design of infiltration basins for stormwater, Ch. 9, *Stormwater Collection Systems Design Handbook,* L. W. Mays, ed., McGraw-Hill, New York, 2001.

28. Hantush, M. S., Growth and decay of groundwater-mounds in response to uniform percolation, *Water Resources Research,* v. 3, pp. 227–234, 1967.

29. Harmeson, R. H., et al., Coarse media filtration for artificial recharge, *Jour. Amer. Water Works Assoc.,* v. 60, pp. 1396–1403, 1968.

30. Hauser, V. L., and F. B. Lotspeich, Artificial groundwater recharge through wells, *Jour. Soil and Water Conservation,* v. 22, pp. 11–15, 1967.

31. Ho, G. E., K. Mathew, and P. W. G. Newman, Water quality improvement of treated wastewater by soil percolation, in Proceedings of seminar, Water Quality, Its Significance in Western Australia, Water Research Foundation of Australia, Perth, 1983.

32. Huisman, L., and T. N. Olsthoorn, *Artificial Groundwater Recharge,* Pitman Publishing, Mansfield, MA, 1983.

33. Idelovitch, E., R. Terkeltoub, and M. Michall, The role of ground-water recharges in wastewater reuse: Israel's Dan Region Project, *Jour. Am. Water Works Assn.,* v. 72, no. 7, pp. 391–400, 1980.

34. Johnson, A. I., and D. J. Finlayson, eds., *Artificial Recharge of Groundwater,* Amer. Soc. Civil Engrs., 1988.

35. Kelly, T. E., Artificial recharge at Valley City, North Dakota, 1932 to 1965, *Ground Water,* v. 5, no. 2, pp. 20–25, 1967.

36. Kool, J. B., and M. Th. Van Genuchten, *HYDRUS: One dimensional variably saturated flow and transport model, including hystersis and root uptake, Version 3.31,* U.S. Salinity Laboratory, U.S. Department of Agriculture, Pineside, CA, 1991.

37. Kumar, A., and O. K. Kimbler, Effect of dispersion, gravitational segregation, and formation stratification on the recovery of freshwater stored in saline aquifers, *Water Resources Research,* v. 6, pp. 1689–1700, 1970.

38. Lappala, E. G., R. W. Healy, and E. P. Weeks, *Documentation of Computer Program VS2D to Solve the Equations of Fluid Flow in Variably Saturated Porous Media, USGS Water Resources Investigations Report,* No. 83-409, U.S. Geological Survey, Lakewood, CO, 1987.

39. Marino, M. A., Artificial ground water recharge, I. Circular recharging area, *Jour. Hydrology,* v. 25, pp. 201–208, 1975a.

40. Marino, M. A., Artificial ground water recharge, II. Rectangular recharging area, *Jour. Hydrology,* v. 26, pp. 29–37, 1975b.

41. McWhorter, D. B., and J. A. Brookman, Pit recharge influenced by sub-surface spreading, *Ground Water,* v. 10, no. 5, pp. 6–11, 1972.

42. Molden, D. J., D. K. Sunada, and J. W. Warner, Microcomputer model of artificial recharge using Glover's solution, *Ground Water,* v. 22, no. 1, pp. 73–79, 1984.

43. Moore, I. D., Infiltration equations modified for surface effects, *Jour. Irrig. Div., Trans. ASAE,* v. 107, no. IR 1, pp. 71–79, 1981.

44. Morel-Seytoux, H. J., C. Miracapillo, and M. J. Abdulrazzak, A reductionist physical approach to unsaturated aquifer recharge from a circular spreading basin, *Water Resources Research,* v. 26, pp. 771–777, 1990.

45. Muckel, D. C., *Replenishment of Ground Water Supplies by Artificial Means,* Tech. Bull. 1195, Agric. Research Service, U.S. Dept. Agric., 51 pp., 1959.

46. Mushtaq, H., L. W. Mays, and K. E. Lansey, Optimum operation of recharge basins, *Jour. Water Resources Plan. Mgt.,* v. 129, no. 6, pp. 927–943, 1994.

47. Nellor, M. H., R. B. Baird, and J. R. Smyth, *Health Effects Study: Final Report.* County Sanitation District of Los Angeles County, Whittier, CA, 1984.

48. Nightingale, H. I., and W. C. Bianchi, Ground-water recharge for urban use: Leaky Acres project, *Ground Water,* v. 11, no. 6, pp. 36–43, 1973.

49. Orange County Water District, *The Groundwater Management Plan,* Fountain Valley, CA, 1994.

50. Ortiz, N. V., D. Q. Zachmann, D. B. McWhorter, and D. K. Sunada, Effects of in-transit water on groundwater mounds beneath circular and rectangular recharge areas, *Water Resources Research,* v. 15, no. 3, pp. 577–582, 1979.

51. Rahman, M. A., et al., Effect of sediment concentration on well recharge in a fine sand aquifer, *Water Resources Research,* v. 5, pp. 641–646, 1969.

52. Rao, N. H., and P. B. S. Sarma, Ground water recharge from rectangular areas, *Ground Water,* v. 19, no. 3, pp. 270–274, 1981.

53. Rastogi, A. K., and S. N. Pandy, Modeling of artificial recharge basins of different shapes and effect on underlying aquifer system, *Jour. Hydrol. Engrg.,* v. 123, no. 3, pp. 62–68, Jan. 1998.

54. Rebhun, M., and J. Schwartz, Clogging and contamination processes in recharge wells, *Water Resources Research,* v. 4, pp. 1207–1217, 1968.

55. Ripley, D. P., and Z. A. Saleem, Clogging in simulated glacial aquifers due to artificial recharge, *Water Resources Research,* v. 9, pp. 1047–1057, 1973.

56. Sceva, J. E., *Liquid waste disposal in the lava terranes of Central Oregon,* U.S. Federal Water Pollution Control Admin., Corvallis, OR, 2 vols., 162 pp. 1968.

57. Schicht, R. J., Feasibility of recharging treated sewage effluent into a deep sandstone aquifer, *Ground Water,* v. 9, no. 6, pp. 29–35, 1971.

58. Scott, V. H., and G. Aron, Aquifer recharge efficiency of wells and trenches, *Ground Water,* v. 5, no. 3, pp. 6–14, 1967.

59. State Water Resources Control Board, Dept. Water Resources, and Dept. Health, *A "State of the Art" Review of Health Aspects of Wastewater Reclamation for Groundwater Recharge,* Water Information Center, Huntington, NY, 240 pp. 1978.

60. Sumner, D. M., D. E. Rolston, and M. A. Marino, Effects of unsaturated zone on groundwater mounding, *Jour. of Hydrol. Engrg.,* v. 4, no. 1, pp. 65–69, 1999.

61. Suter, M., and R. H. Harmeson, *Artificial ground-water recharge at Peoria, Illinois,* Bull. 48, Illinois State Water Survey, Urbana, 48 pp., 1960.

62. Swamee, P. K., and C. S. P. Ojha, Ground-water mound equation for rectangular recharge area, *Jour. Irrig. and Drain. Engrg.,* v. 123, no. 3, pp. 215–217, May/June 1997.

63. Tang, Z., G. Li, L. W. Mays, and P. Fox, Development of methodology for the optimal operation of soil aquifer treatment systems, Proceed-ings, Second International Symposium on Wastewater Reclamation and Reuse, Iraklio, Greece, pp. 925–935, 1995.

64. Tang, Z., Li, G., L. W. Mays, and P. Fox, New methodology for optimal operation of soil aquifer treatment systems, *Water Resources Management,* Kluwer Acadmic Publishers, European Water Resources Association, The Netherlands, Vol. 14, no. 1, pp. 13–33, Feb. 2000.

65. Taylor. S. W., and P. R. Jaffe, Subsurface and biomass transfer in a porous medium, *Water Resources Research,* v. 26, no. 2, pp. 2181–2194, 1990.

66. U.S. Environmental Protection Agency (EPA), *Arid Water Reclamation Reuse Guidelines,* Cincinnati, OH, 1991.

67. Valliant, J., Artificial recharge of surface water to the Ogallala Formation in the High Plains of Texas, *Ground Water,* v. 2, no. 2, pp. 42–45, 1964.

68. Warner, J. W., D. Molden, M. Chehata, and D. K. Sunada, Mathematical analysis of artificial recharge from basins, *Water Resources Bull.,* v. 25, no. 2, pp. 401–411, April 1989.

69. Wilson, L. G., Water quality changes during soil aquifer treatment of tertiary effluents, *Water Environment Research,* v. 67, pp. 371–376, 1995.

70. Zomorodi, K., Optimal artificial recharge in intermittent multibasin system, *Jour. Water Resources Plan. Mgt.* v. 116, no. 5, pp. 639–651, Sept./Oct. 1990.

Chapter 14

Saline Water Intrusion in Aquifers

Saline water is the most common pollutant in fresh groundwater. Intrusion of saline water occurs where saline water displaces or mixes with freshwater in an aquifer. The phenomenon can occur in deep aquifers with the upward advance of saline waters of geologic origin, in shallow aquifers from surface waste discharges, and in coastal aquifers from an invasion of seawater. The interrelations of two miscible fluids in porous media have been studied extensively both theoretically and under field conditions. Management techniques that enable development of fresh water and at the same time control of saline intrusion are discussed in this chapter.

14.1 OCCURRENCE OF SALINE WATER INTRUSION

Saltwater intrusion into fresh groundwater formations generally results inadvertently from human activities.[*] Most large sources of fresh groundwater are in close proximity to the sea, to natural bodies of saline groundwater, or to salts from effluent wastes released by human activities.[44] Typically, shallow freshwater overlies saline water because the flushing action during recent times removes salts from ancient marine deposits. But at greater depths groundwater movement is much less, so that displacement of saline water is slower. Furthermore, at depths of several thousand meters, brines are normally encountered.

Saline water in aquifers may be derived from any of several sources:[59]

1. Encroachment of seawater in coastal areas
2. Seawater that entered aquifers during past geologic time
3. Salt in salt domes, thin beds, or disseminated in geologic formations
4. Water concentrated by evaporation in tidal lagoons, playas, or other enclosed areas
5. Return flows to streams from irrigated land
6. Human saline wastes

The mechanisms responsible for saline water intrusion fall into three categories.[19] (1) Reduction or reversal of groundwater gradients, which permits denser saline water to displace fresh water. This situation commonly occurs in coastal aquifers in hydraulic continuity with the sea when pumping of wells disturbs the natural hydrodynamic balance. (2) Destruction of natural barriers that separate fresh and saline waters. An example would be the construction

---

[*]It should be noted that saline groundwater can represent a valuable resource, particularly in arid inland regions. Potential uses include industrial processes such as cooling; irrigation, where the mineral content is moderate; and desalination for local domestic purposes.

of a coastal drainage canal that enables tidal water to advance inland and to percolate into a freshwater aquifer. (3) Subsurface disposal of waste saline water, such as into disposal wells, landfills, or other waste repositories (see Chapter 8).

The occurrence of saline water intrusion is extensive and represents a special category of groundwater pollution. A comprehensive listing of examples of intrusion has been prepared;[3, 4, 59] this reveals that the problem exists in localities of most parts of the United States. Seawater intrusion along coasts has received the most attention.[32, 35, 39, 67] Internationally the problem has received attention in populated coastal areas.[7, 31, 63, 72, 73, 74] Many small oceanic islands are completely underlain with aquifers containing sea water; these pose special problems in meeting water supply demands. Reilly and Goodman[50] presented a historical perspective on the quantitative analysis of saltwater–freshwater relationships.

14.2 GHYBEN–HERZBERG RELATION BETWEEN FRESH AND SALINE WATERS

More than 100 years ago two groups of investigators,[18, 27] working independently along the European coast, found that salt water occurred underground, not at sea level but at a depth below sea level of about 40 times the height of the fresh water above sea level. This distribution was attributed to a hydrostatic equilibrium existing between the two fluids of different densities. The equation derived to explain the phenomenon is generally referred to as the *Ghyben–Herzberg relation*, after its originators.*

The hydrostatic balance between fresh and saline water can be illustrated by the U-tube shown in Fig. 14.2.1 Pressures on each side of the tube must be equal; therefore,

$$\rho_s g z = \rho_f g (z + h_f) \tag{14.2.1}$$

where ρ_s is the density of the saline water, ρ_f is the density of the fresh water, g is the acceleration of gravity, and z and h_f are as shown in Figure 14.2.1. Solving for z yields

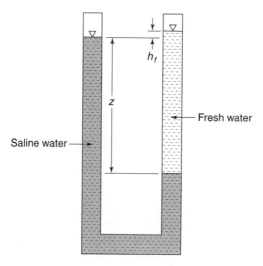

Figure 14.2.1. Hydrostatic balance between fresh water and saline water illustrated by a U-tube.

---

*Almost unnoticed in the hydrologic literature is the much earlier contribution of Joseph DuCommun, a French teacher at West Point Military Academy. He first stated clearly and correctly the fresh–salt water balance existing in coastal aquifers (Du-Commun, J., On the cause of fresh water springs, fountains, etc., *Amer. Jour. Science and Arts*, v. 14, pp. 174-176, 1828).

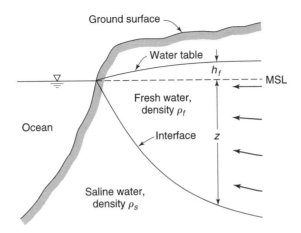

Figure 14.2.2. Idealized sketch of occurrence of fresh and saline groundwater in an unconfined coastal aquifer.

$$z = \frac{\rho_f}{\rho_s - \rho_f} h_f \qquad (14.2.2)$$

which is the Ghyben–Herzberg relation. For typical seawater conditions, let $\rho_s = 1.025$ g/cm^3 and $\rho_f = 1.000$ g/cm^3, so that

$$z = 40 h_f \qquad (14.2.3)$$

Translating the U-tube to a coastal situation, as shown in Figure 14.2.2, h_f becomes the elevation of the water table above sea level and z is the depth to the fresh–saline interface below sea level. This is a hydrodynamic rather than a hydrostatic balance because fresh water is flowing toward the sea. From density considerations alone, without flow, a horizontal interface would develop with fresh water everywhere floating above saline water. Where the flow is nearly horizontal, the Ghyben–Herzberg relation[28] gives satisfactory results. Only near the shoreline, where vertical flow components become pronounced (see Figure 14.2.2), do significant errors in the position of the interface occur.

For confined aquifers, the above derivation can also be applied by replacing the water table by the piezometric surface. It is important to note from the Ghyben–Herzberg relation that fresh–salt water equilibrium requires that the water table, or piezometric surface (1) lie above sea level and (2) slope downward toward the ocean. Without these conditions, seawater will advance directly inland.

Starting from the work of Hubbert,[28] the Ghyben–Herzberg relation has been generalized by Lusczynski and others[40, 41] for situations where the underlying saline water is in motion with heads above or below sea level. The result for nonequilibrium conditions has the form

$$z = \frac{\rho_f}{\rho_s - \rho_f} h_f - \frac{\rho_f}{\rho_s - \rho_f} h_s \qquad (14.2.4)$$

where h_f is the altitude of the water level in a well filled with fresh water of density ρ_f and terminated at depth z, while h_s is the altitude of the water level in a well filled with saline water of density ρ_s and also terminated at depth z (see Figure 14.2.3). When $h_s = 0$, the saline water is in equilibrium with the sea, and Equation 14.2.4 reduces to Equation 14.2.2.

EXAMPLE 14.2.1

The steady fresh–saline interface is located 43 m below sea level 500 m inland from the shoreline in an unconfined aquifer. Determine the elevation of the water table above this point.

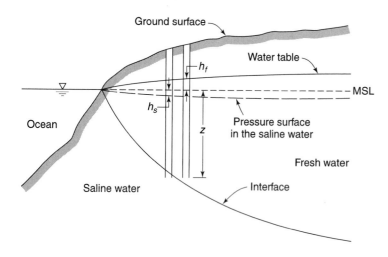

Figure 14.2.3. Diagram illustrating different heads for nonequilibrium conditions in fresh and saline water in an unconfined coastal aquifer.

SOLUTION

Using the fresh and saline water densities for typical seawater conditions, we have $\rho_s = 1.025$ g/cm^3 and $\rho_f = 1.000$ g/cm^3. Substituting into Equation 14.2.2 yields Equation 14.2.3: $z = 40\ h_f$.

$$\text{Given } z = 43 \text{ m,} \qquad h_f = \frac{43 \text{ m}}{40} = 1.075 \text{ m}$$

Thus, the groundwater table is located 1.075 m above sea level at this location. ∎

14.3 SHAPE OF THE FRESH–SALT WATER INTERFACE

Recognizing the approximations inherent in the Ghyben–Herzberg relation, more exact solutions for the shape of the interface have been developed from potential flow theory.[14, 16] The result by Glover (see Cooper et al.[16]) has the form

$$z^2 = \frac{2\rho q x}{\Delta \rho\ K} + \left(\frac{\rho q}{\Delta \rho\ K}\right)^2 \tag{14.3.1}$$

where z and x are as shown in Figure 14.3.1, $\Delta \rho = \rho_s - \rho_f$, ρ is the density of freshwater, K is the hydraulic conductivity of the aquifer, and q is the freshwater flow per unit length of shoreline. The corresponding shape for the water table is given by

$$h_f = \left(\frac{2\Delta \rho\ q x}{(\rho + \Delta \rho)\ K}\right)^{1/2} \tag{14.3.2}$$

The width x_0 of the submarine zone through which fresh water discharges into the sea can be obtained for $z = 0$, yielding

$$x_0 = \frac{\rho q}{2\ \Delta \rho\ K} \tag{14.3.3}$$

The depth of the interface beneath the shoreline z_o occurs where $x = 0$ (see Figure 14.3.1) so that

$$z_0 = \frac{\rho q}{\Delta \rho\ K} \tag{14.3.4}$$

EXAMPLE 14.3.1

The fresh-water flow from a coastal aquifer into the sea per unit length of shoreline is estimated as 1.5 (m^3/day)/m. The hydraulic conductivity of the unconfined aquifer is 13.6 m/day. Compute the water table

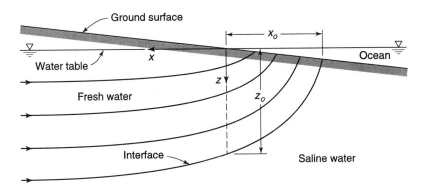

Figure 14.3.1. Flow pattern of fresh water in an unconfined coastal aquifer.

profile and shape of the fresh–saline interface between the shoreline and 1 km inland. Use both the Ghyben–Herzberg relation and the solution by Glover. Assume steady conditions.

SOLUTION

The solution by Glover has the following form (Equation 14.3.1):

$$z = \sqrt{\frac{2\rho q x}{\Delta\rho\, K} + \left(\frac{\rho q}{\Delta\rho\, K}\right)^2}$$

Note that the constant second term inside the square bracket has been added to the Ghyben–Herzberg model to account for the missing seepage face that allows for the vertical components of flow and discharge of the fresh water onto the sea floor.[22] The fresh–saline interface intercepts the water table at the coastline in the Ghyben–Herzberg model.

Thus, the Ghyben–Herzberg model has the following form of Equation 14.3.1:

$$z = \sqrt{\frac{2\rho q x}{\Delta\rho\, K}}$$

The resulting water table profile and the fresh–saline interface estimated by the two models are summarized in Table 14.3.1. Note that the water table profile for Ghyben–Herzberg model is estimated using

Table 14.3.1 Example 14.3.1 Results

x	z (Ghyben–Herzberg)	z (Glover)	h_f (Ghyben–Herzberg)	h_f (Glover)
20	13.28	14.00	0.33	0.33
40	18.79	19.30	0.47	0.46
60	23.01	23.43	0.58	0.57
80	26.57	26.93	0.66	0.66
100	29.70	30.03	0.74	0.73
200	42.01	42.24	1.05	1.04
300	51.45	51.64	1.29	1.27
400	59.41	59.57	1.49	1.47
500	66.42	66.57	1.66	1.64
600	72.76	72.89	1.82	1.80
700	78.59	78.71	1.96	1.94
800	84.02	84.13	2.10	2.07
900	89.11	89.22	2.23	2.20
1000	93.93	94.04	2.35	2.32

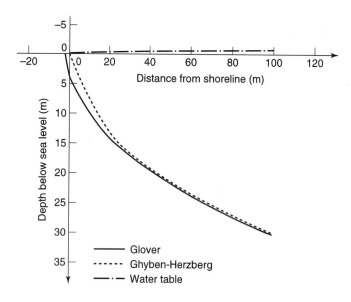

Figure 14.3.2. Example 14.3.1 results showing interface.

$$h_{\mathrm{f}} = \frac{z}{40}$$

The Ghyben–Herzberg model, which does not allow for an outflow face, underestimates the depth to salt-water interface compared to the more exact solution by Glover. The difference becomes more significant closer to the shoreline, especially less than 20 m from the shoreline. On the other hand, the water table profile does not change significantly between the two models. The shape of the interface and the water table profile between the shoreline and 100 m inland are shown in Figure 14.3.2.

The width of the submarine zone through which fresh water discharges into the sea can be computed using Equation 14.3.3:

$$x_0 = \frac{\rho q}{2\,\Delta\rho\,K} = -\frac{\left(1.000\ \mathrm{g/cm}^3\right)\left(1.5\left(\mathrm{m}^3/\mathrm{day}\right)/\mathrm{m}\right)}{2\left(0.025\ \mathrm{g/cm}^3\right)\left(13.6\ \mathrm{m/day}\right)} = 2.21\ \mathrm{m} \qquad \blacksquare$$

EXAMPLE 14.3.2

The salt-water interface 80 m from the shoreline is located 27.5 m below sea level in an unconfined coastal aquifer. If the hydraulic conductivity of the aquifer is 11.8 m/day, determine the fresh water discharge from the aquifer into the sea using Ghyben–Herzberg model and Glover's model. Assume steady-state conditions.

SOLUTION

Ghyben-Herzberg model:

$$z = \sqrt{\frac{2\rho q x}{\Delta\rho\,K}}$$

$$27.5\ \mathrm{m} = \sqrt{\frac{2\left(1.000\ \mathrm{g/cm}^3\right)(q)(80\ \mathrm{m})}{\left(0.025\ \mathrm{g/cm}^3\right)(11.8\ \mathrm{m/day})}}$$

$$q = 1.394\left(\mathrm{m}^3/\mathrm{day}\right)/\mathrm{m}$$

Glover's relation:

$$z = \sqrt{\frac{2\rho q x}{\Delta\rho\,K} + \left(\frac{\rho q}{\Delta\rho\,K}\right)^2}$$

$$27.5 \text{ m} = \sqrt{\frac{2\left(1.000 \text{ g/cm}^3\right)(q)(80 \text{ m})}{\left(0.025 \text{ g/cm}^3\right)(11.8 \text{ m/day})} + \left(\frac{\left(1.000 \text{ g/cm}^3\right)(q)}{\left(0.025 \text{ g/cm}^3\right)(11.8 \text{ m/day})}\right)^2}$$

$$(11.491)q^2 + (542.373)q - 756.25 = 0$$

$$q = 1.355 \left(\text{m}^3/\text{day}\right)/\text{m}$$

Glover's relation provides a more exact answer, although the result from the Ghyben–Herzberg model does not differ significantly from that of Glover's in this problem. It should be noted that the difference between the two models becomes more significant as one gets closer to the shoreline.   ∎

EXAMPLE 14.3.3

The steady fresh-water discharge from an unconfined coastal aquifer into the sea is about 1.6 (m^3/day)/m along the shoreline. The hydraulic conductivity of the aquifer is 8.6 m/day and the horizontal impervious stratum is located 57 m below sea level. Determine the shape of the fresh–saline interface and the location of the toe of the interface from the shoreline.

SOLUTION

First, the exact location of the toe of the interface (where the interface meets the bedrock), $z = 57$ m, must be substituted into Equation 14.3.1, yielding

$$z = \sqrt{\frac{2\rho q x}{\Delta\rho \, K} + \left(\frac{\rho q}{\Delta\rho \, K}\right)^2}$$

$$57 \text{ m} = \sqrt{\frac{2(40)\left(1.6 \text{ m}^3/\text{day/m}\right)(x)}{(8.6 \text{ m/day})} + \left(\frac{(40)\left(1.6 \text{ m}^3/\text{day/m}\right)}{(8.6 \text{ m/day})}\right)^2}$$

$$x = 214.57 \text{ m} \approx 215 \text{ m}$$

where $\rho/\Delta\rho = 40$ assuming typical fresh and salt water densities. Equation 14.3.1 is used to solve for z for x ranging from the shoreline ($x = 0$) to the toe at $x = 215$ m. Glover's method can be solved through using a spreadsheet solution. The resulting shape of the interface is shown in Figure 14.3.3.   ∎

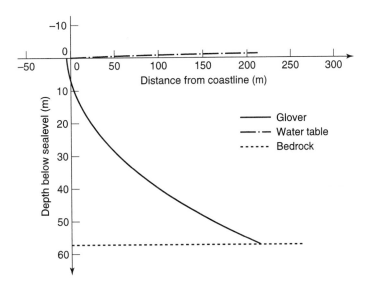

Figure 14.3.3. Example 14.3.3 results showing interface.

14.4 STRUCTURE OF THE FRESH–SALT WATER INTERFACE

The sharp interfacial boundary between fresh and saline water as described above does not occur under field conditions. Instead, a brackish transition zone of finite thickness separates the two fluids. This zone develops from dispersion by flow of the fresh water plus unsteady displacement of the interface by external influences such as tides, recharge, and pumping of wells. In general, the greatest thicknesses of transition zones are found in highly permeable coastal aquifers subject to heavy pumping. Observed thicknesses vary from less than 1 m to more than 100 m.* Figure 14.4.1 shows a transition zone in a highly permeable limestone aquifer in Miami, Florida. Note that the isochlors approach the base of the aquifer perpendicularly; this results because the flow parallels the base of the aquifer, thereby restricting vertical mixing.[16]

An important consequence of the transition zone and its seaward flow is the transport of saline water to the sea. This water originates from the underlying saline water; hence, from continuity considerations, there must exist a small landward flow in the saline water region.[60] Figure 14.4.2 schematically illustrates the flow patterns in the three subsurface zones. Field measurements at Miami[16] and experimental studies[11] have confirmed the landward movement of the saline water body. Where tidal action is the predominant mixing mechanism, fluctuations of groundwater—and hence the thickness of the transition zone—become greatest near the shoreline.

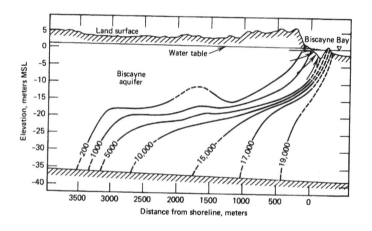

Figure 14.4.1. Cross section through the transition zone of the Biscayne aquifer near Miami, Florida. Numbered lines are isochlors in mg/l (after Cooper et al.[16]).

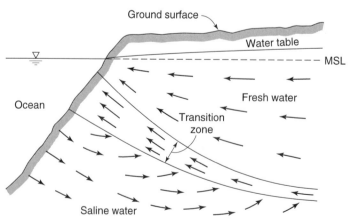

Figure 14.4.2. Vertical cross section showing flow patterns of fresh and saline water in an unconfined coastal aquifer (after Todd[61]; reprinted from the *Journal American Water Works Association,* vol. 66, by permission of the Association, copyright © 1974 by American Water Works Association, 6666 West Quincy Avenue, Denver, CO, 80235).

*As an extreme case, concentrated pumping in the Honolulu–Pearl Harbor area of Hawaii has created localized transition zones more than 300 m thick; these occupy essentially the entire vertical extent of the aquifer.[67]

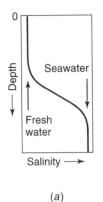

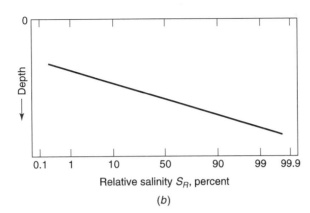

Figure 14.4.3. Increase in salinity with depth through the transition zone. (*a*) Linear scale (*b*) Probability scale

Within the transition zone, the salinity of the groundwater increases progressively with depth from that of the fresh water to that of the saline water. Typically, the distribution of salinity with depth varies as an error function, as shown in Figure 14.4.3*a*.[16] It then becomes advantageous to calculate the relative salinity S_R as a percentage by

$$S_R = 100 \left(\frac{c - c_f}{c_s - c_f} \right) \tag{14.4.1}$$

where c is the salinity* at a particular depth within the transition zone, and c_f and c_s are the salinities of the fresh and saline waters, respectively.

Plotting values of S_R versus depth on probability paper will generally yield a straight line, as demonstrated in Figure 14.4.3*b*. Because detailed data on the structure of a transition zone are difficult to obtain in the field, this graphic technique enables the zone to be estimated from any two point measurements of salinity The 50 percent value of S_R, representing the midline of the transition zone, defines the position of the interface without mixing.

Where inhomogeneities occur in coastal aquifers, stratifications and irregularities in the distribution of fresh and saline waters occur.[25, 47] Considerable research has been done on seawater intrusion in layered aquifers[18, 20, 21] and on unsteady movement of the transition zone.[33, 34]

Anderson et al.,[2] Bear et al.,[6] Bear and Verruijt,[8] Guvanasen et al.,[24] Huyakorn et al,[29, 30] Ledoux et al.,[37] McElwee,[43] Padilla and Cruz-Sanjulian,[46] Segol and Pinder,[55] Person et al.,[48] Reichard,[49] Verruijt,[66] and Wilson and Sada Costa[70] are just a few of the references relating to the simulation of saltwater intrusion in groundwater.

14.5 EFFECT OF WELLS ON SEAWATER INTRUSION

Strack[56] developed a technique for evaluating the effect of wells on seawater intrusion.[5] As a demonstration of the application of the technique, Strack considered a case of a single well located at a distance x_w from the shoreline as illustrated in Figure 14.5.1. For an isotropic, homogeneous aquifer with a horizontal impervious base and with the assumptions of essentially horizontal fresh water flow and no flow in the salt water zone, the following relationship is used for determining the position of the toe of an interface under steady conditions:[5]

$$\frac{1}{2}(1+\delta)\frac{B^2}{\delta^2} = \frac{Q'_{0x}}{K}x + \frac{Q_w}{4\pi K}\ln\left[\frac{(x-x_w)^2 + y^2}{(x+x_w)^2 + y^2}\right] \tag{14.5.1}$$

---

*Salinity can be measured as total dissolved solids, chloride, or electrical conductivity.

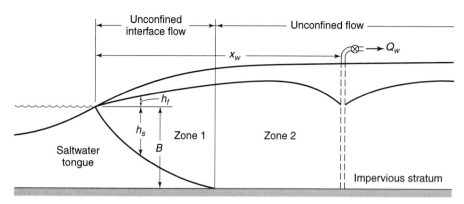

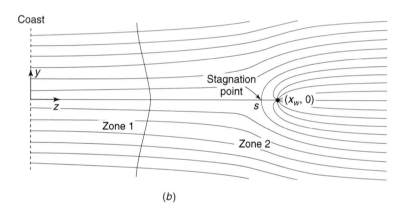

Figure 14.5.1. Shallow coastal aquifer with a well (after Strack[56]).

where

$$\delta = \frac{\rho_f}{\rho_s - \rho_f} \tag{14.5.2}$$

B: depth to bedrock below mean sea level

Q'_{0x}:   freshwater flow per unit length of shoreline (assumed uniform from infinity to the coast)

Q_w: constant pumping rate of the well superimposed on Q'_{0x}

K: hydraulic conductivity of the aquifer

x_w: distance between the well and the shoreline (see Figure 14.5.1); and

(x,y): x–y coordinates of the toe of the interface (see Figure 14.5.1).

For the stagnation point (x_s, y_s), Strack[56] obtained

$$x_s = x_w \left[1 - \frac{Q_w}{\pi Q'_{0,x} x_w} \right]^{1/2} \; ; \quad y_s = 0 \tag{14.5.3}$$

Assuming that an unstable critical situation occurs when the toe of the interface passes through the stagnation point, he obtained the following expression for computing the critical well discharge, Q_w, for unconfined aquifers:

$$\lambda = 2\left(1-\frac{\mu}{\pi}\right)^{1/2} + \frac{\mu}{\pi} \ln\left[\frac{1-(1-\mu/\pi)^{1/2}}{1+(1-\mu/\pi)^{1/2}}\right] \tag{14.5.4}$$

where

$$\lambda = \left(\frac{KB^2}{Q'_{0x}x_w}\right)\left(\frac{1+\delta}{\delta^2}\right) \tag{14.5.5}$$

and

$$\mu = \frac{Q_w}{Q'_{0x}x_w} \tag{14.5.6}$$

The corresponding expressions for confined conditions were presented by Strack.[56]

EXAMPLE 14.5.1   If a production well pumping at a constant rate of 5 L/s is constructed 800 m from the shoreline of the coastal aquifer as described in Example 14.3.3, determine the new location of the toe of the fresh–saline interface under steady conditions. If the well is fully penetrating the aquifer, does brackish water occur?

SOLUTION   Strack's solution estimates the location of the toe of the interface using Equation 14.5.1 given the following: $B = 57$ m, $Q'_{0x} = 1.6$ (m^3/day)/m, $Q_w = 5$ L/s $= 432$ m^3/day, $K = 8.6$ m/day, $x_w = 800$ m.
Substituting $\delta = 40$ for typical sea water conditions and $y = 0$ to find the location of the tip only yields

$$\frac{1}{2}(1+\delta)\frac{B^2}{\delta^2} = \frac{Q'_{0x}}{K}x + \frac{Q_w}{4\pi K}\ln\left[\frac{(x-x_w)^2+y^2}{(x+x_w)^2+y^2}\right]$$

$$\frac{1}{2}(1+40)\frac{(57\text{ m})^2}{40^2} = \frac{(1.6\,(\text{m}^3/\text{day})/\text{m})}{8.6\text{ m/day}}x + \frac{(432\text{ m}^3/\text{day})}{4\pi(8.6\text{ m/day})}\ln\left[\frac{(x-800\text{ m})^2+(0\text{ m})^2}{(x+800\text{ m})^2+(0\text{ m})^2}\right]$$

$$x = 252\text{ m}$$

Note that the toe of the interface was located 215 m from the shoreline (see Example 14.3.3). Thus, the toe of the interface moves 252 m – 215 m = 37 m inland because of the well pumping at a constant rate of 5 L/s.

To determine if brackish water occur in the well or not, we must determine the critical well discharge beyond which the interface would advance rapidly until a new equilibrium is reached, with the interface toe landward of the well. Using Strack's solution for the critical situation (Equation 14.5.5 is used to compute)

$$\lambda = \left(\frac{KB^2}{Q'_{0x}x_w}\right)\left(\frac{1+\delta}{\delta^2}\right) = \left(\frac{(8.6\text{ m/day})(57\text{ m})^2}{(1.6\text{ m}^3/\text{day/m})(800\text{ m})}\right)\left(\frac{1+40}{40^2}\right) = 0.559$$

Substituting $\lambda = 0.559$ into Equation 14.5.4 and solving for μ yields

$$0.559 = 2\left(1-\frac{\mu}{\pi}\right)^{1/2} + \frac{\mu}{\pi} \ln\frac{1-(1-\mu/\pi)^{1/2}}{1+(1-\mu/\pi)^{1/2}}$$

$$\mu = 1.524$$

Finally, the critical well discharge is computed using Equation 14.5.6:

$$1.524 = \frac{Q_w}{(1.6\text{ m}^3/\text{day/m})(800\text{ m})}$$

$$Q_w = 1951\text{ m}^3/\text{day}$$

Since the proposed pumping rate, 5 L/s = 432 m³/day, is below the critical pumping rate, the toe of the interface will stabilize some distance away from the *stagnation point* or the *water divide* defining the region of water supply to the well. Thus, there won't be brackish water in the pumping well of this problem.   ■

14.6 UPCONING OF SALINE WATER

When an aquifer contains an underlying layer of saline water and is pumped by a well penetrating only the upper freshwater portion of the aquifer, a local rise of the interface below the well occurs. This phenomenon, known as *upconing*, is illustrated by Figure 14.6.1. Here the interface is horizontal at the start of pumping when $t = t_0$. With continued pumping the interface rises to successively higher levels until eventually it can reach the well. This generally necessitates the well having to be shut down because of the degrading influence of the saline water. When pumping is stopped, the denser saline water tends to settle downward and return to its former position.

Upconing is a complex phenomenon. Only in recent years has significant headway been made in research studies to enable criteria to be formulated for the design and operation of wells for skimming fresh water from above saline water.[1, 33, 51, 52, 69, 71] From a water-supply standpoint it is important to determine the optimum location, depth, spacing, pumping rate, and pumping sequence that will ensure production of the largest quantity of fresh groundwater, while at the same time striving to minimize any underground mixing of the fresh saline water.

Most investigations of upconing have assumed an abrupt interface between the two fluids,[17, 26] as shown in Figure 14.6.1. This situation would pertain between immiscible fluids, but for miscible fluids such as fresh and saline groundwater, a mixing, or transition, zone having a finite thickness occurs. Although an abrupt interface neglects the physical reality of a transition zone found in groundwater, the assumption has the advantage of simplicity. Furthermore, an interface can be considered as an approximation to the position of the 50-percent relative salinity in a transition zone.

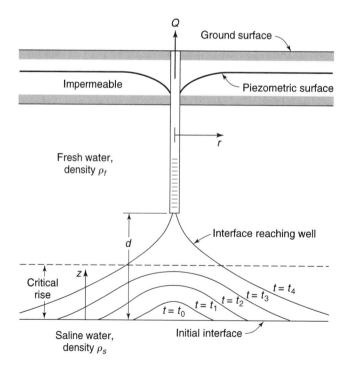

Figure 14.6.1. Diagram of upconing of underlying saline water to a pumping well (after Schmorak and Mercado[54]).

An approximate analytic solution for the upconing directly beneath a well, based on the Dupuit assumptions and the Ghyben–Herzberg relation, is given by[54]

$$z = \frac{Q}{2\pi d K \left(\Delta \rho / \rho_f \right)}$$

(14.6.1)

where $\Delta \rho = \rho_s - \rho_f$, K is the hydraulic conductivity, and all other quantities are defined by Figure 14.6.1. This equation indicates an ultimate rise of the interface to a new equilibrium position that is directly proportional to the pumping rate Q.

Hydraulic model experiments have revealed that the relation in Equation 14.6.1 holds only if the rise is limited. If the upconing exceeds a certain critical rise (see Figure 14.6.1), it accelerates upward to the well. The critical rise has been estimated to approximate $z/d = 0.3$ to 0.5. Thus, adopting an upper limit of $z/d = 0.5$, it follows that the maximum permissible pumping rate without salt entering the well is

$$Q_{max} \leq \pi d^2 K (\Delta \rho / \rho_f)$$

(14.6.2)

For anisotropic aquifers where the vertical permeability is less than the horizontal, a maximum well discharge larger than that for the isotropic case is possible.[13]

In the real-world situation, a transition zone with a finite thickness of brackish water occurs above the body of undiluted saline water. The water at the upper edge of the zone is essentially fresh water and moves accordingly. Upward movement of the almost-fresh water occurs readily along with the adjoining fresh water; consequently, even with a relatively low pumping rate, no limiting critical rise exists above which saline water will not rise. It follows that with any rate of continuous pumping, some saline water must sooner or later reach a well.

Examination of Equation 14.6.1 supports the previous statements. In a transition zone, the salinity changes gradually; hence, $\Delta \rho$ in an incremental width at the top of the zone approaches zero. As a result, z must tend toward infinity, indicating that there can be no finite limit to z. Similarly, in Equation 14.6.2, as $\Delta \rho$ approaches zero, so also must Q_{max} approach zero.

A comparison of the arrivals of salinity at a pumping well for an abrupt interface and for a transition zone is shown qualitatively in Figure 14.6.2. With an abrupt interface, if $Q > Q_{max}$, the salinity appears later and increases more rapidly than with a transition zone. For $Q < Q_{max}$, there will be no salinity reaching the well in the abrupt case; however, a gradual invasion of saline water will occur from a rising transition zone. The ultimate well-water salinity with upconing approaches an intermediate value between the extremes of the fresh and saline waters; empirical data indicate this lies in the range of 5 to 8 percent of the salt concentration in the saline water.[54] Reilly and Goodman[51] presented an analysis of saltwater upconing beneath a pumping well.

It follows from the above analysis that upconing can be minimized by the proper design and operation of wells and galleries. For given aquifer conditions, wells should be separated as far as

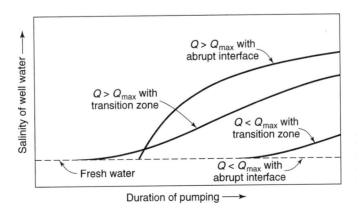

Figure 14.6.2. Well water salinity curves for upconing of an abrupt interface and a transition zone (after Schmorak and Mercado[54]).

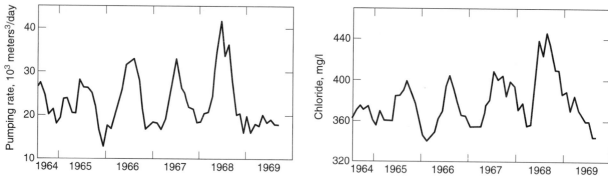

Figure 14.6.3. Effect of annual variation in pumping rate on upconing of underlying saline water in Honolulu, Hawaii. Chloride contents were measured in an observation well at a depth of 128 m and located 430 m from the pumping well (after Todd and Meyer[63]).

possible vertically from the saline zone and pumped at a low uniform rate. Well and gallery designs for regions where a thin layer of fresh water overlies saline water are discussed in the section in this chapter on oceanic islands. Field tests have shown that *scavenger wells*, which pump brackish or saline water from below the fresh water, can also successfully counteract upconing.[38]

The correspondence of pumping rate and salinity in Figure 14.6.3 dramatically illustrates the effects of upconing caused by increased pumping to meet annual midsummer demands in Honolulu.

EXAMPLE 14.6.1

A pumping well partially penetrates a confined aquifer with the bottom of the well screen at an elevation of +25 m. The impervious top and bottom elevations of the aquifer are +64 m and +8 m, respectively. If a horizontal fresh–salt water interface exists at an elevation of +12 m prior to pumping, determine the maximum allowable pumping rate in order to prevent salt water from entering the well. The aquifer's hydraulic conductivity is $K = 25$ m/day.

SOLUTION

Given is $d = 25$ m $- 12$ m $= 13$ m. The critical rise is approximately $z/d = 0.3$ to 0.5 and $\rho_f/\Delta\rho = 40$ for typical fresh–salt water densities. Substituting into Equation 14.6.2 yields

$$Q_{max} = \frac{\pi d^2 K}{40} = \frac{\pi (13 \text{ m})^2 (25 \text{ m/day})}{40} = 332 \text{ m}^3/\text{day} \quad \text{for} \quad \frac{z}{d} = 0.5$$

or

$$Q_{max} = \frac{0.6\pi d^2 K}{40} = \frac{0.6\pi (13 \text{ m})^2 (25 \text{ m/day})}{40} = 199 \text{ m}^3/\text{day} \quad \text{for} \quad \frac{z}{d} = 0.3$$

■

EXAMPLE 14.6.2

If the desired pumping rate is 450 m³/day in Example 14.6.1 and the interruptions of the well pumping are to be minimized, determine the maximum permissible length of the well screen and compare it to the given well screen length.

SOLUTION

Since the desired pumping rate exceeds the critical pumping rates found in the solution to Example 14.6.1, the critical rise of the interface will often be experienced and the pumping will have to be stopped for the interface to return to its former position. However, the interruptions can be minimized if the well screen length is adjusted such that

$$z/d = 0.3 \text{ (assuming this is the limit for critical rise)}$$

Substituting into Equation 14.6.1 yields

$$0.3d = \frac{Q}{2\pi dK\left(\Delta\rho/\rho_f\right)}$$

$$d = \sqrt{\frac{Q}{0.6\pi K\left(\Delta\rho/\rho_f\right)}}$$

$$d = \sqrt{\frac{450 \text{ m}^3/\text{day} \times 40}{0.6\pi\left(25 \text{ m/day}\right)}}$$

$$d = 19.5 \text{ m}$$

Since the interface is initially located at +12 m, the bottom of the screen must be at +12 m + 19.5 m = +31.5 m or above.

This results in a maximum permissible screen length of L_s = +64 m – 31.5 m = 32.5 m ∎

EXAMPLE 14.6.3

A partially penetrating pumping well is installed in an unconfined coastal aquifer for which the horizontal saltwater interface lies 90 ft below the bottom of the well screen. The hydraulic conductivity of the aquifer is 40 ft/day. If the well pumps from the aquifer at a constant rate of 1,200 ft³/hr, how close can the saltwater edge approach the bottom of the screen before the water quality in the pumping well is affected?

SOLUTION

First we must determine the ultimate rise of the interface by substituting $\Delta\rho/\rho_f$ = 1/40, Q = 1200 ft³/hr = 28800 ft³/day, K = 40 ft/day, and d = 90 ft into Equation 14.6.1:

$$z = \frac{Q}{2\pi dK\left(\Delta\rho/\rho_f\right)}$$

$$= \frac{28800 \text{ ft}^3/\text{day}}{2\pi\left(90 \text{ ft}\right)\left(40 \text{ ft/day}\right)\left(1/40\right)}$$

$$= 51 \text{ ft}$$

and

$$\frac{z}{d} = \frac{51 \text{ ft}}{90 \text{ ft}} = 0.57$$

Since the ultimate rise of the interface exceeds the critical rise estimated to approximate z/d = 0.3 to 0.5, the pumping should be stopped when z = 0.3 × 90 ft = 30 ft (assuming an upper limit of z/d = 0.3 for the critical rise). In other words, the pumping should be stopped when the interface is located 60 ft below the bottom of the well screen. ∎

14.7 FRESH–SALT WATER RELATIONS ON OCEANIC ISLANDS

Most small oceanic islands are relatively permeable, consisting of sand, lava, coral, or limestone, so that seawater is in contact with groundwater on all sides. Because fresh groundwater originates entirely from rainfall, only a limited quantity is available. A fresh-water lens, shown schematically in Figure 14.7.1, is formed by the radial movement of the fresh water toward the coast. This lens floats on the underlying salt water; its thickness decreases from the center toward the coast.[15, 23]

From the Dupuit assumptions and the Ghyben–Herzberg relation, an approximate fresh-water boundary can be determined. Assume a circular island of radius R, as shown in Figure 14.7.1, receiving an effective recharge from rainfall at a rate W, the outward flow Q at radius r is

$$Q = 2\pi rK(z+h)\frac{dh}{dr} \tag{14.7.1}$$

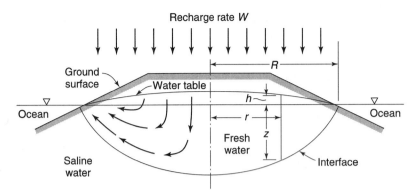

Figure 14.7.1. Freshwater lens in an oceanic island under natural conditions.

where K is the hydraulic conductivity and h and z are defined in Figure 14.7.1. Noting that $h = (\Delta\rho/\rho)z$ and that from continuity $Q = \pi r^2 W$, we have

$$z\,dz = \frac{Wr\,dr}{2K\left[1+\dfrac{\Delta\rho}{\rho}\right]\left[\dfrac{\Delta\rho}{\rho}\right]}$$

(14.7.2)

Integrating and applying the boundary condition that $h = 0$ when $r = R$ yields

$$z^2 = \frac{W\left(R^2 - r^2\right)}{2K\left[1+\dfrac{\Delta\rho}{\rho}\right]\left[\dfrac{\Delta\rho}{\rho}\right]}$$

(14.7.3)

Thus, the depth to salt water at any location is a function of the rainfall recharge, the size of the island, and the hydraulic conductivity. For almost all island conditions, it can be shown that this approximate solution is indistinguishable from more exact solutions by potential theory.[16]

Tidal, atmospheric, and rainfall fluctuations (see Chapter 6), together with dispersion, create a transition zone along the interface in an oceanic island. The close proximity of this boundary zone to the water table can introduce saline water into a well by upconing. Therefore, care must be exercised in development of underground water supplies so that pumping causes a minimal disturbance to the fresh–salt water equilibrium. To avoid the danger of entrainment of saline water, island wells should be designed for minimum drawdown, just skimming fresh water from the top of the lens. If small diameter wells are employed, they should be shallow, dispersed, and pumped at low uniform rates.

In areas where water tables are shallow, an infiltration gallery (see Chapter 5), consisting of a horizontal collecting tunnel at the water table, is advantageous.[53] Drawdowns of a few centimeters can in many instances furnish plentiful water supplies. Installations of infiltration galleries for local water supplies were developed on Bermuda and the Bahamas in the Atlantic Ocean and on the Gilbert and Marianas islands in the Pacific Ocean.[42, 45] Where water tables are deep, dug wells or shafts are sunk to the water table with horizontal tunnels (adits) extending outward to intercept the uppermost layer of fresh water.

Illustrative of the occurrence of fresh water on an oceanic island is Barbados (see Figure 14.7.2). Here a thin highly permeable coral limestone layer serves as the aquifer. Recharge from rainfall percolates through solution channels along the bottom of the limestone until it reaches sea level. From there to the coast the fresh water floats in a layer (known locally as *sheet water*) above the underlying saline water. Water is extracted from large-diameter dug wells connected to horizontal adits at the water table.

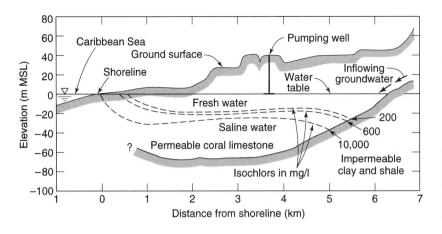

Figure 14.7.2. Geologic profile showing interrelations of fresh and saline water in an unconfined aquifer at Bridgetown, Barbados (courtesy Waterworks Department, Government of Barbados).

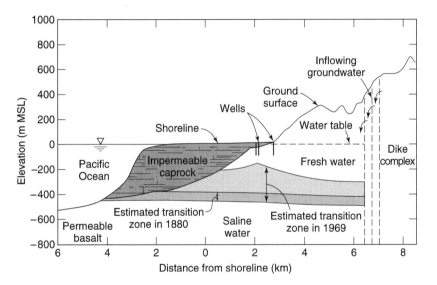

Figure 14.7.3. Geologic profile showing interrelations of fresh and saline water at Honolulu, Hawaii. The thickening of the transition zone resulted from pumping begun in 1880 (after Todd and Meyer[63]).

Another example of the balance between fresh and saline groundwater on an oceanic island is Honolulu (see Figure 14.7.3). Permeable basalt forms the aquifer, but impermeable caprock (erosional material and marine deposits) acts as a groundwater dam. Before wells were first drilled in 1880, groundwater discharged as springs either at the inland or at submarine boundaries of the caprock. At that time the transition zone was narrow and nearly horizontal. Subsequent development of water by wells has lowered the water table and expanded the transition zone so that the volume of fresh water has been substantially reduced. Depths of production wells have decreased gradually from 450 m to 85 m as upconing has progressively increased.[63]

14.8 SEAWATER INTRUSION IN KARST TERRAINS

Coastal aquifers consisting of karstic limestone pose special problems of seawater intrusion. Irregular fissures and solution openings enable seawater to enter the aquifer in configurations that may differ appreciably from those for more homogeneous aquifers.[57, 58] Unique features sometimes found in karst are intermittent brackish springs, which can result from channels connecting with the sea or where saline heads under high tides exceed inland freshwater heads, causing seawater to be discharged inland.

A common feature in the karst regions surrounding the Mediterranean Sea are solutions channels, which discharge fresh water as submarine springs. But pumping of these channels to prevent wastage of fresh water often yields saline water within hours because seawater can freely enter the channel if the freshwater flow is reduced.

14.9 CONTROL OF SALINE WATER INTRUSION

Methods for controlling intrusion vary widely depending on the source of the saline water, the extent of intrusion, local geology, water use, and economic factors. Table 14.9.1 summarizes the generally recognized methods for controlling intrusion from various sources. Because as little as two percent of seawater in fresh water can render water unpotable, considerable attention has been focused on methods to control seawater intrusion. Alternative methods are discussed in the following paragraphs.[9, 61] Measures for coping with upconing were described in earlier sections of this chapter.

Modification of Pumping Pattern

Changing the locations of pumping wells, typically by dispersing them in inland areas, can aid in reestablishing a stronger seaward hydraulic gradient. Also, reduction in pumping of existing wells can produce the same beneficial effect.

Artificial Recharge

Groundwater levels can be raised and maintained by artificial recharge, using surface spreading for unconfined aquifers and recharge wells for confined aquifers. This necessitates development of a supplemental water source.

Extraction Barrier

An extraction barrier is created by maintaining a continuous pumping trough with a line of wells adjacent to the sea. Seawater flows inland from the ocean to the trough, while fresh water within the basin flows seaward toward the trough, as shown in Figure 14.9.1. The water pumped is brackish and normally is discharged into the sea.

Table 14.9.1 Methods for Controlling Saline Water Intrusion

Source or cause of intrusion	Control methods
Seawater in coastal aquifer	Modification of pumping pattern
	Artificial recharge
	Extraction barrier
	Injection barrier
	Subsurface barrier
Upconing	Modification of pumping pattern
	Saline scavenger wells
Oil field brine	Elimination of surface disposal
	Injection wells
	Plugging of abandoned wells
Defective well casings	Plugging of faulty wells
Surface infiltration	Elimination of source
Saline water zones in freshwater aquifers	Relocation and redesign of wells

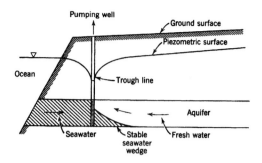

Figure 14.9.1. Control of seawater intrusion by an extraction barrier forming a pumping trough parallel to the coast.

Injection Barrier

This method maintains a pressure ridge along the coast by a line of recharge wells.[10, 62] Injected fresh water flows both seaward and landward, as indicated in Figure 14.9.2. High-quality imported water is required for recharge into wells. A combination of injection and extraction barriers is feasible; this reduces both recharge and extraction rates but requires a larger number of wells.

Subsurface Barrier

Construction of an impermeable subsurface barrier parallel to the coast and through the vertical extent of the aquifer can effectively prevent the inflow of seawater into the basin (see Figure 14.9.3). Materials to construct a barrier might include sheet piling, puddled clay, emulsified asphalt, cement grout, bentonite, silica gel, calcium acrylate, or plastics. Chief problems are construction cost and resistance to earthquakes and chemical erosion.

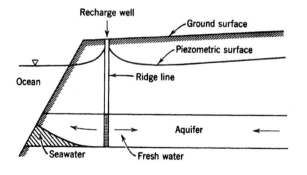

Figure 14.9.2. Control of seawater intrusion by an injection barrier forming a pressure ridge parallel to the coast.

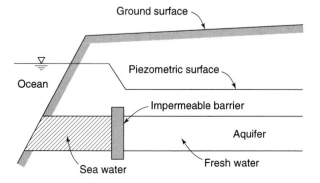

Figure 14.9.3. Control of seawater intrusion by an impermeable subsurface barrier parallel to the coast.

14.10 EXAMPLES OF SEAWATER INTRUSION

Barlow and Wild[4] developed a bibliography on the occurrence and intrusion of saltwater in aquifers along the U.S. Atlantic coast. Of the many localities along the coast of the United States experiencing seawater intrusion, those on Long Island and Southern California are perhaps most significant. Brief descriptions of each of these situations follow.

14.10.1 Long Island, New York

Long Island is underlain by a wedge-shaped mass of unconsolidated sediments that extends more than 600 m in depth. Major intrusion has occurred in western Long Island as a result of development of groundwater and of a decrease in recharge by improved drainage and sewer systems. By the mid-1930s, pumping had lowered water levels at the western end to as much as 10 m below sea level and caused extensive intrusion. Along the southern coast, similar development and reduction of recharge caused active intrusion at depth.[41]

Intrusion has been largely controlled by abandoning pumping and using an imported water supply and by requiring that water pumped from industrial wells be recharged back into the ground after use. Farther east in Nassau County, intrusion is not yet a serious problem; however, to prevent this tendency an extensive program to salvage storm runoff and recharge it through infiltration basins has been initiated (see Chapter 13). In addition, experimental work is underway to reclaim wastewater for recharge to form an injection barrier. Figure 14.10.1 shows a cross section of the distribution of saline water in southwestern Nassau County; the advance of the saline wedge inland varies from 3 to 60 m per year depending on local pumping conditions.

14.10.2 Southern California

Los Angeles, California

Intrusion appeared in the early 1930s along the west coast of Los Angeles County and the rate accelerated rapidly with development of the area in the 1940s. The first and largest injection barrier project has been constructed along a 11-km portion of this shoreline. Some 94 recharge wells form a pressure ridge in the confined aquifers so that seawater is effectively separated from the overpumped basin inland.[10, 36]

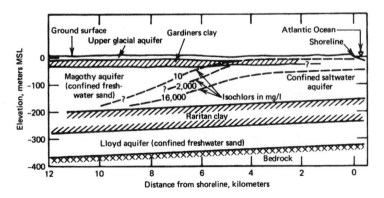

Figure 14.10.1. Geologic profile showing interrelations of fresh and saline water at Far Rockaway, Nassau County, Long Island (after Perlmutter and Geraghty[47]).

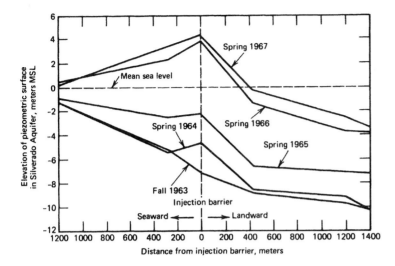

Figure 14.10.2. Piezometric surface profiles perpendicular to the seawater intrusion injection barrier, Los Angeles County, after various time intervals. Note establishment of the pressure ridge following initiation of the barrier at the end of 1963 (courtesy Los Angeles County Flood Control District).

Figure 14.10.2 shows a profile of the piezometric surface perpendicular to the barrier during the years that this portion of the barrier was established. It can be seen how the landward gradient in the injected Silverado Aquifer was transformed into a pressure ridge extending above sea level.

A total of 267 observation wells are monitored to study effects of the injection barrier.

Oxnard Plain, Ventura, California

In Section 10.11, a case study is presented of using simulation–optimization modeling to look the Santa Clara–Calleguas Basin in the Oxnard Plain in southern California. Reichard[49] presented an application of a response matrix model for groundwater–surface water management to identify efficient strategies for meeting water demand (with uncertain surface water supplies) and controlling regional water quality (seawater intrusion) in the Santa Clara–Calleguas Basin in southern California, shown in Figures 10.11.6–10.11.8.

Figure 14.10.3 illustrates the chloride concentrations in water from wells in the upper aquifer system in the Oxnard Plain from 1955 to 1989. Seawater entered the aquifers through outcrop areas in the Hueneme and Mugu submarine canyons (see Figure 14.10.4) in the mid-1950s and advanced inland in response to changes in the amount and distribution of pumping, as illustrated in Figure 14.10.3. By 1989, approximately 23 square miles of the upper aquifer system was believed to be intruded by seawater.[31] Because of the increasing chloride concentrations, pumping was shifted to the lower system from the upper system. As a result the water levels in the lower system declined to below sea level. A combination of groundwater management strategies and increased availability of water from the Santa Clara River for groundwater recharge, by 1993, caused water levels in wells near the coast to rise above sea level and above water levels in the perched aquifer.[31]

The studies by Reichard[49] showed that a significant reduction in pumping, particularly in the lower aquifer system, combined with a large quantity of additional water (either recharged to the aquifers or used in place of groundwater pumping), would be required to control the seawater intrusion.

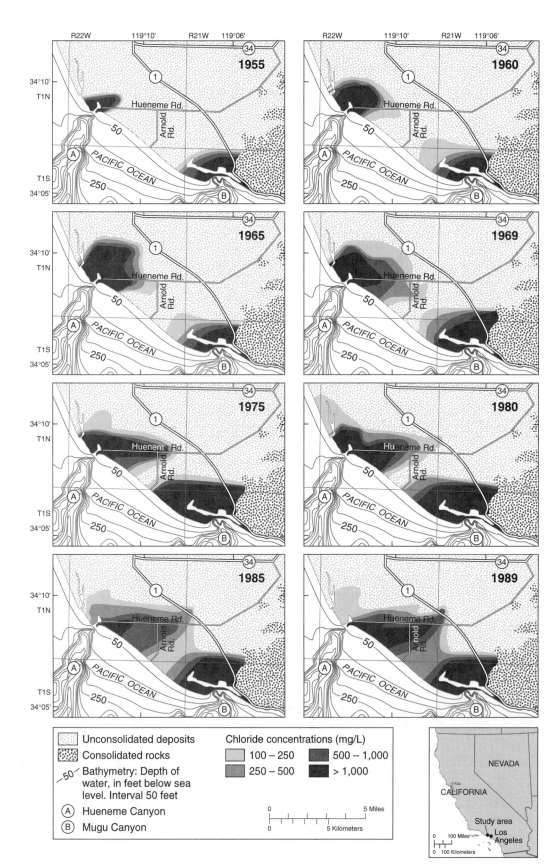

Figure 14.10.3.
Chloride concentrations in water from wells in the upper aquifer system in the Oxnard Plain, 1955–89. Data from California Department of Water Resources and County of Ventura Public Works Agency (Izbicki[31]).

Legend:

Unconsolidated deposits

Consolidated rocks

$-50-$ Bathymetry: Depth of water, in feet below sea level. Interval 50 feet

(A) Hueneme Canyon

(B) Mugu Canyon

Chloride concentrations (mg/L)
100 – 250
250 – 500
500 – 1,000
> 1,000

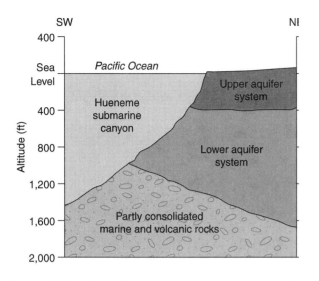

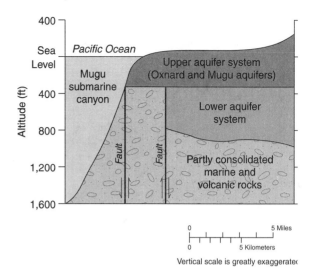

Figure 14.10.4. Generalized geologic sections of the two submarine canyons in Figure 14.10.3. Hueneme Canyon is shown as A and Mugu Canyon is shown as B in Figure 14.10.3 (Izbicki[31]).

PROBLEMS

14.3.1 The water table 30 m from the shoreline is located 0.65 m above sea level in an unconfined coastal aquifer. Determine the depth to the saltwater interface at this location using the Ghyben–Herzberg relation.

14.3.2 Rework Problem 14.3.1 using Glover's relation. The hydraulic conductivity of the aquifer is 8.1 m/day. Which method provides a more exact solution?

14.3.3 The saltwater interface 40 ft from the shoreline is found 48.5 ft below sea level in a coastal aquifer. The hydraulic conductivity of the subsurface medium is 68 ft/day. Determine the rate of fresh water input from the aquifer into the sea and the width of the submarine zone through which fresh water discharges.

14.3.4 Determine the depth of the interface beneath the shoreline in Example 14.3.2 using the Ghyben–Herzberg model and Glover's model.

14.5.1 The seaward specific discharge in an unconfined coastal aquifer is 2.2 (m^3/day)/m. The aquifer is underlain by a horizontal impervious layer 38 m below sea level. The hydraulic conductivity of the aquifer is estimated as 20 m/day. If a fully penetrating well with a constant pumping rate of 850 m^3/day is installed 350 m from the shoreline, determine if brackish water occurs in the pumping well.

14.5.2 The freshwater discharge into the sea in an unconfined coastal aquifer is 50 (ft^3/day)/ft. The horizontal impermeable bottom lies 100 ft below sea level and the hydraulic conductivity of the aquifer is 80 ft/day. If a new well pumping at a constant rate of 750 ft^3/hr is installed 1,000 ft from the shoreline, determine
(a) the location of the toe of the interface before the well installation;
(b) if brackish water occurs in the pumping well;

(c) the new location of the toe after the well installation;
(d) the location of the water divide.

14.5.3 Freshwater flows from an unconfined coastal aquifer into the sea at a steady rate of 30 (ft^3/day)/ft. The impermeable bedrock underlying the aquifer is horizontal and found at a depth of 140 ft. The hydraulic conductivity of the aquifer is 45 ft/day. If a new fully penetrating pumping well were constructed 500 ft from the shoreline, what is the maximum allowable pumping rate in order to prevent brackish water from occurring in the pumping well?

14.5.4 Determine the location of the toe of fresh–saline interface in Problem 14.5.1 without the well pumping using

(a) the Ghyben–Herzberg relation;
(b) Glover's relation;
(c) Strack's solution.

14.6.1 A well pumps at a constant rate of 15 L/s in an unconfined coastal aquifer. The aquifer is 40 m thick and has a hydraulic conductivity of 68 m/day. An initial horizontal saltwater interface lies 14 m below the top of well screen. If the local rise of the interface below the well is monitored continuously, when should the pumping be stopped before the quality of the pumped water is affected?

14.6.2 A new pumping well for potable water extraction will be constructed in a 35-m thick confined aquifer that has a hydraulic conductivity of 50 m/day. A 6-m thick horizontal layer of saline water is detected at the bottom of the aquifer before the well installation. If the desired pumping rate is 800 m^3/day, determine the maximum permissible length of the well screen in order to minimize the interruption of pumping. Assume that the well screen starts from the upper confining layer.

REFERENCES

1. Ackermann, N. L., and Y. Y. Chang, Salt water interface during groundwater pumping, *Jour. Hydraulics Div.*, Amer. Soc. Civil Engrs., v. 97, no. HY2, pp. 223–232, 1971.

2. Anderson, P. F., J. W. Mercer, and H. O. White, Jr., Numerical modeling of salt-water intrusion at Hallandale, Florida, *Ground Water*, v. 26, no. 5, pp. 619–630, 1988.

3. Atkinson, S. F., G. D. Miller, D. S. Curry, and S. B. Lee, *Salt Water Intrusion—Status and Potential in the Contiguous United States,* Lewis Publishers, Chelsea, MI, 1986.

4. Barlow, P. M., and E. C. Wild, Bibliography on the Occurrence and Intrusion of Saltwater in Aquifers along the Atlantic Coast of the United States, *U.S. Geological Survey Open-File Report* 02-235, Northborough, MA, 2002.

5. Bear, J., *Hydraulics of Groundwater,* McGraw-Hill, New York, 1979.

6. Bear, J., A. H. D. Cheng, S. Sorek, D. Ouazar, and Ismael, eds., *Seawater Intrusion in Coastal Aquifers—Concepts, Methods, and Practices,* Kluwer Academic, Dordrecht, The Netherlands, 1999.

7. Bear, J., and G. Dagan, Intercepting freshwater above the interface in a coastal aquifer, *Intl. Assoc. Sci. Hydrology Publ.* 64, pp. 154–181, 1964.

8. Bear, J., and A.Verruijt, *Modeling Groundwater Flow and Pollution,* Reidel, Dordrecht, The Netherlands, 1987.

9. Bruington, A. E., Control of salt-water intrusion in a ground-water aquifer, *Ground Water*, v. 7, no. 3, pp. 9–14, 1969.

10. Bruington, A. E., and F. D. Seares, Operating a sea water barrier project, *Jour. Irrig. Drain. Div.*, Amer. Soc. Civil Engrs., v. 91, no. IR1, pp. 117–140, 1965.

11. Cahill, J. M., Hydraulic Sand-Model Study of the Cyclic Flow of Salt Water in a Coastal Aquifer, *U.S. Geological Survey Professional Paper* 575-B, pp. 240–244, 1967.

12. Calvache, M. L., and A. Pulido-Bosch, Modeling the effects of salt-water intrusion dynamics for a coastal karstified block connected to a detrital aquifer, *Ground Water*, v. 32, no. 5, pp. 767–777, 1994.

13. Chandler, R. A., and D. B. McWhorter, Upconing of the salt-water–fresh-water interface beneath a pumping well, *Ground Water*, v. 13, pp. 354–359, 1975.

14. Charmonman, S., A solution of the pattern of fresh-water flow in an unconfined coastal aquifer, *Jour. Geophysical Research*, v. 70, pp. 2813–2819, 1965.

15. Chidley, T. R. E., and J. W. Lloyd, A mathematical model study of freshwater lenses, *Ground Water*, v. 15, pp. 215–222, 1977.

16. Cooper, H. H., Jr. et al., Sea Water in Coastal Aquifers, *U.S. Geological Survey Water-Supply Paper* 1613-C, 84 pp., 1964.

17. Dagan, G., and J. Bear, Solving the problem of local interface upconing in a coastal aquifer by the method of small perturbations, *Jour. Hydr. Research*, v. 6, pp. 15–44, 1968.

18. Drabbe, J., and W. Badon Ghyben, Nota in verband met de voorgenomen putboring nabij Amsterdam, *Tijdschrift van het Koninklijk Instituut van Ingenieurs*, The Hague, Netherlands, pp. 8–22, 1888–1889.

19. Ernest, L. F., Groundwater flow in the Netherlands delta area and its influence on the salt balance of the future Lake Zeeland, *Jour. Hydrology*, v. 8, pp. 137–172, 1969.

20. Essaid, H. I., A multilayered sharp interface model of coupled freshwater and saltwater flow in coastal systems: Model development and application, *Water Resources Research*, v. 26, no. 7, pp. 1431–1454, 1990a.

21. Essaid, H. I., The Computer Model Sharp, A Quasi-Three-Dimensional Finite Difference Model to Simulate Freshwater and Saltwater Flow in Layered Coastal Aquifer Systems, *U.S. Geological Survey, Water Resources Investigations Report* 90-4130, 1990b.

22. Fetter, C. W., *Applied Hydrogeology,* 4th ed., Prentice Hall, Upper Saddle River, NJ, 2001.

23. Fetter, C. W., Jr., Position of the saline water interface beneath oceanic islands, *Water Resources Research*, v. 8, pp. 1307–1315, 1972.

24. Guvanasen, V., S. C. Wade, and M. D. Barcelo, Simulation of regional groundwater flow and salt water intrusion in Hernando County, Florida, *Ground Water*, v. 39, no. 5, 2000.

25. Harris, W. H., stratification of fresh and salt water on barrier islands as a result of differences in sediment permeability. *Water Resources Research*, v. 3, pp. 89–97, 1967.

26. Haubold, R. G. Approximation for steady interface beneath a well pumping fresh water overlying salt water, *Ground Water*, v. 13, pp. 254–259, 1975.

27. Herzberg, B., Die Wasserversorgung einiger Nordseebader, *Jour. Gasbeleuchtung und Wasserversorgung*, v. 44, pp. 815–819, 842–844, Munich, 1901.

28. Hubbert, M. K., The theory of ground-water motion, *Jour. Geol.*, v. 48, pp. 785–944, 1940.

29. Huyakorn, P. S., P. F. Anderson, J. W. Mercer, and H. O. White, Jr., Saltwater intrusion in aquifers: Development and testing of a three-dimensional finite element model, *Water Resources Research,* v. 23, no. 2, pp. 293–312, 1987.

30. Huyakorn, P. S., Y. S.Wu, and N. S. Park, Multiphase approach to the numerical solution of a sharp interface saltwater intrusion problem, *Water Resources Research,* v. 32, no. 1, pp. 93–102, 1996.

31. Izbicki, J. A., Seawater Intrusion in a Coastal California Aquifer, *U.S. Geological Survey Fact Sheet* 125-96, http://ca.water.usgs.gov/archive/fact_sheets/b07/ref.html, 1996.

32. Kashef, A. I., On the management of ground water in coastal aquifers, *Ground Water*, v. 9, no. 2, pp. 12–20, 1971.

33. Kemblowski, M., Saltwater-freshwater transient upconing—An implicit boundary-element solution, *Jour. Hydrology*, v. 78, pp. 35–47, 1985.

34. Kemblowski, M., The impact of the Dupuit-Forcheimer approximation on salt-water intrusion simulation, *Ground Water*, v. 25, no. 3, pp. 331–336, 1987.

35. Lau, L. S., and J. F. Mink, A step in optimizing the development of the basal water lens of Southern Oahu, Hawaii, *Intl. Assoc. Sci. Hydrology Publ.* 72, pp. 500–508, 1967.

36. Laverty, F. B., and H. A. van der Goot, Development of a fresh-water barrier in southern California for the prevention of sea-water intrustion, *Jour. Amer. Water Works Assoc.,* v. 47, pp. 886–908, 1955.

37. Ledoux, E., S. Sauvagnac, and A. Rivera, A compatible single-phase/two-phase numerical model: Modeling the transient saltwater/fresh-water interface motion, *Ground Water*, v. 28, no. 1, pp. 79–87, 1990.

38. Long, R. A., *Feasibility of a Scavenger-Well System as a Solution to the Problem of Vertical Salt-Water Encroachment,* Water Resources Pamphlet 15, Louisiana Geological Survey, Baton Rouge, 27 pp., 1965.

39. Louisiana Water Resources Research Inst., *Salt-water encroachment into aquifers*, Bull. 3, Louisiana State Univ., Baton Rouge, 192 pp., 1968.

40. Lusczynski, N. J., Head and flow of ground water of variable density, *Jour. Geophysical Research,* v. 66, pp. 4247–4256, 1961.

41. Lusczynski, N. J., and W. V. Swarzenski, Salt-Water Encroachment in Southern Nassau and Southeastern Queens Counties, Long Island, New York, *U.S. Geological Survey Water-Supply Paper* 1613-F, 76 pp., 1966.

42. Mather, J. D., Development of the groundwater resources of small limestone islands, *Quarterly Jour. Engrng. Geol.,* v. 8, pp. 141–150, 1975.

43. McElwee, C. D., A model study of salt-water intrusion to a river using the sharp interface approximation, *Ground Water,* v. 23, no. 4, pp. 465–475, 1985.

44. Newport, B. D., *Salt Water Intrusion in the United States,* Rept. EPA-600/8-77-011, U.S. Environmental Protection Agency, Ada, OK, 30 pp., 1977.

45. Ohrt, F., Water development and salt-water intrusion on Pacific Islands, *Jour. Amer. Works Assoc.,* v. 39, pp. 979–988, 1947.

46. Padilla, F., and J. Cruz-Sanjulian, Modeling sea-water intrusion with open boundary conditions, *Ground Water,* v. 35, no. 4, pp. 704–712, 1997.

47. Perlmutter, N. M., and J. J. Geraghty, Geology and Ground-Water Conditions in Southern Nassau and Southeastern Queens Counties, Long Island, N.Y., *U.S. Geological Survey Water-Supply Paper* 1613-A, 205 pp., 1963.

48. Person, M., J. Z. Taylor, and S. L. Dingman, Sharp interface models of salt water intrusion and wellhead delineation on Nantucket Island, Massachusetts, *Ground Water,* v. 36, no. 4, pp. 731–742, 1998.

49. Reichard, E. G., Groundwater-surface water management with stochastic surface water supplies: A simulation optimization approach, *Water Resources Research,* v. 31, no. 11, pp. 2845–2865, Nov. 1995.

50. Reilly, T. E., and A. S. Goodman, Quantitative analysis of saltwater—freshwater relationships in groundwater systems—A historical perspective, *Jour. Hydrology,* v. 80, pp. 125–160, 1985.

51. Reilly, T. E., and A. S. Goodman, Analysis of saltwater upconing beneath a pumping well, *Jour. Hydrology,* v. 89, no. 3–4, pp. 169–204, 1987.

52. Reilly, T. E., M. H. Frimpter, D. R. LeBlanc, and A. S. Goodman, Analysis of steadystate salt-water upconing with application at Truro well field, Cape Cod, Massachusetts, *Ground Water,* v. 25, no. 2, pp. 194–206, 1987.

53. Rochester, E. W., Jr., and G. J. Kriz, Potable water availability on long oceanic islands, *Jour. San. Engng. Div.,* Amer. Soc. Civil Engrs., v. 96, no. SA5, pp. 1235–1248, 1970.

54. Schmorak, S., and A. Mercado, Upconing of fresh water–sea water interface below pumping wells, field study, *Water Resources Research,* v. 5, pp. 1290–1311, 1969.

55. Segol, G., and G. F. Pinder, Transient simulation of saltwater intrusion in southeastern Florida, *Water Resources Research,* v. 12, no. 1, pp. 65–70, 1976.

56. Strack, O. D. L., A single-potential solution for regional interface problems in coastal aquifers, *Water Resources Research,* v. 12, no. 6, Dec. 1976.

57. Stringfield, V. T., and H. E. LeGrand, Relation of sea water to fresh water in carbonate rocks in coastal areas, with special reference to Florida, USA., and Cephalonia, Greece, *Jour. Hydrology,* v. 9, pp. 387–404, 1969.

58. Stringfield, V. T., and H. E. LeGrand, Effects of karst features on circulation of water in carbonate rocks in coastal areas, *Jour. Hydrology,* v. 14, pp. 139–157, 1971.

59. Task Committee on Salt Water Intrusion, Saltwater intrusion in the United States, *Jour. Hydraulics Div.,* Amer. Soc. Civil Engrs., v. 95, no. HY5, pp. 1651–1669, 1969.

60. Todd, D. K., Salt water intrusion of coastal aquifers in the United States, *Intl. Assoc. Sci. Hydrology Publ.* 52, pp. 452–461, 1960.

61. Todd, D. K., Salt water intrusion and its control, *Jour. Amer. Water Works Assoc.,* v. 66, pp. 180–187, 1974.

62. Todd, D. K., and L. Huisman, Ground water flow in the Netherlands coastal dunes, *Jour. Hydraulics Div.,* Amer. Soc. Civil Engrs., v. 85, no. HY7, pp. 63–81, 1959.

63. Todd, D. K., and C. F. Meyer, Hydrology and geology of the Honolulu aquifer, *Jour. Hydraulics Div.,* Amer. Soc. Civil Engrs., v. 97, no. HY2, pp. 233–256, 1971.

64. Vacher, H. L., Dupuit–Ghyben–Herzberg analysis of strip-island lenses, *Geol. Soc. Amer. Bull.,* v. 100, pp. 580–591, 1988.

65. Vacher, H. L., and T. N. Wallis, Comparative hydrogeology of freshwater lenses of Bermuda and Great Exuma Island, Bahamas, *Ground Water,* v. 30, no. 1, pp. 15–20, 1992.

66. Verruijt, A., A finite element model for interface problems in groundwater flow, *Microcomputers in Engineering Applications,* B. Schrefler and R. W. Lewis, eds., John Wiley, New York, pp. 251–271, 1987.

67. Visher, F. N., and J. F. Mink, Ground-Water Resources in Southern Oahu, Hawaii, *U.S. Geological Survey Water-Supply Paper* 177B, 133 pp. 1964.

68. Wallis, T. N., H. L. Vacher, and M. T. Stewart, Hydrogeology of freshwater lens beneath a Halocene strandplain, Great Exuma, Bahamas, *Jour. Hydrology,* v. 125, pp. 93–109, 1991.

69. Wang, F. C., Approximate theory for skimming well formulation in the Indus Plain of West Pakistan, *Jour. Geophysical Research,* v. 70, pp. 5055–5063, 1965.

70. Wilson, J. L., and A. Sada Costa, Finite element simulation of a saltwater–freshwater interface with interface toe tracking, *Water Resources Research,* vol. 18, pp. 1078–1080, 1982.

71. Wiroganagud, P., and R. J. Charbeneau, Saltwater upconing in unconfined aquifers, *Jour. Hydr. Engrg.,* v. 111, no. 3, pp. 417–434, 1985.

72. Wu, P., Y. Xue, J. Liu, J. Wang, Q. Jiang, and H. Shi, Sea-water intrusion in the coastal area of Laizhou Bay, China: 2. Sea-water intrusion monitoring, *Ground Water,* v. 31, no. 5, pp. 740–745, 1993.

73. Xue, Y., J. Wu, P. Liu, J. Wang, Q. Jiang, and H. Shi, Sea-water intrusion in the coastal area of Laizhou Bay, China: 1. Distribution of seawater intrusion and its hydrochemical characteristics, *Ground Water,* v. 31, no. 4, pp. 532–537, 1993.

74. Yechiele, Y., Fresh-saline ground water interface in the western Dead Sea area, *Ground Water,* v. 38, no. 3, pp. 615–623, 2000.

Appendix A

METRIC UNITS AND ENGLISH EQUIVALENTS

Metric Units

Quantity	Unit	Symbol	Formula
Base units			
Length	meter	m	
Mass	kilogram	kg	
Time	second	s	
Electric current	ampere	A	
Thermodynamic temperature	kelvin	K	
Plane angle	radian	rad	
Solid angle	steradian	sr	
Derived units			
Acceleration	meter per second squared	—	m/s^2
Activity (of a radioactive source)	disintegration per second	—	(disintegration)/s
Angular acceleration	radian per second squared	—	rad/s^2
Angular velocity	radian per second	—	rad/s
Area	square meter	—	m^2
Density	kilogram per cubic meter	—	kg/m^3
Electric capacitance	farad	F	$A \cdot s/V$
Electric conductance	siemens	S	A/V
Electric field strength	volt per meter	—	V/m
Electric inductance	henry	H	W/A
Electric potential difference	volt	V	W/A
Electric resistance	ohm	Ω	V/A
Electromotive force	volt	V	W/A
Energy	joule	J	$N \cdot m$
Force	newton	N	$kg \cdot m/s^2$
Power	watt	W	J/s
Pressure	pascal	Pa	N/m^2
Quantity of electricity	coulomb	C	$A \cdot s$
Quantity of heat	joule	J	$N \cdot m$
Radiant intensity	watt per steradian	—	W/sr
Specific heat	joule per kilogram-kelvin	—	$J/kg \cdot K$
Stress	pascal	Pa	N/m^2
Thermal conductivity	watt per meter-kelvin	—	$W/m \cdot K$

Metric Units

Quantity	Unit	Symbol	Formula
Velocity	meter per second	—	m/s
Viscosity, dynamic	pascal-second	—	Pa·s
Viscosity, kinematic	square meter per second	—	m^2/s
Voltage	volt	V	W/A
Volume	cubic meter	—	m^3
Work	joule	J	N·m

Common Metric Equivalents

Length

$1\ m = 10^3\ mm$
$= 10^2\ cm$
$= 10^{-3}\ km$

Area

$1\ hectare\ (ha) = 10^4\ m^2$
$= 10^{-2}\ km^2$

Volume

$1\ liter\ (l) = 10^{-3}\ m^3$

Mass

$1\ kg = 10^3\ g$

Metric Prefixes

Prefix	Symbol	Multiplication Factor
tera	T	10^{12}
giga	G	10^9
mega	M	10^6
kilo	k	10^3
hecto	h	10^2
deka	da	10^1
deci	d	10^{-1}
centi	c	10^{-2}
milli	m	10^{-3}
micro	μ	10^{-6}
nano	n	10^{-9}
pico	p	10^{-12}
femto	f	10^{-15}
atto	a	10^{-18}

Metric-English Equivalents

Length
1 cm = 0.3937 in.
1 m = 3.281 ft
1 km = 0.6214 mi

Area
1 cm² = 0.1550 in.²
1 m² = 10.76 ft²
1 ha = 2.471 acre
1 km² = 0.3861 mi²

Flow Rate
1 l/s = 15.85 gpm
= 0.02282 mgd
= 0.03531 cfs
$1\ m^3/s = 1.585 \times 10^4$ gpm
= 22.82 mgd
= 35.31 cfs
$1\ m^3/day = 0.1834$ gpm
$= 2.642 \times 10^{-4}$ mgd
$= 4.087 \times 10^{-4}$ cfs

Metric-English Equivalents

Volume
1 cm^3 = 0.06102 in.^{3}
1 l = 0.2642 gal
   = 0.03531 ft^3
1 m^3 = 264.2 gal
   = 35.31 ft^3
   = 8.106 × 10^{-4} acre-ft

Mass
1 g = 2.205 × 10^{-3} lb (mass)
1 kg = 2.205 lb (mass)
   = 9.842 × 10^{-4} long ton

Pressure
1 Pa = 9.872 × 10^{-6} atmosphere
   = 1.000 × 10^{-5} bar
   = 0.01000 millibar
   = 10.00 dyne/cm^2
   = 3.346 × 10^{-4} ft H$_2$O (4°C)
   = 2.953 × 10^{-4} in. Hg (0°C)
   = 0.1020 kg (force)/m^2
   = 0.02089 lb (force)/ft^2

Heat
1 J/m^2 = 8.806 × 10^{-5} BTU/ft^2
   = 2.390 × 10^{-5} calorie/cm^2
1 J/kg = 4.299 × 10^{-4} BTU/lb (mass)
   = 2.388 × 10^{-4} calorie/g

Hydraulic Conductivity
1 m/day = 24.54 gpd/ft^2
1 cm/s = 2.121 × 10^4 gpd/ft^2
1 m/day = 1.198 darcy (for water
      at 20°C)
1 cm/s = 1035 darcy (for water at
      20°C)

Water Quality
1 mg/l = 1 ppm
   = 0.0584 grain/gal
Equivalent weight of ion = atomic weight of ion/valence of ion
meq/l of ion = mg/l of ion/equivalent wt of ion
1 meq/l = 1 me/l
   = 1 epm
1 μS/cm = 1 μmho/cm
1 μS/cm = 0.65 mg/l   } Approximations for most natural
                            waters in the range of 100 to
   = 0.10 meq/l of cations } 5000 μS/cm at 25°C

Velocity
1 m/s = 3.281 ft/s
   = 2.237 mi/hr
1 km/hr = 0.9113 ft/s
   = 0.6214 mi/hr

Temperature
degree Celsius
   = kelvin − 273.15
   = (degree
     Fahrenheit − 32)/1.8

Viscosity
1 Pa·s = 1.000 × 10^3 centistoke
   = 10.00 poise
   = 0.02089 lb (force)·s/ft^2
1 m^2/s = 1.000 × 10^6 centistoke
   = 10.76 ft^2/s

Force
1 N = 1.000 × 10^5 dyne
   = 0.1020 kg (force)
   = 0.2248 lb (force)

Power
1 W = 9.478 × 10^{-4} BTU/s
   = 0.2388 calorie/s
   = 0.7376 ft-lb (force)/s

Energy
1 J = 9.478 × 10^{-4} BTU
   = 0.2388 calorie
   = 0.7376 ft-lb (force)
   = 2.778 × 10^{-7} kw-hr

Numerical Values for Physical Properties

Quantity	Metric	English
Gravitational acceleration, g (std., free fall)	$9.807 \ \text{m/s}^2$	$32.2 \ \text{ft/s}^2$
Density of water, ρ @ 50° F/10° C	$1000 \ \text{kg/m}^3$	$1.94 \ \text{slugs/ft}^3$
Specific weight of water, γ @ 50° F/10° C	$9.807 \times 10^3 \text{N/m}^3$	$62.4 \ \text{lb/ft}^3$
Dynamic viscosity of water, μ @ 50° F/10° C	$1.30 \times 10^{-3} \text{Pa} \cdot \text{s}$	$2.73 \times 10^{-5} \text{lb} \cdot \text{s/ft}^2$
@ 68° F/20° C	$1.00 \times 10^{-3} \text{Pa} \cdot \text{s}$	$2.05 \times 10^{-5} \text{lb} \cdot \text{s/ft}^2$
Kinematic viscosity of water, ν @ 50° F/10° C	$1.30 \times 10^{-6} \text{m}^2/\text{s}$	$1.41 \times 10^{-5} \text{ft}^2/\text{s}$
@ 68° F/20° C	$1.00 \times 10^{-6} \text{m}^2/\text{s}$	$1.06 \times 10^{-5} \text{ft}^2/\text{s}$
Atmospheric pressure, p (std.)	$1.013 \times 10^5 \text{Pa}$	$14.70 \ \text{psia}$

Appendix B

FOLLOWING IS A LIST OF U.S. GOVERNMENT WEB SITES THAT ARE AVAILABLE:

U.S. Geological Survey	www.usgs.gov
Environmental Protection Agency (EPA)	www.epa.gov
Office of Groundwater and Drinking Water	www.epa.gov/ogwdw
U.S. Department of Agriculture	www.usda.gov
Natural Resources Conservation Service	www.nrcs.usda.gov
U.S. Department of Interior	www.doi.gov
Bureau of Reclamation	www.usbr.gov
U.S. Army Corps of Engineers	www.usace.army.mil
Fish and Wildlife Service	www.fws.gov
Federal Legislation	http://thomas.loc.gov
Library of Congress	http://lcweb.loc.gov
National Environmental Methods Index	www.nemi.gov
National Oceanic and Atmospheric Administration	www.noaa.gov

FOLLOWING IS A LIST OF NONGOVERNMENTAL ORGANIZATIONS:

American Geophysical Union	www.agu.org
American Institute of Hydrology	www.aihydro.org
American Society of Civil Engineers	www.asce.org
American Water Resources Association	www.awra.org
American Water Works Association	www.awwa.org
Geological Society of America	www.geosociety.org
Groundwater Management Districts Association	www.gmdausa.org
Groundwater Foundation	www.groundwater.org
International Association of Hydrogeologists	www.iah.org
International Association of Hydrological Sciences	www.cig.ensmp.fr/~iahs
International Water Association	www.iwahq.org.uk
International Water Resources Association	www.iwra.siu.edu
National Ground Water Association	www.ngwa.org
National Institutes for Water Resources	http://wrri.nmsu.edu/niwr
National Water Resources Association	www.nwra.org
Universities Council on Water Resources	www.uwin.siu.edu/ucowr
Water Education Foundation	www.water-ed.org
Water Environment Federation	www.wef.org

Appendix C

Table 9.9.1 Input Data for Example 9.9.1 (part of Example 9.9.1)

INPUT FILE
Basic package

steady state problem

```
     1        7        7        1        4
FREE
     0        1
     1        1(7I3)                          -1  6. DATA IBOUND Array for Layer 1
 -1 -1 -1 -1 -1 -1 -1
  1  1  1  1  1  1 -1
  1  1  1  1  1  1 -1
  1  1  1  1  1  1 -1
  1  1  1  1  1  1 -1
  1  1  1  1  1  1 -1
  1  1  1  1  1  1 -1
     0
     1        1(7G14.0)                       -1  8. DATA Starting Heads in Layer 1
           10        9        8        6        4        2        0
           10       10       10       10       10       10        3
           10       10       10       10       10       10        6
           10       10       10       10       10       10        8
           10       10       10       10       10       10       12
           10       10       10       10       10       10       15
           10       10       10       10       10       10       20
    365        1        1
```

BLOCK CENTERED FLOW PACKAGE

```
     1       50        0        0        0        0        0
 0
    11        1(1G14.0)                       -1  3. DATA TRPY(): Anisotropy factor of layers
         1
    11        1(7G14.0)                       -1  4.DATA: DELR(NCOL)
      500      500      500      500      500      500      500
    11        1(7G14.0)                       -1  5.DATA: DELC(NROW)
      500      500      500      500      500      500      500
    11        1(7G14.0)                       -1  7. DATA Tran() = user specified Transmissivity of layer 1
      500      500      500      500      500      500      500
      500      500      500      500      500      500      500
      500      500      500      500      500      500      500
      500      500      500      500      500      500      500
      500      500      500      500      500      500      500
      500      500      500      500      500      500      500
      500      500      500      500      500      500      500
```

WELL PACKAGE

```
     1       50
     1
     1        5        3    -8000
```

SIP PACKAGE

```
    50        5
     1      .01        1        0        1
```

RECHARGE PACKAGE

1	50		0
1	0		
18	1(7G14.0)		-1 Recharge

.001	.001	.001	.001	.001	.001	.001
.001	.001	.001	.001	.001	.001	.001
.001	.001	.001	.001	.001	.001	.001
.001	.001	.001	.001	.001	.001	.001
.001	.001	.001	.001	.001	.001	.001
.001	.001	.001	.001	.001	.001	.001
.001	.001	.001	.001	.001	.001	.001

OUTPUT CONTROL PACKAGE

0	0	51	52		
1	1	1	1	PERIOD: 1	TIME STEP: 1
0	0	1	1	LAYER: 1	

Table 9.9.2 Output Data for Example 9.9.1

OUTPUT FILE

11 ITERATIONS FOR TIME STEP 1 IN STRESS PERIOD 1

VOLUMETRIC BUDGET FOR ENTIRE MODEL AT END OF TIME STEP 1 IN STRESS PERIOD 1

CUMULATIVE VOLUMES	$L^{**}3$	RATES FOR THIS TIME STEP	$L^{**}3/T$
IN:		IN:	
CONSTANT HEAD =	1269498.8800	CONSTANT HEAD =	3478.0791
WELLS =	0.0000	WELLS =	0.0000
RECHARGE =	3285000.2500	RECHARGE =	9000.0010
TOTAL IN =	4554499.0000	TOTAL IN =	12478.0801
OUT:		OUT:	
CONSTANT HEAD =	1635180.3800	CONSTANT HEAD =	4479.9463
WELLS =	2920000.0000	WELLS =	8000.0000
RECHARGE =	0.0000	RECHARGE =	0.0000
TOTAL OUT =	4555180.5000	TOTAL OUT =	12479.9463
IN - OUT =	-681.5000	IN - OUT =	-1.8662
PERCENT DISCREPANCY =	-0.01	PERCENT DISCREPANCY =	-0.01

TIME SUMMARY AT END OF TIME STEP 1 IN STRESS PERIOD 1

	SECONDS	MINUTES	HOURS	DAYS	YEARS
TIME STEP LENGTH	3.15360E+07	5.25600E+05	8760.0	365.00	0.99932
STRESS PERIOD TIME	3.15360E+07	5.25600E+05	8760.0	365.00	0.99932
TOTAL TIME	3.15360E+07	5.25600E+05	8760.0	365.00	0.99932

HEAD IN LAYER 1 AT END OF TIME STEP 1 IN STRESS PERIOD 1

7	7					
10	9	8	6	4	2	0
9.582833	9.00734	8.179432	7.097559	5.900103	4.558105	3
9.241158	8.766907	8.112735	7.810666	7.444777	6.832418	6
8.873644	8.206147	7.193868	8.087639	8.736045	8.826828	8
8.673501	7.489902	3.868852	8.109984	10.085	11.2388	12
9.156809	8.710836	8.181492	9.898345	11.7551	13.54326	15
9.585927	9.514864	9.747688	11.04657	12.99357	15.67894	20

Table 9.9.3 Input Data for Example 9.9.2

INPUT FILE
Basic Package

```
        1        19        19         1          1
FREE
        0         1
        1                              1(19I3)                    -1    6. DATA IBOUND Array for Layer 1
 -1  -1  -1  -1  -1  -1  -1  -1  -1  -1  -1  -1  -1  -1  -1  -1  -1  -1  -1
 -1   1   1   1   1   1   1   1   1   1   1   1   1   1   1   1   1   1  -1
 -1   1   1   1   1   1   1   1   1   1   1   1   1   1   1   1   1   1  -1
 -1   1   1   1   1   1   1   1   1   1   1   1   1   1   1   1   1   1  -1
 -1   1   1   1   1   1   1   1   1   1   1   1   1   1   1   1   1   1  -1
 -1   1   1   1   1   1   1   1   1   1   1   1   1   1   1   1   1   1  -1
 -1   1   1   1   1   1   1   1   1   1   1   1   1   1   1   1   1   1  -1
 -1   1   1   1   1   1   1   1   1   1   1   1   1   1   1   1   1   1  -1
 -1   1   1   1   1   1   1   1   1   1   1   1   1   1   1   1   1   1  -1
 -1   1   1   1   1   1   1   1   1   1   1   1   1   1   1   1   1   1  -1
 -1   1   1   1   1   1   1   1   1   1   1   1   1   1   1   1   1   1  -1
 -1   1   1   1   1   1   1   1   1   1   1   1   1   1   1   1   1   1  -1
 -1   1   1   1   1   1   1   1   1   1   1   1   1   1   1   1   1   1  -1
 -1   1   1   1   1   1   1   1   1   1   1   1   1   1   1   1   1   1  -1
 -1   1   1   1   1   1   1   1   1   1   1   1   1   1   1   1   1   1  -1
 -1   1   1   1   1   1   1   1   1   1   1   1   1   1   1   1   1   1  -1
 -1   1   1   1   1   1   1   1   1   1   1   1   1   1   1   1   1   1  -1
 -1   1   1   1   1   1   1   1   1   1   1   1   1   1   1   1   1   1  -1
 -1  -1  -1  -1  -1  -1  -1  -1  -1  -1  -1  -1  -1  -1  -1  -1  -1  -1  -1
      -999.99
        1                              1(19G14.0)                 -1    8. DATA Starting Heads in Layer 1
        0      .000E+00
        1               1              1
```

BLOCK CENTERED FLOW PACKAGE

```
        1               50      -1E+30         0         .1         1          0
0
       11                              1(1G14.0)                   -1    3. DATA TRPY(): Anisotropy factor of layers
        1
       11                              1(19G14.0)                  -1    4.DATA: DELR(NCOL)
      63.75 40   40   40   40   40   20   10    5   2.5   5   10   20   40   40   40   40   40 63.75
       11                              1(19G14.0)                  -1    5.DATA: DELC(NROW)
      63.75 40   40   40   40   40   20   10    5   2.5   5   10   20   40   40   40   40   40 63.75
       11                              1(19G14.0)                  -1    7. DATA Tran() = user specified Transmissivity of layer 1
          164.3  ...  164.3
          164.3  ...  164.3
          164.3  ...  164.3
          164.3  ...  164.3
          164.3  ...  164.3
          164.3  ...  164.3
          164.3  ...  164.3
          164.3  ...  164.3
          164.3  ...  164.3
          164.3  ...  164.3
          164.3  ...  164.3
          164.3  ...  164.3
          164.3  ...  164.3
          164.3  ...  164.3
          164.3  ...  164.3
          164.3  ...  164.3
          164.3  ...  164.3
          164.3  ...  164.3
          164.3  ...  164.3
```

WELL PACKAGE

```
1      50
1
1      10     10     -425
```

SIP PACKAGE

```
50      5
 1     .01     1      0      1
```

OUTPUT CONTROL PACKAGE

```
0      0      51     52
1      1      1      1       PERIOD: 1      TIME STEP: 1
0      0      1      1       LAYER: 1
```

Table 9.9.4 Grid Scheme for Example 9.9.2

Row number, i = Column number, j	DELC (i) = DELR (j) (m)
1	63.75
2	40
3	40
4	40
5	40
6	40
7	20
8	10
9	5
10	2.5
11	5
12	10
13	20
14	40
15	40
16	40
17	40
18	40
19	63.75

Table 9.9.5 Output Data for Example 9.9.2

OUTPUT FILE

11 ITERATIONS FOR TIME STEP 1 IN STRESS PERIOD 1

VOLUMETRIC BUDGET FOR ENTIRE MODEL AT END OF TIME STEP 1 IN STRESS PERIOD 1

CUMULATIVE VOLUMES	L**3	RATES FOR THIS TIME STEP	L**3/T
IN:		IN:	
CONSTANT HEAD =	424.5620	CONSTANT HEAD =	424.5620
WELLS =	0.0000	WELLS =	0.0000
TOTAL IN =	424.5620	TOTAL IN =	424.5620
OUT:		OUT:	
CONSTANT HEAD =	0.0000	CONSTANT HEAD =	0.0000
WELLS =	425.0000	WELLS =	425.0000
TOTAL OUT =	425.0000	TOTAL OUT =	425.0000
IN - OUT =	-0.4380	IN - OUT =	-0.4380
PERCENT DISCREPANCY =	-0.10	PERCENT DISCREPANCY =	-0.10

Table 9.9.6 Results for Example 9.9.2

r (m)	Drawdown (Thiem)	Drawdown (MODFLOW)	r (m)	Drawdown (Thiem)	Drawdown (MODFLOW)
0.5	2.634	2.737	70	0.599	0.596
2	2.063	2.349	80	0.544	0.541
4	1.777	1.879	100	0.452	0.438
8	1.492	1.605	120	0.377	0.365
12	1.325	1.363	140	0.314	0.296
16	1.207	1.263	160	0.259	0.242
20	1.115	1.163	180	0.210	0.189
24	1.040	1.063	200	0.167	0.144
30	0.948	0.965	220	0.128	0.099
35	0.884	0.909	240	0.092	0.058
40	0.830	0.853	260	0.059	0.017
45	0.781	0.797	280	0.028	0.000
50	0.738	0.741	300	0.000	0.000
60	0.663	0.651			

Table 9.9.7 Two Grid Schemes for Example 9.9.3.

Row number, i = Column number, j	DELC (i) = DELR (j) (m) Grid-1	Grid-2	Row number, i = Column number, j	DELC (i) = DELR (j) (m) Grid-1	Grid-2
1	300	300	16	100	30
2	200	300	17	150	30
3	150	300	18	200	40
4	100	300	19	300	60
5	80	300	20		80
6	60	300	21		100
7	40	200	22		150
8	30	150	23		200
9	30	100	24		300
10	20	80	25		300
11	30	60	26		300
12	30	40	27		300
13	40	30	28		300
14	60	30	29		300
15	80	20			

Index

Index note: The f or t following a page number refers to a figure or table, respectively, on that page.